XX Chilean Physics Symposium 2016

Journal of Physics: Conference Series Volume 1043

Santiago, Chile
30 November - 2 December 2016

Editors:

Rafael Correa
Luis Huerta
Luis Roa
Leopoldo Soto

ISBN: 978-1-5108-6442-9
ISSN: 1742-6588

Printed from e-media with permission by:

Curran Associates, Inc.
57 Morehouse Lane
Red Hook, NY 12571

Some format issues inherent in the e-media version may also appear in this print version.

Printed by Curran Associates, Inc. (2018)

Institute of Physics
Dirac House, Temple Back
Bristol BS1 6BE UK

Phone: 44 1 17 929 7481
Fax: 44 1 17 920 0979

techtracking@iop.org

Additional copies of this publication are available from:

Curran Associates, Inc.
57 Morehouse Lane
Red Hook, NY 12571 USA
Phone: 845-758-0400
Fax: 845-758-2633
Email: curran@proceedings.com
Web: www.proceedings.com

Table of contents

Volume 1043

XX Chilean Physics Symposium

30 November to 2 December 2016, Santiago, Chile

Accepted papers received: 25 May 2018
Published online: 25 June 2018

Gravitation and Cosmology

Education

IOP Conf. Series: Journal of Physics: Conf. Series **1043** (2018) 011001 doi:10.1088/1742-6596/1043/1/011001

XX Chilean Physics Symposium
November 30th to December 2nd, 2016
Universidad Tecnológica Metropolitana, Santiago of Chile.

Rafael Correa, Luis Huerta, Luis Roa and Leopoldo Soto

Editors

Preface

The Chilean Physics Symposium is the main gathering of physics researchers in Chile, and its organization is one of the central activities of the Chilean Physical Society (Sociedad Chilena de Física, SOCHIFI). The Symposium assembles the largest number of Chilean and foreign physicists resident in the country. Recent advances in the various areas in Physics are presented, by researchers from universities and national research centres. At the same time this is an occasion for the participation of physics students from both the pre- and post-graduate programs.

The Symposium has been organized continuously every two years, since 1978. The organization of the XX symposium was in charge of the Physics Department of the "Universidad Tecnológica Metropolitana", and it took place at the Central Campus, in the city of Santiago, Chile. During this symposium the relation of research in Physics with education and with the society in the country was also analyzed.

During the Symposium, 263 abstracts were submitted, from 273 authors; 143 were oral presentations and 120 posters. In addition, 3 invited plenary talks from outstanding Latin American researchers were presented:

- "Development of accelerator technologies for medical and nuclear applications", by Andrés Kreiner, Head of the Department of Accelerator Technology Research and Applications Area, Constituyentes Atomic Center, National Commission of Atomic Energy CNEA, Argentina and National University of San Martin, Argentina.
- "Multi-energy X-ray cameras for magnetically confinement fusion plasmas", by Luis Delgado-Aparicio, U.S. Department of Energy's (DOE) Princeton Plasma Physics Laboratory (PPPL), USA. He received the Early Career Award of the US Energy Department in 2015. Currently he designs and builds SXR diagnostics for NSTX-U, MST at University of Wisconsin - Madison, DIII-D at General Atomics, JT60SA in Japan, and WEST and ITER in France.
- "Domain walls, black holes entropy and the dynamics of gravitation", by Claudio Bunster, Director of the Centro de Estudios Científicos, CECs, Chile. Bunster was elected to the Natural Academy of Sciences of the United States in 2005 and was made Honorary Member of the Solvay Institute in 2007. He received the National Science Prize of Chile in 1995, the Humboldt Award in 2011 and the TWAS-Lenovo Science Prize in 2013.

  These three plenary talks can be seen in the following link:
  https://xxsimposiofisica.utem.cl/videos/

All authors were invited to submit articles for publication in the Symposium Proceedings. The articles received were reviewed by the Symposium Scientific Committee and by invited peers. The criteria for review focused on the demand for a consistent piece of research, and a clear statement of results. Most of the articles received report the work of research groups where advanced students and young investigators are prominent. Thanks to their enthusiasm, 67 articles are presented in this issue. We would like to express our appreciation to their authors.

All of the oral presentations, including the plenary talks were transmitted via streaming, being this the first time for the Chilean Physics Symposium.

The XX Chilean Physics Symposium was held in the frame of the 50th anniversary of the Chilean Physical Society, that was founded in 1965. In this context, a special recognition was granted to the Chilean poet Nicanor Parra, for his pioneering contribution to physics in Chile from anti-poetry and teaching. In an intimate ceremony, on November 8, the physicists Luis Huerta (Secretary of the SOCHIFI), Rafael Correa (President of the Organizing Committee of the symposium, member of the Board of the SOCHIFI and academic of the Universidad Tecnológica Metropolitana), and Leopoldo Soto (member of the Organizing Committee of the symposium and former President of the SOCHIFI) were received by Nicanor Parra at his home in Las Cruces. On behalf of the physics community of Chile, SOCHIFI expressed to Nicanor Parra his recognition for being an inspirer of several generations of students and many professional physicists in the country. The Chilean Physical Society thanks the family of Nicanor Parra for making this intimate and inspiring encounter possible. On January 23rd, 2018, Nicanor Parra died at 103 years old. The Government of Chile declared 2 days of national mourning in its memory.

In addition, during the symposium, there was a recognition ceremony to three Chilean Physicists for their outstanding trajectory and contribution in both, research and formation of new generations of scientists. The Prize to the Trajectory 2016 of the Chilean Physical Society was granted to the professors: Francisco Claro, Patricio Cordero and Carlos López. The first time in which this prize was granted was in 2014, being the winners the professors: Ivan Schmidt, Jaime Roessler and Jorge Zanelli. In addition, a special session in memory of Prof. Igor Saavedra, pioneer in the constitution of the Chilean physics community, who passed away in 2016, took place during the symposium.

Finally, we would like to express our depth acknowledgement to Mr. Luis Pinto Faverio, Rector of the "Universidad Tecnológica Metropolitana" and to the workers from the university for their important support.

Scientific Committee

Physicists from various research institutions, specialty areas, and regions of the country were invited by the Directive Board of SOCHIFI to join the Symposium Scientific Committee, which was formed by:

Rafael Correa, Universidad Tecnológica Metropolitana, Santiago
Marcia García, Universidad de la Frontera, Temuco
Patricio Häberle, Universidad Técnica Federico Santa María, Valparaíso.
Carla Hernández, Universidad de Santiago, Santiago
Luis Huerta, Universidad de Talca, Talca
María Loreto Ladrón de Guevara, Universidad Católica del Norte, Antofagasta
Edmundo Lazo, Universidad de Tarapacá, Arica
Fernando Lund, Universidad de Chile, Santiago
Eduardo Menéndez, Universidad de Chile, Santiago
Luis Roa, Universidad de Concepción, Concepción
Leopoldo Soto, Comisión Chilena de Energía Nuclear, Santiago
Edgardo Stockmeyer, P. Universidad Católica de Chile, Santiago
Jorge Zanelli (Chairman), Centro de Estudios Científicos, Valdivia

Organizing Commitee

Rafael Correa (President), Universidad Tecnológica Metropolitana
Sofía Camilla, Universidad Tecnológica Metropolitana

IOP Conf. Series: Journal of Physics: Conf. Series **1043** (2018) 011001 doi:10.1088/1742-6596/1043/1/011001

Hugo Durney, Universidad Tecnológica Metropolitana
Luis Huerta, Universidad de Talca
Pamela Quintuprai, Comisión Chilena de Energía Nuclear
Luis Roa, Universidad de Concepción
Leopoldo Soto, Comisión Chilena de Energía Nuclear
Mario Torres, Universidad Tecnológica Metropolitana
José Urbano, Universidad Tecnológica Metropolitana
Javier Wachter, Universidad Tecnológica Metropolitana

Executive Board, Chilean Physical Society April 2015 – April 2017
Gonzalo Gutiérrez, President
Luis Huerta, Secretary
Julio Yañez: Treasurer
Rafael Correa: Director
Pedro Labraña: Director

Peer review statement

All papers published in this volume of *Journal of Physics: Conference Series* have been peer reviewed through processes administered by the proceedings Editors. Reviews were conducted by expert referees to the professional and scientific standards expected of a proceedings journal published by IOP Publishing.

Generation of soliton bubbles in a sine-Gordon system with localised inhomogeneities

Juan F. Marín

Instituto de Física, Pontificia Universidad Católica de Valparaíso, Avenida Brasil, Valparaíso, Casilla 2950, Chile.

E-mail: juanfmarinm@gmail.com

Abstract. Nonlinear wave propagation plays a crucial role in the functioning of many physical and biophysical systems. In the propagation regime, disturbances due to the presence of local external perturbations, such as localised defects or boundary interphase walls have gained great attention. In this article, the complex phenomena that occur when sine-Gordon line solitons collide with localised inhomogeneities are investigated. By a one-dimensional theory, it is shown that internal modes of two-dimensional sine-Gordon solitons can be activated depending on the topological properties of the inhomogeneities. Shape mode instabilities cause the formation of bubble-like and drop-like structures for both stationary and travelling line solitons. It is shown that such structures are formed and stabilised by arrays of localised inhomogeneities distributed in space. Implications of the observed phenomena in physical and biological systems are discussed.

1. Introduction

The propagation of nonlinear waves in inhomogeneous media has captured much recent interest in physics and biophysics [1]. Nonlinear pulse excitations regulate the functioning in many physical models of biological systems, such as the Purkinje's fibres of the cardiac muscles [2, 3], neural fibres [4], DNA chains [5,6] and muscular networks [7]. The study of the effects that can be produced on these regulatory signals by the presence of spatial inhomogeneities, such as localised injuries and necrotic sites, has gained major interest in recent works [3, 8, 9]. For instance, injuries in the heart muscular tissue may produce annihilation, transmission or reflection of incident waves, depending on the geometrical size and shape of the defects. These disturbances may produce a disorder in the rhythmic contractions of the muscular walls, leading to cardiac arrhythmias and different pathologies [3]. A similar situation occurs in the propagation of topological one-dimensional solitons in DNA chains, which has been considered as a suitable model to explain the formation of open states or *bubbles* in the double helix [8]. This bubble formation is an essential mechanism for the regulation and control of gene transcription [5, 6]. Structure heterogeneities in the DNA chain and the presence of boundary interphase walls may affect the propagation regime of these regulatory signals during replication processes.

In the scattering of waves by localised defects, a crucial point to consider is that nonlinear waves exhibit a highly complex behaviour. Transmission, reflection, and annihilation are far from being the only possible phenomena that may occur. Indeed, solitons with an internal structure, such as sine-Gordon (sG) kinks and breathers, may produce many remarkable phenomena when colliding with localised inhomogeneities. The main phenomenon is the internal mode instability

[10, 11], which may cause the breaking of solitons, the formation of multikinks, the creation of kink-antikink pairs and the formation of localised structures [12, 13]. The formation of such structures in sG systems may have important implications not only in biological systems [1], but also in physical systems such as Josephson Junctions (JJs) [14] and ferromagnetic materials [15].

In this article, we investigate the complex phenomena exhibited by a two-dimensional sG system with localised inhomogeneities. Formation of localised structures with bubble-like and drop-like shape are produced in two situations: (a) for line solitons trapped by a single localised inhomogeneity, and (b) for travelling line solitons colliding with an array of localised inhomogeneities. The structure formation is understood using the one-dimensional theory of activation of internal modes in sG solitons. The outline of the article is as follows: Section 2 gives an overview of the one-dimensional theory of activation of internal modes of sG kinks under the action of inhomogeneous external forces. In Sec. 3 the two-dimensional model is introduced, and the internal mode instabilities of line solitons are studied. To avoid any scattering effect, we consider steady line solitons. In Sec. 4, the scattering of line solitons with a spatial array of localised inhomogeneities is investigated. Finally, Sec. 5 gives some concluding remarks.

2. Preliminaries

In this section, an overview of well-known preliminary results of sG kinks under the action of external forces is given. These results provide the theoretical framework for the understanding of the phenomena reported in the subsequent sections.

2.1. Kinks driven by homogeneous external forces

The driven and damped two-dimensional sG system is a particular case of the more general nonlinear Klein-Gordon (KG) system

$$\partial_{tt}\phi(\mathbf{r},t) - \nabla^2\phi(\mathbf{r},t) + \gamma\partial_t\phi(\mathbf{r},t) - G(\phi) = f, \tag{1}$$

where $\mathbf{r} = (x, y)$, γ is a linear-damping coefficient, f is a constant external force and $G(\phi) = -\partial U(\phi)/\partial\phi$, with the nonlinear potential $U(\phi)$ possessing at least two minima separated by a barrier. It is well known that these two minima are fixed points of the associated dynamical system, and that kink solutions are heteroclinic trajectories joining such fixed points [16]. For sG systems, the potential function is given by $U(\phi) = 1 - \cos\phi$ [17], and the equation (1) for $f = 0$ has the well known stationary kink-antikink solutions (line solitons), given by $\phi_\pm(\mathbf{r}) = 4\arctan\exp(\pm x)$. Here, ϕ_+ and ϕ_- reads for the kink and antikink, respectively.

A space independent force with a constant value f leads to an effective potential of the form $U_{eff}(\phi) = U(\phi) - f\phi$. In this case, kink solutions exist for equation (1) only if this effective potential still have at least two minima separated by a barrier. That condition ceases to be fulfilled for a certain critical value f_c of the external force. When $f > |f_c|$ kink solutions do not exist, which means that an initially at rest kink will be destroyed by the external force [18, 19]. The exact value of f_c depends indeed on the particular potential function of the system. For instance, using the sG potential, minima ϕ_j ($j = 0, 1 \ldots$) of the effective potential satisfy the equation $\sin\phi_j = f$, which has no real solutions for ϕ_j if $|f| > 1$. Thus, the critical value is $f_c = 1$ for sG systems.

If $f < |f_c|$, kink solutions in perturbed sG systems exists, and its dynamics in the neighbourhood of fixed points and separatrices can be investigated using the so-called qualitative theory of dynamical systems [20–25]. Based on this, it is possible to generalise the results to other equations that are topologically equivalent to those with the exact solutions [11, 26].

2.2. Soliton breaking by inhomogeneous forces

In the sense of the qualitative theory of dynamical systems, for a general x-dependent force $F(\mathbf{r}) = f(x)$, it is known that a zero of $f(x)$ at $x = x_*$ is an equilibrium point for the centre-of-mass position of the kink [11, 18, 26, 27]. The equilibrium is stable if $df(x)/dx|_{x=x_*} > 0$, and unstable if $df(x)/dx|_{x=x_*} < 0$. If the kink is on a stable equilibrium point, the shape and position of the kink will be recovered for any initial perturbation [11]. On the contrary, if the kink is in an unstable equilibrium position, it is stretched by the inhomogeneous force that is acting on its body in opposite directions. The value f_c is the limit of $|f(x)|$ that the kink can resist the stretching without being destroyed [19]. If $|f(x)| > f_c \forall x$, the force will destroy the line soliton.

Notwithstanding, there is a more subtle mechanism for the kink destruction when $|f(x)| > f_c$ only in a localised region in space. This mechanism is the internal mode instability [25]. In one-dimensional systems, the stability analysis of internal modes of sG kinks has been solved exactly [10, 11]. The procedure starts by introducing the ansatz

$$\phi(x) = 4 \arctan \exp(\pm Bx), \tag{2}$$

where B is a parameter that controls the width of the kink. Solving an inverse problem, one obtains that equation (2) is a solution of the one-dimensional sG equation if the external force is given by

$$F(x) = 2(B^2 - 1)\sinh(Bx)\text{sech}^2(Bx), \tag{3}$$

which is an antisymmetric spatial function that vanishes exponentially for $x \to \pm\infty$. For increasing (decreasing) values of B, the force is less (more) localised in space and has decreasing (increasing) extreme values. Thus, the parameter B also controls the extreme values and the extension of $F(x)$ in space. Interactions of DNA special sites with ligands or perturbation of fluxons by dipole currents in JJs are typical examples of strong and local external influences with the same topological form as equation (3) [5, 28]. From the stability analysis follows that the first stable internal shape mode arises for $1/6 < B^2 < 1/3$, while for $B^2 < 1/6$ many other internal modes can appear. For $B^2 < 2/[\Lambda_*(\Lambda_* + 1)]$, where $\Lambda_* = (5 + \sqrt{17})/2$, the first internal mode becomes unstable, leading to soliton breakups and shape instabilities [10, 11, 18]. These instabilities play a central role in the formation of structures reported in the following sections.

3. Stationary line solitons in the presence of a single localised inhomogeneity

Following the one-dimensional system, consider the two-dimensional perturbed sG equation

$$\partial_{tt}\phi(\mathbf{r}, t) = \nabla^2\phi(\mathbf{r}, t) - \sin\phi(\mathbf{r}, t) - \gamma\partial_t\phi(\mathbf{r}, t) + F(\mathbf{r}), \tag{4}$$

where the inhomogeneous force $F(x, y)$ is given by [13]

$$F(x, y) = 2(B^2 - 1)\text{sech}(Bx)\text{tanh}(Bx)e^{-y^2/\sigma^2}, \tag{5}$$

and a real parameter σ has been introduced to control the spread of the force in the y-direction. Figure 1(a) shows the force of equation (5), which represents a single localised inhomogeneity centred at the origin. The point $x_* = 0$ will be an equilibrium point for two-dimensional kinks. The external force of equation (5) is physically realisable in real experiments. Indeed, the sG equation (4) with this kind of external perturbation models the propagation of fluxons in two-dimensional JJs with a current dipole device inserted into one of the superconducting electrodes [13, 28, 29].

This section shows how the inhomogeneity of equation (5) affects the dynamics of a line soliton, having a special focus on the dynamics of the soliton internal structure. To avoid any scattering effect, consider an initially at rest kink with its centre-of-mass located at the equilibrium point. This initial condition is depicted in the inset of figure 1(a) and is given by $\phi(x, y, 0) = 4 \arctan \exp(x/L)$ with $\partial_t\phi(x, y, 0) = 0$, where L is an arbitrary parameter. We

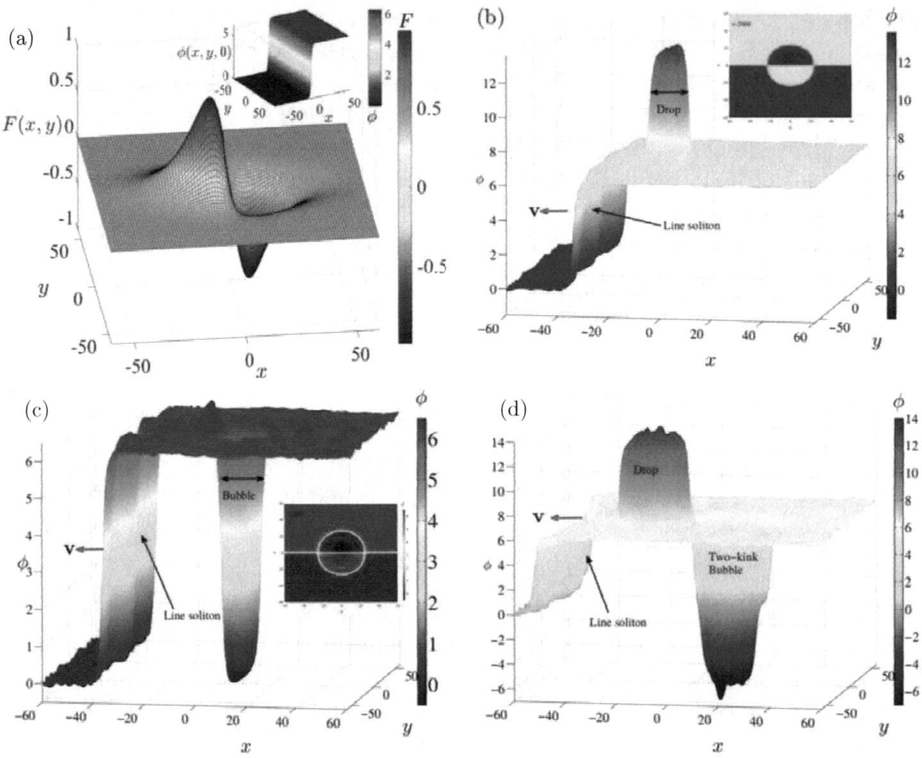

Figure 1. (Colour online) (a) Inhomogeneous force of equation (5) for $B = 0.1$ and $\sigma = 15$. The inset shows the static line soliton used as initial condition in the numerical simulations of section 3. (b) A drop-like structure after 10 000 time iterations for $B = 0.28$ and $\sigma = 8.9$. The inset shows the elliptic-like structure that precedes the drop formation. The value of the field ϕ is indicated with the same colour scale. (c) A bubble-like structure after 10 000 time iterations for $B = 0.27$ and $\sigma = 10$. The inset shows the elliptic-like structure that precedes the bubble formation. (d) A multi-structure formed after 8 300 time iterations for $B = 0.1$ and $\sigma = 6$.

have used an explicit finite difference scheme of the second order of accuracy for the numerical solution of equation (4), with $\gamma = 0.01$ and homogeneous Neumann boundary conditions. The used regular mesh size is $1\,200 \times 1\,200$. The space intervals are $\Delta x = \Delta y = 0.1$, and the time increment is $\Delta t = 0.02$.

Figure 1 shows the main results obtained from numerical simulations. Figure 1(b) shows a drop-like structure created for $B = 0.28$ and $\sigma = 8.9$. The local instability of the shape mode leads to a local break-up of the initially at rest line soliton. This initial breaking consists in a local creation of a pair kink-antikink, giving an elliptic-like structure that grows in time [see the inset of figure 1(b)]. This kink zone propagates until the localised force is weak enough, and the contribution of curvature effects and kink interactions are important. The elliptic form eventually collapses due to the attractive interaction forces between the kink and antikink, developing a final stable drop-like structure in the negative region of x. Notice that the drop is sustained by the positive part of the external force. A kink is still present in the system due to the conservation of the topological charge and propagates towards the negative x-direction with a velocity $\mathbf{v}$, as showed in figure 1(b).

If we proceed to diminish the value of B, the preferred structure is one with the reverse topological charge: a bubble-like structure. Figure 1(c) shows a bubble-like structure created for $B = 0.27$ and $\sigma = 10$. In this case, the force is extended enough in the x-direction to prevent the collapse and sustain the walls of the elliptic form, forming a bubble sustained by the negative part of the force. The mechanism of bubble formation is similar to the drop formation, giving that both are preceded by an elliptic-like form that grows in time [see inset of figure 1(c)]. Over a critical value of the elliptical area, the kink-antikink interaction is too weak to produce the collapse of the bubble. In this case, the external force supports the structure. Otherwise, the external force cannot sustain the walls of the bubble and the structure collapses by kink-antikink interaction, triggering the formation of a drop-like structure. Therefore, the interplay of both parameters B and σ will determine which kind of structure will be formed. Indeed, further numerical simulations have show that these instabilities in sG kinks can be activated varying any of both parameters B or σ [13].

If we continue further diminishing the values of parameters B and σ, it is expected that the line soliton will exhibit more complex phenomena due to the local destabilisation of more internal modes. Indeed, there is a coexistence region in the parameter space where a bubble and a drop merge in a stable structure, like the bubble-drop bound state depicted in figure 1(d) for $B = 0.1$ and $\sigma = 6$. From the one-dimensional theory, it is known that for $B \lesssim 0.2564$ the formation of two-kink solitons occurs [12]. The later explains the two-kink profile of the bubble showed in figure 1(d). More complex multi-structures can be observed for smaller values of B and σ, such as multi-kink bubbles and drops.

Notice that the localised structures showed in Fig. 1 are called *bubbles* and *drops* giving that the equation (4) also appears in the description of structural phase transitions [13, 30, 31]. The results of this article can be interpreted as the driving of a system to a new phase by the activation of a local instability.

4. Scattering of solitons in an array of localised inhomogeneities

In order to simulate some real situations in physical and biological systems, this section deals with travelling line solitons that collide with a distribution of localised inhomogeneities. Consider in equation (4) the following spatial distribution of inhomogeneous forces

$$F(x,y) = F_o + 2\sum_{i=1}^{4}(B_i^2 - 1)\text{sech}\,[B_i(x-x_i)]\tanh\,[B_i(x-x_i)]\,e^{-(y-y_i)^2/\sigma^2}, \tag{6}$$

where $F_o = 0.1$ is a constant homogeneous contribution to the total external force, (x_i, y_i) is the coordinate-position of the core of the i-th inhomogeneity, B_i is its respective control parameter, $\sigma = 10$, and $i \in \{1, 2, 3, 4\}$. Notice that the inhomogeneous contribution of force (6) is given by the superposition of forces of the same form as equation (5).

The position of the inhomogeneities are fixed for all simulations in this section, and are given by $\mathbf{r}_1 = (-15, 25)$, $\mathbf{r}_2 = (15, 25)$, $\mathbf{r}_3 = (-15, -25)$ and $\mathbf{r}_4 = (15, -25)$. With these coordinates, the inhomogeneities are placed at the vertices of a rectangle centred at the origin. The initial condition is given by $\phi(x, y, 0) = 4\arctan\,\exp[(x - xo)/L]$ with $\partial_t\phi(x, y, 0) = 0$, which is an initially at rest line soliton placed some distance away from the array of inhomogeneities. The initial x-coordinate of the centre-of-mass of the kink is given by $x_o = 50$. Due to the homogeneous part of the force (6) and the damping term of equation (4), the line soliton accelerates towards the negative x-direction before reaching an asymptotically constant velocity. When the soliton collides with the inhomogeneities, many remarkable phenomena may occur due to possible internal mode instabilities.

Figure 2(a) shows the results from numerical simulations for $B_1 = 0.27$, $B_2 = 3.0$, $B_3 = B_4 = 0.3$, and $\gamma = 0.5$. For this combination of parameters, the 1st inhomogeneity activates the internal modes of kinks, while the 2nd inhomogeneity tends to trap them at

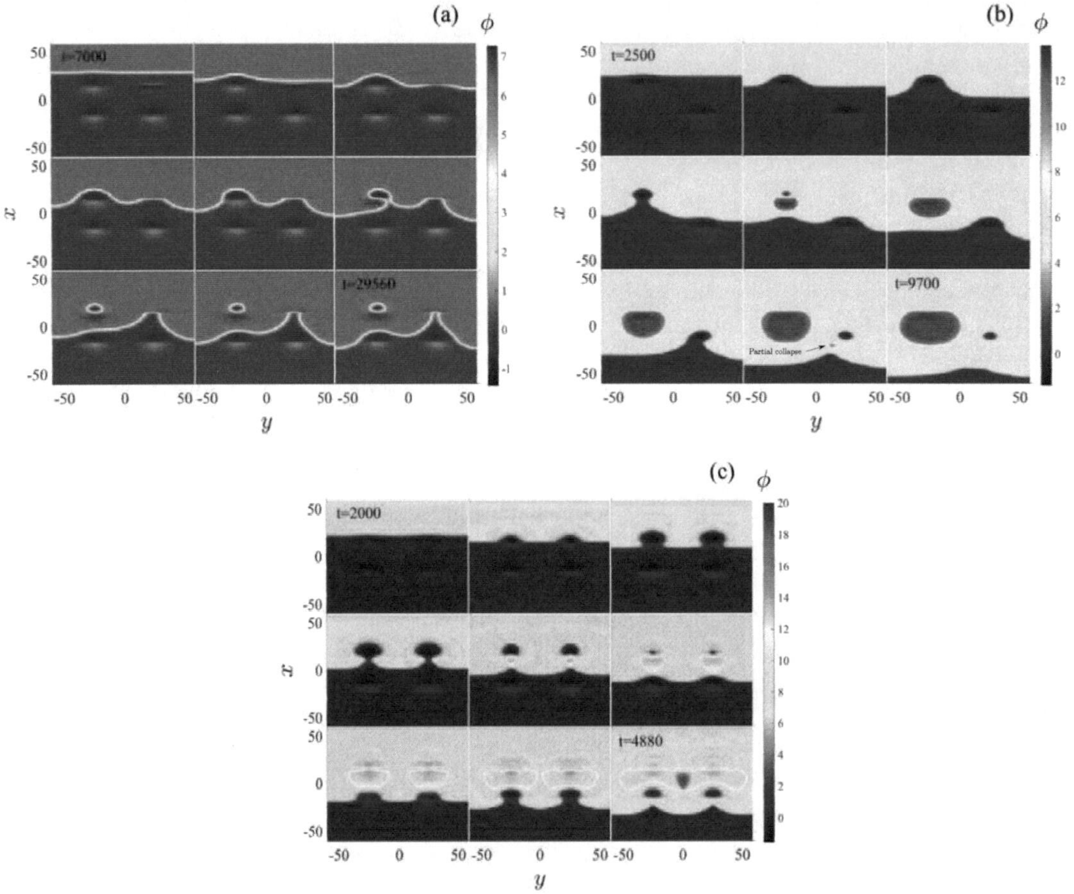

Figure 2. (Colour online) Scattering of travelling line solitons after collisions with a distribution of localised inhomogeneities. (a) Generation of a bubble induced by the collapse of an echo-wave, for $B_1 = 0.27$, $B_2 = 3.0$, $B_3 = B_4 = 0.3$, and $\gamma = 0.5$. (b) Kink transmission for $\gamma = 0.1$, $B_1 = B_4 = 0.27$ and $B_2 = B_3 = 1.1$. A bubble and a drop remains trapped at the inhomogeneities after the transmission of the soliton. (c) Kink transmission and formation of complex interacting structures (bubbles and drops) for $\sigma = 10$, $\gamma = 0.01$, and $B_i = 0.27$ $(i = 1, 2, 3, 4)$.

position $\mathbf{r}_2$. When the line soliton reaches the first line of inhomogeneities, part of the kink is deformed due to the collision with the 1st inhomogeneity, while another part is trapped by the 2nd inhomogeneity. Notice that this initial bent of the soliton is the so-called *echo effect* reported in biophysical systems and is considered as responsible for certain types of disorders and cardiac arrhythmias [3]. This echo-wave is eventually stabilised in size, and forms a stable bubble-like structure sustained by the 1st inhomogeneity. Notice that the echo-waves observed in this section have some resemblance with the elliptic-like structures observed in section 3 [see insets of figure 1(b) and (c)].

For smaller values of γ the soliton reaches the inhomogeneities with higher terminal speeds, and we would expect the enhancing of the transmission of the soliton across the inhomogeneities. Figure 2(b) shows the results from numerical simulations for $\gamma = 0.1$, $B_1 = B_4 = 0.27$ and

$B_2 = B_3 = 1.1$. In this case, the 1st and 4th inhomogeneities activate the internal modes of the soliton. The 2nd and 3rd inhomogeneities tend weakly to trap the soliton. Indeed, when the kink collides with the first pair of inhomogeneities, the 2nd cannot capture the kink, while the 1st inhomogeneity produces an echo-effect. This echo-wave eventually collapse and forms a stable drop, as showed in figure 2(b). A similar mechanism occurs when the line soliton reaches the second pair of inhomogeneities. However, the speed of the line soliton has diminished due to previous collisions. The echo-wave generated at the 4th inhomogeneity partially collapse, as indicated in figure 2(b), but is eventually stabilised and forms a bubble. Thus, the relative speed of collision also plays an important role in determining which kind of structure will be formed.

A highly complex dynamics is observed diminishing the parameters B_i in the array of inhomogeneities. Figure 2(c) shows the results from numerical simulations when $B_i = 0.27$ for all i. After the collision, there is an echo effect followed by the formation of bubbles. The inhomogeneities cannot sustain such bubbles, and they collapse to form drops eventually. However, during this drop formation, another pair of bubbles are created in the neighbour inhomogeneities. In figure 2(c) is clearly seen repulsive interactions between these new bubbles and the previously formed drops. Notice that these drops also collide, enhancing the formation of new structures with the topological charge of three around the origin. More complex dynamics can be observed diminishing the values of B_i or σ.

5. Conclusion

In this article, the formation of localised structures in a two-dimensional sG system has been reported considering two different scenarios: (a) the presence of a single localised inhomogeneity, and (b) the presence of an array of localised inhomogeneities. These inhomogeneities are modelled by a family of topologically equivalent external forces. The localised structures are formed with a bubble-like and drop-like shape and are sustained by the same inhomogeneities that create them. The structure creation is entirely due to the dynamics of the internal modes of the soliton, and not to any scattering effects. However, the internal mode instabilities can also occur during scattering processes, and exhibit a highly complex dynamics.

As a final remark, notice that the reported phenomena may have important implications in biological systems. The formation of bubbles, and the trapping and breaking of such structures by localised inhomogeneities may be related to different functional disorders in similar systems. Moreover, the study of the stability and transport of such bubbles may give some light on how to introduce shock perturbations, and produce the desired effects in a controlled way. Gene transcription in DNA chains can also be regulated by the insertion of this kind of external perturbations.

Acknowledgments

This article was written after a conference at the XX Symposium of the Physical Society of Chile (SOCHIFI), held in Santiago de Chile on November 2016. The author thanks the organising committee for this occasion. The author also thanks Prof. Mónica A. García-Ñustes and Prof. Jorge A. González for many fruitful discussions and the active collaboration on the subject of this article. This work was supported by CONICYT doctorado nacional No. 21150292.

References

[1] Ivancevic V and Ivancevic T 2013 *J. Geom. Symmetry Phys.* **31** 1–56
[2] Aslanidi O V and Mornev O A 1999 *J. Biol. Phys.* **25** 149–164 ISSN 0092-0606
[3] Mornev O A, Tsyganov L M, Aslanidi O, Ordanovich A E and Chailakhyan L M 2000 *Dokl. Biophys.* **373** 32–36 ISSN 0012-4974
[4] Scott A C 1975 *Rev. Mod. Phys.* **47**(2) 487–533
[5] Polozov R V and Yakushevich L V 1988 *J. Theor. Biol.* **130** 423–430 ISSN 0022-5193
[6] Yakushevich L V 2004 *Nonlinear Physics of DNA* (Weinheim: Wiley–VCH)

[7] Gojkovic Z and Ivancevic T 2016 *Nonlinear Dynamics* **86** 2071–2080 ISSN 1573-269X
[8] Grinevich A A, Ryasik A A and Yakushevich L V 2015 *Chaos Soliton. Fract.* **75** 62–75 ISSN 0960-0779
[9] Grinevich A A and Yakushevich L V 2016 *Biophysics* **61** 539–546 ISSN 1555-6654
[10] González J A, Bellorín A and Guerrero L E 2002 *Phys. Rev. E* **65**(6) 065601
[11] González J A, Bellorín A and Guerrero L E 2003 *Chaos Soliton. Fract.* 907–919 ISSN 0960-0779
[12] García-Ñustes M A and González J A 2012 *Phys. Rev. E* **86**(6) 066602
[13] García-Ñustes M A, Marín J F and González J A 2017 *Phys. Rev. E* **95**(3) 032222
[14] Barone A and Paternò G 1982 *Physics and Applications of the Josephson Effect* (New York: Wiley)
[15] Mikeska H J 1978 *J. Phys. C: Solid State Phys.* **11** L29–32
[16] van Saarloos W and Hohenberg P C 1990 *Phys. Rev. Lett.* **64**(7) 749–752
[17] Peyrard M and Dauxois T 2004 *Physique des Solitons* (Paris: Savoirs Actuels)
[18] González J A and Holyst J A 1992 *Phys. Rev. B* **45**(18) 10338–10343
[19] González J, Bellorín A and Guerrero L 2007 *Chaos Soliton. Fract.* **33** 143–155 ISSN 0960-0779
[20] Guckenheimer J and Holmes P 1986 *Nonlinear Oscillations, Dynamical Systems, and Bifurcations of Vector Fields* (New York: Springer-Verlag)
[21] McLaughlin D W and Scott A C 1978 *Phys. Rev. A* **18**(4) 1652–1680
[22] Kivshar Y S and Malomed B A 1989 *Rev. Mod. Phys.* **61**(4) 763–915
[23] Sánchez A and Bishop A R 1998 *SIAM Rev.* **40** 579–615
[24] Chacón R, Bellorín A, Guerrero L E and González J A 2008 *Phys. Rev. E* **77**(4) 046212
[25] González J A, Cuenda S and Sánchez A 2007 *Phys. Rev. E* **75**(3) 036611
[26] González J A and Mello B A 1996 *Phys. Scr.* **54** 14
[27] González J A, García-Ñustes M A, Sánchez A and McClintock P V E 2008 *New J. Phys.* **10** 113015
[28] Malomed B A and Ustinov A V 2004 *Phys. Rev. B* **69**(6) 064502
[29] Ustinov A V 2002 *Appl. Phys. Lett.* **80** 3153–3155
[30] González J A and Oliveira F A 1999 *Phys. Rev. B* **59**(9) 6100–6105
[31] González J, Marcano A, Mello B and Trujillo L 2006 *Chaos Soliton. Fract.* **28** 804–821 ISSN 0960-0779

A new method to obtain analytic approximations applied to the $J_1(x)$ function

Fernando Maass [1] and Pablo Martin [1]

[1] Depto. de Fisca, Universidad de Antofagasta, Av. Angamos 601, Antofagasta, Chile.

E-mail: fernando.maass@uantof.cl; pablo.martin@uantof.cl

Abstract. In this work an analytic approximation for $J_1(x)$ has been found, which is simple and precise, and good for most of the applications of this functions in Physics. The technique here used remarks a little the Pade method, since rational functions are used, but now this type of function are combined in a efficient way with elementary functions, furthermore series power and asymptotic expansions are used simultaneously, as in the Multipoint Quasi-rational Approximation, MPQA method. However here important improvements have been introduced. Though the form of the approximate is built considering the above two expansions, however the parameters of the approximations are determinates by two methods, one similar to the minimum square error method and the other using the coefficients of two expansions, power and asymptotic. Better results are obtained with the first procedure.

1. Introduction

In this work a very efficient and precise technique is used to obtain analytic approximated functions defined by its differential equation or by other mathematical way, wherein its analytic solution is not known. In the present technique power series and asymptotic expansions are used as well as some statistic way of error minimization, someway similar to the minimun square error technique. The approximants here found are a combination of elementary functions with rational functions. In this way there is some semblance with the Pade method, but the actual method is rather an improvement of the Multipoint Quasirational Approximation, MPQA, used in several previous works as quadratic Zeeman effect, Morse potentials with centrifugal terms, an-harmonic potentials, Plasma Dispersion functions, partition functions in Quantum Field Theory and others [1-3]. In the present paper, this technique has been applied to the Bessel functions of first order, $J_1(x)$, in this way we show how to use it. Two technique have been used here, and the most simple approximant is function of 4 parameters. In the first procedure, those parameters are determined directly from the coefficients of the power series and asymptotic expansion, and the maximum absolute error was 0.004 for every positive value of the variable x. In the second procedure both power series and asymptotic expansions are used to find the form of the approximations, but later the 4 parameters are determine using a small number of coefficients of the two expansions than in the first case, as well as some statistic method. Using the second method the maximum error decreases for the same number of parameters.

2. Theoretical Treatment

The power series of $J_1(x)$ is well known and its convergence radius is infinite, thus

$$J_1(x) = \frac{x}{2}\sum_{j=0}^{\infty}\frac{(-1)^j}{j!(j+1)!}\Big(\frac{x}{2}\Big)^{2j} = \frac{x}{2}\Big(1 - \frac{x^2}{8} + \frac{x^4}{192} - \cdots\Big) \tag{1}$$

The asymptotic expansion is also well known

$$J_1(x) \sim \sqrt{\frac{2}{\pi x}}\Big[\Big(1+\frac{35}{128}\frac{1}{x^2}+\cdots\Big)\cos\Big(x-\frac{\pi}{2}-\frac{\pi}{4}\Big)+\Big(\frac{3}{8x}-\frac{3521}{683}\frac{1}{x^3}+\cdots\Big)\sin\Big(x-\frac{\pi}{2}-\frac{\pi}{4}\Big)\Big] \tag{2}$$

This expansion is useful to obtain $J_1(x)$ for large values of x, because if the power series is used, the number of necessary terms to obtain some accuracy is very large, sometimes more than a hundred terms. The procedure to follow is to build a bridge between both expansions, power and asymptotic, by using rational functions combined with elementary functions as fractional powers and trigonometric functions. The word efficiency is taken here in the sense that the power series of the approximate function $\tilde{J}_1(x)$ has only odd powers as the actual function $J_1(x)$, in this way no additional parametres have to be used to cancel the even powers, as it was the case of previous works [6-7]. With this criterion the general forms of the approximate function has been chosen as

$$\tilde{J}_1(x) = \frac{1}{(1+\lambda^2x^2)^{\frac{1}{4}}}\Big[\frac{\sum_{i=0}^{n} p_i x^{2i}}{1+\sum_{j=1}^{n} q_j x^{2j}}\sin(x) + \frac{x}{(1+\lambda^2x^2)^{\frac{1}{2}}}\frac{\sum_{i=0}^{n}\tilde{p}_i x^{2i}}{1+\sum_{j=1}^{n} q_j x^{2j}}\cos(x)\Big] \tag{3}$$

As it can be seen in this equation, there are two rational functions instead of one. Furthermore these two rational functions have the same denominator, since it is desired to determine the parameters of the approximant through linear algebraic equations, avoiding non-linear algebraic equations. The initial fractional function is $(1+\lambda^2x^2)^{-1/4}$, instead of $x^{-1/2}$, because $J_1(x)$ has this ramification point at infinite, but not at zero. As it is known the ramification point of fractional powers come by pairs, and therefore the second ramification point has to be out of the region of interest, which now is the positive real axis. A second requirement is that this auxiliary function must have only even powers in its power series, because of the previous analysis about the power series of $\tilde{J}_1(x)$. The auxiliary trigonometric functions in $\tilde{J}_1(x)$ are just the right ones because of the leading term of the asymptotic expansion of $J_1(x)$, and also in this way the parameters p_n and $\tilde{p}_n$ are related to q_n by

$$p_n = \frac{\lambda^{1/2}}{\sqrt{\pi}}q_n \quad ; \quad \tilde{p}_n = -\frac{\lambda^{3/2}}{\sqrt{\pi}}q_n \tag{4}$$

The simplest approximant to determine is for n=1

$$\tilde{J}_1(x) = \frac{1}{(1+\lambda^2x^2)^{\frac{1}{4}}}\Big[\frac{p_0+p_1x^2}{1+q_1x^2}\sin(x) + \frac{x}{(1+\lambda^2x^2)^{\frac{1}{2}}}\frac{\tilde{p}_0+\tilde{p}_1x^2}{1+q_1x^2}\cos(x)\Big] \tag{5}$$

leaving aside the parameter λ, the parameters to be determined are reduced to tree, p_0, $\tilde{p}_0$ and $\tilde{q}_1$; since the parameters p_1 and p_1 are determined by Eq.(4), once q_1 is obtained. In this way, equalising three terms of the power series of $J_1(x)$, and $\tilde{J}_1(x)$, all the parameters will be determined as a function of λ. Previous to equalize terms from both power series, it is convenient to rationalize, multiplying for the denominators of $\tilde{J}_1(x)$, and in this way avoiding non-linear algebraic equations. To determine λ, one important condition is that q_1 should be positive to avoid undesirable poles. A second condition is to look for the minimum absolute error of the approximation. In this calculation, the relative error can not be considered, because of the zeros of $J_1(x)$, the relative error is always infinite. Following this procedure, the parameters of the function $\tilde{J}_1(x)$ are determined, this function is given by

$$\tilde{J}_1(x) = \frac{\sqrt{1+0.12138x^2}(46.68634+5.82514x^2)\sin(x)}{(57.70003+17.49211x^2)(1+0.12138x^2)^{\frac{3}{4}}} - \frac{x(17.83632+2.02948x^2)\cos(x)}{(57.70003+17.49211x^2)(1+0.12138x^2)^{\frac{3}{4}}} \tag{6}$$

The absolute error, $|\tilde{J}_1(x) - J_1(x)|$, of the function $J_1(x)$ is show in Figure 1.

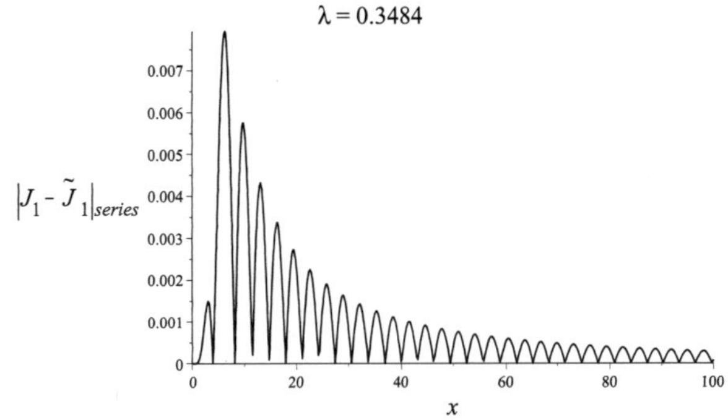

Figure 1. Absolute error of the approximation as a function of variable x, with the parameters determined with the best value of $\lambda = 0.3484$ for x=6.3

The maximum absolute error here found, is abouth 0.008, very small considering the small number of parameters.
The second procedure here followed is that once the form of the approximate function $\tilde{J}_1(x)$ has been defined as it was done in the first procedure, then the value of the parameters are determined as follows. First, a certain number of values for x are selected, for instance, a thousand values between cero and 100, with the same difference between consecutive values. Then the parameters p's, q's and λ are determined with the requirement of a minimum standard deviation. In this way the new approximant, $\hat{J}_1(x)$, for $J_1(x)$ is obtained and given by

$$\hat{J}_1(x) = \frac{(0.1601\,x^2+0.8660)\sin(x)}{(1+0.3489\,x^2)\sqrt[4]{1+0.4181\,x^2}} - \frac{x\,(0.1007\,x^2+0.3718)\cos(x)}{(1+0.4181\,x^2)^{3/4}\,(1+0.3489\,x^2)} \tag{7}$$

In Figure 2, the absolute error, $|\hat{J}_1(x) - J_1(x)|$, is shown as a function of the variable x.

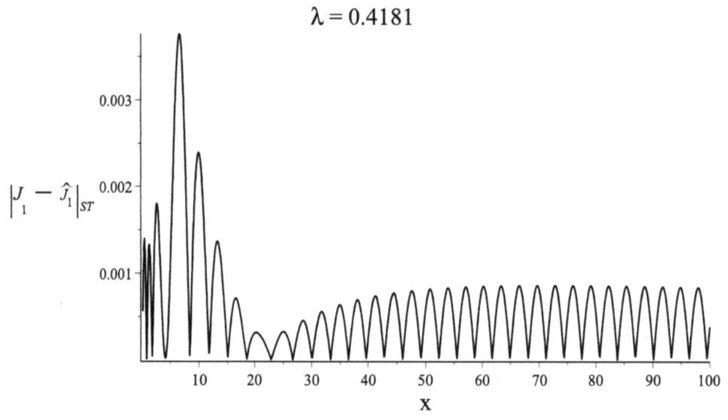

Figure 2. Absolute error of the approximation as a function of the variable x, with the parameters determined with the statistical method and the form of approximant with the MPQA method.

As it can be seen in this Figure, the maximun absolute error decreases comparing with previous maximum error. Now the maximum absolute error is 0.004, abauth a half of previus one 0.008. The point where this maximun occurs is x=6.6, different to previous one of x=6.3. The zeros $x_1, x_2, ..., x_n$ of both approximations $\tilde{J}_1(x)$ and $\hat{J}_1(x)$ are also almost coincident with those of $J_1(x)$. In this case, the relative error is more appropiated to valuate the approximation and the maximum relative errors is 16×10^{-4} and 30.2×10^{-4} for $\tilde{J}_1(x)$ and $\hat{J}_1(x)$ in the zeros x_1 and x_2.

3. Conclusion

Two new methods have been described to obtain analytic approximations of differential equations, where no analytic solution is known. The method is efficient and it allows to obtain precise approximation to the required solution. The methos have been applied to the Bessel function of order one, $J_1(x)$, obtaining approximate functions $\tilde{J}_1(x)$ and $\hat{J}_1(x)$ valids for every positive value of the variable x, with maximum absolute error de 0.008 and 0.004 receptively, depending og the alternative procedure here used. Both methods required to use simultaneously the power series and asymptotic expansion of the solution. However the first one uses also the coefficients of both series, and the second one, the parameters are determined by minimizing the the standard deviation of a selected number of points. The form of the approximate function in both methos is a combination of rational functions with certain type of elementary functions determine by the structure of the power series and asymptotic expansion. The approximations here obtained are very simple, not with standing that the accuracy of these is high and enough for most of the applications of this function in Physics. The relative errors of the approximate function are also with very small relative error, and the maximum of these errors is 0.008 for x = 6.3 in $\tilde{J}_1(x)$ and 0.004 for x = 6.6 in $\hat{J}_1(x)$.

References

[1] P. Martin, E. Castro, J. L. Paz and A. DeFreitas "Multipoint quasi-rational Approximations in Quantum Chemistry", Chapt. 3, in "Developments in Quantum Chemistry", Ed. J. L. Paz and A. J. Hernandez (Transworld Research Network, Kerala, India, 2009) pp 57-77.
[2] P. Martin, E. Castro and J. L. Paz, Rev. Mexicana Fisica, 58, 301-307 (2012).
[3] P. Martin, Rodriguez-Nuez and J. L. Marquez, Phys. Rev. B. 26, 8359 (1992).
[4] J. D. Jackson, "Classical Electrodynamics", 2nd. edition (John Wiley & Sons, New York, USA, 1975) pp.102-106.
[5] M. Abramowitz and I. A. Stegun, "Handbook of Mathematical Functions", Ninth Printing (Dover Publications Inc. New York, 1970) pp.373-379.

[6] P. Martin and L. Guerrero, J. Math. Phys. **26**, 705 (1985).
[7] P. Martin and L. Guerrero, J. Comp. Phys. **85**, 487 (1989).

IOP Conf. Series: Journal of Physics: Conf. Series **1043** (2018) 012003 doi:10.1088/1742-6596/1043/1/012003

A simple approximation for the modified Bessel function of zero order $I_0(x)$

J. Olivares[1], **P. Martin**[2] **and E. Valero**[3]
[1] *Mathematics department, Universidad de Antofagasta, Chile*
[2] *Physic department, Universidad de Antofagasta, Chile*
[3] *Mathematics department, Universidad de Tarapacá, Chile*
e-mail: pablo.martin@uantof.cl; jorge.olivares@uantof.cl; evalero@uta.cl

Abstract. New efficient analytic approximations have been found for the modified zero-order Bessel functions $I_0(x)$. The method, used herein, improves prior techniques, such as multipoint quasi-rational approximations, MPQA. The present work is also an improvement of previous works, in the sense that the form of the approach is more efficient, that is, a smaller relative error is obtained using a lower number of parameters. As in the Pade method rational functions are used with coefficients determined from the powers series, but now also asymptotic expansions are also used simultaneously with that series, and consequently the rational functions have to be combined with elementary functions, in such a way, that the approach is a bridge between both expansions. The approximation now found is a rational function combined with a hyperbolic function. Only three parameters are needed to obtain a relative error smaller than 0.6 percent.

1. Introduction

The Bessel functions are important in Physics and in particular the modified Besel function $I_0(x)$ [1,3]. Its power series and asymptotic expansion are known, the first is good for small values of the variable x, and the second for large values. Although the power series is entire, this is, with an infinite radius of convergence, however, many terms need to be used for large and intermediate values of x. It is the object of the present work to find an analytic approximation of this function for every value of x, which can be derived and integrated, as it is required in Thermodynamics, Optics and other areas of Physics. This will be simple and wich a small relative error, so that, it can be used in almost all cases where this function is of interest in Physics. For this purpose, the Multipoint Quasirational Aproximation, MPQA, has been used and improved. This technique has been successfully used in several physics problems such as Plasma dispersion function, Zeemman Quadratic effect in $2D$, and others [4]. The method uses rational functions, as in Pade, but now combined with elementary functions. The reason for introducing these types of functions is that in addition to use potential series as in Pade, asymptotic series are also used simultaneously with the power series. Asymptotic series are usually expressed in terms of sometimes fractional negative powers combined with non-potential elementary functions. What is desired with the quasirational approximations is to have a function like a bridge, between the power series and the asymptotic expansion.

The usual fact that asymptotic series are not valid in the whole circle of infinity, but only in one sector, does not present problems, because although the approximation contains all the singularities of the exact function, this is only true in the sector of the complex plane wherein the asymptotic expansion is valid. It is not desired that the approximation and the exact function have the same singularities in the whole complex plane, because in that case, as demonstrated in complex variable theory, both functions would differ only in one constant.

2. Theoretical development

The potential series of $I_0(x)$ is

$$I_0(x) = \sum \frac{(\frac{x^2}{4})^k}{k\,\Gamma(k+1)} \tag{1}$$

and the asymptotic series is

$$I_0(x) = \frac{e^x}{\sqrt{2\pi x}}\left(1 + \frac{1}{8x} + \frac{9}{128x^2} + \cdots\right) \tag{2}$$

The approximant used here will be

$$\widetilde{I}_0(x) = (1+\lambda^2x^2)^{-1/4}\cosh x\,\frac{p_0+p_1x^2}{1+q_1x^2} \tag{3}$$

The form of this approximant is selected by the following reasons:

(i) It is necessary that when x tends to infinity the factor that accompanies the rational function has the factor $x^{-1/2}$. Choosing this factor as an auxiliary function does not work, because since fractional powers have two branching points, the second point would be a branch point at $x = 0$, which the function $I_0(x)$ does not have. Thus, the second branching point must be placed outside the region of interest, which in this case has been chosen to be such that the approximant is good for every positive value of x. One way to do this would be to select the auxiliary function as $(1+x)^{1/2}$, so that the second branch point would be in $x = -1$, which clearly is outside the region of interest. While this selection is good and has been done well in other problems, however is not the best, as in this paper is shown. The reason for this is that we also have to observe the potential series that has only paired powers, so that, if we want to have more efficiency we should select the auxiliary function in order to obtain only even powers, such as the function $(1+x^2)^{1/4}$ which in infinity behaves as $x^{-1/2}$, and at the same time by expanding it around x we would have only even functions. In this way we make more efficient the approximate in its totality. Using an additional λ parameter is also a way of introducing a degree of freedom that allows us to obtain more accurate approximations as shown below.

(ii) The asymptotic expansion has the factor e^x, which must also be included as an auxiliary function; but this function is not even, and since we want the function to be even for the reasons given in the previous paragraph, we choose $\cosh x$, instead of e^x, as an auxiliary function, so that the behavior at infinite is correct and at the same time, its power series will have are even powers as desired.

(iii) Since the auxiliary functions are even and $I_0(x)$ is also even, the rational function of x, will have only even powers.

3. Results and Discussion

The determination of the parameters p_0, p_1 and q_1, are now made using the coefficients of the power series and the asymptotic expansion. From the coefficient in the leading term of the asymptotic expansion we obtain that

IOP Conf. Series: Journal of Physics: Conf. Series **1043** (2018) 012003 doi:10.1088/1742-6596/1043/1/012003

$$\frac{1}{\sqrt{2\pi}} = \frac{p_1}{2\lambda^{1/2} q_1} \tag{4}$$

Since we should to have equal numbers of equations than of parameters, then we need only two equations to determine all the parameters, that is, we only need two coefficients of the power series, so the series for $I_0(x)$ will be written as

$$I_0(x) = 1 + \frac{x^2}{4} + O(x^4). \tag{5}$$

Rationalizing equation (3), it is obtained

$$\begin{aligned} &(1+\lambda^2 x^2)^{1/4}(1+q_1 x^2)(1+\frac{1}{4}x^2) = \\ &\cosh x (p_0 + \sqrt{\frac{2}{\pi}}\lambda^{1/2} q_1 x^2) + O(x^4) \end{aligned} \tag{6}$$

Replacing $\cosh x$ by its series expansion $(1+\frac{x^2}{2})$, using only two terms, and matching the coefficients in x^0 and x^2 in both sides of the equation two equations are obtained to find p_0 and q_1 as a function of λ. In this way, the parameters p_0 and q_1 are

$$p_0 = 1\ ;\ q_1 = \frac{\frac{1}{4} - \frac{1}{4}\lambda^2}{1 - \sqrt{\frac{2}{\pi}}\lambda^{1/2}} \tag{7}$$

So the approximant is

$$\widetilde{I}_0(x) = \frac{\cosh x}{(1+\lambda^2 x^2)^{\frac{1}{4}}} \frac{1 + \sqrt{\frac{2}{\pi}} \left[\frac{\frac{1-\lambda^2}{4}}{1-\sqrt{\frac{2}{\pi}}\lambda^{\frac{1}{2}}}\right] \lambda^{\frac{1}{2}} x^2}{1 + \left[\frac{\frac{1-\lambda^2}{4}}{1-\sqrt{\frac{2}{\pi}}\lambda^{\frac{1}{2}}}\right]} \tag{8}$$

The value of λ must be defined using the condition that there are no poles of $\widetilde{I}_0(x)$ in the region of interest, so that q_1 is required to be positive. A graph of q_1 as a function of λ is shown in Figure 1 adn 2.

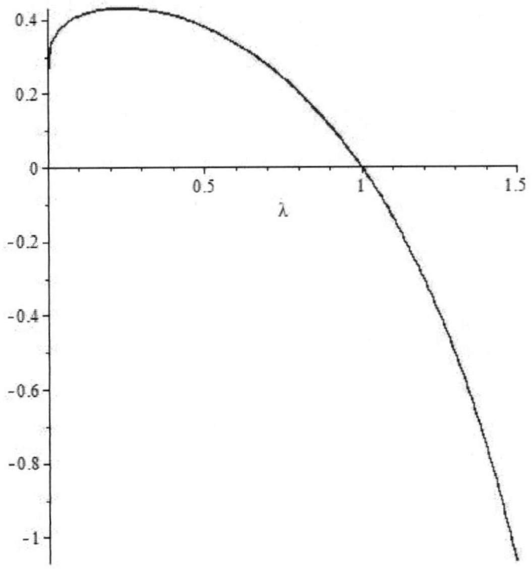

Figure 1. The q_1 value is shown as function of λ

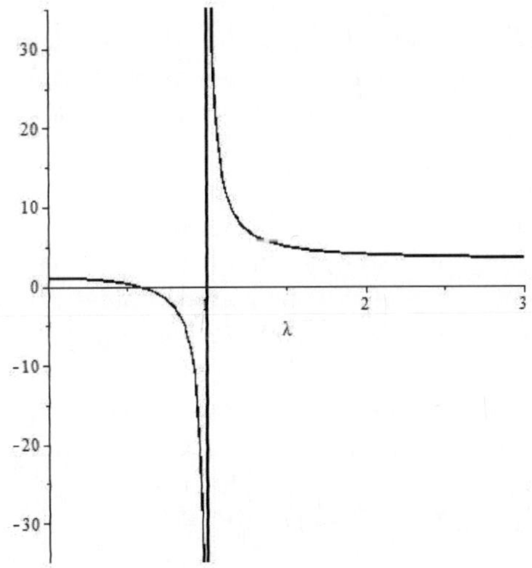

Figure 2. The q_1 values as function of λ is shown in the interval[0, 1]

From the first graph we obtain that λ must be $0 \leq \lambda \leq 1$ and $\frac{1}{2}\pi \leq \lambda \leq \infty$.

Of all these values the minimum relative error is obtained for $\lambda = \frac{1}{2}$. In this way the approximant is

$$\widetilde{I}_0(x) = \frac{\cosh x}{(1+\frac{1}{2}^2 x^2)^{\frac{1}{4}}} \frac{1+\sqrt{\frac{2}{\pi}}\left[\frac{\frac{1-\frac{1}{2}^2}{4}}{1-\sqrt{\frac{2}{\pi}}\frac{1}{2}^{\frac{1}{2}}}\right]\frac{1}{2}^{\frac{1}{2}}x^2}{1+\left[\frac{1-\frac{1}{2}^2}{1-\sqrt{\frac{2}{\pi}}\frac{1}{2}^{\frac{1}{2}}}\right]} \qquad (9)$$

A graph of the relative error is shown in Figure 3. The maximum relative error is 0.00564, but if the region near this value is discarded, the relative error of the approximant is always less than 0.002. When calculating each term of the approach it is convenient to use only the digits necessary, so that, the relative error does not increase. With this criterion the final approximation is obtained as

$$\widetilde{I}_0(x) = \frac{\cosh x}{(1+\frac{1}{2}^2 x^2)^{\frac{1}{4}}} \frac{1+0.24273x^2}{1+0.43023x^2} \tag{10}$$

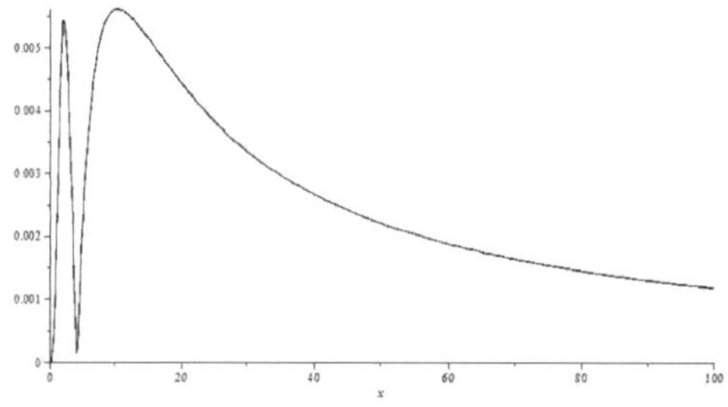

Figure 3. Relative error as function of the variable x

4. Conclusions

A quasi-rational approximation has been found for the modified Bessel $\widetilde{I}_0(x)$ function, very simple with only four parameters λ, p_0, p_1, q_1, but with a great precision lower than 0.005 for almost all x value, except for those close to the value 11, where the relative error can be 0.002, also quite small value. To find the approximation, rational functions combined with the hyperbolic cosine function and a fractional power have been used. In the present technique, the power series and the asymptotic expansion of $I_0(x)$ are used simultaneously, so that the Is a bridge between the two expansions. The validity of the approximant is for positive values of the variable x and in a broader sense for the right half-plane of the complex plane. In the present work the MPQA approximation technique has been improved, in the sense of obtaining approximations with a parity required in order to make more efficient the obtaining of the parameters of the approver, as well as to use less number of equations and of parameters for get good approximations with a small relative error. The approximate here obtained analytical and valid for all positive value of x, which can be derived in integrated with confidence, and we think that the function $I_0(x)$ in the different applications in different areas of Physics.

References

[1] G. N. Watson, " A treatise on the theory of Bessel functions", 2nd. ed., Cambrigde, England: Cambrigde University Press, (1966).
[2] J. D. Jackson, "Clasical Electrodynamics", 2nd. ed., John Wiley and Sons, Inc., Ch.3 (1975).
[3] M. Abramovitz and I. A. Stegun, "Handbook of mathematical Functions" (Dover Books New York, 1972) pp. 692-697.
[4] P. Martin, E. Castro, J. L. Paz and A. De Freitas, "Multipoint quasi-rational approximants in Quantum Chemistry", Chapter 3 of "New Developments in Quantum Chemistry by Jose Luis Paz and Antonio J. Hernan" (Transworld Research Network, Kerala, India, 2009) pp. 55-78.
[5] P. Martin, J. Olivares, L. Cortés-Vega and A. Sotomayor,"Multi-point quasi-rational approximants for the midified Bessel function $I_1(x)$", J. Phys Conf. **738**: 012066 (2016).
[6] D. Kh. Morozov, V.V. Voilsekhovich. A new wide-rauge approximation of modified Besscl functions iu terms of elementary functions. Revista Mexicana de Física 44(**3**) 231-234 (1998).

IOP Conf. Series: Journal of Physics: Conf. Series **1043** (2018) 012004 doi:10.1088/1742-6596/1043/1/012004

On the integrability of the octonionic Korteweg-de Vries equation

Alvaro Restuccia[1], **Adrián Sotomayor**[2], **Jean Pierre Veiro**[3]

[1]Physics Department, University of Antofagasta, Chile
[2] Department of Mathematics, University of Antofagasta, Chile
[3] Department of Mathematics, Universidad Simón Bolívar, Caracas, Venezuela

E-mail: arestu@usb.ve, adrian.sotomayor@uantof.cl, jpveiro@usb.ve

Abstract. We introduce the octonionic Korteweg-de Vries (KdV) equation, starting with a Lagrangian formulation and presenting some of its symmetries. We introduce also the associated octonionic Gardner equation and deduce from it the infinite sequence of conserved quantities for octonionic KdV itself. We finally give a master Lagrangian from which we can get both the octonionic KdV and also the corresponding octonionic Miura KdV.

1. Introduction

Since the work of Miura and Gardner et al [1, 2, 3, 4, 5], carried out in the context of the Korteweg-de Vries (KdV) equation, a lot of interesting results about the so-called integrable systems have been obtained.

A class of extensions of KdV equation arises by introducing supersymmetry. Several integrable supersymmetric extensions were given in [7, 8, 9, 10, 11, 12].

In this work we consider the octonionic Korteweg-de Vries equation, that is, the defining field has values in the octonion algebra and the product being also the product in the same algebra.

A famous theorem by Hurwitz establishes that the only real normalized division algebras are the reals $\mathbb{R}$, the complex $\mathbb{C}$, the quaternions $\mathbb{H}$ and the octonions $\mathbb{O}$. In particular, these division algebras are directly related to the existence of Super Yang-Mills in several dimensions: three, four, six and ten dimensions [13]. The octonion algebra may be explicitly used in the formulation of superstring theory in ten dimensions and in the Supermembrane theory in eleven dimensions, relevant theories in the search for a unified theory of all the known fundamental forces in nature. The extension of the KdV equation to a partial differential equation for a field valued on a octonion algebra is then an interesting goal [14].

In [15] was showed that an extension of the KdV equation valued on a Clifford algebra gives rise to a coupled system with Liapunov stable soliton solutions but without an infinite sequence of local conserved quantities. In distinction, the system we will consider has an infinite sequence of conserved quantities.

The octonion algebra contains as subalgebras all other division algebras, hence our construction may be reduced to any of them.

The KdV equation on the octonion algebra can be seen as a coupled KdV system, as we will see it has some similarities to the construction in the previous sections. However, it is invariant under the exceptional Lie group G_2, the automorphisms of the octonions, and under the Galileo transformations. Those symmetries characterize the octonionic system. In particular, they are not present on the Clifford algebra valued system in [15].

In section 2 we analyze the KdV extension where the field is valued on the octonion algebra. The system shares several properties of the original real KdV equation. It has soliton solutions and also has an infinite sequence of local conserved quantities derived from a Bäcklund transformation and a bi-Lagrangian and bi-Hamiltonian structure [14]. We also will show in section 2 a Lagrangian formulation for the octonion KdV equation.

In sections 3 and 4 we give the Gardner formulation and present the first conserved quantities (corresponding to the infinite sequence of conserved quantities) for the octonionic KdV, presenting also the master Lagrangian from which we can deduce both octonionic KdV and Miura equations.

In section 5 we give the conclusions.

2. Lagrangian formulation and symmetries for the octonionic Korteweg-de Vries equation

We denote $u = u(x,t)$ a function with domain in $\mathbb{R} \times \mathbb{R}$ valued on the octonion algebra. If we denote $e_i, i = 1, \ldots, 7$ the imaginary basis of the octonions, u can be expressed as

$$u(x,t) = b(x,t) + \overrightarrow{B}(x,t) \tag{1}$$

where $b(x,t)$ is the real part and $\overrightarrow{B} = \sum_{i=1}^{7} B_i(x,t)e_i$ its imaginary part.

The KdV equation formulated on the algebra of octonions , or simply the octonion KdV equation, is given by

$$u_t + u_{xxx} + \frac{1}{2}(u^2)_x = 0, \tag{2}$$

when $\overrightarrow{B} = \overrightarrow{0}$ it reduces to the scalar KdV equation. In terms of b and $\overrightarrow{B}$ the equation can be re-expressed as

$$b_t + b_{xxx} + bb_x - \sum_{i=1}^{7} B_i B_{ix} = 0, \tag{3}$$

$$(B_i)_t + (B_i)_{xxx} + (bB_i)_x = 0. \tag{4}$$

A Lagrangian for (2) is given by

$$L(w) = \int_{t_i}^{t_f} dt \int_{-\infty}^{+\infty} dx \mathbb{R}\mathrm{e} \left[-\frac{1}{2} w_x w_t - \frac{1}{6}(w_x)^3 + \frac{1}{2}(w_{xx})^2 \right],$$

where $u = w_x$. Independent variations with respect to w yields the octonionic KdV equation (2).

Equation (2) is invariant under the Galileo transformation given by

$$\widetilde{x} = x + ct,$$
$$\widetilde{t} = t,$$
$$\widetilde{u} = u + c$$

where c is a real constant.

Additionally, equation (2) is invariant under the automorphisms of the octonions, that is, under the group G_2. If under an automorphism

$$u \to \phi(u)$$

then

$$u_1u_2 \to \phi(u_1u_2) = \phi(u_1)\phi(u_2)$$

and consequently

$$[\phi(u)]_t + [\phi(u)]_{xxx} + \frac{1}{2}\left([\phi(u)]^2\right)_x = 0.$$

3. The Gardner formulation for the octonion valued algebra KdV equation

Associated to the real KdV equation there is a Gardner ε-transformation and a Gardner equation which allows to obtain in a direct way the corresponding infinite sequence of conserved quantities. There exists a generalization of this approach for the KdV valued on the octonion algebra. The generalized Gardner transformation, expressed in terms of a new field $r(x,t)$ valued on the octonion is given by

$$u = r + \varepsilon r_x - \frac{1}{6}\varepsilon^2 r^2. \tag{5}$$

The generalized Gardner equation is then

$$r_t + r_{xxx} + \frac{1}{2}(rr_x + r_x r) - \frac{1}{12}\left((r^2)r_x + r_x(r^2)\right)\varepsilon^2 = 0 \tag{6}$$

where ε is a real parameter.

If $r(x,t)$ is a solution of the generalized Gardner equation (6), then $u(x,t)$ is a solution of the octonion algebra valued KdV equation (2).

It has been shown in [14] that $\int_{-\infty}^{+\infty} \mathbb{R}e\,[r(x,t)]\,dx$ is a conserved quantity of (6). We note that this fact is a non-trivial one because of the non-associativity (and non-conmutativity) of the algebra.

We can then invert (5), assuming a formal ε-expansion of the solution $r(x,t)$, to obtain an infinite sequence of conserved quantities for the KdV equation valued on the octonion algebra.

The first few of them are

$$H_1 = \int_{-\infty}^{+\infty} \mathbb{R}e(u)dx,$$
$$H_2 = \int_{-\infty}^{+\infty} \left((\mathbb{R}e(u))^2 - \|\mathbb{I}m(u)\|^2\right)dx,$$
$$H_3 = \int_{-\infty}^{+\infty} \left(\frac{1}{3}(\mathbb{R}e(u))^3 - (\mathbb{R}e(u_x))^2 + \|\mathbb{I}m(u_x)\|^2 - \mathbb{R}e(u)\|\mathbb{I}m(u)\|^2\right)dx.$$

4. The master Lagrangian for the KdV equation valued on the octonion algebra

We may now use the Helmholtz procedure to obtain a Lagrangian density for the generalized Gardner equation. The master Lagrangian formulated in terms of the casimir potential $s(x,t)$,

$$r(x,t) = s_x(x,t),$$

is

$$L_\varepsilon(s) = \int_{t_i}^{t_f} dl \int_{-\infty}^{+\infty} \mathcal{L}_\varepsilon(s)dx$$

where the Lagrangian density is given by

$$\mathcal{L}_\varepsilon(s) = \mathbb{R}e\left[-\frac{1}{2}s_x s_t - \frac{1}{6}(s_x)^3 + \frac{1}{2}(s_{xx})^2 + \frac{1}{72}\varepsilon^2(s_x)^4\right].$$

The Lagrangian density $\mathcal{L}_\varepsilon(s)$ is invariant under the action of the exceptional Lie group G_2.

Independent variations with respect to s yields

$$\delta\mathcal{L}_\varepsilon(s) = \mathbb{R}e\left[-\frac{1}{2}(\delta s)_x s_t - \frac{1}{2}s_x(\delta s)_t - \frac{1}{6}\left((\delta s)_x(s_x)^2 + s_x(\delta s)_x s_x + (s_x)^2(\delta s)_x\right)\right] +$$
$$+\mathbb{R}e\left[\frac{1}{2}\left((\delta s)_{xx}s_{xx} + (\delta s)_{xx}s_{xx}\right) + \frac{1}{72}\varepsilon^2\left((\delta s)_x(s_x)^3 + s_x(\delta s)_x(s_x)^2 + (s_x)^2(\delta s)_x s_x + (s_x)^3(\delta s)_x\right)\right].$$

Using properties of the octonion algebra we obtain from the stationary requirement $\delta L_\varepsilon(s) = 0$ the generalized Gardner equation (6).

In the calculation the property to be a division algebra of the octonions is explicitly used.

If we take the limit $\epsilon \to 0$, we obtain a first Lagrangian for the KdV equation valued on the octonion algebra,

$$L(w) = \int_{t_i}^{t_f} dt \int_{-\infty}^{+\infty} dx \mathbb{R}e\left[-\frac{1}{2}w_x w_t - \frac{1}{6}(w_x)^3 + \frac{1}{2}(w_{xx})^2\right].$$

If we consider the following redefinition

$$s \to \hat{s} = \epsilon s$$
$$\mathcal{L}_\epsilon(s) \to \epsilon^2\mathcal{L}_\epsilon(\hat{s})$$

and take the limit $\epsilon \to \infty$ we obtain

$$\lim_{\epsilon\to\infty} \epsilon^2\mathcal{L}_\epsilon(\hat{s}) = \mathcal{L}^M(\hat{s}),$$

where

$$\mathcal{L}^M(\hat{s}) = \mathbb{R}e\left[-\frac{1}{2}\hat{s}_x\hat{s}_t + \frac{1}{2}(\hat{s}_{xx})^2 + \frac{1}{72}(\hat{s}_x)^4\right].$$

We get in this limit the generalized Miura Lagrangian

$$L^M(\hat{s}) = \int_{t_i}^{t_f} dt \int_{-\infty}^{+\infty} dx \mathcal{L}^M(\hat{s}).$$

The Miura equation is then obtained by taking variations with respect to $\hat{s}$, we get

$$\hat{r}_t + \hat{r}_{xxx} - \frac{1}{18}(\hat{r})^3_x = 0, \quad \hat{r} \equiv \hat{s}_x,$$

while the Miura transformation arises after the redefinition process, it is $u = \hat{r}_x - \frac{1}{6}\hat{r}^2$.

Any solution of the Miura equation, through the Miura transformation, yields a solution of the KdV equation valued on the octonion algebra. Since $L_\epsilon(s)$ is invariant under G_2, the same occurs for $L(w)$ and $L^M(\hat{s})$ and consequently for the equations arising from variations of them.

The Lagrangian formulation of the octonionic KdV equation may be used as the starting step to obtain the Hamiltonian structure of the octonion algebra valued KdV equation.

5. Conclusions

We introduced the octonionic Korteweg-de Vries equation, starting with a Lagrangian formulation and presenting some of its symmetries. We analyzed the integrable properties of this equation using a generalized Gardner equation. With this tool we obtained the infinite conserved quantities for the octonionic KdV equation. We obtained also a master Lagrangian from which we can get both the octonionic KdV and the octonionic Miura KdV. We expect, in view of the connection of the octonion algebra with string theories, to consider a supersymmetic formulation for the octonionic KdV equation. We will report on this extension elsewhere.

Acknowledgments

A. R. and A. S. are partially supported by Project Fondecyt 1161192, Chile.

IOP Conf. Series: Journal of Physics: Conf. Series **1043** (2018) 012004 doi:10.1088/1742-6596/1043/1/012004

References

[1] C. S. Gardner, J. M. Greene, M. D. Kruskal and R. M. Miura, Phys. Rev. Lett. 19, 1096 (1967).
[2] R. M. Miura, J. Math. Phys. 9, 1202 (1968).
[3] R. M. Miura, C. S. Gardner and M. D. Kruskal, J. Math. Phys. 9, 1204 (1968).
[4] C. Gardner, J. Math. Phys. 12, 1548 (1971).
[5] R. M. Miura, Siam Review 18, 412 (1976).
[6] B. A. Kupershmidt, Phys. Lett. A 102, 449 (1984).
[7] P. Mathieu, J. Math. Phys. 29, 2499 (1988).
[8] P. Labelle and P. Mathieu, J. Math. Phys. 32, 923 (1991).
[9] S. Bellucci, E. Ivanov and S. Krivonos, J. Math. Phys. 34, 3087 (1993).
[10] F. Delduc and E. Ivanov, Phys. Lett. B 309, 312-319 (1993).
[11] F. Delduc and E. Ivanov and S. Krivonos, J. Math. Phys. 37, 1356 (1996).
[12] Z. Popowicz, Phys. Lett. B 459, 150-158 (1999).
[13] J. M. Evans, Nucl. Phys. B 298, 92-108 (1988).
[14] A. Restuccia, A. Sotomayor and J. P. Veiro, arXiv: 1609.05410 [math-ph] (2016).
[15] A. Restuccia and A. Sotomayor, J. Math. Phys. 54, 113510 (2013).

Vortex and glass bead interaction in nematic liquid crystal layer

V. Zambra

Departamento de Física, Facultad de Ciencias, Universidad de Chile, Santiago, Chile

E-mail: `valeska.zambra@ug.uchile.cl`

Abstract. An homeotropically aligned liquid crystal layer driven by external electric or magnetic fields exhibit an intricate network of defects called *umbilics* or vortexes. Here, we report an experimental characterization of vortexes-glass bead interaction in nematic liquid crystal layer. The glass spheres, embedded in the liquid crystal, are found to be of two type: some of them attractive, with different strength of attraction; and most of them non-attractive. The attractive glass spheres pull the vortexes so that the distance decays approximately with the square root of time. The glass bead sphere induces an anisotropic force on the vortex. This force can be well approximate by an inverse power law of the distance between vortex and glass bead with exponent 3/4. The vortexes are attracted to a polar region of the attractive glass bead.

1. Introduction

Nematic liquid crystal layers with negative anisotropy and homeotropic anchoring under the influence of an external voltage can exhibit complex spatiotemporal dynamics textures [1]. The critical voltage from which one observes this intriguing dynamics is usually denominated as the Fréedericksz voltage [1, 2, 3]. Above the Fréederickz voltage the liquid crystal molecules change their orientation. This phenomenon is a result of the imbalance between the intrinsic elastic torque and electric torque resulting from the applied field. Indeed, the electric field forces the molecules to reorient in orthogonal directions. In opposition, the elastic coupling tends to restore the molecules in the electric field direction. Hence, the molecules can be oriented with different direction with the similar angle with respect to the electric field, generating an overall highly nonuniform texture.

To monitor this texture, one can consider crossed linear polarizers. Figure 1 shows the typical *Schlieren*-like texture, where the dark brushes account for directions where the optical axis, to be precise the averaged direction of the molecule in the plane of the cell, parallel or orthogonal to either the polarizer or the analyzer. The dark brushes meet at the defects, *umbilical defects* [1, 2]. Around this defect, when one circles clockwise, the molecules rotate clockwise or anti-clockwise. From the mathematical point of view, these defects correspond to topological singularities of the averaged 2-D projection of the molecules onto the plane of the cell, which usually are named vortexes [4, 5]. Defects with molecules rotating in clockwise (anti-clockwise) sense are termed positive (negative) charge vortexes. Figure 1 depicts the umbilical defects with different topological charge. Indeed, the the spatiotemporal dynamics of the texture is governed by the dynamics of the vortexes. In the last decades the vortex dynamics have been study theoretical and experimentally in details [4, 5, 6, 7]. The vortex dynamics are characterized by attraction (repulsion) between opposite (equal) charged vortexes. When the vortexes with opposite charge collide, they are annihilate and disappear [4, 6, 8]. An unified theory of these defects is based on the Ginzburg-Landau equation with real coefficients [4, 5, 9]. Based on this description, one can show that the pair interaction law of vortex is governed by a force, inversely proportional to the distance between the vortexes at a dominant order. A thorough study of the vortex interaction is performed in Ref. [8].

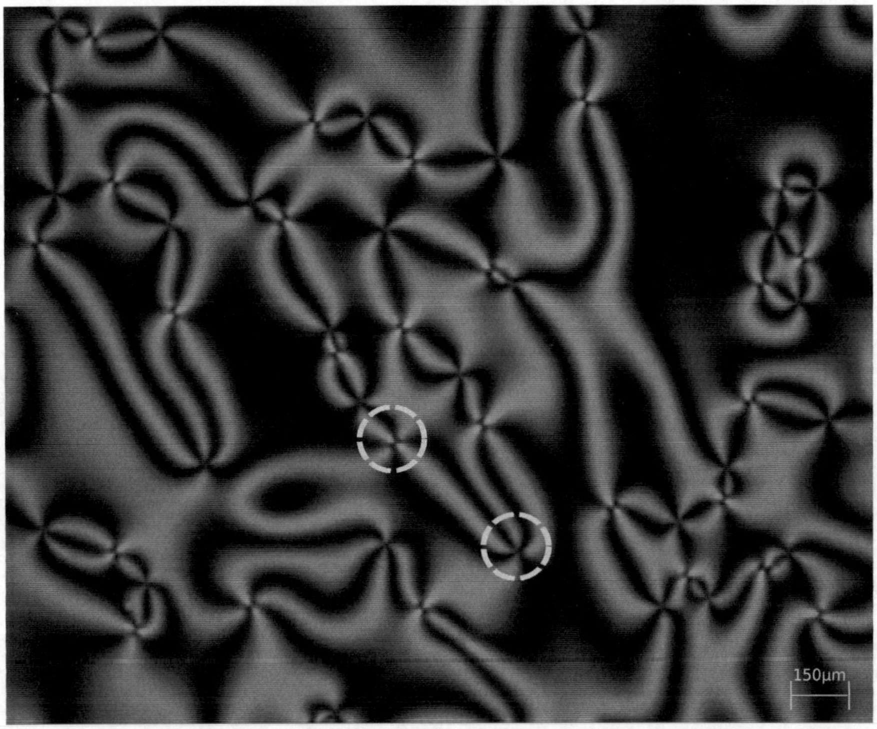

Figure 1. Gas of defects obtained from a biased uniform homeotropically aligned nematic liquid crystal layer. The dashed circumferences account for opposite topological charges. The optical micrograph is obtained using an Olympus BX51 microscope with crossed linear polarizers.

To study experimentally the umbilical defects dynamics, one can consider a nematic liquid crystal layer with negative anisotropy dielectric constant and homeotropic anchoring. To keep the thickness of the cell and avoid cell plates buckling, often one can use randomly distributed glass spheres between cell plates. These spheres are mono-disperse and have approximately the same diameter as the thickness of cell. It could be expected that the presence of glass beads does not affect the vortex dynamics. However, recently it has been shown that these glass beads change the average defects dynamics [10].

In this manuscript, the vortex-glass bead interaction is experimentally characterized. The glass bead have been found to be attractive or non-attractive. We observed that not all glass beads are attractive, at least for the sample and the regions of the sample so far observed, there are more non-attractive beads than attractive ones. The attractive glass bead approximatively pull the vortexes so that the distance decays with the square root of time. Hence, the glass bead induces a force on the vortex proportional to an inverse power law of the distance between them with exponent 3/4. This dynamics is not isotropic, that is, the vortexes are attracted and annihilated to a polar region of the sphere.

2. Experimental Setup

We have considered a cell, 5B100A150UT180, manufactured by Instec. The cell is filled by capillarity with MLC − 6608 nematic liquid crystal that has negative anisotropy. The inner walls of this cell are chemically treated so that the molecules of liquid crystal are orthogonally aligned to the cell plates. This configuration is usually named *homeotropic anchoring* [1]. The glass plates of the cell have 15 μm of separation between them. To keep the thickness of the inner gap between the glass plates fixed, the cell is filled with monodisperse glass beads (clear/transparent ceramics) which are randomly distributed in the sample. The sample of liquid crystal is subjected to an external electric field in the vertical direction (z-axis), using a low frequency (100kHz) sinusoidal voltage (9V) produced by function generator (Agilent 33521A) with a high voltage amplifier (Tabor Electronics 9200).

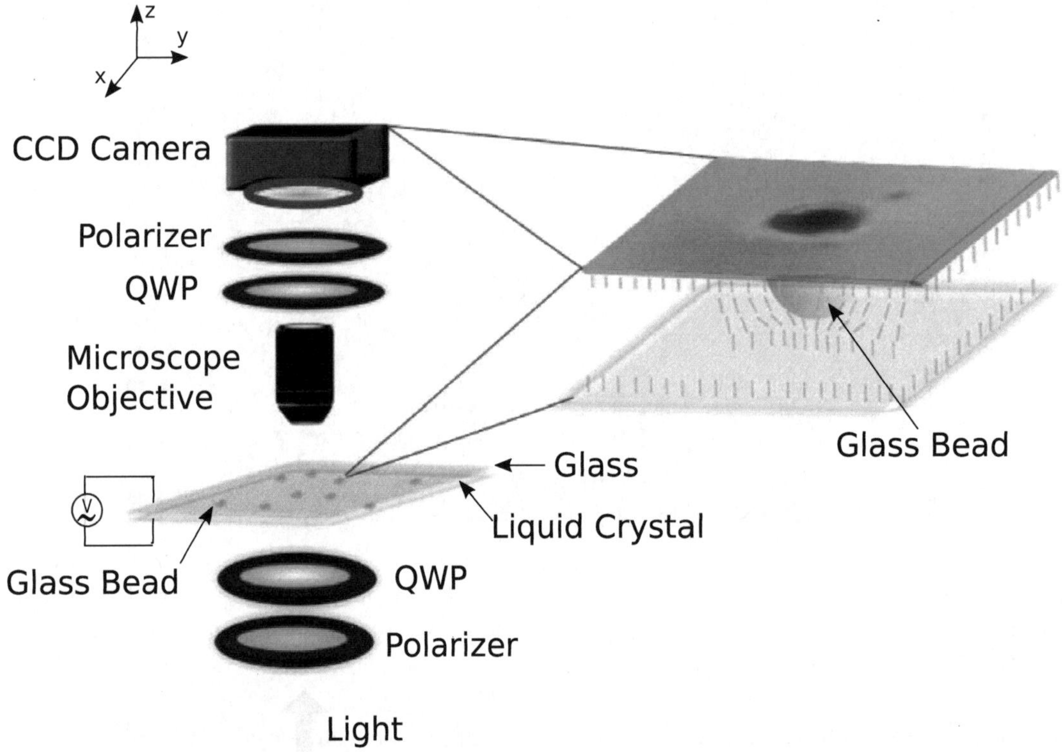

Figure 2. Sketch of the setup. The microscope, here represented by the objective, and accommodates a circular polarizer, bottom quarter-wave plate (QWP) and polarizer, and a circular analyzer, top polarizer and quarter-wave plate. The image of the cell is recorded via a CCD camera. The inset sketches the glass bead and a configuration of liquid crystal directors.

The imaging system used, see Figure 2, is a modified Motic 310 microscope equipped with a circular polarizer (bottom linear polarizer and quarter-wave plate) and a circular analyzer (top quarter-wave plate and polarizer). The light from the microscope condenser, that illuminate the cell mounted on the microscope stage, is filtered with a band-pass filter since the QWPs are designed to operate at $\lambda = 532$nm. Then, the texture of the sample is captured by a charge-coupled device (CCD camera). The images obtained by the CCD camera are recorded and then analyzed with an image processing and analysis software Fiji (ImageJ)[12]. The position of the glass beads and the vortexes are extracted for further analysis.

3. Results

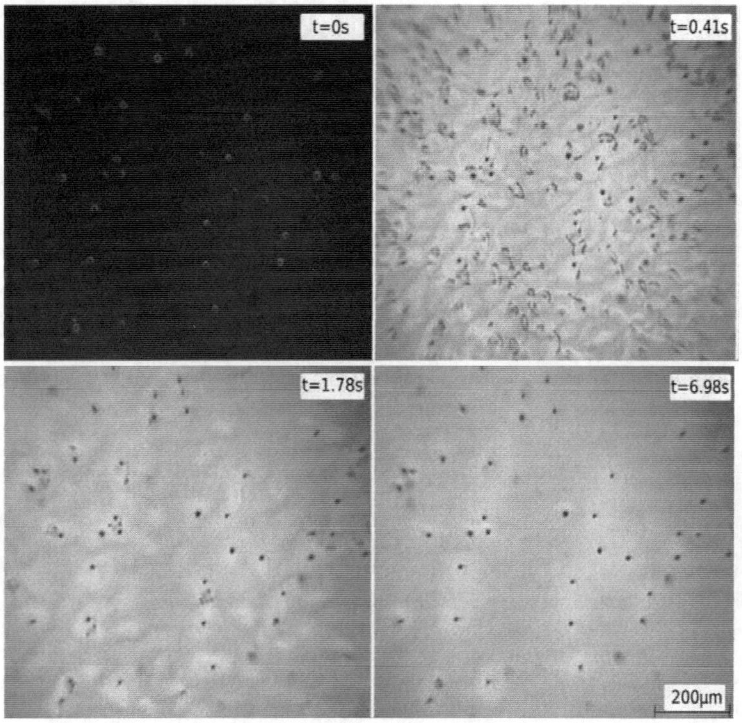

Figure 3. Vortexes dynamics in inhomogeneous medium. Sequence of temporal snapshots. Top left panel accounts for the liquid crystal layer without voltage. The gray and green-marked gray spheres correspond to glass beads. The green-marked spheres stand for attractive glass beads. Once the electric voltage is applied, a large number of vortexes are observed. Each vortex is identified as a black spot, which has a typical size of 5 μm, as illustrated in the top right panel. Subsequently, the vortexes of opposite charge are attracted to each other and annihilated, decreasing the density of vortexes (cf. bottom left panel). In this regime of dilute gas of vortexes, some of them are attracted by some glass beads. Finally, for a sufficiently long time all the vortexes disappear, see bottom right panel.

When a low frequency voltage with a sufficiently large amplitude greater than the Freédericksz voltage is applied to the nematic liquid crystal layer—which was previously described—the creation of vortexes is observed. Due to the presence of the glass beads in the sample, this process occurs in an inhomogeneous medium. The dynamics of vortexes is strongly affected by the glass beads [10]. We observe that there are attractive glass beads. Indeed, the attractive spheres pull the near vortexes. Figure 3 shows a typical observed vortex dynamics.

In the monitored region, there are 39 glass beads, only five of them are attractive. The density of glass non-attractive beads and attractive ones are 27 beads/mm^2 and 4 beads/mm^2, respectively. Hence, the proportion of attractive glass beads is 15%. Initially, a large number of vortexes are created, then the vortexes with opposite topological charges are attracted to each

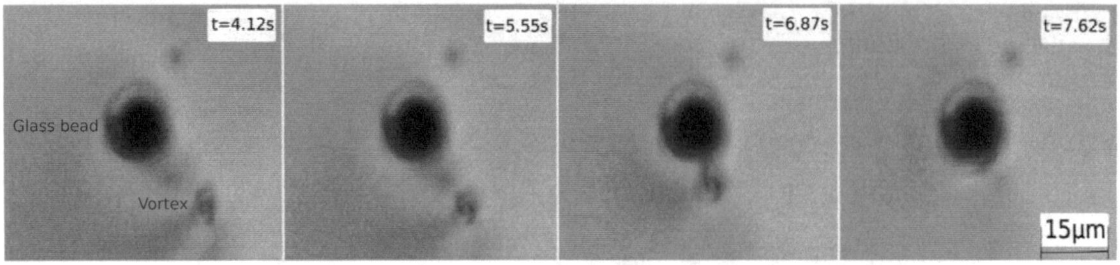

Figure 4. Vortex glass bead interaction. From the left to the right is presented a temporal sequence of snapshots. Initially, an attractive glass bead pulls a vortex. When a vortex collides with the attractive glass bead, it disappears (cf. right panel).

other and eventually annihilate and those with same charges are repel each other. As result of the vortex interaction, the density of vortex decreases with time. In this regime of dilute gas of vortexes, some of the defects are attracted by some glass beads. When a vortex collides with the attractive glass bead, it disappears. Figure 4 illustrates the vortex and glass bead interaction, showing a sequence of snapshots. To characterize this process we have considered a region around an attractive bead embedded in the nematic liquid crystal layer, region that is magnified 20 times. Notice that the vortex trajectory is not radial. In addition, the typical time of collision between vortexes is of the order 2.33 s. On the other hand, the typical time of collision between a vortex and glass bead is of the order of 13.0 s.

To illustrate the non attractive glass beads, we have identified a pair of glass beads, one attractive and one unattractive (see Fig. 5). In this figure is illustrated how a sphere attracts three vortexes. Note that also this attractive glass bead modifies the dynamics between vortexes. The non attractive glass bead presumably weakly repels the vortexes, however experimentally we have not been able to verify this property. The above assertion is based on the fact that vortexes motion close to this glass bead have never been observed in our case. Even, the vortexes that are observed close to this sphere move away. A detailed study of this repulsive phenomenon is in progress. In the next section, we will present a detailed study of the vortex attraction towards a glass bead.

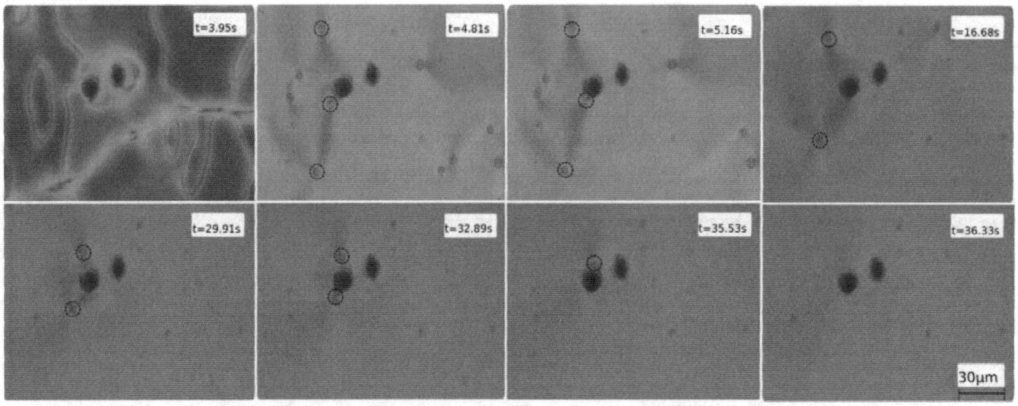

Figure 5. Attractive and non-attractive glass beads. From the left to the right is presented a temporal sequence of snapshots. The attractive sphere pulls three vortex. The vortexes that which are pulled by the attractive glass bead are highlighted with blue circular lines.

3.1. Experimental analysis of vortex and glass bead interaction

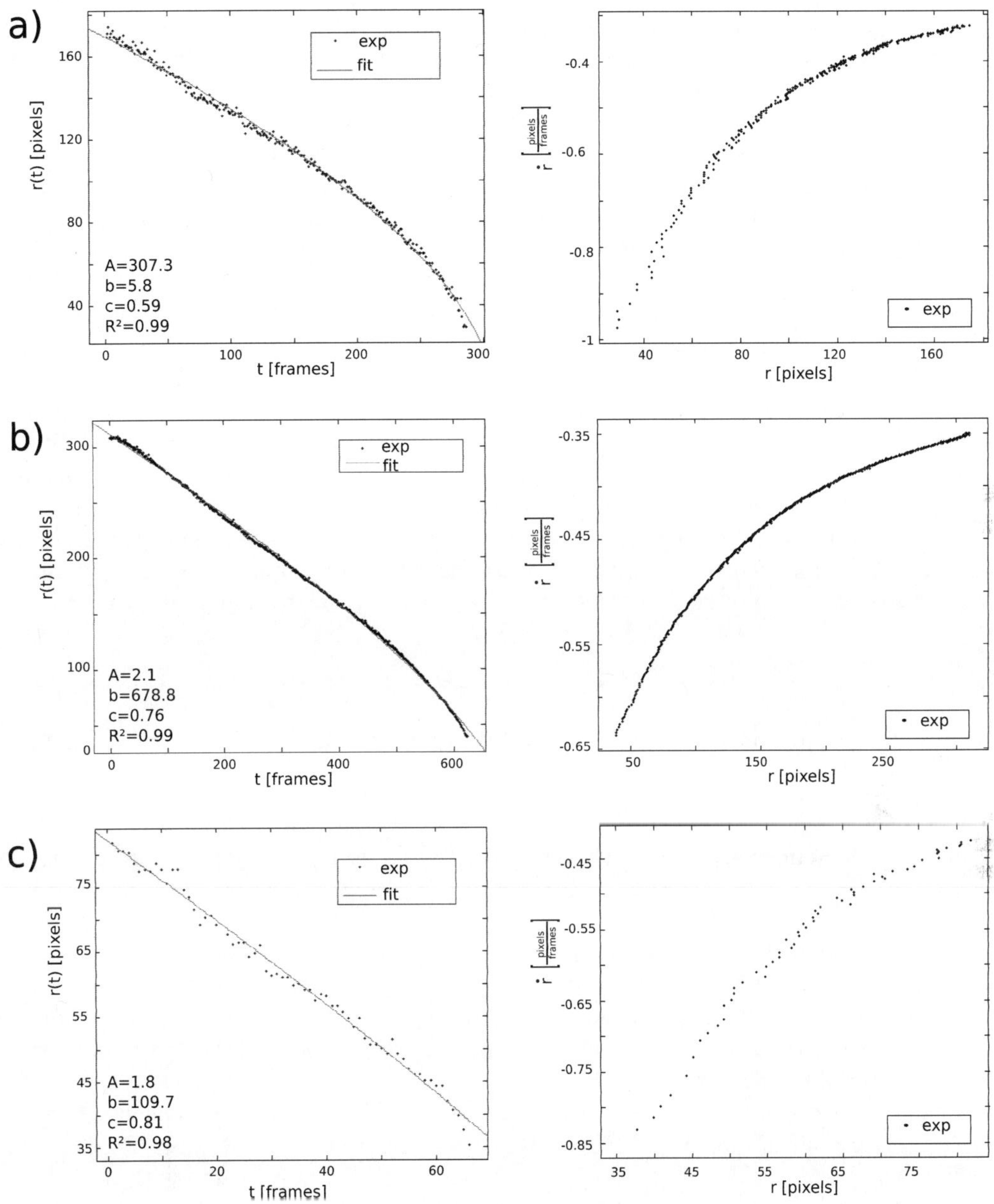

Figure 6. Interaction dynamics of vortex and glass bead. Points and continuous curves in the left panels, respectively, account for the experimental evolution of the distance between vortex and the glass bead as function of time for three experimental realizations and fitted curves using formula (2) with their respective adjustment R-square. Points and continuous curves in the right panels, respectively, stand for the experimental evolution and fit curves using formula (1) of the force as function of the vortex radial position (r) with respect to the attractive glass bead, where 1 pixel $=0.254$ μm and 1 frame$=$ 0.06 s.

To characterize the interaction dynamics of vortex and glass bead, we have measured and monitored the distance between them. This distance was obtained by processing the glass beads and the vortexes positions, the latter determined from the analysis of a sequence of frames in the acquired video (fixed frame rate) using Fiji (ImageJ).

To carry on with this process, first an attractive spheres was identified via testing. Subsequently, an electric voltage is applied to the liquid crystal sample and the evolution of the vortexes attracted by the glass bead is studied. Later on, the external electrical voltage is turned off and the system is allowed to relax at equilibrium for a period of about 10 s. All this procedure is repeated several times to achieve an adequate statistical description. Left panels of figure 6 shows the experimental evolution of the distance between vortex and the glass bead as function of time for different experimental realizations.

The measurements are represented by the points cloud. Clearly from these measurements, one observes that the distance as a function of time has a well define evolution law. It is well-known that the vortex-pair interaction is characterized by an over-damped force that goes as the inverse of the distance between the vortexes [7]. Namely, the distance between opposite vortexes, $d(t)$, decreases as the square root of time, $d(t) = \kappa\sqrt{t_o - t}$ where t_o accounts for collision time and κ is an appropriated dimensional parameter. It is worthy to note that vortexes are characterized by the induced molecular distortion around vortex position [1]. Likewise, glass beads can generate molecular distortion in their own surrounding [11]. This molecular distortion generates interaction with the vortexes. Therefore, one would expect the attractive glass bead and the umbilical vortex interaction to be an over-damped force, $f(r)$, which is inversely proportional to a power of the distance, $r(t)$, between vortex and glass bead, that is,

$$\dot{r} = f(r) = -\frac{\gamma}{r^n}, \tag{1}$$

where γ is a dimension parameter and n is a real positive number. After straightforward calculations, one gets

$$r(t) = A(b - t)^c, \tag{2}$$

where b accounts for collision time, $A \equiv ((1+n)\gamma)^{1/(1+n)}$ is a dimensional parameter, and $c \equiv 1/(1+n)$.

In order to determine A, b and c from the experimental data, we will use formula (2) as a fit law. The continuous curves in left panels of figure 6 are the best fits of the experimental data for different realization. Note that there is a quite fair agreement between the experimental data and the fit law, formula (2). Note that the main source of fluctuations in the data is the detection algorithm of the vortexes.

Table 1 summarizes the fit parameters, using formula (2), of the vortex-glass bead interaction for 14 experimental realizations. We note that there is a trend between A and c. Indeed, when the c coefficient increases the A coefficient decreases. The b coefficient is related with the collision time of the vortex and the glass bead. The collision time is of the order of 13 seconds.

The above analyzes assume that the interaction is of a central nature. However, the trajectories of the vortexes to the glass bead are not radial.

Figure 7 shows the trajectories of different vortexes that move towards to glass bead, and the vortex glass bead interaction appears to be anisotropic. Experimentally, we observe that the vortexes collide with the glass bead in certain well define region which we termed as *south pole*. Probably, this anisotropic behavior is due to the fact that the deformation around the glass bead is not cylindrically symmetric, which could generate a deformation equivalent to a dipole. However, we have not detected vortexes that collide with the bead at its *north pole*. The detailed structure of the interaction is an open problem, and its understanding is in progress. Likewise, when the liquid crystal samples are not subjected to an external voltage and between to circular crossed polarizers, we observe that attractive and non-attractive glass beads exhibit

Table 1. Fit values, using formula (2), of the vortex-glass bead interaction for 14 experimental realizations.

cases	A	b	c	R^2
I	30.6	1027.0	0.26	0.98
II	22.2	29.2	0.29	0.99
III	16.5	74.1	0.35	0.99
IV	8.8	64.8	0.49	0.99
V	8.5	79.1	0.53	0.99
VI	7.4	127.8	0.54	0.99
VII	8.0	71.3	0.57	0.99
VIII	6.5	124.6	0.58	0.99
IX	5.8	307.3	0.59	0.99
X	5.0	108.7	0.63	0.99
XI	5.1	82.5	0.67	0.99
XII	2.1	678.8	0.76	0.99
XIII	1.8	109.7	0.81	0.99
XIV	0.4	369.6	0.91	0.99
Average	9.19	232.46	0.57	0.99

an asymmetrical and symmetric halo of light, respectively. The inset of Fig. 7 depicts the typical observed halo of light.

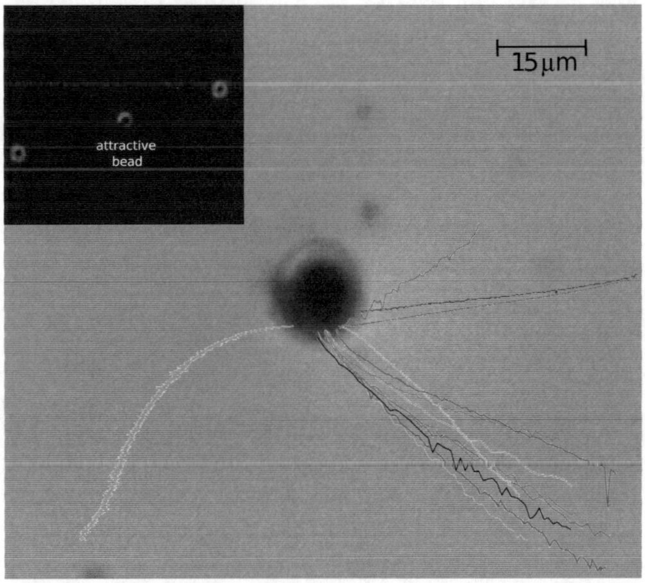

Figure 7. Anisotropic interaction between vortex and glass bead. Different colors account for different vortex trajectories. Inset accounts for the liquid crystal sample without external voltage and between to circular crossed polarizers. The attractive (glass bead in the middle) and non-attractive glass beads (glass beads outside) exhibit an asymmetrical and symmetric halo of light, respectively.

3.2. Vortex-glass bead force

Although the interaction is not isotropic, as a first approximation it will be considered isotropic. From the experimental data on the evolution of the distance between the vortex and the glass bead, one can deduce an evolution law using formula (2), see left panels of Fig. 6. Similarly, from the experimental data and fit curve one can reconstruct the interaction force between the vortex and the glass bead.

The points and the continuous curves in right panels of figure (6) show, respectively, the vortex speed ($\dot{r}$) as function of the distance (r) obtained from experimental data and theoretical fit, formula (1) for different experimental realization. From figure 6, one can infer that as the distance (r) increases, the interaction force is more abrupt. From this figure we can conclude that the force of interaction as function of the distance between the vortex and the glass bead is an inverse power law in good approximation.

To understand how the force changes as a function of time, from the fit expression of the temporal evolution of the distance between vortex and glass bead, formula (2), we can rebuild the interaction force as function of time,

$$f(t) = \dot{r} = -Ac\,(b-t)^{c-1}. \tag{3}$$

Rewriting this expression $f(t) = -C_1\,(b-t)^{(c-1)}$, where $C_1 = cA$. Considering the average value c from table 1, $\langle c\rangle = 0.57$, we obtain the average force of vortex-glass bead

$$\langle f(t)\rangle = -\frac{C_1}{(b-t)^{0.43}}. \tag{4}$$

Hence, the attractive glass bead spheres pull the vortexes so that the distance decays approximately as the inverse of the square root of the time to collision. On the other hand, the interaction force as a function of distance has the form given by formula (1), where average exponent is $\langle n\rangle = 0.75$. Therefore from these results, we conclude that when the distance between vortex pair and the distance between vortex and glass sphere are the same, the interaction between vortex and glass bead ($\approx 1/r^{3/4}$) is greater than the vortex pair interaction ($\approx 1/r$). However, we note that after the vortexes are created, the distance between vortex pair is less than distance between vortex and glass bead, hence, the main interaction is between vortex pair. Subsequently, when the remaining number of vortexes is reduced (diluted gas), they begin to interact with the attractive glass beads. This scenario is the one observed experimentally (cf. Fig. 3).

4. Conclusions and Remarks

Liquid crystal driven by external electric and magnetic fields exhibit an intricate network topological defects. These types of defects have attracted the attention of physicists, mathematicians, and engineers for their fundamental properties as macroscopic topological particles and for their potential applications in particular in communications, image processing and quantum computation [4, 5]. Here, we have reported an experimental characterization of vortex-glass bead interaction in nematic liquid crystal layer. The glass bead spheres can be classified in two type from the interaction point of view: attractive and non-attractive.

The attractive glass bead spheres pull the vortexes so that the distance decays as the square root of the time to collision. The glass bead sphere induces an anisotropic force on the vortex. This force can be well approximate by an inverse power law of the distance between vortex and glass bead with exponent 3/4. The vortexes are attracted to a polar region of the attractive glass bead. After the vortexes are created the main interaction is mediated by vortex pair interaction. Subsequently, when a reduced number of vortexes is left (diluted limit), they begin to interact with the attractive glass beads. This scenario is the one observed experimentally.

Still unknown the origin of these attractive force between vortex and glass bead. Certainly, the presence of glass bead generate deformation around itself. This deformation is responsible of vortex-glass bead interaction, and its characterization is in progress. Likewise, in the present study we have neglected the effect of distant vortexes (vortex pair interaction). In addition, mean field theory is not valid as consequence of long range interaction between vortexes. Hence, the presence of distant vortex can play a role. A systematic study of this issue is in progress. The role of non-attractive glass beads has not been performed, they are probably of repulsive type.

Acknowledgments
V.Z. thanks for fruitful comments and observations given by M.G. Clerc, R. Barboza, G. Gonzalez-Cortes, and M. Ferré.

5. References

[1] S. Chandrasekhar, 1977, *Liquid crystals of disc-like molecules* (Cambridge University press, Cambridge).
[2] P. G. de Gennes and J. Prost, 1974, *The Physics of Liquid Crystals* (Oxford University press, New York).
[3] P. Oswald, and P. Pieranski, 2005, *Nematic and cholesteric liquid crystals: concepts and physical properties illustrated by experiments* (CRC press, Boca Raton).
[4] L.M. Pismen, 1999, *Vortices in Nonlinear Fields: From Liquid Crystals to Superfluids, From Non-Equilibrium Patterns to Cosmic Strings* (Clarendon press, Oxford).
[5] F. Bethuel, H. Brezis, and F. Hélein, *Ginzburg-Landau Vortices* (Springer Science & Business Media, 2012).
[6] I. Dierking, *Textures of liquid crystals* (Wiley-VCH, Darmstadt, 2003).
[7] R. Barboza, U. Bortolozzo, G. Assanto, E. Vidal-Henriquez, M.G. Clerc, and S. Residori (2012), Vortex induction via anisotropy stabilized light-matter interaction, Phys. Rev. Lett. **109**, 143901.
[8] R. Barboza, T. Sauma, U. Bortolozzo, G. Assanto, M.G. Clerc, and S. Residori, (2013), *Characterization of the vortex-pair interaction law and nonlinear mobility effects*. New J. Phys. **15**, 013028.
[9] M.G. Clerc, E. Vidal-Henriquez, J.D. Davila, J.D., and M. Kowalczyk, *Symmetry breaking of nematic umbilical defects through an amplitude equation*, Phys. Rev. E, **90**, 012507 (2014).
[10] R. Barboza, U. Bortolozzo, M.G. Clerc, R. Residori, and V. Zambra, (2016), *Coarsening Dynamics of Umbilical Defects in Inhomogeneous Medium*, In Nonlinear Dynamics: Materials, Theory and Experiments (pp. 31-43). Springer International Publishing.
[11] M.A. Gharbi, M. Nobili, G. Prvot, P. Galatola, J.B. Fournier, and C. Blanc, (2011), *Behavior of colloidal particles at a nematic liquid crystal interface*, Soft Matter, **7**, 1467.
[12] J. Schindelin, I. Arganda-Carreras, E. Frise, et al., (2012), *Fiji: an open-source platform for biological-image analysis*, Nature methods, **9(7)**, 676-682.

High accuracy approximation for the modified Bessel function of fractional order $I_{1/3}(x)$

P. Martin [1], J. Olivares [2], F. Maass [1] and E. Valero [3]

[1] Physics Dept., Universidad de Antofagasta, Antofagasta, Chile
[2] Mathematics Dept., Universidad de Antofagasta, Antofagasta, Chile
[3] Physics Dept., Universidad de Tarapaca, Arica, Chile.

E-mail: pablo.martin@uantof.cl, jorge.olivares@uantof.cl, fernando.maass@uantof.cl, e.valero@ucn.cl

Abstract. The modified Bessel functions of fractional order $I_{1/3}$(x) has been approximated by an analytic function containing rational and hyperbolic fumctions. A technique using both power series and assymptotic expansion has been used. The accuracy of the approximation is very high using only three parameters., and the largest relative error is smaller than 0,004.

1. Introduction

A very precise approximation has been found for the modified Bessel functions of fractional order $I_{1/3}$(x)[1, 2]. The method here applied, improves prior techniques such as quasi-rational multipoint approximations, MPQA [3-4]. The Bessel functions the order 1/3 are very important, because its connection with the Airy functions [2]. The approximations presented here are a combination of a rational function, as Pad method, with hyperbolic functions. The idea is, as in the MPQA method, to use simultaneously power series and asymptotic expansions. To do this, it is neccesary to use auxiliary functions as hyperbolic functions [5]. The MPQA method has also been improved, seeking greater efficiency. Only three parameters are needed to obtain a relative error of less than 0.4%. In the next Section, the description of the method will be presented, as well as, the results, The last section will be devoted to the Conclusion.

2. Theoretical development

The potential series for $I_{\frac{1}{3}}(x)$ is known and it is given by

$$I_{1/3}(x) = \frac{x^{\frac{1}{3}}}{2^{\frac{1}{3}}\Gamma(\frac{3}{4})}\left[1 + \frac{3x^2}{4} + \frac{9x^4}{112} + \cdots\right] \tag{1}$$

Asymptotic expansion is also known and it is

$$I_{1/3}(x) \sim \frac{e^x}{\sqrt{2\pi x}}\left[1 + \frac{5}{72x} + \frac{77}{1152x^2} + \cdots\right] \tag{2}$$

Considering both expansions we have chosen the form of the approximant $\tilde{I}_{1/3}(x)$ as

$$\tilde{I}_{1/3}(x) = \cosh x \left[\frac{p_0 + \sum_{i=0}^{N} x^{2i} p_i}{1 + \sum_{j=0}^{n} q_j x^{2j}} \right] \cdot \frac{x^{\frac{1}{3}}}{\Gamma(\frac{4}{3}) 2^{\frac{1}{3}} (1 + \lambda^2 x^2)^{\frac{5}{12}}} \tag{3}$$

The criterion followed for this choice is:

(i) First, it should be noted that the asymptotic series has a singularity $x^{-1/2}$, which is a fractional power of x. It is necessary to take this singularity into account, but it is also important to note that the singular points of fractional powers or branching points come in pairs, and in this case $x^{-1/2}$, has singularities in zero and in infinity. The second singularity in this case does not exist in $I_{1/3}(x)$, and this auxiliary function should not be included with the rational functions, whose singularities are only poles. The way to treat this problem is to divide the complex plane into regions, and to pay attention only to the region of interest, which in this case are the positive values of x, that is the right half-plane of the complex plane. With this criterion in mind the unwanted singularities should be placed outside of the region of interest, that is the left halfplane ($x < 0$) of the complex plane. Following these considerations the auxiliary function $(1+x)^{-1/2}$ would fulfill this condition since now the branch points would be infinite and -1.

(ii) The previous type of reasoning was also followed in previous approximations to integer order Bessel functions, although acceptable, but the analysis is incomplete and leads to forms of inefficient approximation. To complement this analysis, consider the potential series. This series apart from the initial factor, has only even powers, so that it would be desirable that the power series of the approximate function would have only even powers. In this way the determination of the parameters of the potential function could be done more efficiently, because in this way only even powers are equalized, avoiding the inconvenience of equations between parameters equalized to zero because of odd powers. By including these considerations in our analysis we can obtain better approximations with the same number of parameters. In this way the highest efficiency is obtained. This last auxiliary function can be put in a more general form by entering a parameter λ, to be defined later. Thus the selected auxiliary function was $\frac{1}{\sqrt[4]{1+\lambda^2 x^2}}$.This is transformed into $(1 + \lambda^2 x^2)^{-\frac{5}{12}}$,because at infinite the fractional power $x_{1/3}$ must be canceled. Thus this factor is because of the power series of the $I_{1/3}$(x).. The second selected auxiliary function was $\cosh(x)$. The reason for this selection is that in the asymptotic expansion appears the exponential function e^x. This could be the auxiliary function, but it has the drawback that its powerl series has odd and even powers, so by considering as auxiliary function $cosh(x)$, only even, power series are obtained and the exponential behavior at infinite is also reached since the auxiliary function has only even powers..

(iii) Now the selected rational function has only even powers, in so satisfying the efficiency conditions required by the potential series of $I_{1/3}(x)$ according to the previous analyses.

The parameters p_0, p_1, and q_1 are now determined by the coefficients of the power and asymptotic series. Let is take two terms from the potential series and the leading term of the asymptotic expansion. From this last condition you get

$$p_1 = \Gamma(\frac{4}{3}) 2^{\frac{1}{3}} \sqrt{\frac{2}{\pi}} \lambda^{\frac{5}{6}} q_1 \tag{4}$$

In this way Eq.(3) is transformed into

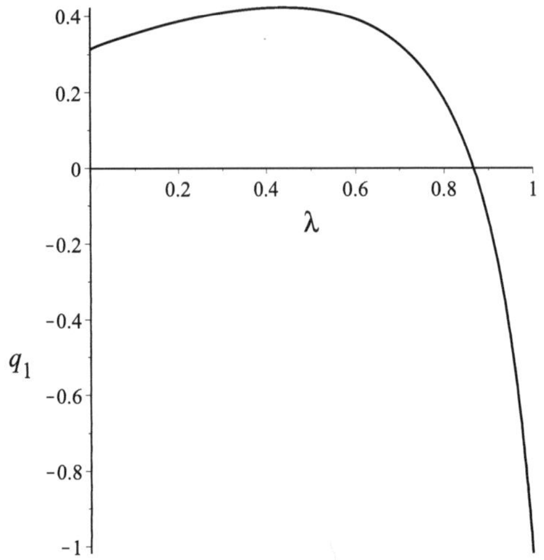

Figure 1. The value of q_1 as a function of λ, is shown

$$I_{\frac{1}{3}}(x) = x^{\frac{1}{3}}\cosh(x)\frac{\frac{p_0+\Gamma(\frac{4}{3})2^{\frac{1}{3}}\sqrt{\frac{2}{\pi}}\lambda^{\frac{5}{6}}q_1x^2}{1+q_1x^2}}{\Gamma(\frac{4}{3})2^{\frac{1}{3}}(1+\lambda^2x^2)^{\frac{5}{12}}} \tag{5}$$

This new equation is rationalized by multiplying both sides by $(1+q_1x^2)(1+\lambda^2x^2)^{5/12}$, and the first two coefficients of these expressions on both sides are equalized. In this way two linear algebraic equations are obtained for p_0 and q_1, and they can be determined, resulting

$$p_0 = 1, q_1 = \frac{5}{12}\frac{\lambda^2-\frac{3}{4}}{1-2^{\frac{1}{3}}\Gamma(\frac{4}{3})\sqrt{\frac{2}{\pi}}\lambda^{\frac{5}{6}}} \tag{6}$$

Entering this value into $\tilde{I}_{\frac{1}{3}}(x)$, it is obtained

$$\tilde{I}_{\frac{1}{3}}(x) = \frac{x^{\frac{1}{3}}\cosh x}{2^{\frac{1}{3}}\Gamma(\frac{4}{3})(1+\lambda^2x^2)^{\frac{5}{12}}}\frac{1+2^{\frac{1}{3}}\Gamma(\frac{4}{3})\frac{5}{12}\sqrt{\frac{2}{\pi}}\frac{\lambda^2-\frac{3}{4}}{1-2^{\frac{1}{3}}\Gamma(\frac{4}{3})\sqrt{\frac{2}{\pi}}\lambda^{\frac{5}{6}}}x^2}{1+\frac{5}{12}\frac{\lambda^2-\frac{3}{4}}{1-2^{\frac{1}{3}}\Gamma(\frac{4}{3})\sqrt{\frac{2}{\pi}}\lambda^{\frac{5}{6}}}x^2} \tag{7}$$

A graph of q_1 in function of λ is shown in Figure 1, and
q_1 is positive for the interval $0 < \lambda < \frac{\sqrt{3}}{2}$ and $0 < \lambda < 0.32914$.
When computing the maximum relative error as function of λ the best value for λ is obtained when $\lambda = 0.3$. In Figure 2 the relative error as a function of x is shown for this value of λ

The maximum relative error is 0.0038, at the value of $x = 9$. Doing the calculations and taking only the number of appropriate digits so that the relative error does not change, finally it is got

$$I_{\frac{1}{3}}(x) = \frac{x^{\frac{1}{3}}\cosh x}{1.12508(1+0.009x^2)^{\frac{5}{12}}}\frac{1+0.13492x^2}{1+0.40552x^2} \tag{8}$$

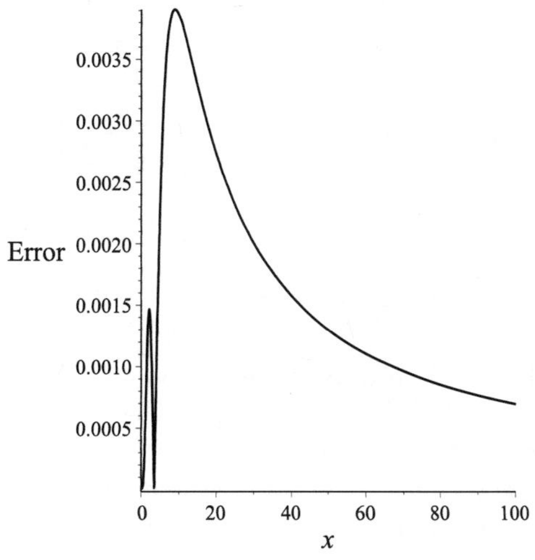

Figure 2. Relative error as a function of x for $\lambda = 0.3$

3. Conclusion

An approximation has been found for the modified Bessel function $I_{1/3}(x)$. This function is related to the Airy function, so it is also important for this property. The approximate function found is good for all positive values of the independent variable x. The error is less than a thousandth for almost all x, except for a region around $x = 9$, where the maximum relative error is 0.004. Thus the analytic approximation found can be used with confidence for almost all of its applications in Physics. The function can also be integrated and derived although the approximation may vary slightly. The approximant found here is a combination of a rational function with simple auxiliary functions. The parameters and the form of the approximant were found using simultaneously the power series and the asymptotic expansion of $I_{1/3}(x)$, the approximant being a bridge between both expansions.

References

[1] G. N. Watson, " A treatise on the theory of Bessel functions", 2nd. ed.(Cambrigde University Press, Cambrigde (1966).
[2] M. Abramovitz and I. A. Stegun, "Handbook of mathematical Functions" (Dover Books, New York, 1972) pp. 692-697.
[3] P. Martin, E. Castro, J. L. Paz and A. De Freitas, "Multipoint quasi-rational approximants in Quantum Chemistry", Chapter 3 of "New Developments in Quantum Chemistry" by J. A. Paz and J. Hernandez (Transworld Research Network, Kerala, India, 2009) pp. 55-78.
[4] P. Martin, E. Castro and J. L. Paz, Rev. Mexicana Fisica, **58** , 301 (2012).
[5] P. Martin, J. Olivares, and A. Sotomayor, Rev. Mexicana Fisica **63**,130 (2017).

Spiky ice and penitente tilting

Pablo Guilleminot and Rodrigo Olea
Departamento de Física, Universidad Andres Bello,
Sazié 2212, Piso 7, Santiago, Chile.

E-mail: yemheno@gmail.com rodrigo.olea@unab.cl

Abstract. Under certain conditions, at high altitude, the surface of snow develops spike-like structures known as *penitentes*. This is a rather counterintuitive phenomenon, which is a consequence of surface sublimation at a given point as a result of the incidence of light scattered by the surrounding region.

Following the existing literature, we model the time evolution of the phenomenon described above as a 1D diffusion equation with a non-local source term, as it represents the light coming from all the line of sight defined for a point of the curve. For small initial perturbations in the surface, the system undergoes a thermodynamic instability which triggers the formation of spikes.

For sunlight coming in at a given angle, numerical simulations account for a feature observed in the real system: penitentes get tilted in the direction of the sunlight.

1. Introduction

Structure formation in snow is controlled by many factors: surface reflectance, wind, daily melt-freeze cycle on the surface, incidence angle of the sunlight, etc. This paper models penitente formation in snow fields as a purely radiation-driven phenomenon. The key condition for the differential ablation is sublimation, what requires that the dew point is lower than the freezing point [1]. On the contrary, if the snow melts, the water weighs down the surface and smooths out any spiky features.

2. Toy model for penitente formation

The ablation at any point is proportional to the heat absorbed due to either direct or scattered sunlight. Part of the heat is lost by convection to the air, while another fraction is transmitted to the neighbouring points. We will restrict ourselves to heat transmission within the surface which is governed by the diffusion equation

$$\frac{\partial U}{\partial t} = \Upsilon(x) + D\frac{\partial^2 U}{\partial x^2} \tag{1}$$

$$\frac{\partial h}{\partial t} = -K\frac{\partial U}{\partial t} \tag{2}$$

where $U(x,t)$ is the thermal energy at a point x for the time t, $h(x,t)$ is the profile height, $D\frac{\partial^2 U}{\partial x^2}$ is the diffusion term and $\Upsilon(x,t)$ is the total energy received at a given point.

2.1. Geometry of the model
The fact that the snow looks white is a consequence of a diffuse reflection of incident sunlight. The light is partially reflected beneath the surface such that all colors are scattered roughly equally in all directions. Every single point of the curve acts as a source of equal intensity per unit of length, which is spread out over the tangent line to that point. This assumption simplifies the discussion, as surface geometry is the only input required to compute the amount of absorbed energy on a particular point of the curve.

The contribution of an infinitesimal length ds in the emitting surface (at a point x') to the absorbed energy on another point x is proportional to the sustained angle $d\theta$ (See figure).

Figure 1. Line of sight.

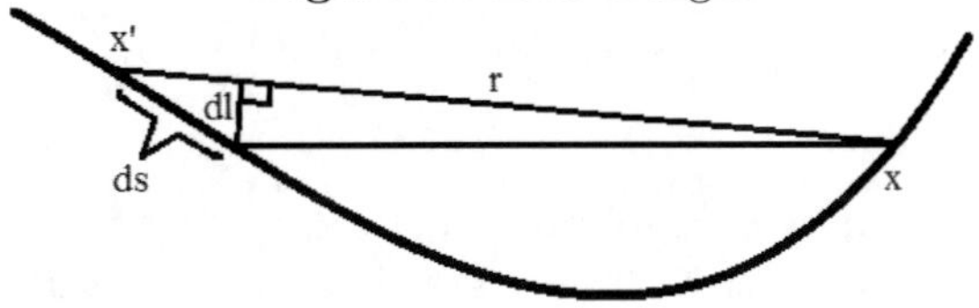

Thus, the heat source at any point $\Delta(x)$ comes as the sum over all infinitesimal contributions coming from the *line of sight* (LoS)

$$\Delta(x) = \frac{1}{\pi}\int_{LoS} d\theta = \frac{1}{\pi}\int_{LoS} \frac{\|\vec{r}\times d\vec{s}\|}{r^2}. \tag{3}$$

The arc element $d\theta$ in terms of x and $h(x)$, from the above figure, is simply given by $d\theta = \frac{dl}{r}$. At the same time, dl is the height of the parallelogram formed by $\vec{r}$ and $d\vec{s}$. Plugging this in eq.(3)

$$\Delta(x) = \frac{1}{\pi}\int_{LoS} \frac{xh'(x) - h(x)}{x^2 + h(x)^2} dx \tag{4}$$

Here $\Delta(x)$ is a portion of the angle π that measures the amount of light scattered within the surface. The total energy absorbed is then

$$\Upsilon(x) = I\alpha\,(1 + (1-\alpha)\Delta(x)) \tag{5}$$

Where α is the portion of energy absorbed by the surface in a single bounce and I is the energy of the beam. This is a non-local expression that acts as a source of the diffusion eq. (1).

2.2. Line of sight
For given point $P = (x, h(x))$ on the curve, the line of sight in the neighboring peaks is defined as the interval $x' \in [a, b]$ such that $Q = (x', h(x'))$ can be connected by a straight line to P without intersecting the surface. The extreme points in $x' = a, b$ one can see from P are given by the condition that the slope of the connecting segment coincides with the derivative there.

The line of sight must include portions of more distant peaks to P, if they are tall enough. The numerical resolution of the PDE (1) must take this issue into account in the non-local source term Υ by summing over the visible points of all peaks around.

IOP Conf. Series: Journal of Physics: Conf. Series **1043** (2018) 012007 doi:10.1088/1742-6596/1043/1/012007

2.3. Shady zones

At intermediate latitudes, the sunlight does not fall in a vertical line. The geometric description below is a model to account for the tilting of the icy spikes in the direction of the sun, which can be observed in penitente fields.

If the light comes in at a given angle ϕ, we suppress the contribution to the scattered light from the portion of the surface which is under the shadow of a peak.

The last lightened point, namely $x = c$, must satisfy

$$h'(c) = \tan(\phi) \tag{6}$$

and such that the curve is convex at that point, i.e., in other words $h''(c) < 0$.
Thus, the shady zone is given as an interval $[c, d]$, where the other end $x = d$ obeys

$$h(d) = \tan(\phi)(d - c) + h(c)\,. \tag{7}$$

3. Results

The set of partial differential equations is solved numerically, using a code in C++ language that is developed to show the time evolution of the system. For an initial surface that contains micro-features, the surface will develop spikes as a result of an instability of the PDEs. The program uses forward difference at t and second-order central difference for the position x corresponding to a finite difference method.

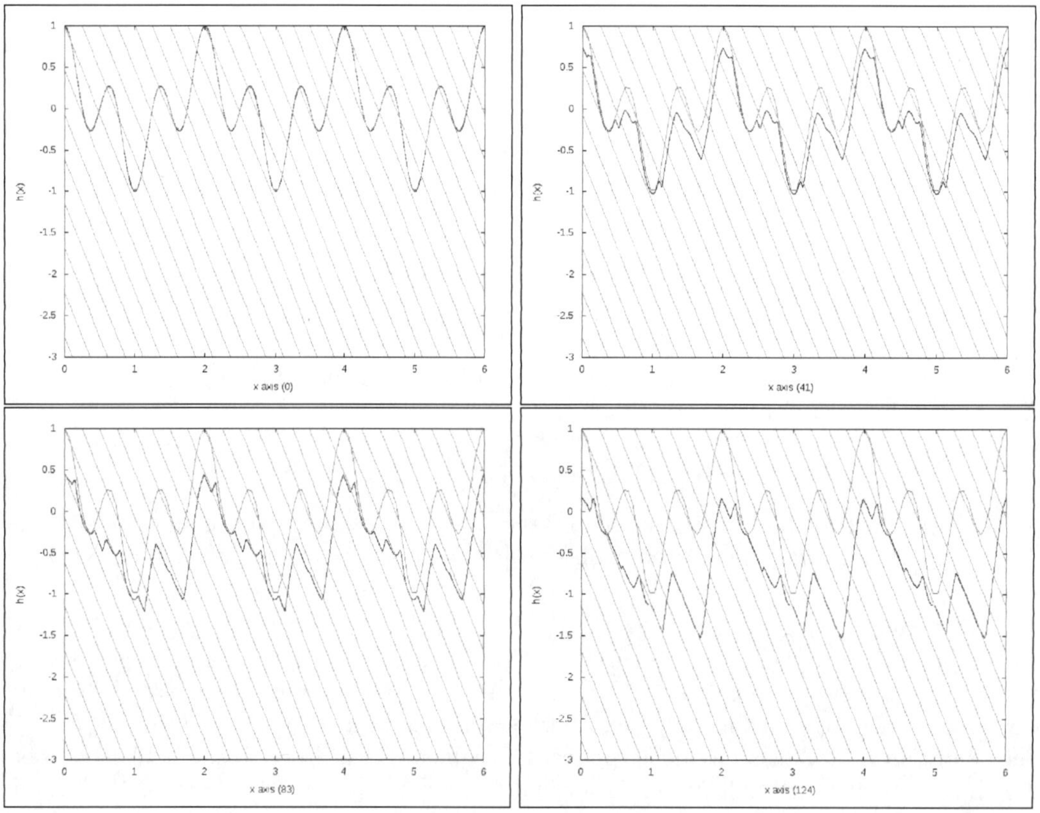

Figure 2. Time evolution

In the images, the time evolution of the penitentes is depicted. The parameters $D = 1$ and $K = 1$ in eq.(1) and eq.(2), respectively, have been taken. We consider $\alpha = 0.5$ in eq.(5), as well. The transversal lines correspond to the incident light rays. Initial and final shapes of the surface profile are given as a reference. Notice that the zones which are lower undergo more ablation than the upper ones. It is clear that, due to the presence of shady zones in the line of sight, the structures tilt in the direction of the coming sunlight. We will provide an extended version reporting on this issue in a forthcoming publication [4].

4. Acknowledgments
We thank R. Rojas and R. Soto for interesting discussions. P.G. was financially supported by Facultad de Ciencias Exactas, UNAB, to attend SOCHIFI 2016 Meeting.

References
[1] Lliboutry L 1954 *The origin of penitentes* Jour. Glaciol **2** 331
[2] Betterton M 2001 *Theory of structure formation in snowfields motivated by penitentes, suncups, and dirt cones*, Phys. Rev. E **63** 056129
[3] Bergeron V, Berger C and Betterton M 2006 *Controlled Irradiative Formation of Penitentes*, Phys. Rev. Lett. **96** 098502
[4] Guilleminot P and Olea R, manuscript in preparation.

IOP Conf. Series: Journal of Physics: Conf. Series **1043** (2018) 012008 doi:10.1088/1742-6596/1043/1/012008

Expectation values of general observables in the Vlasov formalism

Diego González[1,2], Abiam Tamburrini[1], Sergio Davis[2], Jalaj Jain[2], Gonzalo Gutiérrez[1]

[1]Grupo de Nanomateriales, Departamento de Física, Facultad de Ciencias, Universidad de Chile.

[2]Comisión Chilena de Energía Nuclear, Casilla 188-D, Santiago.

E-mail: dgonzalez@gnm.cl, abiamtamburrini@gmail.com, sdavis@cchen.cl

Abstract. Collisionless plasmas in an arbitrary dynamical state are described by the Vlasov equation, which gives the time evolution of the probability density $\rho(\boldsymbol{x}, \boldsymbol{v})$. In this work we introduce a new analytical procedure to generate particular partial differential equations (PDEs) for an arbitrary macroscopic observable $w(\boldsymbol{x}, \boldsymbol{v})$ that can be expressed as a function of positions and velocities, without solving the time evolution of the probability itself. This technique, which we will call the "Ehrenfest procedure" (as it produces relations that are analogous to Ehrenfest's theorem in Quantum Mechanics), is based on the iterative application of the fluctuation-dissipation theorem and the recently proposed conjugate variables theorem (CVT) in order to eliminate the explicit dependence on ρ. In particular, we show how this formalism is applied to the Vlasov equation for collisionless plasmas, and derive a general evolution equation for the fluctuations of any macroscopic property w in this kind of plasma.

1. Introduction

The standard treatment of nonequilibrium systems in Statistical Mechanics involves the solution of a partial differential equation (PDE) for the time-dependent probability density of microstates of the system. This is a formidable problem for systems with large number of degrees of freedom, and most of the time it is approximated using numerical methods on a discretized time and space grid. After one solves for the probability density, it is possible to take expectations to compute the nonequilibrium values of macroscopic quantities.

An interesting problem in plasma physics approached from the point of view of Statistical Mechanics is to obtain properties of collisionless plasmas [1] which are described by the Vlasov equation. This equation is a particular case of the Liouville theorem of conservation of the volume of phase space,

$$\frac{\partial \rho}{\partial t} + \{\rho, H\} = 0. \tag{1}$$

where $\rho(\boldsymbol{x}, t)$ is the time-dependent probability density of microstates $\boldsymbol{x}$. In the particular case of collisionless plasma, the Hamiltonian involves kinetic terms and the interaction of charged particles with electromagnetic fields, and Liouville's theorem becomes the Vlasov equation

$$\frac{\partial \rho}{\partial t} + \boldsymbol{v} \cdot \nabla_r \rho + \frac{q}{m}(\boldsymbol{E} + \frac{1}{c}\boldsymbol{v} \times \boldsymbol{B}) \cdot \nabla_v \rho = 0. \tag{2}$$

This equation is used to model collisionless plasmas out of equilibrium, where the force on each particle is given by the Lorentz force, with time-dependent electric and magnetic fields $\boldsymbol{E}(\boldsymbol{r}, t)$ and $\boldsymbol{B}(\boldsymbol{r}, t)$ respectively, and where those $\boldsymbol{E}$ and $\boldsymbol{B}$ are consistent with Maxwell's equations.

There is no general analytical solution to this problem for arbitrary electromagnetic fields (or arbitrary charge and current densities) and the numerical methods are extremely expensive in terms of computational cost, as they require the discretization of space and time with a grid size as small as the desired resolution, which depends of the highest frequency of modes of oscillation in the plasma, typically the frequency associated with the electrons [2].

In this work we show a general formalism that allows the construction of the particular PDE associated to an arbitrary, time-dependent macroscopic property $w(\boldsymbol{r}, \boldsymbol{v}, t)$ from the original PDE describing the time evolution of the probability density ρ. The development of this procedure, which we will call the "Ehrenfest procedure" (in reference to the Ehrenfest theorem in Quantum Mechanics), requires the application of two mathematical identities of probability theory, namely the fluctuation-dissipation theorem and the conjugate variable theorem, which we will introduce in the next section.

2. Mathematical Tools

As we have explained in the previous section, the objective of this procedure is to move from the problem of solving a PDE for the probability density function to the (usually much simpler) problem of solving a PDE for one of its macroscopic properties. In order to achieve this, we will employ two theorems of probability theory for continuous variables, that are of general validity. These are the Fluctuation-Dissipation Theorem (FDT) and the recently introduced Conjugate Variable Theorem (CVT), which will be used to eliminate the explicit dependence of the probability density function from the PDEs to be solved. In the following we will review the mathematical statement of these theorems.

2.1. Fluctuation-Dissipation Theorem

According to the definition of expectation, any time-dependent macroscopic property W_t can be written as

$$W_t = \big\langle w(x,t) \big\rangle_t = \int dx \rho(x,t) w(x,t), \tag{3}$$

where $w(x,t)$ is a macroscopic property, $\rho(x,t)$ is the probability density function and $x = (x^1, x^2, ..., x^n)$ is an instantaneous state of the system. The subindex t in the expectation $< \cdot >_t$ and W_t indicates the temporal dependence of those quantities. Taking partial derivatives of the definition of expectation with respect to time we get

$$\frac{\partial}{\partial t}\big\langle w(x,t) \big\rangle_t = \int dx [w(x,t)\frac{\partial}{\partial t}\rho(x,t) + \rho(x,t)\frac{\partial}{\partial t}w(x,t)],$$

and rewriting $\frac{\partial \rho(x,t)}{\partial t}$ as $\rho(x,t)\frac{\partial}{\partial t}\ln\rho(x,t)$ is possible to rewrite each term as an expectation. This leads to the so-called fluctuation-dissipation theorem,

$$\frac{\partial}{\partial t}\big\langle w(x,t) \big\rangle_t - \Big\langle \frac{\partial}{\partial t} w(x,t) \Big\rangle_t = \Big\langle w(x,t)\frac{\partial}{\partial t}\ln\rho(x,t) \Big\rangle_t. \tag{4}$$

This identity is valid for any expectation that depends parametrically on variables, as in this case the time t.

2.2. Conjugate Variable Theorem
In a similar way to the FDT, we can construct an identity involving spatial derivatives of an arbitrary observable $w(\boldsymbol{x},t)$ and of the probability density $\rho(\boldsymbol{x},t)$, the recently introduced conjugate variables theorem [3]. Its N-dimensional version reads

$$\Big\langle \nabla\cdot\boldsymbol{w}(\boldsymbol{x},t)\Big\rangle = -\Big\langle \boldsymbol{w}(\boldsymbol{x},t)\cdot\nabla\ln\rho(\boldsymbol{x},t)\Big\rangle, \tag{5}$$

where $\nabla = \frac{\partial}{\partial x_i}\hat{x}_i$ and $\boldsymbol{w}(\boldsymbol{x},t)$ is a macroscopic vector property dependent of the state of the system and time.

Both theorems (FDT and CVT) are expectation identities which are valid for states arbitrarily far from equilibrium, and they connect arbitrary expectations involving $w(\boldsymbol{x},t)$ and their derivatives, with expectations that include derivatives of $\ln\rho(\boldsymbol{x},t)$. In the following we will use this interesting feature of these theorems in order to produce differential equations for the expectation of $w(\boldsymbol{r},\boldsymbol{v},t)$, where we consider the state space as $\boldsymbol{x}=\boldsymbol{r}\otimes\boldsymbol{v}$.

3. Ehrenfest Procedure

Now we are able to present the analytical procedure that we will call the "Ehrenfest procedure", first employed in Ref. [4]. The idea is to obtain a PDE for the time-dependent expectation $\big\langle w\big\rangle_t$ of a macroscopic quantity $w(\boldsymbol{x},t)$ from the original PDE for the probability density $\rho(\boldsymbol{x},t)$.

In order to fix ideas, we will demonstrate the procedure on the Vlasov equation, which is a PDE for $\rho(\boldsymbol{r},\boldsymbol{v},t)$ of the form

$$\frac{\partial\rho}{\partial t}+\boldsymbol{v}\cdot\nabla_r\rho+\frac{\boldsymbol{F}}{m}\cdot\nabla_v\rho=0, \tag{6}$$

where $\rho=\rho(\boldsymbol{x},t)$ with $\boldsymbol{x}=\boldsymbol{r}\otimes\boldsymbol{v}$ and $\boldsymbol{F}$ is the Lorentz force, given by $\boldsymbol{F}=q(\boldsymbol{E}+\frac{1}{c}\boldsymbol{v}\times\boldsymbol{B})$.

The procedure is divided in three stages. First, we rewrite the PDE for ρ as a PDE for $\ln\rho$ by using the relations

$$\frac{\partial}{\partial t}\rho=\rho\,\frac{\partial}{\partial t}\ln\rho, \tag{7}$$

$$\nabla_x\rho=\rho\nabla_x\ln\rho. \tag{8}$$

Once we have put the original PDE in logarithmic form,

$$\rho\,\frac{\partial}{\partial t}\ln\rho+\rho\,\boldsymbol{v}\cdot\nabla_x\ln\rho+\rho\,\frac{\boldsymbol{F}}{m}\cdot\nabla_v\ln\rho=0. \tag{9}$$

we introduce expectations by applying the operator $\int d\boldsymbol{r}d\boldsymbol{v}\;w\cdot$ with the test function $w(\boldsymbol{x},\boldsymbol{v},t)$ defined over the state space) and using the definition of expectation,

$$\Big\langle w(\boldsymbol{r},\boldsymbol{v},t)\Big\rangle_t=\int d\boldsymbol{r}d\boldsymbol{v}w(\boldsymbol{r},\boldsymbol{v},t)\rho(\boldsymbol{r},\boldsymbol{v},t), \tag{10}$$

we obtain

$$\int d\boldsymbol{r}d\boldsymbol{v}\;\rho\left[w\frac{\partial}{\partial t}\ln\rho+w\boldsymbol{v}\cdot\nabla_x\ln\rho+w\frac{\boldsymbol{F}}{m}\cdot\nabla_v\ln\rho\right]=0,$$

that is,

$$\Big\langle w\frac{\partial}{\partial t}\ln\rho+w\boldsymbol{v}\cdot\nabla_x\ln\rho+w\frac{\boldsymbol{F}}{m}\cdot\nabla_v\ln\rho\Big\rangle=0. \tag{11}$$

Finally, we use CVT and FDT to eliminate $\ln\rho$ from each expectation, in order to reduce them to expectations involving derivatives (spatial or temporal) of w. More precisely, we eliminate $\partial_t \ln\rho$ using the FDT, and $\nabla \ln\rho$ using CVT. In this case, we obtain

$$\frac{\partial}{\partial t}\big\langle w\big\rangle_t = \Big\langle \frac{\partial w}{\partial t} + \partial_{\mathbf{x}}(w\mathbf{v}) + \partial_{\mathbf{v}}(w\frac{\boldsymbol{F}}{m})\Big\rangle,$$

which can be expanded as

$$\frac{\partial}{\partial t}\big\langle w\big\rangle_t = \Big\langle \frac{\partial w}{\partial t}\Big\rangle_t + \Big\langle \boldsymbol{v}\partial_x w + w\partial_x \boldsymbol{v}\Big\rangle_x + \Big\langle \frac{\boldsymbol{F}}{m}\partial_v w + w\partial_v \frac{\boldsymbol{F}}{m}\Big\rangle_v. \tag{12}$$

Here we notice that we can simplify this result even further. As $\boldsymbol{v}$ is not dependent of $\boldsymbol{x}$ ($\nabla_{\boldsymbol{x}}\boldsymbol{v} = 0$) and for the case of the Lorentz force $\nabla_{\boldsymbol{v}} \cdot \boldsymbol{F} = 0$, we arrive at our main result, which is a PDE for an arbitrary property $\big\langle w\big\rangle_t$,

$$\frac{\partial}{\partial t}\big\langle w\big\rangle_t = \Big\langle \frac{\partial w}{\partial t}\Big\rangle_t + \Big\langle \boldsymbol{v}\cdot\partial_{\boldsymbol{x}} w\Big\rangle_t + \Big\langle \frac{\boldsymbol{F}}{m}\cdot\partial_{\boldsymbol{v}} w\Big\rangle_t. \tag{13}$$

This resembles a classical analog of Ehrenfest's theorem in Quantum Mechanics [5],

$$\frac{d}{dt}\Big\langle \hat{\Omega}\Big\rangle = \Big\langle \frac{\partial\hat{\Omega}}{\partial t}\Big\rangle - \frac{i}{\hbar}\Big\langle \big[\hat{\Omega}, \hat{H}\big]\Big\rangle. \tag{14}$$

where the Poisson bracket has to be used instead of the quantum-mechanical commutator.

4. Results

Using the main result of this work, namely Eq. 13, it is possible to obtain the PDE for the fluctuations of an arbitrary quantity $w(\boldsymbol{r}, \boldsymbol{v}, t)$. We will illustrate this by considering the velocity fluctuations $\big\langle(\delta v)^2\big\rangle_t$, and for this we will use $w = v$ and $w = v^2$ separately in Eq. 13. First, we obtain the time evolution of $\boldsymbol{v}$ as

$$\frac{\partial}{\partial t}\big\langle \boldsymbol{v}\big\rangle_t - \big\langle\frac{\partial \boldsymbol{v}}{\partial t}\big\rangle_t + \Big\langle \boldsymbol{v}\partial_{\boldsymbol{x}}\boldsymbol{v}\Big\rangle_t + \Big\langle \frac{\boldsymbol{F}}{m}\partial_v \boldsymbol{v}\Big\rangle_t, \tag{15}$$

but $\partial_t \boldsymbol{v} = 0$ and $\partial_{\boldsymbol{x}}\boldsymbol{v} = 0$, so we simplify it as

$$\frac{\partial}{\partial t}\big\langle \boldsymbol{v}\big\rangle_t = \Big\langle \frac{\boldsymbol{F}}{m}\Big\rangle_t. \tag{16}$$

This has a clear meaning: the change in the mean value of the velocity is given by the mean acceleration. Now, we obtain the evolution equation for $w = v^2$,

$$\frac{\partial}{\partial t}\big\langle v^2\big\rangle_t = \frac{2}{m}\Big\langle \boldsymbol{F}\cdot\boldsymbol{v}\Big\rangle_t. \tag{17}$$

We can join Eqs. 16 and 17 by taking

$$\frac{\partial}{\partial t}\Big\langle (\delta v)^2\Big\rangle_t = \frac{\partial}{\partial t}\Big\langle v^2\Big\rangle_t - 2\Big\langle \boldsymbol{v}\Big\rangle_t \cdot \frac{\partial}{\partial t}\Big\langle \boldsymbol{v}\Big\rangle_t, \tag{18}$$

and we can write a PDE for $\big\langle(\delta \boldsymbol{v})^2\big\rangle_t = \big\langle v^2\big\rangle_t - \big\langle \boldsymbol{v}\big\rangle_t^2$ of the form

$$\frac{\partial}{\partial t}\big\langle(\delta v)^2\big\rangle_t = \frac{2}{m}\Big\{\big\langle q\boldsymbol{E}\cdot\boldsymbol{v}\big\rangle_t - \big\langle \boldsymbol{v}\big\rangle\big\langle q\boldsymbol{E}\big\rangle - \frac{q}{c}\big\langle \boldsymbol{v}\times\boldsymbol{B}\big\rangle\big\langle \boldsymbol{v}\big\rangle_t\Big\}. \tag{19}$$

Here is interesting to note that, for the case where electromagnetic fields $\boldsymbol{E}$ and $\boldsymbol{B}$ are spatially homogeneous, the right-hand side vanishes and we have that the size of the velocity fluctuations is kept constant in time.

5. Concluding remarks

We have shown the application of our "Ehrenfest procedure" to the Vlasov equation for collisionless plasmas. Our main result here was a general relation, Eq. 13, which gives the time evolution of any macroscopic property in a Vlasov system. From the general Ehrenfest relation we have derived the partial differential equation for the velocity fluctuations, Eq. 19, which indicates that for a collisionless plasma under spatially homogeneous electromagnetic fields, its initial velocity distribution does not widen with time. We can see that the Ehrenfest procedure is a general tool applicable whenever we have a system that follows a continuity equation for probability. It allows to quickly obtain dynamical evolution equations for particular properties, and thus appears as a powerful addition to the study of dynamical systems and non-equilibrium Statistical Mechanics.

Testing of Eq. 13 and 19 on different configurations of fields remains as future work, to be connected with numerical simulations of Vlasov systems.

Acknowledgements

SD and DG gratefully acknowledge funding from FONDECYT grant 1140514. DG acknowledges funding from CONICYT PhD Fellowship 21140914.

References

[1] Bellan P M 2006 *Fundamentals of Plasma Physics* (Cambridge University)
[2] Birdsall C and Langdon A 1991 *Plasma Physics via Computer Simulation*
[3] Davis S and Gutiérrez G 2012 *Phys. Rev. E* **86** 051136
[4] Davis S and González D 2015 *J. Phys. A: Math. Theor.* **48** 425003
[5] Shankar R 1994 *Principles of Quantum Mechanics* (Plenum Press)

IOP Conf. Series: Journal of Physics: Conf. Series **1043** (2018) 012009 doi:10.1088/1742-6596/1043/1/012009

Dynamics of the d-HMF model: Sensitive dependence on size and initial conditions

Boris Atenas and Sergio Curilef

E-mail: scurilef@ucn.cl
Departamento de Física, Universidad Católica del Norte, Av. Angamos 0610, Antofagasta, Chile.

Abstract. We characterize the dynamics of systems with type-dipole interactions included in a new Hamiltonian mean field model [proposed by Atenas and Curilef, Phys. Rev. E (2017) 022110]. The sensitivity on the initial conditions is a particular behavior that we try to describe properly using molecular dynamics. We carry out simulations to focus our attention in the primary evolution before reaching the equilibrium. The current anomalies seem to characterize the general behavior of systems with long-range interactions. The dynamics of the system remains in quasi-stationary states before arriving at equilibrium.

1. Introduction

The main motivation in this work is the potential future applications in nanoscience and nanotechnology, where systems of rotors are mounted on surfaces or inside solids of particular interest. Until now, by far most molecular rotors have been studied in solution. After all, the rotor molecules need to be synthesized and their basic characteristics, such as rotational barriers, were established before mounting them on surfaces or examining them inside solids, and such synthesis and characterization are nearly always performed in solution[1]. Besides, to many authors, the investigation of molecular rotors freely floating in a solution is in itself a fascinating concern[1]. A variety of molecular systems have been artificially created, which can exhibit controlled rotational motion [2]. In the development of such systems, the key step is the addition of communication between molecules in a network. Therefore, the total energy of the rotor network, for the observed synchronized rotation mechanism, is analyzed through a model that is described by energy terms composed by: the internal rotational kinetic energy, the dipole energy, the external field energy, the structural energy including the dispersion interactions in the network, and the thermal energy corresponding to a substrate special temperature [2]. In our perspective, we approach the model taking into account just two terms in a Hamiltonian, namely the classical rotational energy and dipole potential in the mean field approximation.

In this regard, families of Hamiltonian mean field models may give a favorable outlook. Such models are related to systems with long-range interactions, which are very common in nature. They are observed from the atomic to the astronomical scale and exhibit some anomalies, such as inequivalence of ensembles, negative heat capacity, ergodicity breaking, non-equilibrium phase transitions, quasi-stationary states and anomalous diffusion [3, 4]. These anomalies are exacerbated when special initial conditions are imposed [4]. This is a topic that we also discuss in the course of the current paper. Canonical ensemble in statistical mechanics does not explain the molecular dynamics on a short-time scale, though it proves to be correct on a larger time scale; i.e., canonical and microcanonical, whose phase diagrams are overlapped only at equilibrium.

As early concluded by Lynden-Bell statistics[6], a description based on Vlasov dynamics that preserves the hypervolume of phase-space density levels and uses the distribution functions as the solutions of Vlasov equation, that systems with long-range interactions can be trapped in quasi-stationary states (QSS), before going to equilibrium. However, the limitation of this theory is that properties similar to the usual statistical mechanics, as ergodicity and mixing, are required, which are not valid for systems with long-range interactions, where those properties are broken.

The Hamiltonian mean field (HMF) model [4, 5] is axiomatic and useful to characterize general properties of systems with long-range interactions. The existence of quasi-stationary states (QSS) before reaching equilibrium constitutes a fingerprint of this kind of systems, whose behavior of the kinetic energy and other thermodynamic observables are partly used to characterize the QSS [5, 6] in physical models and systems.

Some interesting systems related to charges, spins, rotors and dipoles are previously discussed in literature; for instance, charged particles[7] and noninteracting dipoles[8] in an external magnetic field, long range interactions in a type of Ising model and other generalized models [9, 10].

2. Basic equations

We consider N identical coupled dipole-type particles with a mass equal to 1, whose dynamics evolves in a periodic cell described by a 1D, dipole-type Hamiltonian mean field (d-HMF) model as previously proposed[11], given by

$$H=\sum_{i=1}^{N}\frac{p_i^2}{2}+\frac{\epsilon}{2N}\sum_{i\neq j}^{N}[\cos(\theta_i-\theta_j)-3\cos\theta_i\cos\theta_j-\Delta_{i,j}], \quad (1)$$

where the variable p_i represents the momentum of the particle i and θ_i represents its corresponding orientation angle. With the integer $i \in [1, N]$, N is the size of the system. In Figure 1 we illustrate two dipoles in a ring, whose angles are defined by the position the ring. Initial condition are defined in $t = 0$ and the illustration in $t \neq 0$ is arbitrary. The parameters ϵ and $\Delta_{i,j}$ denote the coupling and initial conditions, respectively. The parameter $\Delta_{i,j}$ suitably establishes, as shown[11], the zero of the potential energy.

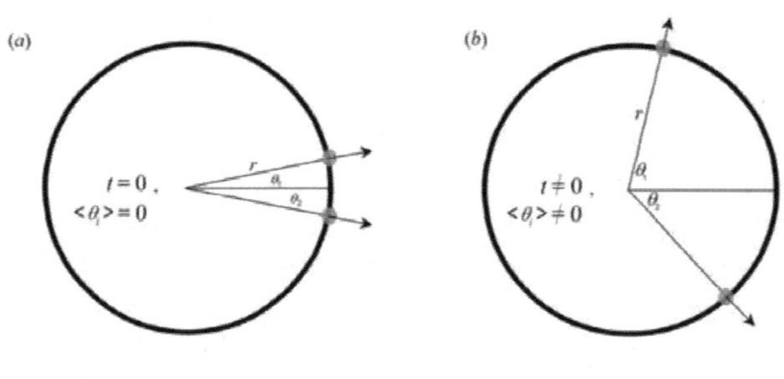

Figure 1. Two dipoles are shown in a ring. a) For $t = 0$, initial conditions are represented. b) For $t \neq 0$, an arbitrary state is illustrated.

2.1. Initial distribution

The interaction coupling is rescaled by the number of particles, to make the potential thermodynamically extensive[9, 10, 11, 12]. We take a ferromagnetic system with a coupling,

ϵ, positive. The application of numerical methods has become an acceptable tool for the study and characterization of anomalies nonequilibrium. Following a previous line of research, some details related to these interactions are shown. Therefore, we can introduce the total spin vector

$$\vec{M} = \left(\frac{1}{N}\sum_{i=1}^{N}\cos\theta_i, \frac{1}{N}\sum_{i=1}^{N}\sin\theta_i\right) \tag{2}$$
$$= (M_x, M_y) \tag{3}$$
$$= M\exp(i\phi), \tag{4}$$

where (M_x, M_y) are the components of the vector $\vec{M}$, and $M = |\vec{M}|$. In addition, ϕ denotes the phase of the order parameter. The equation of motion is denoted by the following set of coupled equations given by

$$\dot{p}_i = -\frac{\epsilon}{2N}\left(2M_x\sin\theta_i + M_y\cos\theta_i\right), \tag{5}$$

where M_x and M_y represent the coupling terms . Then, the potential can be written as

$$V = -\frac{\epsilon}{2N}(2M_x^2 - M_y^2 - \Delta), \tag{6}$$

where $\Delta = \sum_{i,j}\Delta_{i,j}$ as aforementioned defines the zero of the energy[11]. Moreover, we see that $M_y \approx 0$ for all time of the evolution. Therefore, $M \approx M_x$ that it allows us writing the internal energy per particle can be written as:

$$U = \langle K\rangle - \frac{\epsilon}{2}(2M^2 - \Delta), \tag{7}$$

where$\langle K\rangle$ represents the mean value of the kinetic energy.

3. Discussion of Results

In this section, we introduce and discuss some results that characterize the present model. Numerical simulations in the microcanonical ensemble are carried out through molecular dynamics. The evolution of the system is characterized by two time intervals of QSS.

3.1. Initial distribution

The evolution of thermodynamical properties of the system is discussed by carrying out molecular dynamics simulation using a symplectic integrator[13]. At this stage, the relevant consideration is the choose of the initial distribution that we know as the water bag initial conditions (WBIC). With this respect, let us take the particles uniformly distributed in phase space. Thus, the general definition of the WBIC is related with the probability distribution function, $f(\theta, p)$, as

$$f(x,p)\,\mathrm{d}\theta\,\mathrm{d}p = \frac{\mathrm{d}\theta\,\mathrm{d}p}{4\,\theta_0\,p_0}, \tag{8}$$

where $-p_0 \leq p \leq p_0$ and $-\theta_0 \leq \theta \leq \theta_0$, which is properly normalized in all the space. If $\theta_0 = \pi$ the function $f(x,p)$ may be called "homogeneous in space" and it is a stationary solution of the Vlasov equation[14]. In the inhomogeneous case, $0 < \theta_0 < \pi$, it is expected that the dynamics of the system evolves in a self-consistent way to satisfy its pertinent thermodynamic properties. As will see, we mainly discuss in the present contribution, the case where θ_0 is close to zero and it corresponds to the case where the dipoles are aligned almost in the same direction, preparing the samples with external fields and disconnecting them in $t = 0$ to make the effort of characterizing the evolution of the properties of the system with this kind of WBIC because the anomalous behavior is exacerbated.

In Figure 2, we represent the average dynamics of the system by the evolution of $2\langle K\rangle/N$ in 100 samples with $N = 25.000$ particles. The difference among curves comes from three different initial conditions that we consider. We take the WBIC given by Eq.(8) with a typical value of p_0 and three different values of θ_0, which typically correspond to the extremely inhomogeneous case, $\theta_0 \approx 0$, an intermediate inhomogeneous case, $\theta_0 = \pi/7$, and the completely homogeneous cases, $\theta_0 = \pi$. In the inset of Figure 2, the first evolution is known as violent relaxation, which is clearly shown when WBIC is inhomogeneous; nevertheless, for the homogeneous case the quantity $2\langle K\rangle/N = 0.76$ coincides with its numerical value for the second QSS. While the dynamics of the HMF model does not present evident differences if variations on the WBIC are considered, the d-HMF shows two different plateaus when the WBIC departs from the homogeneous case. Therefore, the results, shown in Figure 2, emphasize the relevance of the initial conditions in the dynamics of the problem. The appearance of two plateaus, before reaching the equilibrium, is due to considering the inhomogeneous WBIC.

3.2. Quasi-stationary states

We remark here some thermodynamic properties that we obtain from the simulations, related to effects of finite size and finite time, where QSS have been observed. In the dynamics of the model, it is important to observe where the kinetic energy is constant. As far as we know, this is relevant for determining the general behavior of other thermodynamic observables.

In Figure 3 (a), the time evolution of $2\langle K\rangle/N$ is reported for several sizes of the system, this is $N = 8.000$, 16.000, 32.000, 70.000 and 170.000. The limiting cases are indicated in the figure, these are $N = 8.000$ and $N = 170.000$. The quantity $2\langle K\rangle/N$ shows two plateaus, whose value and duration clearly depend on N. If N increases, the value decreases, however, the duration increases. States defined by these plateaus, known as the QSS, are notoriously lower than the canonical temperature. It is known that WBIC stimulates the time that the systems remain in states of non-equilibrium, which are the QSS in the current case. They need times that

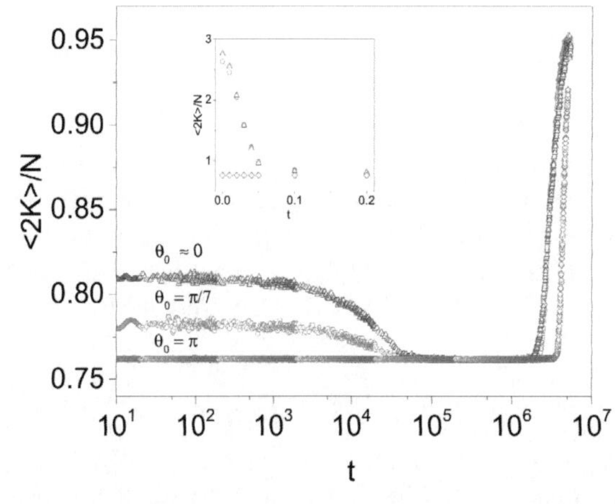

Figure 2. The evolution of the average over 100 samples of the kinetic energy with $N = 25.000$ as a function of the time. We see the evolution from three different initial condition $\theta_0 = 10^{-4}, \pi/7$ and π. In the homogeneous case, this is $\theta_0 = \pi$, the behavior is similar to the HMF model. In the extremely inhomogeneous case, $\theta_0 \approx 0$, the anomaly is the greatest and this is te case that we use to characterize the two QSS that are found in this dynamics. In the inset, the first evolution is shown, which is known as violent relaxation in linear scale.

increase with N to transit from the first QSS to the second QSS, finally to relax to the canonical equilibrium. Averages represented in Figure 3(a) are taken in 100 samples for $N = 8.000$ and in 32 samples for $N = 16.000, 32.000, 70.000$ and 170.000.

As discussed before, the behavior of systems is sensitive to the WBIC, which is remarked in Figure 3 (a). In addition, these results evoke the behavior of systems that strongly depend on the size of systems. Properties slowly reach the thermodynamic limit. Results seem to indicate that the difference, between the two plateaus, tends to disappear as $N \to \infty$. Therefore, we indicate the first plateau as a QSS-type, contrary the second plateau that seems to define a true QSS. Hence, it is expected that the dynamics of the system has only one plateau when the thermodynamics limit is reached. Thus, the two plateaus is a peculiar effect induced by applying the inhomogeneous WBIC and by considering finite size of systems.

For simulation with $N = 32.000$ and greater, it is not possible to observe when the mean kinetic energy goes to the equilibrium in the time interval that we have carried out our molecular dynamics simulation, because of the computational limitations. Hence, a detailed characterization of the duration of the second QSS as a function of N, the size of the system, is a good challenge for future works. However, it is expected that the behavior becomes similar to the QSS observed for the HMF model. Nevertheless, if we focus our attention in the first plateau, this is the duration of the first QSS as $\tau(N)$, we observe that such a value depends on N according a power law given by $\tau(N) \propto N^{\mu}$. In Figure 3 (b) we obtain the approximate slope in log-log scale, that gives $\mu \approx 0.7$. This slope, μ, is less than the μ known for the unique QSS that is observed in the HMF model[14]. The duration of the second QSS will be studied in the future, but we expect that becomes similar to the unique QSS that shows the HMF model[14].

4. Summary and Concluding Remarks

In summary, the d-HMF model(1) gives a good perspective for studying systems with long-range interactions.

It is generally accepted that two phases appear in the phase diagram that depend on the energy and, consequently, the temperature. At low energy, a phase identified by the presence of a single cluster of particles arises by floating in a diluted homogeneous background. At high energy a homogeneous phase is recovered; the cluster disappears and the particles move (almost) freely. The system is characterized by the microcanonical ensemble with negative specific heat and the resulting instability is extremely relevant[4] because of its strong implications on both experimental and theoretical features. It is expected that the non-equilibrium distribution related to the single cluster evolves into another non-equilibrium distribution, before reaching equilibrium. This behavior corresponds to an apparent thermodynamical inconsistency as explained by Hertel and Thirring[15], who proposed that the canonical and microcanonical ensembles are not equivalent near the transition region. We have successfully confirmed this proposal at least in an interval of time.

To describe the evolution of the model proposed here, we performed molecular dynamics simulations to obtain the QSS before arriving at equilibrium. Due to the non-equilibrium states, we can define two different QSS, as depicted in Figure 3. This observation is novel compared with other models that are known in the literature. We characterize the QSS as a constant value of the average kinetic energy. The second QSS has a value less than the first QSS.

Finally, we remark that the initial conditions and the size of the systems are two relevant considerations that we use in the study of the present topic. Thermodynamic and dynamic properties strongly depend on N the size of systems that we partially characterize for the d-HMF model. We focus the analysis considering three possible initial distributions that correspond to the homogeneous WBIC and two cases of inhomogeneous WBIC. We see that if the homogeneous WBIC is considered, the behavior of the dynamic from the d-HMF model seems to be similar to the standard HMF model. After the process of violent relaxation, the system remains trapped

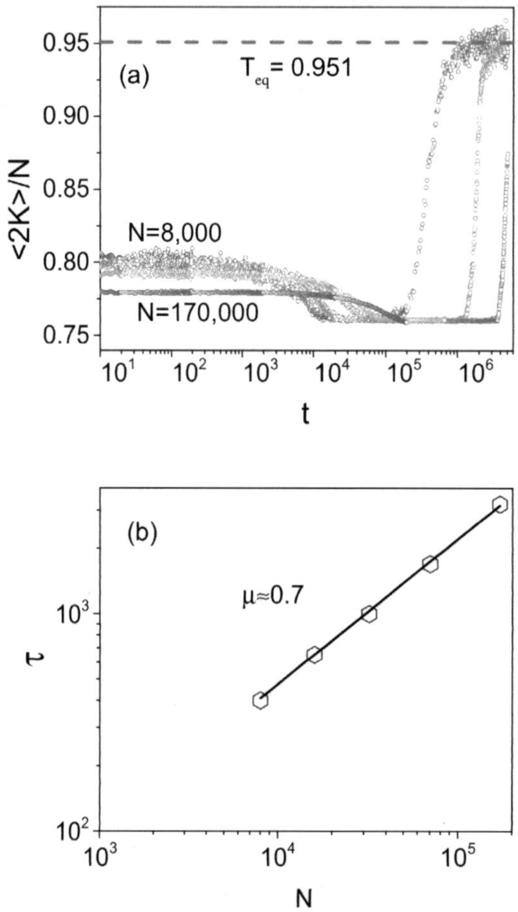

Figure 3. The dependence on N, the size of the system, of the twice mean kinetic energy is illustrated. The quantity $2\langle K\rangle/N$ is depicted as a function of the time. In (a), several sizes of the systems are considered. The cases indicated in the figure are the limiting ones: $N = 8.000$ and 170.000. Intermediate cases correspond to $N =$ 16.000, 32.000, 70.000. In (b), we depict τ, the relaxation time of the first QSS that goes to second QSS, as a function of $1/N$ in log-log scale to obtain a power law of the duration of the first QSS in term of the size of the system. Thus, we expect that the lifetime observed in the in the first QSS behaves as $\tau(N) \propto N^{0.7}$.

in a unique QSS before reaching the equilibrium. However, if the system is prepared in an inhomogeneous WBIC, the dynamics of the system shows two different plateaus before arriving at equilibrium as illustrated in Figure 2. The duration τ of the first QSS depend on a power law of N, typically this is $\tau(N) \propto N^{0.7}$ as depicted in Figure 3. Due to the cost of the computational time, the description of the second QSS will be made in future works.

References

[1] S. Erbas-Cakmak, D. A. Leigh, C. T. McTernan, and A. L. Nussbaumer, Chem. Rev., **115** (2015) 10081-10206.

[2] Y. Zhang, H. Kersell, R. Stefak, J. Echeverria, V. Iancu, U. G. E. Perera, Y. Li, A. Deshpande, K.-F. Braun, C. Joachim, G. Rapenne and S.-W. Hla, Nature Nanotechnology **11** (2016) 706-712.

[3] Y. Levin, R. Pakter, F.B. Rizzato, T.N. Teles and F.P. da C. Benetti, Physics Reports **535** (2014) 160.
[4] M. Antoni and S. Ruffo, Phys. Rev. **E 52** (1995) 2361.
[5] A. Pluchino, V. Latora and A. Rapisarda, Continuum Mech. Thermodyn. **16** (2004) 245-255.
[6] D. Lynden-Bell, Monthly Notices of the Royal Astronomical Society **136** (1967) 101.
[7] S. Curilef and F. Claro, Am. J. Phys. **65** (1997) 244-250
[8] B. Atenas, L. A. del Pino and S. Curilef, Annals of Physics **350** (2014) 605-614
[9] L.A. del Pino, P. Troncoso and S. Curilef, Phys. Rev. **B 76** (2007) 172402;
S. Curilef, L. A. del Pino and P. Orellana, Phys. Rev. **B 72** (2005) 224410.
[10] S. Curilef, Phys. Lett. **A 299** (2002) 366-370.
[11] B. Atenas, S. Curilef, Phys. Rev. **E 95** (2017) 022110.
[12] M. Kac, G. Uhlenbeck, P.C. Hemmer, J. Math. Phys. **4** (1963) 216.
[13] H. Yoshida, Physics Letters **A 150** (1990) 262.
[14] Y. Yamagushi, J. Barré, F. Bouchet, T. Dauxois and S. Ruffo, Physica **A 337** (2004) 36-66.
[15] P. Hertel and W. Thirring, Ann. Phys. **63** (1970) 520.

Thermal conductivity of argon: Modeling and comparison with experimental data

Henrry Revollo and Sergio Curilef

Departamento de Física,Universidad Catolica del Norte, Angamos 610, Antofagasta

E-mail: hrz002@aulumnos.ucn.cl

Abstract. The thermal conductivity of argon in solid state phase is modeled by studying the physics property of atom vibrations given by different potentials in crystalline networks at temperature less than 80 $[K]$, in the canonical ensemble where volume, mass and temperature are held constant. Mfolecular dynamics simulations are carried out taking into account the Verlet algorithm and the Green-Kubo formalism. We consider some interatomic interactions as Lennard-Jones, Mie and Kihara potential to obtain results that we compare to experimental data. We show that the present modeling based on the Kihara potential is the best, showing a precision (8.8%), which is not found in recent literature.
Keywords: Thermal conductivity; Interatomic interaction potential; Molecular dynamics.

1. Introduction

In physics, the thermal conductivity, λ, is the property that allows the heat conduction[1] in materials, which is possible to be assessed via the Fourier law, given by

$$\mathbf{J} = -\lambda \nabla T, \tag{1}$$

where $\mathbf{J}$ is the heat flux in units $\left[\frac{W}{m^2}\right]$, ∇T is the gradient of the temperature inside of the material in $\left[\frac{K}{m}\right]$ and λ is the thermal conductivity in units $\left[\frac{W}{mK}\right]$. In materials of low conductivity, the heat transfer is less than in materials of high conductivity. In one case, the materials of high conductivity are widely used as heat dissipators. In the contrary case, materials with low conductivity are used as thermal insulators. The thermal conductivity of materials depends on the temperature, and its inverse is the thermal resistivity[2].

There are proper tools for determining the thermal conductivity for steady and transient states. In general, techniques for the steady states are useful when the temperature is unchanged in time [2]. At atomic level, there are no simple expressions to phrase the thermal conductivity. Then, there are two perspectives to make this calculation, those are:

(i) Green-Kubo relations, which are employed in the present paper. Although these are analytical, the calculation is carried out by molecular dynamics (MD) for dense fluid and solids[2].

(ii) Relaxation time approximation. The heat transfer is because of the anharmonicity of the potential in the crystal, which is known as phonon conductivity, studied by the Monte Carlo method[2].

The MD simulation is for now a well-established method to calculate several transport and thermal properties in fluids [3]. The Lennard-Jones potential has been considered a pivotal model[4], which takes the molecules as hard spheres containing two main characteristics: repulsive at short distance and attractive at intermediate and long distances[5, 6]. There are other models related to potentials of interacting molecules[7]. In this work, we use the following potentials:

$$\phi_{L-J}(r_{ij}) = 4\epsilon\left[\left(\frac{\sigma}{r_{ij}}\right)^{12} - \left(\frac{\sigma}{r_{ij}}\right)^{6}\right] \quad \text{Lennard-Jones,} \tag{2}$$

$$\phi_{M14-7}(r_{ij}) = 4\epsilon\left[\left(\frac{\sigma}{r_{ij}}\right)^{14} - \left(\frac{\sigma}{r_{ij}}\right)^{7}\right] \quad \text{Mie (14-7),} \tag{3}$$

$$\phi_{M10-5}(r_{ij}) = 4\epsilon\left[\left(\frac{\sigma}{r_{ij}}\right)^{10} - \left(\frac{\sigma}{r_{ij}}\right)^{5}\right] \quad \text{Mie (10-5,)} \tag{4}$$

$$\phi_{K}(r_{ij}) = 4\epsilon\left[\left(\frac{\sigma(1+a)}{r_{ij}+a\sigma}\right)^{12} - \left(\frac{\sigma(1+a)}{r_{ij}+a\sigma}\right)^{6}\right] \quad \text{Kihara} \tag{5}$$

where r_{ij} is the distance between the i-th and the j-th atom, the parameters L-J, Kihara and Mie are: $\epsilon/k_B = 119.8[K]$ and $\sigma = 3.405[\text{Å}]$, and additionally for the Kihara potential we use $a = 0.01$. k_B is the Boltzmann constant. The Kihara potential is defined by Eq.(5). For its parameters we use $a << 1$. The Kihara potential is expanded in Taylor series in terms of a that we show explicitly at first order of approximation, as

$$\phi_K(r_{ij}) = \phi_{L-J}(r_{ij}) - 24\epsilon\sigma\left[2\frac{\sigma^{12}}{r_{ij}^{13}} - \frac{\sigma^6}{r_{ij}^7} - 2\frac{\sigma^{11}}{r_{ij}^{12}} + \frac{\sigma^5}{r_{ij}^6}\right] a + \mathcal{O}(a^2). \tag{6}$$

Therefore, the Kihara potential can be written as the L-J potential plus correction terms. Furthermore, the thermal conductivity λ is calculated as [5, 8, 9, 10]:

$$\lambda = \frac{1}{DVk_BT^2}\int_0^\infty \langle \mathbf{j}(0)\,\mathbf{j}(t)\rangle\, dt, \tag{7}$$

where V is the volume, T is the temperature, the symbol $\langle ... \rangle$ represents the time average and the factor $D = 3$ is related to the dimensions of the systems. The microscopic current is given by[5, 8, 9, 10]:

$$\mathbf{j}(t) = \sum_i \mathbf{v}_i e_i + \frac{1}{2}\sum_{i,j,i\neq j} \mathbf{r}_{ij}\left[\mathbf{F}_{ji}\cdot \mathbf{v}_i\right], \tag{8}$$

where $\mathbf{v}_i$ is the velocity of the particle i and $\mathbf{F}_{ij}$ is the force over the particle i owing he particle j. The energy is estimated by[5, 8, 9, 10].

$$e_i = \frac{1}{2}\left[m_i\left|\mathbf{v}_i\right|^2 + \sum_j \phi(\mathbf{r}_{ij})\right], \tag{9}$$

where $\phi(\mathbf{r}_{ij})$ stands for different potentials defined by the Eqs.(2)-(5). The Green-Kubo formalism allows us assessing other dynamic properties as the viscosity, electric conductivity, etc., from simulations of atomic systems. Anyway, calculations are made in thermodynamic equilibrium conditions because of results converge for large sizes of systems and large scales of time. The size of samples are considered as 108, 256, 500, 864, 1372, 2048, 2916 and 4000 particles, with computation time as 10^6, with temperatures less than 80 $[K]$.

There are reported previous calculations through MD simulation with precision greater than 10% compared with experimental data [11, 12, 13]. Some variations of the L-J potential are proposed in the literature, for instance L-J potential of Kihara-type [13] and Mie (14,7) potential[7, 14].

The main aim in the current contribution is to compare the thermal conductivity of argon obtained from intermolecular interaction proposed in Eqs.(2)-(5) to experimental data[15, 16]. We expect to perform MD simulations in the NVT ensemble to find out which model gives the best precision related to referred experimental data.

2. The model

At the beginning of simulation, atoms in fluids are settled on face-centered cubic (FCC) [1, 17]. The argon at environment temperature is at gaseous state phase.

2.1. Simulation

The MD simulation is a method which solves numerically the equations of motion for a system of N interacting particles and derive dynamical, thermodynamical and structural properties of the system[17, 18].

2.2. Initial conditions

The system evolves keeping the particle number (N), the volume (V) and the average temperature (T) fixed. This outlook is known as canonical ensemble or NVT ensemble. According to statistical mechanics we know that the translation velocities of atoms in a classical system are distributed by the Maxwell-Boltzmann distribution. Then, if the temperature of the system T, the probability of velocity of an atom is between v_r and $(v_r + dv_r)$ is:

$$f(v_r)dv_r = \sqrt{\frac{m_i}{2\pi k_B T}}\, e^{\left(\frac{m_i}{2k_B T} v^2\right)}\, dv_r. \tag{10}$$

The initial values of velocities are related to a distributions given by

$$T = \frac{2}{N_{df} k_B} \langle K \rangle , \tag{11}$$

where N_{df} corresponds to the freedom degree of the system and $\langle K \rangle$ is the mean kinetic energy. The following expression can be instantaneously used

$$T(t) = \frac{2}{N_{df} k_B} K(t). \tag{12}$$

This equation allows us relating the instantaneous temperature to velocities after starting the simulation[19].

2.3. Periodic boundary conditions

The simulation is frequently carried out using a finite and small number N of particles because of the time spent to obtain results increases with N^2. In the current contribution we consider a sample of N particles where $108 \leq N \leq 8000$, which is found in a central cube. To apply periodic boundary conditions means to make infinite replicas of the original cube. Image particles move in the same way as the original particle does. If a particle goes out of the original cube, other particle comes in by the opposite size. Thus, the total number of particles inside the original cube and its replicas remains constant[20].

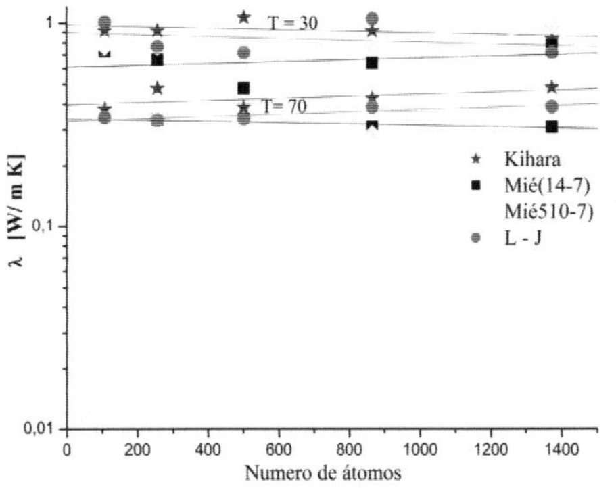

Figure 1. Thermal conductivity, λ, as a function of the size of systems, N, is depicted for L-J, Mie (14-7), Mie (10-5) and Kihara potential.

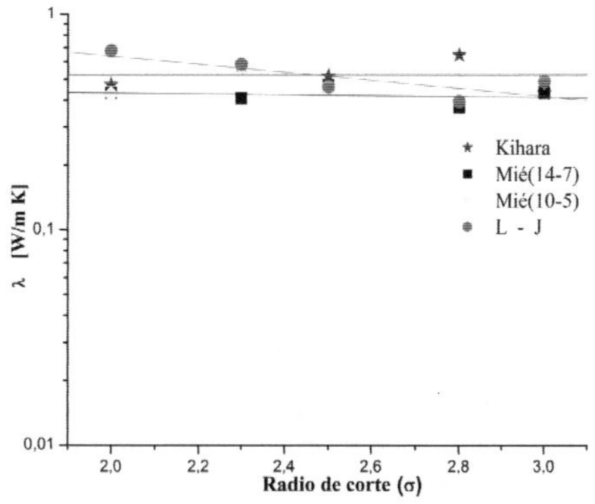

Figure 2. Thermal conductivity, λ, as a function of the cutoff, r_c, is depicted for potentials as L-J, Mie (14-7), Mie(10,5) and Kihara.

2.4. Force calculation

If we consider a system of N interacting particles in a volume V. The internal energy of the system is given by

$$E \equiv K + U, \qquad \text{with} \qquad K \equiv \sum_{i=1}^{N} \frac{1}{2} m_i \left|\dot{\mathbf{r}}_i(t)\right|^2, \tag{13}$$

where K is the kinetic energy; $U = U(\mathbf{r}_1, \mathbf{r}_2, \mathbf{r}_3, ..., \mathbf{r}_N)$ is the potential energy of interacting particles and $\mathbf{r}_i(t) = (x_i(t), y_i(t), z_i(t))$ is the position vector of the i-th particle. In the NVT ensemble, the temperature T remains around its expected average and the evolution of the system is governed by a conservative outlook given by equations of motion

$$\mathbf{F}_i = -\nabla_{r_i} U, \tag{14}$$

where $\mathbf{F}_i$ is the net force over the i-th particle due to other, which is derived from the potential energy U [18].

Other relevant aspect is to take into account the reduced units in the simulation, which are: $r \to r\sigma$, $E \to E\epsilon$, $t \to t/\sigma \left(\frac{m}{\epsilon}\right)^{\frac{1}{2}}$, $T \to T\left(\frac{\epsilon}{k_B}\right)$, $P \to P\left(\frac{\epsilon}{\sigma^3}\right)$, $v \to v\sqrt{\frac{\sigma}{m}}$ and $\mathbf{F} \to \mathbf{F}\left(\frac{\epsilon}{\sigma}\right)$; where r is the position, E is the energy, m is the mass, T is the temperature, P is the pressure, v is the velocity and $\mathbf{F}$ is the force. When $r_{ij} < r_{min}$, the interaction is repulsive and when $r_{ij} > r_{min}$ the potential is attractive. The potential Mie (14-7) has a range shorter than Mie (10-5) and this latter larger than L-J potential. The Kihara potential compared to L-J potential has additional terms analytically given by Eq.(6). Therefore, the interaction force between i-th and j-th atoms is defined by $\mathbf{F} = -\nabla\phi$.

2.5. Thermal conductivity

The thermal conductivity, λ, graphically corresponds to the area under the curve of the autocorrelation heat current as a function of the time, where its discrete form is given by[9, 13]

$$\lambda = \frac{\Delta t}{3Vk_BT^2} \sum_{m=1}^{M} \frac{1}{N-m} \sum_{n=1}^{N-m} \mathbf{j}(n) \cdot \mathbf{j}(m+n). \tag{15}$$

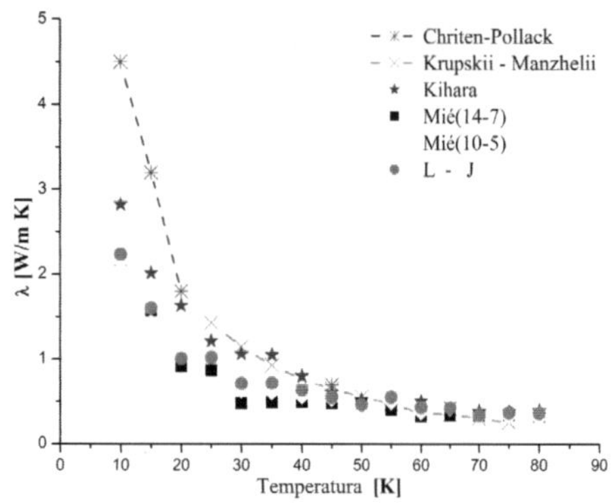

Figure 3. Thermal conductivity, λ, as a function of temperature, T, for potentials as L-J, Mie (14-7), Mie (10,5) and Kihara.

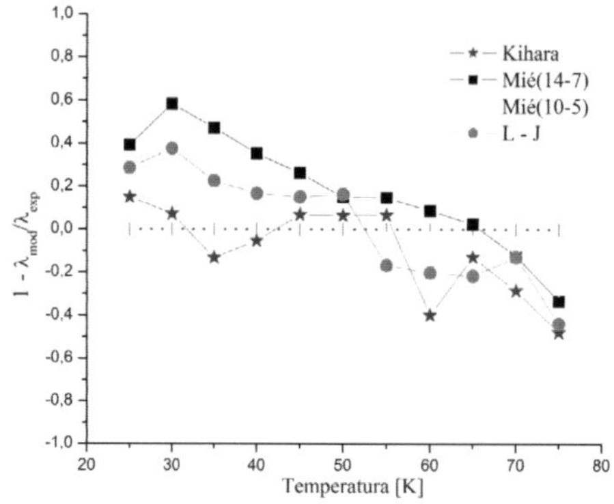

Figure 4. Relative difference of the thermal conductivity related to experimental data as a function of temperature every potential.

2.6. Other parameters

We carry out simulations of samples using equations defined before, with times 10^5 and 10^6 in simulation units. The useful temperatures are T = 20, 30, 40, 50 and 60 $[K]$. We use in each calculation the defined potential of interaction as Lennard-Jones, Mie (14-7), Mie(10-5) and Kihara defined in Eqs.(2)-(5) We use the NVT ensemble taking samples with N = 108, 256, 500, 864 and 1372. Our test was made with reduced density $\rho^* = 1.043$ and cutoff given by 2σ, 2.3σ, 2.5σ, 2.8σ and 3σ at temperature T =50 $[K]$ for each defined potential Results are depicted in Figure 1 and Figure 2 to show that the dependence on cutoff r_c and the size of samples N are practically negligible. Thus, we choose the cutoff as $r_c = 2.5\sigma$ and $N = 500$.

3. Discussion of results and concluding remarks

The values of thermal conductivity are calculated by MD simulation using the Green-Kubo formula and the Verlet algorithm for each potential of interaction. The experimental data were measured by D.K.Christen and G.L.Pollack [15] and I.N.Krupskii and Manzhelii [16].

In Figure 3, the thermal conductivity as a function of temperature is depicted for L-J, Mie(14-7), Mie(10,5) and Kihara potentials. In Figure 4, the relative difference of the thermal conductivity related to experimental data, $\lambda/\lambda_{exp} - 1$, as a function of temperature, T, is shown for L-J, Mie(14-7), Mie(10,5) and Kihara potentials. The correlation and the precision

Potential	Correlation	Precision
Kihara	0.967 Kr	8.8%
Mie (14,7)	0.894 Kr	22%
Mie (10,5)	0.939 Kr	16.7%
L - J	0.962 Kr	15.9%
Kihara	0.983 Ch	28.7%
Mie (14,7)	0.999 Ch	46.7%
Mie (10,5)	0.985 Ch	44.4%
L - J	0.999 Ch	43.7%

Table 1. Correlation and precision among experimental data and calculated values.

of the theoretical values related to experimental data are presented in the Table 1, where the correlations related to D.K.Christen G.L.Pollack is denoted as **Ch**, and related to I.N.Krupskii and Manzhelii is denoted as **Kr**.

We notice that the Kihara potential in the way defined in Eq.(5) is the model that presents the best results based on the Figure 4 and the Table 1. Kihara presents the lowest relative difference with the data and the best values of precision related to range measured by each authors. We think other properties can be calculated by Kihara potential in the future for the argon and other elements in its solid phase as krypton and xenon.

Acknowledments
We would like to thank partial financial support by "Beca-UCN" and VRIDT/UCN-105/2016.

References

[1] Hiebert E N, *Historical Remarks on the Discovery of Argon.* Ed. H. Hyman. 3-20.Chicago, University of Chicago Press (1963).
[2] Shind S L, and Goela J, . *High Thermal Conductivity Materials.*(2006) 10.1007/b106785.
[3] Ackland G J, DMellowa K, Daraszewicza S L, Hepburna D J, Uhrina M, Stratford K, *The MOLDY Short-Range Molecular Dynamics Package.* Computer Physics Communications 182, 25872604. (2011)
[4] Shankar P Das, *Statistical Physics of Liquids at Freezing and Beyond* Editorial Cambridge University Press (2011).
[5] Cheong, Seng k, *Thermal Conductivity of Argon from the Green-Kubo Method.*, 1011 (2004).
[6] Shimuzu F, Kimizuka H, Kaburaki H, Li J, Yip S, *Parallel Molecular Dynamics Simulation on Elastic Properties of Solid Argon.* Proceedings of the fourth International Conference on Supercomputing in Nuclear Aplications September 4-7, Tokyo, Japan (2000).
[7] Eskandari A, *Theory and Atomistic Simulation of Krypton Fluid.* Journal of Chemical Physics 129, 18 (2008).
[8] Stackhouse S., and Stixrude L., *Theoretical Methods for Calculating the Lattice Thermal Conductivity of Minerals.* Reviews in Mineralogy & Geochemistry 71, 25369 (2010).
[9] Tretiakov K. V., and Scandolo S. *Thermal Conductivity of Solid Argon from Molecular Dynamics Simulations.* The Journal of Chemical Physics 120, 376569 (2004).
[10] Trokhymchuk A, Alejandre J, *Computer Simulations of Liquid-Vapor Interface in Lennard-Jones Fluids.*, Journal of Chemical Physics 111 (1999).
[11] Hanley H J M, *The Viscosity and Thermal Conductivity Coefficients of Dilute Argon, Krypton, and Xenon.*, Journal of Physical and Chemical Reference Data 2, 61942 (1973).
[12] MacDowell L G, Garzon B, Calero S, and Lago S, *Dynamical Properties and Transport Coefficients of Kihara Linear Fluids.*, The Journal of Chemical Physics. 106, 475367 (1997).
[13] Kaburaki H., Li J., and Yip S., *Thermal Conductivity of Solid Argon by Classical Molecular Dynamics.* Materials Research Society Symposium Proceedings. Proc. 538, 503 (1999).
[14] Eskandari A, Nader O, and Behzad H, *Transport Properties of Mie (14,7) Fluids: Molecular Dynamics Simulation and Theory.* The Journal of Chemical Physics 129, 024507 (2008).
[15] Christen D K, Pollack G L, *Thermal conductivity of solid argon* Phys. Rev. B 12, 3380 (1975).
[16] Krupskii I N and Manzhelii V G, *Multiphonon interactions and the thermal conductivity of crystalline argon, krypton, and xenon.* Sov.Phys.JETP 28,1097 (1969).
[17] Li J, *Basic Molecular Dynamics, Handbook of Materials Modeling,* 565-588 (2005).
[18] Allen M P, and Tildesley D J, *Computer Simulation of Liquids*, Clarendon, Oxford, (1987).
[19] Leach A R, *Molecular Dynamics: Simulation Methods, in Molecular Modelling Principles and Applications*, Pearson Prentice Hall.
[20] Allen M P, *Introduction to Molecular Dynamics Simulation*, 1-28 (2004).

IOP Conf. Series: Journal of Physics: Conf. Series **1043** (2018) 012011 doi:10.1088/1742-6596/1043/1/012011

Implications of Superstatistics for steady-state plasmas

Sergio Davis[1], Jalaj Jain[1], Diego González[1,2], Gonzalo Gutiérrez[2]

[1] Departamento de Ciencias Nucleares, Comisión Chilena de Energía Nuclear
[2] Grupo de Nanomateriales, Departamento de Física, Facultad de Ciencias, Universidad de Chile

E-mail: sdavis@cchen.cl

Abstract. Colissionless plasmas are non-extensive systems which, due to the long-range interaction between their components, are incapable of reaching thermal equilibrium, even in a steady state. These systems cannot, therefore, be described statistically by a single canonical distribution with well-defined inverse temperature $\beta = 1/k_B T$ and are commonly described via two alternative approaches: namely Tsallis statistics and Superstatistics. The use of Superstatistics in describing steady-state plasmas has been proposed by several authors, more recently Ourabah *et al.* In this work we study the consequences of this assumption of Superstatistics for steady-state plasmas. We explicitly show that only the ensembles characterized by the condition $P(\boldsymbol{x}|\rho) = \rho(H(\boldsymbol{x}))$ are consistent with a generalized definition of temperature introduced recently. We show how this formalism is employed in the case of plasma, considering interaction with external electromagnetic fields as well as between particles in the plasma. Our results clearly illustrate why low-energy particles tend to the Maxwellian distribution of velocities, while high-energy particles contribute to the long tails of the velocity distribution.

1. Introduction

In the plasma state, the long-range interactions between different ions and electrons are dominant, and this causes these systems to be found in highly non-equilibrium states. Even when these systems are able to reach a steady state, they do not follow Boltzmann-Gibbs statistics [1]. A pressing question then arises: Do these steady-state plasmas have a well-defined temperature? According to equilibrium Thermodynamics, temperature T is defined through the relation

$$\frac{1}{T} = \frac{dS(E)}{dE}, \tag{1}$$

where $S(E) = k_B \ln \Omega(E)$ is the Boltzmann entropy. For a system in thermal equilibrium at temperature T, the velocity distribution of its components is given by the Maxwell-Boltzmann distribution,

$$P(\boldsymbol{v}|\beta) = \frac{1}{Z(\beta)} \exp\left(-\beta \frac{mv^2}{2}\right) \tag{2}$$

with $\beta = 1/(k_B T)$ the inverse temperature. The equipartition theorem connects the parameter β with the average kinetic energy of the system,

$$\frac{3N}{2}k_B T = \Big\langle K \Big\rangle_\beta = \Big\langle \frac{1}{2}\sum_{i=1}^{N} m_i v_i^2 \Big\rangle_\beta. \tag{3}$$

In general, we can classify systems into three classes:

(a) Systems in thermodynamical equilibrium, where β is well-defined,

(b) Systems with non-constant β where it makes sense to introduce a *temperature distribution function*,

(c) Other systems where a temperature cannot even be defined. In this last case, usually transient states, we can only speak of average kinetic energy.

For systems in classes (a) and (b) we will denote the probability of having an inverse temperature β as $P(\beta|S)$, in accordance with the formalism known as Superstatistics, which we will explain next.

2. Superstatistics

An interesting proposal for the treatment of non-equilibrium, steady-state systems was introduced in 2003 by Beck and Cohen, called *Superstatistics* [2, 3]. In this framework, one goes from the canonical ensemble

$$P(\boldsymbol{r},\boldsymbol{p}|\beta) = \frac{\exp(-\beta H(\boldsymbol{r},\boldsymbol{p}))}{Z(\beta)}, \tag{4}$$

to a superposition of infinite canonical ensembles at different inverse temperatures β,

$$P(\boldsymbol{r},\boldsymbol{p}|S) = \int_0^\infty d\beta P(\beta|S)\left[\frac{\exp(-\beta H(\boldsymbol{r},\boldsymbol{p}))}{Z(\beta)}\right] \tag{5}$$

each of them weighted by its probability of occurrence, $P(\beta|S)$. Two cases are of interest. The trivial case when $P(\beta|S)$ is a Dirac delta function recovers the canonical ensemble, and the case when $P(\beta|S)/Z(\beta)$ is a Gamma distribution

$$P(\beta|k,\theta) = \frac{\exp(-\beta/\theta)\beta^{k-1}}{\Gamma(k)\theta^k}, \tag{6}$$

generates the so-called q-canonical ensemble,

$$P(\boldsymbol{r},\boldsymbol{p}|\beta_0,q) = \frac{1}{\eta(\beta_0,q)}\left[1-(1-q)\beta_0 H(\boldsymbol{r},\boldsymbol{p})\right]_+^{\frac{1}{1-q}} \tag{7}$$

also known as Tsallis distributions. In space plasmas, the Kappa distribution [4] is an instance of the q-canonical (Tsallis) distribution and, accordingly, has been derived by the use of Superstatistics [5].

3. The fundamental inverse temperature

A "thermal" system, in the sense that temperature can be defined, is characterized by the condition $P(\boldsymbol{r},\boldsymbol{p}|S) = \rho(\mathcal{H}(\boldsymbol{r},\boldsymbol{p}))$. In order to completely describe these systems, it is no longer enough to have a single number β; we need a continuous function. This can be $P(\beta|S)$ in the framework of Superstatistics or the ensemble function $\rho(E)$. There is a more convenient possibility, the fundamental inverse temperature function [6, 7], defined by

Ensemble	$\rho(E)$	$\beta_F(E)$
Canonical	$\exp(-\beta_0 E)/Z(\beta_0)$	β_0
q-canonical	$\frac{1}{\eta(\beta_0,q)}\left[1-(1-q)\beta_0 E\right]_+^{\frac{1}{1-q}}$	$\beta_0/(1-(1-q)\beta_0 E)$
Gaussian	$\frac{1}{Z(a,E_t)}\exp(-a(E-E_t)^2)$	$2a(E-E_t)$

Table 1. Generalized Boltzmann factor $\rho(E)$ and fundamental inverse temperature function $\beta_F(E)$ for the canonical, q-canonical and Gaussian [8, 9] ensembles.

$$\beta_F(E) = -\frac{d}{dE}\ln\rho(E). \tag{8}$$

In the case of Superstatistics, $\beta_F(E)$ corresponds to $\langle\beta\rangle_E$, i.e., the average inverse temperature at fixed energy E.

By employing the statistical identity presented in Ref. [10] as the conjugate variables theorem (CVT),

$$\Big\langle \nabla\cdot\boldsymbol{v}\Big\rangle_S = -\Big\langle \boldsymbol{v}\cdot\nabla\ln\rho(H)\Big\rangle_S, \tag{9}$$

we can obtain the generalization of the equipartition theorem for a steady-state plasma,

$$\Big\langle \beta K\Big\rangle_S = \Big\langle \beta_F(E)K\Big\rangle_S = \frac{3N}{2}, \tag{10}$$

where the role of the parameter β is played by the fundamental inverse temperature function $\beta_F(E)$.

4. Thermodynamical description of plasma

In a plasma we have N particles, each with a mass m_i and a charge q_i, with position and velocity vectors $\boldsymbol{r_i}$ and $\boldsymbol{v_i}$ respectively. The probability density of microstates of the plasma can be written as $P(\boldsymbol{r}_1,\ldots,\boldsymbol{r}_N,\boldsymbol{v}_1,\ldots,\boldsymbol{v}_N|t) = P(\boldsymbol{R},\boldsymbol{V}|S)$ such that Vlasov equation holds,

$$\frac{dP}{dt} = \frac{\partial P}{\partial t} + \sum_i \boldsymbol{v_i}\cdot\frac{\partial P}{\partial \boldsymbol{r}_i} + \sum_i \frac{\boldsymbol{F}_i}{m_i}\cdot\frac{\partial P}{\partial \boldsymbol{v}_i} = 0 \tag{11}$$

where $\boldsymbol{F}_i = q_i\Big(\boldsymbol{E}(\boldsymbol{r}_i) + \boldsymbol{v}_i\times\boldsymbol{B}(\boldsymbol{r}_i)\Big)$ is the Lorentz force. Vlasov equation is equivalent to Liouville's theorem of conservation of phase space volume for plasma, and due to the presence of the Lorentz force, must be solved self-consistently with Maxwell's equations,

$$\nabla\cdot\boldsymbol{E} = \frac{\rho}{\epsilon_0}, \tag{12}$$

$$\nabla\times\boldsymbol{B} = \mu_0\Big(\boldsymbol{J} + \epsilon_0\frac{\partial\boldsymbol{E}}{\partial t}\Big) = \mu_0\boldsymbol{J}, \tag{13}$$

$$\nabla\cdot\boldsymbol{B} = 0, \tag{14}$$

$$\nabla\times\boldsymbol{E} = -\frac{\partial\boldsymbol{B}}{\partial t} = 0. \tag{15}$$

Here $\rho(\boldsymbol{r},t)$ and $\boldsymbol{J}(\boldsymbol{r},t)$ are the charge and current densities, respectively, which in a steady state can be expressed as expectations over the ensemble as

$$\rho = \rho(\boldsymbol{r}) = \langle \sum_i q_i \delta(\boldsymbol{r}_i - \boldsymbol{r}) \rangle_S$$
$$\boldsymbol{J} = \boldsymbol{J}(\boldsymbol{r}) = \langle \sum_i q_i \boldsymbol{v}_i \delta(\boldsymbol{r}_i - \boldsymbol{r}) \rangle_S. \tag{16}$$

We see then that, in a steady state, $P(\boldsymbol{R}, \boldsymbol{V}|S)$ is uniquely determined either by $\{\Phi(\boldsymbol{r}), \boldsymbol{A}(\boldsymbol{r})\}$ or $\{\rho(\boldsymbol{r}), \boldsymbol{J}(\boldsymbol{r})\}$. The non-relativistic Hamiltonian of the system is $\mathcal{H} = \sum_i \mathcal{H}_i$, with

$$\mathcal{H}_i(\boldsymbol{r}, \boldsymbol{p}) = \frac{1}{2m_i}(\boldsymbol{p} - q_i \boldsymbol{A}(\boldsymbol{r}))^2 + q_i \Phi(\boldsymbol{r}), \tag{17}$$

and where the momentum $\boldsymbol{p}_i = m_i \boldsymbol{v}_i + q_i \boldsymbol{A}(\boldsymbol{r}_i)$. We can write the energy for the i-th particle in terms of its position and velocity as

$$\mathcal{E}_i(\boldsymbol{r}, \boldsymbol{v}) = \frac{1}{2} m_i \boldsymbol{v}^2 + q_i \Phi(\boldsymbol{r}), \tag{18}$$

which does not depend on the vector potential $\boldsymbol{A}$. In order to define temperature, the probability density $P(\boldsymbol{R}, \boldsymbol{V}|S)$ must have the form

$$P(\boldsymbol{R}, \boldsymbol{V}|S) = \rho(\mathcal{E}) = \rho\Big(\sum_i \mathcal{E}_i(\boldsymbol{r}_i, \boldsymbol{v}_i)\Big), \tag{19}$$
$$\mathcal{E}_i(\boldsymbol{r}, \boldsymbol{v}) = \frac{1}{2} m_i \boldsymbol{v}^2 + q_i \Phi(\boldsymbol{r}). \tag{20}$$

We can verify that this *Ansatz* automatically solves the steady-state Vlasov equation,

$$\sum_i \boldsymbol{v}_i \cdot \frac{\partial}{\partial \boldsymbol{r}_i} \ln P(\boldsymbol{R}, \boldsymbol{V}|S) + \frac{\boldsymbol{F}_i}{m_i} \cdot \frac{\partial}{\partial \boldsymbol{v}_i} \ln P(\boldsymbol{R}, \boldsymbol{V}|S) =$$
$$\rho'(\mathcal{E}) \sum_i \left[\boldsymbol{v}_i \cdot q_i \nabla \Phi(\boldsymbol{r}_i) + \frac{q_i}{m_i} \left(-\nabla \Phi(\boldsymbol{r}_i) + \boldsymbol{v}_i \times \boldsymbol{B}(\boldsymbol{r}_i)\right) \cdot m_i \boldsymbol{v}_i \right] = 0. \tag{21}$$

By imposing consistency with Gauss law we find that any such "thermal" plasma must be found in an ensemble $\rho(\mathcal{H})$ for which

$$-\epsilon_0 \nabla^2 \Phi(\boldsymbol{r}) = \int_0^\infty dK \cdot K^{3N/2} \int d\boldsymbol{R} \left[\sum_{i=1}^N q_i \delta(\boldsymbol{r}_i - \boldsymbol{r}) \right] \rho\left(K + \sum_i q_i \Phi(\boldsymbol{r}_i) \right), \tag{22}$$

holds. There cannot be any net current inside such a plasma, because by requiring consistency with Ampère's law,

$$\boldsymbol{J}(\boldsymbol{r}) = \sum_{i=1} q_i \int d\boldsymbol{R} \delta(\boldsymbol{r}_i - \boldsymbol{r}) \int d\boldsymbol{V} \cdot \boldsymbol{v}_i \rho \left(\sum_i \frac{1}{2} m_i v_i^2 + \sum_i q_i \Phi(\boldsymbol{r}_i) \right) = 0, \tag{23}$$

which is zero by symmetry. For a given electrostatic potential $\Phi(\boldsymbol{r})$, the steady-state ensemble must be a solution of

$$-\epsilon_0 \nabla^2 \Phi(\boldsymbol{r}) = \int_0^\infty dK \cdot K^{3N/2} \int d\boldsymbol{R} \left[\sum_{i=1}^N q_i \delta(\boldsymbol{r}_i - \boldsymbol{r}) \right] \rho\left(K + \sum_i q_i \Phi(\boldsymbol{r}_i) \right) \tag{24}$$

and this constrains the form of the fundamental inverse temperature compatible with Φ. By defining $\rho_1(\epsilon)$ such that

$$\frac{\partial}{\partial \epsilon_i} \ln \rho(\epsilon) = \frac{\partial}{\partial \epsilon_i} \ln \rho_1(\epsilon_i), \tag{25}$$

we can greatly simplify our consistency equation and write it as

$$-\epsilon_0 \nabla^2 \Phi(\boldsymbol{r}) = \sum_{i=1}^{N} q_i \rho_1(q_i \Phi(\boldsymbol{r})). \tag{26}$$

We can also write the single-particle distributions of position and velocity in terms of ρ_1. They are given by

$$P_i(\boldsymbol{r}|S) = \rho_1(q_i \Phi(\boldsymbol{r})), \tag{27}$$

$$P_i(\boldsymbol{v}|S) = \rho_1\Big(\frac{m_i \boldsymbol{v}^2}{2}\Big). \tag{28}$$

In the same way, the single-particle fundamental inverse temperature is then

$$\beta_{1F}(\mathcal{E}_i) = -\frac{\partial}{\partial \mathcal{E}_i} \ln \rho_1(\mathcal{E}_i). \tag{29}$$

An interesting consequence of this is that ions or electrons in different ranges of energy will be seen as different populations described by different velocity distributions. This is commonly observed in plasmas. Furthermore, different spatial regions can appear as having different temperatures.

Assuming $\mathcal{E}_i \ll \mathcal{E}$ we can approximate the effective single-particle temperature by using the following Taylor expansion,

$$\beta_{1F}(\mathcal{E}_i) \approx \Big\langle \hat{\beta}_F(\mathcal{E} + \mathcal{E}_i) \Big\rangle_S = \Big\langle \hat{\beta}_F(\mathcal{E}) \Big\rangle_S + \mathcal{E}_i \Big\langle \hat{\beta}'_F(\mathcal{E}) \Big\rangle_S + \frac{1}{2}\mathcal{E}_i^2 \Big\langle \hat{\beta}''_F(\mathcal{E}) \Big\rangle_S + \ldots \tag{30}$$

This means that, in a first approximation, the single-particle fundamental temperature is a constant, equal to the average fundamental temperature of the system, therefore the particles are approximately Maxwellian for low kinetic energies, and can have other, more "exotic" distributions for higher energies.

5. Conclusions

In this work we have shown that, for any steady-state plasma in which $P(\boldsymbol{R}, \boldsymbol{V}|S) = \rho(\mathcal{H})$, we can provide a "fundamental temperature" that completely describes the system in Superstatistics. A thermal plasma in this sense is fully described by either its single-particle ensemble function ρ_1 or its single-particle fundamental inverse temperature β_{1F} . These plasmas can naturally have a segregation in their energy distributions between slow (approximately Maxwellian) and fast particles.

Acknowledgments

SD acknowledges funding from FONDECYT grant 1140514. DG acknowledges funding from CONICYT PhD Fellowship grant 21140914.

References

[1] Bellan P M 2006 *Fundamentals of Plasma Physics* (Cambridge University)
[2] Beck C and Cohen E G D 2003 *Physica A* **322** 267–275
[3] Beck C 2004 *Continuum Mech. Thermodyn.* **16** 293–304
[4] Vasyliunas V M 1968 *J. Geophys. Res.* **73** 7519–7523
[5] Ourabah K, Gougam L A and Tribeche M 2015 *Phys. Rev. E* **91** 4–7
[6] Palma G, Gutiérrez G and Davis S 2016 *Phys. Rev. E* **94** 062113
[7] Loguercio H and Davis S 2016 *J. Phys.: Conf. Series* **720** 12007
[8] Challa M S S and Hetherington J H 1988 *Phys. Rev. A* **38** 6324–6337
[9] Johal R S, Planes A and Vives E 2003 *Phys. Rev. E* **68** 056113
[10] Davis S and Gutiérrez G 2012 *Phys. Rev. E* **86** 051136

IOP Conf. Series: Journal of Physics: Conf. Series **1043** (2018) 012012 doi:10.1088/1742-6596/1043/1/012012

The Role of Dissipation in the Two–Color Second–Order Coherence Function for Light Sources

I. C. Andrade[1] and B. A. Rodríguez[2]

Grupo de Física Atómica y Molecular, Universidad de Antioquia UdeA, Calle 70 No. 52-21, Medellín, Colombia

E-mail: [1]isabel.andrade@udea.edu.co

E-mail: [2]boris.rodriguez@udea.edu.co

Abstract. We discuss the effect of the dissipation and pump rates in the two-photon spectrum (2PS) for different light sources. In this paper we show the second-order coherence function for the different frequencies given of the systems, we show how this measure reveals more information about the light source, and we show how it changes with dissipation and pump rates. We emphasize the importance of modelling system as an open quantum system for a more realistic description of the spectrum $S(\omega)$ and 2PS of field light

1. Introduction

The study of light sources with well-defined emission statistics and coherence properties has always been of scientific interest due to its possible technological and fundamental applications. Research has focused on thermal light sources [1, 2], quantum emissions [3] and coherent emissions [4, 5], among others. The photon statistics and coherence properties are the main tools for discussing the nonclassical properties of light fields. This can be described in the framework of Glauber's theory of coherence function [6]. We distinguish between the various orders of coherence, according to the number of conditions satisfied. The first-order coherence define the emission spectrum which, gives us the intensity of photons emitted by the system $S(\omega)$ at a certain frequency in the steady state $\hat{\rho}_{ss}$:

$$S(\omega) = \frac{1}{2\pi} \frac{\int_0^\infty d\tau e^{i\omega\tau} \langle \hat{a}^\dagger \hat{a}(\tau) \rangle}{\langle \hat{a}^\dagger \hat{a} \rangle}, \tag{1}$$

where $\hat{a}$ ($\hat{a}^\dagger$) is the annihilation (creation) field operator, τ is the delay time with $\tau = 0$ the time in which the system reaches the steady state and the expected values $\langle \ldots \rangle$ are calculated in $\hat{\rho}_{ss}$. The quantum theory of optical second-order coherence was developed by Glauber [6] in an attempt to understand the Hanbury Brown & Twiss (HBT) experiment [7], and both are cornerstones in the research of quantum light. The second-order coherence function $g^{(2)}(\tau)$ is defined by [8]:

$$g^{(2)}(\tau) = \frac{\langle \hat{a}^\dagger \hat{a}^\dagger(\tau) \hat{a}(\tau) \hat{a} \rangle}{\langle \hat{a}^\dagger(\tau) \hat{a}(\tau) \rangle \langle \hat{a}^\dagger \hat{a} \rangle}. \tag{2}$$

Where again the expected values are calculated at the steady state. At zero delay time $\tau = 0$, the second-order coherence function $g^{(2)}(0)$, for some quantum states, is able to take values that are excluded in the classical optical coherence theory in which holds the inequalities $g^{(2)}_{classic}(0) \geq g^{(2)}_{classic}(\tau) \geq 1$. In the quantum case we can have:

$$g^{(2)}(0) \begin{cases} > 1 & \textit{bunching} \\ = 1 & \textit{coherent} \\ < 1 & \textit{antibunching} \end{cases} \tag{3}$$

Therefore the antibunching regime is a strong signature of non–classical light.

A generalization of the coherence theory is a measure of the coherence simultaneously in times and frequencies [9]. Mathematically, this amounts to adding exponential decays in the Fourier transform of the time–autocorrelation. To first-order, the output field allows to compute the time-dependent power spectrum of emission as the density of output photons with frequency ω_1 at time T_1, i.e.,

$$S^{(1)}_{\Gamma_1}(\omega_1, T_1) = \frac{\Gamma_1}{2\pi} \int\int_{-\infty}^{T_1} dt'_1 dt'_4 e^{-\frac{\Gamma_1}{2}(T_1 - t'_1)} e^{-\frac{\Gamma_1}{2}(T_1 - t'_4)} e^{i\omega_1(t'_4 - t'_1)} \langle \hat{a}^\dagger(t'_1)\hat{a}(t'4)\rangle , \tag{4}$$

where Γ_1 is interpreted as the detector's line width. This so-called physical spectrum reduces to the Wiener–Khinchin theorem in the steady state and in the limit $\Gamma_1 \to 0$ [9].

The generalization at second-order is the so-called *two–color second-order coherence function*, which measure the intensity correlation by two detectors with delay time τ at frequencies ω_1 and ω_2 [8, 10]. These photon correlations combining both their frequency and time information are now routinely measured in the laboratory. These experiments have proven extremely powerful in characterising quantum systems such as a resonantly driven emitter [11], the strong coupling of light and matter [12], to perform quantum state tomography [13], to monitor heralded single photon sources [14] or to access spectral diffusion of single emitters [15]. This correlator can be defined as:

$$S^{(2)}_{\Gamma_1\Gamma_2}(\omega_1, T_1; \omega_2, T_2) = \frac{\Gamma_1\Gamma_2}{(2\pi)^2} \int\int_{-\infty}^{T_1} dt'_1 dt'_4 e^{-\frac{\Gamma_1}{2}(T_1 - t'_1)} e^{-\frac{\Gamma_1}{2}(T_1 - t'_4)} \times \tag{5}$$

$$\int\int_{-\infty}^{T_2} dt'_2 dt'_3 e^{-\frac{\Gamma_2}{2}(T_2 - t'_2)} e^{-\frac{\Gamma_2}{2}(T_2 - t'_3)} e^{i\omega(t'_4 - t'_1)} e^{i\omega(t'_3 - t'_2)} \langle \mathcal{T}_-[\hat{a}^\dagger(t'_1)\hat{a}^\dagger(t'_2)] \mathcal{T}_+[\hat{a}(t'_3)\hat{a}(t'_4)]\rangle ,$$

where we have defined $\mathcal{T}_+$ ($\mathcal{T}_-$) to order the operators in a product with the latest time to the far left (far right) [8]. Normalising this expression yields the sought two–color second–order coherence function

$$g^{(2)}_{\Gamma_1\Gamma_2}(\omega_1, T_1; \omega_2, T_2) = \frac{S^{(2)}_{\Gamma_1\Gamma_2}(\omega_1, T_1; \omega_2, T_2)}{S^{(1)}_{\Gamma_1}(\omega_1, T_1) S^{(1)}_{\Gamma_2}(\omega_2, T_2)}. \tag{6}$$

This is an increasingly popular experimental observable by inserting filters in the arms of a standard HTB setup [3]. We compute efficiently and easily $g^{(2)}_{\Gamma_1\Gamma_2}(\omega_1, T_1; \omega_2, T_2)$ through the coupling of two 'sensors' to the open quantum system of interest, using a new theory of photo-detection described in [16], through the Lindblad Master Equation (LME) [17]

$$\frac{d}{dt}\hat{\rho}(t) = -i\left[\hat{H}_{system} + \hat{H}_{sensor}, \hat{\rho}(t)\right] + \left(\frac{\kappa}{2}\mathcal{L}_{\hat{a}} + \frac{P}{2}\mathcal{L}_{\hat{a}^\dagger} + \sum_{i=1}^{2}\frac{\Gamma_i}{2}\mathcal{L}_{\hat{\zeta}_i}\right)\hat{\rho}(t);$$

$$\hat{H}_{sensor} = \sum_{i=1}^{2}\left[\omega_i \hat{\zeta}_i^\dagger \hat{\zeta}_i + \varepsilon_i(\hat{a}^\dagger\hat{\zeta}_i + \hat{\zeta}_i^\dagger\hat{a})\right]; \quad \hat{n}_i = \hat{\zeta}_i^\dagger\hat{\zeta}_i; \tag{7}$$

$$\mathcal{L}_{\hat{c}}\hat{\rho}(t) = 2\hat{c}\hat{\rho}(t)\hat{c}^\dagger - \hat{c}^\dagger\hat{c}\hat{\rho}(t) - \hat{\rho}(t)\hat{c}^\dagger\hat{c},$$

where $\hat{H}_{system}$ describes the system with $\kappa(P)$ the dissipation(pumping) rate. $\hat{H}_{sensor}$ is the coupling between the system and sensors represented by two-level operators ζ_i with $i = 1, 2$, with ε_i the coupling strength between sensors and system. The sensors are two level systems characterised by transition frequencies ω_i, and decay rates Γ_i. The two–color second–order coherence function is given by [16]:

$$g^{(2)}_{\Gamma_1\Gamma_2}(\omega_1, T_1; \omega_2, T_2) = \lim_{\varepsilon_1\varepsilon_2 \to 0} \frac{\langle \hat{n}_1(T_1)\hat{n}_2(T_2)\rangle}{\langle \hat{n}_1(T_1)\rangle \langle \hat{n}_2(T_2)\rangle}. \tag{8}$$

The simultaneous measure of the two–color second–order coherence function—$T_1 = T_2 = 0$, define the two-photon spectrum (2PS), which describe the statistic of the light sources in function of the frequency. In this work we compute the 2PS for the different systems, under the theory of the generation of light states.

2. Generation of the light sources

In this section we briefly review the generation theory of some relevant quantum states of light: coherent, thermal and photon number states. We focus in the dissipation effects on the second-order coherence function. We consider the following three cases that model optical cavities for the generation of light states:

$$\hat{H}_{th} = \omega_a a^\dagger a \qquad \kappa = 0.1 \quad P = 0.01, \tag{9}$$

$$\hat{H}_{coh} = \omega_a \hat{a}^\dagger \hat{a} + \omega_a(\hat{a}^\dagger + \hat{a}) \qquad \kappa = 0.1 \quad P = 0.01, \tag{10}$$

$$\hat{H}_{SPS} = \omega_a \hat{a}^\dagger \hat{a} + \omega_a(\hat{a}^\dagger + \hat{a}) - \omega_a(\hat{a}^\dagger \hat{a}^2 + \hat{a}^{\dagger 2}\hat{a}) \qquad \kappa = 0.1 \quad P = 0, \tag{11}$$

where ω_a is the frequency of the free field. Their respective quantum algebra is given by the bosonic commutation rule ($\hat{a}\hat{a}^\dagger - \hat{a}^\dagger\hat{a} = 1$). The first case (9), is the single mode electromagnetic field Hamiltonian. The steady state of (9) under a LME evolution (7) is a mixed thermal state. Physically is the incoherent superposition of many uncorrelated sources. Note that to generate a thermal state, it is always necessary to interact with the cavity walls ($\kappa \neq 0$ and $P \neq 0$). The second one (10), is the Hamiltonian to generate a coherent state by displacing the vacuum state $|0\rangle$ [18]. To generate a coherent state of light we need an externally forced-driven harmonic oscillator, for example a classical oscillating current that is resonantly coupled to the electromagnetic field in a coherent process ($\kappa = 0$ and $P = 0$). But real systems are not perfectly isolated and therefore we take dissipation effects into account: $\kappa \neq 0$ and $P \neq 0$. Finally, the third case (11), is the spontaneous parametric down conversion (SPDC) effective Hamiltonian to produce a single photon source (SPS) [19]. As in the previous model, we simulate a more realistic SPDC source introducing a dissipative rate.

3. Results and discussions

The second-order coherence function at zero delay $g^{(2)}(0)$, can be calculated analytically in the above systems restricting ourselves to the steady state and the $N = 6$ photon manifold. For thermal and SPS sources we have $g^{(2)}(0)|_{th} = 2$–bunching and $g^{(2)}(0)|_{SPS} = 0$–antibunching, respectively. In the coherent source we obtain:

$$g^{(2)}(0)|_{coh} = 2 - \frac{16(P-\kappa)^2}{\left(P(P-\kappa)^2 + 4\kappa\right)^2}, \tag{12}$$

where we have taken $\omega_a = 1$. This expression reveals several interesting features: (i) When $P \to 0$ and $\kappa \to 0$, then $g^{(2)}(0) \to 1$–coherent, (ii) if $P = \kappa$, or $P \to \infty$ ($\kappa \to \infty$), then the system is bunching $g^{(2)}(0) \to 2$, and (iii) there exist values for which the emission is coherent

IOP Conf. Series: Journal of Physics: Conf. Series **1043** (2018) 012012 doi:10.1088/1742-6596/1043/1/012012

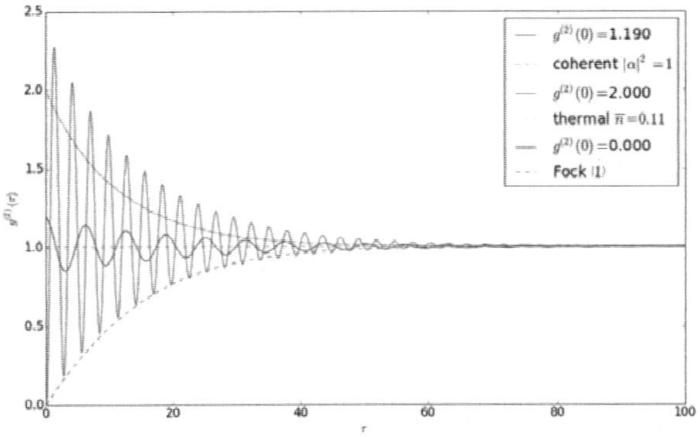

Figure 1. Steady state second-order coherence function in the time domain $g^{(2)}(\tau)$ for the different sources: $\hat{H}_{coh}$ (blue line), $\hat{H}_{th}$ (green line) and $\hat{H}_{SPS}$ (purple line). Exact theoretical states: Coherent state $|\alpha|^2 = 1$, $\kappa = 0.1$, $P = 0.01$ (blue dashed line), thermal state $\bar{n} = 0.11$, $\kappa = 0.1$, $P = 0.01$ (green dashed line) and Fock state $|1\rangle$, $\kappa = 0.15$, $P = 0.06$ (purple dashed).

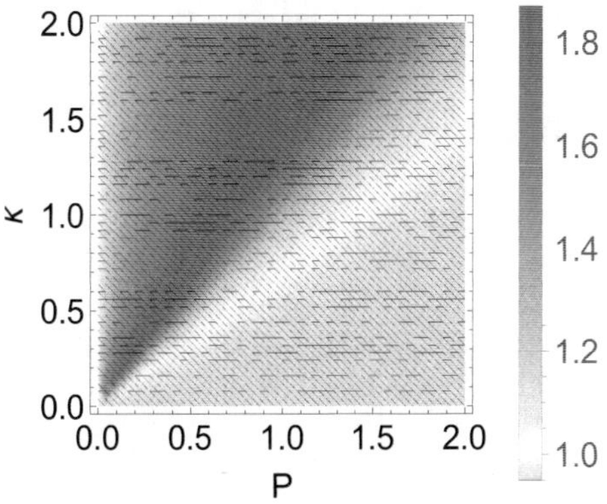

Figure 2. $g^{(2)}(0)$ for the coherent source (see (12)). In the white region the system is coherent, in the red region is bunched and blue region is antibunched. Note that the system can not has $g^{(2)}(0) > 2$.

with $P \neq 0$ and $\kappa \neq 0$.

We have solved numerically (9, 10 and 11) under the dynamic given by (7) in the steady state limit, on a Hilbert space truncated to $N_{max} = 20$ photons that ensure convergence. We used different decay κ and pump P rates, in units of ω_a. In the figure 1 we show the $g^{(2)}(\tau)$ behaviour for the different sources and compared with the exact theoretical states. Unsurprisingly, the thermal source $\hat{H}_{th}$ has a $g^{(2)}(\tau)$ that coincides exactly with the theoretical thermal state. The SPS $\hat{H}_{SPS}$ produces antibunching emission and its second-order coherence function oscillates with an envelope given by the theoretical single-photon state. These oscillations are due to the decay is smaller than the fundamental frequency of the system, but for some delay times the emission is bunched. In the case of the coherent source $g^{(2)}(\tau)$ oscillates around the theoretical value $g^{(2)}(\tau) = 1$. The $g^{(2)}(0)$ of the coherent source $\hat{H}_{coh}$ is described by (12), and it depends on the decay and incoherent pump rates. In the figure 2, we show that the source is able to have bunched or coherent emission controlling the dissipation and pump rates. The antibunching region is an effect of the truncation in the solution. In figure 3 we have studied $g^{(2)}(\tau)$ for $\hat{H}_{coh}$ with different decay and pump parameters. For $P < \kappa$ the second-order coherence function oscillates around 1 and the detection is bunched, for $P > \kappa$ the second-order coherence function is damped and the detection can be coherent or bunched. This behaviour resides in the limited range of validity of (12).

We have studied the impact of the dissipation in the 2PS using the sensing method (see (8)) for all three sources. The results are presented in figure 4. The spectrum and the 2PS for the thermal source is showed in the panel (a). Its spectrum corresponds to a Lorentzian lineshape centred in one. The detection in the 2PS is bunched for all frequency pairs correlated. The 2PS of the $\hat{H}_{th}$ is structureless and evidences than the emission process is indistinguishable in frequency [16]. The panel (b) shows the spectrum and 2PS for a coherent source with $\kappa = 0.1$ and $P = 0.01$. The spectrum shows two peaks centred in $\omega = 0$ and $\omega = 1$, and the 2PS

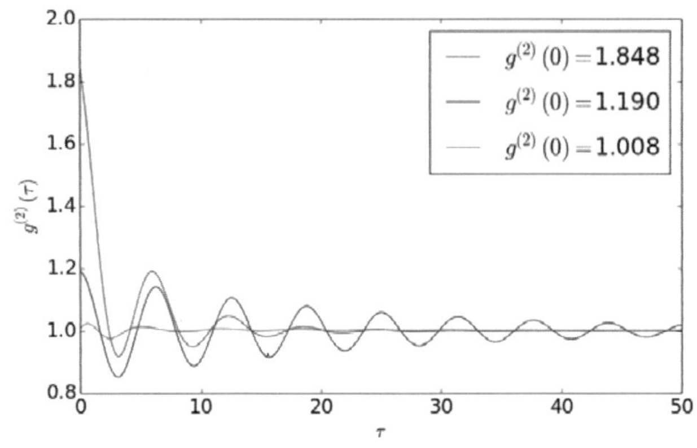

Figure 3. Coherent source $g^{(2)}(\tau)$ for different dissipation and pump regimes. Purple line: $P = 0.67$ and $\kappa = 1$; blue line: $P = 0.01$ and $\kappa = 0.1$; green line: $P = 0.1$ and $\kappa = 0.069$. In the inset the zero delay value are shown.

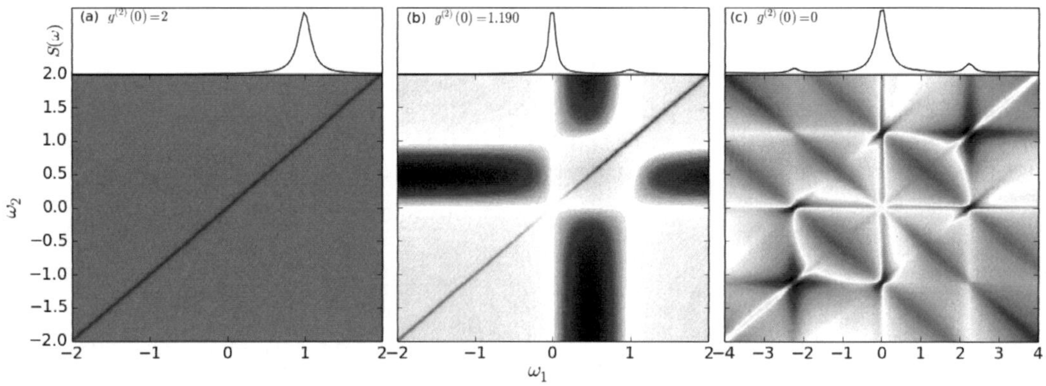

Figure 4. Spectrum (top panel) and 2PS (bottom panel) for the different sources. $\hat{H}_{th}$ in (a), $\hat{H}_{coh}$ in (b), and $\hat{H}_{SPS}$ in (c). The figure colour code is in a logarithmic scale with: blue $<$ log(1)–antibunching, white $=$ log(1)–coherent and red $>$ log(2)–bunching. In the top panel the values of $g^{(2)}(0)$ are shown.

shows bunching and antibunching in the emission. Panel (c) shows the spectrum and the 2PS when we have the SPS. Its spectrum exhibits a Mollow triplet [20], and the 2PS exhibits a richer and complex pattern around the peaks. The correlations of the central peak with its side-peaks produce antibunched detection, which corresponds to distinguishable emissions occurring in different paths. The correlation between the side-peaks produce bunched detection due to the indistinguishable frequencies of emissions with opposite signs. The correlation of the peaks with itself can be indistinguishable bunching if the system has a cascade emission (central peak) or it is antibunching (side peaks).

The figure 5 shows the spectrum and the 2PS for the coherent source $\hat{H}_{coh}$ at different values of $g^{(2)}(0)$. The spectrum exhibits a narrow peak centred in $\omega = 0$ and a lateral Lorentzian lineshape centred in $\omega = 1$, with variable width that changes with pumping and dissipation rates. The 2PS shows than the coherent source has two frequency modes relevant in the cascade emission process at $\omega = 0$ and $\omega = 1$ and their correlations can be distinguishable or indistinguishable.

In the figure 5(a) we show the highly coherent emission regimen, i.e., $g^{(2)}(0) \approx 1$. In this case the correlation at $\omega = 0$ is distinguishable in frequency and the correlation at $\omega = 1$ is indistinguishable. In this regime, the spectrum is composed by more than one frequency mode and the 2PS structure is more complex. The detection filtering around $\omega = 1$ is highly bunched.

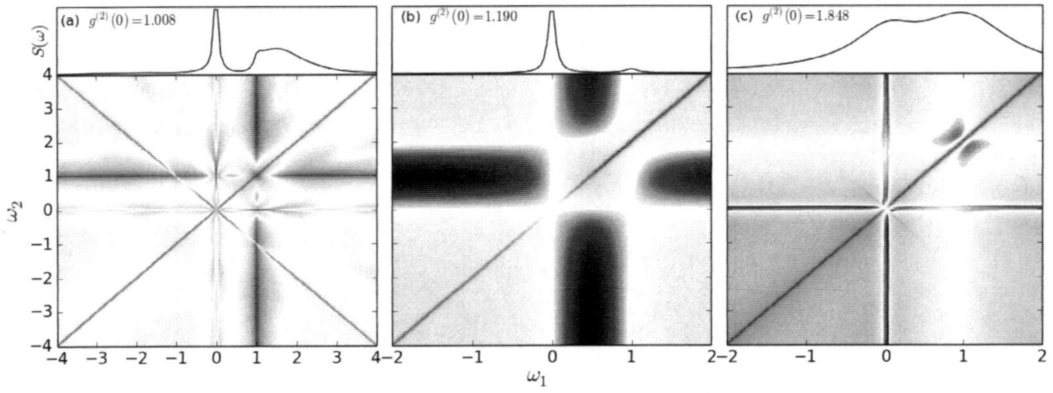

Figure 5. Spectrum (top panel) and 2PS (bottom panel) for the coherent source $\hat{H}_{coh}$. Different regimens in $g^{(2)}(0)$ are shown. Colour code same as in figure 4.

The opposite situation occurs when $g^{(2)}(0) \approx 2$ –see figure 5(c). In this case the correlation at $\omega = 0$ is indistinguishable and at $\omega = 1$ is distinguishable in frequency. The spectrum widens and the 2PS structure disappears, leaving an antibunching region non-localised. Far from the coherent condition (see figure 5(b)), the coherent source produces antibunched detection filtering distinguishable emissions around $\omega = 1$.

4. Conclusions

In this work, the effects of the dissipation and pump rates on the second order coherence function for three models of quantum light sources were studied. We have identified a new condition $\kappa \neq 0$ and $P \neq 0$ where the coherent source $\hat{H}_{coh}$ has incoherent emission. We have simulated, for different parameter regimes the steady state under an open quantum systems dynamic, and calculated the spectrum, the second-order coherence function and the 2PS. Our results emphasize the importance of modelling the quantum source as an open quantum system for a more realistic description of the $S(\omega)$ and 2PS of light field. The values of the dissipation κ and pumping P rates can be used as control parameters to modify the spectral or correlation properties of the sources. Finally, we have showed the importance that has the filtering of the emitted light over the bunching or antibunching properties, particularly in the case of single photon and coherent sources.

Acknowledgments

The authors acknowledge partial financial support from Departamento Administrativo de Ciencia, Tecnología e Innovación (COLCIENCIAS), and from Universidad de Antioquia, under Projects No. 2014-989 (CODI-UdeA) and Estrategia de Sostenibilidad del Grupo de Física Atómica y Molecular. We thank Elena del Valle for enlightened discussions.

References

[1] Scarcelli G, Valencia A and Shih Y 2004 *Europhys. Lett.* **68**, 618
[2] Zhai Y H, Chen X H, Zhang D and Wu L A 2005 *Phys. Rev. A* **72**, 043805
[3] Peiris M, Petrak B, Konthasinghe K, Yu Y, Niu Z C, and Muller A 2015 *Phys. Rev. B* **91**, 195125
[4] Javanainen J 1992 *Europhys. Lett.* **17**, 407
[5] Gmachl C, Capasso F, Sivco D L and Cho A Y 2001 *Rep. Prog. Phys.* **64**, 1533
[6] Glauber R J 2007 *Quantum Theory of Optical Coherence: Selected Papers and Lectures* (Berlin: Wiley-VCH)
[7] Silva I and Freire O 2013 *Hist. Stud. Nat. Sci.*, **43**, 453

IOP Conf. Series: Journal of Physics: Conf. Series **1043** (2018) 012012 doi:10.1088/1742-6596/1043/1/012012

[8] Vogel W and Welsch D G 2006 *Quantum Optics* (Berlin: Wiley-VCH)
[9] Eberly J H and Wddkiewicz K 1977 *J. Opt. Soc. Am.* **67**, 1252
[10] Knoll L, Weber G and Schafer T 1984 *J. Phys. B: At. Mol. Phys.* **17**, 4861
[11] Aspect A, Roger G, Reynaud S, Dalibard J and Cohen-Tannoudji C 1980 *Phys. Rev. Lett.* **45**, 617
[12] Press D, Gotzinger S, Reitzenstein S, Hofmann C, Loffler A, Kamp M, Forchel A and Yamamoto Y 2007 *Phys. Rev. Lett.* **98**, 117402
[13] Akopian N, Lindner N H, Poem E, Berlatzky Y, Avron J, Gershoni D, Gerardot B D and Petroff P M 2006 *Phys. Rev. Lett.* **96**, 130501
[14] Moreau E, Robert I, Manin L, Thierry-Mieg V, Gerard J M and Abram I 2001 *Phys. Rev. Lett.* **87**, 183601
[15] Sallen G, Tribu A, Aichele T, Andre R, Besombes L, Bougerol C, Richard M, Tatarenko S, Kheng K and Poizat J P 2010 *Nat. Photon.* **4**, 696
[16] del Valle E, Gonzalez-Tudela A, Laussy F P, Tejedor C and Hartmann M J 2012 *Phys. Rev. Lett* **109**, 183601
[17] Breuer H P and Petruccione F 2002 *The Theory of Open Quantum Systems* (Oxford: Oxford University Press)
[18] Yamamoto Y and Imamoglu A 1999 *Mesoscopic Quantum Optics* (New York: Wiley-Interscience)
[19] Shih Y 2011 *An Introduction to Quantum Optics: Photon and Biphoton Physics.* Taylor & Francis
[20] Mollow B R 1969 *Phys. Rev.* **188**

Non-local interactions in quantum mechanics modelled by shifted Dirac delta functions

Alvaro Restuccia[1], **Adrián Sotomayor**[2], **Vladimir Strauss**[3]

[1] Physics Department, University of Antofagasta, Chile
[2] Department of Mathematics, University of Antofagasta, Chile
[3] Department of High Mathematics, Ulyanovsk State Pedagogical University, Russia

E-mail: arestu@usb.ve,adrian.sotomayor@uantof.cl,vstrauss@mail.ru

Abstract. We introduce a new interaction formulated in terms of two shifted delta functions with non-local boundary conditions. For this purpose we use a method of self-adjoint extensions for a particular simmetric operator. We obtain bound states of the associated quantum mechanical problem and study its symmetry properties. We compare the results with the ground state of the quantum mechanical problem with a local interaction in terms of two delta functions.

1. Introduction

The Dirac delta distribution as a potential in quantum physics allows to construct approximate models with great historical and physical interest [1, 2, 3, 4, 5, 6, 7, 8, 9]. One of them is an approximate model of an hydrogen molecule ion, H_2^+ . The ion is commonly formed in molecular clouds in space. The first successful quantum mechanical treatment of H_2^+ was published by O. Burrau in 1927 [1]. The electron of the molecule moves in an attractive potential generated by the two protons. The quantum analysis of the electron arises from the solution of the Schrödinger equation, where the potential can be approximated by two delta "functions" with negative coefficients describing the attractive interaction.

Another interesting application arises from an approximation to the Kronig-Penney [3] model, which describes some basic quantum effects in the conduction of electrical charges in metals. The potential is expressed as an infinite sequence of Dirac deltas with a finite separation between them, a combination of Dirac deltas. The delta interactions have been also used in toy models of quantum wires or nanowires.

In this work we introduce a new non-local interaction of the type of two delta functions. We will determine the bound states, in particular the ground state of the quantum mechanical hamiltonian and study its symmetry properties. A main motivation for this work is the formulation of non-local field theory, based on string field theory [10].

2. Preliminaries

Let us consider the differential operator $D = -d^2\cdot/dt^2$ acting on the real line $\mathbb{R}$ with the domain

$$\mathcal{D}(D) = \{y(t)|\ y(t), y'(t), y''(t) \in L^2(\mathbb{R}),\ y(0) = y'(0) = 0\}.$$

Then the adjoint operator D^* has the domain

$$\mathcal{D}(D^*) = \{y(t)|\ y(t), y'(t), y''(t)|_{\mathbb{R}_+} \in L^2(\mathbb{R}_+),\ y(t), y', (t)y''(t)|_{\mathbb{R}_-} \in L^2(\mathbb{R}_-)\},$$

where $\mathbb{R}_+ = \{t|\ t \geq 0\}$, $\mathbb{R}_- = \{t|\ t \leq 0\}$. Since $D \subset D^*$, for every self-adjoint extension $\widetilde{D}$ of D we have $D \subset \widetilde{D} \subset D^*$ and an extension of D can be reduced to a restriction of D^*. For D^* we have $(D^*y, z) = -y'(-0)\bar{z}(-0) + y'(+0)\bar{z}(+0) + y(-0)\bar{z}'(-0) - y(+0)\bar{z}'(+0) + (y, D^*z)$. So, we obtain a self-adjoint restriction if $-y'(-0)\bar{z}(-0)+y'(+0)\bar{z}(+0)+y(-0)\bar{z}'(-0)-y(+0)\bar{z}'(+0) = 0$. The space of boundary values in our case is four-dimensional, so for a self-adjoint restriction we need two linear homogenous conditions. One of these conditions have the form (the same for y and z) $y(-0) = y(+0)$ and $\frac{y'(+0)-y'(0)}{y(0)} = c = const$. In this case the first generalized derivative of $y(t)$ has a jump and, so, the second one has a generalized summand with the delta-function. Thus the corresponding extension $\widetilde{D}$ can be naturally presented in the form $\widetilde{D}y(t) = -d^2y(t)/dt^2 + c \cdot y(0)\delta(t) = -d^2y(t)/dt^2 + c \cdot y(t)\delta(t)$. If $c < 0$, then $\widetilde{A}$ has the negative eigenvalue $\lambda = -c^2$ that corresponds to the eigenfunction $y(t) = e^{-c|t|}$. These facts are well known and can be find in the book of Albeverio et al [11].

In the present paper we study some generalization of the described above scheme for two boundary problems for points $-h$, h and a behaviour of the corresponding extension if $h \to 0$. We shall show that this behaviour involves the derivatives of the delta-function.

Note that even for the one-point problem there are some self-adjoint extensions with one or two negative eigenvalues, naturally involve not only the delta-function but its first derivative. One of such extension is given by the boundary conditions ($\alpha > 0,\ \beta > 0$)

$$\begin{aligned} y(+0) &= \frac{1}{2}\left\{-\left(\frac{1}{\alpha}+\frac{1}{\beta}\right)y'(+0) + \left(\frac{1}{\alpha}-\frac{1}{\beta}\right)y'(-0)\right\}, \\ y(-0) &= \frac{1}{2}\left\{\left(-\frac{1}{\alpha}+\frac{1}{\beta}\right)y'(+0) + \left(\frac{1}{\alpha}+\frac{1}{\beta}\right)y'(-0)\right\}, \end{aligned} \tag{1}$$

where $\int_{-\infty}^{-0}(|y(t)|^2+|y(t)'|^2+|y(t)''|^2)dt+\int_{+0}^{+\infty}(|y(t)|^2+|y(t)'|^2+|y(t)''|^2)dt < \infty$. The eigenvalues for this extension are $-\alpha^2$ and $-\beta^2$, the corresponding eigenfunctions are $e^{-\alpha|t|}$ and $\mathrm{Sgn}\,(t)e^{-\beta|t|}$ respectively. The extension $\widetilde{D}$ has the representation

$$\widetilde{D}y(t) = -y''(t) - \frac{1}{\beta}\cdot\delta'(t)\left(y'(-0)+y'(+0)\right) - \alpha\cdot\delta(t)\left(y(-0)+y(+0)\right). \tag{2}$$

3. Main constructions

3.1. Continuous function

Let the differential operator $D = -d^2\cdot/dt^2$ act on the real line $\mathbb{R}$ and have the domain $\mathcal{D}(D) =$

$$\{y(t)|\ y(t), y'(t)y''(t) \in L^2(\mathbb{R}),\ y(-h) = y'(-h) = y(h) = y'(h) = 0\}.$$

Then for the adjoint operator D^* we have

$$\begin{aligned} (D^*y, z) - (y, D^*z) = \;& -y'(-h-0)\bar{z}(-h-0) + y'(-h+0)\bar{z}(-h+0) \\ & -y'(h-0)\bar{z}(h-0) + y'(h+0)\bar{z}(h+0) \\ & +y(-h-0)\bar{z}'(-h-0) - y(-h+0)\bar{z}'(-h+0) \\ & +y(h-0)\bar{z}'(h-0) - y(h+0)\bar{z}'(h+0). \end{aligned} \tag{3}$$

Seeking self-adjoint extensions of D let us suppose that

$$y(-h-0) = y(-h+0), \quad y(h-0) = y(h+0) \tag{4}$$

and the same for $z(t)$. Then the conditions of self-adjointnity for extensions of D convert to $(y'(-h+0)-y'(-h-0))\bar{z}(-h)+(y'(h+0)-y'(h-0))\bar{z}(h)-y(-h)(\bar{z}'(-h+0)-\bar{z}'(-h-0))-y(h)(\bar{z}'(h+0)-\bar{z}'(h-0))=0$. Let us seek self-adjoint extensions such that

$$\begin{pmatrix}(y'(-h+0)-y'(-h-0))\\ (y'(h+0)-y'(h-0))\end{pmatrix}=\begin{pmatrix}a_{11} & a_{12}\\ a_{21} & a_{22}\end{pmatrix}\begin{pmatrix}y(-h)\\ y(h).\end{pmatrix} \tag{5}$$

It easy to check that the corresponding extension will be self-adjoint if and only if the matrix

$$A=\begin{pmatrix}a_{11} & a_{12}\\ a_{21} & a_{22}\end{pmatrix} \tag{6}$$

is symmetric. Note that the first derivative of $y(t)$ can have jumps in the points $-h$ and h, so the second one (if $y(t)$ is considered as a generalized function) two shifted δ-summands:

$$y''(t)=y''_{cl}(t)+(a_{11}y(-h)+a_{12}y(h))\delta(t+h)+(a_{21}y(-h)+a_{22}y(h))\delta(t-h)\ ,$$

where

$$f''_{cl}(t)=\begin{cases}f''(t) & \text{if } f''(t) \text{ exists in the classical sense,}\\ 0 & \text{in the opposite case.}\end{cases}$$

Thus, the corresponding extension $\widetilde{D}_h$ of the operator D can be re-written as

$$\begin{aligned}\widetilde{D}_h y(t)= & -y''(t)+(a_{11}y(-h)+a_{12}y(h))\delta(t+h)\\ & +(a_{21}y(-h)+a_{22}y(h))\delta(t-h)\ .\end{aligned} \tag{7}$$

Note that due to presence of α_{12} and α_{21} in (7) the interaction in question is not local.

Proposition 1 *Let an extension $\widetilde{D}$ satisfies the conditions (4) and (5). Then $\widetilde{D}$ has no non-negative eigenvalues and the number of its negative eigenvalues doesn't exceed the number of negative eigenvalues (taking in account the multiplicity) of matrix A from (6).*

Proof of Proposition 1 *Let us suppose that $\widetilde{D}$ has a positive eigenvalue ω^2, $\omega>0$. Then a corresponding eigenfunction $y(t)$ has a form*

$$y(t)=\begin{cases}a_1\cos\omega t+b_1\sin\omega t, & t\le h,\\ a_2\cos\omega t+b_2\sin\omega t, & t\in(-h,h),\\ a_3\cos\omega t+b_3\sin\omega t, & t\ge h.\end{cases}$$

Since $y(t)\in L^2(R)$, $a_1=b_1=a_3=b_3=0$. Due to the condition (4) it means that $y(-h)=y(h)=0$ and $y'(-h-0)=y'(h+0)=0$, therefore (see (5)) $y'(-h+0)=y'(h-0)=0$. From here it is easy to obtain $a_2=b_2=0$, i.e. $y(t)\equiv 0$. Thus ω^2 cannot be an eigenvalue for $\widetilde{D}$. In the same way one can show that 0 is not an eigenvalue for $\widetilde{D}$ too. The last assertion of Proposition 1 follows from the equality $(\widetilde{D}y(t),y(t))=(y(-h),y(h))\cdot A\cdot(y(-h),y(h))^T+\|y'\|^2$.

Let us find the extension $\widetilde{D}_h$ of D such that it has two eigenvalues $\lambda=-\alpha^2$ and $\mu=-\beta^2$. In order to do it we construct a matrix A such that for every positive h the function

$$f_h(t)=\begin{cases}e^{\alpha t}, & t\le -h\\ \dfrac{e^{-\alpha h}(e^{-\alpha t}+e^{\alpha t})}{e^{\alpha h}+e^{-\alpha h}}, & t\in(-h,h)\\ e^{-\alpha t}, & t\ge h\end{cases} \tag{8}$$

would be an eigenfunction of the operator $\widetilde{D}_h$.

According to (5) we have

$$\begin{aligned} -\frac{2\alpha \cdot e^{-\alpha h}}{1+e^{-2\alpha h}} &= e^{-\alpha h}\cdot(a_{11}+a_{12}), \\ -\frac{2\alpha \cdot e^{-\alpha h}}{1+e^{-2\alpha h}} &= e^{-\alpha h}\cdot(a_{21}+a_{22}). \end{aligned} \tag{9}$$

In the same way let us demand that for every positive h the function

$$g_h(t) = \begin{cases} e^{\beta t}, & t \le -h \\ \dfrac{e^{-\beta h}(e^{\beta t}-e^{-\beta t})}{e^{-\beta h}-e^{\beta h}}, & t \in (-h,h) \\ -e^{-\beta t}, & t \ge h \end{cases} \tag{10}$$

would be an eigenfunction of the operator $\widetilde{D}_h$. According to (5) we have

$$\begin{aligned} -\frac{2\beta \cdot e^{-\beta h}}{1-e^{-2\beta h}} &= e^{-\beta h}\cdot(a_{11}-a_{12}), \\ \frac{2\beta \cdot e^{-\beta h}}{1-e^{-2\beta h}} &= e^{-\beta h}\cdot(a_{21}-a_{22}). \end{aligned} \tag{11}$$

The system (9), (11) has a solution and we have

$$\begin{aligned} a_{11}=a_{22} &= -\left(\frac{\alpha}{(1+e^{-2\alpha h})}+\frac{\beta}{(1-e^{-2\beta h})}\right), \\ a_{12}=a_{21} &= \left(-\frac{\alpha}{(1+e^{-2\alpha h})}+\frac{\beta}{(1-e^{-2\beta h})}\right). \end{aligned} \tag{12}$$

Then the expression (7) is converted to

$$\begin{aligned} \widetilde{D}_h y(t) = & -y''(t) - \tfrac{\alpha(\delta(t-h)+\delta(t+h))(y(-h)+y(h))}{(1+e^{-2\alpha h})} \\ & + \tfrac{\beta(\delta(t+h)-\delta(t-h))(y(h)-y(-h))}{(1-e^{-2\beta h})}. \end{aligned} \tag{13}$$

Let us consider some particular cases. The interaction in (5) is not, generally speaking, local, because the jump of the derivative y' in the point $-h$ depends of the values y not only in the point $-h$ but also in the point h, etc. This interaction is local if and only if $a_{12} = a_{21} = 0$, that means

$$\frac{\alpha}{(1+e^{-2\alpha h})} = \frac{\beta}{(1-e^{-2\beta h})}.$$

The latter brings $\beta < \beta\cdot(1+e^{-2\alpha h}) = \alpha\cdot(1-e^{-2\beta h}) < \alpha$, so in the case of locality the bound state is given by (8) . If $\alpha = \beta$ then $a_{12} > 0$ and the bound state is a linear combination of (10) (8). It can be asymmetric. If we consider (8)+(10) we obtain a lefthanded valued eigenfunction and if we take (8)-(10) we get a right handed valued eigenfunction.

Note that $a_{12} \to 0$ and $a_{11} \to \infty$ if $\alpha \to \infty$, so for relatively big α a violence of locality in (5) is weak.

Let us consider the behaviour of $\widetilde{D}_h$ if $h \to 0$. For this aim we calculate the resolvent of $\widetilde{D}_h$ for a fixed real point $-\gamma^2$, $\gamma > \alpha$, $\gamma > \beta$. Let

$$G(t) = \frac{e^{-\gamma|t|}}{2\gamma}.$$

IOP Conf. Series: Journal of Physics: Conf. Series **1043** (2018) 012013 doi:10.1088/1742-6596/1043/1/012013

Then for $z(t) \in L^2(\mathbb{R})$, such that

$$\int_{-\infty}^{+\infty} z(\tau)G(-h-\tau)d\tau = 0, \quad \int_{-\infty}^{+\infty} z(\tau)G(h-\tau)d\tau = 0\,,$$

and

$$y(t) = \int_{-\infty}^{+\infty} z(\tau)G(t-\tau)d\tau$$

we have (see (13))

$$(\gamma^2 I + \widetilde{D}_h)y(t) = z(t)$$

thus,

$$(\gamma^2 I + \widetilde{D}_h)^{-1}z(t) = y(t) = \int_{-\infty}^{+\infty} z(\tau)G(t-\tau)d\tau. \tag{14}$$

Let us introduce some notations. We put

$$u_h(t) = G(-h-t) + G(h-t), \;\; v_h(t) = G(-h-t) - G(h-t),$$

$$M_h = \{f(t) : \, f(t) \in L^2(\mathbb{R}), \, f(t) \perp u_h(t), \, f(t) \perp v_h(t)\}$$

and denote by N_h the linear span of $f_h(t)$ and $g_h(t)$. Then (note that this decomposition is not orthogonal)

$$L^2(\mathbb{R}) = M_h \dot{+} N_h. \tag{15}$$

Let $\mu_h = \int_{-\infty}^{+\infty} u_h(t)f_h(t)dt$ and $\nu_h = \int_{-\infty}^{+\infty} v_h(t)g_h(t)dt$. Then for every $f(t) \in L^2(\mathbb{R})$ the decomposition (15) the form

$$\begin{aligned} f(t) = \Big(f(t) - (\frac{f_h(t)}{\mu_h}\int_{-\infty}^{+\infty} f(t)u_h(t)dt + \frac{g_h(t)}{\nu_h}\int_{-\infty}^{+\infty} f(t)v_h(t)dt)\Big) + \\ (\frac{f_h(t)}{\mu_h}\int_{-\infty}^{+\infty} f(t)u_h(t)dt + \frac{g_h(t)}{\nu_h}\int_{-\infty}^{+\infty} f(t)v_h(t)dt)\,. \end{aligned}$$

Thus (see (14)),

$$\begin{aligned} &(\gamma^2 I + \widetilde{D}_h)^{-1}f(t) = \Big(\textstyle\int_{-\infty}^{+\infty} f(\tau)G(t-\tau)d\tau \\ &-(\frac{\int_{-\infty}^{+\infty} f_h(\tau)G(t-\tau)d\tau}{\mu_h}\textstyle\int_{-\infty}^{+\infty} f(t)u_h(t)dt + \\ &\frac{\int_{-\infty}^{+\infty} g_h(\tau)G(t-\tau)d\tau}{\nu_h}\textstyle\int_{-\infty}^{+\infty} f(t)v_h(t)dt)\Big) + \\ &(\frac{f_h(t)}{(\gamma^2-\alpha^2)\mu_h}\textstyle\int_{-\infty}^{+\infty} f(t)u_h(t)dt + \frac{g_h(t)}{(\gamma^2-\beta^2)\nu_h}\int_{-\infty}^{+\infty} f(t)v_h(t)dt)\,. \end{aligned} \tag{16}$$

Note that

$$u_h(t) = \begin{cases} \dfrac{e^{\gamma(t+h)} + e^{\gamma(t-h)}}{2\gamma}, & t \le -h \\ \dfrac{e^{-\gamma(t+h)} + e^{\gamma(t-h)}}{2\gamma}, & t \in (-h, h) \\ \dfrac{e^{-\gamma(t+h)} + e^{-\gamma(t-h)}}{2\gamma}, & t \ge h \end{cases} \tag{17}$$

IOP Conf. Series: Journal of Physics: Conf. Series **1043** (2018) 012013 doi:10.1088/1742-6596/1043/1/012013

and

$$v_h(t) = \begin{cases} \dfrac{e^{\gamma(t+h)} - e^{\gamma(t-h)}}{2\gamma}, & t \le -h \\ \dfrac{e^{-\gamma(t+h)} - e^{\gamma(t-h)}}{2\gamma}, & t \in (-h, h) \\ \dfrac{e^{-\gamma(t+h)} - e^{-\gamma(t-h)}}{2\gamma}, & t \ge h\ . \end{cases} \tag{18}$$

The direct calculation brings

$$\mu_h = 2\frac{\cosh(\gamma h)}{\gamma} \cdot \frac{e^{-(\alpha+\gamma)h}}{\alpha+\gamma} + \frac{e^{-(\alpha+\gamma)h}}{\gamma \cosh(\alpha h)} \cdot \{\frac{\sinh(\alpha+\gamma)h}{(\alpha+\gamma)} + \frac{\sinh(\gamma-\alpha)h}{(\gamma-\alpha)}\}\ ,$$

$$\nu_h = 2\frac{\sinh(\gamma h)}{\gamma} \cdot \frac{e^{-(\beta+\gamma)h}}{\beta+\gamma} + \frac{e^{-(\beta+\gamma)h}}{\gamma \sinh(\beta h)} \cdot \{\frac{\sinh(\beta+\gamma)h}{(\beta+\gamma)} - \frac{\sinh(\gamma-\beta)h}{(\gamma-\beta)}\}\ ,$$

so for small h we have $\nu_h = \frac{2h}{\beta+\gamma} - \frac{2h^2}{3} + \dots$.

The latter equalities bring

$$\lim_{h\to 0} \frac{u_h(t)}{\mu_h} = \gamma(\alpha+\gamma)G(t),\ \lim_{h\to 0} \frac{v_h(t)}{\nu_h} = -\gamma(\beta+\gamma)G(t)\,\mathrm{Sgn}\,(t) \tag{19}$$

From (16) and (19) we have

$$\begin{aligned} &\lim_{h\to 0}(\gamma^2 I + \widetilde{D}_h)^{-1} f(t) = \Big(\int_{-\infty}^{+\infty} f(\tau)G(t-\tau)d\tau \\ &-\gamma\big((\alpha+\gamma)\int_{-\infty}^{+\infty} f_0(\tau)G(t-\tau)d\tau \int_{-\infty}^{+\infty} f(t)G(t)dt - \\ &(\beta+\gamma)\int_{-\infty}^{+\infty} g_0(\tau)G(t-\tau)d\tau \int_{-\infty}^{+\infty} f(t)G(t)\,\mathrm{Sgn}\,(t)dt\big)\Big) + \\ &\gamma(\frac{f_0(t)}{(\gamma-\alpha)}\int_{-\infty}^{+\infty} f(t)G(t)dt - \frac{g_0(t)}{(\gamma-\beta)}\int_{-\infty}^{+\infty} f(t)G(t)\,\mathrm{Sgn}\,(t)dt)\ , \end{aligned} \tag{20}$$

where $f_0(t) = e^{-\alpha|t|}$ and $g_0(t) = -e^{-\beta|t|}\,\mathrm{Sgn}\,(t)$. It is easy to check that the right part in (20) the resolvent of the operator (2) calculated in the point $-\gamma^2$, thus this operator is a limit of $\widetilde{D}_h$ and, respectively, the boundary conditions (1) be treated as a limit of the boundary conditions (5), (12) for $h \to 0$. Let us note that this limit pass eliminates a possibility of a spontaneous symmetry breaking model because the condition $\alpha = \beta$ transforms (1) to the conditions

$$\alpha \cdot y(-0) = y'(-0) \text{ and } \alpha \cdot y(+0) = -y'(+0)\ .$$

The latter means that waves on $\mathbb{R}_-$ and $\mathbb{R}_+$ are independent, so it is a model of a non transitable barrier.

Remark 1 *There is an extension $\widetilde{D}$ that doesn't satisfy the conditions (4), (5) and has a positive eigenvalue.*

3.2. Continuous derivative

Seeking self-adjoint extensions of D let us suppose that

$$y'(-h-0) = y'(-h+0), \quad y'(h-0) = y'(h+0) \tag{21}$$

and the same for $z(t)$. Then the conditions of self-adjointness for extensions of D convert to

$$y'(-h)\left(-\bar{z}(-h-0)+\bar{z}(-h+0)\right)+y'(h)\left(-\bar{z}(h-0)+\bar{z}(h+0)\right)+$$
$$\left(y(-h-0)-y(-h+0)\right)\bar{z}'(-h)+\left(y(h-0)-y(h+0)\right)\bar{z}'(h)=0.$$

Let us seek self-adjoint extensions such that

$$\begin{pmatrix}(y(-h+0)-y(-h-0))\\(y(h+0)-y(h-0))\end{pmatrix}=\begin{pmatrix}b_{11}&b_{12}\\b_{21}&b_{22}\end{pmatrix}\begin{pmatrix}y'(-h)\\y'(h)\end{pmatrix}. \tag{22}$$

It easy to check that the corresponding extension will be self-adjoint if and only if the matrix

$$B=\begin{pmatrix}b_{11}&b_{12}\\b_{21}&b_{22}\end{pmatrix} \tag{23}$$

is symmetric.

Let us consider a function $y(t)$ under Conditions (22) as a generalized function (distribution). Then

$$y''(t)=y''_{cl}(t)+(b_{11}y'(-h)+b_{12}y'(h))\delta'(t+h)+(b_{21}y'(-h)+b_{22}y'(h))\delta'(t-h)\ ,$$

where

$$f''_{cl}(t)=\begin{cases}f''(t) & \text{if } f''(t) \text{ exists in the classical sense,}\\ 0 & \text{in the opposite case,}\end{cases}$$

as before. Thus, the corresponding extension $\widetilde{D}_h$ of the operator D can be re-written as

$$\begin{aligned}\widetilde{D}_h y(t)= &\ -y''(t)+(b_{11}y'(-h)+b_{12}y'(h))\delta'(t+h)\\ &+(b_{21}y'(-h)+b_{22}y'(h))\delta'(t-h)\ .\end{aligned} \tag{24}$$

Let us construct a matrix B such that for every positive h the function

$$f_h(t)=\begin{cases}e^{\alpha t}, & t\le h\\ -\dfrac{e^{-\alpha h}(e^{-\alpha t}+e^{\alpha t})}{e^{\alpha h}-e^{-\alpha h}}, & t\in(-h,h)\\ e^{-\alpha t}, & t\ge h\end{cases} \tag{25}$$

would be an eigenfunction of the operator $\widetilde{D}$. It is clear that the corresponding eigenvalue is $\lambda=-\alpha^2$. Note that $\int_{-h}^{h}f_h(t)dt=-\frac{2e^{-\alpha h}}{\alpha}$, so in the sense of distributions $\lim_{h\to+0}f_h(t)=f_0(t)-\frac{2}{\alpha}\delta(t)$, where $f_0(t)=e^{-\alpha^2|t|}$, therefore in this case the limit generates a new boundary problem, that directly involves $\delta(t)$. Let us show that there is an extension $\widetilde{D}$ that corresponds to $f_h(t)$. According to (5) we have

$$-\frac{2e^{-\alpha h}}{1-e^{-2\alpha h}}=\alpha\cdot e^{-\alpha h}\cdot(b_{11}-b_{12}),$$

$$\frac{2e^{-\alpha h}}{1-e^{-2\alpha h}}=\alpha\cdot e^{-\alpha h}\cdot(b_{21}-b_{22}).$$

The last system implies that the matrix elements $b_{12}=b_{21}$ are real and $b_{11}=b_{22}$. The second type of conditions we obtain by the hypothesis that $\widetilde{D}$ has another eigenvalue $\mu=-\beta^2$. If an eigenfunction $g_h(t)$ corresponds to μ, then it must be orthogonal to $f_h(t)$. This condition, in particular, would be fulfilled if $g_h(t)$ is odd. Put

$$g_h(t) = \begin{cases} e^{\beta t}, & t < h \\ \dfrac{e^{-\beta h}(-e^{-\beta t} + e^{\beta t})}{e^{\beta h} + e^{-\beta h}}, & t \in (-h, h) \\ -e^{-\beta t}, & t \geq h \end{cases} \tag{26}$$

According to (5) we have

$$-\frac{2e^{-\beta h}}{1 + e^{-2\beta h}} = \beta \cdot e^{-\beta h} \cdot (b_{11} + b_{12}),$$

$$-\frac{2e^{-\beta h}}{1 + e^{-2\beta h}} = \beta \cdot e^{-\beta h} \cdot (b_{21} + b_{22}).$$

So

$$\begin{aligned} b_{11} &= -\left(\frac{1}{\alpha(1 - e^{-2\alpha h})} + \frac{1}{\beta(1 + e^{-2\beta h})}\right), \\ b_{12} &= \left(\frac{1}{\alpha(1 - e^{-2\alpha h})} - \frac{1}{\beta(1 + e^{-2\beta h})}\right). \end{aligned} \tag{27}$$

Then the expression (24) is converted to

$$\begin{aligned} \widetilde{D}_h y(t) = \quad & -y''(t) + \tfrac{(\delta'(t-h) - \delta'(t+h))(y'(-h) - y'(h))}{\alpha(1 - e^{-2\alpha h})} \\ & - \tfrac{(\delta'(t+h) + \delta'(t-h))(y'(-h) + y'(h))}{\beta(1 + e^{-2\beta h})}. \end{aligned}$$

Strictly speaking in the latter expression we cannot pass to limit for $h \to 0$ because the domain of $\widetilde{D}_h$ depends of h and, moreover, the eigenfunction $f_h(t)$ does not converge to any function of $L^2(\mathbb{R})$, but from some heuristic point of view we can say that $\widetilde{D}_h \to \widetilde{D}_0$, where

$$\widetilde{D}_0 y(t) = -y''(t) - \frac{\delta''(t)(y'(-0) - y'(+0))}{\alpha^2} - \frac{\delta'(t)(y'(-0) + y'(+0))}{\beta}.$$

The constructed operator $\widetilde{D}_0$ is not well defined because its domain is not evident.

4. Conclusions

We introduced a new interaction of the type of two delta functions or their derivatives.

For this aim we used a method of self-adjoint extensions of a given symmetric operator. We determined the associated bound states. There exists a symmetric and an antisymmetric eigenfunction. We analized the dependence on the intensity of the interaction of the corresponding eigenvalues as well as the limit of degeneracy of the ground state. We compared the results with the corresponding ones for the interaction given by a potential with only two delta functions. We expect to extend this exotic non-local interaction to non-local field theory arising from string theory.

Acknowledgments

A. R. and A. S. are partially supported by Project Fondecyt 1161192, Chile.

References

[1] Ø. Burrau, Naturwissenschaften, Volume 15, Issue 1, 16-17 (1927).
[2] C. N. Yang, Phys. Rev. Lett. 19:13121315 (1967).
[3] R. de L. Kronig and W. G. Penney, Proc. Roy. Soc. (London) 130A, 499-513 (1931).
[4] A. A. Frost, J. Chem. Phys. 25, 1150-1154 (1956).
[5] A. A. Frost and F. E. Leland J. Chem. Phys. 25, 1154-1160 (1956).
[6] R. Ahlrichs and P. Claverie, J. Quant. Chem. 6, 1001-1009 (1972).
[7] P. R. Certain and W. Byers Brown, J. Quant. Chem. 6, 131-142 (1972).
[8] S. Albeverio, F. Gesztesy, R. Hoegh-Krohn and W. Kirsch, Operator Theory 12, 101-126. Erratum 13, 419 (1985).
[9] F. Gesztesy and H. Holden, J. Phys. A 20, 5157-5177 (1987).
[10] "Strings Theory and Fundamental Interactions", Edited by M. Gasparini and J. Maharana, Lecture Notes in Physics 737, Springer, Berlin Heidelberg (2008).
[11] S. Albeverio, F. Gesztesy, R. Hoegh-Kron and H. Holden, "Solvable models in quantum mechanics", AMS Chelsea Publishing, American Mathematical Society-Providence, Rhode Island (2004).

Natural Inflation on the brane with a TeV-scale gravity: Parameter constraints after Planck 2015

Nelson Videla [1], **Grigorios Panotopoulos** [2]

[1] Departamento de Física, FCFM, Universidad de Chile, Blanco Encalada 2008, Santiago, Chile

[2] CENTRA, Instituto Superior Técnico, Universidade de Lisboa, Av. Rovisco País 1, Lisboa, Portugal

E-mail: nelson.videla@ing.uchile.cl; grigorios.panotopoulos@tecnico.ulisboa.pt

Abstract. In the present work we have studied Natural Inflation in the framework of the Randall-Sundrum II brane model (RS-II) in the light of the latest Planck results. Adopting the Randall-Sundrum fine-tuning, the model is characterized by 3 parameters in total, namely the 5-dimensional Planck mass M_5 and the two mass scales of the inflaton potential f and Λ. We show in the $n_s - r$ plane the theoretical predictions of the model together with the allowed contour plots, and we conclude that the model is viable. By using the Planck results only it is possible to determine the two mass scales of the inflaton potential in terms of M_5, which remains undetermined. However, there are several good theoretical reasons to consider a higher-dimensional Planck mass of the order of 10 TeV, which is compatible with primordial nucleosynthesis. If we insist on considering a M_5 of this order of magnitude all parameters are known.

1. Introduction

The inflationary universe has become one of the central paradigms in modern cosmology. This is due to the fact that many long-standing problems of the Big Bang model, such as the horizon, flatness, homogeneity and monopole problems, find a natural explanation in the framework of the inflationary universe [1]. However, the essential feature of inflation is that it generates a mechanism to explain the Large-Scale Structure (LSS) of the universe [2] and provides a causal interpretation of the origin of the anisotropies observed in the Cosmic Microwave Background (CMB) radiation[3], since primordial density perturbations may be produced from quantum fluctuations during the inflationary era.

Despite its success, a theory of inflation is still missing, all we have is a large collection of inflationary models (see e.g.[4], and for a classification of inflationary models see[5]). The inflaton potential cannot be derived from a fundamental theory in a unique way. Moreover, the potential must be specially designed to be flat, and this is the fine-tuning problem of inflation. Natural Inflation (NI) with a pseudo-Nambu Goldstone boson (pNGB) as the inflaton[6] is provided in certain particle physics model[7]. The scalar potential, which is flat due to shift symmetries, has the form

$$V(\phi) = \Lambda^4 \left[1 - \cos\left(\frac{\phi}{f}\right)\right]. \quad (1)$$

and it is characterized by two mass scales f and Λ with $f \gg \Lambda$. It is assumed that a global symmetry is spontaneously broken at some scale f, with a soft explicit symmetry breaking at a lower scale Λ. Natural inflation has been already studied in standard cosmology based on Einstein's General Relativity[8]. Upon comparison to the current cosmological and astronomical observations, specially those related with the CMB temperature anisotropies, it is possible to constrain several inflation models. Particularly, the constraints in the $n_s - r$ plane give us the predictions of a number of representative inflationary potentials. Recently, the Planck collaboration has published new data of enhanced precision of the CMB anisotropies [9]. Here, the Planck full mission data has improved the upper bound on the tensor-to-scalar ratio $r_{0.002} < 0.11(95\%\ \mathrm{CL})$. In particular, Natural Inflation is consistent with current data [9] for trans-Planckian values of the symmetry breaking scale f, for which it may be expect the low-energy effective theory, on which (1) is based, to break down [10]. Regarding the last data of Planck, this model is consistent for $\log_{10}(f/M_p) > 0.84$, where M_p is the reduced Planck mass $M_p \equiv 1/\sqrt{8\pi G_N} = 2 \times 10^{18}$ GeV. Nevertheless, several mechanisms which yield the potential (1) with super-Planckian values for the symmetry breaking scale f consistent with a low energy description, have been proposed recently [11, 12].

On the other hand, it would be challenging to study natural inflation in non-standard cosmologies. Considering non-standard cosmologies is motivated by at least two facts, namely i) deviations from the standard Friedmann equation arise in higher-dimensional theories of gravity, and ii) there is no observational test of the Friedmann equation before the primordial big-bang nucleosynthesis (BBN) epoch. A well-studied example of a novel higher-dimensional theory is brane-world models. Brane models are inspired from M/superstring theory and although they are not yet derivable from the fundamental theory, at least they contain the basic ingredients, such as extra dimensions, higher-dimensional objects (branes), higher-curvature corrections to gravity (Gauss-Bonnet) etc. Since superstring theory claims to give us a fundamental description of nature it is important to study what kind of cosmology it predicts. Regarding the realization of Natural Inflation in non-standard cosmologies, some works have been put forward in the literature so far [13, 14], achieving sub-Planckian values for the symmetry breaking scale f, being consistent with the data available at that time.

The main goal of the present work is to study the realization of NI in the high-energy regime of the RS-II brane model, in the light of the recent Planck results. Later on we will show that our results are modified significantly compared to[13] using the Planck results. By comparing the theoretical predictions of the model together with the allowed contour plots, and we conclude that the model is viable. Using the latest Planck results only the inflaton potential mass scales f, Λ are given in terms of the five-dimensional Planck mass, which remains unconstraint though. However, we insist on considering a higher-dimensional Planck mass of the order of, say, 10 TeV, since there are several good theoretical reasons for that.

We organize our work as follows: After this introduction, in the next section we summarize the basics of the brane model as well as the dynamics of inflation. In the third section we analyze natural inflation in the framework of RS-II model and present our results, and in the last section we finish with our conclusions. We choose units so that $c = \hbar = 1$.

2. Basics of Braneworld inflation

2.1. Braneworld cosmology

In the brane-world scenario the main idea is that our four-dimensional world and the standard model of particle physics are confined to live on a 3-dimensional brane, while gravity lives in the higher-dimensional bulk. Since the higher-dimensional Plank mass M_D is the fundamental mass scale instead of the usual four-dimensional Planck mass M_4, the brane concept has been used to address the hierarchy problem of particle physics, initially in the simple framework of a flat (4+n) spacetime with 4 large dimensions and n small compact dimensions[15], and later refined by Randall and Sundrum[16, 17]. For an introduction to brane cosmology see e.g.[18]. In the RS-II model [17], the four-dimensional Einstein equations may be written as [19]

$$^{(4)}G_{\mu\nu} = -\Lambda_4 g_{\mu\nu} + \frac{8\pi}{M_4^2}\tau_{\mu\nu} + \left(\frac{8\pi}{M_5^3}\right)^2 \pi_{\mu\nu} - E_{\mu\nu}, \tag{2}$$

where Λ_4 is the four-dimensional cosmological constant, $\tau_{\mu\nu}$ is the energy-momentum tensor of matter on the brane, $\pi_{\mu\nu} = (1/12)\tau\tau_{\mu\nu} + (1/8)g_{\mu\nu}\tau_{\alpha\beta}\tau^{\alpha\beta} - (1/4)\tau_{\mu\alpha}\tau^{\alpha}_{\nu} - (1/24)\tau^2 g_{\mu\nu}$, and $E_{\mu\nu} = C^{\alpha}_{\beta\rho\sigma}n_\alpha n^\rho g^\beta_\mu g^\sigma_\nu$ is the projection of the five-dimensional Weyl tensor $C_{\alpha\beta\rho\sigma}$ on the brane, where n^α is the unit vector normal to the brane. $E_{\mu\nu}$ and $\pi_{\mu\nu}$ encode the information about the bulk. The four-dimensional quantities can be computed in terms of the five-dimensional ones as follows[20]

$$M_4 = \sqrt{\frac{3}{4\pi}}\left(\frac{M_5^2}{\sqrt{\lambda}}\right) M_5 \tag{3}$$

$$\Lambda_4 = \frac{4\pi}{M_5^3}\left(\Lambda_5 + \frac{4\pi\lambda^2}{3M_5^3}\right) \tag{4}$$

The Friedmann equation for a flat FRW background is given by [21]

$$H^2 = \frac{\Lambda_4}{3} + \frac{8\pi}{3M_4^2}\rho\left(1 + \frac{\rho}{2\lambda}\right) + \frac{\mathcal{E}}{a^4}. \tag{5}$$

where $\mathcal{E}$ is an integration constant coming from $E_{\mu\nu}$. The term $\frac{\mathcal{E}}{a^4}$ is known as the dark radiation, since it decays in the same way as radiation. However, during inflation this term will be rapidly diluted, and we can neglect it. The five-dimensional Planck mass is constraint by the standard bib-bang nucleosynthesis (BBN) to be $M_5 \geq 10$ TeV [22]. In the following, we will take the four-dimensional cosmological constant Λ_4 to be zero, or in other words we adopt the RS fine tuning $\Lambda_5 = -4\pi\lambda^2/(3M_5^3)$ so that model can explain the cosmic acceleration without cosmological constant, and neglecting the term $\frac{\mathcal{E}}{a^4}$ the Friedmann equation (5) becomes

$$H^2 = \frac{8\pi}{3M_4^2}\rho\left(1 + \frac{\rho}{2\lambda}\right), \tag{6}$$

which becomes the basis of our study on brane-world inflation.

2.2. *Inflationary dynamics*

At low energies, i.e., when $\rho \ll \lambda$, inflation in the brane-world scenario behaves in exactly the same way as standard inflation. But at higher energies we would expect the dynamics of inflation to be changed.

We consider slow-roll inflation driven by a scalar field ϕ, for which the energy density ρ and the pressure P are given by $\rho = \frac{\dot{\phi}^2}{2} + V(\phi)$ and $P = \frac{\dot{\phi}^2}{2} - V(\phi)$, respectively, where $V(\phi)$ is the scalar potential. We assume that the scalar field is confined to the brane, so the four-dimensional Klein-Gordon equation still holds

$$\ddot{\phi} + 3H\dot{\phi} + V' = 0, \tag{7}$$

where prime indicates derivative with respect to ϕ, and dot a derivative with respect to cosmic time. We can use the slow-roll approximation to write (6) and (7) as

$$H^2 \simeq \frac{8\pi}{3M_4^2}V\left(1 + \frac{V}{2\lambda}\right), \tag{8}$$

and

$$3H\dot{\phi} \simeq -V'. \tag{9}$$

In this way, using these two equations, it is possible to write the slow-roll parameters on the brane as[20]

$$\epsilon_V \equiv \frac{M_4^2}{16\pi}\left(\frac{V'}{V}\right)^2 \frac{1+V/\lambda}{(1+V/2\lambda)^2}, \tag{10}$$

$$\eta_V \equiv \frac{M_4^2}{8\pi}\frac{V''}{V}\frac{1}{1+V/2\lambda}. \tag{11}$$

Slow-roll inflation implies that $\epsilon_V \ll 1$ and $|\eta_V| \ll 1$, as in standard cosmology. These reduce to standard slow-roll parameters at the the low-energy limit $V \ll \lambda$. On the other hand, in the high-energy limit, i.e., $V \gg \lambda$, these expressions become

$$\epsilon_V \simeq \frac{M_4^2\lambda}{4\pi}\frac{V'^2}{V^3}, \tag{12}$$

$$\eta_V \simeq \frac{M_4^2\lambda}{4\pi}\frac{V''}{V^2}. \tag{13}$$

The deviations from standard slow-roll inflation can be seen in the high-energy as both the parameters are suppressed by a factor V/λ.

The number of e-folds in the slow-roll approximation, using (6) and (7), yields

$$N \simeq -\frac{8\pi}{M_4^2}\int_{\phi_*}^{\phi_{end}} \frac{V}{V'}\left(1+\frac{V}{2\lambda}\right) d\phi, \tag{14}$$

where ϕ_* and ϕ_{end} are the values of the scalar field when the cosmological scales cross the Hubble-radius and at the end of inflation, respectively. As it can be seen, the number of e-folds is increased due to an extra term of V/λ. This implies a more amount of inflation, between these two values of the field, compared to standard inflation.

2.3. Perturbations

In the following, we will give a review of cosmological perturbations in brane-world inflation. We consider the gauge invariant quantity $\zeta = -\psi - H\frac{\delta\rho}{\dot{\rho}}$. Here, ζ is defined on slices of uniform density and reduces to the curvature perturbation at super-horizon scales. A fundamental feature of ζ is that it is nearly constant on super-horizon scales[23], and in fact this property does not depend on the gravitational field equations[24]. Therefore, for the spatially flat gauge, we have $\zeta = H\frac{\delta\phi}{\dot{\phi}}$, where $|\delta\phi| = H/2\pi$. In this way, using the slow-roll approximation, the spectra of scalar perturbations is given by[20]

$$\mathcal{P}_{\mathcal{R}} = \frac{H^2}{\dot{\phi}^2}\left(\frac{H}{2\pi}\right)^2 \simeq \frac{128\pi}{3M_4^6}\frac{V^3}{V'^2}\left(1+\frac{V}{2\lambda}\right)^3. \tag{15}$$

On the other hand, the tensor perturbations are more involved since the gravitons can propagate in the bulk. The amplitude of tensor perturbations is given by[20]

$$\mathcal{P}_g = \frac{64\pi}{M_4^2}\left(\frac{H}{2\pi}\right)^2 F^2(x), \tag{16}$$

where

$$F(x) = \left[\sqrt{1+x^2} - x^2 \ln\left(\frac{1}{x} + \sqrt{1+\frac{1}{x^2}}\right)\right]^2$$
$$= \left[\sqrt{1+x^2} - x^2 \sinh^{-1}\left(\frac{1}{x}\right)\right]^{-1/2}, \quad (17)$$

and x is given by

$$x = HM_4\sqrt{\frac{3}{4\pi\lambda}}. \quad (18)$$

The expressions for the spectra are, as always, to be evaluated at Hubble radius crossing $k = aH$. As expected, in the the low-energy limit the expressions for the scalar and tensor spectra become the same as those derived without considering the brane effects. However, in the high-energy limit, these expressions become

$$\mathcal{P}_{\mathcal{R}} \simeq \frac{16\pi}{3M_4^6\lambda^3}\frac{V^6}{V'^2}, \quad (19)$$
$$\mathcal{P}_g \simeq \frac{32V^3}{M_4^4\lambda^2}. \quad (20)$$

The scale dependence of the scalar power spectra is determined by the scalar spectral index, which under the slow-roll approximation, obeys the usual relation

$$n_s = 1 + \frac{d\ln\mathcal{P}_{\mathcal{R}}}{d\ln k}$$
$$n_s \simeq 1 - 6\epsilon_V + 2\eta_V. \quad (21)$$

The amplitude of tensor perturbations can be parameterized by the tensor-to-scalar ratio, defined to be

$$r \equiv \frac{\mathcal{P}_g}{\mathcal{P}_{\mathcal{R}}}, \quad (22)$$

which implies that in the low-energy limit this expression becomes $r \simeq 16\epsilon_V$, where ϵ_V is the standard slow-roll parameter, whereas in the high-energy limit we have that $r \simeq 24\epsilon_V$, with ϵ_V corresponding to Eq.(12).

As we have seen, at late times the brane-world cosmology is identical to the standard one. During the early universe, particularly during inflation, there may be changes to the perturbations predicted by the standard cosmology, if the energy density is sufficiently high compared with the brane tension. In the following, we will obtain the predictions for the natural inflationary model in the brane-world scenario in the high-energy limit, and to try to ascertain whether these predictions are compatible with current observational constraints.

3. Natural inflation on the brane

3.1. Dynamics of inflation

Natural inflation in the Randall-Sundrum brane-world scenario is characterized by 3 parameters in total, namely the 5-dimensional Planck mass and the mass scales of the inflaton potential (1), Λ and f, respectively. In the high-energy limit, the slow-roll parameters ϵ_V and η_V, using Eqs.(12) and (13) are given by

$$\epsilon_V = \alpha\frac{(1+\cos(y))}{(1-\cos(y))^2}, \quad (23)$$

$$\eta_V = \alpha\frac{\cos(y)}{(1-\cos(y))^2}, \quad (24)$$

where $y \equiv \phi/f$ and α is a dimensionless parameter defined as

$$\alpha \equiv \frac{M_4^2 \lambda}{4\pi f^2 \Lambda^4}. \tag{25}$$

For this model, the condition for the end of inflation is found to be $\epsilon(y_{end}) = 1$, leading to

$$\cos(y_{end}) = \cos\left(\frac{\phi_{end}}{f}\right) = \frac{1}{2}\left(2 + \alpha - \sqrt{\alpha}\sqrt{8+\alpha}\right). \tag{26}$$

The number of inflationary e-folds that occur between the values of the scalar field when a given perturbation scale leaves the Hubble-radius and at the end of inflation, can be computed from (1) and the high-energy limit of (14), yielding

$$N = \frac{1}{\alpha}\left[\cos(y_*) - \cos\left(y_{end}\right) - 2\ln\left(\frac{1+\cos(y_*)}{1+\cos(y_{end})}\right)\right], \tag{27}$$

with $\cos(y_{end})$ given by (26). Solving Eq.(27) for $\cos(y_*)$, after replacing Eq.(26), we may obtain the value of the scalar field at the time of Hubble-radius crossing, given by

$$\cos(y_*) = \cos\left(\frac{\phi_*}{f}\right) = -1 - 2\,W\left[z(N,\alpha)\right], \tag{28}$$

where

$$z(N,\alpha) \equiv -\frac{\sqrt{e^{-1-\frac{\sqrt{\alpha}}{2}(2N\sqrt{\alpha}+\sqrt{\alpha}-\sqrt{\alpha+8})}\left(\alpha^2+8\alpha+8-\alpha^{3/2}\sqrt{\alpha+8}-4\sqrt{\alpha}\sqrt{\alpha+8}\right)}}{2\sqrt{2e}}, \tag{29}$$

and W denotes the Lambert W function [25].

3.2. Cosmological perturbations

Regarding the cosmological perturbations, the amplitude of scalar perturbations, using Eqs.(1) and (19), is found to be

$$\mathcal{P}_{\mathcal{R}} = \frac{1}{12\pi^2\alpha^3}\gamma^4\frac{(1-\cos(y))^5}{(1+\cos(y))}, \tag{30}$$

where $\gamma \equiv \frac{\Lambda}{f}$ is the ratio between both mass scales of the inflaton potential (1), Λ and f. The scalar spectral index, using Eqs.(21), (23), and (24), becomes

$$n_s = 1 - 2\alpha\frac{(3+2\cos(y))}{(1-\cos(y))^2}. \tag{31}$$

Finally, the tensor-to-scalar ratio can be obtained from the high-energy limit of Eq.(22), yielding

$$r = 24\alpha\frac{(1+\cos(y))}{(1-\cos(y))^2}. \tag{32}$$

After evaluating these inflationary observables at the value of the scalar field when a given perturbation scale leaves the Hubble-radius, given by (28), we may compare the theoretical predictions of our model with the observational data in order to obtain constraints on the parameters that characterize it.

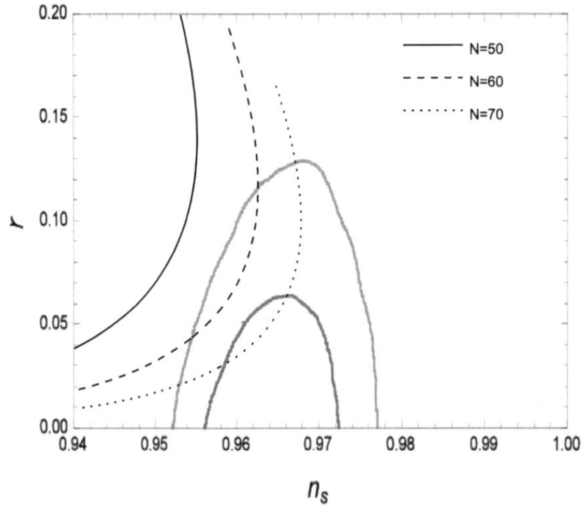

Figure 1. Plot of the tensor-to-scalar ratio r versus the scalar spectral index n_s for natural inflation in the high-energy limit of braneworld scenario. Here, we have considered the two-dimensional marginalized joint confidence contours for (n_s, r), at the 68% and 95% CL, from the latest Planck data [9]. In this plot we have used 3 different values for the number of e-folds N: the solid, dashed, and dotted lines correspond to $N = 50$, 60, and 70, respectively.

The amplitude of the scalar perturbations, the scalar spectral index, and the tensor-to-scalar ratio, evaluated at the Hubble-radius crossing $k = aH$, become

$$\mathcal{P}_{\mathcal{R}} = \frac{4}{3\pi\alpha^3}\gamma^4\frac{(1+W[z(N,\alpha)])^5}{(-W[z(N,\alpha)])}, \tag{33}$$

$$n_s = 1 - \frac{\alpha}{2}\frac{(-W[z(N,\alpha)])}{(1+W[z(N,\alpha)])^2}, \tag{34}$$

$$r = 12\alpha\frac{(-W[z(N,\alpha)])}{(1+W[z(N,\alpha)])^2}. \tag{35}$$

The trajectories in the n_s - r plane for the model studied here may be generated by plotting Eqs.(34) and (35) parametrically. In particular, we have obtained three different curves by fixing the number of e-folds to $N = 50$, 60, and 70, and plotting with respect to the parameter α in a wide range. The Fig.(1) shows the plot of the tensor-to-scalar ratio r versus the scalar spectral index n_s for natural inflation in the high-energy limit of brane-world scenario. Here, we have considered the two-dimensional marginalized joint confidence contours for (n_s, r), at the 68% and 95% CL, from the latest Planck data [9]. The corresponding allowed range for the dimensionless parameter α defined by Eq.(25), for each $r(n_s)$ curve, may be inferred by finding the points when the trajectory enters and exits the 95% CL region from Planck. We can determinate numerically from Eqs.(34) and (35) that, by fixing N, the tensor-to-scalar ratio decreases as the parameter α is increasing. On the other hand, after to reach a maximum value, the scalar spectral index begins to decrease as α is increasing. In this way, we may obtain a lower limit as well an upper limit for α. The trajectory for $N = 50$ lies well outside of the joint 95% CL region in the n_s - r plane, so it is ruled out by the data. For $N = 60$, the trajectory lies inside the 95% CL region, obtaining the constraint on α given by $3.1 \times 10^{-2} \lesssim \alpha \lesssim 7.1 \times 10^{-2}$.

N	constraint on α	constraint on γ
60	$3.1 \times 10^{-2} \lesssim \alpha \lesssim 7.1 \times 10^{-2}$	$6.6 \times 10^{-3} \lesssim \gamma \lesssim 7.4 \times 10^{-3}$
70	$4.8 \times 10^{-3} \lesssim \alpha \lesssim 8.5 \times 10^{-2}$	$4.3 \times 10^{-3} \lesssim \gamma \lesssim 6.5 \times 10^{-3}$

Table 1. Results for the constraints on the parameters α and γ for natural inflation in the high-energy of Randall-Sundrum brane model, using the last data of Planck.

Finally, for $N = 70$ the trajectory lies inside the the joint 95% CL region as well the 68% CL region, obtaining the constraint on α given by $4.8 \times 10^{-3} \lesssim \alpha \lesssim 8.5 \times 10^{-2}$. On the other hand, by combining the scalar power spectrum (33), the constraints on α already obtained, and the observational value for amplitude of the scalar power spectrum $\mathcal{P}_{\mathcal{R}} \simeq 2 \times 10^{-9}$ [9], me may obtain the allowed range for the ratio $\gamma \equiv \Lambda/f$ for each value of N. For $N = 60$, this constraint becomes $6.6 \times 10^{-3} \lesssim \gamma \lesssim 7.4 \times 10^{-3}$, and for $N = 70$ we have that $4.3 \times 10^{-3} \lesssim \gamma \lesssim 6.5 \times 10^{-3}$. Table (1) summarizes the constraints obtained on α and γ using the last data of Planck.

As we can see, using the latest Planck results we find that the γ parameter must be one order of magnitude larger than that found in[13], while the ratio $f/M_5 \sim 10$ (see equation (36) below) must be one order of magnitude lower compared to that obtained in[13].

Clearly, we were not able to obtain the allowed range for all three parameters of the model, namely the 5-dimensional Planck mass M_5 and the mass scales of the inflaton potential Λ and f, only by considering the Planck data. We have found the allowed range for certain combinations of these parameter, α and γ, instead. However, using the definitions of α, γ as well as the formulas relating the four-dimensional quantities with the five-dimensional ones, we can express everything in terms of the fundamental Planck mass M_5 as follows

$$f = \left(\frac{3}{16\pi^2\alpha\gamma^4}\right)^{1/6} M_5 \tag{36}$$

$$\Lambda = \gamma f = \gamma \left(\frac{3}{16\pi^2\alpha\gamma^4}\right)^{1/6} M_5 \tag{37}$$

However, a string scale/higher-dimensional Planck mass of the order of a few TeV is very attractive from the theoretical point of view for several reasons, as it addresses the hierarchy problem, provides an alternative to gauge coupling unification in D-brane constructions of the Standard Model[26], and the evaporation a la Hawking of TeV mini-black holes can be seen at the colliders[27], and possibly explain anomalies related to cosmic ray observations[28]. Furthermore, non-standard cosmologies that at high energies predict a novel expansion law of the universe of the form $H(T) = H_{st}(T)A(T)$, where $A(T)$ is a certain function of the temperature depending on the model at hand and H_{st} is the Hubble parameter in standard cosmology, can rescue electroweak baryogenesis. In particular, in the framework of the RS-II brane model the window for the five-dimensional Planck mass is a few tens of TeV, $M_5 = (15 - 50)$ TeV [29], compatible with the bound from BBN $M_5 \geq 10$ TeV. Therefore, we shall take it seriously and assume for M_5 a value of 10 TeV. Then, by combining the constraints found earlier on α and γ, and $M_5 \geq 10$ TeV, Eqs.(36) and (37) give us the allowed range for the spontaneous symmetry breaking scale f and the soft explicit symmetry breaking scale Λ, yielding

(i) $211.4\,\text{TeV} \lesssim f \lesssim 261.9\,\text{TeV}$ and $1.5\,\text{TeV} \lesssim \Lambda \lesssim 1.8\,\text{TeV}$ $(N = 60)$

(ii) $223.7\,\text{TeV} \lesssim f \lesssim 475.6\,\text{TeV}$ and $1.3\,\text{TeV} \lesssim \Lambda \lesssim 2.3\,\text{TeV}$ $(N = 70)$

However, a value of a few TeV for M_5 is in strong conflict with with phenomenology in the RS-II model. A stronger constraint on M_5 results from current tests for deviation from Newton's gravitational law ($r \sim 1$ mm) [30]

$$V(r) \approx \frac{G_N m_1 m_2}{r}\left(1+\frac{2\ell^2}{3r^2}+\mathcal{O}(r^{-3})\right), \ (r \gg \ell)$$

where $\ell = \frac{M_4^2}{M_5^3} \lesssim 1$ mm, implying that $M_5 \gtrsim 10^5$ TeV. Then, by considering $M_5 = 10^5$ TeV, the new allowed ranges for the spontaneous symmetry breaking scale f and the soft explicit symmetry breaking scale Λ become

(i) 2.1×10^6 TeV $\lesssim f \lesssim 2.6\times10^6$ TeV and 1.5×10^4 TeV $\lesssim \Lambda \lesssim 1.7\times10^4$ TeV ($N = 60$)
(ii) 2.2×10^6 TeV $\lesssim f \lesssim 4.7\times10^6$ TeV and 1.4×10^4 TeV $\lesssim \Lambda \lesssim 2.1\times10^4$ TeV ($N = 70$)

These results imply a hierarchy between the mass scales consistent with $f \gg \Lambda$. On the other hand, the constraints found on α and Eqs.(26) and (28) imply that, during inflation $\phi \sim f$, therefore natural inflation in the framework of the high-energy regime of the RS-II brane model takes place at sub-Planckian values of the scalar field.

4. Conclusions

To summarize, in this article we have studied natural inflation in the framework of the Randall-Sundrum II brane model in the light of the recent Planck results. Adopting the Randall-Sundrum fine-tuning, the brane model is characterized only by the 5-dimensional Planck mass, while the inflationary model is characterized by the two mass scales f, Λ of the inflaton potential. We have used the COBE normalization as well as the allowed contour plots in the $n_s - r$ plane. First, in the $n_s - r$ plane we show the theoretical predictions of the model for three different values of e-folds $N = 50, 60, 70$. According to the plot the $N = 50$ case is excluded, while for the $N = 60, 70$ cases the model is viable for a certain range of the α parameter (a combination of the three parameters of the model f, M_5, Λ) defined in the text. After that, using the constraint for the amplitude of scalar perturbations we determined the ratio Λ/f. We have expressed the mass scales of the inflaton potential in terms of the five-dimensional Planck mass which remains unconstraint using the Planck results only. It is known, however, that there are good theoretical reasons, such us hierarchy problem, alternative to gauge coupling unification, mini black hole evaporation etc., to believe that M_5 is at the TeV scale. If we take it seriously and insist on a M_5 10 TeV all parameters of the model are known. However, a stronger constraint on M_5 comes from current tests for deviation from Newton's gravitational law, but consistent with a sub-Planckian excursion of the inflaton scalar field.

Acknowledgments

N.V. was supported by Comisión Nacional de Ciencias y Tecnología of Chile through FONDECYT Grant N_o 3150490. G.P. was supported by Comisión Nacional de Ciencias y Tecnología of Chile through Anillo project ACT1122.

References

[1] A. Guth , Phys. Rev. D **23**, 347 (1981)
[2] V.F. Mukhanov and G.V. Chibisov , JETP Letters **33**, 532(1981)
[3] P. A. R. Ade *et al.* [Planck Collaboration], Astron. Astrophys. **594**, A13 (2016).
[4] D. H. Lyth and A. Riotto, Phys. Rept. **314** (1999) 1.
[5] W. H. Kinney, E. W. Kolb, A. Melchiorri and A. Riotto, Phys. Rev. D **69** (2004) 103516.
[6] K. Freese, J. A. Frieman and A. V. Olinto, Phys. Rev. Lett. **65** (1990) 3233.
[7] F. C. Adams, J. R. Bond, K. Freese, J. A. Frieman and A. V. Olinto, Phys. Rev. D **47** (1993) 426.

[8] K. Freese and W. H. Kinney, JCAP **1503** (2015) 044.
[9] P. A. R. Ade *et al.* [Planck Collaboration], arXiv:1502.02114 [astro-ph.CO].
[10] T. Banks, M. Dine, P. J. Fox and E. Gorbatov, JCAP **0306**, 001 (2003)
[11] X. Gao, T. Li and P. Shukla, JCAP **1410**, 048 (2014)
[12] T. Higaki and F. Takahashi, Phys. Lett. B **744**, 153 (2015)
[13] R. G. Felipe, Phys. Lett. B **618**, 7 (2005)
[14] I. P. Neupane, Phys. Rev. D **90**, no. 12, 123502 (2014)
[15] I. Antoniadis, N. Arkani-Hamed, S. Dimopoulos and G. R. Dvali, Phys. Lett. B **436** (1998) 257.
[16] L. Randall and R. Sundrum, Phys. Rev. Lett. **83** (1999) 3370.
[17] L. Randall and R. Sundrum, Phys. Rev. Lett. **83** (1999) 4690.
[18] D. Langlois, Prog. Theor. Phys. Suppl. **148** (2003) 181.
[19] T. Shiromizu, K. i. Maeda and M. Sasaki, Phys. Rev. D **62** (2000) 024012.
[20] R. Maartens, D. Wands, B. A. Bassett and I. Heard, Phys. Rev. D **62** (2000) 041301.
[21] P. Binetruy, C. Deffayet and D. Langlois, Nucl. Phys. B **565** (2000) 269.
[22] J. M. Cline, C. Grojean and G. Servant, Phys. Rev. Lett. **83** (1999) 4245.
[23] A. Riotto, hep-ph/0210162.
[24] D. Wands, K. A. Malik, D. H. Lyth and A. R. Liddle, Phys. Rev. D **62** (2000) 043527.
[25] D. Veberic, Comput. Phys. Commun. **183** (2012) 2622.
[26] I. Antoniadis, E. Kiritsis and T. N. Tomaras, Phys. Lett. B **486** (2000) 186.
[27] A. Casanova and E. Spallucci, Class. Quant. Grav. **23** (2006) R45.
[28] A. Mironov, A. Morozov and T. N. Tomaras, Int. J. Mod. Phys. A **24** (2009) 4097.
[29] G. Barenboim and J. Rasero, JHEP **1207** (2012) 028.
[30] T. Clifton, P. G. Ferreira, A. Padilla and C. Skordis, Phys. Rept. **513**, 1 (2012).

On the null energy condition in scale dependent frameworks with spherical symmetry

Ángel Rincón, Benjamin Koch

Physics Institute, Pontifical Catholic University of Chile, Av. Vicuña Mackenna 4860, Santiago, Chile

E-mail: `arrincon@uc.cl`

Abstract. The effect of a scale dependent newton coupling, which is a crucial ingredient of many quantum gravity theories, is investigated in this paper. For case of non-rotating black holes, and using the null energy condition, we show that Newtons scale dependent coupling can actually be obtained without solving the full quantum gap equations.

1. Introduction

Nowadays, different alternatives exist to obtain an unified theory of quantum gravity. Loop quantum gravity [1] as well as string theory [2,3] are famous candidates to achieve this unification. Besides, a self-consistent theory of quantum gravity is still lacking, being studied assuming different points of view [4–20]. Many quantum extensions of general relativity present a common feature: a scale dependence appearing at the level of the effective action of gravity. The classical couplings are not constants any more, instead, they depend on arbitrary scale k. The physics of a black hole strongly depends on the scale setting e.g. the connection between k and r. For the case of spherically symmetric black holes, it can be expected that $k = k(r)$ and thus $\{G_k, \Lambda_k, e_k, \mathcal{O}_k\} \rightarrow \{G(r), \Lambda(r), e(r), \mathcal{O}(r)\}$ [21, 22]. Those r-dependent couplings can be obtained directly from the quantum gap equations, if one assumes certain energy conditions. Recent papers have shown that the so-called "Null Energy Condition" (hereafter NEC) plays a crucial role in this context [23–25]. Thus, in this conference proceedings we analyze the link between the so-called Schwarzschild ansatz and the NEC, in the light of scale dependent couplings. We show farther the important role played by the NEC in determining the gravitational running coupling $G(r)$. The present work is organized as follows: after this short introduction, we explain the main idea of this letter in Section 2, whereas the technique and results are collected at Section 3. Finally, the conclusions are given at Section 4.

2. Main Idea

The so-called Null Energy Condition is one of the usual energy conditions (dominant, weak, strong, and null) and, besides, the least restrictive of them. It helps to obtain suitable solutions of the Einstein field equation and it is usually applied to discern whether a solution has physical validity or not. We will explore this condition in the context of spherical symmetry where the line element reads

$$ds^2 = -f(r)dt^2 + g(r)dr^2 + r^2 d\Omega, \tag{1}$$

which have to be determined by theorems. Here $f(r)$ and $g(r)$ are arbitrary functions, and $d\Omega$ is the solid angle (which depends on the studied dimension). In many cases the task of solving the EOM's is simplified by the so called Schwarzschild ansatz $f(r) \cdot g(r) = 1$. As was explained by Jacobson [26], the aforementioned ansatz can be seen as a consequence of NEC. This finding follows from the fact that the Ricci tensor is proportional to the metric in the $t - r$ subspace. In addition, and regarding the application of this condition, one should note that validity of NEC was guaranteed at Ref. [26], and it applies in spherical symmetry with either Maxwell electrodynamics (i.e. the Reissner-Nordstrom solution) or Born-Infeld non-linear electrodynamics [27], and persists in the presence of a cosmological constant.
In the present work, Jacobson's argument is applied to black holes in the light of the running couplings, analysing the effect of this on the underlying physics as well as the implications associated with the specific form of gravitational coupling $G(r)$. Thus, for an effective Einstein-Hilbert action with cosmological constant, one has the effective Einstein field equations which read [23, 25]

$$G_{\mu\nu} + \Lambda(r) g_{\mu\nu} = 8\pi G(r) T^{\text{effec}}_{\mu\nu}, \tag{2}$$

and the effective energy momentum tensor is defined according to

$$8\pi G(r) T^{\text{effec}}_{\mu\nu} = 8\pi G(r) T^{m}_{\mu\nu} - \Delta t_{\mu\nu}. \tag{3}$$

Here, $T^m_{\mu\nu}$ is the matter energy momentum tensor (which for simplicity is taken to be zero) and the additional term $\Delta t_{\mu\nu}$ is related to the scale dependence of the gravitational running coupling $G(r)$ as shown in next section.

3. Technique and Results

In spherical symmetry 1 and according to Jacobson's idea [26], the Null Energy Condition reads

$$T^{\text{effec}}_{\mu\nu} \ell^\mu \ell^\nu = 0, \qquad \therefore \qquad R_{\mu\nu} \ell^\mu \ell^\nu = 0, \tag{4}$$

where, in three dimensional spacetime, ℓ^μ is a null vector given by

$$\ell^\mu \equiv \Big(\sqrt{g(r)}, \ \sqrt{f(r)}, \ 0 \Big), \tag{5}$$

For arbitrary dimension, one always can define appropriate a null vector such as $\ell^\mu = (g^{1/2}, f^{1/2}, \mathbf{0})$, where $\mathbf{0}$ encoded the zero component for an arbitrary dimension of this null vector. Using the aforementioned ℓ^μ combined with Eq. 4, it is ever possible to obtain a simple differential equation which is straightforward to solve.
Besides, note that in the Einstein-Hilbert truncation [28] the field equation of motion are consistent with Eq. 2. Thus, from 4 one sees that

$$\left[R_{\mu\nu} - \left(\frac{1}{2} R - \Lambda(r) \right) g_{\mu\nu} \right] \ell^\mu \ell^\nu = \left[8\pi G(r) T^m_{\mu\nu} - \Delta t_{\mu\nu} \right] \ell^\mu \ell^\nu, \tag{6}$$

where the additional object $\Delta t_{\mu\nu}$ is defined as follows

$$\Delta t_{\mu\nu} \equiv G_k \Big(g_{\mu\nu} \Box - \nabla_\mu \nabla_\nu \Big) G_k^{-1}. \tag{7}$$

Note that, by definition, a null vector satisfies that $g_{\mu\nu}\ell^\mu \ell^\nu = 0$, which allows getting the simple equation

$$R_{\mu\nu} \ell^\mu \ell^\nu = 0, \qquad \therefore \qquad \Delta t_{\mu\nu} \ell^\mu \ell^\nu = 0, \tag{8}$$

which produces the following ordinary differential equation:

$$2\left[\frac{dG(r)}{dr}\right]^2 - G(r)\frac{d^2G(r)}{dr^2} = -\frac{1}{2}G(r)\frac{dG(r)}{dr}\left[\Big(f(r)\cdot g(r)\Big)^{-1}\frac{d}{dr}\Big(f(r)\cdot g(r)\Big)\right], \tag{9}$$

thereby, one finds without even solving the set of equations 2 that

$$G(r) = a\left[\int_{r_0}^{r}\sqrt{f(r')\cdot g(r')}\,dr'\right]^{-1}, \tag{10}$$

independent of the actual form of $f(r)$ and $g(r)$. In addition, a is a real value which is taken such as one recovering $G(r) \to G_0$ in some limit. Particular attention must be dedicated to case when the Schwarzschild ansatz is used. Under this assumption, the aforementioned differential equation is simplified to be

$$2\left[\frac{dG(r)}{dr}\right]^2 = G(r)\frac{d^2G(r)}{dr^2}, \tag{11}$$

which gives the following scale dependent gravitational coupling

$$G(r) = \frac{G_0}{1+\epsilon r}. \tag{12}$$

The running coupling found in [23, 25, 29, 30] corresponds to $f(r)\cdot g(r) = 1$, which is consistent with Eq. 12. Please, note that the relation $f(r)\cdot g(r) = 1$ is actually independent of the truncation and form of E.O.M.'s since by 8 alone one finds the solution.

4. Conclusion

In the present work, the role of Null Energy Condition [26] is analyzed in the context of running couplings. It is found that the NEC allows to obtain a justification of the Schwarzschild ansatz (by the virtue of [23]). By imposing non-generation of stress energy tensor due to scale dependent gravitational coupling [23, 25] it allows to determine the form of $G(r)$.

Acknowledgments

The author A.R. was supported by the CONICYT-PCHA/Doctorado Nacional/2015-21151658 whereas the work of B.K. was supported by the Fondecyt 1161150.

References

[1] Rovelli C 2007 *Quantum gravity* Cambridge Monographs on Mathematical Physics (Cambridge University Press)
[2] Polchinski J 2007 *String theory. Vol. 1: An introduction to the bosonic string* (Cambridge University Press)
[3] Polchinski J 2007 *String theory. Vol. 2: Superstring theory and beyond* (Cambridge University Press)
[4] Wheeler J A 1957 *Annals Phys.* **2** 604–614
[5] Deser S and Zumino B 1976 *Phys. Lett.* **B62** 335
[6] Rovelli C 1998 *Living Rev. Rel.* **1** 1 (*Preprint* gr-qc/9710008)
[7] Bombelli L, Lee J, Meyer D and Sorkin R 1987 *Phys. Rev. Lett.* **59** 521–524
[8] Ashtekar A 2005 *New J. Phys.* **7** 198 (*Preprint* gr-qc/0410054)
[9] Sakharov A D 1968 *Sov. Phys. Dokl.* **12** 1040–1041 [Gen. Rel. Grav.32,365(2000)]
[10] Jacobson T 1995 *Phys. Rev. Lett.* **75** 1260–1263 (*Preprint* gr-qc/9504004)
[11] Verlinde E P 2011 *JHEP* **04** 029 (*Preprint* 1001.0785)
[12] Reuter M 1998 *Phys. Rev.* **D57** 971–985 (*Preprint* hep-th/9605030)
[13] Litim D F 2004 *Phys. Rev. Lett.* **92** 201301 (*Preprint* hep-th/0312114)

[14] Horava P 2009 *Phys. Rev.* **D79** 084008 (*Preprint* 0901.3775)
[15] Charmousis C, Niz G, Padilla A and Saffin P M 2009 *JHEP* **08** 070 (*Preprint* 0905.2579)
[16] Ashtekar A 1981 *Phys. Rev. Lett.* **46** 573–576
[17] Penrose R and Rindler W 1988 *SPINORS AND SPACE-TIME. VOL. 2: SPINOR AND TWISTOR METHODS IN SPACE-TIME GEOMETRY* (Cambridge University Press) ISBN 9780521347860, 9780511868429
[18] Connes A 1996 *Commun. Math. Phys.* **182** 155–176 (*Preprint* hep-th/9603053)
[19] Nicolini P 2009 *Int. J. Mod. Phys.* **A24** 1229–1308 (*Preprint* 0807.1939)
[20] Gambini R and Pullin J 2005 *Phys. Rev. Lett.* **94** 101302 (*Preprint* gr-qc/0409057)
[21] Koch B, Rioseco P and Contreras C 2015 *Phys. Rev.* **D91** 025009 (*Preprint* 1409.4443)
[22] Koch B and Ramirez I 2011 *Class. Quant. Grav.* **28** 055008 (*Preprint* 1010.2799)
[23] Koch B, Reyes I A and Rincón A 2016 *Class. Quant. Grav.* **33** 225010 (*Preprint* 1606.04123)
[24] Contreras C, Koch B and Rioseco P 2013 *Class. Quant. Grav.* **30** 175009 (*Preprint* 1303.3892)
[25] Rincón A, Koch B and Reyes I 2017 *J. Phys. Conf. Ser.* **831** 012007 (*Preprint* 1701.04531)
[26] Jacobson T 2007 *Class. Quant. Grav.* **24** 5717–5719 (*Preprint* 0707.3222)
[27] Demianski M 1986 *Foundations of Physics* **16** 187–190 ISSN 1572-9516 URL http://dx.doi.org/10.1007/BF01889380
[28] Lauscher O and Reuter M 2002 *Phys. Rev.* **D66** 025026 (*Preprint* hep-th/0205062)
[29] Rincón A, Contreras E, Bargueño P, Koch B, Panotopoulos G and Hernández-Arboleda A 2017 (*Preprint* 1704.04845)
[30] Koch B and Rioseco P 2016 *Class. Quant. Grav.* **33** 035002 (*Preprint* 1501.00904)

IOP Conf. Series: Journal of Physics: Conf. Series **1043** (2018) 012016 doi:10.1088/1742-6596/1043/1/012016

Pontryagin Term and Magnetic Mass in 4D AdS Gravity

René Araneda[1], Rodrigo Aros[2], Olivera Miskovic[3] and Rodrigo Olea[2]

[1] Instituto de Física, Pontificia Universidad Católica de Chile, Av. Vicuña Mackenna 4860, Santiago, Chile.

[2] Departamento de Ciencias Físicas, Universidad Andres Bello, Sazié 2212, Piso 7, Santiago, Chile.

[3] Instituto de Física, Pontificia Universidad Católica de Valparaíso, Casilla 4059, Valparaíso, Chile.

E-mail: rraraneda@uc.cl, raros@unab.cl, olivera.miskovic@pucv.cl, rodrigo.olea@unab.cl

Abstract. In the context of the anti-de Sitter/Conformal Field Theory (AdS/CFT) correspondence, 4D AdS gravity is suitably renormalized by adding the Gauss Bonnet term to the Einstein-Hilbert action. The subsequent addition of the Pontryagin term, with a specific coupling, allows to write the on-shell action in terms of the Weyl tensor and its dual, such that the action becomes stationary for asymptotic (anti) self-dual solutions in the Weyl tensor. The addition to the action of both topological invariants mentioned above does not modify the bulk dynamics, but it does modify the expression of the Noether current and, therefore, the conserved quantities of the theory. Here, we show that the method of Iyer and Wald leads to a fully-covariant Noether charge, which contains both the electric and magnetic parts of the Weyl tensor. For configurations which are globally (anti) self-dual in the Weyl tensor, both the action and the Noether charge identically vanish. This means that, for such spacetimes, the *magnetic mass* is equal to the *electric mass.*

1. Introduction

It has been shown that in asymptotically anti-de Sitter (AAdS) gravity in four dimensions the infrared divergences are cancelled by the Gauss-Bonnet (GB) term with a given coupling [1]. The addition of this topological invariant cancels divergent terms in the asymptotic region, such as those present in the Euclidean gravity action, only for the specific value of coupling constant mentioned above [2],

$$I_4 = \frac{1}{16\pi G}\int\limits_M d^4x\sqrt{g}\left[R + \frac{6}{\ell^2} + \frac{\ell^2}{4}\left(R_{\mu\nu\alpha\beta}R^{\mu\nu\alpha\beta} - 4R_{\mu\nu}R^{\mu\nu} + R^2\right)\right]. \qquad (1)$$

With the addition of this term, the action has an extremum for AAdS spacetimes. On the other hand, the renormalized action (1) can be rewritten in terms of the Weyl tensor [3],

$$I_4 = \frac{\ell^2}{64\pi G}\int_M d^4x\ \sqrt{g}\, W^{\mu\nu}_{\alpha\beta}W^{\alpha\beta}_{\mu\nu}\,, \qquad (2)$$

for any Einstein space. The GB invariant is not the only topological invariant of the Lorentz group in four dimensions. There exists yet another topological term, of opposite parity, known as Pontryagin term,

$$\mathcal{P}_4 = -\frac{1}{4}\,\epsilon^{\mu\nu\alpha\beta} R^{\sigma\lambda}_{\mu\nu} R_{\sigma\lambda\alpha\beta}\,, \tag{3}$$

which is locally equivalent to the derivative of the gravitational Chern-Simons term in three dimensions. The addition of $\mathcal{P}_4$ to the gravity action does not induce additional divergences in the asymptotic region. To our best knowledge, the only way to fix the Pontryagin coupling is to consider (in the Euclidean sector) an asymptotic (anti) self-duality condition in the Weyl tensor,[1]

$$W_{\alpha\beta\mu\nu} = \pm\,{}^{*}W_{\alpha\beta\mu\nu}\,, \tag{4}$$

such that the total action is stationary under arbitrary variations of the fields.

Then, the renormalized action (1) acquires an additional term which can be written in terms of the Weyl tensor and its dual,

$$I = I_4 \pm \frac{\ell^2}{32\pi G}\int_M d^4x\,\sqrt{g}\,W^{\mu\nu}_{\alpha\beta}\,{}^{*}W^{\alpha\beta}_{\mu\nu}. \tag{5}$$

This is analogous to the form in which the Pontryagin density for $U(1)$ group may be added to the Euclidean action in the Maxwell electromagnetism,

$$I^E = \frac{1}{4}\int\limits_M \left(F^{\mu\nu}F_{\mu\nu} \mp {}^{*}F^{\mu\nu}F_{\mu\nu}\right) d^4x\,, \tag{6}$$

where the $F_{\mu\nu}$ is the Abelian field strength. Due to the Bianchi identity ($\partial_\mu\,{}^{*}F^{\mu\nu} = 0$), the second term in the Lagrangian does not modify the equations of motion. The Pontryagin coupling in eq.(6) is a consequence of an asymptotic (anti) self-dual condition for F,

$$F^{\mu\nu} = \pm{}^{*}F^{\mu\nu} \quad \text{at } \partial M\,. \tag{7}$$

The analogy between AdS gravity and self-dual electromagnetism can be pushed forward, as both theories portray striking similarities at the level of the conserved charges.

The Noether theorem allows to obtain an expression of conserved quantities in a gauge theory and gravity, as well, though the symmetries present in each case are of different nature. As long as the Noether current is written (globally) as a total derivative, the conserved charges are given by surface integrals in the asymptotic region, without the need of introducing a source at the origin.

Gauge symmetry $U(1)$ in electromagnetism is realized by the transformation in the gauge field $\delta_\lambda A_\nu = \partial_\nu\lambda$. As this is an internal symmetry ($\delta x^\mu = 0$), the Lagrangian density (6) produces a Noether current

$$J^\mu = \partial_\nu\left[\left(F^{\mu\nu} + \gamma\,{}^{*}F^{\mu\nu}\right)\lambda\right]. \tag{8}$$

Due to the conservation law, $\partial_\mu J^\mu = 0$, and the fact that J^μ can be written as a total derivative, the conserved charge is an integral on a two-dimensional sphere

$$Q[\lambda] = -\int\limits_{S^2} \left(F^{\mu\nu} \mp {}^{*}F^{\mu\nu}\right)\lambda\, d\Sigma_{\mu\nu}\,, \tag{9}$$

[1] The Hodge star dual is defined by ${}^{*}W_{\alpha\beta\mu\nu} = \frac{1}{2}\sqrt{g}\,\epsilon_{\alpha\beta\gamma\delta}\,W^{\gamma\delta}{}_{\mu\nu}$.

in terms of the electromagnetic tensor and its dual. The expression (9), when λ is covariantly constant, leads to total charge which is the sum of the electric and the magnetic charges. Therefore, the magnetic charge is understood as a topological charge which is the Noether charge associated to a topological invariant. The total charge, for a global (anti) self-dual condition in F, will be identically zero, and so the Euclidean action. This defines a subclass of Euclidean instantons which can be taken as ground states of the theory.

In what follows, we apply the method developed by Wald and Iyer [4] to the 4D AdS gravity action supplemented by the topological invariants discussed above. The resulting charge features striking similarities to the case of self-dual electromagnetism.

2. Noether-Wald charges and Topological Invariants

In Riemannian gravity, the action $\int d^4xL$ depends on the metric and its derivatives combined in the form of the Christoffel symbol. Invariance under coordinate transformations of the type $(\delta x^\mu = \xi^\mu(x))$ leads to a conserved current

$$\sqrt{g}\, J^\mu[\xi] = \Theta(\delta_\xi g)^\mu + \Theta(\delta_\xi \Gamma)^\mu + L\xi^\mu ,$$

where we have assumed that the surface term Θ which appears in the arbitrary variation of the action is separable in contributions along the Lie derivative of both the metric and Christoffel connection.

When ξ^μ is a Killing vector, the diffeomorphic variation of the metric, $\delta_\xi g_{\mu\nu} = -\left(\nabla_\mu \xi_\nu + \nabla_\nu \xi_\mu\right)$, vanishes.

In an arbitrary gravity theory with a Lagrangian L given in terms of the metric, the Riemann tensor and its contractions,

$$L = L(g_{\mu\nu}, R_{\mu\nu\alpha\beta}, R_{\mu\nu}, R) , \tag{10}$$

the conserved current turns out to be

$$\sqrt{g}\, J^\mu[\xi] = E^{\mu\nu\alpha\beta} g_{\alpha\lambda} \delta_\xi \Gamma^\lambda_{\nu\beta} + L\xi^\mu, \tag{11}$$

where the tensor $E^{\mu\nu\alpha\beta}$ corresponds to the variation of L with respect to the Riemann tensor $R_{\mu\nu\alpha\beta}$, and the Lie derivative of the Christoffel symbol takes the form

$$\delta_\xi \Gamma^\lambda_{\nu\beta} = -\frac{1}{2}\left(\nabla_\nu \nabla_\beta \xi^\lambda + \nabla_\beta \nabla_\nu \xi^\lambda\right) + \frac{1}{2}\left(R^\lambda{}_{\beta\nu\sigma} + R^\lambda{}_{\nu\beta\sigma}\right)\xi^\sigma .$$

In addition, we can use the following identities for the Riemann tensor

$$[\nabla_\beta, \nabla_\nu]\xi_\alpha = R_{\beta\nu\alpha\sigma}\xi^\sigma \quad \text{and} \quad E^{\mu\nu\alpha\beta}\left(R_{\alpha\beta\nu\sigma} - 2R_{\alpha\nu\beta\sigma}\right) = 0 , \tag{12}$$

where the last one is obtained by contracting the first type Bianchi identity with $E^{\mu\nu\alpha\beta}$ tensor. In doing so, one can isolate the contribution from the field equations, such that the Noether current can be finally rewritten as

$$\sqrt{g} J^\mu = 2E^{\mu\nu\alpha\beta} \nabla_\nu \nabla_\alpha \xi_\beta . \tag{13}$$

This is the key formula for the conserved current in any gravitational theory defined by the Lagrangian (10).

For the particular case of the AdS gravity action in eq.(5), the Noether current is

$$J^\mu[\xi] = \frac{\ell^2}{64\pi G} \nabla_\nu \left[\delta^{[\mu\nu\sigma\lambda]}_{[\alpha\beta\gamma\delta]} \nabla^\alpha \xi^\beta \left(W^{\gamma\delta}_{\sigma\lambda} \mp {}^*W^{\gamma\delta}_{\lambda\sigma}\right)\right] , \tag{14}$$

as it was shown in ref.[5]. It is an analog of the current (8) found in electromagnetism. As it is a total derivative, the corresponding Noether charge is a surface integral on a two-dimensional surface, given in terms of the Weyl tensor and its dual. The spacetime topology is $M \simeq \mathbb{R} \times \Sigma$, where Σ is the space section with unitary normal $u^\mu = (-\tilde{N}, 0, 0, 0)$, thus we can write the conserved charge as

$$Q[\xi] = \frac{\ell^2}{64\pi G} \int\limits_{\partial\Sigma} d^2x \, \sqrt{\sigma} \, n_\mu u_\nu \delta^{[\mu\nu\lambda\sigma]}_{[\alpha\beta\gamma\delta]} \nabla^\alpha \xi^\beta \left(W^{\gamma\delta}_{\lambda\sigma} \mp {}^*W^{\gamma\delta}_{\lambda\sigma} \right) . \tag{15}$$

Here, $n^\mu = (0, N, 0, 0)$ is the unitary normal at the boundary $\partial M \simeq \mathbb{R} \times \partial\Sigma$ and σ is the determinant of the induced metric at the surface $\partial\Sigma$, which satisfies the relation $\sqrt{g} = N\sqrt{h} = N\tilde{N}\sqrt{\sigma}$.

As the first term in the Noether charge is a function of the Weyl tensor, it allows to connect the addition of the topological invariant of GB to the notion of conformal mass in AAdS gravity defined by Ashtekar, Magnon and Das [6], given in terms of the electric part of the Weyl tensor. In analogous way to the electromagnetic case, the form of eq. (15) makes manifest the fact that configurations globally (anti) self-dual in the Weyl tensor have a vanishing Noether charge. For these solutions, the magnetic mass is equal to the conformal mass, as it is the case for Taub-NUT-AdS solution [7, 8, 9].

3. Magnetic Mass

The discussion above is unnecessary for Schwarzschild-AdS or Kerr-AdS solutions as they have a vanishing Pontryagin density. Some of simplest examples to consider are Taub-NUT-AdS and Taub-Bolt-AdS solutions. Taub-NUT-AdS has an (anti) self-dual Weyl tensor and Taub-Bolt-AdS meets that condition in the asymptotic region. Indeed, both solutions present topological defects and have a non-vanishing magnetic mass [5].

Both Taub-NUT-AdS and Taub-Bolt-AdS solutions can be written, in the coordinates (τ, r, θ, ϕ), as

$$ds^2 = f(r)(d\tau + 2n\cos\theta d\phi)^2 + \frac{dr^2}{f(r)} + (r^2 - n^2)(d\theta^2 + \sin^2\theta d\phi^2),$$

where

$$f(r) = \frac{r^2 - 2Mr + n^2 - \frac{3}{\ell^2}\left(n^4 + 2n^2r^2 - \frac{r^4}{3}\right)}{r^2 - n^2} . \tag{16}$$

Here, n is a parameter related to a topological defect present in the geometry [11], and M is a mass parameter [12].

For either NUT or Bolt solution the largest root of the equation of $f(r) = 0$, namely $r = r_*$, defines a horizon. The difference between these two geometries is that, for NUT solution, the relation $r_* = |n|$ holds exactly while, for Bolt solution, it is the inequality $r_* = r_b > |n|$. Moreover, it must be satisfied $f'(r_*) = \frac{1}{2n}$ to avoid a singularity at r_*. The mass for the NUT solution, computed in ref.[1], is

$$M_{\mathrm{NUT}} = \pm n \left(1 - 4\ell^{-2}n^2\right) , \tag{17}$$

while for the Bolt solution

$$M_{\mathrm{Bolt}} = \frac{r_b^2 + n^2}{2r_b} - \frac{3}{2\ell^2}\left(\frac{n^4}{r_b} + 2n^2r_b - \frac{r_b^3}{3}\right) .$$

As mentioned above, we are interested in analyzing the conserved charges as defined by eq.(15). For NUT solution, since its Weyl tensor is self-dual, we have

$$Q_{\mathrm{NUT}}[\xi] = 0, \tag{18}$$

for any Killing vector $\xi = \xi^\mu \partial_\mu$. For Bolt solution, the only non-vanishing conserved charge in eq.(15) is the one associated to a timelike Killing vector

$$Q_{\mathrm{Bolt}}[\partial_\tau] = M_{\mathrm{Bolt}} \pm M_{\mathrm{NUT}}. \tag{19}$$

4. Conclusions

In this article, we have shown that the addition of topological invariants (GB and Pontryagin terms), with specific couplings, modifies the value of the Noether charges of the theory. While the electric part of the Weyl tensor gives rise to the energy of AAdS black holes, the magnetic part produces a non-vanishing magnetic mass for gravitational instantons in AdS. Furthermore, we have proved that solutions with (anti) self-dual Weyl tensors are suitable candidates for ground states. Since the mass $Q[\partial\tau]$ obtained from eq.(15) is a thermodynamic variable, we may expect a change in the thermodynamics of the system when solutions with a non vanish Pontryagin term, as Taub-NUT-AdS and Taub-Bolt-AdS, are considered.

Acknowledgements

This work was funded in part by FONDECYT Grants No. 1131075, 1140296 and 1151107, CONICYT Grant DPI 2014-0115, UNAB Grants DI-735-15/R and DI-1336-16/RG and VRIEA-PUCV Grant 039.345/2016. R.Araneda received financial support from Facultad de Física, Pontificia Universidad Católica de Chile to participate in SOCHIFI Congress.

References

[1] R. Aros, M. Contreras, R. Olea, R. Troncoso, and J. Zanelli 2000, Phys. Rev. Lett. **84**, 1647
[2] R. Olea 2005, J. High Energy Phys. **06**, 023
[3] O. Miskovic and R. Olea 2009, Phys. Rev. **D79**, 124020
[4] V. Iyer and R. Wald 1994, Phys. Rev. D **50**, 846
[5] R. Araneda, R. Aros, O. Miskovic and R. Olea 2016, Phys. Rev. D **93**, 084022
[6] A. Ashtekar and A. Magnon 1984, Classical Quantum Gravity **1**, L39; A. Ashtekar and S. Das 2000, Classical Quantum Gravity **17**, L17
[7] E. Newman, L. Tamburino and T. Unti 1963, J. Math. Phys. **4**, 915
[8] D. N. Page and C. N. Pope 1987, Classical Quantum Gravity **4**, 213
[9] D. Kramer, H. Stephani, M. MacCallum and E. Herlt, *Exact solutions of Einstein's field equations*, Cambridge University Press, 1980
[10] A. Chamblin, R. Emparan, C.V. Johnson and R.C. Myers 1999, Phys. Rev. D **59**, 064010
[11] T. Eguchi, P. B. Gilkey and A. J. Hanson 1980, Phys. Rept. **66**, 213
[12] S. W. Hawking, C. J. Hunter and D. N. Page 1999, Phys. Rev. D **59**, 044033

IOP Conf. Series: Journal of Physics: Conf. Series **1043** (2018) 012017 doi:10.1088/1742-6596/1043/1/012017

Inflation in $\tilde{\delta}$ Gravity.

P González.

Departamento de Física, FCFM, Universidad de Chile. Blanco Encalada 2008, Santiago, Chile.

E-mail: pgonzalez@ing.uchile.cl

Abstract. $\tilde{\delta}$ Gravity is a gravitational fields model based on two symmetric tensors, $g_{\mu\nu}$ and $\tilde{g}_{\mu\nu}$, and new matter fields given by $\tilde{\phi}_I = \tilde{\delta}\phi_I$. This model has great properties at the quantum level. For example: It lives at one loop only, the classical equations of motion of the original fields are preserved and it is a finite quantum theory in the vacuum. Besides, massive particles follow an anomalous geodesic, while trajectories of massless particles are given by a null geodesic of the effective metric $\mathbf{g}_{\mu\nu} = g_{\mu\nu} + \kappa_2 \tilde{g}_{\mu\nu}$. In previous works, the accelerated expansion of the universe was explained without dark energy. Additionally, we saw that $\tilde{\delta}$ matter helps to explain dark matter. All these are motivations to study the effect of $\tilde{\delta}$ Gravity in Cosmic Inflation. For this, we introduce one traditional inflaton, φ, plus his $\tilde{\delta}$ partner, $\tilde{\varphi}$. Additionally, we have the contribution of $g_{\mu\nu}$ and $\tilde{g}_{\mu\nu}$.

1. Introduction

In the physics word, the Quantum Mechanics (QM) and General Relativity (GR) are the principal theories used to explain the universe. On one side, QM explain the sub-atomic word, and on the other side, GR describe the physics from scales larger than a millimeter to Cosmological scales [1]. Both theories are really successful, but when we try to mix them, some problems appear. That is because GR is non-renormalizable [2]. Moreover, recent discoveries in cosmology have revealed that most part of matter is in the form of unknown matter, dark matter [3], and that the dynamics of the expansion of the Universe is governed by a mysterious component that accelerates its expansion, the so called dark energy [4, 5]. This is the Dark Sector. Although GR is able to accommodate the Dark Sector, its interpretation in terms of fundamental theories of elementary particles is problematic [5]. Some candidates exist that could play the role of dark matter, but none have been detected yet. On the other side, dark energy can be explained if a small cosmological constant (Λ) is present. However Λ is too small to be generated in quantum field theory (QFT) models, because Λ is the vacuum energy, which is usually predicted to be very large [6]. For these reasons, there has been various proposals to explain the observed acceleration of the Universe.

In [7], a model of gravitation that is very similar to GR is presented, but works different at the quantum level. In that paper, we considered two different points. The first is that GR is finite on shell at one loop in vacuum [2], so renormalization is not necessary in this case. The second is a special kind of modification, called the $\tilde{\delta}$ gauge theories (DGT), originally presented in [8], where the main properties are: (a) A new kind of field $\tilde{\phi}_I$ is introduced, different from the original set ϕ_I. (b) The classical equations of motion of ϕ_I are satisfied even in the full quantum theory. (c) The model lives at one loop. (d) The action is obtained through the extension of the original gauge symmetry of the model, introducing an extra symmetry that we call $\tilde{\delta}$ symmetry,

IOP Conf. Series: Journal of Physics: Conf. Series **1043** (2018) 012017 doi:10.1088/1742-6596/1043/1/012017

since it is formally obtained as the variation of the original symmetry. When we apply this modification to GR, we obtain $\tilde{\delta}$ Gravity.

The original motivation was to develop the quantum properties of this model (See [7]) and then we studied the classical properties of $\tilde{\delta}$ Gravity (See [9]), where we saw that: (a) it agrees with GR, far from the sources. In particular, the causal structure of $\tilde{\delta}$ Gravity in vacuum is the same as in general relativity. (b) The necessary quantity of dark matter could be considerably less that we expected. (c) When we study the evolution of the Universe, it predicts an accelerated expansion without a cosmological constant or additional scalar fields. The Universe ends in a Big-Rip, similar to the scenario considered in [10]. (d) The scale factor agrees with the standard cosmology at early times and shows acceleration only at late times. Therefore we expect that primordial density perturbations should not have large corrections.

Now, Inflation is a brief period of exponential expansion before the Radiation era [11], governed by a scalar field ϕ called Inflaton. Some inflationary parameters can be defined to show the expansion behavior. The more important is ϵ, that represent the acceleration of the expansion. As long as $\epsilon \ll 1$, the expansion is exponential and it will be ended when $\epsilon = 1$. In that moment, the Inflaton decay on matter and radiation in a process called Reheating [12]. In the same way, the quantum fluctuation of Inflaton produce density fluctuations and they will give rise to the large structures of the universe, including dark matter distribution. The anisotropy produced by these fluctuation is represented by the Power Spectrum of Inflation, including information about the evolution of the universe produced by dark energy. Therefore, it is useful to study the Dark Sector. Motivated by this, it is important to study the effect of $\tilde{\delta}$ Gravity in Cosmic Inflation, that is $\tilde{\delta}$ Inflation.

Unlike traditional Inflation, in $\tilde{\delta}$ Inflation the accelerated expansion could be affected by $\tilde{\delta}$ Gravity, just like the expansion is accelerated in the present era. Besides, we will have an additional inflationary field $\tilde{\phi}$ and new inflationary parameters. In the same way, $\tilde{\delta}$ Gravity will affect the perturbative equation and then the Power Spectrum. With this, we can understand much better how $\tilde{\delta}$ Gravity explain the Dark Sector and verify if it is phenomenologically viable. In this work, we will present an introduction to $\tilde{\delta}$ Inflation, just like it was presented in XX Simposio SOCHIFI 2016. This work is in progress and a complete article will be published soon.

2. Cosmology in $\tilde{\delta}$ Gravity.

From [9], we know that the action of $\tilde{\delta}$ Gravity is:

$$S = \int d^4x\sqrt{-g}\left(\frac{R}{2\kappa} + L_M - \frac{\kappa_2}{2\kappa}\left(G^{\alpha\beta} - \kappa T^{\alpha\beta}\right)\tilde{g}_{\alpha\beta} + \kappa_2 \tilde{L}_M\right), \tag{1}$$

where $L_M = L_M(\phi_I, \partial_\mu\phi_I)$ is the lagrangian of the matter fields ϕ_I, $\tilde{g}_{\mu\nu} = \tilde{\delta} g_{\mu\nu}$, $T^{\mu\nu} = \frac{2}{\sqrt{-g}}\frac{\delta}{\delta g_{\mu\nu}}\left[\sqrt{-g}L_M\right]$ and $\tilde{L}_M = \tilde{\phi}_I \frac{\delta L_M}{\delta \phi_I} + (\partial_\mu\tilde{\phi}_I)\frac{\delta L_M}{\delta(\partial_\mu\phi_I)}$, where $\tilde{\phi}_I = \tilde{\delta}\phi_I$ are the $\tilde{\delta}$ matter fields. From this action, we can obtain the equations of motion of $g_{\mu\nu}$ and $\tilde{g}_{\mu\nu}$. It is easy to see that the Einstein's equations are still valid. Besides, the equation for $\tilde{g}_{\mu\nu}$ is:

$$G^{\mu\nu} = \kappa T^{\mu\nu} \quad \wedge \quad F^{(\mu\nu)(\alpha\beta)\rho\lambda} D_\rho D_\lambda \tilde{g}_{\alpha\beta} + \frac{1}{2} g^{\mu\nu} R^{\alpha\beta}\tilde{g}_{\alpha\beta} - \frac{1}{2}\tilde{g}^{\mu\nu} R = \kappa \tilde{T}^{\mu\nu}. \tag{2}$$

with $F^{(\mu\nu)(\alpha\beta)\rho\lambda} = P^{((\rho\mu)(\alpha\beta))}g^{\nu\lambda} + P^{((\rho\nu)(\alpha\beta))}g^{\mu\lambda} - P^{((\mu\nu)(\alpha\beta))}g^{\rho\lambda} - P^{((\rho\lambda)(\alpha\beta))}g^{\mu\nu}$, $P^{((\alpha\beta)(\mu\nu))} = \frac{1}{4}\left(g^{\alpha\mu}g^{\beta\nu} + g^{\alpha\nu}g^{\beta\mu} - g^{\alpha\beta}g^{\mu\nu}\right)$, where $(\mu\nu)$ denotes that μ and ν are in a totally symmetric combination and $\tilde{T}_{\mu\nu} = \tilde{\delta} T_{\mu\nu}$ is the $\tilde{\delta}$ energy momentum tensor, related to $\tilde{\delta}$ matter fields. An important fact to notice is that our equations are of second order in derivatives which is needed to preserve causality. Besides, we have two conservation rules:

$$D_\nu T^{\mu\nu} = 0 \quad \wedge \quad D_\nu \tilde{T}^{\mu\nu} = \frac{1}{2} T^{\alpha\beta} D^\mu \tilde{g}_{\alpha\beta} - \frac{1}{2} T^{\mu\beta} D_\beta \tilde{g}^\alpha_\alpha + D_\beta(\tilde{g}^\beta_\alpha T^{\alpha\mu}). \tag{3}$$

Particulary, in the cosmological case, we have:

$$g_{\mu\nu} dx^\mu dx^\nu = -N^2(t) c^2 dt^2 + a^2(t) \left(dx^2 + dy^2 + dz^2\right) \tag{4}$$

$$\tilde{g}_{\mu\nu} dx^\mu dx^\nu = -\tilde{F}_0(t) N^2(t) c^2 dt^2 + \tilde{F}_1(t) a^2(t) \left(dx^2 + dy^2 + dz^2\right), \tag{5}$$

where $N(t)$ and $\tilde{F}_0(t)$ will be determined fixing the gauge. In $\tilde{\delta}$ Gravity, the proper time is given by $g_{\mu\nu}\dot{x}^\mu\dot{x}^\nu = -1$, so we will use $N(t) = 1$ such that t is the proper time. In this case, we can use $\tilde{F}_0(t) = 3\tilde{F}_1(t)$ (See **Appendix B** of [9]). We will impose the gauge below.

On the other side, in our model, the light (or any massless particle) move in a effective null geodesic given by $(g_{\mu\nu} + \kappa_2 \tilde{g}_{\mu\nu})\,\dot{x}^\mu\dot{x}^\nu = 0$. For this reason, the expansion of the universe is produced by an effective scale factor $a_{eff}(t) = a(t)\sqrt{\frac{1+\kappa_2\tilde{F}_1(t)}{1+\kappa_2\tilde{F}_0(t)}}$, such that $N(t)dt = -a_{eff}(t)dr$ now. This means that all definitions must be modified such that $a(t) \to a_{eff}(t)$. For example, the redshift and luminosity distance:

$$1 + z(t_1) = \frac{a_{eff}(t_0)}{a_{eff}(t_1)} \quad \wedge \quad d_L = \frac{a^2_{eff}(t_0)}{a_{eff}(t_1)} \int_{t_1}^{t_0} \frac{N(t)dt}{a_{eff}(t)}. \tag{6}$$

With all these, we explained the accelerated expansion of the universe without dark energy in [9]. Now, in this paper, we will developed inflation with $\tilde{\delta}$ Gravity, producing the expansion in the same way. But first, in the next chapter, we will analyze the solution for a perfect fluid using the Cosmological case.

3. Analysis of the Perfect Fluid.

In this chapter, as an example of how the expansion works in $\tilde{\delta}$ Gravity, we will study the perfect fluid solution. The Energy-Momentum Tensors for a perfect fluid are $T_{\mu\nu} = p g_{\mu\nu} + (\rho + p) U_\mu U_\nu$ and $\tilde{T}_{\mu\nu} = p\tilde{g}_{\mu\nu} + \tilde{p} g_{\mu\nu} + (\tilde{\rho} + \tilde{p}) U_\mu U_\nu + (\rho + p)\left(\tilde{U}_\mu U_\nu + U_\mu \tilde{U}_\nu\right)$, where $U_\mu = (1, 0, 0, 0)$ and $\tilde{U}_\mu = \left(\frac{3}{2}\tilde{F}_1(t), 0, 0, 0\right)$. Besides, ρ, p, $\tilde{\rho}$ and $\tilde{p}$ just depend on time in a Background level. With all these, the equations in (2) are reduced to:

$$H^2(t) = \frac{\kappa}{3}\rho(t) \quad \wedge \quad \dot{H}(t) = -\frac{\kappa}{2}\left(p(t) + \rho(t)\right) \tag{7}$$

$$3H(t)\left(\dot{\tilde{F}}_1(t) - 3H(t)\tilde{F}_1(t)\right) = \kappa\tilde{\rho}(t) \quad \wedge \quad \ddot{\tilde{F}}_1(t) + 3\kappa p(t)\tilde{F}_1(t) = -\kappa\tilde{p}(t). \tag{8}$$

Additionally, when we have a fluid, equations of state are required to solve the system. In general, they are $p(t) = \omega(t)\rho(t)$ and $\tilde{p}(t) = \omega(t)\tilde{\rho}(t) + \tilde{\omega}(t)\rho(t)$, where $\omega(t)$ is generally assumed to be constant and $\tilde{\omega}(t)$ go to zero in that case. So, when we use these equations of state in (7-8), we obtain:

$$\dot{H}(t) + \frac{3}{2}\left(\omega(t) + 1\right) H^2(t) = 0 \quad \wedge \quad \ddot{\tilde{F}}_1(t) + 3\omega(t) H(t)\dot{\tilde{F}}_1(t) + 3\tilde{\omega}(t) H^2(t) = 0. \tag{9}$$

In order to understand the behavior of these equations, we will solve the case where $\omega(t)$ is constant and $\tilde{\omega}(t) = 0$. That is:

$$a(t) = \begin{cases} a_0 \left(\frac{t}{t_0}\right)^{\frac{2}{3(1+\omega)}} & \text{, with: } \omega \neq -1 \\ a_0 e^{\frac{t-t_0}{t_0}} & \text{, with: } \omega = -1 \end{cases} \quad \wedge \quad \tilde{F}_1(t) = \begin{cases} \left(\frac{a(t)}{a_1}\right)^{\frac{3}{2}(1-\omega)} + b_1 & \text{, with: } \omega \neq 1 \\ b_2 \ln\left(\frac{a(t)}{a_2}\right) & \text{, with: } \omega = 1 \end{cases} \tag{10}$$

where t_0, a_0, a_1, a_2, b_1 and b_2 are integration constants. From this solution and (7), we obtain the well-known result:

$$H^2(X) = \begin{cases} \frac{\kappa\rho_0}{3X^{3(1+\omega)}} & \text{, with: } \omega \neq -1 \\ H_0^2 & \text{, with: } \omega = -1 \end{cases} \quad \wedge \quad t(X) = \begin{cases} \frac{2X^{\frac{3}{2}(1+\omega)}}{\sqrt{3\kappa\rho_0}(1+\omega)} & \text{, with: } \omega \neq -1 \\ \frac{\ln(X)}{H_0} & \text{, with: } \omega = -1, \end{cases} \tag{11}$$

where the new independent variable is now the normalized scale factor, $X \equiv \frac{a(t)}{a_0}$. (11) correspond to the usual GR solution, however we have new contribution given by $\tilde{F}_1(t)$ and $\tilde{\delta}$ fluid, $\tilde{\rho}(t)$. Additionally, we can rewrite (10) such that:

$$\tilde{F}_1(X) = \begin{cases} b_0 X^{\frac{3}{2}(1-\omega)} + b_1 & \text{, with: } \omega \neq 1 \\ b_2 \ln(X) + b_3 & \text{, with: } \omega = 1 \end{cases}. \tag{12}$$

Then, using the modified scale factor and equations (8,11-12), we obtain:

$$\mathcal{X}(X) \equiv \frac{a_{eff}(t)}{a_0} = \begin{cases} X\sqrt{\frac{c_1+2c_2-X^{(3-\epsilon)}}{3\left(c_1-X^{(3-\epsilon)}\right)}} & \text{, with: } \epsilon \neq 3 \\ X\sqrt{\frac{d_1+2d_2-\ln(X)}{3(d_1-\ln(X))}} & \text{, with: } \epsilon = 3 \end{cases} \tag{13}$$

$$\mathcal{H}(X) \equiv \frac{\dot{a}_{eff}(t)}{a_{eff}(t)} = \begin{cases} H(X)\left(1+\frac{c_2(3-\epsilon)X^{(3-\epsilon)}}{\left(c_1+2c_2-X^{(3-\epsilon)}\right)\left(c_1-X^{(3-\epsilon)}\right)}\right) & \text{, with: } \epsilon \neq 3 \\ H(X)\left(1+\frac{d_2}{(d_1+2d_2-\ln(X))(d_1-\ln(X))}\right) & \text{, with: } \epsilon = 3 \end{cases} \tag{14}$$

$$\epsilon_{eff}(X) = \begin{cases} \epsilon - \frac{c_2(3-\epsilon)X^{(3-\epsilon)}\left((2\epsilon-3)X^{2(3-\epsilon)}-\epsilon(2c_1+(\epsilon-1)c_2)X^{(3-\epsilon)}+3c_1(c_1+2c_2)\right)}{\left(X^{2(3-\epsilon)}-(2c_1+(\epsilon-1)c_2)X^{(3-\epsilon)}+c_1(c_1+2c_2)\right)^2} & \text{, with: } \epsilon \neq 3 \\ 3 - \frac{d_2\left(3\ln^2(X)-2(1+3d_1+3d_2)\ln(X)+d_1(3d_1+6d_2+2)+5d_2\right)}{\left(\ln^2(X)-2(d_1+d_2)\ln(X)+d_1(d_1+2d_2)+d_2\right)^2} & \text{, with: } \epsilon = 3, \end{cases} \tag{15}$$

where we used $\epsilon = \frac{3}{2}(1+\omega)$, $c_1 = -\frac{1+3\kappa_2 b_1}{3\kappa_2 b_0}$, $c_2 = -\frac{1}{3\kappa_2 b_0}$, $d_1 = -\frac{1+3\kappa_2 b_3}{3\kappa_2 b_2}$, $d_2 = -\frac{1}{3\kappa_2 b_2}$, $\mathcal{X}(X)$ is the modified scale factor, $\mathcal{H}(X)$ is the effective Hubble parameter and $\epsilon_{eff}(X) \equiv -\frac{\dot{\mathcal{H}}(X)}{\mathcal{H}^2(X)} = \frac{3}{2}(1+\omega_{eff}(X))$ is the effective inflationary parameter. This parameter is used in inflation to impose slow-roll conditions. For this reason, we will use it in this paper.

In conclusion, $\epsilon_{eff}(X)$ represent the acceleration of the expansion in $\tilde{\delta}$ Gravity. When inflation is ended, $\epsilon_{eff}(X)$ has to be closed to one. Unfortunately, that is impossible with a perfect fluid: As was developed in [9], the expansion is produced by a pole in the modified scale factor in (13), producing a Big-Rip. Additionally, the numerator of $\mathcal{X}(X)$ must be positive when

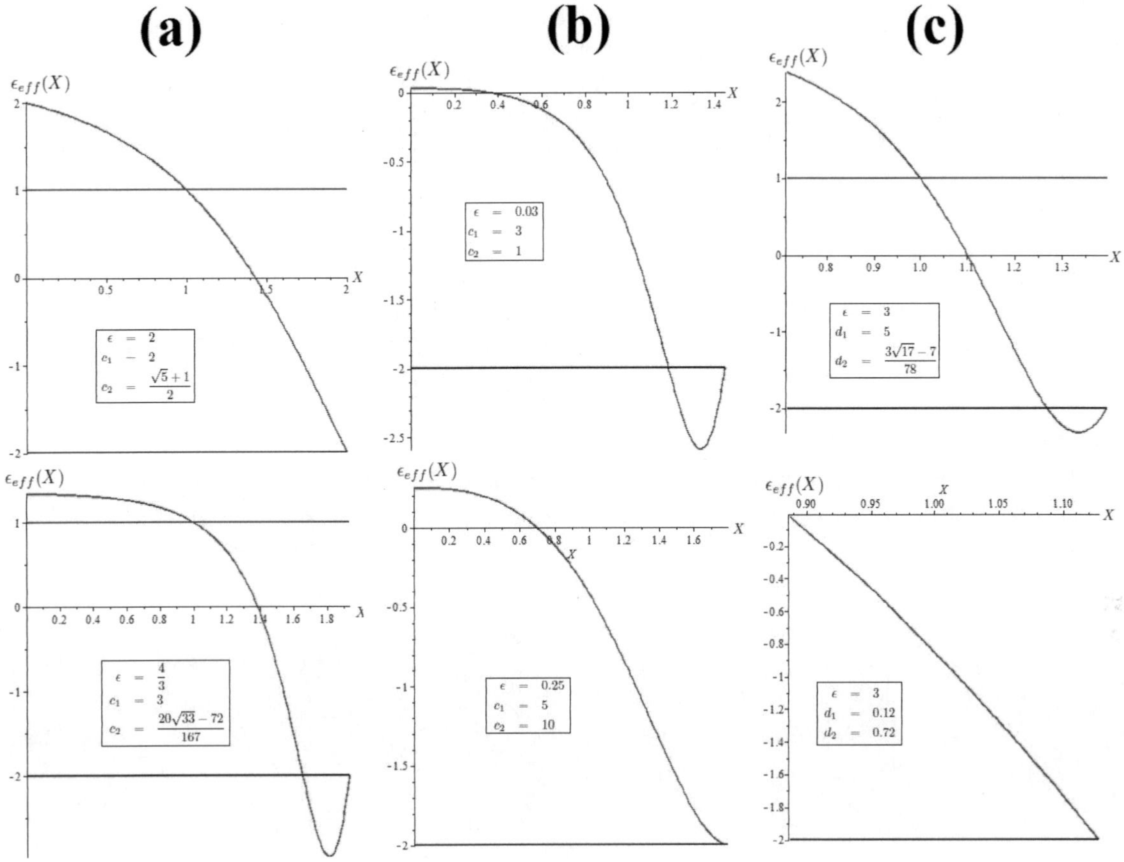

Figure 1. $\epsilon_{eff}(X)$ vs X. Behavior of equation (15) for different values of the parameters. **(a)** We have two cases where $\epsilon_{eff}(1) = 1$, so c_2 is given by (16). We can see that the accelerated expansion is produced very late and it is permanent. **(b)** We have two cases where $\epsilon_{eff}(X \sim 0)$ is small. Clearly, the expansion is too strong. **(c)** We have two cases with $\epsilon = 3$. In the first place, $\epsilon_{eff}(1) = 1$, so d_2 is given by (17). Then, we have a case where $\epsilon_{eff}(X \sim 0)$ is small. We obtain similar result to (a) and (b).

$X < X_{BR}$. So, defining the end of inflation when $X = 1$, then $c_1 > 1$, $c_2 > 0$, $d_1 > 0$ y $d_2 > 0$. These conditions are enough to obtain an accelerated expansion, however it is too strong. We can prove that, if $\epsilon_{eff}(1) = 1$, then:

$$c2 = \frac{(c_1-1)}{2}\left(\frac{(3-\epsilon)\left((5-2\epsilon)c_1 + \epsilon - 2 + \sqrt{(4\epsilon^2-24\epsilon+29)c_1^2 + (4\epsilon+2)c_1 - 4\epsilon+5}\right)}{1-\epsilon^2+2(2\epsilon^2-7\epsilon+11)c_1 - 4(\epsilon-1)c_1^2} - 1\right), \quad (16)$$

when $1 < \epsilon < 3$ and $1 < c_1 < \frac{2\epsilon^2-7\epsilon+11+(3-\epsilon)\sqrt{4\epsilon^2-8\epsilon+13}}{4(\epsilon-1)}$, or:

$$d2 = \frac{d_1\left(2 - d_1 - 8d_1^2 - \sqrt{4-4d_1-7d_1^2}\right)}{2(4d_1+3)(2d_1-1)}, \quad (17)$$

when $\epsilon = 3$ and $0 < d_1 < \frac{1}{2}$. For $0 \leq \epsilon \leq 1$, the condition is impossible to satisfy. Besides, we can prove, in all these cases, that $\epsilon_{eff}(X < 1) > 1$ and $\epsilon_{eff}(X > 1) < 1$. This means that $\epsilon_{eff}(X)$ takes bigger values during "inflation" and the expansion is begun later. This result is a contradiction, so inflation does not work. On the other side, if we redefine X such that $\epsilon_{eff}(X \sim 0) \sim 0$ (Exponential expansion) or < 0 (Phantom expansion), then we obtain that the accelerated expansion is perfectly produced, but inflation will never end. In fact, $\epsilon_{eff}(X_{BR}) = -2$. Few example are presented in **Figure 1**. In conclusion, inflation does not work with a single perfect fluid. A possible solution is to use two or more fluids, just like [9] where we explain the accelerated expansion of the universe with radiation ($\omega = \frac{1}{3}$) and non-relativistic matter ($\omega = 0$), so that the condition can be satisfied. However, the behavior of $\epsilon_{eff}(X)$ ($\omega_{eff}(X)$) for two fluids is really similar to one fluid, so explain inflation could be difficult (See **Appendix A**). A best option to explain inflation is a fluid with a dynamic $\omega(t)$ in (7-9), so that we have a new component represented by $\tilde{\omega}(t) \neq 0$. In the next chapter, we will focus on this option using a scalar field, the inflaton φ, to produce inflation.

4. $\tilde{\delta}$ Inflation.

In inflation, the fluid is usually represented by a single scalar field with a lagrangian $L_M = -\frac{1}{2}g^{\alpha\beta}(\partial_\alpha\varphi)(\partial_\beta\varphi) - V(\varphi)$, where φ is the inflaton and $V(\varphi)$ is the potential. The modified Action is obtained using (1). So, with (4) and (5), the action in the background level is:

$$S^{(0)} = \int d^4x a^3 \left(\frac{3\left(1+\kappa_2\tilde{F}_1\right)}{\kappa a}\frac{d}{dt}\left(\frac{\dot{a}}{N}\right) + \frac{3\left(2+\kappa_2\tilde{F}_1+\kappa_2\tilde{F}_0\right)\dot{a}^2}{2\kappa N a^2} + \frac{1}{2N}\left(1+\frac{\kappa_2}{2}\left(3\tilde{F}_1-\tilde{F}_0\right)\right)\dot{\varphi}_0^2 \right.$$
$$\left. + \frac{\kappa_2}{N}\dot{\varphi}_0\dot{\tilde{\varphi}}_0 - N\mathbf{V}(\varphi_0,\tilde{\varphi}_0)\right), \quad (18)$$

where $\mathbf{V}(\varphi_0,\tilde{\varphi}_0) = \left(1+\frac{\kappa_2}{2}\left(\tilde{F}_0+3\tilde{F}_1\right)\right)V(\varphi_0) + \kappa_2 V_{,\varphi}(\varphi_0)\tilde{\varphi}_0$ is the effective potential, with $V_{,\varphi}(\varphi) = \frac{\partial V}{\partial\varphi}(\varphi)$. Additionally, we have another scalar field, $\tilde{\varphi}$, called $\tilde{\delta}$ inflaton. In a background level, they are $\varphi = \varphi_0(t)$ and $\tilde{\varphi} = \tilde{\varphi}_0(t)$. Now, calculating the equations of motions and fixing the gauge, we obtain:

$$3H^2 - \kappa\left(\frac{1}{2}\dot{\varphi}_0^2 + V(\varphi_0)\right) = 0 \quad \wedge \quad \dot{H} + \frac{\kappa}{2}\dot{\varphi}_0^2 = 0 \quad (19)$$

$$\ddot{\tilde{F}}_1 - 3H\dot{\tilde{F}}_1 + 2\kappa\dot{\varphi}_0\dot{\tilde{\varphi}}_0 = 0 \quad \wedge \quad 3H\dot{\tilde{F}}_1 - \kappa\left(\dot{\varphi}_0\dot{\tilde{\varphi}}_0 + 3\tilde{F}_1 V(\varphi_0) + V_{,\varphi}(\varphi_0)\tilde{\varphi}_0\right) = 0 \quad (20)$$

$$\ddot{\varphi}_0 + 3H\dot{\varphi}_0 + V_{,\varphi}(\varphi_0) = 0 \quad \wedge \quad \ddot{\tilde{\varphi}}_0 + 3H\dot{\tilde{\varphi}}_0 + V_{,\varphi\varphi}(\varphi_0)\tilde{\varphi}_0 + 3\tilde{F}_1 V_{,\varphi}(\varphi_0) = 0. \quad (21)$$

We can verify that (19-21) are respectively (7-9) when $\omega(t) = \frac{\dot{\varphi}_0^2-2V(\varphi_0)}{\dot{\varphi}_0^2+2V(\varphi_0)}$ and $\tilde{\omega}(t) = \frac{4\left(V(\varphi_0)\left(2\dot{\varphi}_0\dot{\tilde{\varphi}}_0-3\tilde{F}_1\dot{\varphi}_0^2\right)-\dot{\varphi}_0^2 V_{,\varphi}(\varphi_0)\tilde{\varphi}_0\right)}{\left(\dot{\varphi}_0^2+2V(\varphi_0)\right)^2}$. In the next section, we will rewrite them using the field space formalism and we will define the effective inflationary parameters.

4.1. Fields Space.

In this formalism (For instance see [13]), we have to define a vector in the fields space given by $\phi_0^a = \begin{pmatrix} \varphi_0 \\ \tilde{\varphi}_0 \end{pmatrix}$. So, the background action (18) can be rewritten as:

$$S^{(0)} = \int d^4x a^3 N\left(L_{GR} + \frac{1}{2}\gamma_{ab}\dot{\phi}_0^a\dot{\phi}_0^b - \mathbf{V}(\phi_0^a)\right), \quad (22)$$

where L_{GR} is the Background of General Relativity lagrangian, $\mathbf{V}(\phi_0^a)$ and γ_{ab} are the effective potential and the metric in the fields space. After to fixing the gauge, they are given by $\mathbf{V}(\phi_0^a) = \left(1 + 3\kappa_2\tilde{F}_1\right) V(\varphi_0) + \kappa_2 V_{,\varphi}(\varphi_0)\tilde{\varphi}_0$ and $\gamma_{ab} = \begin{pmatrix} 1 & \kappa_2 \\ \kappa_2 & 0 \end{pmatrix}$. In this notation, the equations in (21) are reduced to:

$$\ddot{\phi}_0 + 3H\dot{\phi}_0 + \mathcal{V}_T = 0 \quad \wedge \quad \dot{T}^a - \frac{\mathcal{V}_N}{\dot{\phi}_0} N^a = 0, \tag{23}$$

with $\mathcal{V}^1 = V_{,\varphi}(\varphi_0)$, $\mathcal{V}^2 = V_{,\varphi\varphi}(\varphi_0)\tilde{\varphi}_0 + 3\tilde{F}_1 V_{,\varphi}(\varphi_0)$, $\dot{\phi}_0 \equiv \sqrt{\gamma_{ab}\dot{\phi}_0^a\dot{\phi}_0^b} = \sqrt{\dot{\varphi}_0^2 + 2\kappa_2\dot{\varphi}_0\dot{\tilde{\varphi}}_0}$, $T^a = \frac{\dot{\phi}_0^a}{\dot{\phi}_0}$ and $N_a = \sqrt{-\gamma}\epsilon_{ab}T^b$. To obtain (23), we use that $T_a T^a = 1$, $T_a\dot{T}^a = 0$, $N_a N^a = -1$ and $N_a T^a = 0$. This means that T^a and N^a is a complete base in the field space of φ_0 and $\tilde{\varphi}_0$. At this point, we must to say that our model is anomalous, because it has phantom fields. For this reason the determinant of the fields space metric is negative, specifically $\gamma = -\kappa_2^2$. For this reason, we have that $\mathcal{V}^a = \mathcal{V}_T T^a - \mathcal{V}_N N^a$ with $\mathcal{V}_T = T^a\mathcal{V}_a$ and $\mathcal{V}_N = N^a\mathcal{V}_a$. Additionally, we have the condition $\gamma_{ab} = \begin{pmatrix} 1 & \kappa_2 \\ \kappa_2 & 0 \end{pmatrix} = T_a T_b - N_a N_b$. To satisfy this, we need $T^a = \begin{pmatrix} T \\ \frac{1-T^2}{2\kappa_2 T} \end{pmatrix}$ and $N^a = \begin{pmatrix} -T \\ \frac{1+T^2}{2\kappa_2 T} \end{pmatrix}$, with $\dot{T} + \frac{\mathcal{V}_N}{\dot{\phi}_0}T = 0$.

In conclusion, T represent the direction of ϕ^a in the field space in $\tilde{\delta}$ Inflation, where $\dot{\varphi}_0 = T\dot{\phi}_0$ and $\dot{\tilde{\varphi}}_0 = \frac{1-T^2}{2\kappa_2 T}\dot{\phi}_0$. Now, we can define the inflationary parameters and simplify the equations of motion.

4.2. Inflationary Parameters and Background Equations.

Usually, slow-roll parameters are defined to study inflation. In this model, we can define similar parameters. In first place, we have the usual parameters: $\epsilon \equiv -\frac{\dot{H}}{H^2}$ and $\eta^a \equiv -\frac{\ddot{\phi}_0^a}{H\dot{\phi}_0}$. Then, using $a(t) \propto e^N$, we have:

$$\epsilon = \frac{\kappa}{2}T^2\phi_0'^2 \ , \ \eta_T \equiv \eta_a T^a = 3 + \frac{\mathcal{V}_T}{H^2\phi_0'} \text{ and } \eta_N \equiv \eta_a N^a = \frac{\mathcal{V}_N}{H^2\phi_0'}. \tag{24}$$

Besides, the equation of motion say $H^2 = \frac{\kappa V(\varphi_0)}{3 - \frac{\kappa}{2}T^2\phi_0'^2}$. On the other side, we have a effective scale factor, given now by $a_{eff}(t) \equiv a(t)\sqrt{\frac{1+\kappa_2\tilde{F}_1(t)}{1+3\kappa_2\tilde{F}_1(t)}}$. As we said in **Chapter 3**, an accelerated growth can be produced by a pole in $a_{eff}(t)$. In [9], we fixed the parameters to produce a pole to explain the accelerate expansion of the universe without dark energy. However, a Big-Rip is generated. For this reason, the dynamic of the scalar fields have to produce a $\omega(t)$ such that the expansion slow down when inflation ends. To consider this effect in inflation, we need to define additional parameters, like the effective Hubble parameter and the effective expansion parameter:

$$\mathcal{H} \equiv \frac{\dot{a}_{eff}(t)}{a_{eff}(t)} = H\left(1 - \frac{\kappa_2\tilde{F}_1'}{\left(1+\kappa_2\tilde{F}_1\right)\left(1+3\kappa_2\tilde{F}_1\right)}\right) \tag{25}$$

$$\epsilon_{eff} \equiv -\frac{\dot{\mathcal{H}}}{\mathcal{H}^2} = \frac{H^2}{\mathcal{H}^2}\left(\epsilon + \frac{\kappa_2\left(3\tilde{F}_1' - 2\tilde{\epsilon}\right)}{\left(1+\kappa_2\tilde{F}_1\right)\left(1+3\kappa_2\tilde{F}_1\right)} - \frac{2\left(2+3\kappa_2\tilde{F}_1\right)\kappa_2^2\tilde{F}_1'^2}{\left(1+\kappa_2\tilde{F}_1\right)^2\left(1+3\kappa_2\tilde{F}_1\right)^2}\right), \tag{26}$$

where $()' = \frac{d}{dN}$ and we use the background equations (19-21) and $\tilde{\epsilon} \equiv \varphi_0'\tilde{\varphi}_0' = \frac{\kappa}{2\kappa_2}\left(1-T^2\right)\phi_0'^2$. In GR, inflation needs $\epsilon \ll 1$, because it must obey $V(\varphi_0(t)) \gg \dot{\varphi}_0^2(t)$ to obtain an exponential expansion. On the other side, in $\tilde{\delta}$ Gravity, the accelerated expansion could be produced by a pole in $a_{eff}(t)$, such that $\mathcal{H}$ is the effective expansion rate. For this reason, the condition to guarantee inflation in $\tilde{\delta}$ Gravity is $\epsilon_{eff} \ll 1$. On the other side, the background equations of motion, (19-21), can be reduced to:

$$\phi_0'' + (3-\epsilon)\left(\phi_0' + \frac{\mathcal{V}_T}{\kappa V(\varphi_0)}\right) = 0 \quad (27)$$

$$T' + \frac{\mathcal{V}_N}{H^2\phi_0'}T = 0 \quad (28)$$

$$\tilde{F}_1'' + (3-\epsilon)\left(\tilde{F}_1' - 6\tilde{F}_1 - \frac{2V_{,\varphi}(\varphi_0)}{V(\varphi_0)}\tilde{\varphi}_0\right) = 0, \quad (29)$$

where $\mathcal{V}_T = \frac{1+T^2}{2T}V_{,\varphi}(\varphi_0) + \kappa_2 T\left(3V_{,\varphi}(\varphi_0)\tilde{F}_1 + V_{,\varphi\varphi}(\varphi_0)\tilde{\varphi}_0\right)$ and $\mathcal{V}_N = \frac{1-T^2}{2T}V_{,\varphi}(\varphi_0) - \kappa_2 T\left(3V_{,\varphi}(\varphi_0)\tilde{F}_1 + V_{,\varphi\varphi}(\varphi_0)\tilde{\varphi}_0\right)$. From (25), we can see that the simplest case is $\tilde{F}_1 \to cte$, because $\mathcal{H} \to H$. In fact, the effective scale factor is $a_{eff}(t) \propto a(t)$. This means that $\epsilon_{eff} \to \epsilon$ and $\tilde{\epsilon} \to 0$. Now, from equation (29), we have $\tilde{\varphi}_0 \to -3\tilde{F}_1\frac{V(\varphi_0)}{V_{,\varphi}(\varphi_0)}$, so $\tilde{F}_1\epsilon\left(\frac{V(\varphi_0)V_{,\varphi\varphi}(\varphi_0)}{V_{,\varphi}^2(\varphi_0)} - 1\right) \to 0$. This means that $\epsilon \to 0$, unless $\tilde{F}_1 = 0$ or $V(\varphi) = V_0 e^{\lambda\varphi}$. Then, in any other case, if $\tilde{\epsilon} \ll 1$, the inflation expansion is totally slow-rolling and the evolution is just like the usual inflation with GR. A similar situation is happened when $\tilde{F}_1 \gg 1$, such that $a_{eff}(t) \to \frac{a(t)}{\sqrt{3}}$.

In conclusion, if $\tilde{F}_1 \to cte$ or $\tilde{F}_1 \gg 1$, the $\tilde{\delta}$ Gravity effect is negligible and the expansion is governed by the original inflaton φ_0. Otherwise $\tilde{\delta}$ Gravity is important for the expansion and we have to take it into account for a specifical model, given by $V(\varphi_0)$. This work is in progress and we will publish an article very soon studying different models and we will compute the power spectrum.

5. Conclusions.

In this work, we have introduced basics concept and an initial analysis of $\tilde{\delta}$ Gravity in inflation. $\tilde{\delta}$ Gravity is a modified model with a new gravitational field $\tilde{g}_{\mu\nu}$. A quantum field theory analysis of $\tilde{\delta}$ Gravity has been developed in [7], where we proved that it lives at just one loop. Besides, in a classical level, we saw that the necessary quantity of dark matter could be considerably less that we expected and it predicts an accelerated expansion without a cosmological constant or additional scalar fields [9]). Additionally, the scale factor agrees with the standard cosmology at early times and shows acceleration only at late times. Therefore we expect that primordial density perturbations should not have large corrections. All these are the motivation to study inflation with $\tilde{\delta}$ Gravity.

In first place, we studied the equation of motion for just one perfect fluid with equations of state given by $p(t) = \omega\rho(t)$ and $\tilde{p}(t) = \omega\tilde{\rho}(t)$. In that case, we conclude that an accelerated expansion is possible, but inflation never ends. Therefore, a perfect fluid with a $\omega = cte$ can not explain inflation. A natural second option is to use two or more fluids. In [9] we used two fluid, radiation ($\omega = \frac{1}{3}$) and non-relativistic matter ($\omega = 0$), to explain the accelerated expansion of the universe. However, the behavior of $\omega_{eff}(X)$ for one and two fluid are very similar. For that reason, probably the best option is to use a fluid with a non-constant $\omega(t)$. In that case, we have a new element given by a $\tilde{\omega}(t) \neq 0$. For this, we used a scalar field to produce inflation.

In second place, we developed $\tilde{\delta}$ inflation. It is a model with a scalar field called inflaton and its $\tilde{\delta}$ partner, called $\tilde{\delta}$ inflaton. In traditional inflation with GR, the exponential expansion is produced by the inflaton and its potential $V(\varphi)$, in a slow-roll approximation. In $\tilde{\delta}$ inflation, the expansion could be produced by a $\tilde{\delta}$ Gravity effect, given by the modified scale factor. To describe the dynamic of the expansion, we defined effective inflationary parameters. For example, if $\epsilon_{eff} \sim 0$ or $\epsilon_{eff} < 0$, the expansion is exponential or phantom-like respectively, and if $\epsilon_{eff} = 1$, inflation is finished. Then, ϵ_{eff} represent the effects in the expansion by inflatons and $\tilde{\delta}$ Gravity. In fact, the expansion could be accelerated and $\epsilon \sim 1$.

In resume, $\tilde{\delta}$ Gravity could be important for the expansion in inflation for a specifical potential. In a future work, different potentials will be elaborated, including a calculation of the power spectrum to understand much better how $\tilde{\delta}$ Gravity explain the Dark Sector.

Acknowledgements.
This work has been financed by CONICYT Programa de Postdoctorado FONDECYT N° 3150398. The author wants to thank G Palma and N Videla for useful remarks.

Appendix A: Cosmological Case.
From [9], we know that the effective scale factor in the cosmological case is:

$$\mathcal{Y}(Y) = Y\sqrt{\frac{1 - L_2\frac{Y}{3}\sqrt{Y+C} + L_1\frac{Y}{C}\left(\sqrt{\frac{Y}{C}+1}\ln\left(\frac{\sqrt{\frac{Y}{C}+1}+1}{\sqrt{\frac{Y}{C}+1}-1}\right) - 2\right)}{1 - L_2Y\sqrt{Y+C} + 3L_1\frac{Y}{C}\left(\sqrt{\frac{Y}{C}+1}\ln\left(\frac{\sqrt{\frac{Y}{C}+1}+1}{\sqrt{\frac{Y}{C}+1}-1}\right) - 2\right)}}, \quad \text{(A.1)}$$

where $Y = \frac{a(t)}{a_0}$ is the usual scale factor normalized in the present and L_1, L_2 and $C = \frac{\Omega_R}{\Omega_M}$ are parameters in the model. The effective equation of state parameter is given by $\omega_{eff}(Y) \equiv -1 - \frac{2YH(Y)\mathcal{H}'(Y)}{3\mathcal{H}^2(Y)}$, with $\mathcal{H}(Y) \equiv \frac{YH(Y)\mathcal{Y}'(Y)}{\mathcal{Y}(Y)}$ and $()' = \frac{d}{dY}$. The expression of this parameter is very complicated, however we can prove that:

$$\omega_{eff}(Y \to 0) = \frac{1}{3} + O(C) \quad \text{(A.2)}$$

$$\omega_{eff}(Y \to 1) = -\frac{3L_2\left(16L_1^2 - 8L_1L_2 + 32L_1 - 5L_2 + 12\right)}{\left(8L_1^2 - 8L_1L_2 + 2L_2^2 + 16L_1 - 5L_2 + 6\right)^2} + O(C) \quad \text{(A.3)}$$

$$\omega_{eff}(Y \to Y_{Rip}) = -\frac{7}{3} + O(C), \quad \text{(A.4)}$$

where $Y_{Rip} = \left(\frac{1+2L_1}{L_2}\right)^{\frac{2}{3}}$ and $C \ll 1$. Therefore, the effective equation of state evolve from a radiation fluid to a phantom-like fluid ($\omega_{eff} < -1$). In **Figure A1**, we can see the behavior of $\omega_{eff}(Y)$ using the parameters values obtained with the supernovae data [9]. They are $L_1 = 1.565$, $L_2 = 2.262$ and $C = 1.82 \times 10^{-4}$. In this case, we used two different fluids to explain the accelerated expansion of the universe, non-relativistic matter ($\omega = 0$) and radiation ($\omega = \frac{1}{3}$). With these parameters, we have that $\omega_{eff}(1) = -0.926$.

References

[1] Turyshev S 2008 Experimental tests of general relativity *Ann. Rev. of Nuc. and Part. Science* **58** pp 207-48 doi:10.1146/annurev.nucl.58.020807.111839.

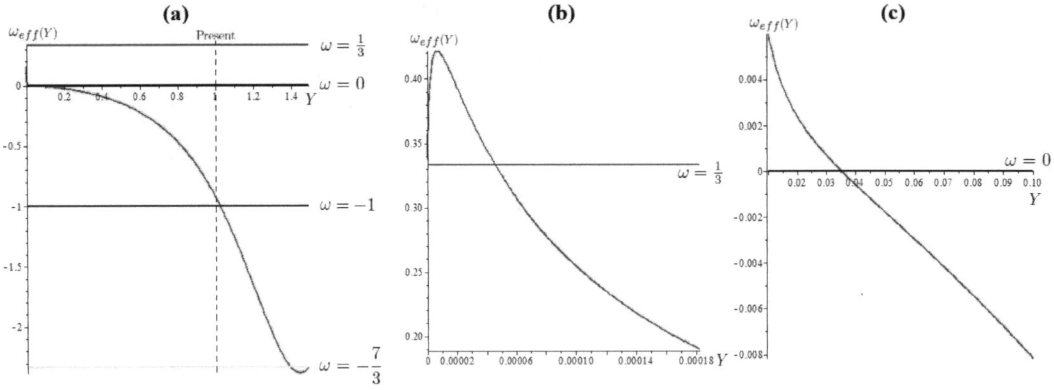

Figure A1. $\omega_{eff}(Y)$ vs Y with non-relativistic matter ($\omega = 0$) and radiation ($\omega = \frac{1}{3}$), using the parameters values obtained in [9] to explain the accelerated expansion of the universe. **(a)** Behavior of $\omega_{eff}(Y)$ from the beginning of the universe to the Big-Rip. **(b)** Behavior of $\omega_{eff}(Y)$ in the early universe until the matter-radiation equality epoch, where $Y = C = 1.82 \times 10^{-4}$. **(c)** Behavior of $\omega_{eff}(Y)$ close to the moment where the effective fluid is non-relativistic ($\omega = 0$), like cold dark matter (CDM). Clearly, we can see how the effective equation of state evolve from a radiation fluid ($\omega_{eff} = \frac{1}{3}$) to a phantom-like fluid ($\omega_{eff} < -1$) close to Big-Rip epoch.

[2] 't Hooft G and Veltman M 1974 One-loop divergencies in the theory of gravitation *Annales de l'institut Henri Poincaré, Section A, Physique théorique)* **20** 1 pp 69-94, http://www.numdam.org/item?id=AIHPA_1974__20_1_69_0.

[3] Hooper D and Baltz E 2008 Strategies for determining the nature of dark matter *Annual Review of Nuclear and Particle Science* **58** pp 293-314 doi:10.1146/annurev.nucl.58.110707.171217; Salucci P, Lapi A, Tonini C, Gentile G, Yegorova I, and Klein U 2007 The universal rotation curve of spiral galaxies II. The dark matter distribution out to the virial radius *MNRAS* **378** 1 pp 41-7; Ashman K 1992 Dark matter in galaxies *Astronomical Society of the Pacific* **104** 682 pp 1109-38 doi:10.1086/133099.

[4] Caldwell R and Kamionkowski M 2009 The physics of cosmic acceleration *Ann. Rev. of Nuc. and Part. Science* **59** pp 397-429 doi:10.1146/annurev-nucl-010709-151330; Perlmutter S, et al. 1999 Measurements of Ω and Λ from 42 high-redshift supernovae *Astrophysical Journal* **517** 2 pp 565-586 doi:10.1086/307221.

[5] Frieman J, Turner M and Huterer D 2008 Dark energy and the accelerating universe *Annual Review of Astronomy and Astrophysics* **46** pp 385-432 doi:10.1146/annurev.astro.46.060407.145243.

[6] Martin J 2012 Everything you always wanted to know about the cosmological constant problem (but were afraid to ask) *Comptes Rendus Physique* **13** 6 pp 566-665 doi:10.1016/j.crhy.2012.04.008.

[7] Alfaro J, González P and Ávila R 2011 A finite quantum gravity field theory model *Class. Quantum Grav* **28** 215020 (25pp) doi:10.1088/0264-9381/28/21/215020.

[8] Alfaro J and Labraña P 2002 Semiclassical gauge theories *Phys. Rev.* D **65** 045002 doi:http://dx.doi.org/10.1103/PhysRevD.65.045002.

[9] Alfaro J and González P 2017 Classical $\tilde{\delta}$ gravity: the dark sector and more *Preprint* arXiv:1704.02888.

[10] Cline J, Jeon S and Moore G 2004 The phantom menaced: constraints on low-energy effective ghosts *Phys. Rev.* D **70** 043543 doi:http://dx.doi.org/10.1103/PhysRevD.70.043543; Caldwell R, Kamionkowski M and Weinberg N 2003 Phantom energy: dark energy with $\omega < -1$ causes a cosmic doomsday *Phys. Rev. Lett.* **91** 7 071301 doi:http://dx.doi.org/10.1103/PhysRevLett.91.071301.

[11] Guth A 1981 Inflationary universe: a possible solution to the horizon and flatness problem *Phys. Rev.* D **23** 347 doi:https://doi.org/10.1103/PhysRevD.23.347; Starobinsky A 1980 A new type of isotropic cosmological models without singularity *Phys. Lett.* B **91** 99 doi:https://doi.org/10.1016/0370-2693(80)90670-X.

[12] Kofman L, Linde A and Starobinsky A 1994 Reheating after inflation *Phys. Rev. Lett* **73** pp 3195-98 doi:https://doi.org/10.1103/PhysRevLett.73.3195.

[13] Céspedes S, Atal V and Palma G 2012 On the importance of heavy fields during inflation *Journal of Cosmology and Astroparticle Physics* JCAP05(2012)008 doi:10.1088/1475-7516/2012/05/008.

IOP Conf. Series: Journal of Physics: Conf. Series **1043** (2018) 012018 doi:10.1088/1742-6596/1043/1/012018

Hamiltonian formulation of the linearized hypersurface-orthogonal Einstein-aether theory

J. Bellorín[1], A. Restuccia[1,2] and F. Tello[3]

[1] Department of Physics, Universidad de Antofagasta, Antofagasta, Chile.
[2] Department of Physics, Universidad Simón Bolívar, Caracas, Venezuela.
[3] Department of Physics, Universidad Católica del Norte, Antofagasta, Chile.

E-mail: jbellori@gmail.com, arestu@usb.ve, ftello@ucn.cl

Abstract. We perform the Hamiltonian formulation of the Einstein-aether theory subject to the condition of hypersurface-orthogonality on the aether vector. Our main aim is to obtain the bulk Hamiltonian together with its constraints and their algebra. We implement a perturbative approach, studying the theory at the linearized level. Under the condition of hypersurface orthogonality the aether vector can be represented by a scalar field. The theory has two first-class constraints, which are associated to the symmetry of general diffeomorphisms over the spacetime. One of the constraints depends on the perturbative aether field, generating linear-order diffeomorphisms on it. We also present the reduced bulk Hamiltonian depending on the physical degrees of freedom. We find conditions on the coupling constants enusring the positiveness of the reduced Hamiltonian and the hyperbolicity of the propagation equations of the independent modes.

1. Introduction

There is a great interest in studying theories that break the Lorentz symmetry, either by quantum motivations or even at the classical level with theories that are modifications of general relativity (GR). One of such classical theories is the Einstein-aether theory (EA-theory) [1]. In esence, this theory modifies GR by coupling to it a dynamical vector field u^μ, called the aether, which is regarded as a fundamental field rather than a matter source. The dynamical aether vector is restricted to be timelike and unitary, $u^\alpha u_\alpha = -1$. On any configuration, the presence of the vector u^μ breaks the local Lorentz symmetry. On the other hand, since the vector u^μ is dynamical and the action is fully written in term of spacetime tensors, the action is manifestly invariant under general diffeomorphisms on the spacetime. The action of the standard theory, which is the one we study here, is of second order in derivatives of u^μ, although higher order couplings can be considered as well.

In the general case there are three independent propagating modes associated to the aether vector. This number can be reduced by imposing the geometrical condition of hypersurface orthogonality on the aether vector u^μ. Locally this condition can be defined by means of a function T whose gradient is everywhere timelike, $\partial_\mu T \partial^\mu T < 0$. Then the condition of hypersurface orthogonality is

$$u_\alpha = \frac{\partial_\alpha T}{(-g^{\mu\nu}\partial_\mu T \partial_\nu T)^{1/2}}. \tag{1}$$

With this condition the normalization constraint on the aether field is satisfied. We call the theory with this condition the restricted EA-theory (other used names are the khronometric theory and the T theory). Intrerestingly, the Lagrangian of the restricted EA-theory is related [2, 3] to the Lagrangian of the nonprojectable Hořava theory [4, 5] truncated up to second order in spatial derivatives.

Regardless of the condition of hypersurface orthogonality, the full Hamiltonian formulation of the EA-theory has not been successfully achieved due to the mathematical complications that arise in the Legendre transformation (boundary terms that determine the energy and other conserved charges for asymptotically flat configurations have been previously studied in Refs. [6, 7]).

In this paper we give an answer to the problem of finding the bulk Hamiltonian of the Einstein-aether theory. Specifically, by means of a perturbative analysis we obtain the bulk Hamiltonian of the linearized restricted EA-theory with its corresponding constraints. This analysis allows us to determine rigorously the dynamics of the true propagating modes of the theory. We also study the relation of the constraints with symmetries.

2. Lagrangian and Hamiltonian formulations

2.1. The perturbative Lagrangian

The action of the (unrestricted) EA-theory, omiting a cosmological-constant term and matter couplings, is [1]:

$$S_{ae} = \int \sqrt{-g}\,(R + \mathcal{L}_{ae})\, d^4x. \tag{2}$$

In this action R is the four-dimensional Ricci scalar and $\mathcal{L}_{ae}$ the Lagrangian density of the aether vector, which is given by

$$\mathcal{L}_{ae} = -\mathcal{M}^{\alpha\beta\lambda\gamma}\nabla_\alpha u_\lambda \nabla_\beta u_\gamma, \tag{3}$$

where $\mathcal{M}^{\alpha\beta\lambda\gamma}$ a matrix defined as

$$\mathcal{M}^{\alpha\beta\lambda\gamma} = c_1 g^{\alpha\beta}g^{\lambda\gamma} + c_2 g^{\alpha\lambda}g^{\beta\gamma} + c_3 g^{\alpha\gamma}g^{\beta\lambda} + c_4 u^\alpha u^\beta g^{\lambda\gamma}. \tag{4}$$

$c_{1,2,3,4}$ are dimensionless coupling constants. The action (2) is manifestly invariant under general diffeomorphism over the spacetime.

The restricted EA-theory is defined by imposing the hypersurface-orthogonality condition (1) at the level of the action (2), i. e., before deriving the equation of motion. We assume this restriction from now on. Thus, the action (2) becomes a functional of $g_{\mu\nu}$ and T.

In order to address the Hamiltonian formulation of the restricted EA-theory, we first decompose the spacetime metric $g_{\mu\nu}$ in the standard Arnowitt-Deser-Misner (ADM) variables γ_{ij}, N and N_i. With these variables the action (2) takes the form

$$S_{ae} = \int N\sqrt{\gamma}\left(K^{ij}K_{ij} - K^2 + {}^{(3)}R - \mathcal{M}^{\alpha\beta\lambda\gamma}\nabla_\alpha u_\lambda \nabla_\beta u_\gamma\right) d^3x dt, \tag{5}$$

where ${}^{(3)}R$ is the spatial Ricci scalar, K_{ij} the extrinsic curvature defined by

$$K_{ij} = \frac{1}{2N}\left(\dot{\gamma}_{ij} - \nabla_i N_j - \nabla_j N_i\right), \tag{6}$$

and $K = \gamma^{ij}K_{ij}$ is the trace of the extrinsic curvature. The dot denotes time derivative, $\dot{\gamma}_{ij} = \partial\gamma_{ij}/\partial t$. It is supposed that the ADM variables are substituted in the aether Lagrangian as well. We comment that there are contributions to the kinetic terms of γ_{ij} coming from the aether sector. Below we shall see this explicitly.

Now we implement the perturbations. We choose as the background the Minkowski spacetime together with the condition $T = t$, which constitute a solution of the field equations ($T = t$ gives a zero energy-momentum tensor). We introduce the perturbative variables n, n_i, h_{ij} and τ in the following way

$$N = 1 + \epsilon n, \tag{7}$$

$$N^i = \epsilon n_i, \tag{8}$$

$$\gamma_{ij} = \delta_{ij} + \epsilon h_{ij}, \tag{9}$$

$$T = t + \epsilon\tau. \tag{10}$$

For the sake of simplify computations, in the whole perturbative analysis we impose the transverse gauge

$$\partial_i h_{ij} = 0. \tag{11}$$

After introducing the perturbations (7 - 10) in (5) and (1), we obtain the action of quadratic order in perturbations. It is

$$S = \frac{1}{\beta}\int dtd^3x \left[\kappa_{ij}\kappa_{ij} - \lambda\kappa^2 + \alpha\left(\partial_i\dot{\tau} - \partial_i n\right)^2 + \frac{\beta}{4}\left(h_{ij}\Delta h_{ij} - h\Delta h\right) - \beta n\Delta h \right. \\ \left. -2\left(1-\beta\right)\kappa_{ij}\partial_{ij}\tau - 2\left(\beta - \lambda\right)\kappa\Delta\tau + (1-\lambda)(\Delta\tau)^2\right], \tag{12}$$

where κ_{ij} is the linear-order perturbation of the extrinsic curvature,

$$\kappa_{ij} = \frac{1}{2}\left(\dot{h}_{ij} - \partial_i n_j - \partial_j n_i\right), \tag{13}$$

$\kappa = \kappa_{kk}$, $h = h_{kk}$ and $\Delta = \partial_{kk}$ is the flat Laplacian. α, β and λ are combinations of the coupling constants defined by

$$\alpha = \frac{c_1 - c_4}{1 - c_1 - c_3}, \quad \beta = \frac{1}{1 - c_1 - c_3}, \quad \lambda = \frac{1 + c_2}{1 - c_1 - c_3}. \tag{14}$$

These combinations, which are the ones the perturbative theory naturally adopts, can be identified with the coupling constants of the $z = 1$ nonprojectable Hořava theory [3]. As we anticipated, there are contributions to the kinetic term of h_{ij} coming from the aether sector; their presence can be recognized by the fact that the coupling constant β^{-1} in front of the action (12) is equal to $1 - c_1 - c_3$.

2.2. *The Legendre transformation*

From the Lagrangian (12) we obtain the following conjugate momenta

$$\phi_0 = \frac{\partial\mathcal{L}}{\partial\dot{n}} = 0, \quad \phi_i = \frac{\partial\mathcal{L}}{\partial\dot{n}^i} = 0, \tag{15}$$

$$p_\tau = \frac{\partial\mathcal{L}}{\partial\dot{\tau}} = -\frac{2\alpha}{\beta}\Delta\left(\dot{\tau} - n\right), \tag{16}$$

$$\pi^{ij} = \frac{\partial\mathcal{L}}{\partial\dot{h}_{ij}} = \frac{1}{\beta}\left[\kappa_{ij} - \lambda\kappa\delta_{ij} - (1-\beta)\left(\partial_{ij}\tau\right) - \left(\beta - \lambda\right)\delta_{ij}\Delta\tau\right]. \tag{17}$$

Equations (15) are primary constraints. Equations (16) and (17) allow us to solve the variables $\dot{\tau}$ and $\dot{h}_{ij}$ in terms of the canonical variables,

$$\dot{\tau} = -\frac{\beta}{2\alpha\Delta}p_\tau + n, \tag{18}$$

$$\kappa_{ij} = \beta\pi^{ij} + \frac{\beta\lambda}{1-3\lambda}\delta_{ij}\pi + (1-\beta)\partial_{ij}\tau + \frac{\beta(1-\lambda)}{1-3\lambda}\delta_{ij}\Delta\tau. \tag{19}$$

IOP Conf. Series: Journal of Physics: Conf. Series **1043** (2018) 012018 doi:10.1088/1742-6596/1043/1/012018

Notice that this inversion cannot be performed when the constant λ takes the value $\lambda = 1/3$. In the nonprojectable Hořava theory there is an analog to this particular case. Its Hamiltonian formulation has been studied in Refs. [8, 9]. In that theory two additional second-class constraints emerge and the extra scalar mode is eliminated. Throughout this paper we assume $\lambda \neq 1/3$, hence the inversion (19) is consistent. In addition, Eq. (18) requires $\alpha \neq 0$.

With these formulas we may build the Hamiltonian, which takes the form

$$H = \int d^3x \left[\frac{1}{\beta} \left(\kappa_{ij}\kappa_{ij} - \lambda \kappa^2 \right) - \frac{\beta}{4\alpha} p_{\mathcal{T}} \frac{1}{\Delta} p_{\mathcal{T}} - \frac{1}{4} (h_{ij} \Delta h_{ij} - h \Delta h) + \frac{1}{\beta} (\lambda - 1)(\Delta \mathcal{T})^2 \right. \\ \left. + n \mathcal{H}^{(1)} + n_i \mathcal{H}^{i(1)} \right] , \tag{20}$$

where

$$\mathcal{H}^{(1)} \equiv \Delta h + p_{\mathcal{T}} , \tag{21}$$

$$\mathcal{H}^{j(1)} \equiv -2 \partial_i \pi^{ij} . \tag{22}$$

We have put the superscript $^{(1)}$ on these objects to remark that they are of linear order in perturbations. In the Hamiltonian (20) it is understood that the variable κ_{ij} must be substituted by its expression (19) in terms of the canonical variables.

We observe that there is a nonlocal term in the Hamiltonian density (20). This nonlocality is a mere effect of using a phase space with nonreduced variables. Below we show that, when the Hamiltonian is reduced to the physical phase space (the reduced Hamiltonian), the resulting Hamiltonian density is completely local.

Our next step is to implement the Dirac procedure for preserving the constraints in time. The preservation in time of the ϕ_0 and ϕ_i constraints leads, respectively, to the secondary constraints

$$\mathcal{H}^{(1)} = 0 , \tag{23}$$

$$\mathcal{H}^{i(1)} = 0 . \tag{24}$$

Subsequently, we impose the preservation in time of these last constraints. We obtain that they are automatically preserved by the Hamiltonian (20),

$$\dot{\mathcal{H}}^{(1)} = \{ \mathcal{H}^{(1)} , H \} = 0 . \tag{25}$$

$$\dot{\mathcal{H}}^{i(1)} = \{ \mathcal{H}^{i(1)} , H \} = 0 . \tag{26}$$

This closes Dirac's procedure. Therefore, we have that the set of all constraints is given by ϕ_0, ϕ_i, $\mathcal{H}^{(1)}$ and $\mathcal{H}^{i(1)}$, and all these constraints are preserved. The structure of the Hamiltonian (20) allows us to make the usual simplification of this kind of generally-covariant theories: n and n_i can be regarded as Lagrange multipliers, such that the conjugated pairs (n, ϕ_0) and (n_i, ϕ_i) can be dropped out from the phase space.

We comment that the Hamiltonian (20) is composed of two parts: a part that is explicitly the sum of constraints and the other one that is not. On the other hand, this is a theory with general convariance, which implies the freedom of transforming arbitrary the time. For this kind of symmetry one expects the bulk Hamiltonian is a total sum of constraints, such that it vanishes on the constrained phase space. Actually, there is no fundamental contradiction between this and the fact that H in (20) has terms outside the constraints; the discrepancy is only an effect of the perturbative theory. The terms in H that do not explicitly belong to constraints arise as a consequence of truncating the Hamiltonian up to a given order in perturbations. We expect that the nonperturbative Hamiltonian is effectively a sum of constraints, but this nonperturbative Hamiltonian is not known for this theory.

2.3. Symmetries and the reduced Hamiltonian

The two constraints of the theory, $\mathcal{H}^{(1)}$ and $\mathcal{H}^{i(1)}$, are first-class constraints (this can be explicitly checked using the full expressions of the constraints without fixing the gauge (11)). This implies that $\mathcal{H}^{(1)}$ and $\mathcal{H}^{i(1)}$ are generators of gauge symmetries. As in linearized GR, $\mathcal{H}^{i(1)}$ is the generator of purely spatial diffeomorphisms on h_{ij}, that is, it is the momentum constraint. The symmetry associated to this constraint has been already fixed in (11). Interestingly, $\mathcal{H}^{(1)}$ has a double action. It generates transformations on the metric variables associated to time transformations, as does the Hamiltonian constraint of linearized GR; but it also generates redefinitions of τ,

$$\left\{\tau, \int d^3z\, \varphi\mathcal{H}^{(1)}\right\} = \varphi\,. \tag{27}$$

Since φ is an arbitrary test function, this transformation is a total redefinition of τ. This transformation can also be regarded as a diffeomorphism associated to the time coordinate. Indeed, coming back to the nonperturbative field T, we have that it is an scalar under a general diffeomorphism given by $\delta x^\mu = \epsilon^\mu(x^\alpha)$,

$$\delta T = -\epsilon^\mu \partial_\mu T. \tag{28}$$

If we now substitute the perturbative expansion (10) in this transformation rule, at linear order we obtain

$$\delta\tau = -\epsilon^0. \tag{29}$$

Thus, at linear order the variable τ transforms under a diffeomorphism only with the time component (at linear order it is invariant under a diffeomorphism affecting only the spatial directions), and it is totally redefined by the component ϵ^0. As we may see from (27), this transformation on τ is generated by $\mathcal{H}^{(1)}$. The fact that this is a symmetry of the theory means that the dynamics is independent of the particular τ or, in other words, τ is a pure-gauge variable and can be gauged fixed. In the exact theory, this feature can be traced back to the fact that the equation of motion corresponding to the variations of the action with respect to the T field is implied by the other field equations [3].

We are now ready to study the reduced (physical) Hamiltonian, which requires to solve all the constraints and fix the gauge symmetries. The purely spatial diffeomorphisms have been fixed by (11). To fix the diffeomorphisms along the time coordiante we choose the gauge in which τ vanishes,

$$\tau = 0\,. \tag{30}$$

To further advance we decompose the perturbative spatial metric into transverse and longitudinal parts,

$$h_{ij} = h_{ij}^{TT} + \frac{1}{2}\left(\delta_{ij} - \frac{\partial_{ij}}{\Delta}\right) h^T + \partial_j h_i^L + \partial_i h_j^L, \tag{31}$$

where these variables are restricted by $\partial_k h_{kl}^{TT} = h_{kk}^{TT} = 0$. We also decompose π_{ij} in the same way as (31). The transverse gauge (11) eliminates the longitudinal sector h_i^L. By virtue of the constraint (24) the longitudinal sector of π_{ij} is also eliminated. Constraint (23) can be solved for p_τ in favour of h^T,

$$p_\tau = -\Delta h^T. \tag{32}$$

After these considerations, we obtain the reduced Hamiltonian for the true physical degrees of freedom. It takes the form

$$H_{\text{RED}} = \int d^3x \left[\beta\pi_{ij}^{TT}\pi_{ij}^{TT} - \frac{1}{4}h_{ij}^{TT}\Delta h_{ij}^{TT} + \frac{\beta(1-\lambda)}{2\,(1-3\lambda)}(\pi^T)^2 - \frac{2\beta-\alpha}{8\alpha}h^T\Delta h^T\right]. \tag{33}$$

Of course, the variables h_{ij}^{TT}, π_{ij}^{TT} are still subject to the conditions of transversality and vanishing trace. As we anticipated, the reduced Hamiltonian density is completely local.

In order to ensure the positiveness of the reduced Hamiltonian (33), the following conditions on the coupling constants are required

$$\beta \geq 0, \quad \frac{1-\lambda}{1-3\lambda} \geq 0, \quad \frac{2\beta-\alpha}{\alpha} \geq 0. \tag{34}$$

The second condition can be satisfied only in the ranges $\lambda \in (-\infty, 1/3)$ and $\lambda \in [1, +\infty)$. The third condition can be satisfied only in $\alpha \in (0, 2\beta]$. We recall that we have assumed $\lambda \neq 1/3$ and $\alpha \neq 0$ in all the analysis.

From the reduced Hamiltonian (33) the evolution equations of the several modes can be deduced. From these equations we obtain the second-order equations for h_{ij}^{TT} and h^T,

$$\ddot{h}_{ij}^{TT} - \beta \Delta h_{ij}^{TT} = 0, \tag{35}$$

$$\ddot{h}^T - c_s^2 \Delta h^T = 0, \tag{36}$$

where

$$c_s = \sqrt{\frac{\beta(2\beta-\alpha)(1-\lambda)}{\alpha(1-3\lambda)}}. \tag{37}$$

The conditions (34) on the coupling constants ensure the hyperbolicity of the Eqs. (35) and (36), hence they are wave equations. The speed of the tensorial modes is $\sqrt{\beta}$ whereas the speed of the scalar mode is c_s.

3. Conclusions

We have successfully performed the Hamiltonian formulation of the linearized Einstein-aether theory with the condition of hypersurface orthogonality on the aether vector, under which the aether vector can be modelled via a scalar field. We have focused the problem on a perturbative approach in order to overcome the difficulties arising in the nonperturbative theory, whose bulk Hamiltonian is still not known. We have found that the theory possesses two first-class constraints associated to the symmetry of general diffeomorphisms over the spacetime. This is in agreement with the fact that the Einstein-aether theory is a theory with general covariance. In particular, one of the constraints generates diffeomorphisms along the time coordinate on the scalar field corresponding to the aether. This symmetry transformation is reinterpreted as this scalar field is a pure-gauge variable. The reduced Hamiltonian is positive if certains bounds on the coupling constants are imposed. These conditions match with the conditions found in Ref. [5] in the context of the nonprojectable Hořava theory. We have found that the restricted EA-theory has three canonical propagating modes, the two transverse-traceless tensorial modes and a scalar mode. They propagate with wave equations with different speeds. This confirms that the condition of hypersurface orthogonality eliminates two of the three modes associated to the aether vector.

It would be interesting to investigate if an iterative process on the perturbative theory can shed some light on the Hamiltonian of the nonperturbative theory. Another interesting application of our results is the comparison with the Hamiltonian formulation of the nonprojectable Hořava theory [10, 11, 12]. The idea is to ellucidate whether both Hamiltonian formulations are physically equivalent. It is of particular interest to perform the analysis with general boundary conditions, such that the relation can be determined for general configurations, not only for asymptotically flat ones.

Acknowledgments
F. T. thanks financial support from U. C. N. graduate program. A. R. and F. T. are partially supported by grant Fondecyt No. 1161192, Chile.

References
[1] Jacobson T and Mattingly D 2001 *Phys. Rev.* D **64** 024028.
[2] Blas D, Pujolas O and Sibiryakov S 2010 *Phys. Lett.* B **688** 350.
[3] Jacobson T 2010 *Phys. Rev.* D **82** 129901.
[4] P. Hořava 2009 *Phys. Rev.* D **79** 084008.
[5] Blas D, Pujolas O and Sibiryakov S 2010 *Phys. Rev. Lett.* **104** 181302.
[6] Eling C 2006 *Phys. Rev.* D **73** 084026 [Erratum: 2009 *Phys. Rev.* D **80** 129905].
[7] Foster B 2006 *Phys. Rev.* D **73** 024005.
[8] Bellorín J, Restuccia A and Sotomayor A 2013 *Phys. Rev.* D **87** 084020.
[9] Bellorín J and Restuccia A 2016 *Phys. Rev.* D **94** 064041.
[10] Kluson J 2010 *JHEP* **1007** 038.
[11] Donnelly W and Jacobson T 2011 *Phys. Rev.* D **84** 104019.
[12] Bellorín J and Restuccia A 2011 *Phys. Rev.* D **84** 104037.

IOP Conf. Series: Journal of Physics: Conf. Series **1043** (2018) 012019 doi:10.1088/1742-6596/1043/1/012019

Asymmetry and dark energy evolution

Víctor H. Cárdenas

Instituto de Física y Astronomía, Facultad de Ciencias, Universidad de Valparaíso, Av. Gran Bretaña 1111, Valparaíso, Chile.

E-mail: `victor.cardenas@uv.cl`

Abstract. The well known asymmetry of type Ia supernova data is used to infer general features of an alternative model for dark energy. Using a model that allows DE to vary with redshift as a probe, the results indicate the low redshift transition of the deceleration parameter $q(z)$ is evident only taken data points from one hemisphere of the sky. A related effect is a reconstructed DE density $X(z)$ that shows evolution (it decreases as redshift increases) only towards the hemisphere where $q(z)$ manifest the transition. We discuss on possible theoretical frameworks where these effects can be explained.

1. Introduction

So far the ΛCDM has been a very successful model in confronting the data, not only using type Ia supernova (SNIa), from which it was proposed for the first time as a model [1],[2], but also using a myriad of different observational probes, both geometric and dynamical. In this setup Λ drives the cosmic acceleration. However, although successful, the model forces us to accept uncomfortable facts: we do not know the origin of Λ, and with this we do not know why it has the value we measure, and also, we must accept that we live in a very unique time, just that time where the contribution of Λ and matter have the same order of magnitude.

Among the models that have been proposed as alternatives to the ΛCDM we can mention the dark energy (DE) [3], a fluid that fills all the space and has negative pressure. The source of this DE can be a scalar field as in quintessential models [4]. Another group of alternative models explore a modification of the Einstein - Hilbert action to explain what we observe [5].

Also, there are models that relax the homogeneity assumption of the standard cosmological model, by assuming that we live near the center of a void. Using this assumption it is possible to fit the SNIa data without invoking DE [7]. Although it is a seductive idea, the initial program testing this against the observations seems to indicate that we need a void of a size never observed (gigaparsec in size) [8]. Research in these areas continues [9].

In the context of the homogeneous and isotropic scheme, direct explorations of possible departures from the ΛCDM model have been focused mainly through parameterizing specific physical quantities. We can parametrize the equation of state (EoS) parameter $w = p/\rho$ (where p is the pressure and ρ the energy density) as a function of redshift [10, 11, 12, 13], or we can parametrize the DE density itself [14].

Using the Chevallier-Polarski-Linder (CPL) parametrization [47], [46] $w(z) = w_0 + w_1 z/(1+z)$, the authors of [10] found that the reconstructed deceleration parameter $q(z)$ shows a low-redshift transition leading to the cosmic slow down of acceleration scenario. Later, this phenomena was found using new SNIa data [11], [12], also using data from gas mass fraction in galaxy clusters [13], and using different parameterizations [15], showing that this effect – the low redshift transition of the reconstructed $q(z)$ – is telling us something about our local universe, because the evidence for such transition disappears once data from BAO and CMB (high z physics) are added.

A similar tension between low-redshift with high-redshift data manifest in measurements of the Hubble constant H_0 [18]. This tension emerges using the ΛCDM model as the "base" model. Although simple extensions of the ΛCDM model were studied, not all of them alleviated the tension. In [19] the authors showed that the assumption of a large local under-density on radial scales of a few Mpc it is enough to alleviate such a tension.

Regarding the low redshift transition, it was noticed that such a transition is a expected output if we are positioned near the center of a void [16]. Actually, it was claimed that such a feature have to be considered a signal of the existence of a void (underdensity) in our neighborhood [17]. Then the fact that both, the tension between low and high redshift data and the inhomogeneous connection for this feature of the reconstructed $q(z)$, make sense only if we assume a local underdensity, means that something in the local distribution of matter is affecting the data, something that we would like to characterize.

In this context, it is interesting to cite the case of a preferred direction of expansion, as for example, the large scale bulk-flow [20, 22, 21, 24, 23], the low multipoles alignment in the CMB angular power spectrum [25], the large scale alignment of quasar polarization [26], and studies using SNIa [27, 32, 33, 28, 34, 35, 30, 31, 29]. So the evidence points towards the necessity to a study further the effects on the cosmological parameters of a inhomogeneous local distribution of matter. In [38] using two different SNIa samples, it was shown that this low-redshift transition of $q(z)$ is evident only towards one hemisphere and is absent in the opposite one. If we interpret the low-redshift feature as a indication of a local void, this last results implies that such a void is only apparent toward one hemisphere, and so not only implying a local inhomogeneous distribution of matter but also having a dipolar anisotropy.

In this work we perform a study of our local universe, using two samples of SNIa, and using parametrizations for the EoS parameter $w(z)$ and also for the DE density $X(z)$. We also discuss the first steps towards to enhance the cosmological model to take into account these local effects.

2. The method

We begin recalling that the co-moving radial coordinate is

$$r(z) = \frac{c}{H_0}\int_0^z \frac{dz'}{E(z')}, \tag{1}$$

in a flat universe, where $E(z) = H(z)/H_0$ is the reduced Hubble function. Assuming an EoS parameterized by [47, 46]

$$w(z) = w_0 + \frac{w_1 z}{1+z}, \tag{2}$$

the reduced Hubble function takes the form

$$E^2(z) = \Omega_m(1+z)^3 + (1-\Omega_m)X(z), \tag{3}$$

where Ω_m comprises both the baryonic and non baryonic DM, and the reconstructed DE density is

$$X(z) = e^{-\frac{3w_1 z}{1+z}}(1+z)^{3(1+w_0+w_1)}. \tag{4}$$

From this, following previous works [11, 12], we can reconstruct the deceleration parameter function

$$q(z) = (1+z)\frac{1}{E(z)}\frac{dE(z)}{dz} - 1. \tag{5}$$

As was shown in [14], the low redshift transition of $q(z)$ at 2σ C.L is observed in both the more recent SNIa data and gas mass fraction data from galaxy clusters. The idea here is to study how this picture changes once the position of the SNIa data is taken into account, and restudy both the effects on $q(z)$ and $X(z)$.

3. Using the Union 2 sample

The Union 2 dataset [40] consist of 557 SNIa in the range $0.015 < z < 1.4$. Union 2 join the Union dataset [43], six SNIa at high redshift found by the Hubble Space Telescope, and low and intermediate redshift data from [44] and [45], respectively.

In this section we use the result of [27] using the hemispherical asymmetry technique, were they found a dipolar asymmetry with a maximum acceleration towards $(l, b) = (309^o, 18^o)$ and a minimum acceleration towards $(l, b) = (129^o, -18^o)$, in galactic coordinates.

Given these points, we separate the data into two hemispheres each one with center in the points of maximal (minimal) acceleration. In the hemisphere corresponding to the maximum acceleration we get 116 supernovae and in the other the remaining 441. This huge asymmetry in the number is essentially due to the fact that in the minimum acceleration hemisphere we found the SDSS stripe of SNIa data. The best fit values in each hemisphere are displayed in Table 1.

	χ^2_{min}/dof	Ω_m	ω_0	ω_1
A	$103.65/113$	0.21 ± 0.64	-1.3 ± 1.2	1.5 ± 3.4
B	$427.57/438$	0.46 ± 0.06	-0.41 ± 0.53	-9.2 ± 7.3

Table 1. Best fit values of the cosmological parameters for the two hemispheres. Row A for the 116 SNIa in the maximum acceleration hemisphere. Row B the 441 SNIa in the opposite direction. See also Fig.(1).

From inspection of Figure 1 we notice that the transition feature at low redshift is only evident in the direction of the minimum acceleration, meanwhile in the opposite direction the best fit is in agreement with ΛCDM. Following standard procedure (see [38] for details) we can estimate the confidence for each set of best fit parameters. In the case of the 116 SNIa in the maximum acceleration hemisphere we get a distance of 0.31σ away from the ΛCDM point. In the other hemisphere – in which we observe the low redshift transition of the reconstructed $q(z)$ – we get a distance of 1.76σ away from the ΛCDM best fit.

Further, the reconstructed DE density Eq.(4) for the Union 2 sample in hemispheres is display in Fig. 2. As we can see, the reconstructed DE density towards the hemisphere of maximum acceleration does not show any deviation with respect to a cosmological constant, however towards the minimum acceleration zone, this function decrease as a function of the redshift, which is what we expect if we have a void in that direction.

Taking the result of [16] as our main hypothesis – that a low redshift transition of $q(z)$ can be considered a signal for a local under-density – the result we have found indicates the data (Union 2 set) is able to detect such an under-density only in one (hemisphere) direction of the sky. This implies that the assumption of a spherically symmetric model for such under-density is

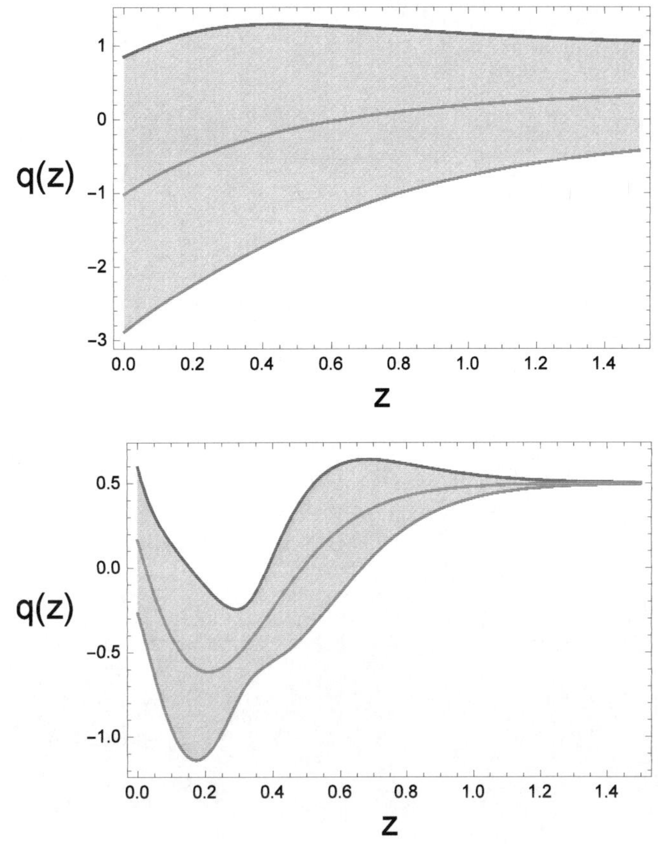

Figure 1. Here we show the reconstructed deceleration parameter $q(z)$ using data from two hemispheres. The upper panel showing the one around the maximum acceleration point, and the bottom the one in the opposite direction. See Table 1 for the best fit values for each case.

not a good idea (as is usual working with LTB models). Moreover, this result probably suggest the use of a metric with a dipolar asymmetry.

The low redshift transition apparent in Figure 1 and the decrease in $X(z)$ in Figure 2, disappear once data from baryon acoustic oscillation (BAO) and cosmic microwave background radiation (CMBR) are taken into account.

It is also worth notice the effect of the number of points used in each hemisphere. The set pointing towards the maximum acceleration has only 116 SNIa. Although the best fit is only 0.3σ away from the ΛCDM values, indicating these data points are in agreement with the concordance model, the sparsity of data points increases the errors considerably. The actual redshift distribution is also worth mentioned in this context. In this hemisphere most of the data, around 65 points, have $z < 0.2$, with no data in the range $0.2 < z < 0.3$, with 45 points distributed around redshift $z \simeq 0.5$, and almost one point for every redshift bin ($\triangle z = 0.1$) for $z > 0.8$. The other hemisphere has a greater number of data points (441) with a redshift distribution that, although decreasing with redshift, it still maintain a large number of data points per redshift bin ($\triangle z = 0.1$) until $z \simeq 1.4$.

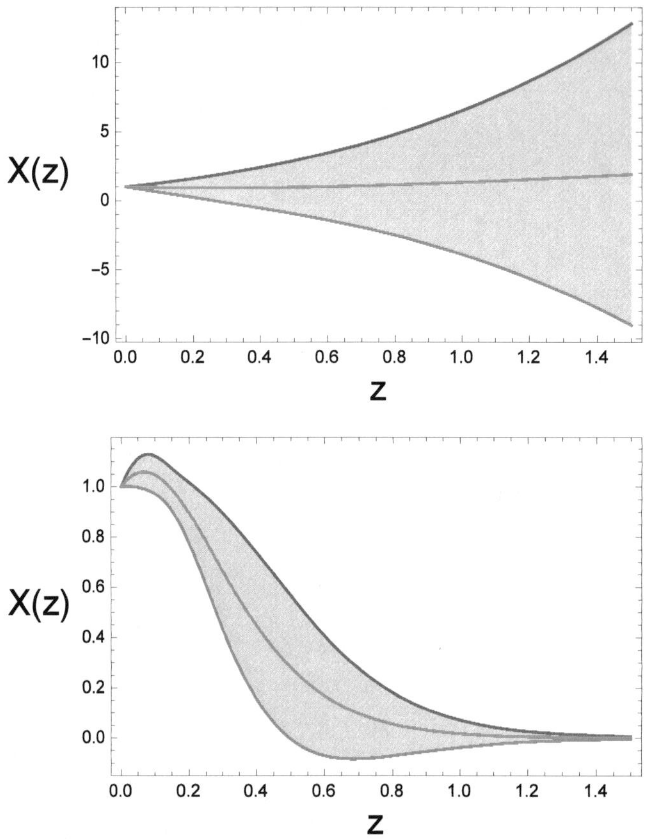

Figure 2. Here we show the reconstructed DE density $X(z)$ using data from two hemispheres. The upper panel showing the one around the maximum acceleration point, and the bottom the one in the opposite direction. See Table 1 for the best fit values for each case.

4. Analysis using the Loss sample

We have also performed the study of the consequences, in the reconstructed deceleration parameter, of the existence of an axis of maximal asymmetry using the Loss data set [42]. In order to find this axis, we follow the procedure of [27], using a flat ΛCDM as a base model, and start a scan of random axes looking for the maximum variation in the best fit of Ω_m. Using the Loss sample we found that the direction of maximal acceleration points towards $(l, b) = (309^o, 31^o)$.

In the hemisphere corresponding to the maximum acceleration, there are 216 supernovae and in the other the remaining 370. Again we observe the asymmetry in the distribution, although less severe than in the case of the Union 2 sample. In fact, using the Union 2 set we have verified the value $(\Delta\Omega_m)\mathrm{max}/\Omega_m = 0.43 \pm 0.06$ from [27]. Using the same procedure with the Loss sample we get $(\Delta\Omega_m)\mathrm{max}/\Omega_m = 0.30 \pm 0.06$.

The best fit values in each hemisphere are displayed in Table 2:

In agreement with the previous analysis using the Union 2 data set, the result using the hemisphere towards the maximum acceleration is consistent with ΛCDM, showing no more than a 0.1σ departure from it (see Fig. 3).

Our results are also in agreement with the previous case, the analysis with data from the hemisphere of minimum acceleration shows a sharp low redshift transition. In this case the best

	χ^2_{min}/dof	Ω_m	ω_0	ω_1
A	$197.28/213$	0.2 ± 1.1	-0.9 ± 0.9	0.2 ± 5.6
B	$374.76/367$	0.39 ± 0.07	-0.47 ± 0.43	-7.6 ± 6.2

Table 2. Best fit values of the cosmological parameters using the LOSS sample separating the sky in two hemispheres. Row A for the 216 SNIa in the maximum acceleration hemisphere. Row B the 370 SNIa in the opposite direction. See also Fig.3.

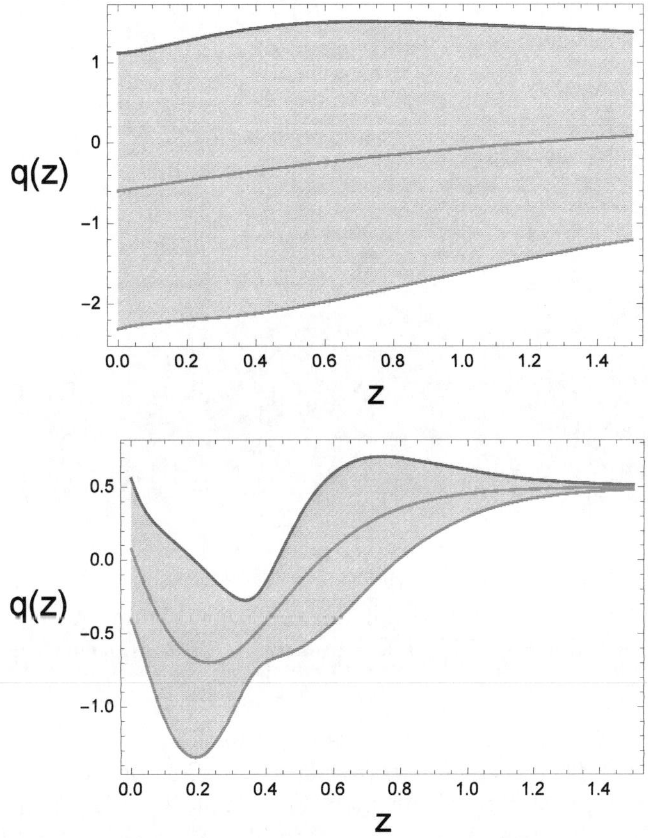

Figure 3. Reconstructed deceleration parameter $q(z)$ using data from two hemispheres using the Loss sample. The panels show the one around the maximum (upper) and minimum (lower) acceleration points respectively. The upper panel show the one around the maximum acceleration point, and the bottom one that around the minimum acceleration point. See Table 2 for the best fit values for each case.

fit is around 1.5σ away from the ΛCDM value.

It would be interesting to study this effect in the newest data set, as the JLA set [48] comprising more than 740 data points, and also using different parameterizations, as we have performed in [15]. These issues are under study.

There remains to check if this asymmetry in the behavior of the reconstructed deceleration parameter is not due to the anisotropic distribution of data. This issue as been discussed in the past. As we have mentioned before, using the Union 2 dataset, the authors of [27] found

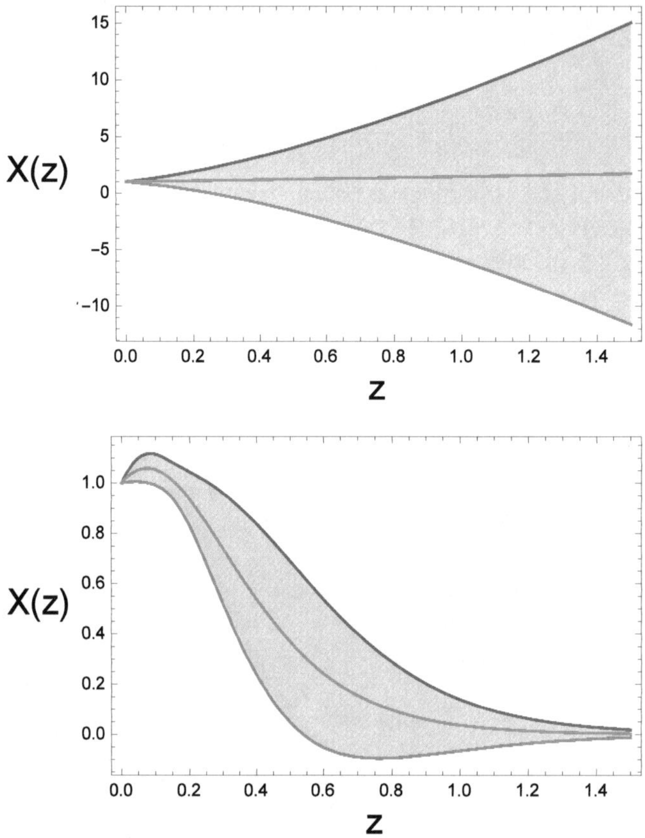

Figure 4. Here we show the reconstructed DE density $X(z)$ using data from two hemispheres. The upper panel showing the one around the maximum acceleration point, and the bottom the one in the opposite direction. See Table 2 for the best fit values for each case.

a preferred axis in the data, but they also checked if such anisotropy can be obtained using simulated isotropic data. Their method suggest that only a 30% of the simulated data can reach such anisotropy, implying that the anisotropy of the Union 2 set is consistent with statistical isotropy. However, the coincidence of the axes (a total of six phenomenological axes identified in [27]) within a small angular region in the sky, makes this dipolar anisotropy a feature that needs further scrutiny. Using the Loss sample, the authors of [31] found a hint for anisotropy – most pronounced in the range $0.015 < z < 0.045$ – but once they take into account large scale velocity perturbations, the results shows no evidence for any anomalous deviation from the isotropic ΛCDM. In [29] the authors studied this issue using the Union 2.1 and JLA set. Although they found that both datasets are statistically anisotropic, the correlation with the anisotropic distribution found turns out to be different: using the Union 2.1 they found a small correlation but using the JLA sample they found a stronger correlation. A similar study [37] using the JLA dataset also showed that by ignoring the velocity covariance may produce a hint of anisotropy from the data. In this context, it is interesting to notice here the result of the work [49] where the authors studied the JLA set, considering all the information from possible systematics (encoded in the covariance) in the analysis, finding a marginal (less than 3σ) evidence for the accepted cosmic acceleration.

5. Theoretical ideas

We have already mentioned a possible theoretical framework where this behavior of $q(z)$ and $X(z)$ can be observed. A model with a void towards one hemisphere. In fact, notice that Figures 2 and 4 shows that $X(z)$ decreases with z towards one hemisphere and into the other, the DE density is consistent with a constant lambda, as in the ΛCDM model. Actually, given the errors, this last solution is also consistent with *zero* cosmological constant for most of the range of the SNIa data.

Another way to get a decreasing DE density with redshift is considering an explicit interaction term between DE and DM. In the simplest case we have

$$\dot{\rho}_m + 3H\rho_m = Q, \tag{6}$$
$$\dot{\rho}_x + 3H(1+w_x)\rho_x = -Q, \tag{7}$$

where Q is the interaction and let us assume that $Q = 3H\alpha\rho_x$, as is usual in this kind of models. It is easy to find that $\rho_x(z) = \rho_{x0}(1+z)^{3(1+w_x+\alpha)}$ and

$$\rho_m(z) = \rho_{m0}(1+z)^3 + \frac{\alpha}{\alpha+w}\rho_{x0}\left[1-(1+z)^{3(w_x+\alpha)}\right](1+z)^3, \tag{8}$$

so finally the Hubble function can be written as

$$H^2 = H_0^2\left[\Omega_m(1+z)^3 + \Omega_x\frac{(1+z)^3}{\alpha+w_x}\left(\alpha + w_x(1+z)^{3(w_x+\alpha)}\right)\right]. \tag{9}$$

If someone give us this Hubble function, we could interpreted as the DE density the second term inside the square brackets,

$$X(z) = \frac{(1+z)^3}{\alpha+w_x}\left(\alpha + w_x(1+z)^{3(w_x+\alpha)}\right). \tag{10}$$

As the redshift increase, this function can decrease (even getting negative values after some redshift point) just by requiring that for example, $w_x = -0.9$ and $\alpha = 0.01$, which are typical values obtained from observations. In general, given a positive α (which ensure energy transfer from DE to DM) this behavior is found, sooner or later, for negative values of w_x, with a transition redshift $z_c = -1+(\alpha/|w_x|)^{(1/(\alpha-|w_x|))}$. However, it is not possible to get a deceleration parameter $q(z)$ – obtained from Eq.(5) using $H(z)$ from Eq. (9) – that shows the low redshift transition. This is very interesting, because this means that not every $X(z)$ that decrease with z implies a low redshift transition in their related $q(z)$.

6. Discussion

In this paper we have study the effects in the reconstructed deceleration parameter and dark energy density using the well known dipolar asymmetry present in SNIa data. We have used the Union 2 set and the result of [27], a dipolar asymmetry with a maximum acceleration towards $(l,b) = (309^o, 18^o)$ and a minimum acceleration towards $(l,b) = (129^o, -18^o)$ is found. The low reshift transition, previously recognized in [10, 11, 12, 13, 14, 15], appears only towards the direction of minimum acceleration, and it is absent towards the opposite direction. This result was obtained separating the data in two hemispheres constructed along the maximal asymmetry axis.

We have also performed a similar analysis using the Loss dataset [42] finding the results displayed in Fig. 3, which are in agreement with those using the Union 2 dataset (see Fig. 1). Assuming the result of [16] as our main hypothesis – i.e., a low redshift transition of $q(z)$ can be considered as a signal of a local under-density – our work indicates that the supernova data is able to detect such an under-density only towards one direction in the sky. This result suggest that a better description of the data would be a background inhomogeneous (because the dark energy density varies with z) and also with at least a dipolar asymmetry.

As we discussed in previous section, there remains to check if this asymmetry in the behavior of the reconstructed deceleration parameter is not due to the anisotropic distribution of data. However, it seems difficult to explain why all the evidence points towards the CMB dipole axes.

Altogether, the apparent detection of a local under density can be considered as a possible cause of the original tension between low and high redshift observational constraints discussed in [10, 11, 12, 13, 15], which as a solution also agree with the analysis of [19] where they proposed the existence of a local under-density to alleviate the tension in the determination of H_0.

Further study on this subject would be interesting. For example, the effects of using different parameterizations, and also the use of different SNIa data sets. Given that the SNIa data used in this work is obtained after the calibration of the whole set (by fitting globally also the α, β and M parameters, being the last one connected with the value of H_0) it would be interesting to study the impact of the hemispherical asymmetry on such a calibration, and then on the cosmological parameters. All these issues are under study.

Acknowledgments

The author want to thank V. Motta, C. Bernal and J. Magaña for useful discussions. This work was supported through DIUV project No. 50/2013.

[1] A. G. Riess et al., Astron. J. 116, 1009 (1998).
[2] S. Perlmutter et al., Astrophys. J. 517, 565 (1999).
[3] J. Frieman, M. Turner and D. Huterer, Ann. Rev. Astron. Astrophys. **46**, 385 (2008).
[4] J. Martin, Mod. Phys. Lett. A **23**, 1252 (2008) [arXiv:0803.4076 [astro-ph]].
[5] S. Tsujikawa, Lect. Notes Phys. **800**, 99 (2010) [arXiv:1101.0191 [gr-qc]].
[6] J.F. Pascual-Sanchez, Mod. Phys. Lett. A **14**, 1539 (1999).
[7] M.N. Celerier, Astron. Astrophys. **353**, 63 (2000).
[8] K. Bolejko, M.N. Celerier, & A. Krasinski, Classical and Quantum Gravity, **28**, 164002 (2011).
[9] V. Marra, & A. Notari, Classical and Quantum Gravity, **28**, 164004 (2011).
[10] Shafieloo A., Sahni V. and Starobinsky A.A., Phys. Rev. D 80, 101301, (2009).
[11] Z. Li , P. Wu and H. Yu, Phys. Lett. B 695, 1 , (2011).
[12] V.H. Cardenas and M. Rivera, Phys. Lett. B 710, 251 , (2011).
[13] V.H. Cardenas , C. Bernal and A. Bonilla, MNRAS, 433, 3534, (2013).
[14] V.H. Cardenas, Phys. Lett. B **750**, 128 (2015) [arXiv:1405.5116 [astro-ph.CO]].
[15] J. Magana, V.H. Cardenas and V. Motta, JCAP **1410**, no. 10, 017 (2014) [arXiv:1407.1632 [astro-ph.CO]].
[16] R.A. Vanderveld , E.E. Flanagan and I. Wasserman, Phys. Rev. D 74, 023506 , (2006).
[17] S. February , J. Larena , M. Smith and C. Clarkson, MNRAS, 405, 2231, (2010).
[18] L. Verde, P. Protopapas and R. Jimenez, Phys. Dark Univ. **2**, 166 (2013) [arXiv:1306.6766 [astro-ph.CO]].
[19] R. C. Keenan, A. J. Barger and L. L. Cowie, Astrophys. J. **775**, 62 (2013) [arXiv:1304.2884 [astro-ph.CO]].
[20] R. Watkins, H. A. Feldman and M. J. Hudson, Mon. Not. Roy. Astron. Soc. **392**, 743 (2009) [arXiv:0809.4041 [astro-ph]].
[21] A. Kashlinsky, F. Atrio-Barandela, D. Kocevski and H. Ebeling, Astrophys. J. **686**, L49 (2009) [arXiv:0809.3734 [astro-ph]].
[22] A. Kashlinsky, F. Atrio-Barandela, D. Kocevski and H. Ebeling, Astrophys. J. **691**, 1479 (2009) [arXiv:0809.3733 [astro-ph]].
[23] F. Atrio-Barandela, A. Kashlinsky, H. Ebeling, D. J. Fixsen and D. Kocevski, Astrophys. J. **810**, 143 (2015)
[24] U. Feindt *et al.*, Astron. Astrophys. **560**, A90 (2013) [arXiv:1310.4184 [astro-ph.CO]].
[25] Y. Akrami, Y. Fantaye, A. Shafieloo, H. K. Eriksen, F. K. Hansen, A. J. Banday and K. M. Grski, Astrophys. J. **784**, L42 (2014) [arXiv:1402.0870 [astro-ph.CO]].
[26] D. Hutsemékers, L. Braibant, V. Pelgrims, & D. Sluse, Astron. Astrophys. **572**, A18 (2014)
[27] I. Antoniou and L. Perivolaropoulos., JCAP 12, 012A (2010), [arXiv:1007.4347v3 [astro-ph.CO].
[28] R. Cai and Z. Tuo, JCAP 1202, 004 (2012).
[29] C. A. P. Bengaly, A. Bernui and J. S. Alcaniz, Astrophys. J. **808**, 39 (2015) [arXiv:1503.01413 [astro-ph.CO]].
[30] S. Appleby and A. Shafieloo, JCAP **1410**, no. 10, 070 (2014) [arXiv:1405.4595 [astro-ph.CO]].

[31] S. Appleby, A. Shafieloo and A. Johnson, Astrophys. J. **801**, no. 2, 76 (2015) [arXiv:1410.5562 [astro-ph.CO]].
[32] D. C. Dai, W. H. Kinney and D. Stojkovic, JCAP **1104**, 015 (2011) [arXiv:1102.0800 [astro-ph.CO]].
[33] J. Colin, R. Mohayaee, S. Sarkar and A. Shafieloo, Mon. Not. Roy. Astron. Soc. **414**, 264 (2011) [arXiv:1011.6292 [astro-ph.CO]].
[34] G. J. Mathews, B. Rose, P. Garnavich, D. Yamazaki and T. Kajino, arXiv:1412.1529 [astro-ph.CO].
[35] J. Beltran Jimenez, V. Salzano and R. Lazkoz, Phys. Lett. B **741**, 168 (2015) [arXiv:1402.1760 [astro-ph.CO]].
[36] B. Javanmardi, C. Porciani, P. Kroupa, & J. Pflamm-Altenburg, Astrophys. J. **810**, 47 (2015).
[37] D. Huterer, D. L. Shafer and F. Schmidt, arXiv:1509.04708 [astro-ph.CO].
[38] C. Bernal, V. H. Cardenas and V. Motta, Phys. Lett. B **765**, 163 (2017) [arXiv:1606.07333 [astro-ph.CO]].
[39] M. Hicken *et al.*, Astrophys. J. **700**, 1097 (2009) [arXiv:0901.4804 [astro-ph.CO]].
[40] R. Amanullah et al., ApJ, 716, 712, (2010).
[41] N. Suzuki, D. Rubin, C. Lidman, G. Aldering, R. Amanullah, K. Barbary, L. F. Barrientos and J. Botyanszki *et al.*, Astrophys. J. **746**, 85 (2012) [arXiv:1105.3470 [astro-ph.CO]].
[42] M. Ganeshalingam, W. Li, and A.V. Filippenko, MNRAS, 433, 2240 (2013).
[43] M. Kowalski *et al.* [Supernova Cosmology Project Collaboration], Astrophys. J. **686**, 749 (2008) [arXiv:0804.4142 [astro-ph]].
[44] M. Hicken, , W. M. Wood-Vasey, S. Blondin, P. Challis, S. Jha, P. L. Kelly, A. Rest and R. P. Kirshner Astrophys. J. 700, 1097H, (2009) [arXiv:0901.4804 [astro-ph.CO]]
[45] J. A. Holtzman et al. Astron. J. 136, 2306-2320, (2008) [arXiv:0908.4277 [astro-ph.CO]]
[46] Linder E. V., Phys. Rev. Lett. 90, 091301, (2003) [astro-ph/0311403].
[47] M. Chevallier and D. Polarski, Int. J. Mod. Phys. D **10**, 213 (2001) [gr-qc/0009008].
[48] M. Betoule *et al.* [SDSS Collaboration], Astron. Astrophys. **568**, A22 (2014) [arXiv:1401.4064 [astro-ph.CO]].
[49] J. T. Nielsen, A. Guffanti and S. Sarkar, Scientific Reports, 6, 35596, (2016) [arXiv:1506.01354 [astro-ph.CO]].

IOP Conf. Series: Journal of Physics: Conf. Series **1043** (2018) 012020 doi:10.1088/1742-6596/1043/1/012020

Quantum phase transition of the static extremal black hole coupled to a Stückelberg scalar

Paula Quezada Leon, Olivera Miskovic

Abstract. We show that a static extremal black hole in 4D gravity, non-linearly coupled to a massive Stückelberg scalar, can exhibit a quantum phase transition due to electric charge variations. The critical point exists only if the spacetime possesses cosmological constant (positive or negative) and the mass of the scalar is non-vanishing. In that case, around the critical point, two solutions coexist: Reissner-Nordström (A)dS black hole and the hairy (A)dS black hole. A near-critical expansion reveals that the entropy and its derivative are continuous functions at the critical point. Obtained results are analytical and based on the entropy function formalism.

Instituto de Física, Pontificia Universidad Católica de Valparaíso, Casilla 4059, Valparaíso.

E-mail: pquezada.l@gmail.com, olivera.miskovic@pucv.cl

1. Introduction

An equilibrium state of a thermodynamic system corresponds to a minimum of the internal energy in the energy representation of states, or a maximum of the entropy in the entropy representation of states. The latter one is suitable for studying equilibria of extremal black holes because they have zero temperature and are not described by conventional thermodynamics. The entropy of extremal black hole arises due to a degenerate quantum ground state. This phenomenon is known in the physics of condensed matter (spin glasses).

The macroscopic entropy of extremal black hole can be calculated using the entropy function formalism [1, 2]. It is based on a variational principle applied to a generic class of entropy functions of the charges, scalar fields and the parameters of the near-horizon geometry, $AdS_2 \times \mathbb{S}^{D-2}$. The formalism can be generalized to other extremal geometries, such as warped ones [3]. Extremization of the entropy function determines all the near-horizon parameters, enabling to extract the information about the black holes without knowledge of a particular solution.

On the other hand, phase transitions due to thermal or, in our case, quantum fluctuations, are related to instabilities of the system around a critical point. In particular, it has been noticed that a massless scalar field produces an instability at the horizon of an extreme Reissner-Nordström black hole [4] and that the axisymmetric extremal horizons are unstable under linear scalar perturbations [5]. Similar instability also occurs for a massive scalar field [6]. Recently, these instabilities were studied from the extremal limit point of view through analysis of charged scalar perturbations for Reissner-Nordström and Kerr solutions [7].

We are interested in studying quantum phase transitions of a 4D charged black hole in General Relativity, when it is non-minimally coupled to a Stückelberg scalar [8]. Namely, non-linear Stückelberg interaction has been known to describe both first and second order thermal phase transitions [9, 10]. A question we address is whether a similar change would also occur at the zero temperature.

2. Entopy of the extremal black hole

The extremal black hole is the smallest mass black hole for a given electric charge and angular momentum. Geometrically, it has at least two horizons that overlap, what implies that its near-horizon geometry has the topology $\mathbb{H} \simeq \mathrm{AdS}_2 \times \mathbb{S}^2$ in four space-time dimensions, where AdS_2 is two-dimensional anti-de Sitter space and $\mathbb{S}^2$ is a two-sphere. Let v_1 and v_2 be the radii of the AdS_2 and 2-sphere, respectively. Then the horizon of the black hole is described by the following metric in the spherical coordinates $x^\mu = (t, r, \theta, \varphi)$,

$$\left. ds^2 \right|_{\mathbb{H}} = g_{\mu\nu}(x)\, dx^\mu dx^\nu = v_1 \left(-r^2\, dt^2 + \frac{dr^2}{r^2} \right) + v_2 \left(d\theta^2 + \sin^2\theta\, d\varphi^2 \right) , \tag{1}$$

where r is a radial distance from the horizon.

The metric (1) corresponds to a hear-horizon limit of some *static* black hole coupled to the electromagnetic field $A_\mu(x)$ and the scalar field $\phi(x)$, described by the action

$$I = \int d^4x \sqrt{-g}\, \mathcal{L}(g, A, \phi) \,. \tag{2}$$

Near the horizon, the metric behaves as (1), the electric field is e and the scalar field, due to the attractor mechanism, depends only on it asymptotic value u. Thus, on the horizon, the fields are represented by a set of parameters

$$\mathbb{H}: \qquad g_{\mu\nu} \to (v_1\,,\, v_2)\,, \quad A_\mu \to e\,, \quad \phi \to u\,. \tag{3}$$

The action (2) evaluated on the horizon is given by an auxiliary function

$$f(v, e, u) = \int\limits_{\mathbb{H}} d\theta d\phi \sqrt{-g}\, \mathcal{L}(v, e, u)\,, \tag{4}$$

which satisfies the action principle, that is, has an extremum on the equations of motion, for given boundary conditions. The entropy function $\mathcal{E}(v, e, u)$ is a Legendre transformation of the function f with respect to the electric field,

$$\mathcal{E}(v, e, u) = 2\pi \left[eq - f(v, e, u) \right] , \tag{5}$$

where q is the asymptotic electric charge. The parameters near the horizon are calculated as an extremum of the entropy function [1, 2],

$$\frac{\partial \mathcal{E}}{\partial v_i} = 0 \quad (i = 1, 2)\,, \qquad \frac{\partial \mathcal{E}}{\partial u} = 0\,, \qquad \frac{\partial \mathcal{E}}{\partial e} = 0\,, \tag{6}$$

and the black hole entropy S is its extremal value,

$$S = \mathcal{E}_{\mathrm{extr}}\,. \tag{7}$$

Therefore, finding the entropy function $\mathcal{E}(v, e, u)$ and its maximum, one can calculate the entropy, electric field and AdS_2 and $\mathbb{S}^2$ radii of the extremal black hole, independently on a particular static solution considered.

IOP Conf. Series: Journal of Physics: Conf. Series **1043** (2018) 012020 doi:10.1088/1742-6596/1043/1/012020

3. Black hole coupled to a Stückelberg scalar

Consider General Relativity with a cosmological constant Λ coupled to electromagnetic and scalar fields,

$$I=\int d^4x\sqrt{-g}\left\{\frac{1}{16\pi G_N}(R-2\Lambda)-\frac{1}{4}F_{\mu\nu}F^{\mu\nu}-\frac{1}{2}\left[(\partial\phi)^2+m^2\phi^2+\mathcal{F}(\phi)(\partial\sigma-A)^2\right]\right\}, \quad (8)$$

where $R=g^{\mu\nu}R^{\alpha}{}_{\mu\alpha\nu}$ is the scalar curvature and $F_{\mu\nu}=\partial_\mu A_\nu-\partial_\nu A_\mu$ is the EM field strength. Stückelberg complex scalar $\hat{\phi}(x)=\phi(x)\,e^{i\sigma(x)}$ (with ϕ and σ real) is minimally coupled when the Stückelber function is quadratic, $\mathcal{F}(\phi)=\phi^2$. Non-minimal interaction has to satisfy $\mathcal{F}(\phi)>0$, $\phi\neq 0$, and $\mathcal{F}(0)=0$ [8]. Such couplings preserve an $U(1)$ invariance and lead to first and second order phase transitions in non-extremal cases [9, 10]. In the extremal case, non-linear terms in $\mathcal{F}$ might produce instabilities at the horizon [4], so we choose

$$\mathcal{F}(\phi)=\phi^2-\frac{a}{4}\phi^4\geq 0\,, \quad (9)$$

where a is some coupling constant of dimension (lenghth)^2 in natural units.

On the horizon $\mathbb{H}$, spherically symmetric static field configurations are replaced by four parameters (v_1,v_2,e,u), where the electric field $F_{rt}=e$ can be obtained from the electric potential $A_\mu=\delta^t_\mu\,er$. The Lagrangian density on $\mathbb{H}$ reads

$$\mathcal{L}=\frac{1}{8\pi G_N}\left(\frac{1}{v_2}-\frac{1}{v_1}-\Lambda\right)+\frac{e^2}{2v_1^2}-\frac{1}{2}\left(m^2u^2-\frac{e^2}{v_1}\mathcal{F}(u)\right), \quad (10)$$

where for the scalar field we assumed that its value on the horizon is $\phi=u$, $\partial_\mu\phi=0$. Furthermore, the equation of motion of the field $\sigma(x)$ is not independent on other equations due to a $U(1)$ gauge freedom, so it can be gauge-fixed to $\sigma=0$.

Gathering all the terms together, the auxiliary function $f=\int d\theta d\phi\, v_1v_2\sin\theta\,\mathcal{L}$ is obtained using Eq.(10) and the entropy function (5) reads

$$\mathcal{E}=2\pi eq-8\pi^2v_1v_2\left[\frac{1}{8\pi G_N}\left(\frac{1}{v_2}-\frac{1}{v_1}-\Lambda\right)+\frac{e^2}{2v_1^2}-\frac{1}{2}\left(m^2u^2-\frac{e^2}{v_1}\mathcal{F}(u)\right)\right]. \quad (11)$$

An extremum of the above function gives rise to the following equations of motion,

$$\begin{aligned}
0&=\frac{\partial\mathcal{E}}{\partial v_1} &\Rightarrow\quad & 1-\Lambda v_2=v_2\left(\frac{e^2}{v_1^2}+m^2u^2\right),\\
0&=\frac{\partial\mathcal{E}}{\partial v_2} &\Rightarrow\quad & 1+\Lambda v_1=\frac{e^2}{v_1}-v_1\,m^2u^2+e^2\mathcal{F}(u)\,,\\
0&=\frac{\partial\mathcal{E}}{\partial e} &\Rightarrow\quad & Q=v_2e\left(\frac{1}{v_1}+\mathcal{F}(u)\right),\\
0&=\frac{\partial\mathcal{E}}{\partial u} &\Rightarrow\quad & 0=v_2\left(2v_1\,m^2u-e^2\frac{d\mathcal{F}}{du}\right).
\end{aligned} \quad (12)$$

We set $4\pi G_N=1$ for the sake of simplicity and also denote $Q=\frac{q}{4\pi}$. Eqs.(12) are invariant under the reflection $(e,Q)\rightarrow(-e,-Q)$, and e and Q have the same sign (see third equation) so we can choose, without loss of generality, e, $Q>0$.

Equations (12) have different branches of solutions that have to be discussed independently.

4. Entropy of a non-hairy black hole

$u = 0$ is always a particular solution of the scalar equation in (12), thus we first focus to that case. A general solution for fixed Q exists provided $\Delta = 1 - 4\Lambda Q^2 > 0$, and it reads

$$\begin{aligned} v_1^{\pm}(Q) &= \frac{2Q^2}{1 - 4\Lambda Q^2 \pm \sqrt{1 - 4\Lambda Q^2}}\,, \\ v_2^{\pm}(Q) &= \frac{2Q^2}{1 \pm \sqrt{1 - 4\Lambda Q^2}}\,, \\ e^{\pm}(Q) &= Q\,\frac{1 \pm \sqrt{1 - 4\Lambda Q^2}}{1 - 4\Lambda Q^2 \pm \sqrt{1 - 4\Lambda Q^2}}\,. \end{aligned} \tag{13}$$

Other inequalities that have to be fulfilled are

$$\frac{1}{v_1} + \Lambda > 0\,, \qquad 2\Lambda + \frac{1}{v_1} > 0\,, \tag{14}$$

what is satisfied if $1 \pm \sqrt{\Delta} > 0$. We conclude:

(i) When $\Lambda < \frac{1}{4Q^2}$, there is only one black hole solution, given by Eqs.(13) with the sign '+';

(ii) When $0 < \Lambda < \frac{1}{4Q^2}$, there are two solutions, given by Eqs.(13) with the signs '+' and '−'.

(iii) When $\Lambda = 1/4Q^2$, there is no a finite solution for v_1.

(iv) The case $\Lambda = 0$ can be reproduced from the limit $\Lambda \to 0$ of the positive branch of the solution (13).

Under these conditions, the extremum of the entropy function (11) for the fixed charge and the values (13) of the black hole parameters is

$$S_{\pm}(Q) = \frac{8\pi^2 Q^2}{1 \pm \sqrt{1 - 4\Lambda Q^2}}\,. \tag{15}$$

Note that $S_+(Q) > 0$ is always fulfilled while, for the negative branch, $S_-(Q) > 0$ is true only for the negative values of the cosmological constant. Thus, spaces with positive cosmological constant have only one solution, corresponding to the positive branch in (13).

When $\Lambda = 0$, then the positive branch reproduces the well-known result

$$S_+(Q)|_{\Lambda=0} = 4\pi^2 Q^2 = \frac{q^2}{4}\,. \tag{16}$$

5. Hairy extremal black hole and the critical point

For the study of phase transitions, the most interesting cases involve non-linear Stückelberg interaction $\mathcal{F}$ of the form (9). The last equation in (12) implies that, for the minimal coupling $a = 0$, the only solution for the scalar field is $u = 0$, for which the black hole entropy was calculated in the previous section. When $a \neq 0$, then there are three solutions for the scalar parameter,

$$u = 0\,, \qquad u = \pm\sqrt{\frac{2}{a}\left(1 - \frac{v_1 m^2}{e^2}\right)}\,. \tag{17}$$

The case $u = 0$ was analyzed in Section 4. When $u \neq 0$, the equations of motion (17) are invariant under the replacement $u \to -u$, so we can chose $u > 0$.

Now we address the following question: Is there a critical point for which two solutions $u = 0$ and $u \geq 0$ co-exist? If the answer is yes, this would be a possible branch point where a quantum phase transition from one configuration to another might occur.

To answer this question, let us analyze the critical point limit $u \to 0$ of the non-trivial solution in (17). The critical point exists only for the massive fields ($m \neq 0$). Then Eq.(17) implies

$$v_{1c} = \frac{e_c^2}{m^2}\,. \tag{18}$$

First two equations of (12) give that the critical point exist only for $\Lambda \neq 0$ and $m^2 \neq 1, \frac{1}{2}$, leading to

$$\begin{aligned} v_{1c} &= \frac{m^2-1}{\Lambda}\,, \\ e_c &= \sqrt{\frac{m^2\left(m^2-1\right)}{\Lambda}}\,, \\ v_{2c} &= \frac{m^2-1}{\Lambda\left(2m^2-1\right)}\,. \end{aligned} \tag{19}$$

The result does not depend on the strength of the scalar coupling $a \neq 0$. Non-vanishing and positive v_{ic} and e_c exist only if

$$m^2 > \frac{1}{2}\,, \qquad \frac{m^2-1}{\Lambda} > 0\,. \tag{20}$$

Note that the above conditions have to be satisfied for both positive and negative values of the cosmological constant. In addition, in case of negative cosmological constant, the Breitenlohner-Freedman bound that ensures the stability of scalar field in AdS_4 imposes $m^2 > \frac{3\Lambda}{4}$ to the scalar mass, what is weaker than the inequalities (20), so it is always satisfied.

The last unsolved equation (third in Eq.(12)) shows that the critical values of the parameters can be reached only for the critical charge

$$Q_c = \frac{v_{2c}e_c}{v_{1c}} = \frac{1}{2m^2-1}\sqrt{\frac{m^2\left(m^2-1\right)}{\Lambda}}\,. \tag{21}$$

Compared to the original two $u = 0$ solutions given by Eqs.(13) where Q is replaced by Q_c we find that, for the positive branch '+', we reproduce the known critical results of the parameters,

$$v_i^+(Q_c) = v_{ic}\,, \qquad e^+(Q_c) = e_c\,, \tag{22}$$

whereas the branch with the sign '−' is inconsistent because it gives negative values of parameters. The critical entropy reads

$$S_c = \frac{4\pi^2\left(m^2-1\right)}{\Lambda\left(2m^2-1\right)} > 0\,, \tag{23}$$

and it is always positive. Thus, the entropy is a continuous function at Q_c because

$$S_+(Q_c) = S_c\,. \tag{24}$$

When $\Lambda = 0$, then $m^2 = 1$ (and vice versa), and the solution is $v_{1c} = v_{2c} = Q^2$, $e_c = Q$ with the charge Q which remains arbitrary. The case $m^2 = \frac{1}{2}$ does not have solutions.

We will discuss only the spacetimes with $\Lambda \neq 0$ when the critical point is well-defined and analyze a behavior of the hairy black hole in its vicinity. We will also explore whether the quantum phase transition would occur in this point using the second thermodynamics law ($dS > 0$).

6. Near-critical behaviour of the entropy

Knowing the critical point given by Eqs.(19) and (21), we study a near-critical behavior of the solution by introducing a small parameter

$$\epsilon = Q - Q_c\,, \tag{25}$$

where ϵ can be either positive or negative. We assume that $u \to 0$ when $Q \to Q_c$ so that, near Q_c, it behaves as

$$u^2 = A\,\epsilon^\beta + \cdots > 0\,, \qquad \beta > 0\,, \tag{26}$$

where β is a critical exponent. We also seek for the expansions

$$\begin{aligned} e &= e_c + B\,\epsilon^\beta + \cdots\,, \\ v_2 &= v_{2c} + C\,\epsilon^\beta + \cdots\,, \\ v_1 &= v_{1c} + V\,\epsilon^\beta + \cdots\,. \end{aligned} \tag{27}$$

Expanding the third equation in (12), all finite terms cancel out since they correspond to the critical values and, at the lowest order of ϵ, we obtain

$$\epsilon = \left(e_c v_{2c}\,A + \frac{v_{2c}}{v_{1c}}\,B + \frac{e_c}{v_{1c}}\,C - \frac{e_c v_{2c}}{v_{1c}^2}\,V \right) \epsilon^\beta + \cdots\,, \tag{28}$$

leading to the critical exponent $\beta = 1$. Thus, the order parameter shows a typical mean-field behavior near the critical point, namely

$$u = \sqrt{A\,(Q - Q_c)} + \mathcal{O}((Q - Q_c)^2)\,. \tag{29}$$

With this result at hand, all equations at the linear order become

$$\begin{aligned} 0 &= v_{1c}v_{2c}e_c^2\,A + 2e_c v_{2c}\,B + \left(\Lambda v_{1c} + m^2\right) v_{1c}\,C - 2m^2 v_{2c}\,V\,, \\ 0 &= 2e_c\,B - \left(\Lambda + m^2\right)\,V\,, \\ v_{1c}^2 &= e_c v_{1c}^2 v_{2c}\,A + v_{1c}v_{2c}\,B + v_{1c}e_c\,C - e_c v_{2c}\,V\,, \\ 0 &= ae_c^2\,A - 4e_c B + 2m^2 V\,, \end{aligned} \tag{30}$$

where Eq.(18) was used to cancel out some terms. The above system is algebraic in A, B, C and V, with a general solution

$$\begin{aligned} A &= \sqrt{\frac{m^2-1}{\Lambda m^2}}\,\frac{4\Lambda^2\left(2m^2-1\right)^2}{4\left(m^2-1\right)^3 + \Lambda a}\,, \\ B &= \frac{\Lambda a\,\left(2m^2-1\right)^3}{4\left(m^2-1\right)^3 + \Lambda a}\,, \\ C &= \sqrt{\frac{m^2\left(m^2-1\right)}{\Lambda}}\,\frac{2\left[-2\left(m^2-1\right)^2 + \Lambda a\right]}{4\left(m^2-1\right)^3 + \Lambda a}\,, \\ V &= \sqrt{\frac{m^2\left(m^2-1\right)}{\Lambda}}\,\frac{2\Lambda a\,\left(2m^2-1\right)^2}{4\left(m^2-1\right)^3 + \Lambda a}\,. \end{aligned} \tag{31}$$

Replacing the above solutions in the expression for the entropy function (11), we obtain the entropy near the critical point in the form

$$S|_{u\neq 0} = S_c + 8\pi^2 e_c\,(Q - Q_c) + \mathcal{O}((Q - Q_c)^2)\,. \tag{32}$$

We conclude that, near the critical value of the charge Q_c, there are two extremal black hole solutions – the hairy one whose entropy is given by Eq.(32)., and a non-hairy one (Reissner-Nordström (A)dS) with the entropy given by $S_+(Q)$ in Eq.(15). When expanded around the critical point, the non-hairy solution becomes

$$\begin{aligned} S|_{u=0} &= S_+(Q) = S_c + \left.\frac{dS_+}{dQ}\right|_{Q_c} (Q - Q_c) + \mathcal{O}((Q - Q_c)^2) \\ &= S_c + 8\pi^2 e_c\,(Q - Q_c) + \mathcal{O}((Q - Q_c)^2)\,, \end{aligned} \quad (33)$$

what exactly matches the expression (32) up to quadratic terms. Surprisingly, two different solutions that extremize the entropy, $u = 0$ and $u > 0$, glue almost smoothly at the critical point such that both S and its first derivative are continuous functions around Q_c. This indicates that the hair develops in the black hole in such a way that there is a quantum phase transition of the order higher than two. Higher-power terms in the expansions (32) and (33) are needed to determine the exact order of the phase transition.

7. Conclusions

We studied quantum phase transitions of black holes in 4D at zero temperature that arise due to variations of electric charge. We showed analytically that the static extremal black hole coupled to the Stückelberg scalar possesses two different solutions, one with scalar hair and another without it, which indicates that a phase transition might be possible at some critical electric charge. If so, a change in the phases is such that S and $\partial S/\partial Q$ are continuous at the critical point, so the discontinuity should appear at higher order. This type of a phase transition would be possible only for spacetimes with non-vanishing cosmological constant and non-vanishing mass of the scalar, which interact non-linearly with gravitational and electromagnetic fields. Higher-power terms in the expansions of two branches of solutions are necessary to gain more insight in the process.

One should also include in this study the AdS black holes with planar and hyperbolic horizons, magnetic charge with the magnetic field and ungauged U(1) freedom through the field $\sigma(x)$. Another open question is how to embed the model in the framework of supergravity.

Acknowledgments

The authors thank Alessio Marani for useful discussions about the SUGRA context of the results. This work was supported by the VRIEA-PUCV grant N°039.345/2016.

References

[1] A. Sen 2005, J. High Energy Phys. **0509**, 038.
[2] A. Sen 2006, J. High Energy Phys. **0603**, 008.
[3] D. Astefanesei, O. Miskovic and R. Olea 2012, Phys. Lett. B **714**, 331.
[4] S. Aretakis 2013, Phys. Rev. D **87**, 084052.
[5] S. Aretakis 2015, Adv. Theor. Math. Phys. **19**, 507.
[6] J. Lucietti, K. Murata and H.S. Reall 2013, J. High Energy Phys. **1303**, 035.
[7] P. Zimmerman, *Horizon instability of extremal black holes to charged perturbations*, arXiv:1612.03172 [gr-qc].
[8] E. C. G. Stückelberg 1938, Helv. Phys. Acta **11**, 299. (in German)
[9] S. A. Hartnoll, C. P. Herzog and G. T. Horowitz 2008, J. High Energy Phys. **0812**, 015.
[10] S. Franco, A. Garcia-Garcia and D. Rodriguez-Gomez 2010, J. High Energy Phys. **1004**, 092.

IOP Conf. Series: Journal of Physics: Conf. Series **1043** (2018) 012021 doi:10.1088/1742-6596/1043/1/012021

Tunneling and the Emergent Universe scheme in a JBD Theory

Hobby Cossio and Pedro Labraña

Departamento de Física, Universidad del Bío-Bío and Grupo de Cosmología y Gravitación-UBB, Casilla 5-C, Concepción, Chile.

E-mail: hobbycosio@ubiobio.cl

Abstract.
In this work it is proposed an alternative scheme for an Emergent Universe scenario in the context of a Jordan-Brans-Dicke theory, where the universe is initially in a static state supported by a scalar field located in a false vacuum. The universe begins to evolve when, by quantum tunneling, the scalar field decays into a state of true vacuum.

In particular, in this work we study the stability of the static universe, under the aforementioned characteristics, and the probability of creating a true vacuum bubble. It is shown that in this model, unlike what happens in previous works, the static universe is stable under small classical perturbations.

1. Introduction

The standard cosmological model (SCM) is shown as a satisfactory description of our universe [1, 2, 3]. However, despite its great success, the SCM presents several problems related to its initial conditions at the time of the Big Bang which can be solved through the scenario of Inflationary Universe [4, 5, 6, 7]. Independent of the success of SCM and the inflationary paradigm to explain the cosmological data there are still important open questions to be answered. One of these questions is whether the universe had a definite origin, characterized by an initial singularity or if, on the contrary, it did not have a beginning, that is, it extends infinitely to the past. Theorems about space-time singularities have been developed in the context of inflationary universes, proving that the universe necessarily has a beginning. In other words, according to these theorems, the existence of an initial singularity can not be avoided even if the inflationary period occurs. We can refer to the BGV Theorem. This theorem was developed in 2003 by Arvind Borde, Alan Guth and Alex Vilenkin [8].Vilenkin's arguments demonstrate that null and time-like geodesics are generally incomplete in inflationary models, regardless of whether energy conditions are maintained, provided that the average expansion condition ($H > 0$) is maintained throughout of these geodesics directed towards the past. The search for cosmological models without initial singularities has led to the development of emergent universes models (UE) [9, 10, 11, 12, 13, 14, 15, 16]. However, most simple models developed in a context of General Relativity suffer from instabilities, associated with the instability of Einstein's static universe [17].

On the other hand, Tensor-Scalar theories can give us the possibility to explore the viability of a stable UE model in which there is no initial singularity [19, 18]. The theory normally associated

with Scalar-Tensor models is the Jordan-Brans-Dicke (JBD) theory of gravitation [20, 21]. In this work we propose a scheme for an EU scenario, where the universe is initially in a stable static state. This state is supported by a scalar field located in a false vacuum. The universe begins to evolve when, by quantum tunneling, the scalar field decays to a state of true vacuum. Then, a bubble of a new phase is formed and expands as it converts the energy of vacuum and feeds the energy released to the kinetic energy of the bubble wall.
This is a preliminary report of our results in this respect, where we focus on the study of the existence of a static universe solution and its stability conditions. We also calculate the probability of creation of the bubble in this context, leaving for later the report of the study of the classic evolution of the bubble once materialized [22].

2. Emergent Universes on Jordan-Brans-Dicke Theories

In this work we consider a scheme for EU scenario where the universe is initially in a stable static state. This state is supported by a scalar field located in a false vacuum. The universe begins to evolve when, by quantum tunneling, the scalar field decays to a state of true vacuum. For this reason we will start by studying the possibility of obtaining a static and stable solution in this theory when the scalar field is in the false vacuum. Following the EU scheme we consider a closed Friedmann-Robertson-Walker metric:

$$ds^2 = dt^2 - a(t)^2 \left[\frac{dt^2}{1-r^2} + r^2 \left(d\theta^2 + \sin^2\theta d\varphi^2\right)\right], \tag{1}$$

where $a(t)$ is the scale factor and t represents the cosmological time. The matter content is modeled by a perfect fluid with an effective state equation given by $P_f = (\gamma - 1)\,\rho_f$, with γ constant, and a scalar field (inflaton) for which

$$P_\psi = \frac{1}{2}\dot{\psi}^2 - U(\psi), \quad \rho_\psi = \frac{1}{2}\dot{\psi}^2 + U(\psi). \tag{2}$$

By assuming that the momentum energy tensor of ordinary matter takes the form corresponding to a perfect fluid, we obtain the field equations of our model.

$$H^2 + \frac{1}{a^2} + H\frac{\dot{\phi}}{\phi} = \frac{\rho}{3\phi} + \frac{\omega}{6}\left(\frac{\dot{\phi}}{\phi}\right)^2 + \frac{V}{3\phi}. \tag{3}$$

Using the Ricci's scalar $R = 6\left(\dot{H} + 2H^2 + 1/a^2\right)$ we have

$$2\frac{\ddot{a}}{a} + H^2 + \frac{1}{a^2} + \frac{\ddot{\phi}}{\phi} + 2H\frac{\dot{\phi}}{\phi} + \frac{\omega}{2}\left(\frac{\dot{\phi}}{\phi}\right)^2 - \frac{V}{\phi} = -\frac{P}{\phi}, \tag{4}$$

and the field equation for the JBD field is

$$\ddot{\phi} + 3H\dot{\phi} = \frac{\rho - 3P}{2\omega + 3} + \frac{2}{2\omega + 3}\left[2V - \phi V'\right], \tag{5}$$

where $V' = dV(\phi)/d\phi$, $\rho = \rho_f + \rho_\psi$ and $P = P_f + P_\psi$. The points mean derivatives with respect to cosmological time. In the context of JBD theory solutions are closed universes characterized by the conditions $a = a_0 = Const.$, $\dot{a}_0 = \ddot{a}_0 = 0$ y $\phi = \phi_0 = Const.$, $\dot{\phi}_0 = \ddot{\phi}_0 = 0$. Then by replacing these conditions in the field equations (3), (4) and (5) the static solution for a universe dominated by a general perfect fluid is obtained if the following conditions are satisfied:

$$a_0^2 = \frac{3}{V_0'}, \tag{6}$$

$$\rho_0 = -U_0 - V_0 + V_0'\phi_0, \tag{7}$$

and

$$\gamma = 2\frac{\phi_0}{a_0^2\rho_0}, \tag{8}$$

where $V_0 = V(\phi_0)$ y $V_0' = (dV(\phi)/d\phi)_{\phi=\phi_0}$, $\rho_{f0} = \rho_0$ y $\rho_{\psi 0} = U_0$. These equations connect the equilibrium values of the scale factor and the JBD field with the energy density and the JBD potential at the equilibrium point. We now study the stability of this solution against small homogeneous and isotropic perturbations. In order to do this, we consider small perturbations around the static solution for the scaling factor and the JBD field. We set

$$a(t) = a_0\left[1 + \alpha(t)\right], \tag{9}$$

and

$$\phi(t) = \phi_0\left[1 + \beta(t)\right]. \tag{10}$$

Then, we have

$$\rho(t) = \rho_0 + \delta\rho \approx \rho_0 - 3\gamma\rho_0\alpha(t), \tag{11}$$

where $\alpha \ll 1$ y $\beta \ll 1$ are small perturbations.
By introducing the expressions (9), (10) and (11) into equations (4) and (5), and retaining terms at a linear order in α and β, we obtain the following coupled equations

$$\ddot{\alpha} - \left[\frac{1}{a_0^2} + 3\frac{(\gamma-1)}{a_0^2}\right]\alpha - \frac{\ddot{\beta}}{2} - \frac{\beta}{a_0^2} = 0, \tag{12}$$

and

$$(3+2\omega)\ddot{\beta} - \left(\frac{6}{a_0^2} - 2\phi_0 V_0''\right)\beta + (4-3\gamma)\frac{6}{a_0^2}\alpha = 0, \tag{13}$$

where $V_0'' = \left(d^2V(\phi)/d\phi^2\right)_{\phi=\phi_0}$.
From this system of coupled differential equations we can obtain the frequencies for small oscillations

$$w_\pm^2 = \frac{1}{a_0^2(3+2\omega)}\left[a_0^2\phi_0 V_0'' - 6 + \omega(2-3\gamma)\right. \tag{14}$$

$$\left.\pm\sqrt{\left[-6 + a_0^2\phi_0 V_0'' + 2\omega - 3\omega\gamma\right]^2 + 2(3+2\omega)\left(-6 + a_0^2\phi_0 V_0''\left[3\gamma-2\right]\right)}\right]. \tag{15}$$

The static solution is stable if $w_\pm^2 > 0$. Then assuming that the parameter ω satisfies the constraint, $(3+2\omega) > 0$, we find that the following inequalities must be fulfilled in order to have a stable static solution

$$\frac{2}{3} < \gamma < \frac{4}{3}, \tag{16}$$

$$-\frac{3}{2} < \omega < -18\frac{(\gamma-1)}{(2-3\gamma)^2}, \tag{17}$$

and

$$2(6+\omega) - 3(3+\omega)\gamma + \sqrt{3}\,|4-3\gamma|\sqrt{3+2\omega} < a_0^2\phi_0 V_0'' < \frac{6}{3\gamma-2}. \tag{18}$$

From these inequalities we can conclude that for a universe dominated by a perfect standard fluid (with $\gamma > 2/3$) and a scalar field localized in a false vacuum, it is possible to find a solution where the universe is static and stable.

IOP Conf. Series: Journal of Physics: Conf. Series **1043** (2018) 012021 doi:10.1088/1742-6596/1043/1/012021

3. Bubble Nucleation

We calculate the probability of creation of the bubble in a context of JBD Theory. The calculation of the decay of the false vacuum was solved by Coleman and Callan [23, 24] and the incorporation of gravitational effects was studied by Coleman and De Luccia [25]. However, the latter analysis is not applicable here because of the non-standard gravitational action. In order to avoid these difficulties, we reformulate the theory by perform a Weyl rescaling in order to remove the coupling of ϕ with $\mathcal{R}$. We consider the following action

$$S = \frac{1}{16\pi}\int d^4x\sqrt{-g}\left[\phi\mathcal{R} - \frac{\omega}{\phi}\nabla_\mu\phi\nabla^\mu\phi + V(\phi) + \mathcal{L}_m\right]. \tag{19}$$

In particular, we define the new fields $\tilde{g}$ y $\tilde{\phi}$ as

$$g_{\mu\nu} = \Omega^{-2}(x)\tilde{g}_{\mu\nu}, \tag{20}$$

$$\tilde{\phi} = \sqrt{\frac{2\omega+3}{16\pi G}}\ln\left(\phi/\phi_0\right), \tag{21}$$

where $\phi_0^{-1} = G$, here G is the value of the Newton constant observed in the present. We choose the conform factor $\Omega(x)$ so that $\Omega = \sqrt{G\phi}$. Finally we see that the action in the rescaled theory can be expressed as:

$$S[\tilde{g},\tilde{\phi},\psi] = \int d^4x\sqrt{-\tilde{g}}\left[\frac{1}{16\pi G}\tilde{\mathcal{R}} - \frac{1}{2}\tilde{g}^{\mu\nu}\tilde{\nabla}_\mu\tilde{\phi}\tilde{\nabla}_\mu\tilde{\phi} + W(\tilde{\phi})\right. \tag{22}$$

$$\left. + \exp\left(-4\sqrt{\frac{\pi G}{2\omega+3}}\tilde{\phi}\right)\frac{1}{2}\tilde{g}^{\mu\nu}\nabla_\mu\psi\nabla_\mu\psi - \exp\left(-8\sqrt{\frac{\pi G}{2\omega+3}}\tilde{\phi}\right)U(\psi)\right]. \tag{23}$$

We note that we have a model where we removed the non-standard coupling between ϕ and $\mathcal{R}$, where the gravitational coupling constant is independent of time, also have a minimally coupled field $\tilde{\phi}$, And an inflaton field now coupled to $\tilde{\phi}$.

Now that we have removed the coupling of ϕ to $\mathcal{R}$ we can calculate the nucleation rate in this rescaled theory. Due to Weyl's rescaling the two line elements $d\bar{s}^2 = d\bar{t}^2 - \bar{a}^2(\bar{t})d\bar{l}^2$ and $ds^2 = dt^2 - a^2(t)dl^2$ are related by $ds = \Omega d\bar{s}$, we also have $d\bar{l} = dl$, hence

$$\frac{dt}{d\bar{t}} = \frac{a(t)}{\bar{a}(\bar{t})} = \Omega(t). \tag{24}$$

In the rescaled theory the probability of nucleation per unit of physical volume per time is given by

$$\bar{\Gamma}(\bar{t}) \equiv \frac{dP(\bar{t})}{d\bar{V}(\bar{t})d\bar{t}}, \tag{25}$$

where $P(\bar{t})$ is the bubble probability of nucleation and $\bar{V}(\bar{t})$ is the physical volume measured in the rescaled system at time $\bar{t}$. Therefore, the equivalent rate in the original theory is

$$\Gamma(t) \equiv \frac{dP(t)}{dV(t)dt} = \frac{dP(\bar{t})}{\left(\Omega^3 d\bar{V}\right)\left(\Omega d\bar{t}\right)} = \Omega^{-4}(t)\bar{\Gamma}(\bar{t}), \tag{26}$$

where we have used the equation (24) and the fact that $P(t) = P\left[\bar{t}(t)\right]$.

In JBD theories, the gravitational coupling and the time dependence of the JBD field might cause the bubble nucleation rate to deviate from its value to a flat space. In order to eliminate any time dependence coming directly from JBD field, we consider the limit $G \to 0$ in the theory of the equation (22) adopted by [26]. At this limit the term gravitational in action can be omitted. The term, which contains the kinetic energy of the JBD field, should also be neglected. To make this we replace $\tilde{\phi}$ with

$$\tilde{\Phi} = \sqrt{\frac{16\pi G}{2\omega + 3}}\tilde{\phi}. \tag{27}$$

The term containing the kinetic energy of the JBD field now has the form

$$\frac{1}{2}\left(\frac{2\omega + 3}{16\pi G}\right)\tilde{g}^{\mu\nu}\tilde{\nabla}_{\mu}\tilde{\Phi}\tilde{\nabla}_{\mu}\tilde{\Phi}. \tag{28}$$

Then this term is scaled precisely in the same way as the term containing the Ricci scalar as $G \to 0$.
Finally we are left with the last two terms of the equation (22). Only the field ψ remains as a dynamic quantity. Without their derivative terms, the metric and JBD field now enter essentially as externally specified non-dynamic quantities. While it has been shown that if it is assumed that the background gravitational field remains classical, static and is not perturbed during the decay, Coleman-De Luccia's prescription can be justified as a direct extension of the formalism of flat space, the problem general of back-reaction of the geometry against a quantum (non-causal)decay has not been answered.
In order to eliminate this problem, we neglect during this calculation the gravitational back-reaction of the bubble on the geometry of space-time.
Since gravitational effects may be small only if the bubble size is much smaller than the curvature of space, we take the metric as flat, and write the action of this theory as

$$\bar{S}[\psi] = \int d^4x \left[\frac{1}{2}\, b\, \tilde{g}^{\mu\nu}\nabla_{\mu}\psi\nabla_{\mu}\psi - b^2 U(\psi)\right], \tag{29}$$

where we have defined

$$b \equiv \exp\left(-4\sqrt{\frac{\pi G}{2\omega + 3}}\tilde{\phi}\right). \tag{30}$$

Now we would like to calculate the rate of nucleation of a bubble in this theory, or more precisely, to determine its dependence on the value b. Without the present gravitational interaction we can use the methods of the references [23, 24], although with some necessary modifications due to the non-standard form of the kinetic energy term.
The nucleation rate $\bar{\Gamma}$ is given by

$$\bar{\Gamma}(t) = A\exp\left(-S_E\right), \tag{31}$$

where S_E is Euclidean action. It is shown in [26] that the prefactor A has a net factor of b^2 with respect to a theory in which $b = 1$, this entails that $\bar{\Gamma}(t) = b^2\Gamma_0$, where Γ_0 is the nucleation rate for a normal scalar field theory with potential $U(\psi)$.
Since b is a function of the time-dependent JBD field, then we have a time-dependent nucleation rate in the rescaled theory. However, this time dependence disappears if we return to the original theory. The equations (20) and (21), together with the definition of ϕ_0 and Ω imply that $b = \Omega^2$. Using the equation (26) we find that the nucleation rate of the original theory is

$$\Gamma(t) = \Omega^{-4}(t)\bar{\Gamma}(\bar{t}) = \left(b^{-2}\right)\left(b^2\Gamma_0\right) = \Gamma_0, \tag{32}$$

and it is then independent of time on the approach with which we are working.
From the results obtained by [27] we have the imaginary part of the action for a universe with a closed geometry $k = 1$ is

$$Im[S] = \frac{27\sigma^4\pi}{4\varepsilon^3}\left[1 - \frac{1}{2}r_0^2\right], \tag{33}$$

where ϵ is the difference of energy density between the two phases (latent heat), σ the energy density of the wall and $r_0 = 3\sigma/(\varepsilon a_0)$ s the nucleation radius of the bubble when the space is flat and static.
Then, the nucleation rate Γ_0 for our model will be

$$\Gamma_0 \approx \exp\left(-2Im\ [S]\right) \approx \exp\left[-\frac{27\sigma^4\pi}{2\varepsilon^3}\left(1-\frac{81\gamma^2\rho_0^2\sigma^4}{8\phi_0^2\varepsilon^4}\right)\right]. \tag{34}$$

We can conclude that the bubble nucleation rate, within this approximation, is essentially that of the theory used by Coleman-De Luccia to which a correction of the order of r_0^2 has been added.

Once the nucleation rate was obtained, we have to study the evolution of the true vacuum bubble after its nucleation. To calculate this evolution we will consider the response of the background. In particular, we will follow the scheme developed by [28] regarding the conditions of joining the formalism of Darmois-Israel ([29, 30]) in JBD Theories. At this respect, we note that the bubble is formed in a region of the initial static universe, which has a perfect fluid as content of matter. As the bubble expands, this matter accumulates on the outer side of its wall. In this sense, the bubble could collapse or, on the contrary, continue to expand until it fills the entire space. On the other hand, inside the bubble it constitutes our universe, according to the scheme of open inflation, so it is important to determine whether the bubble once materialized collapses or expands in this model. We have postponed the report of the study of the classical evolution of the bubble, but preliminary results show that once the bubble materializes it grows without collapsing [22].

4. Conclusions

In this work we propose an scheme for an Emergent Universe scenario, where the universe is initially in a truly static state. This state is supported by a scalar field which is located in a false vacuum. The universe leaves the static state when, by quantum tunneling, the scalar field decays into a state of true vacuum.

In our model the matter content, in the initial static universe, is modeled by a perfect fluid and by an inflaton scalar field located in the false vacuum.

In particular, in Section 2, we search for static universe solutions for our emergent universe model. Then we study the stability of these solutions, finding the conditions that must be satisfied in order to obtain a static universe solution that is stable under small isotropic and homogeneous perturbations. These results show that our emergent universe model, based on Jordan-Brans-Dicke theory, presents static-universe solutions that do not suffer from classical instabilities, improving the results of [27].

In the section 3 we calculate the nucleation rate of the true vacuum bubble using the approximation developed by [26]. According to this approach, the calculation of the bubble nucleation rate in a context of Jordan-Brans-Dicke theories, can be reduced to the case of a coupling constant independent of time. This allows us to use the results obtained by [27] in our case. The calculation of the nucleation rate of the true vacuum bubble was performed applying these results to our model.

We have left for later the report of the study of the classical evolution of the bubble once materialized [22], but preliminary results show that once the bubble materializes it grows without collapsing. Then this model is viable and an interesting alternative for EU to be studied in detail.

5. Acknowledgments

P. L. is supported by Dirección de Investigación de la Universidad del Bío-Bío through Grants N^0 166907 2/R, and GI 150407/VC. H. C. is partially supported by Dirección de Postgrado de la Universidad del Bío-Bío.

References

[1] S. Weinberg, Gravitation and Cosmology: Principle and Application of the General Relativity, Wiley, NY, 1972; Ch. W. Misner, K. S. Turner, J. A. Wheeler, Gravitation, W. H.: Freeman and Company, SF 1973.
[2] P. J. E. Peebles, Principles of Physical Cosmology, Princeton University Press 1993; J. A. Peacock, Cosmological Physics, Cambridge University Press, 1998); S. Weinberg, Cosmology, Oxford University Press, USA, 2008.
[3] E. Kolb and M. Turner, The Early Universe, Addison- Wesley Publishing(1989).
[4] Guth A., The inflationary universe: A possible solution to the horizon and flatness problems, 1981 Phys. Rev. D **23** 347.
[5] Albrecht A. and Steinhardt P. J., Cosmology for grand unified theories with radiatively induced symmetry breaking, 1982 Phys. Rev. Lett. **48** 1220.
[6] Linde A. D., A new inflationary universe scenario: A possible solution of the horizon, flatness, homogeneity, isotropy and primordial monopole problems, 1982 Phys. Lett. **108B** 389.
[7] Linde A. D., Chaotic inflation, 1983 Phys. Lett. **129B** 177.
[8] Borde A., Guth A. H. and Vilenkin A., Inflationary space-times are incompletein past directions, 2003 Phys. Rev. Lett. **90** 151301.
[9] Ellis G. F. R. and Maartens R., The emergent universe: Inflationary cosmology with no singularity, 2004 Class. Quant. Grav. **21** 223.
[10] Ellis G. F. R., Murugan J. and Tsagas C. G., The emergent universe: An explicit construction, 2004 Class. Quant. Grav. **21** 233.
[11] Mulryne D. J., Tavakol R., Lidsey J. E. and Ellis G. F. R., An emergent universe from a loop, 2005 Phys. Rev. D **71** 123512.
[12] Mukherjee S., Paul B. C., Maharaj S. D. and Beesham A., Emergent universe in Starobinsky model, arXiv:gr-qc/0505103.
[13] Mukherjee S., Paul B. C., Dadhich N. K., Maharaj S. D. and Beesham A. , Emergent universe with exotic matter, 2006 Class. Quant. Grav. **23** 6927.
[14] A. Banerjee, T. Bandyopadhyay and S. Chakraborty, Emergent universe in brane world scenario, Grav. Cosmol. **13**, 290 (2007).
[15] Nunes N. J., Inflation: A graceful entrance from loop quantum cosmology, 2005 Phys. Rev. D **72** 103510.
[16] Lidsey J. E. and Mulryne D. J., A graceful entrance to braneworld inflation, 2006 Phys. Rev. D **73** 083508.
[17] J. D. Barrow, G. F. R. Ellis, R. Maartens and C. G. Tsagas, "On the stability of the Einstein static universe," Class. Quant. Grav. **20**, L155 (2003) doi:10.1088/0264-9381/20/11/102 [gr-qc/0302094].
[18] S. del Campo, R. Herrera and P. Labrana, JCAP **0711**, 030 (2007) doi:10.1088/1475-7516/2007/11/030 [arXiv:0711.1559 [gr-qc]].
[19] S. del Campo, R. Herrera and P. Labrana, JCAP **0907**, 006 (2009) doi:10.1088/1475-7516/2009/07/006 [arXiv:0905.0614 [gr-qc]].
[20] P. Jordan, Z. Phys. **157** (1959) 112. doi:10.1007/BF01375155
[21] C. Brans and R. H. Dicke, Phys. Rev. **124** (1961) 925. doi:10.1103/PhysRev.124.925
[22] P. Labrana and H. Cossio, Work in progress.
[23] S. R. Coleman, "The Fate of the False Vacuum. 1. Semiclassical Theory," Phys. Rev. D **15**, 2929 (1977) Erratum: [Phys. Rev. D **16**, 1248 (1977)]. doi:10.1103/PhysRevD.15.2929, 10.1103/PhysRevD.16.1248
[24] C. G. Callan, Jr. and S. R. Coleman, "The Fate of the False Vacuum. 2. First Quantum Corrections," Phys. Rev. D **16**, 1762 (1977). doi:10.1103/PhysRevD.16.1762
[25] S. R. Coleman and F. De Luccia, "Gravitational Effects on and of Vacuum Decay," Phys. Rev. D **21**, 3305 (1980). doi:10.1103/PhysRevD.21.3305
[26] R. Holman, E. W. Kolb, S. L. Vadas, Y. Wang and E. J. Weinberg, "False Vacuum Decay in Jordan-brans-dicke Cosmologies," Phys. Lett. B **237**, 37 (1990). doi:10.1016/0370-2693(90)90457-H
[27] P. Labrana, "Emergent Universe by Tunneling," Phys. Rev. D **86**, 083524 (2012) doi:10.1103/PhysRevD.86.083524 [arXiv:1111.5360 [gr-qc]].
[28] N. Sakai and K. i. Maeda, "Bubble dynamics in generalized Einstein theories," Prog. Theor. Phys. **90**, 1001 (1993). doi:10.1143/PTP.90.1001
[29] W. Israel, "Singular hypersurfaces and thin shells in general relativity," Nuovo Cim. B **44S10**, 1 (1966) [Nuovo Cim. B **44**, 1 (1966)] Erratum: [Nuovo Cim. B **48**, 463 (1967)].
[30] Darmois, G."Memorial des Sciences Mathematiques" Fasc. 25, Gauthier-Villars (1927)

IOP Conf. Series: Journal of Physics: Conf. Series **1043** (2018) 012022 doi:10.1088/1742-6596/1043/1/012022

Conserved quantities for a charged rotating black holes in 5D Einstein-Maxwell-Chern-Simons theory

Felipe Díaz-Martínez and Rodrigo Olea

Departamento de Física, Universidad Andres Bello, Sazié 2212, Piso 7, Santiago, Chile

E-mail: fdiaz47@gmail.com, rodrigo.olea@unab.cl

Abstract. In this work, we compute the conserved quantities of a charged rotating black hole which appears as the solution of Einstein-Maxwell action in five dimensions coupled to a Chern-Simons term for $U(1)$ field. The addition of the Chern-Simons term will modify the Maxwell equations and the definition of charge but not the Einstein field equations. Upon the addition of suitable boundary terms for the pure gravity sector of the theory, which depend on the extrinsic and intrinsic curvatures (Kounterterms), we obtain the correct conserved quantities of the solution.

1. Introduction

A black hole is a solution of General Relativity with nontrivial causal structure, which has been focus of study in fundamental physics in the last decades. It is expected that such objects can give better insight on the understanding of Quantum Gravity. It has been recently found in Ref.[3] the metric of Kerr-AdS black hole for an arbitrary dimension D, with maximal number of rotation parameters $N = [(D-1)/2]$. Electrically charged generalizations of rotating AdS black holes can be obtained in the form of Kerr-Newman solution in $D = 4$. No higher-dimensional counterparts have been reported in the literature.

On the other hand, non-extremal rotating charged black holes appear as solutions of the bosonic sector of gauged supergravity in five dimensions, which contains a Chern-Simons term for $U(1)$ on top of the Einstein-Maxwell Lagrangian. This is the closest one may be to the concept of Kerr-Newman solution in $D > 4$. In this case, the action adopts the form

$$I = \frac{1}{16\pi G_5}\int\limits_{\mathcal{M}} d^5x\sqrt{-g}\left(R - 2\Lambda - \frac{1}{4}F_{\mu\nu}F^{\mu\nu}\right) - \frac{\lambda}{12\sqrt{3}}\epsilon^{\mu\nu\alpha\beta\gamma}A_\mu F_{\nu\alpha}F_{\beta\gamma} + c_4\int\limits_{\partial\mathcal{M}} d^4x\ B_4 \quad (1)$$

where the cosmological constant takes the value $\Lambda = -6/\ell^2$, A_μ is the vector potential, $F_{\mu\nu}$ is the field strength and B_4 its the boundary term that regularizes both the Euclidean action and the conserved charges.

As the Chern-Simons term does not depend on the metric, it does not contribute to the Einstein equation, but it does modify the Maxwell equation

$$\nabla_\mu F^{\mu\nu} + \frac{\lambda}{4\sqrt{3}}\varepsilon^{\mu\nu\alpha\beta\gamma}F_{\mu\alpha}F_{\beta\gamma} = 0\,. \quad (2)$$

2. Black Hole Solution

The metric for this black hole solution, with two rotation parameters and electric charge, can be written in BoyerLindquist coordinates $x^\mu = (t, r, \theta, \phi, \psi)$ [1]

$$ds^2 = -\frac{\Delta_\theta\left[\left(1+\frac{r^2}{\ell^2}\right)\rho^2 dt + 2q\nu\right]dt}{\Xi_a\Xi_b\rho^2} + \frac{2q\nu\omega}{\rho^2} + \frac{f}{\rho^4}\left(\frac{\Delta_\theta dt}{\Xi_a\Xi_b} - \omega\right)^2 + \frac{\rho^2 dr^2}{\Delta_r} + \frac{\rho^2 d\theta^2}{\Delta_\theta} + \frac{(r^2+a^2)}{\Xi_a} sin^2\theta d\phi^2 + \frac{(r^2+b^2)}{\Xi_b} cos^2\theta d\psi^2 \,, \tag{3}$$

with

$$\begin{aligned}
\nu &= b sin^2\theta d\phi + a cos^2\theta d\psi \,, \\
\omega &= a sin^2\theta \frac{d\phi}{\Xi_a} + b cos^2\theta \frac{d\psi}{\Xi_b} \,, \\
\Delta_\theta &= 1 - \frac{a^2}{l^2} cos^2\theta - \frac{b^2}{\ell^2} sin^2\theta \,, \\
\Delta_r &= \frac{(r^2+a^2)(r^2+b^2)(1+r^2/l^2) + q^2 + 2qab}{r^2} - 2m \,, \\
\rho^2 &= r^2 + a cos^2\theta + b^2 sin^2\theta \,, \\
\Xi_a &= 1 - \frac{a^2}{\ell^2} \,, \quad \Xi_b = 1 - \frac{b^2}{\ell^2} \,, \\
f &= 2m\rho^2 - q^2 + 2abq\rho^2/\ell^2 \,,
\end{aligned} \tag{4}$$

and the gauge potential

$$A = \frac{q\sqrt{3}}{\rho^2}\left(\frac{\Delta_\theta dt}{\Xi_a\Xi_b} - \omega\right) . \tag{5}$$

The conserved quantities of this solution were given in Ref [1], as obtained from the integration of the First Law of Thermodynamics, and not as surface integrals at infinity. Here, following the Kounterterms method, we intend to provide an alternative derivation of the existing results.

3. Kounterterms

A prescription for the boundary terms required by the regularization of AdS action in odd spacetime dimensions was given in Ref.[2]. The boundary term depends on the intrinsic and extrinsic curvatures and a closed expression exists in any dimension in the form of a polynomial with coefficients given by parametric integrations.

We consider a radial foliation of the spacetime, in such a way that the general form of the line element is written as

$$ds^2 - g_{\mu\nu}dx^\mu dx^\nu - N^2(r)dr^2 + h_{ij}(r,x)dx^i dx^j \tag{6}$$

where h_{ij} is the induced metric, such that (i, j) denote boundary indices.

In this coordinate frame, the extrinsic curvature adopts the simple form

$$K_{ij} = -\frac{1}{2N}\partial_r h_{ij} \,, \tag{7}$$

due to the fact that *shift* functions in the radial ADM-like metric (6) are zero.

In five dimensions, the boundary term needed by the gravitational sector of the theory is

$$B_4 = \sqrt{-h}\delta^{[i_1 i_2 i_3]}_{[j_1 j_2 j_3]} K^{j_1}_{i_1}\left(\mathcal{R}(h)^{j_2 j_3}_{i_2 i_3} - K^{j_2}_{i_2}K^{j_3}_{i_3} + \frac{1}{3\ell^2}\delta^{j_2}_{i_2}\delta^{j_3}_{i_3}\right) , \tag{8}$$

with a fixed coupling constant

$$c_4 = \frac{\ell^2}{128\pi G} \,. \tag{9}$$

As it is a boundary term it does not modify the bulk dynamics, but it does modify the conserved charges, leading to a finite conserved quantity. For the electromagnetic part, we are assuming that the potential vector A_μ is kept fixed at the boundary.
A splitting of the charges of the form

$$Q[\xi] = \int_\Sigma d^{d-1}y\sqrt{\sigma}u_j\xi^i \left(q_i^j + q_{(0)i}^j\right) \tag{10}$$

is a feature that appears naturally in this method. Indeed, it separates the contributions proportional to the Weyl tensor from the ones that give rise to the vacuum energy $q_{(0)i}^j$. Here, ξ^i is a Killing vector associated to boundary isometries and u_j is a unit vector, normal to a constant-time slice.

In five dimensions the explicit expressions for the conserved charges are

$$q_i^j = \frac{\ell^2}{64\pi G}\delta^{[jj_2j_3j_4]}_{[i_1i_2i_4i_4]}K_j^{i_1}\delta^{i_2}_{j_2}\left(R^{i_3i_4}_{j_3j_4} + \frac{1}{\ell^2}\delta^{[j_3j_4]}_{[i_4i_4]}\right) \tag{11}$$

and

$$q_{(0)i}^j = \frac{\ell^2}{128\pi G}\delta^{[jj_2..j_4]}_{[ki_2..i_4]}\left(K_i^k\delta^{i_2}_{j_2} + K^k_{j_2}\delta^{i_2}_i\right)\left(\mathcal{R}^{i_3i_4}_{j_3j_4} - K^{i_3}_{j_3}K^{i_4}_{j_4} + \frac{1}{\ell^2}\delta^{i_3}_{j_3}\delta^{i_4}_{j_4}\right) . \tag{12}$$

4. Conserved quantities

Evaluating the charge formula for the charged rotating black hole metric (3), we obtain

$$E = q(\partial_t) = \frac{m\pi(2\Xi_a + 2\Xi_b - \Xi_a\Xi_b) + 2\pi qab(\Xi_a + \Xi_b)/\ell^2}{4\Xi_a^2\Xi_b^2} \,, \tag{13}$$

$$E_0 = q_0(\partial_t) = \frac{2\pi\ell^2}{32}\left(1 - \frac{(\Xi_a - \Xi_b)^2}{9\Xi_a\Xi_b}\right) , \tag{14}$$

where E_0 is the same as in the case of Kerr-AdS$_5$ black hole. E corresponds to the black hole mass. There is a subtlety regarding this point, though. Commonly, a proper definition of energy for a rotating black hole is associated to a Killing vector that does not rotate at infinity, $\xi = \partial_t - (a/\ell^2)\partial_\phi - (b/\ell^2)\partial_\psi$. As it has been widely discussed in the literature, this is the only notion of energy that satisfies the First Law of Thermodynamics for Kerr-AdS black holes. Being able to reproduce the correct value of the energy from a Killing vector $\xi = \partial_t$ is a reflection of the fact that one does not recover Kerr-AdS metric from the uncharged limit of the solution (3). On the other hand, for the Killing vectors associated to the azimuthal angles ϕ and ψ,

$$J_\phi = q(\partial_\phi) = \frac{\pi}{4\Xi_a^2\Xi_b}\left[2am + qb(1 + a^2/\ell^2)\right] , \tag{15}$$

$$J_\psi = q(\partial_\psi) = \frac{\pi}{4\Xi_b^2\Xi_a}\left[2bm + qa(1 + b^2/\ell^2)\right] , \tag{16}$$

which correspond to the angular momenta of the black hole.
All the conserved quantities computed here are in agreement with Ref.[1], where the charges were obtained from the integration of the First Law.

5. Conclusions

We have calculated the Noether charges associated to the isometries of the charged rotating black hole in five dimensional found in Ref.[1]. When the dilaton fields present in a $5D$ gauged supergravity theory are kept fixed, they act as a cosmological constant. Black holes like the one studied here are solutions to the bosonic sector of this supergravity. The addition of counterterms that depends on the intrinsic and extrinsic curvatures renders the conserved charges both finite and correct. These quantities correspond to the energy of the system, which also involves the electric charge, two angular momenta, and a vacuum energy E_0, which is the same as in Kerr-AdS_5 (computed in Ref.[4] and Ref.[5] using the standard counterterm method).

The finite action defined by Eq.(1) provides the right starting point to obtain thermodynamic relations directly from the Euclidean action. However, depending on thermodynamic ensemble one is working on, the action may require additional boundary terms, as it was discussed in the case of Kerr-Newman-AdS black hole in Ref.[6].

6. Acknowledgments

We thank Giorgos Anastasiou and David Rivera-Betancour for insightful discussions. This work was funded in part by FONDECYT Grant 1131075, UNAB Grant DI-1336-16/RG and CONICYT Grant DPI 20140115.

7. References

[1] Chong Z W, Cvetic M, Lu H and Pope C N 2005 *Phys. Rev. Lett.* **95** 161301
[2] Olea R 2007 *JHEP* **04** 073
[3] Gibbons G W, Lu H, Page D N and Pope C N 2005 *J.Geom.Phys.* **53** 49-73
[4] Papadimitriou I and Skenderis K 2005 *Math. Theor. Phys.* **8** 73-101
[5] Awad A M and Johnson C V 2001 *Phys. Rev. D* **63** 124023
[6] Caldarelli M M, Cognola G and Klemm D 2000 *Class. Quant. Grav.* **17** 399-420

Study of Vainshtein mechanism in Galileon theory

Carla Bernal, Radouane Gannouji

Instituto de Física, Pontificia Universidad Católica de Valparaíso, Casilla 4950, Valparaíso, Chile

E-mail: carlabernalb@gmail.com, radouane.gannouji@pucv.cl

Abstract. In this project, we compute the nonlinear power spectrum for a generalized galileon model. This calculation is a first step to get access to the nonlinear effective gravitational constant and therefore to the scale where the model differs from general relativity.

1. Introduction

Actually, despite the fact there is a cosmological model that is accepted as the standard model, which fits perfectly the observations, there are some problems which are unresolved such as the cosmological constant problem. It can be seen as a fine-tuning problem or why the vacuum energy gravitates so little. Modified gravity models which are often presented as a possible solution to this problem do not actually try to resolve the problem but instead replace the cosmological constant problem by a similar problem. But they can be seen as good laboratories to study gravity models beyond General Relativity (GR). In fact, in various situations such as brane models, an effective theory on the brane provides additional corrections to Einstein's theory.

In higher dimensional theories, the models have more than 2 degrees of freedom. For example, the DGP model has 5 degrees of freedom which can be split into a massless graviton, a massless vector field and a scalar field. But the main interesting parts of the model are contained in the additional scalar field (along with the massless graviton). In fact, integrating out the additional dimension [1] gives rise to an effective theory describing the brane bending mode, known as galileon theory [5]. Higher dimensional models contains often scalar fields which can be very relevant to cosmology. In fact, more generically many theories of high-energy physics, such as string theory and supergravity, predict light gravitationally coupled scalar fields [3, 4].

We see therefore that the path towards modified gravity can be very rich and interesting. But cosmology demands the new degree of freedom to be incredibly light in order to be relevant. Unfortunately light particles are highly constrained by fifth force experiments and solar system observations. A light scalar field is required for cosmology but not from local tests which put severe upper bound on its mass. Therefore this light degree of freedom should hide in local experiments, that is to say a screening mechanism.

One of the most relevant mechanism dubbed Vainshtein mechanism corresponds to a scalar field coupled to matter but becomes weakly coupled because of a large kinetic term. In this context, we would like to study this mechanism in realistic situations. The very interesting object in modified gravity is the evolution of the gravitational constant. If it is suppose to deviate from the Newtonian constant at large scales, it should recover this same constant at small scales because of some screening mechanism. This effective constant gives then naturally access to the radius at which a deviation from GR appears and then to possible observations. In this direction, the Vainshtein mechanism has shown to be very interesting but it is not very well understood in a realistic framework of formation of structures. It is usually derived for an isolated object in a spherically symmetric situation. One of the challenging problem is to understand the elephant problem or how various vainshtein radii interact. In this context, we can try to answer this question, by studying the Vainshtein radius in an N-body problem. Also, the Vainshtein mechanism seems to be less effective in time dependent problem. Therefore, we will try to address also this problem by looking to the time evolution of the Vainshtein radius in this context. The strategy is to be able to extract the non-linear gravitational constant during matter evolution, we will have access to the Vainshtein radius (when it deviates sensibly from the Newtonian constant) and we will have access to its evolution during the cosmological evolution.

To be able to compute the effective gravitational constant in a non-linear regime, we need to have access for this model to the non-linear power-spectrum and bispectrum. In this first project, we will compute the nonlinear power spectrum of this model by using standard perturbation theory and compare to numerical simulations.

2. The model

In this paper, we consider a generalized form of the galileon model [5], for which the action takes the following form

$$S = \int d^4x\sqrt{-g}\left[\frac{M_{Pl}^2}{2}R - \frac{c_2}{2}(\partial\phi)^2 - \frac{c_3}{\Lambda_v^3}\Box\phi(\partial\phi)^2 - \frac{c_4}{4\Lambda_\star^4}(\partial\phi)^4\right] \tag{1}$$

where g is the metric determinant, R is the Ricci scalar, M_{Pl} is the reduced Planck's mass, c_2, c_3 and c_4 are dimensionless constants, ϕ is the galileon scalar field and Λ_v with $\Lambda_\star$ are characteristic scales of the model.

2.1. Background

From this action we obtain the Friedmann equations:

$$3M_{Pl}^2H^2 = \rho_m + \rho_\phi \tag{2}$$
$$-2\dot{H}M_{Pl}^2 = \rho_m + \rho_\phi + P_\phi \tag{3}$$
$$\dot{J} + 3HJ = 0 \tag{4}$$

where

$$\rho_\phi = \frac{c_2\dot{\phi}^2}{2} - \frac{3c_4\dot{\phi}^4}{4\Lambda_\star^4} - \frac{6c_3H\dot{\phi}^3}{\Lambda_v^3} \tag{5}$$
$$P_\phi = \frac{c_2\dot{\phi}^2}{2} - \frac{c_4\dot{\phi}^4}{4\Lambda_\star^4} + \frac{2c_3\dot{\phi}^2\ddot{\phi}}{\Lambda_v^3} \tag{6}$$
$$J = c_2\dot{\phi} - \frac{6c_3H\dot{\phi}^2}{\Lambda_v^3} - \frac{c_4\dot{\phi}^3}{\Lambda_\star^4} \tag{7}$$

To analyze the background, we transform the equations into a dynamical system where we define the variables (x, y) by

$$H = H_{dS}\sqrt{\frac{x}{y}} \tag{8}$$
$$\dot{\phi} = \dot{\phi}_{dS}\sqrt{xy} \tag{9}$$

while $(H_{dS}, \dot{\phi}_{dS})$ are defined during the de Sitter era. We define the scales

$$l_{\rm v} = \frac{2c_3 H_{dS}^2 M_{Pl}}{\Lambda_{\rm v}^3}, \quad l_\star = \frac{c_4 H_{dS}^2 M_{Pl}^2}{4\Lambda_\star^4}, \quad r_{dS} = \frac{\dot{\phi}_{dS}}{M_{Pl} H_{dS}} \tag{10}$$

During the de Sitter phase, we have $x = y = 1$ and $\dot{H} = \ddot{\phi} = \rho_m = 0$ which implies from the equations of motion, the constraints

$$\frac{c_2}{6} r_{dS}^2 - l_{\rm v} r_{dS}^3 - l_\star r_{dS}^4 = 1\,, \qquad -\frac{c_2}{6} r_{dS}^2 + \frac{l_\star}{3} r_{dS}^4 = 1 \tag{11}$$

We notice that these relations are invariant under the transformations $c_i \to \gamma^i c_i$ and $r_{dS} \to r_{dS}/\gamma$ along with $l_{\rm v} \to \gamma^3 l_{\rm v}$ and $l_\star \to \gamma^4 l_\star$. We see therefore that it is convenient to define an invariant parameter $\alpha = l_{\rm v} r_{dS}^3$ and express the others as

$$l_\star r_{dS}^4 = -3 - \frac{3}{2}\alpha \tag{12}$$

$$c_2 r_{dS}^2 = -12 - 3\alpha \tag{13}$$

We have therefore shown, that the dynamics of the model depend on only 1 parameter. Notice that $\alpha = -2$ corresponds to galileon model (from eq.(10,12) we have $c_4 = 0$). The system (eq.(2,3,4)) reduces to a 2D dynamical system of the following form

$$\frac{dx}{d\ln a} = f_\alpha(x, y) \tag{14}$$

$$\frac{dy}{d\ln a} = g_\alpha(x, y) \tag{15}$$

where a is the scale factor, (f_α, g_α) are functions of (x, y) and the only parameter of the model is α.

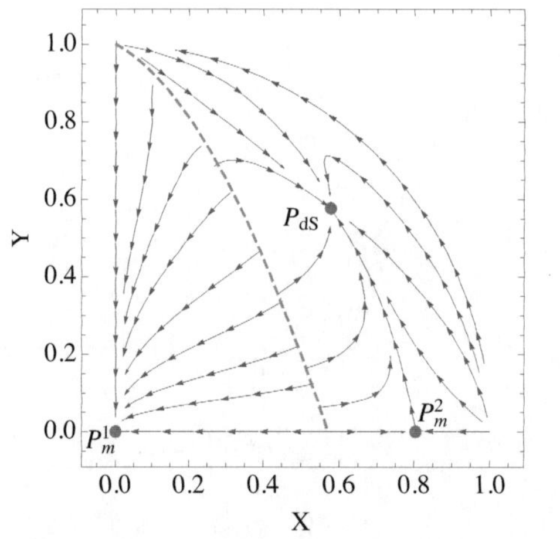

Figure 1. Phase space using the Poincaré sphere $(X, Y) > 0$. The dashed green line is an invariant submanifold which separates two phase spaces. P_{dS} represents the de Sitter phase while P_m^1 and P_m^2 represents 2 matter phases. For definitions of Poincaré sphere and invariant manifold along the way to study a dynamical system we refer to [8].

The stability has been studied by many authors and therefore we will not reproduce it here. We refer to [9] for a full analysis. We found that during the de Sitter phase which is an attractor, the propagation of the perturbed scalar field doesn't have Laplacian instability if $-2 < \alpha < 0$. We have also 2 matter points; the point $P_m^1 = (0,0)$ which is an attractor and therefore can't be followed by an accelerated universe while the point $P_m^2(-1-4/\alpha, 0)$ is saddle with $c_s^2 = 4/3$ (sound speed).

From now, we will consider the case $\alpha = -1$, for which we have

$$\frac{dx}{d\ln a} = -3x\frac{18 - 6x - 20xy - 9y^2 + 9xy^2 - 2x^2y^2 + 21xy^3 - 5x^2y^3 - 6 + x^2y^4}{2(6 - 4x - 12xy - x^2y^2)} \quad (16)$$

$$\frac{dy}{d\ln a} = -3y\frac{6 + 2x + 4xy + 9y^2 - 15xy^2 + 2x^2y^2 - 21xy^3 + 7x^2y^3 + 6x^2y^4}{2(6 - 4x - 12xy - x^2y^2)} \quad (17)$$

The phase space of this system is represented in Fig.(1). It is therefore easy to integrate the system, for which we get (see Fig.(2))

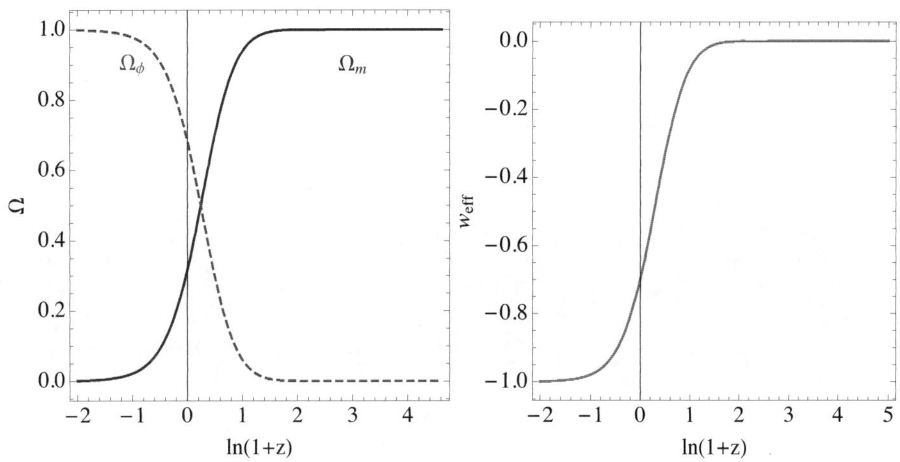

Figure 2. Left: Evolution of the density of matter and scalar field. Right: Evolution of $w_{eff} = -1 + 2\dot{H}/(3H^2)$ as function of $\ln(1+z)$.

2.2. Standard perturbation theory

Considering the metric

$$ds^2 = -(1 + 2\Phi)dt^2 + a^2(1 - 2\Phi)d\mathbf{x}^2 \quad (18)$$

along with $\delta\phi$ the perturbation of the scalar field and $\delta_m = \delta\rho_m/\rho_m$, we found

$$\Delta\delta\phi + \frac{c_3}{a^2\Lambda_{\rm v}^3\beta_1}\left[(\Delta\delta\phi)^2 - (\partial_{ij}\delta\phi)^2\right] = \frac{c_3}{\Lambda_{\rm v}^3\beta_2 M_{Pl}^2}a^2\rho_m\delta_m \quad (19)$$

$$\Delta\Phi = \frac{a^2}{2M_{Pl}^2}\rho_m\delta_m + \frac{c_3}{\Lambda_{\rm v}^3 M_{Pl}^2}\dot{\phi}^2\Delta\delta\phi \quad (20)$$

Along with the Vlasov equation up to the second moment, which represents mass conservation and Euler equation, we have

$$\dot{\delta}_m + \frac{1}{a}\partial_i\left[(1 + \delta_m)u^i\right] = 0 \quad (21)$$

$$\dot{u}_i + Hu_i + \frac{1}{a}u^j\partial_j u_i = -\frac{1}{a}\partial_i\Phi \quad (22)$$

where u^i is the mean density flow and we defined

$$\beta_1 = \frac{c_2}{2} - \frac{2c_3}{\Lambda_{\rm v}^3}(\ddot\phi + 2H\dot\phi) - \frac{c_3^2}{\Lambda_{\rm v}^6 M_{Pl}^2}\dot\phi^4 - \frac{c_4}{2\Lambda_\star^4}\dot\phi^2 \tag{23}$$

$$\beta_2 = \frac{2}{\dot\phi^2}\beta_1 \tag{24}$$

Transforming into Fourier space

$$\delta(t,\mathbf{x}) = \frac{1}{(2\pi)^3}\int d^3\mathbf{k}\delta(t,\mathbf{k})e^{i\mathbf{k}\cdot\mathbf{x}}, \tag{25}$$

$$u^j(t,\mathbf{x}) = \frac{1}{(2\pi)^3}\int d^3\mathbf{k}\frac{-ik^j}{k^2}aH\theta(t,\mathbf{k})e^{i\mathbf{k}\cdot\mathbf{x}}, \tag{26}$$

$$\delta\phi(t,\mathbf{x}) = \frac{1}{(2\pi)^3}\int d^3\mathbf{k}\delta\phi(t,\mathbf{k})e^{i\mathbf{k}\cdot\mathbf{x}}. \tag{27}$$

and defining $\theta = \nabla_i u^i/(aH)$, we have

$$\frac{\dot\delta_m}{H} + \theta + \int\frac{\mathrm{d}^3k_1\mathrm{d}^3k_2}{(2\pi)^3}\delta(\vec k - \vec k_{12})\alpha(k_1,k_2)\delta_m(k_1)\theta(k_2) = 0 \tag{28}$$

$$\frac{\dot\theta}{H} + \left(2 + \frac{\dot H}{H^2}\right)\theta + \frac{3}{2}\Omega_m\delta_m - \frac{c_3\dot\phi^2}{\Lambda_{\rm v}^3 M_{Pl}^2 a^2 H^2}k^2\delta\phi + \int\frac{\mathrm{d}^3k_1\mathrm{d}^3k_2}{(2\pi)^3}\delta(\vec k - \vec k_{12})\beta(k_1,k_2)\theta(k_1)\theta(k_2) = 0 \tag{29}$$

$$k^2\delta\phi + \frac{c_3}{\Lambda_{\rm v}^3\beta_2 M_{Pl}^2}a^2\rho_m\delta_m - \frac{c_3}{a^2\Lambda_{\rm v}^3\beta_1}\int\frac{\mathrm{d}^3k_1\mathrm{d}^3k_2}{(2\pi)^3}\delta(\vec k - \vec k_{12})\gamma(k_1,k_2)\delta\phi(k_1)\delta\phi(k_2) = 0 \tag{30}$$

where $\vec k_{12} = \vec k_1 + \vec k_2$ and

$$\alpha(k_1,k_2) = 1 + \frac{\vec k_1.\vec k_2}{k_2^2} \tag{31}$$

$$\beta(k_1,k_2) = \frac{\vec k_1.\vec k_2|\vec k_1 + \vec k_2|}{2k_1^2k_2^2} \tag{32}$$

$$\gamma(k_1,k_2) = k_1^2k_2^2 - (\vec k_1.\vec k_2)^2 \tag{33}$$

(34)

and modified Poisson equation has been used

$$-k^2\Phi = \frac{a^2}{2M_{Pl}^2}\rho_m\delta_m - \frac{c_3\dot\phi^2}{\Lambda_{\rm v}^3 M_{Pl}^2}k^2\delta\phi \tag{35}$$

These equations can be solved iteratively

$$Y(t,\mathbf{p}) = \sum_{n=1} Y_n(t,\mathbf{p}), \tag{36}$$

where Y represents δ_m, θ or $\delta\phi$, and Y_n is the n-order solution of the expansion.

2.3. First order perturbations

Using these equations to first order, we have

$$\frac{\dot{\delta}_m^{(1)}(t,\mathbf{k})}{H} + \theta^{(1)}(t,\mathbf{k}) = 0 \tag{37}$$

$$\frac{\dot{\theta}^{(1)}(t,\mathbf{k})}{H} + \left(2 + \frac{\dot{H}}{H^2}\right)\theta^{(1)}(t,\mathbf{k}) + \frac{2}{3}\Omega_m\delta_m^{(1)}(t,\mathbf{k}) - \frac{c_3\dot{\phi}^2}{\Lambda_v^3 M_{Pl}^2 a^2 H^2}k^2\delta\phi^{(1)}(t,\mathbf{k}) = 0 \tag{38}$$

$$k^2\delta\phi^{(1)}(t,\mathbf{k}) + \frac{c_3}{\Lambda_v^3\beta_2 M_{Pl}^2}a^2\rho_m\delta_m^{(1)}(t,\mathbf{k}) = 0 \tag{39}$$

Combining the different equations, we can reduce the system to a second order equation for δ_m

$$\ddot{\delta}_m(t,\mathbf{k}) + 2H\dot{\delta}_m(t,\mathbf{k}) - \frac{3}{2}H^2\Omega_m\frac{G_{eff}(t)}{G}\delta_m(t,\mathbf{k}) = 0 \tag{40}$$

where G_{eff} is the linear gravitational constant

$$\frac{G_{eff}}{G} = 1 + \frac{2c_3^2\dot{\phi}^2}{\Lambda_v^6 M_{Pl}^2\beta_2}. \tag{41}$$

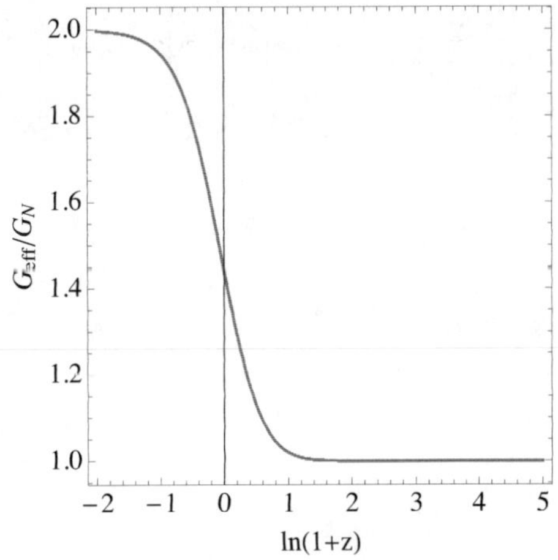

Figure 3. Evolution of the effective gravitational constant for $\alpha = -1$. We see that the model differs from GR only recently and therefore is perfect for a dark energy model.

The eq.(40) has solutions of the following form $\delta_m(t,\mathbf{k}) = D(t)A(\mathbf{k})$ and we can find that D has 2 solutions corresponding to a growing mode (D_+) and decreasing mode (D_-). For the linear solution, it is sufficient to consider only the growing mode, which gives

$$\delta_m^{(1)}(t,\mathbf{k}) = D_+(t)A(\mathbf{k}), \tag{42}$$

$$\theta^{(1)}(t,\mathbf{k}) = -D_+(t)A(\mathbf{k}), \tag{43}$$

$$\delta\phi^{(1)}(t,\mathbf{k}) = -\frac{a^2}{k^2}\frac{\rho_m c_3}{\Lambda_v^3\beta_2 M_{Pl}^2}D_+(t)A(\mathbf{k}). \tag{44}$$

2.4. Second order perturbations

Going to second order, we found

$$\frac{\dot{\delta}_m^{(2)}(t,\mathbf{k})}{H} + \theta^2(t,\mathbf{k}) = -\int \frac{d^3\mathbf{k}_1 d^3 d\mathbf{k}_2}{(2\pi)^3}\delta(\mathbf{k}-\mathbf{k}_{12})\alpha(\mathbf{k}_1,\mathbf{k}_2)\delta_m^{(1)}(t,\mathbf{k}_2)\theta^{(1)}(t,\mathbf{k}_1) \tag{45}$$

$$\frac{\dot{\theta}^{(2)}(t,\mathbf{k})}{H} + \left(2+\frac{\dot{H}}{H^2}\right)\theta^{(2)}(t,\mathbf{k}) + \frac{2}{3}\Omega_m\delta_m^{(2)}(t,\mathbf{k}) - \frac{c_3\dot{\phi}^2}{\Lambda_v^3 M_{Pl}^2 a^2 H^2}k^2\delta\phi^{(2)}(t,\mathbf{k}) = -\int \frac{d^3\mathbf{k}_1 d^3\mathbf{k}_2}{(2\pi)^3}\delta(\mathbf{k}-\mathbf{k}_{12})\beta(\mathbf{k}_1,\mathbf{k}_2)\theta^{(1)}(t,\mathbf{k}_1)\theta^{(1)}(t,\mathbf{k}_2) \tag{46}$$

$$k^2\delta\phi^{(2)}(\mathbf{k}) + \frac{c_3}{\Lambda_v^3\beta_2 M_{Pl}^2}a^2\rho_m\delta_m^{(2)}(t,\mathbf{k}) = \frac{c_3}{a^2\Lambda_v^3\beta_1}\int\frac{d^3\mathbf{k}_1 d^3\mathbf{k}_2}{(2\pi)^3}\delta(\mathbf{k}-\mathbf{k}_{12})\gamma(\mathbf{k}_1,\mathbf{k}_2)\delta\phi^{(1)}(t,\mathbf{k}_1)\delta\phi^{(1)}(t,\mathbf{k}_2) \tag{47}$$

The second order gives an equation similar to eq.(40), but with an inhomogeneous equation

$$\ddot{\delta}_m^{(2)}(t,\mathbf{k}) + 2H\dot{\delta}_m^{(2)}(t,\mathbf{k}) - \frac{3}{2}H^2\Omega_m\frac{G_{eff}}{G}\delta_m^{(2)}(t,\mathbf{k}) = \frac{\dot{a}}{a^2}E(t,\mathbf{k}) + \frac{c_3\dot{\phi}^2}{M_{Pl}^3 H_0^2}I(t,\mathbf{k}) + \frac{1}{a}E(t,\mathbf{k}) + F(t,\mathbf{k}) \tag{48}$$

where

$$\begin{aligned} E(t,\mathbf{k}) &= \int\frac{d^3\mathbf{k}_1 d^3 d\mathbf{k}_2}{(2\pi)^3}\delta(\mathbf{k}-\mathbf{k}_{12})\alpha(\mathbf{k}_1,\mathbf{k}_2)\delta_m^{(1)}(t,\mathbf{k}_2)\theta^{(1)}(t,\mathbf{k}_1) \\ F(t,\mathbf{k}) &= \int\frac{d^3\mathbf{k}_1 d^3\mathbf{k}_2}{(2\pi)^3}\delta(\mathbf{k}-\mathbf{k}_{12})\beta(\mathbf{k}_1,\mathbf{k}_2)\theta^{(1)}(t,\mathbf{k}_1)\theta^{(1)}(t,\mathbf{k}_2) \\ I(t,\mathbf{k}) &= \frac{c_3}{a^2\Lambda_v^3\beta_1}\int\frac{d^3\mathbf{k}_1 d^3\mathbf{k}_2}{(2\pi)^3}\delta(\mathbf{k}-\mathbf{k}_{12})\gamma(\mathbf{k}_1,\mathbf{k}_2)\delta\phi^{(1)}(t,\mathbf{k}_1)\delta\phi^{(1)}(t,\mathbf{k}_2) \end{aligned}$$

The solution can be written in the following form

$$\delta_m^{(2)}(t,\mathbf{k}) = D_+^2(t)\left(\mathcal{W}_\alpha(\mathbf{k}) - \frac{2}{7}\lambda(t)\mathcal{W}_\gamma(\mathbf{k})\right) \tag{49}$$

where

$$\mathcal{W}_\alpha(\mathbf{k}) = \frac{1}{(2\pi)^3}\int d^3\mathbf{k}_1 d^3\mathbf{k}_2\delta(\mathbf{k}_{12}-\mathbf{k})\alpha^{(s)}(\mathbf{k}_1,\mathbf{k}_2)A(\mathbf{k}_1)A(\mathbf{k}_2), \tag{50}$$

$$\mathcal{W}_\gamma(\mathbf{k}) = \frac{1}{(2\pi)^3}\int d^3\mathbf{k}_1 d^3\mathbf{k}_2\delta(\mathbf{k}_{12}-\mathbf{k})\gamma(\mathbf{k}_1,\mathbf{k}_2)A(\mathbf{k}_1)A(\mathbf{k}_2), \tag{51}$$

with

$$\alpha^{(s)}(\mathbf{k}_1,\mathbf{k}_2) = 1 + \frac{\mathbf{k}_1\cdot\mathbf{k}_2(k_1^2+k_2^2)}{2k_1^2k_2^2},\ \gamma(\mathbf{k}_1,\mathbf{k}_2) = 1 - \frac{(\mathbf{k}_1\cdot\mathbf{k}_2)^2}{k_1^2k_2^2}. \tag{52}$$

and $\lambda(t)$ has a complicated expression and for that will not be written but it's evolution is represented below.

Using all these results, it is easy to find $\theta^{(2)}(t,\mathbf{k})$ and $\delta\phi^{(2)}(t,\mathbf{k})$ that we will not reproduce here.

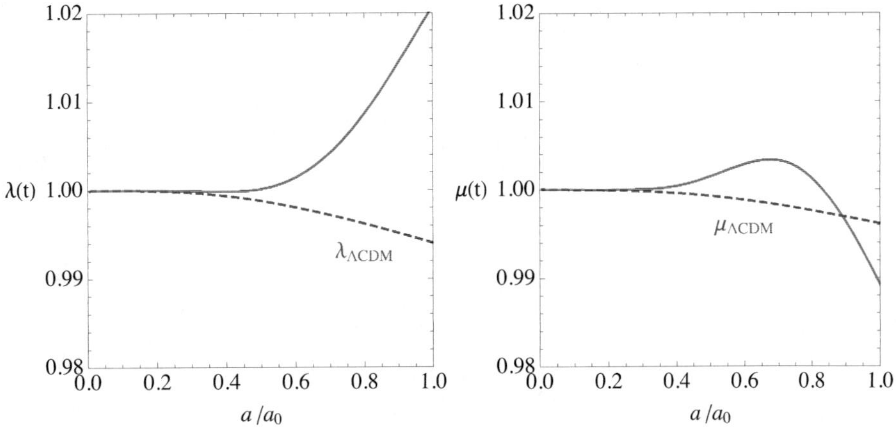

Figure 4. Evolution of the parameters λ and μ as a function of the scale factor. Notice that during matter era $\lambda = \mu \simeq 1$.

2.5. Third order perturbations

Following the same procedure, we can write the third order solution in the following form

$$\delta_m^{(3)}(t,\mathbf{k}) = D_+^3(t)\left(\mathcal{W}_{\alpha\alpha}(\mathbf{k}) - \frac{2}{7}\lambda(t)\mathcal{W}_{\alpha\gamma R}(\mathbf{k}) - \frac{2}{7}\lambda(t)\mathcal{W}_{\alpha\gamma L}(\mathbf{k}) - \frac{2}{21}\mu(t)\mathcal{W}_{\gamma\gamma}(\mathbf{k})\right), \qquad (53)$$

where the function μ is represented in Fig.4, and

$$\begin{aligned}
\mathcal{W}_{\alpha\alpha}(\mathbf{k}) &= \frac{1}{(2\pi)^6}\int d\mathbf{k}_1 d\mathbf{k}_2 d\mathbf{k}_3 \delta^{(3)}(\mathbf{k}_1+\mathbf{k}_2+\mathbf{k}_3-\mathbf{k})\alpha\alpha(\mathbf{k}_1,\mathbf{k}_2,\mathbf{k}_3)A(\mathbf{k}_1)A(\mathbf{k}_2)A(\mathbf{k}_3),\\
\mathcal{W}_{\alpha\gamma R}(\mathbf{k}) &= \frac{1}{(2\pi)^6}\int d\mathbf{k}_1 d\mathbf{k}_2 d\mathbf{k}_3 \delta^{(3)}(\mathbf{k}_1+\mathbf{k}_2+\mathbf{k}_3-\mathbf{k})\alpha\gamma_R(\mathbf{k}_1,\mathbf{k}_2,\mathbf{k}_3)A(\mathbf{k}_1)A(\mathbf{k}_2)A(\mathbf{k}_3),\\
\mathcal{W}_{\alpha\gamma L}(\mathbf{k}) &= \frac{1}{(2\pi)^6}\int d\mathbf{k}_1 d\mathbf{k}_2 d\mathbf{k}_3 \delta^{(3)}(\mathbf{k}_1+\mathbf{k}_2+\mathbf{k}_3-\mathbf{k})\alpha\gamma_L(\mathbf{k}_1,\mathbf{k}_2,\mathbf{k}_3)A(\mathbf{k}_1)A(\mathbf{k}_2)A(\mathbf{k}_3),\\
\mathcal{W}_{\gamma\gamma}(\mathbf{k}) &\equiv \frac{1}{(2\pi)^6}\int d\mathbf{k}_1 d\mathbf{k}_2 d\mathbf{k}_3 \delta^{(3)}(\mathbf{k}_1+\mathbf{k}_2+\mathbf{k}_3-\mathbf{k})\gamma\gamma(\mathbf{k}_1,\mathbf{k}_2,\mathbf{k}_3)A(\mathbf{k}_1)A(\mathbf{k}_2)A(\mathbf{k}_3),
\end{aligned}$$

with

$$\begin{aligned}
\alpha\alpha(\mathbf{k}_1,\mathbf{k}_2,\mathbf{k}_3) &= \frac{1}{3}(\alpha^{(s)}(\mathbf{k}_1,\mathbf{k}_2+\mathbf{k}_3)\alpha^{(s)}(\mathbf{k}_2,\mathbf{k}_3) + 2 \text{ cyclic terms}),\\
\alpha\gamma_R(\mathbf{k}_1,\mathbf{k}_2,\mathbf{k}_3) &= \frac{1}{3}\left(\alpha(\mathbf{k}_1,\mathbf{k}_2+\mathbf{k}_3)\gamma(\mathbf{k}_2,\mathbf{k}_3) + 2 \text{ cyclic terms}\right),\\
\alpha\gamma_L(\mathbf{k}_1,\mathbf{k}_2,\mathbf{k}_3) &= \frac{1}{3}\left(\alpha(\mathbf{k}_1+\mathbf{k}_2,\mathbf{k}_3)\gamma(\mathbf{k}_2,\mathbf{k}_3) + 2 \text{ cyclic terms}\right),\\
\gamma\gamma(\mathbf{k}_1,\mathbf{k}_2,\mathbf{k}_3) &= \frac{1}{3}\left(\gamma(\mathbf{k}_1,\mathbf{k}_2+\mathbf{k}_3)\gamma(\mathbf{k}_2,\mathbf{k}_3) + 2 \text{ cyclic terms}\right).
\end{aligned}$$

3. Power Spectrum

With the third order solution we can compute the second order power spectrum (1-loop correction) of matter density contrast using

$$\langle\delta(t,\mathbf{k}_1)\delta(t,\mathbf{k}_2)\rangle = (2\pi)^3\delta_m^{(3)}(\mathbf{k}_1+\mathbf{k}_2)P_{\delta\delta}(t,k), \qquad (54)$$

with $\delta(t,\mathbf{k}) = \delta_m^{(1)} + \delta_m^{(2)} + \delta_m^{(3)}$, which gives after some simplifications

$$P_{\delta\delta}(t,k) = D_+^2(t)P_L(k) + D_+^4(t)\left(P_{\delta\delta}^{(22)}(t,k) + 2P_{\delta\delta}^{(13)}(t,k)\right), \tag{55}$$

where $P_L(k)$ is the linear matter power spectrum (correlation function of the amplitude $A(k)$) and $P_{\delta\delta}^{(22)}(t,k)$ and $P_{\delta\delta}^{(13)}(t,k)$ are defined by

$$P_{\delta\delta}^{(22)}(t,k) = \frac{k^3}{98(2\pi)^2}\int dr P_{\rm L}(rk)\int_{-1}^{1} dx P_{\rm L}(k(1+r^2-2rx)^{1/2})\frac{((7-4\lambda)r+7x+2(2\lambda-7)rx^2)^2}{(1+r^2-2rx)^2}, \tag{56}$$

$$2P_{\delta\delta}^{(13)}(t,k) = \frac{k^3}{252(2\pi)^2}P_{\rm L}(k)\int dr P_{\rm L}(rk)\Big[12\mu\frac{1}{r^2} - 2(21+36\lambda+22\mu) + 4(84-48\lambda-11\mu)r^2 - 6(21-12\lambda-2\mu)r^4 + \frac{3}{r^3}\left(r^2-1\right)^3\left((21-12\lambda-2\mu)r^2+2\mu\right)\ln\left(\frac{r+1}{|r-1|}\right)\Big]. \tag{57}$$

After integration, we found

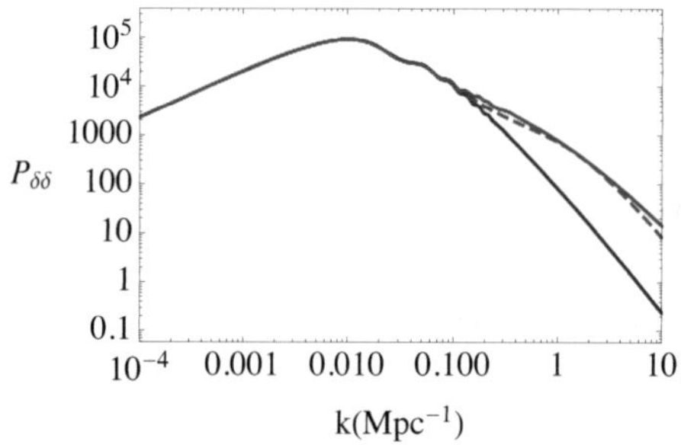

Figure 5. Evolution of the power spectrum as a function of the scale k. In blue, the linear power spectrum, in red the 1-loop correction and in dashed red the halofit model for the nonlinear matter power spectrum.

From Fig.5, we see that our 1-loop correction approximates very well the power spectrum at all scales while of course the linear spectrum fails for $k > 0.1$ Mpc^{-1} (small scales), because of the nonlinearities. This result constitute the first part before getting the nonlinear effective gravitational constant which will be presented in a future project.

References

[1] M. A. Luty, M. Porrati and R. Rattazzi, JHEP **0309** (2003) 029 [hep-th/0303116].
[2] A. Nicolis, R. Rattazzi and E. Trincherini, Phys. Rev. D **79** (2009) 064036 [arXiv:0811.2197 [hep-th]].
[3] P. Binetruy, Oxford, UK: Oxford Univ. Pr. (2006) 520 p
[4] A. D. Linde, Lect. Notes Phys. **738** (2008) 1 [arXiv:0705.0164 [hep-th]].
[5] A. Nicolis, R. Rattazzi and E. Trincherini, Phys. Rev. D **79** (2009) 064036 doi:10.1103/PhysRevD.79.064036 [arXiv:0811.2197 [hep-th]].
[6] Vainshtein A.I., 1972, Phys.Lett.B, 39, 396.
[7] Babichev E. & Deffayet C., 2013, Class.Quantum.Grav, 30, 184001.
[8] L. Perko. 1991. Springer. Differential Equations and Dynamical Systems Berlin
[9] T. Kobayashi, M. Yamaguchi and J. Yokoyama, Prog. Theor. Phys. **126** (2011) 511 doi:10.1143/PTP.126.511 [arXiv:1105.5723 [hep-th]].

IOP Conf. Series: Journal of Physics: Conf. Series **1043** (2018) 012024 doi:10.1088/1742-6596/1043/1/012024

Noether-Wald Charges in Critical Gravity

David Rivera-Betancour[1,2] and Rodrigo Olea[2],

[1]Instituto de Física, Pontificia Universidad Católica de Valparaíso, Casilla 4059, Valparaíso, Chile.

[2]Departamento de Física, Universidad Andres Bello, Sazié 2212, Piso 7, Santiago, Chile

E-mail: d.riverabe@gmail.com, rodrigo.olea@unab.cl

Abstract. Critical Gravity theory is defined by a particular combination of quadratic couplings in the curvature added on top of 4D Einstein-Hilbert action with negative cosmological constant. As the Lagrangian is given by a Weyl-squared term, the asymptotic form of the curvature is not modified. The coupling of the $Weyl^2$ term is such the massive scalar mode is eliminated and the massive spin-2 mode become massless, rendering the theory consistent around the critical point.

In the present work, we construct the Noether-Wald charges for the action of Critical Gravity. Such construction makes manifest a defining property of this theory: both the energy and entropy for Einstein black holes vanish identically.

1. Introduction

Quadratic-curvature gravity can be seen as a correction to Einstein-Hilbert theory, whose action is given by the general expression

$$I = \frac{1}{16\pi G}\int d^4x\sqrt{-g}\left(R - 2\Lambda + \alpha R_{\mu\nu}R^{\mu\nu} + \beta R^2\right) , \tag{1}$$

where α and β are arbitrary couplings, and $\Lambda < 0$ is the cosmological constant. The Riemann-squared term is absent because it can be always traded off by the Gauss-Bonnet (GB) invariant plus the other curvature-squared terms. The GB invariant does not contribute to the bulk dynamics of the theory. Generically speaking, this class of actions produces higher-derivative field equations, which describe a massless spin-2 graviton, a massive spin-2 field and a massive scalar.

For an arbitrary set of couplings, this kind of theories suffers from the appearance of ghosts. That problem can be solved by flipping the sign of the Einstein-Hilbert term. However, this change would produce a negative value for the mass of a Schwarzschild-AdS black hole. Thus, while the energy of the perturbations around a background are negative, the mass of a black hole is positive, or viceversa, what it is clearly inconsistent.

When the couplings take the particular values $\alpha = -3\beta$ and $\beta = -\frac{1}{2\Lambda}$, the massive mode scalar is eliminated and the massive spin-2 mode becomes massless, giving rise to a four-dimensional gravity theory which is free of the inconsistencies mentioned above [1].

Indeed, in [1], it was shown that, using the Ostrogradsky method for higher-order Lagrangians, the on-shell energy of the massive mode in the critical value of the couplings vanishes.

On the other hand, one can obtain the Abbott-Deser-Tekin mass for a Schwarzschild-AdS black hole [2, 3], which is given by

$$M = m\left(1 + 2\Lambda(\alpha + 4\beta)\right) \tag{2}$$

where m is the usual mass parameter in the solution.
It is easy to see that in (2), for the critical condition mentioned above, the mass for Schwarzschild-AdS black hole becomes zero.
The main aim of this note is to arrive at the same conclusion from the point of view of Noether-Wald charges [4].

2. Noether-Wald Charges

For an arbitrary theory of gravity in which the Lagrangian density depends only on the metric and the curvature $L(g_{\mu\nu}, R_{\mu\nu\alpha\beta})$, one can construct a conserved current for a set of Killing vectors $\{\xi^\mu\}$ which has the form [5]

$$\sqrt{-g}J^\mu = \Theta^\mu\left(\delta_\xi g\right) + \Theta^\mu\left(\delta_\xi \Gamma\right) + L\xi^\mu \tag{3}$$

The first term in (3) is zero due to the fact that $\delta_\xi g_{\mu\nu} = \nabla_\mu \xi_\nu + \nabla_\nu \xi_\nu = 0$ which corresponds to the Killing equation.

The rest of the current, for a generic gravity theory, is given by

$$\sqrt{-g}J^\mu = 2E^{\mu\nu}_{\alpha\beta}\left(\nabla_\nu \nabla^\alpha \xi^\beta + R^{\alpha\beta}_{\nu\sigma}\xi^\sigma\right) + \xi^\mu L \tag{4}$$

where $E^{\mu\nu}_{\alpha\beta}$ is the functional derivative with respect to the Riemann tensor $R^{\mu\nu}_{\alpha\beta}$

$$E^{\mu\nu}_{\alpha\beta} = \frac{\partial L}{\partial R^{\alpha\beta}_{\mu\nu}}\,. \tag{5}$$

The last two terms on the right side of (4) are proportional to the equation of motion and, as a consequence, they vanish on-shell.
The first term on the right side of (4) satisfies the Bianchi identity, what implies that the conserved current can be written as

$$\sqrt{-g}J^\mu = 2\nabla_\nu\left(E^{\mu\nu}_{\alpha\beta}\nabla^\alpha \xi^\beta\right) \tag{6}$$

We know that, whenever the current J^μ can be written as $J^\mu = \nabla_\nu Q^{\mu\nu}$, the conserved charge is expressed as an integral on the two-dimensional surface Σ

$$Q = \int_\Sigma d\Sigma_{\mu\nu} Q^{\mu\nu} \tag{7}$$

where $d\Sigma_{\mu\nu} = d^2x\sqrt{\sigma}u_\nu n_\mu$. Here, σ is the determinant of the two-dimensional surface metric, u_ν corresponds to a unit timelike vector, normal to the two-dimensional surface Σ and n_μ is a unit spacelike vector, normal to the boundary ∂M.
Finally the conserved charge is written as

$$Q[\xi] = 2\int_\Sigma d^2x\sqrt{\sigma}u_\nu n_\mu E^{\mu\nu}_{\alpha\beta}\nabla^\alpha \xi^\beta. \tag{8}$$

3. Conserved charges in Critical Gravity

The action for Critical Gravity is given by ($\Lambda = -3/\ell^2$ in terms of the AdS radius)

$$I = \frac{1}{16\pi G}\int d^4x\sqrt{-g}\left[(R+\frac{6}{\ell^2}) - \frac{\ell^2}{2}\left(R_{\mu\nu}R^{\mu\nu} - \frac{1}{3}R^2\right)\right] \tag{9}$$

An alternative form for the action of Critical Gravity considers the difference between $Weyl^2$ and the GB invariant E_4 [6], as the latter term does not contribute to the field equations

$$I = \int d^4x\sqrt{-g}\left[\frac{1}{16\pi G}\left(R+\frac{6}{\ell^2}\right) + \frac{\ell^2}{64\pi G}\left(E_4 - W^{\mu\nu\alpha\beta}W_{\alpha\beta\mu\nu}\right)\right]. \tag{10}$$

Using the Noether-Wald formula for the current (6), we can split the action in two parts, I_{ren} which correspond to the action of Einstein-Hilbert plus GB term and I_{CG} which correspond to the action of Conformal Gravity.
For the first part, the functional derivative with respect to the Riemann tensor of the Lagrangian in I_{ren} produces

$$E^{\mu\nu}_{\alpha\beta} = \sqrt{-g}\frac{\ell^2}{128\pi G}\delta^{[\mu\nu\gamma\lambda]}_{[\alpha\beta\rho\delta]}\left(R^{\rho\delta}_{\gamma\lambda} + \frac{1}{\ell^2}\delta^{[\rho\delta]}_{[\gamma\lambda]}\right), \tag{11}$$

whereas, for the Conformal Gravity part L_{CG}, we get

$$\tilde{E}^{\mu\nu}_{\alpha\beta} = -\sqrt{-g}\frac{\ell^2}{128\pi G}\delta^{[\mu\nu\gamma\lambda]}_{[\alpha\beta\rho\delta]}W^{\rho\delta}_{\gamma\lambda} \tag{12}$$

Using the Noether-Wald formula (8), the total charge for the theory

$$Q = \frac{\ell^2}{64\pi G}\int d^2x\sqrt{\sigma}u_\nu n_\mu\delta^{[\mu\nu\gamma\lambda]}_{[\alpha\beta\rho\delta]}\nabla^\alpha\xi^\beta\left[\left(R^{\rho\delta}_{\gamma\lambda} + \frac{1}{\ell^2}\delta^{[\rho\delta]}_{[\gamma\lambda]}+\right) - W^{\rho\delta}_{\gamma\lambda}\right]. \tag{13}$$

By definition, the Weyl tensor is

$$W^{\mu\nu}_{\alpha\beta} = R^{\mu\nu}_{\alpha\beta} - \frac{1}{2}R^{[\mu}_{[\alpha}\delta^{\nu]}_{\beta]} + \frac{1}{6}R\delta^{[\mu\nu]}_{[\alpha\beta]}. \tag{14}$$

For Einstein spaces, $R_{\mu\nu} = -3/\ell^2 g_{\mu\nu}$, the Weyl tensor adopts the particular form

$$W^{\rho\delta}_{\gamma\lambda} = R^{\rho\delta}_{\gamma\lambda} + \frac{1}{\ell^2}\delta^{[\rho\delta]}_{[\gamma\lambda]}, \tag{15}$$

where the r.h.s, is referred to as AdS curvature.

Using the above fact, the conserved quantity in Critical Gravity is identically zero for Einstein spaces.

4. Conclusions

In this work, we have provided a more explicit proof of the vanishing mass for Einstein black holes in Critical Gravity. Because the black hole entropy can be obtained as the Noether-Wald charge evaluated at the horizon, it is evident that the entropy for this class of solution is zero, as well. The departure from the Einstein condition in a more generic set of solutions for Critical Gravity should link the energy definition to a higher-derivative tensor, e.g., the Bach tensor [7].

IOP Conf. Series: Journal of Physics: Conf. Series **1043** (2018) 012024 doi:10.1088/1742-6596/1043/1/012024

5. Acknowledgments

We thank G. Anastasiou, F. Díaz-Martínez and O. Miskovic for helpful discussions. This work was funded in part by FONDECYT Grant 1131075, CONICYT Grant DPI 2014-0115 and UNAB Grants DI-1336-16/RG. D.R.B was financially supported by Facultad de Ciencias Exactas, UNAB, to attend SOCHIFI 2016 Meeting.

6. References

[1] H. Lu and C.N. Pope 2011, Phys. Rev. Lett. **106**, 181302
[2] S. Deser and B. Tekin 2002, Phys. Rev. Lett. **89**, 101101
[3] S. Deser and B. Tekin 2002, Phys. Rev. D **67**, 084009
[4] V. Iyer and R.M. Wald 1994, Phys. Rev. D **50**, 846
[5] R. Araneda, R. Aros, O. Miskovic and R. Olea 2016, Phys. Rev. D **93**, 084022
[6] O. Miskovic, M. Tsoukalas and R. Olea 2014, JHEP **08**, 108
[7] G. Anastasiou and R. Olea 2016, Phys. Rev. D **94**, 086008

IOP Conf. Series: Journal of Physics: Conf. Series **1043** (2018) 012025 doi:10.1088/1742-6596/1043/1/012025

Observational constraints on the interaction between dark matter and dark energy

Daniela Grandón and Víctor H. Cárdenas

Instituto de Física y Astronomía, Facultad de Ciencias, Universidad de Valparaíso, Av. Gran Bretaña 1111, Valparaíso, Chile.

E-mail: `daniela.grandon@alumnos.uv.cl; victor.cardenas@uv.cl`

Abstract. In this work we use the latest observations on BAO, SNIa, and $H(z)$, to constraint three models, two of them showing an explicit interaction between dark matter and dark energy. We found that using the three observational probes together, one of the interaction model (the IwCDM2) shows evidence for interaction at 2 σ. Although significant, further study is necessary to establish this claim firmly.

1. Introduction

Although the ΛCDM model has been confirmed as the model that best fits all the observational tests, since a few years, the level of precision of measurements and the increase in the number of them, have led to seriously consider a model *extension* [1]. Among the different ways in which we can deform the Model ΛCDM, are:

(i) To propose models where the cosmological constant is dynamic, that is, it changes with time. Among this family are the models of quintessence, for example, and from which comes the name *dark energy*, which is interpreted as a contribution to the matter content of the universe whose nature is unknown.

(ii) Models where the gravitational theory is modified, that is to say, it is expected to account for the effect of the cosmological constant.

(iii) Models where it relaxes one of the fundamental principles of cosmology, which is the homogeneity, openly violating the Copernican principle.

One of the models of the type (i) that has received much attention in recent years is the model of interaction between matter (DM) and dark energy (DE). Since we are unaware of the nature of DM (non baryonic) and DE, it is not unreasonable to assume that both can be related. An exploratory form in which this occurs is by assuming that there is a transfer of energy from DE to DM. In this context, a direct interaction between these dark contributions appears as an observational viable option [2].

The interaction is usually modeled phenomenologically by modifying the conservation equations through a function Q,

$$\dot{\rho} + 3H\rho_m = Q, \tag{1}$$
$$\dot{\rho}_{de} + 3H(1+\omega)\rho_{de} = -Q, \tag{2}$$

in such a way that only the sum of the contributions is conserved but not each one separately. If $Q < 0$ means that there is an energy transfer from DM to DE, and the opposite occurs for $Q > 0$.

2. The models

In this paper we study the restrictions on interaction models imposed by observational data. Here we study two models of interaction and as a reference, also the wCDM model with no interaction. Explicitly we study the cases: $Q_1 = 3\gamma H\rho_{de}$ that defines the model IwCDM1 and $Q_2 = 3\gamma H\rho_{dm}$ defining the model IwCDM2, both already studied in [9]. If γ is zero means there is no interaction. If $\gamma < 0$ indicates that there is transfer of energy from DM to DE.

As is well-known for the wCDM model the reduced Hubble function is

$$E^2(z) = \Omega_m(1+z)^3 + \Omega_r(1+z)^4 + (1-\Omega_m-\Omega_r)(1+z)^{3(1+w)}, \tag{3}$$

where $\Omega_r = 2.469 \times 10^{-5}h^{-2}(1+0.2271N_{\text{eff}})$ and $N_{\text{eff}} = 3.04$. In this case, the free parameters to be constrained by the observational data are: Ω_m, h and w.

For the IwCDM1 model the $E(z)$ function is given by

$$E^2(z) = \Omega_m(1+z)^3 + \Omega_r(1+z)^4 + +\Omega_x\left(\frac{\gamma}{w+\gamma}(1+z)^3 + \frac{w}{w+\gamma}(1+z)^{3(1+w+\gamma)}\right), \tag{4}$$

where γ is the parameter that makes the interaction manifest. Here $\Omega_m = \Omega_c + \Omega_b$ where Ω_c is the non-baryonic part and Ω_b is the baryonic one.

For the IwCDM2 model we obtain

$$\begin{aligned} E^2(z) = {} & \Omega_x(1+z)^{3(1+w)} + \Omega_r(1+z)^4 + \Omega_b(1+z)^3 + \\ & +\Omega_c\left(\frac{\gamma}{w+\gamma}(1+z)^{3(1+w)} + \frac{w}{w+\gamma}(1+z)^{3(1-\gamma)}\right). \end{aligned} \tag{5}$$

Here the free parameters are h, Ω_b, Ω_c, w and γ. It is clear that for $\gamma = 0$ both expressions - those for IwCDM1 and IwCDM2 - reduced to that of the wCDM model.

3. The data

In this work we test the models described in the previous section using 3 types of data: Measurements of $H(z)$, from type Ia supernovae (SNIa), and baryonic acoustic oscillations (BAO). Measurements of the Hubble function, $H(z)$ are taken from [3], which consist of 30 data covering a range of redshift between $z = 0.07$ to $z = 1.95$. The data from SNIa are from the Joint Light-curve Analysis (JLA) [4], where the function to be minimized is

$$\chi^2 = (\mu - \mu_{th})^T C^{-1}(\mu - \mu_{th}). \tag{6}$$

Here C corresponds to the covariance matrix delivered in [4], and the modular distance is assumed to take the shape

$$\mu = m - M + \alpha X - \gamma Y, \tag{7}$$

where m is the maximum apparent magnitude in band B, X is related to the widening of the light curves, and Y corrects the color. In general, cosmology is restricted to the parameters M, X and Y. The author of [4] also released the compressed version of JLA where only M is a free parameter.

In addition, we used data from baryonic acoustic oscillations (BAO). We used the results of the WiggleZ experiment [5], the distance measurements of SDSS DR7 BAO [6] and the data of the 6dFGS BAO [8]. The χ^2 for the BAO data of WiggleZ is given by

$$\chi^2_{WiggleZ} = (\bar{A}_{obs} - \bar{A}_{th})C^{-1}_{WiggleZ}(\bar{A}_{obs} - \bar{A}_{th})^T, \tag{8}$$

Where the data vector is $\bar{A}_{obs} = (0.474.0.442, 0.424)$ for an effective redshift $z = 0.44, 0.6$ and 0.73. The theoretical value corresponding $\bar{A}_{th}$ denotes the acoustic parameter $A(z)$ introduced in [7]. The inverse of the covariance is

$$C^{-1}_{WiggleZ} = \begin{pmatrix} 1040.3 & -807.5 & 336.8 \\ -807.5 & 3720.3 & -1551.9 \\ 336.8 & -1551.9 & 2914.9 \end{pmatrix}. \quad (9)$$

Similarly for SDSS data DR7, χ^2 is expressed as [6]

$$\chi^2_{SDSS} = (\bar{d}_{obs} - \bar{d}_{th}) C^{-1}_{SDSS} (\bar{d}_{obs} - \bar{d}_{th})^T, \quad (10)$$

where $\bar{d}_{obs} = (0.1905.0.1097)$ is the data for $z = 0.2$ and 0.35. $\bar{d}_{th}$ denotes the ratio $d_z = r_s(z_d)/D_V(z)$ where $r_s(z)$ is a comoving sound horizon, and $D_V(z)$ is a distance defined in [6]. The inverse matrix is given by

$$C^{-1}_{SDSS} = \begin{pmatrix} 30124 & -17227 \\ -17227 & 86977 \end{pmatrix}. \quad (11)$$

For the data of 6dFGS [8], there is only one point for $z = 0.106$, and the χ^2 is:

$$\chi^2_{6dFGS} = \left(\frac{d_z - 0.336}{0.015} \right)^2. \quad (12)$$

The total χ^2 for BAO is then $\chi^2_{BAO} = \chi^2_{WiggleZ} + \chi^2_{SDSS} + \chi^2_{6dFGS}$. For settings using only the data of $H(z)$ we have added a prior of Planck [10] for the Hubble constant, $H_0 = 67.8 \pm 0.9$ km s^{-1} Mpc^{-1}. More details of the work with the data see [11].

	χ^2_{min}	h	Ω_m	Ω_b	ω	γ
A	14.36/26	0.71 ± 0.09	0.19 ± 0.35	$--$	-1 ± 1	-0.2 ± 0.5
B	48.51/57	0.70 ± 0.01	0.23 ± 0.08	$--$	-0.9 ± 0.1	-0.1 ± 0.1
C	50.05/62	0.70 ± 0.01	0.30 ± 0.04	0.052 ± 0.018	-1.1 ± 0.1	0.10 ± 0.06

Table 1. Best fit values of the cosmological parameters for the first interaction model IwCDM1 using different set of data. A using only $H(z)$ data. B using $H(z)$+SNIa. C using $H(z)$+SNIa +BAO.

	χ^2_{min}	h	Ω_m	Ω_b	ω	γ
A	14.06/26	0.74 ± 0.29	0.4 ± 0.1	0.04 ± 0.03	-2 ± 3	-0.2 ± 0.2
B	48.50/57	0.70 ± 0.01	0.25 ± 0.05	0.0 ± 0.7	-0.9 ± 0.1	-0.05 ± 0.14
C	50.10/63	0.70 ± 0.01	0.30 ± 0.04	0.08 ± 0.02	-1.0 ± 0.1	0.024 ± 0.016

Table 2. Best fit values of the cosmological parameters for the first interaction model IwCDM2 using different set of data. The meaning of A, B and C is the same as in Table 1.

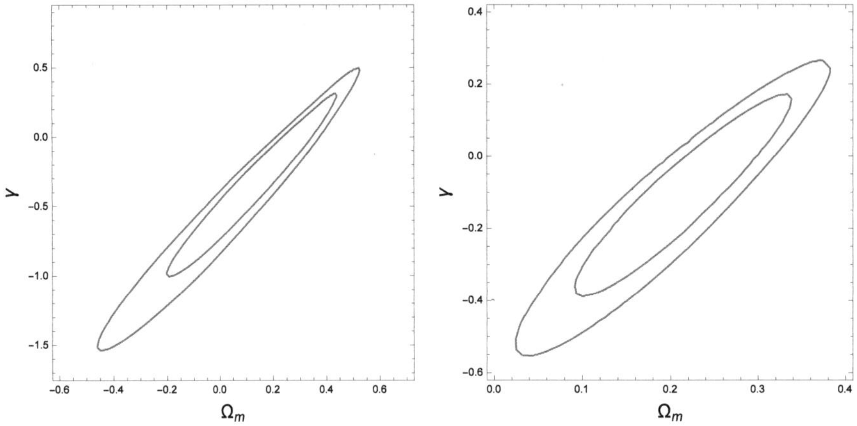

Figure 1. We display confidence boundaries for 1σ and 2σ for the model IwCDM1 in the Ω_m and γ parameters using of $H(z)$ (left panel) and $H(z)$ + SNIa (right panel).

4. Results

In Figure 1, the 1σ and 2σ confidence boundaries for the Ω_m and γ parameters of the IwCDM1 model are shown, using $H(z)$ measurements (left panel) and $H(z)$ + SNIa (right panel). According to the figure, the constraints imposed by the data is consistent with $\gamma = 0$, indicating that there is no preference for an interacting model. Once we add the contribution of SNIa, that is, using simultaneously data of $H(z)$ and SNIa, the confidence boundaries are reduced as shown in Figure 1. In this case, we note that the uncertainty in the determination of the parameters decreases appreciably, however, this time we notice that $\gamma = 0$ is on the edge of the region at 1σ in confidence. To conclude with the IwCDM1 model, we show also the result of the best fit

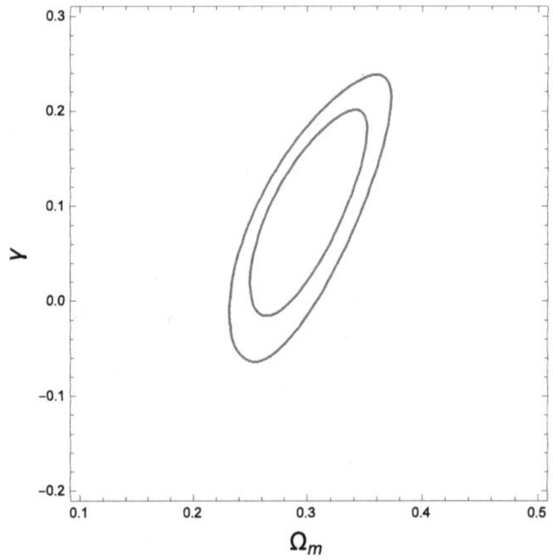

Figure 2. We display confidence boundaries for 1σ and 2σ for the model IwCDM1 in the Ω_m and γ parameters using H(z)+SNIa+BAO.

using the latest measurements of BAO. The 1 and 2 sigma confidence boundaries are shown in Figure 2. Here the effect of the BAO points is clear. The confidence boundaries move along the degeneracy direction, making the best fit for Ω_m increase, and the best fit for γ is now centered on a positive value - indicating energy transfer from dark energy to dark matter - discarding the value $\gamma = 0$ to 1 sigma.

As a comparison with the previous results, we next show the confidence boundaries for the IwCDM2 model using only $H(z)$ data in Figure 3. As we can see, the best fit for this case looks quite similar to that shown in Figure 2, where again $\gamma = 0$ is discarded at 1 sigma and Ω_m is centered around $\Omega_m \simeq 0.3$.

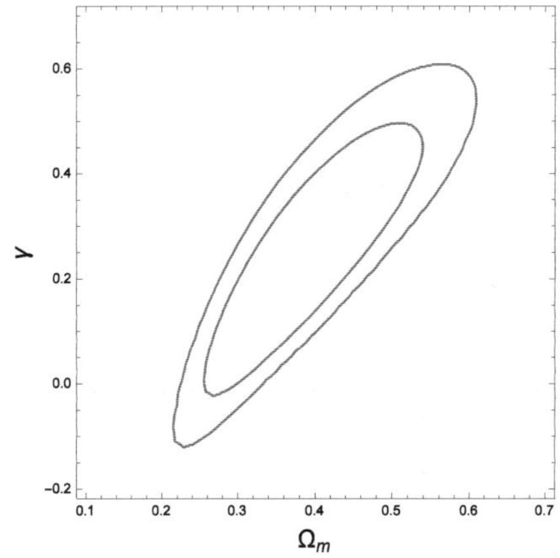

Figure 3. We display confidence boundaries for 1σ and 2σ for the model IwCDM2 in the Ω_m and γ parameters using only the data of $H(z)$.

A direct comparison with Fig. 1 for the previous case, we notice that this time the dispersion is lower, although the best fit value for Ω_m is relatively high (see Table II). We notice that already with only the data of $H(z)$ is seems clear a slight preference for a value $\gamma > 0$.

By plotting the contours for the IwCDM2 model using $H(z)$+SNIa we obtain the contours of Figure 4. Clearly the best fit for this model is consistent with no interaction. As can be notice in Table II, the best fit for H_0 decrease, and Ω_m takes a value closer to the one for the ΛCDM model. In this way, the addition of SNIa leads to $\gamma = 0$.

Now adding the BAO points we find a preference for interaction even valid at 2 σ, as can be see from Table II and Fig. 5. Although this study does not incorporate dynamical constraints, as those from perturbations, our results seems to indicate a chance to find evidence for a non-zero

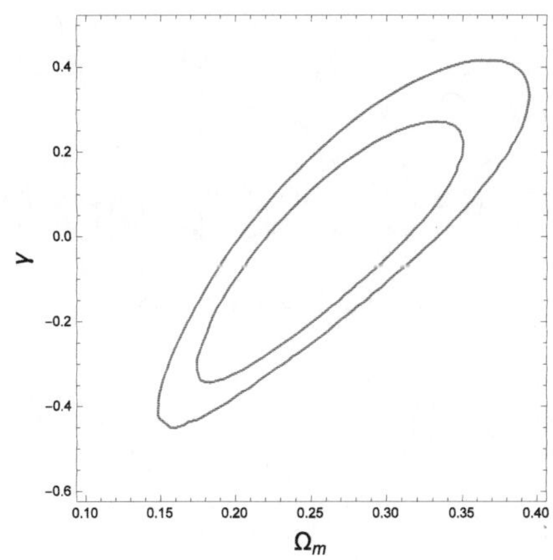

Figure 4. We display confidence boundaries from 1σ to 4σ for the IwCDM2 model in the Ω_m and γ parameters using $H(Z)$+SNIa.

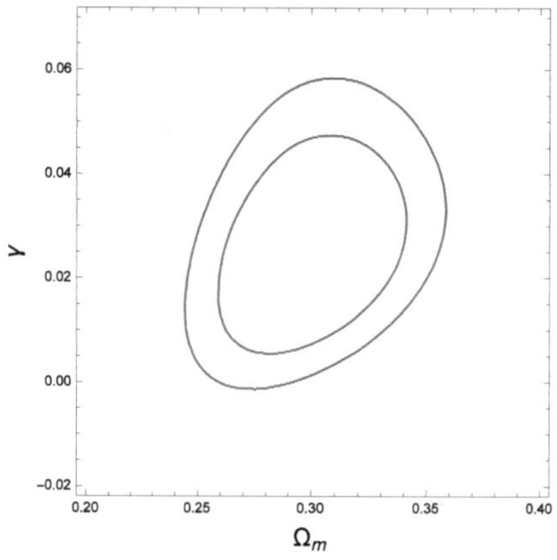

Figure 5. We display confidence boundaries for 1σ and 2σ for the IwCDM2 model in the Ω_m and γ parameters using $H(z)$+SNIa+BAO.

interaction term in the recent cosmological evolution.

Taking into account the results obtained, it is clear that it is necessary to add new observational tests to be more conclusive, mainly because the models have 5 free parameters. This work is in progress.

5. Discussion

Using the data, $H(z)$ + SNIa + BAO, the model IwCDM1 discards the solution $\gamma = 0$ to 1 sigma, whereas in the same case, the model IwCDM2 discards the solution $\gamma = 0$ at 2 sigma, clearly consistent with the existence of interaction. This result is in conflict with what was published in [9] where the authors found that in both models the data they used indicated no interaction. For $H(z)$ + SNIa, the analysis also does not show conclusive positive evidence, however, the evidence for interaction seems slightly larger in the IwCDM2 model than in IwCDM1. Finally, using only $H(z)$ data the evidence of interaction is clearly greater in the IwCDM2 model than in IwCDM1, as can be seen by comparing Figures 1 and 3. Clearly it is necessary to add new observational tests to be more conclusive in this respect.

Acknowledgments

This work has been supported in part by DIUV project No. 50/2013.

References

[1] E. Di Valentino, A. Melchiorri and J. Silk, Phys. Rev. D **92**, no. 12, 121302 (2015).
[2] V. Salvatelli, N. Said, M. Bruni, A. Melchiorri and D. Wands, Phys. Rev. Lett. **113** (2014) 18, 181301.
[3] M. Moresco *et al.*, JCAP **1605**, no. 05, 014 (2016)
[4] M. Betoule *et al.* [SDSS Collaboration], Astron. Astrophys. **568**, A22 (2014)
[5] C. Blake et al., MNRAS, 418, 1707, 2011
[6] W. J. Percival et al., MNRAS, 401, 2148, 2010
[7] D. J. Eisenstein et al., ApJ, 633, 560, 2005
[8] F. Beutler et al., MNRAS, 416, 3017, 2011
[9] D. Xia, S. Wang, MNRAS **463**, 952-956 (2016)
[10] Planck Collaboration, Astron. and Astrophys. 594, A13 (2016).
[11] V. H. Cardenas, Phys. Lett. B **750**, 128 (2015)

IOP Conf. Series: Journal of Physics: Conf. Series **1043** (2018) 012026 doi:10.1088/1742-6596/1043/1/012026

Stable Emergent Universe from Conservation Laws

P Labraña[1], S del Campo[2], R Herrera[2],and E Guendelman[3]

1 Departamento de Física, Universidad del Bío Bío and Grupo de Cosmología y Gravitación-UBB, Avenida Collao 1202, Casilla 5-C, Concepción, Chile.

2 Instituto de Física, Pontificia Universidad Católica de Valparaíso, Avenida Brasil 2950, Casilla 4059, Valparaíso, Chile.

3 Physics Department, Ben Gurion University of the Negev, Beer Sheva 84105, Israel

E-mail: `plabrana@ubiobio.cl`

Abstract. In this paper, based on JCAP 1608 (2016) 049, we show that there is a class of emergent universes models, derived from scale invariant two measures theories with spontaneous symmetry breaking (s.s.b) of the scale invariance, which can have both classical stability and do not suffer the instability pointed out by Mithani-Vilenkin towards collapse [4, 3, 2, 5]. We find that this stability is due to the presence of a symmetry in the "emergent phase", which together with the non linearities of the theory, does not allow that the FLRW scale factor to be smaller that a certain minimum value a_0 in a certain protected region.

The standard hot big-bang model provides us with the description of how the universe evolves, explaining the observational facts, such as the Hubble expansion, the $3K$ microwave background radiation and the abundance of light elements. However, this model presents some problems in its evolution. We will reference some of then; the smoothness or horizon problem, the flatness, the structure or primordial density problem, etc.. These problems can be solved in the context of the inflationary universe [6], where the essential feature of any inflationary model is the rapid but finite period of expansion that the universe underwent at very early times in its evolution. Perhaps the most import feature of the inflationary universe model is that it provides a causal explication for the origin of the observed anisotropy in the cosmic microwave background radiation (CMB), and also to the distribution of large-scale structures, which are consistent with the observations [7, 8].

However, one should point out that even in the context of the inflationary scenario one still encounters the initial singularity problem [9, 10] showing that the universe necessarily had a singular beginning for generic inflationary cosmologies [11].

One interesting way to avoid the initial singularity problem is to consider the emergent universe (EU) scenario [12]. The emergent universe refers to models in which the universe emerges from a past eternal Einstein static state (ES), inflates, and then evolves into a hot big bang era. The EU is an attractive scenario since it avoids the initial singularity and provides a smooth transition towards an inflationary period.

The original proposal for the emergent universe [12] supported an instability at the classical level of the ES state, and various models intended to formulate a stable model have been given [13], in particular the Jordan Brans Dicke models [14]. In this context, Mithani-Vilenkin in Refs. [3]-[5] have shown that certain classically stable static universes could be unstable semiclassically towards collapse. In this work, we show that there is a class of emergent universes derived from scale invariant two measures theories with spontaneous symmetry breaking of the scale invariance, which can have both classical stability and do not suffer the instability pointed out by Mithani-Vilenkin towards collapse of the ES state.

In a series of papers [15]-[18] we have studied a class of EU scenarios which are based on a spontaneously broken scale symmetry induced by the dynamics of a Two Measures Field Theory (TMT)[19]-[24], (see also Ref.[15]). In such model there is a dilaton field ϕ and the EU as the $t \longrightarrow -\infty$ is well described by an Einstein static universe, where t is the cosmological time.

Here we want to consider the detailed analysis of the EU solutions of the model developed in Ref. [16]. The results obtained in this case can also be applied to models studied in Refs. [15]-[18], which present similar symmetries as the model in Ref. [16].

We start by considering the Friedmann-Robertson-Walker closed cosmological solutions of the form

$$ds^2 = dt^2 - a(t)^2\left(\frac{dr^2}{1-r^2} + r^2(d\theta^2 + sin^2\theta d\phi^2)\right), \phi = \phi(t), \tag{1}$$

where $a(t)$ is the scale factor, and the scalar field ϕ is a function of the cosmic time t only, due to homogeneously and isotropy. We will consider a scenario where the scalar field ϕ is moving in the extreme left region $\phi \to -\infty$. In this case, the expressions for the energy density ρ and pressure p are given by

$$\rho = \frac{A}{2}\dot{\phi}^2 + 3B\dot{\phi}^4 + C, \tag{2}$$

and

$$p = \frac{A}{2}\dot{\phi}^2 + B\dot{\phi}^4 - C, \tag{3}$$

where the constants A, B and C are given by,

$$A = 1 - \frac{2\delta b_g V_1}{4(b_g V_1 - V_2)}, \quad B = -\frac{\delta^2 b_g^2}{4(b_g V_1 - V_2)}, \quad \text{and} \quad C = \frac{V_1^2}{4(b_g V_1 - V_2)}. \tag{4}$$

As was discussed in Ref.[16], the emergent universe can turn into inflation only if $C > 0$. On the other hand, in order to have a scenario in which the emergent universe evolves from an static and classically stable universe at $a = a^*$ with

$$a^* = \sqrt{\left(\frac{3}{8\pi G}\right)\frac{12B}{A^2 + 24B\,C - A\sqrt{A^2 + 12B\,C}}}, \tag{5}$$

and then passed to an inflationary phase, the following conditions must to be met, see Ref. [16]:

$$0.5 < y < 0.54\,, \tag{6}$$

$$B < 0\,, \tag{7}$$

$$-\frac{1}{64B} < C < -\sqrt{\frac{3}{B^2}} - \frac{7}{4B}\,. \tag{8}$$

Where $A = 1 - y$ and we have defined $y = \frac{2\delta b_g C}{V_1}$.

Now will turn our attention to possible quantum tunneling from the solution $a = a^*$ to $a = 0$, during the static regimen of the EU scenario. Let us first note that there is a conserved quantity Π_ϕ, due to the fact that from $\phi \to -\infty$, there is other symmetry $\phi \to \phi + c$. Given that in the Einstein frame we can use the action

$$S = \frac{1}{\kappa}\left[\int R\sqrt{-g}\,d^4x + \int p\sqrt{-g}\,d^4x\right], \tag{9}$$

and the symmetry $\phi \to \phi + c$, in which c is a constant, leads to the conservation law

$$a^3(t)\,[A\dot{\phi} + 4\,B\dot{\phi}^3] = \Pi_\phi = const.. \tag{10}$$

Without loss of generality, let us consider $\Pi_\phi > 0$. From conservation equation (10), we can write a as a function of $\dot{\phi}$

$$a(\dot{\phi}) = \left(\frac{\Pi_\phi}{A\dot{\phi} + 4\,B\dot{\phi}^3}\right)^{1/3}. \tag{11}$$

We can note that in this case $-\infty < \dot{\phi} < -\sqrt{\frac{A}{4|B|}}$ or $0 < \dot{\phi} < \sqrt{\frac{A}{4|B|}}$ in order to satisfied $\Pi_\phi > 0$. When $\dot{\phi}$ is in the first region $a(\dot{\phi})$ is a function which approach to zero when $\dot{\phi} \to -\infty$ and diverges when $\dot{\phi} \to -\sqrt{\frac{A}{4|B|}}$. But in this region ρ becomes negative see Eq. (2), then we are not interested in this case.

On the other hand, when $\dot{\phi}$ is in the second region, $a(\dot{\phi})$ has an extremum (minimum) at $\dot{\phi} = \dot{\phi}_0$, where $a(\dot{\phi}_0) = a_0$, with

$$\dot{\phi}_0 = \sqrt{\frac{A}{12|B|}}, \tag{12}$$

$$a_0 = \left(\frac{12|B|}{A}\right)^{1/6}\left[\frac{3\Pi_\phi}{2A}\right]^{1/3}. \tag{13}$$

Also from Eq. (11), we obtain that in this region a diverges when $\dot{\phi}$ approach to zero or to $\sqrt{\frac{A}{4|B|}}$.

Therefore, we can note that a smaller scale factor than a_0 is out of the range where the scale factor is defined for the physical solutions.

As an example, in Fig. 1 we have plotted $a(\dot{\phi})$, where we have considered $B = -1$, $C = 0.016$, $y = 0.505964$ and $\Pi_\phi = 113.41$.

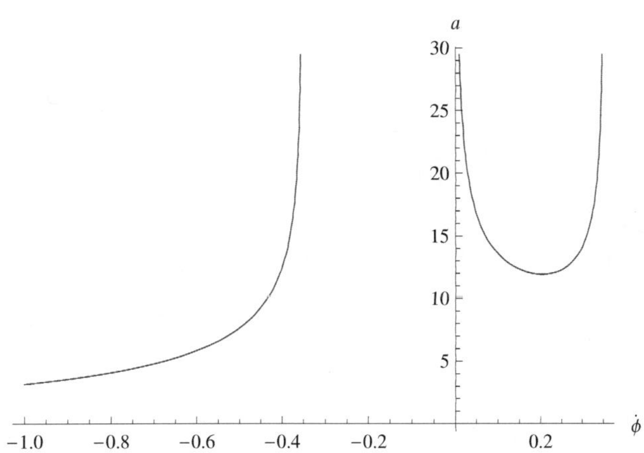

Figure 1. From Eq. (10), the scale factor a as a function of $\dot{\phi}$, when it is consider $\Pi_\phi = 113.41$.

From Eq. (10) we can obtain $\dot{\phi}$ as a function of a. We have three solutions:

$$\dot{\phi}_1 = -\frac{a^3 A}{2\,3^{1/3}\left(9a^6B^2\Pi_\phi + \sqrt{3}\sqrt{a^{18}A^3B^3 + 27a^{12}B^4\Pi_\phi^2}\right)^{1/3}} + \frac{\left(9a^6B^2\Pi_\phi + \sqrt{3}\sqrt{a^{18}A^3B^3 + 27a^{12}B^4\Pi_\phi^2}\right)^{1/3}}{2\,3^{2/3}a^3 B}, \tag{14}$$

$$\dot{\phi}_2 = \frac{\left(1+i\sqrt{3}\right)a^3 A}{4 3^{1/3}\left(9a^6B^2\Pi_\phi + \sqrt{3}\sqrt{a^{18}A^3B^3 + 27a^{12}B^4\Pi_\phi^2}\right)^{1/3}} - \frac{\left(1-i\sqrt{3}\right)\left(9a^6B^2\Pi_\phi + \sqrt{3}\sqrt{a^{18}A^3B^3 + 27a^{12}B^4\Pi_\phi^2}\right)^{1/3}}{4 3^{2/3}a^3 B}, \tag{15}$$

$$\dot{\phi}_3 = \frac{\left(1-i\sqrt{3}\right)a^3 A}{4 3^{1/3}\left(9a^6B^2\Pi_\phi + \sqrt{3}\sqrt{a^{18}A^3B^3 + 27a^{12}B^4\Pi_\phi^2}\right)^{1/3}} - \frac{\left(1+i\sqrt{3}\right)\left(9a^6B^2\Pi_\phi + \sqrt{3}\sqrt{a^{18}A^3B^3 + 27a^{12}B^4\Pi_\phi^2}\right)^{1/3}}{4 3^{2/3}a^3 B}. \tag{16}$$

As an example we plot these solutions in Figs. (2, 3, 4), where we have used the values of Ref. [16] and $\Pi_\phi = 113.41$. We can note that the plots are fully consistent with Fig. 1.

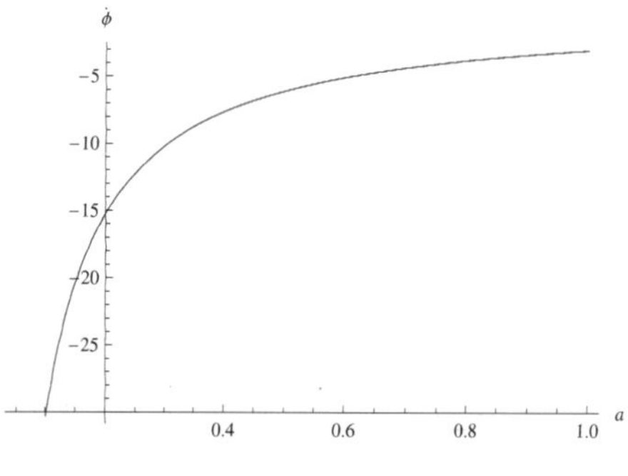

Figure 2. From Eq. (10), $\dot{\phi}_1$ as function of a when it is consider $\Pi_\phi = 113.41$.

The classical theory which describe this universe can be regarded as a constrained dynamical system with a Hamiltonian

$$\mathcal{H} = -\frac{G}{3\pi\,a}\left(p_a^2 + U(a)\right), \tag{17}$$

where

$$p_a = -\frac{3\pi}{2G}\,a\,\dot{a}, \tag{18}$$

is the momentum conjugate to a, and $U(a)$ corresponds to the effective potential given by

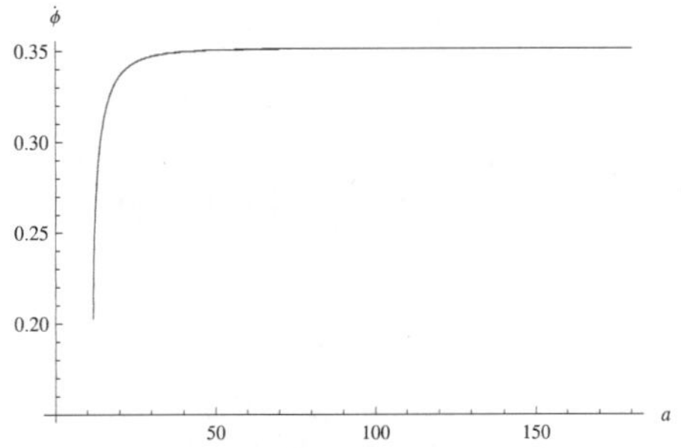

Figure 3. From Eq. (10), $\dot{\phi}_2$ as function of a when it is consider $\Pi_\phi = 113.41$.

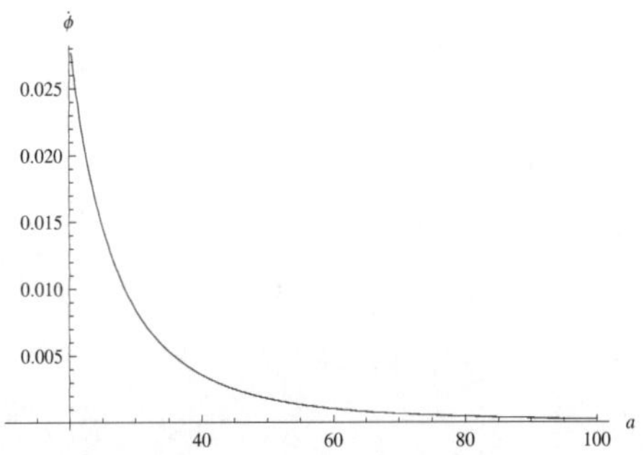

Figure 4. From Eq. (10), $\dot{\phi}_3$ as function of a when it is consider $\Pi_\phi = 113.41$.

$$U(a) = \left(\frac{3\pi}{2G}\right) a^2 \left(1 - \frac{8\pi G}{3} a^2 \rho(a)\right), \tag{19}$$

where we have written ρ as a function of a. It is possible to do that by using the solutions Eqs.(14, 15, 16) and Eq. (2).

The Hamiltonian constraint is $\mathcal{H} = 0$, from where we obtain the Friedmann equation

$$\left(\frac{\dot{a}}{a}\right)^2 = \frac{8\pi G}{3} \rho(a) - \frac{1}{a^2}. \tag{20}$$

Also, we can obtain the equation for $\dot{H}$ given by

$$\dot{H} = -4\pi G(\rho + p) + \frac{1}{a^2}. \tag{21}$$

One of the characteristic of the EU scenario is the period of superinflation after de static regimen and before inflation where $(\dot{H}) > 0$, see [25]. From Eq. (21), we note that in the relevant solutions of our model we do not need to violate the null energy condition, $\rho + p > 0$, in order to have an EU scenario (with a superinflationary phase) and avoid the initial singularity, since we are considering a closed universe. This is different from what happens in models as Ref. [26], which by the way shows that some violations of the energy conditions can be consistent.

In the context of quantum theory, the universe could be described by a wave function $\psi(a)$, the conjugate momentum p_a becomes the differential operator $-id/da$ and the constraint is replaced by the Wheeler-DeWitt (WDW) equation [27]

$$\mathcal{H}\psi(a) = 0, \tag{22}$$

$$\left(-\frac{d^2}{da^2} - \frac{\beta}{a}\frac{d}{da} + U(a)\right)\psi(a) = 0\,, \tag{23}$$

where we have used the minisuperspace approximation, which is appropriated for our model where the universe is homogeneous isotropic and closed during the ES regimen and therefore has a single degree of freedom, the scale factor [28]. The parameter β represents the ambiguity in the ordering of the non-commuting factors a and p_a in the Hamiltonian. The value of β does not affect the wave function in the semiclassical regimen, and usually in the study of semi-classical stability of EU it is chosen to be zero, see [3, 2, 4, 5].

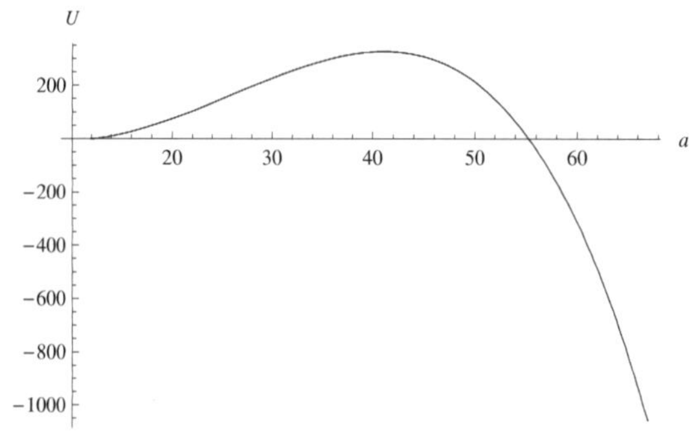

Figure 5. Potential $U(a)$ for $a > a_0$. Here, we have used Eq. (15).

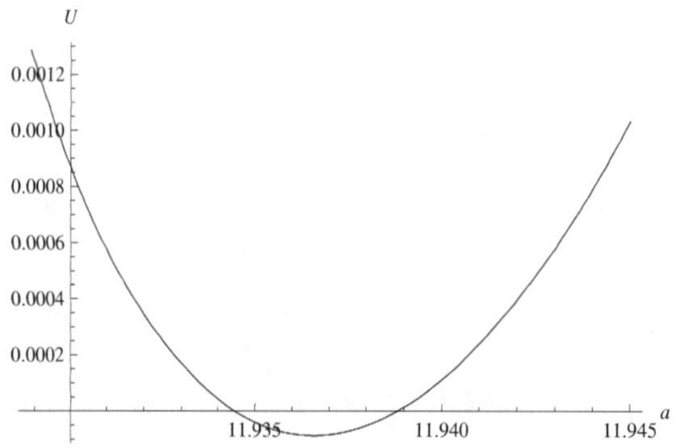

Figure 6. Potential $U(a)$ for $a > a_0$ near the equilibrium point. Here, we have used Eq. (15).

In order to obtain the potential $U(a)$ for the case of the EU, we have to select one of the solutions Eqs.(14, 15, 16) which is related with the static and classically stable solution. This classical solution was discussed in Ref. [16]. In this case this solution is Eq. (15). When we consider this solution the potential $U(a)$ has a local minimum at $a = a^*$, where a^* was defined in Eq. (5) and a local maximum at $a = a'$, where

$$a' = \sqrt{\left(\frac{3}{8\pi G}\right)\frac{12B}{A^2 + 24B\,C + A\sqrt{A^2 + 12B\,C}}}\,. \tag{24}$$

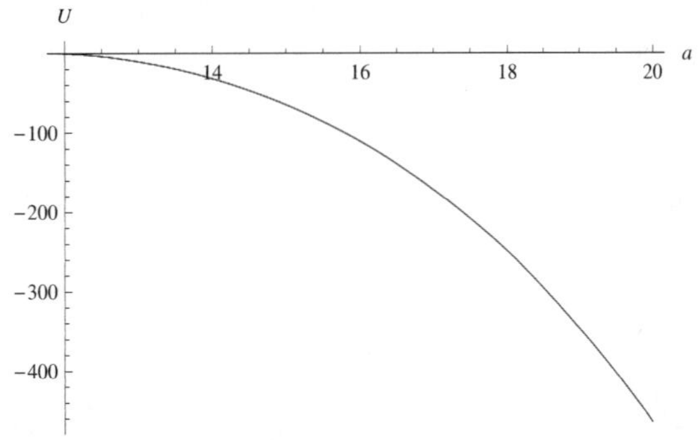

Figure 7. Potential $U(a)$ for $a > a_0$. Here, we have used Eq. (16).

The nature of these two equilibrium points was discussed in Ref.[16], where it is shown that $a = a^*$ is an stable equilibrium point and $a = a'$ is an unstable equilibrium point. Then, the system is classically stable near the static solution, $a \sim a^*$. There is a finite barrier which prevents the scale factor to go from $a \simeq a^*$ to infinity and the potential is not well defined for $a < a_0$ given the discussion above, this can be interpreted as a hard wall at $a = a_0$ for the potential $U(a)$.

As an example, in Fig. 5 it is plot the potential for the values allowed for a, that is $a > a_0$. In Fig. 6 it is plot the potential $U(a)$ near the static point (where also was consider $a > a_0$).

Since a smaller scale factor than a_0 is out of the range where the scale factor is defined for the physical solutions, we find that the possible instability towards a equal zero is not even a logical possibility in this context. Nevertheless, we observe that exist the possibility of tunneling through the finite barrier from the static solution to an expanding universe, see Fig. 5. This is an interesting scenario to study in future works.

As an example, in Fig. 7 it is show the potential for the solution Eq. (16), we can note, as we expected, that in this case there is not a equilibrium point as in Fig. 6.

From the Friedmann equation, Eq. (20), and Eq. (10) we can note that solutions $\dot{\phi}_2$ and $\dot{\phi}_3$ are not connected by the dynamics of the system. Solution Eq. (15) satisfies $\dot{\phi}_2 > \dot{\phi}_0$ and solution Eq. (16) satisfies $\dot{\phi}_3 < \dot{\phi}_0$, and it is not possible to cross the line $\dot{\phi} = \dot{\phi}_0$. At this respect, and by using Eq.(10), we can rewrite Eq. (20) as the following equations for $\dot{\phi}$,

$$\ddot{\phi}^2 + V(\dot{\phi}) = 0\,, \tag{25}$$

where

$$V(\dot{\phi}) = \frac{(A\dot{\phi} + 4B\dot{\phi}^3)^2}{(A + 12B\dot{\phi}^2)^2}\left[\frac{1}{\Pi_\phi^{2/3}}(A\dot{\phi} + 4B\dot{\phi}^3)^{2/3} - \frac{\kappa}{3}\left(\frac{A}{2}\dot{\phi}^2 + 3B\dot{\phi}^4 + C\right)\right]. \tag{26}$$

Then, from Eq. (26) we note that the solutions $\dot{\phi}_2$ and $\dot{\phi}_3$ are classically disconnected since $V \to \infty$ at the value $\dot{\phi} = \dot{\phi}_0 = \sqrt{\frac{A}{12|B|}}$. However, there is the possibility of tunneling through this divergent barrier, see [29]. In this case, the tunneling correspond to a quantum tunneling from the static solution to an expanding universe with initial values $a = a_0$.

As an example in Fig. 8 it is shown the potential $V(\dot{\phi})$, where we have used $\Pi_\phi = 113.41$ and the values of Ref. [16]. We can note that there is an infinite barrier at $\dot{\phi} = \dot{\phi}_0 = 0.20$.

In Fig. 9 we show the dependence of the potential $V(\dot{\phi})$ as a function of $\dot{\phi}$, near the equilibrium point.

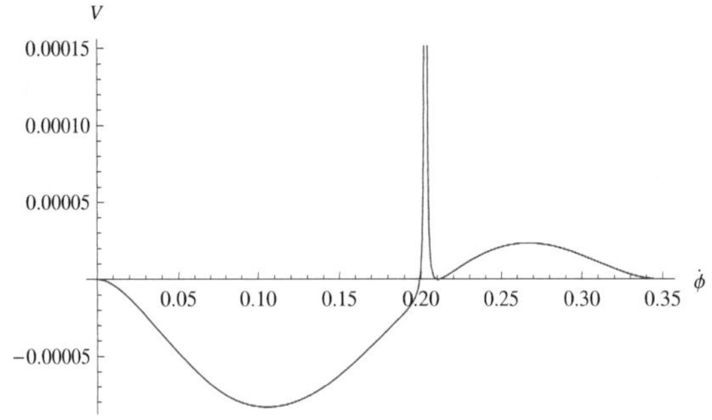

Figure 8. Potential $V(\dot{\phi})$, for $\dot{\phi} > 0$.

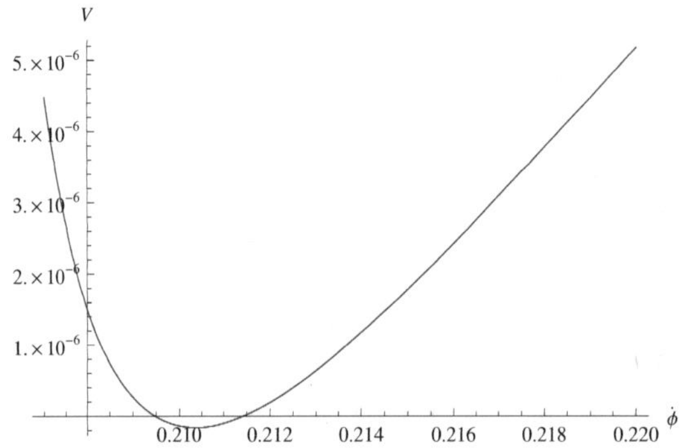

Figure 9. Potential $V(\dot{\phi})$ near the equilibrium point.

Therefore, we note that both tunnelings discussed above do not correspond to a collapse to $a \to 0$, but a creation of an expanding universe.

1. Discussion and Conclusions

It has been recently pointed out by Mithani-Vilenkin [4, 3, 2, 5] that certain emergent universe scenarios which are classically stable are nevertheless unstable semiclassically to collapse. In this work, we shown that there is a class of emergent universes derived from scale invariant two measures theories with spontaneous symmetry breaking of the scale invariance, which can have both classical stability and do not suffer the instability pointed out by Mithani-Vilenkin towards collapse. This stability is due to the presence of a symmetry in the "emergent phase", which together with the non linearities of the theory, does not allow the FLRW scale factor to be smaller that a certain minimum a_0 in a certain protected region.

Since a smaller scale factor than a_0 is out of the range where it is defined for the physical solutions $\dot{\phi}_2$ and $\dot{\phi}_3$ where ρ is positive, we have found that the possible instability towards a scale factor equal zero is not even a logical possibility in this context. Therefore our model is free of the instability towards collapse described in Refs. [4, 3, 2, 5]. The conserved quantity $\Pi_\phi \neq 0$ provides in this case with a protection towards collapse to a equal zero.

It is interesting to observe that exist the possibility of tunneling through the finite barrier of the potentials $U(a)$ and $V(\dot{\phi})$ from the static solution to an expanding universe, but also there is the possibility of tunneling through the divergent barrier of potentials $V(\dot{\phi})$, see Eq. (26). In this case, the tunneling correspond to a quantum tunneling from the static solution to an expanding universe with initial values $a = a_0$. We have noted that both tunnelings process, do not correspond to a collapse to $a \to 0$, instead they correspond to a creation of an expanding universe. This is an interesting scenario to study in future works and correspond to an alternative scheme for an emergent universe scenario, similar to the one studied in Refs. [31].

In particular in this work we studied the model Ref. [16], but the results obtained in this work can also be applied to models studied in Refs. [15]-[18], which present similar symmetries as the model in Ref. [16].

We should mention that, we have considered a closed universe, where the contribution of the curvature term is relevant before the inflationary period. Nevertheless, they are the possibility of contrast the EU with observation by studying the superinflationary period of these models. As it was reported in Ref. [25], during the superinflationary period, the EU scenario produces a suppression of the CMB anisotropies at large scale which could be responsible for the observed lack of power at large angular scales of the CMB. We hope to be able to analyze this suppression and also submitting our model to further test such as CMB temperature anisotropies and density perturbations. This will be the subject of a future work.

2. Acknowledgements

The authors dedicate this article to the memory of Professor Sergio del Campo (R.I.P.).

We thank Professor Alexander Vilenkin for suggesting and encouraging us to study the problem of the quantum stability of the emergent universe scenario and for multiple discussions on this subject. This work was supported by Comisión Nacional de Ciencias y Tecnología through FONDECYT Grants 1110230 (SdC), 1130628 (RH). Also it was supported by Pontificia Universidad Católica de Valparaíso through grants 123.787-2007 (SdC) and 123724 (RH). One of us (E.I.G) would like to thank the astrophysics and cosmology group at the Pontificia Universidad Católica de Valparaíso and the Frankfurt Institute of Advanced Studies of Frankfurt University for hospitality. P. L. is supported by Dirección de Investigación de la Universidad del Bío-Bío through Grants N^0 166907 2/R, and GI 150407/VC.

References

[1] S. del Campo, E. I. Guendelman, R. Herrera and P. Labraña, JCAP **1608** (2016) 049.
[2] A. Mithani and A. Vilenkin, arXiv:1204.4658 [hep-th].
[3] A. T. Mithani and A. Vilenkin, JCAP **1201**, 028 (2012) [arXiv:1110.4096 [hep-th]].
[4] A. T. Mithani and A. Vilenkin, JCAP **1405**, 006 (2014) [arXiv:1403.0818 [hep-th]].
[5] A. T. Mithani and A. Vilenkin, JCAP **1507**, no. 07, 010 (2015) [arXiv:1407.5361 [hep-th]].
[6] A. Guth, Phys. Rev. D23, 347 (1981); A. Linde, Phys Lett. B 129, 177 (1983); A. Linde; Particle physiscs and inflationary cosmology (Gordon and Breach, New York, 1990).
[7] D.Larson *et al.*, Astrophys. J. Suppl. textbf192, 16 (2011); Bennett C. et al., Astrophys. J. Suppl. 192, 17 (2011); Hinshaw G. et al. [WMAP Collaboration], Astrophys. J. Suppl. 208, 19 (2013).
[8] P.A.R. Ade *et al.* [Planck Collaboration], arXiv:1502.02114 [astro-ph.CO]; P.A.R. Ade *et al.*, Phys. Rev. Lett. **114**, 101301 (2015).
[9] R. Penrose and S. W. Hawking, Proc. R. Soc. A**314**, 529 (1970); R. P. Geroch, Phys. Rev. Lett. **17**, 445 (1966).
[10] S. W. Hawking and G. F. R. Ellis, *The Large Scale Structure of Space-time*. Cambridge University Press, New York, 1973.
[11] A. Borde, A. Vilenkin, Phys.Rev.Lett. **72**, 3305 (1994); Int.J.Mod.Phys. **D5**, 813 (1996).
[12] G. F.R. Ellis, Roy Maartens, Class.Quant.Grav. **21**, 223 (2004); G. F.R. Ellis, J. Murugan, C. G. Tsagas, ibid. **21**, 233 (2004).
[13] D.J. Mulryne, R. Tavakol, J.E. Lidsey and G.F.R. Ellis, *Phys. Rev.* **71**, 123512 (2005); A. Banerjee, T. Bandyopadhyay and S. Chaakraborty, Grav. Cosmol., 290 **13** (2007); J.E. Lidsey and D.J. Mulryne *Phys. Rev.* **73**, 083508 (2006); S. Mukherjee, B.C.Paul, S.D. Maharaj and A. Beesham, arXiv:qr-qc/0505103;S. Mukherjee, B.C.Paul, N.K. Dadhich, S.D. Maharaj and A. Beesham, *Class.Quant.Grav.* **23**, 6927 (2006).
[14] S. del Campo, R. Herrera, P. Labrana, *JCAP* **0711**: 030, (2007); S. del Campo, R. Herrera, P. Labrana, *JCAP* **0907**: 006, (2009).
[15] S. del Campo, E.I. Guendelman, R. Herrera, P. Labrana,*JCAP* **1006**: 026, (2010).

IOP Conf. Series: Journal of Physics: Conf. Series **1043** (2018) 012026 doi:10.1088/1742-6596/1043/1/012026

[16] S. del Campo, E. Guendelman, A.B. Kaganovich, R. Herrera and P. Labrana, Phys. Lett. B **699**, 211 (2011).
[17] E. I. Guendelman and P. Labrana, Int. J. Mod. Phys. D **22** (2013) 1330018.
[18] E. Guendelman, R. Herrera, P. Labrana, E. Nissimov and S. Pacheva, Gen. Rel. Grav. **47**, no. 2, 10 (2015) [arXiv:1408.5344 [gr-qc]].
[19] E.I. Guendelman and A.B. Kaganovich, *Phys. Rev.* **D53**, 7020 (1996); *Mod. Phys. Lett.* **A12**, 2421 (1997); *Phys. Rev.* **D55**, 5970 (1997); *Mod. Phys. Lett.* **A12**, 2421 (1997); *Phys. Rev.* **D56**, 3548 (1997); *Mod. Phys. Lett.* **A13**, 1583 (1998).
[20] E.I. Guendelman and A.B. Kaganovich, *Phys. Rev.* **D60**, 065004 (1999).
[21] E.I. Guendelman, *Mod. Phys. Lett.* **A14**, 1043 (1999).
[22] E.I. Guendelman and O. Katz, *Class. Quant. Grav.*, **20**, 1715 (2003).
[23] E.I. Guendelman, A.B. Kaganovich, *Phys.Rev.* **D75**:083505, (2007).
[24] E.I. Guendelman, A.B. Kaganovich, in *Paris 2005, Albert Einstein's century*, p.875, Paris, 2006; hep-th/0603229.
[25] P. Labrana, Phys. Rev. D **91**, no. 8, 083534 (2015) [arXiv:1312.6877 [astro-ph.CO]].
[26] V. A. Rubakov, Phys. Usp. **57**, 128 (2014) doi:10.3367/UFNe.0184.201402b.0137 [arXiv:1401.4024 [hep-th]].
[27] B.S. DeWitt, Quantum Theory of Gravity. 1. The Canonical Theory, Phys. Rev. **160**, 1113 (1967).
[28] A. Vilenkin, Phys. Rev. D **37**, 888 (1988). doi:10.1103/PhysRevD.37.888.
[29] J. Dittrich and P. Exner, J. Math. Phys. **26**, 2000 (1985). J. Dittrich and P. Exner, JINR-E2-84-353. J. Dittrich and P. Exner, JINR-E2-84-352.
[30] E.I. Guendelman, *Class.Quant.Grav.* **17**, 3673 (2000); E.I. Guendelman, *Phys. Rev.* D**63**, 046006 (2002); E.I. Guendelman, A.B. Kaganovich, E. Nissimov, S. Pacheva, *Phys. Rev.* D**66**, 046003 (2002); E.I. Guendelman, A. Kaganovich, E. Nissimov, S. Pacheva, *Phys.Rev.* D**72**, 086011 (2005).
[31] P. Labrana, Phys. Rev. D **86**, 083524 (2012) doi:10.1103/PhysRevD.86.083524 [arXiv:1111.5360 [gr-qc]]; AIP Conf.Proc. 1606 (2014) 38-47; Astrophys.Space Sci.Proc. 38 (2014) 95-106.

Development of a semi-empirical method to determine the efficiency of a gamma radiation detector

Pablo Ortiz-Ramírez, Franco López-Usquiano

Departamento de Física, Facultad de Ciencias, Universidad de Chile, Santiago, Chile

E-mail: rapeitor@ug.uchile.cl, franco.lopez@ug.uchile.cl

Abstract. In the present work we present the development of a new method to calibrate the Full energy peak (FEP) efficiency of a gamma radiation detector. The method consists of the experimentally determination of the detector efficiency for an energy using a monoenergetic gamma source and from this value to extrapolate constructing the complete detector efficiency curve using the physics of first principles of gamma detection theory. For this reason the proposal method corresponds to a semi-empirical method.

In this first work we show the application and the study for the validation of the method by means of simulations based on the Monte Carlo method using FLUKA code. The energies considered in the simulations correspond to those emitted by the radioisotopes ^{137}Cs, for the determination of the efficiency for one energy we will call reference efficiency, and ^{152}Eu, to perform the extrapolation of the intrinsic efficiency from the reference efficiency.

So far, we have calibrated the total instrinsic efficiency of a bare cylindrical detector by applying the proposed method. A systematic difference between the expected values of the intrinsic efficiency obtained by statistical count for the ^{152}Eu gamma energies, and those determined by the proposed method is observed. Despite this, the results show a good agreement with relative biases less than 5% in most cases.

One of the principal advantages of this method is that it let us obtain a calibration curve with all phenomenology involved in the detection, such as the resonance peaks.

1. Introduction

The gamma espectrometry using a Hiper Pure Germaniun (HPGe) detector is the most popular technique for the determination and quantification of the radioactive nuclei present in a radioactive source. When we want to know the quantities of the radioactive nuclei present in a like point-source, we need to have knowledge of the full energy peak (FEP) efficiency of the detector for a point source and for the different energies emitted by the radioactive source under study.

The more common way to determine the FEP efficiency curve is experimentally, using a set of radioactive sources with well known activity (named calibration sources) and covering the complete gamma energy range of interest. Following this, a curve is fitted to the experimental points to construct the FEP efficiency curve. Then, the FEP efficiency for the energy of interest can be determined by interpolation. If we count on a calibration source compounded by the radionuclide under study we can determine the activity of this source directly by the relative method. Due to the uncertainty produced in the interpolation, the best way is using the relative method, but calibration point source of any radionuclide is not easy to obtain.

In this work we are developing a new semi-empirical method to calibrate the FEP efficiency by means of extrapolation from one experimental point of the detector efficiency. The experimental point is used to discover the average traveled path by the photons in the detector and then using the linear attenuation coefficient of the detector material for the interest energy, we determined the FEP efficiency for such energy.

The adventages of the method are: we do not need to use multigamma sources for the calibration, which require complicated corrections in the process as true coincidence correction; we do not need to apply a fit to determine the efficiency curve, and one of the principal advantages of this method is that we can obtain the efficiency curve with all phenomenology involved in the detection, such as the resonance peaks.

In this first aproximation to the validation we have used FLUKA code [1] to determine the efficiencies for the different energies used in this work and XCOM data base [2] to know the attenuation coefficients.

2. Semi-empirical Method to Determine the Efficiency of a Detector

The intrinsic efficiency of a detector for photons with energy E emitted by a gamma source positioned in $\vec{r}$ is defined as,

$$\epsilon_{int}(E,\vec{r}) = 1 - e^{-\lambda(E)d(\vec{r})}, \tag{1}$$

where λ is the linear attenuation coefficient and $d(\vec{r})$ the average traveled path by photons in the detector.

If we know the intrinsic efficiency and the linear attenuation coefficient for an energy E_1, $\epsilon_{int}(E_1,\vec{r})$ and $\lambda(E_1)$, the average traveled path $d(\vec{r})$ by the photons in the detector can be determined by the following equation,

$$d(\vec{r}) = \frac{\ln(1-\epsilon_{int}(E_1))}{-\lambda(E_1)}. \tag{2}$$

Now, if we know the linear attenuation coefficient for another energy E_2, $\lambda(E_2)$, we can determine the intrinsic efficiency for that energy using the next equation,

$$\epsilon_{int}(E_2,\vec{r}) = 1 - e^{-\lambda(E_2)d(\vec{r})}. \tag{3}$$

On the other hand, the intrinsic efficiency is related to the absolute efficiency by the next equation,

$$\epsilon_{int} = \epsilon_{abs}\frac{4\pi}{\Omega}, \tag{4}$$

where Ω is the solid angle that subtends the point source over the detector.

2.1. Hypothesis I: Application of the method to calibrate the FEP efficiency (Direct method)

The linear attenuation coefficient has three important components in the gamma spectrometry, the partial linear attenuation coefficients associated to the photoelectric effect, Compton scattering and pair production. Considering this, we can write the total linear attenuation coefficient as,

$$\lambda(E) = \lambda_{ph}(E) + \lambda_{cs}(E) + \lambda_{pp}(E), \tag{5}$$

where λ_{ph}, λ_{cs} and λ_{pp} represent the partial linear attenuation coefficients related to the photoelectric effect, Compton scattering and pair production respectively. Now, if we consider only the absorptions that deposite all the energy in the detector, i.e. the photons that are recorded in the FEP, wen can define the complete absorption coefficient as,

$$\lambda = \lambda_{ph}(E) + \beta\lambda_{cs}(E) + \gamma\lambda_{pp}(E), \tag{6}$$

where β and γ are parameters that represent the probability of a photon that interacted by Compton scattering or pair production finally deposited all the energy in the detector.

The total linear attenuation coefficient, shown in Equation (5), evaluated in Equation (1) corresponds to the total intrinsic efficiency of the detector, while the evaluation of the Equation (6) in the Equation (1) correponds to the FEP intrinsic efficiency.

Finally, if we can determine the FEP area removing the Compton scattering and Pair production contributions, i.e. if we achieve that β and γ are zero, then we could determine the FEP intrinsic eficiency only using the photoelectric linear attenuation coefficient and we do not need to know β and γ parameters. According to the Equation (3), the FEP intrinsic efficiency would be,

$$\epsilon_{int}(E, \vec{r}) = 1 - e^{-\lambda_{ph}(E)\, d(\vec{r})}. \tag{7}$$

2.2. Hypothesis II: Semi-empirical method to calibrate the FEP efficiency using the total efficiency (Indirect method)

In contrast to the previous method, wich we wanted to determine the FEP efficiency curve by direct extrapolation from the reference FEP efficiency. Now, we want to determine the FEP efficiency curve by indirect method. This means, we will calibrate the total intrinsic efficiency and from this we will determine the FEP intrinsic efficiency. Theoretically, this corresponds to the evaluation of the Equation (5) in the Equation (1), as indicated in

$$\epsilon_{int}(E, \vec{r}) = 1 - e^{-(\lambda_{ph}(E)+\lambda_{cs}(E)+\lambda_{pp}(E))\, d(\vec{r})}. \tag{8}$$

Then, we can determine the average traveled path by the photons in the detector using the total intrinsic efficiency and the total linear attenuation coefficient (both for the same energy) using Equation (2), and to extrapolate the total intrinsic efficiency of the detector for another energy using Equation (3).

Subsequently, from the calibration curve of the total intrinsic efficiency of the detector we can determine the FEP intrinsic efficiency by means of the following equation [3],

$$\epsilon_{int,peak} = \frac{P}{T}\epsilon_{int,total}, \tag{9}$$

where $\epsilon_{int,peak}$ represents the FEP intrinsic efficiency of the detector, $\epsilon_{int,total}$ is the total intrinsic efficiency of the detector and P/T is the Peak-Total ratio. The Peak-Total ratio must be determined using simulations as Moens indicates in his work [3].

IOP Conf. Series: Journal of Physics: Conf. Series **1043** (2018) 012027 doi:10.1088/1742-6596/1043/1/012027

3. Simulation

For simplicity, we have simulated a bare cylindrical detector of Germanium, whose dimensions are 2.5 cm radius and 4.0 cm height.

The gamma sources have been simulated like monoenergetic isotropic sources emitting the interest energies. In Table 1 we show the interest energies and the radionuclides associated to these.

Table 1. Emitted energies by the simulated monoenergetic sources and the radionuclide associated to these.

Energy (MeV)	Radionuclide
0.1218	^{152}Eu
0.2447	^{152}Eu
0.3443	^{152}Eu
0.4111	^{152}Eu
0.6617	^{137}Cs
0.7789	^{152}Eu
0.8674	^{152}Eu
1.086	^{152}Eu
1.333	^{152}Eu
1.408	^{152}Eu

The simulated sources were positioned on the axial axis of the cylindrical photon detector. In all simulations the number of photons emmited by the source was 10^7. In Figure 1 we can see the simulated setup.

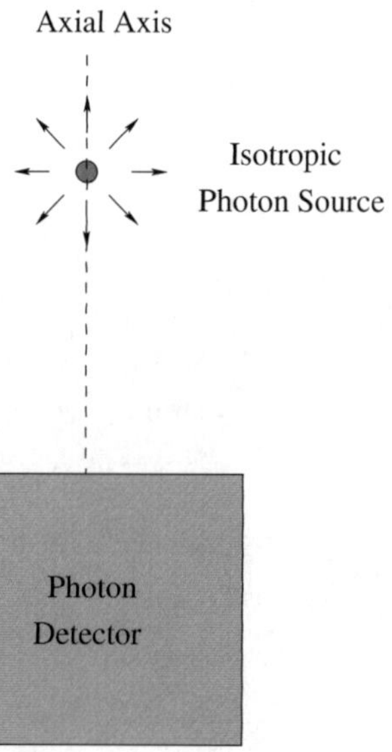

Figure 1. This figure shows the simulated setup. In this we can see the cylindrical gamma detector and the isotropic gamma source positioned on the axial axis of the detector.

4. Validations

4.1. Hypothesis I: Application of the method to calibrate the FEP efficiency (Direct method)

First, we simulated a point isotropic source emmiting photons of 661.65 keV to 10 cm from de detector. From the obtained spechtrum we determined the FEP absolute efficency of the detector and using Equation (4) we obtained the refernce FEP intrinsic efficiency. Afterwards we determined the average traveled path for the photons in the detector using Equation (7). The value of the linear attenuation coeficient was determined using XCOM data base [2].

Then, using the obtained average traveled path and the linear attenuation coefficient for the energies associated to the ^{132}Eu decay from XCOM data base [2], we calculated the efficiencies of the detector for this group of energies using Equation (7).

On the other hand, we simulated a set of mononergetic isotropic sources, one by one, emitting gamma rays with the energies associated to the ^{132}Eu decay and we determined the intrinsic efficiency by statistical count.

The obtained results are shown in Table 2.

Table 2. Comparison between the efficiencies obtained by the statistical count and the direct method.

Energy (MeV)	Photoelectric attenuation coefficient (1/cm)	ϵ_{int} Simulated	ϵ_{int} Method	Δ %
0.1218	1.125×10^{0}	8.410×10^{-1}	1.000×10^{0}	15.9
0.2447	1.385×10^{-1}	4.967×10^{-1}	9.362×10^{-1}	46.9
0.3443	5.174×10^{-2}	3.396×10^{-1}	6.425×10^{-1}	47.1
0.4111	3.168×10^{-2}	2.817×10^{-1}	4.672×10^{-1}	39.7
0.7789	6.446×10^{-3}	1.451×10^{-1}	1.203×10^{-1}	−20.7
0.8674	5.147×10^{-3}	1.295×10^{-1}	9.726×10^{-2}	−33.1
1.086	3.160×10^{-3}	1.054×10^{-1}	6.088×10^{-2}	−73.1
1.333	2.169×10^{-3}	8.794×10^{-2}	4.219×10^{-2}	−108.4
1.408	1.963×10^{-3}	8.196×10^{-2}	3.826×10^{-2}	−114.2

We can see that the results obtained with the direct method are not good. This is because we assume we can remove Compton and Pair production contribution to the peak is false. Although we can discount multiple Compton contribution to the peak we can not discount the interactions that start with a Compton scattering or with a Pair production and follow it by a photoelectric absortion. These contributions to the FEP area can not be separated from the photons that immediately interact by photoelectric effect. Therefore this method can not be applied.

4.2. Hypothesis II: Semi-empirical method to calibrate the FEP efficiency using the total efficiency (Indirect method)

First, we simulated a point isotropic source emitting photons of 661.65 keV to 10, 20 and 30 cm from the detector. From the obtained spectrum we determined the total absolute efficiency of the simulated detector and using Equation (4) we obtained the reference total intrinsic efficiency. Afterwards we determined the average traveled path for the photons in the detector using Equation (8).

Then, using the obtained average traveled path and the linear attenuation coefficient for the energies associated to the ^{132}Eu decay from XCOM data base [2], we calculated the efficiencies of the detector for these energies using Equation (8).

On the other hand, we simulated a set of mononergetic isotropic sources, one by one, emitting gamma rays with the energies associated to the ^{132}Eu decay and we determinate the intrinsic efficiency by statistical count.

The obtained results for the relative bias between the total intrinsic efficiencies determined by statistical count and by extrapolation of the reference efficiency are shown in Table 3.

Table 3. Comparison between the total intrinsic efficiencies obtained by statistical count and by extrapolation. The values correspond to the relative bias between this quantities.

Energy (MeV)	Source-detector distance (cm) 10	20	30
0.1218	−9.44	−6.39	−5.61
0.2447	−5.85	−4.50	−3.64
0.3443	−3.68	−2.89	−2.78
0.4111	−2.31	−2.00	−1.65
0.7789	0.71	0.42	0.29
0.8674	1.62	0.20	0.38
1.086	2.29	1.19	0.59
1.333	3.47	1.65	1.04
1.408	3.15	1.90	1.43

From the results shown in Table 3 we can see that the results are better while we move away from the detector. Another characteristic we can see is that the results are better for energies close to the reference energy (661.65 keV). And finally, we can identify a sistematic error in the method, because the results are greater than the expected values for energies less than the reference energy, and lesser than the expected values in opposite case.

In general the results are in a good agreement, except the result obtained for the 121.8 keV. The obtained results for energies between 344.3 keV and 1408 keV are in an excellent agreement with the expected values.

The results can be improved considering a set of total intrinsic efficiencies along the energy range of interest as references and to do the extrapolation in a short range around these points. Experimentally, such gamma sources can be a set of monoenergetic sources, because the use of a monoenergetic source represents the easier way to determine the efficiency.

The next step to complete the method is to determine the FEP intrinsic efficiency to the energies under study using Equation (9). Due to that this work has been performed using simulaton, the transfer of the total intrinsic efficiency to the FEP intrinsic efficiency can not be carried out, because this needs to be performed by simulation, as Moens indicates [3]. But, as the application of the Equation (9) has already been demostrated we expect it to be successful.

In order to apply the method as proposed, we need a well characterized detector, i.e. a detector with well known all geometric parameters and materials.

5. Conclusions

The semi-empiric method to determine the FEP intrinsic efficiency of the detector can not be directly applied using Equation (7), because when we determine the FEP area of the peak we can not remove the counts on the peak contributed by the Compton and Pair production interactions.

The method to determine the Total intrinsic efficiency by extrapolation, Equation (9) has been successfully applied, except for the energy 121.8 keV.

The next step to improve the method is to use a set of reference efficiency, and to realize a local extrapolation.

The final step to complete this work is to use a well known detector and to apply the complete method.

References

[1] Ferrari A, Sala P R, Fasso A and Ranft J Fluka: a multi-particle transport code *CERN 2005-10 (2005), INFN/TC 05/11, SLAC-R-773*
[2] MJ B, JH H, SM S, J C, JS C, R S, DS Z and K O (2010) Xcom: Photon cross section database (version 1.5).
[3] Moens L, Donder J D, Xi-lei L, Corte F D, Wispelaere A D, Simonits A and Hoste J 1981 *Nuclear Instruments and Methods in Physics Research* **187** 451 – 472 ISSN 0167-5087

IOP Conf. Series: Journal of Physics: Conf. Series **1043** (2018) 012028 doi:10.1088/1742-6596/1043/1/012028

Activity concentration of NORM and ^{137}Cs radionuclide in soil samples from the Andes Cordillera at latitude 33°56' South

R. Correa[a], **P.A. Miranda**[a], **P. Ortiz-Ramirez**[b], **J. Wachter**[a], **S. Camilla**[a], **E. Mera**[a], **E. Piñones**[c].

[a]Departamento de Física, Fac. de Ciencias Naturales, Matemática y del Medio Ambiente, Universidad Tecnológica Metropolitana. Santiago 7800002, Chile

[b]Departamento de Física, Facultad de Ciencias, Universidad de Chile, Casilla 653, Santiago, Chile

[c]Comisión Chilena de Energía Nuclear, Santiago, Chile

E-mail: `pmiranda@utem.cl`

Abstract. Activity concentrations of naturally occurring radioactive material (^{238}U, ^{232}Th, ^{40}K) and ^{137}Cs were determined in surface soil samples collected from four sites at latitude 33°56' South and at an average altitude of 1650 masl in the Andes Cordillera, Chile. The handling of samples and the gamma spectroscopic analyses followed the recommended international protocols for this type of work. The presence of ^{137}Cs due to fallout remains from atomic blasts produced in the Pacific atolls in the seventies was detected only in three sites analyzed in the current study. The activity concentration of this anthropogenic radionuclide showed significant fluctuations ranging from below minimum detection activity to 21.32 Bq/kg. The ranges of activity concentrations of ^{238}U, ^{232}Th, ^{40}K were found to be around 15-30, 22-30 and 411-611 Bq/kg respectively. The weighted mean activities concentrations of NORM radionuclides are within those reported from literature and were then used to calculate the radiological hazard indices. Values of external absorbed dose rate (D), the annual outdoor effective dose ($AEDE$), the radium equivalent activity (Ra_{eq}), and the external radiation hazard (H_{ex}) index are within the accepted international ranges. These results point out that the radiation hazard due to NORM radionuclides found in soil samples from the sites studied in this work is not significant.

1. Introduction

The concern about the amount of environmental radioactivity to which the population is exposed is stimulating new research in this field. Today around the world there is broad interest in studying the radiological characterization of soils, not only to learn about concentration activity levels and associated health risks, but also to know more accurately the contribution of human activity to the environmental radioactivity [1-5]. However, in countries like Chile studies about soil inventories of naturally occurring radionuclide (NORM) and ^{137}Cs are not only scarce but usually their objectives have been rather different. For instance, in Ref. [6] a radon and gamma monitoring program was presented which aimed to estimate population indoor exposure in dwellings and in some Chilean mines. P. Schuller et al., in Ref. [7] evaluated the applicability of a technique using ^{137}Cs concentration to obtain spatial information on mean soil redistribution rates in Central-South Chile, obtaining similar results to those obtained by

IOP Conf. Series: Journal of Physics: Conf. Series **1043** (2018) 012028 doi:10.1088/1742-6596/1043/1/012028

pedological observations. In 2010 a campaign for detection of ^{137}Cs and ^{40}K radionuclides was conducted along the Maipo River at altitudes from 900 to 1800 masl showing that after almost two half life since the nuclear tests were detonated in the atmosphere in the Pacific atolls, between 1966 and 1974, ^{137}Cs is easily detected [8]. Finally, Chamizo et al. in Ref. [9] provides new data on the impact of nuclear tests performed in the Southern Hemisphere to South America soils through the study of ^{239}Pu and ^{240}Pu in different areas of Northern and Southern Chile. Their results showed the presence of debris from the French Tests with homogeneous ^{240}Pu/^{239}Pu ratios ranging from 0.02-0.23. It is worth to mention that the geographical location of Chile, just in front of the Pacific Ocean, extending from 17° latitude South to Cape Horn at 56°, along the Andes Cordillera, favored the reception of fallout debris from those nuclear explosions occurred around 21.8° South. The results presented in this work are the first obtained within a cooperative research project between the University of Chile (UCH), the Metropolitan Technological University (UTEM), and the Chilean Nuclear Energy Commission (CCHEN). This project aims to determine activity concentrations of ^{137}Cs and natural occurring radionuclides in surface soil samples collected in Central Chile along latitude 33° South, from the Andes to the coast of the Pacific Ocean. During the first stage of this project, soil samples from Andes Mountain, near Santiago, Chile, were collected in order to determine the activity concentration of NORM and that of the anthropogenic ^{137}Cs radionuclide. Our long-range objective is to evaluate the contribution of these radionuclides to the radiation background in different zones of the country.

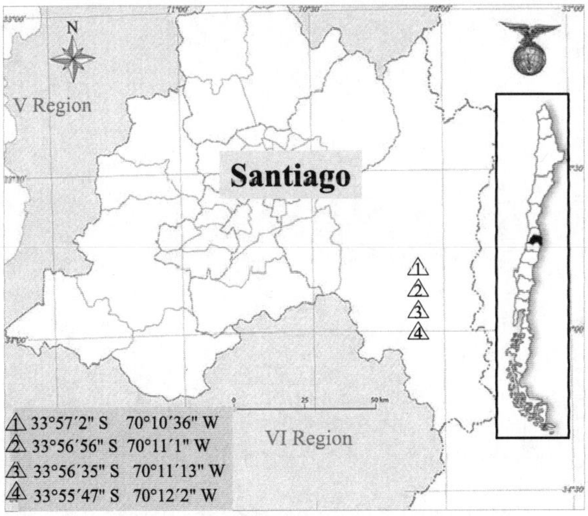

Figure 1. Area of Andes Cordillera where the samples were collected (33° South) during sampling campaign performed in March 2015

2. Experimental

2.1. Study Area

The map of Central Chile shown in figure 1 gives a general view of the locations where the soils samples were collected during a campaign performed in March 2015. In the south western extreme of South America, Chile occupies a narrow (177 km on average) and long (4,270 km) zone that extends roughly from latitude 17° South to Cape Horn at 56°. Other important feature about the geomorphology of continental Chile is the East-West topographical profile. For instance, at 33° it presents four different zones, namely, Marine Terraces, Coastal Cordillera, Longitudinal Central Valley and the Andes Cordillera. In Ref. [10] it has been established that the local relief in the Andes is due to a balance between extended fluvial erosion and glacial erosion acting as opposite agents. On the other hand, and according to the distribution of Soil

Taxonomy system (USDA) orders observed in Chile, the most probable classification of soil at the Central Andes sector would belong to two soil orders, namely Andisols and Inceptisols (see Fig. 2.88 in Ref.[11]). The sampling locations studied in this work are characterized as very thin soils (∼15 cm thick) over rock with poor vegetation and influenced by running water, thermal and gravitational processes, with an average annual precipitation in the range from 1500 mm to 3500 mm, mostly snow.

2.2. Sample collection and preparation

In this study soil samples were collected at four sites in the Andes Mountain near Santiago, Chile, at a distance of about 100 km to South-east, and extracted from holes having a surface of 30×30 cm^2 and depth of 5 cm at an average altitude of 1650 m above sea level. Sites were selected according to their accessibility and by showing no evidence of surface modification by human action. In each case, organic residues and stone fragments were removed from the surface. Soil samples from each site were packed in plastic bags and labeled accordingly. At the laboratory, preparation of samples for gamma spectroscopy measurements followed procedures recommended in [12], i.e., the samples were weighted and dried for 24 hours in an oven at 105 °C. Next, they were sieved with a 0.6 mm mesh screen and transferred to Petri Dishes and finally sealed for one month in order to reach secular equilibrium among the daughter products of ^{222}Ra, ^{220}Ra and their short lived decay products. An average of 0.100 kg of soil was used per sample.

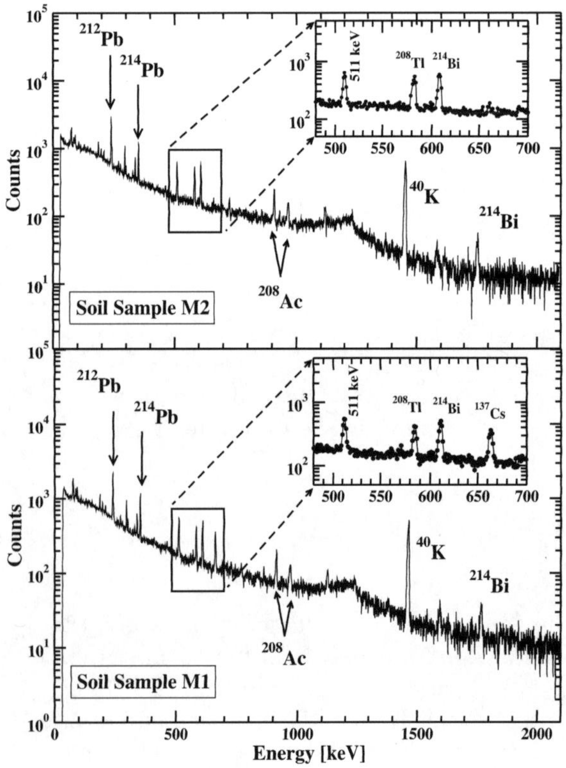

Figure 2. Gamma spectra corresponding to soil samples collected in two different locations at the Andes Cordillera. In the upper part, the gamma spectrum shows that anthropogenic radionuclide ^{137}Cs is not present in site M2 while in the lower part the gamma energy peak at 662 keV is easily detected in sample M1

2.3. Calibration and Gamma spectrometry measurements

Gamma spectrometric measurements were carried out in the Gamma Spectrometry Laboratory at the University of Chile using a HPGe detector ORTEC (GEM-10195) with an energy resolution of 1,95 keV at 1330 keV and with a relative efficiency of 10% reported by the manufacturer. In order to reduce background radiation, the detector is placed inside a 10 cm cylindrical lead shielding with internal covers of cadmium and copper, 1 mm thickness each. Detector efficiency curve was obtained by using Certified Reference Material IAEA 447 [13]. Energy calibration of the acquisition system was frequently checked by using standard reference radionuclide sources. For quality control and quality assurance, three samples were prepared and analyzed for each one of the four sites (M1, M2, M3, and M4). After analyzing these three samples a very good agreement was found since their activity concentrations agreed within 2%. As an example, gamma spectra corresponding to soil samples collected in sites M1 and M2 at the Andes Cordillera are shown in figure 2. The gamma spectrum of M2 shows that anthropogenic radionuclide ^{137}Cs is below the detection limit in this site while gamma energy peak at 662 keV is easily detected in sample M1. Collection times of 48 hours were set for each spectrum. The activity concentration of each radionuclide of our interest, A_n, was calculated from the respective full-energy peak area C_n (background subtracted), after applying appropriate factors such as peak efficiency $\varepsilon(E_\gamma)$ and gamma intensity I_γ. Hence, using the dry weight (M) of the samples, their respective activity concentrations of NORM and ^{137}Cs radionuclide could be expressed in activity per unit weight (Bq/kg) [14]:

$$A_n = \frac{C_n}{M\varepsilon(E_\gamma)I_\gamma \Delta t} \tag{1}$$

Uncertainties were estimated by combining counting statistics, mass and calibration uncertainties. The activity concentration of the single decaying nuclei, ^{137}Cs and ^{40}K were determined directly by the measurement of the gamma-ray transitions at 662 keV and 1461 keV, respectively. On the other hand, the activity concentration of ^{238}U was determined from the 609.3, 1120.3 and 1764.5 keV gamma-ray peaks of ^{214}Bi and from the 295.2 and 351.9 keV gamma-ray peaks of ^{214}Pb. Finally, the activity concentration of ^{232}Th was evaluated by taking into account gamma-ray line of ^{212}Pb at 238.6, gamma-ray lines of ^{228}Ac at 338.2, 911.2 and 969.0 keV, and gamma-ray line of ^{208}Tl at 583.2 keV. The activity concentration obtained from each line was then combined to determine a weighted mean and uncertainty of the weighted mean (k=1) for the corresponding decay chain [14].

3. Results and Discussion

Typical gamma spectra corresponding to soil samples collected at sites M1 and M2 are shown in figure 2. As shown in this figure, the identified radionuclides detected in all samples mostly belong to the ^{232}Th and ^{238}U decay series, though the most prominent gamma-ray energy peak observed in log scale corresponds to ^{40}K. In addition, a significant ^{137}Cs peak to background ratio for gamma emission is clearly observed only in samples M1, M3 and M4, indicating that their ^{137}Cs activity concentrations are greater than the detection limit of 0.1 Bq/kg.

3.1. Natural radionuclides in soil samples

Using gamma spectrometry, radioactivity of materials commonly found in the environment, like water and soil, is frequently studied to determine background levels of radiation or to assess the level of contamination as a consequence of human activity [14]. In this work the latter goal does not apply since in the Andes Cordillera human activities are rather scarce. Therefore, the results presented in this work can be used as a baseline in order to evaluate the contribution of NORM and anthropogenic radionuclides to the radiation background in different zones of the country.

Table 1. Comparison between mean activity concentrations and radiation hazard indices of NORM and ^{137}Cs from soil samples collected at the Andes Cordillera and the mean value worldwide.

Sample	A_U Bq/kg	A_{Th} Bq/kg	A_K Bq/kg	A_{Cs} Bq/kg	D nGy/h	$AEDE$ μSv/y	Ra_{eq} Bq/kg	H_{ex}
M1	30.0±1.6	28.71±0.93	572±26	11.83±0.53	55.1±1,0	67.6±1.3	115,0±2.9	0,3108±0,0078
M2	25.2±1.4	29.95±0.93	611±27	MDA	55.25±0.94	67,8±1.2	115,1±2,8	0,3110±0,0076
M3	17.33±0.96	21.67±0.72	411±19	21.32±0.56	38.22±0,89	46,9±1.1	79,9±2.0	0,2159±0,0055
M4	14.53±0.79	23.23±0.72	429±22	0.97±0,16	38.62±0,79	47.4±1.0	80,7±2.1	0,2181±0,0057
World								
Average	35	30	400		57	70	(<370)	(<1)
Range	17-60	11-64	140-850		18-93			

Table 1 shows that in general, the activity concentration of soil samples collected from four sites at the Andes Mountain are within world range values [15]. Solid lines in figure 3(a) and 3(b) represent the weighted mean activity concentrations of ^{232}Th and ^{238}U respectively as measured for soil sample M3 while the dashed lines correspond to the uncertainties associated with these mean values. In the case of ^{232}Th and ^{238}U and ^{137}Cs our results are generally lower than world average values while for ^{40}K measured activity concentration data are higher. It is worth noting that, regarding NORM radionuclides, samples M1 and M2 show higher values compared to samples M3 and M4. Fallout radionuclide ^{137}Cs does not follow this trend since its activity concentration shows significant fluctuations ranging from below MDA to 21.32 Bq/kg.

3.2. Dose Rates and Hazard Indices

In order to assess radiation hazards associated with soil samples, several quantities have been defined [15]. These quantities are calculated from ^{238}U, ^{232}Th and ^{40}K activity concentration values (A_U, A_{Th} and A_K) and by assuming that in each sample parent radionuclides and their decay progeny are in secular equilibrium.

The external absorbed dose rate (D) in air is defined for terrestrial gamma-radiation at a height of about 1 m above the ground, and it is calculated according to the following relation [16]:

$$D\left[\frac{\text{nGy}}{\text{h}}\right] = 0.462A_U + 0.604A_{Th} + 0.0417A_K \qquad (2)$$

In the present study, the absorbed dose for the four different locations of soil samples varied from 38.22 to 55.25 nGy/h with a mean value of 47 nGy/h, which is within the expected range, and it is lower than the worldwide mean value of 57 nGy/h. From the above dose definition (D), the annual outdoor effective dose ($AEDE$) is obtained by applying a conversion coefficient of 0.7 Sv/Gy. Additionally, an outdoor occupancy factor of 0.2 must be taken into account. Therefore, $AEDE$ values are calculated from [17]:

$$AEDE\left[\frac{\mu\text{Sv}}{\text{y}}\right] = D \times 0.7 \times 0.2 \times 8.766 \qquad (3)$$

The world average annual effective dose equivalent ($AEDE$) from outdoor terrestrial gamma radiation is 70 μSv/y [15]. From Table 1, the effective dose range from 47.4±1.0 to 67.6±1.3 μSv/y with a mean value of around 58 μSv/y.

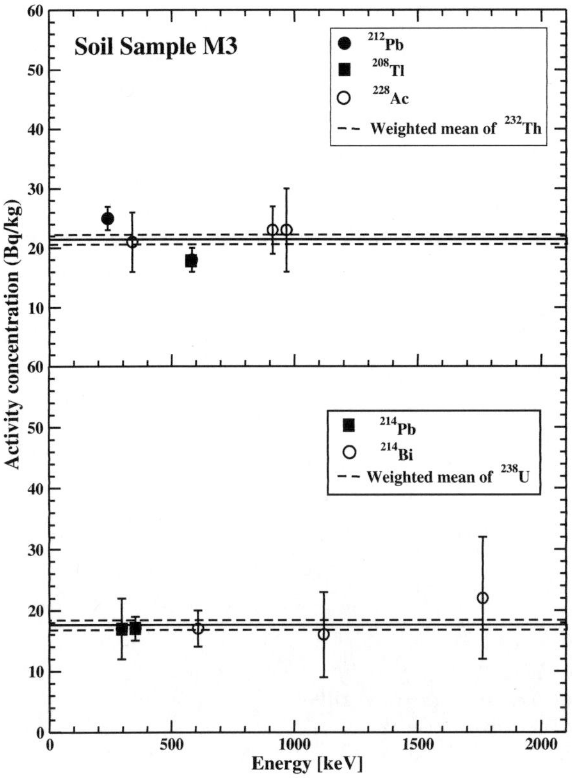

Figure 3. The weighted mean activity concentration of ^{238}U and ^{232}Th measured in sample M3. These values were computed using the individual data points of gamma-ray transitions from ^{214}Bi and ^{214}Pb (^{238}U decay chain) and ^{212}Pb, ^{208}Tl and ^{228}Ac (^{232}Th decay chain). The dashed-dotted lines represent the uncertainties in the calculated weighted mean values

Since the distribution of ^{238}U, ^{232}Th and ^{40}K in soil is not uniform, a common index called Radium Equivalent Activity (Ra_{eq}) is defined in [18] under the assumption that 370 Bq/kg of ^{238}U produces the same gamma dose rate as 259 Bq/kg of ^{232}Th and 4810 Bq/kg of ^{40}K giving rise to the next equation:

$$Ra_{eq}\left[\frac{\text{Bq}}{\text{kg}}\right] = A_U + \frac{370}{259}A_{Th} + \frac{370}{4810}A_K \tag{4}$$

It is observed in Table 1 that the average radium equivalent activity for soil samples studied in this work is lower than the maximum value allowed (370 Bq/kg), ranging from around 80 (M1 and M2) to 115 Bq/kg (M3 and M4), with a mean value of around 98 Bq/kg.

Every human being is continuously exposed to an external radiation field produced by natural radionuclide coming from soil, sediment and rocks. In order to assess this issue, the External Radiation Hazard (H_{ex}) is defined on the basis that the radiation exposure due to natural radionuclides should not surpass the permissible dosed equivalent limit of 1 mSv/yr [18]:

$$H_{ex} = \frac{A_U}{370} + \frac{A_{Th}}{259} + \frac{A_K}{4810} < 1 \tag{5}$$

As can be seen, this value should not exceed unity in order to consider the radiation hazard as insignificant. As listed in Table 1, the H_{ex} values for all soil samples studied ranged from 0.216 to 0.311 with an average value of 0.265, which is well below the limit of unity.

4. Conclusions

In this work gamma radiation from soil samples collected in the Andes Mountains near Santiago, Chile, were analyzed. Activity concentration of natural radioactive isotopes originated in the decay series of ^{232}Th and ^{238}U, together with ^{40}K, were determined. Measured values are below world average values. Average dose rates and calculated hazard index were lower than world recommended values. These results point out that the radiation hazard due to NORM radionuclides found in soil samples from the sites studied in this work is not significant. On the other hand, activity concentration values for ^{137}Cs show significant fluctuations. ^{137}Cs was detected in samples collected in locations M1, M3 and M4 while the activity concentration of ^{137}Cs in samples M2 is below MDA. The intrinsic soil chemical properties and the erosion by water draining at the sites where samples were collected can explain the observed differences in ^{137}Cs activities in samples M1, M2, M3 and M4. It was observed that soils with significant erosion present about 10-20 times less ^{137}Cs than those with minimal erosion. These preliminary results can be used as a baseline for information in the area under study and motivates further research.

5. Acknowledgements

This work was supported by Universidad Tecnológica Metropolitana (UTEM) under grant L2-12 Study on the presence on radioactive substances in urban and rural soils in the Metropolitan Region, Chile. We also acknowledge the support of Centro de Física Experimental (CEFEX), Facultad de Ciencias, Universidad de Chile.

References

[1] Guidotti L, Carini F, Rossi R, Gatti M, Cenci R M, Beone G M 2015 *J. Environ. Radioact.* **142** 36
[2] Agbalagba E O, Avwiri G O, Chad-Umoreh Y E 2012 *J. Environ. Radioact.* **109** 64
[3] Almayahi B A, Tajuddin A A, Jaafar M S 2012 *Appl Radiat Isot.* **70** 2652
[4] Velasco H, Cid A S, Anjos R M, Zamboni C B, Rizzotto M, Valladares D L, Ayub J J 2012 *J. Environ. Radioact.* **104** 64
[5] Song G, Chen D, Tang Z, Zhang Z, Xie W 2012 *J. Environ. Radioact.* **103** 48
[6] Stuardo E 1996 *Radiat. Prot. Dosim.* **67** 129
[7] Schuller P, Sepúlveda A, Trumper R E, Castillo A 2000 *Acta Geol. Hisp.* **35** 285
[8] Correa R, Morales J R, Da Silva L, Donoso S, Godoy S, Mera E 2012 *AIP Conf. Proc.* **1423** 369
[9] Chamizo E, García-León M, Peruchena J I, Cereceda F, Vidal V, Pinilla E, Miró C 2011 *Nucl. Instr. Meth. B.* **269** 3163
[10] Rehak K, Bookhagen B, Strecker M R, Echtler H P 2010 *Earth Surf. Process Landf.* **35** 1516
[11] Casanova M, Salazar O, Seguel O, Luzio W 2013 *The Soils of Chile* (World Soils Book Series) A.E. Hartemink, Ed. (Dordrecht: Springer)
[12] IAEA-TECDOC-1415 2004 *Soil sampling for environmental contaminants* IAEA, Vienna. `http://www-pub.iaea.org/MTCD/publications/PDF/te_1415_web.pdf`
[13] Shakhashiro A, Tarjan S, Ceccatelli A, Kis-Benedek G, Betti M 2012 *Appl. Radiat. Isot.* **70** 1632
[14] Gilmore G R 2008 *Practical Gamma-Ray Spectrometry*, 2nd Edition. (Chichester: John Wiley & Sons Ltd.)
[15] United Nations Scientific Committee on the Effects of Atomic Radiation UNSCEAR 2000 Sources, *Effects and Risks of Ionization Radiation. Report to the General Assembly, with Annexes* (New York)
[16] Chang B U, Koh S M, Kim Y J, Yoon Y Y, Rwo J W, Lee D M 2009 *J. Environ Radioact.* **99** 455
[17] United Nations Scientific Committee on the Effects of Atomic Radiation UNSCEAR 1988 Sources, *Effects and Risks of Ionization Radiation. Report to the General Assembly, with Annexes* (New York)
[18] Beretka J, Mathew P J 1985 *Health Phys.* **48** 87

IOP Conf. Series: Journal of Physics: Conf. Series **1043** (2018) 012029 doi:10.1088/1742-6596/1043/1/012029

Melting temperature for scalar hadrons in AdS / QCD models

Adolfo Ibañez and Alfredo Vega

Instituto de Física y Astronomía, Universidad de Valparaíso,
A. Gran Bretaña 1111, Valparaíso, Chile.

E-mail: alfredo.vega@uv.cl

Abstract.
We consider an analysis to potentials related to Schrödinger like equations for scalar fields in a 5D AdS black hole background with dilaton in order to get melting temperatures for different hadrons in a thermal bath. In approach considered it is not necessary to calculate spectral functions and it is easy to get results for different kind of hadrons. Here we present results for scalar mesons, glueballs, hybrid mesons and tetraquarks.

1. Introduction

Actually in several places are realizing experiments to understand how temperature affect hadron properties, and this topic has attracted interest of theoreticians considering several approaches to describe hadrons under these conditions (e.g see [1, 2]). The main theoretical tools used are lattice QCD and sum rules, and in last 15 years was added techniques based on Gauge / Gravity dualities (e.g see [3, 4, 5, 6, 7]). In the last kind of models, temperature is included considering black hole backgrounds in AdS in $d+1$ dimensions.

With models based on holographic correspondence it is possible to calculate several properties that vary with temperatures as masses and melting temperatures using the spectral function calculated in this approach [5, 6, 7]).

In this work we consider a procedure to calculate melting temperatures for hadrons in a thermal bath without calculate the spectral function. We use a holographic model, and melting temperature it is obtained since an analysis of potential associated to Schrödinger like equations that describe hadrons in AdS side [9, 10]. Here we restrict us to study scalar hadrons, which according to AdS dictionary are related to scalar modes [8], and are shown results for melting temperatures for mesons, glueballs, hybrid mesons and tetraquarks in a thermal bath.

This paper consist of fourth sections and an appendix. The introduction is followed by section 2 where we describe the procedure used to calculate melting temperatures considering an analysis of potentials of equation of motions in AdS side. In section 3 are presented and discussed the results. Section 4 considers conclusions and future perspectives. Additionally we consider an appendix with Liouville substitution, which is the procedure used to get Schrödinger like equations and potentials which are analysed in this work to get melting temperatures.

IOP Conf. Series: Journal of Physics: Conf. Series **1043** (2018) 012029 doi:10.1088/1742-6596/1043/1/012029

2. The model

In AdS / QCD models, scalar hadrons in AdS side are modelled by scalar models in a 5D curved space with dilaton with an action [11, 12, 13, 14]

$$S = \frac{1}{2K}\int d^5x\sqrt{-g}e^{-\phi(z)}\left[g^{MN}\partial_M X\left(x,z\right)\partial_N X\left(x,z\right) + m_5^2 X^2\left(x,z\right)\right] \tag{1}$$

In order to introduce temperature in this description it is necessary to consider a 5D AdS black hole metric [15].

$$ds^2 = e^{2A(z)}\left[-f\left(z\right)dt^2 + \sum_{i=1}^{3}\left(dx^i\right)^2 + \frac{1}{f\left(z\right)}dz^2\right], \tag{2}$$

or

$$g_{MN} = e^{2A(z)}diag\left(-f\left(z\right),1,1,1,\frac{1}{f\left(z\right)}\right). \tag{3}$$

In this case the equation of motion is

$$e^{B(z)}f\left(z\right)\partial_z\left[e^{-B(z)}f\left(z\right)\partial_z\psi\right] - f\left(z\right)e^{2A(z)}m_5^2\psi + \omega^2\psi - f\left(z\right)q^2\psi = 0, \tag{4}$$

where $B(z) = \phi(z) - 3A(z)$.

For particles at rest ($\vec{q} = \vec{0}$), last equation change to

$$\partial_z\left[e^{-B(z)}f\left(z\right)\partial_z\psi\right] + \left[\frac{\omega^2}{e^{B(z)}f\left(z\right)} - e^{-\phi(z)+5A(z)}m_5^2\right]\psi = 0. \tag{5}$$

By using the Liouville substitution (see appendix), it is possible obtain a Schrödinger like equation which is related to the previous one, which potential is,

$$V\left(z\left(\xi\right)\right) = e^{B(z(\xi))}f\left(z\left(\xi\right)\right)e^{-\phi(z(\xi))+5A(z(\xi))}m_5^2 - \left[e^{-2B(z(\xi))}\right]^{-\frac{1}{4}}\frac{d^2}{d\xi^2}\left(e^{-2B(z(\xi))}\right)^{\frac{1}{4}}. \tag{6}$$

This potential at zero temperature ($f(z) = 1$) has normalizable modes ([11, 12, 13, 14]), but at the finite temperature it has the form of quasi bound states, and depth of the well it is reduced when temperature is increased, and at some temperature the well disappear. This temperature it is interpreted as melting temperature in thermal bath [9, 10]. In literature it is possible to find a different way to obtain a potential to extract the melting temperature, but it is necessary to do supposition about how mass vary with temperature [16].

Here we analyse this potential for a specific metric and dilaton. And we consider different values for m_5, which according at AdS / CFT dictionary is related to dimension of operators that create scalar hadrons. Expression that relate AdS mass with operator dimension is [8, 11]

$$m_5^2R^2 = \Delta(\Delta - 4), \tag{7}$$

where R is the AdS radii (without loss of generality can be considered as one) and Δ is the dimension of aperator that create hadrons. In this work are considered mesones ($\Delta = 3$; $m_5^2 = -3$), glueballs ($\Delta = 4$; $m_5^2 = 0$), hybrid mesons ($\Delta = 5$; $m_5^2 = 5$) and tetraquarks ($\Delta = 6$; $m_5^2 = 12$) [11].

In this work was used

$$f(z) = 1 - \frac{z^4}{z_h^4} \quad , \quad 0 < z < z_h, \tag{8}$$

so fore,

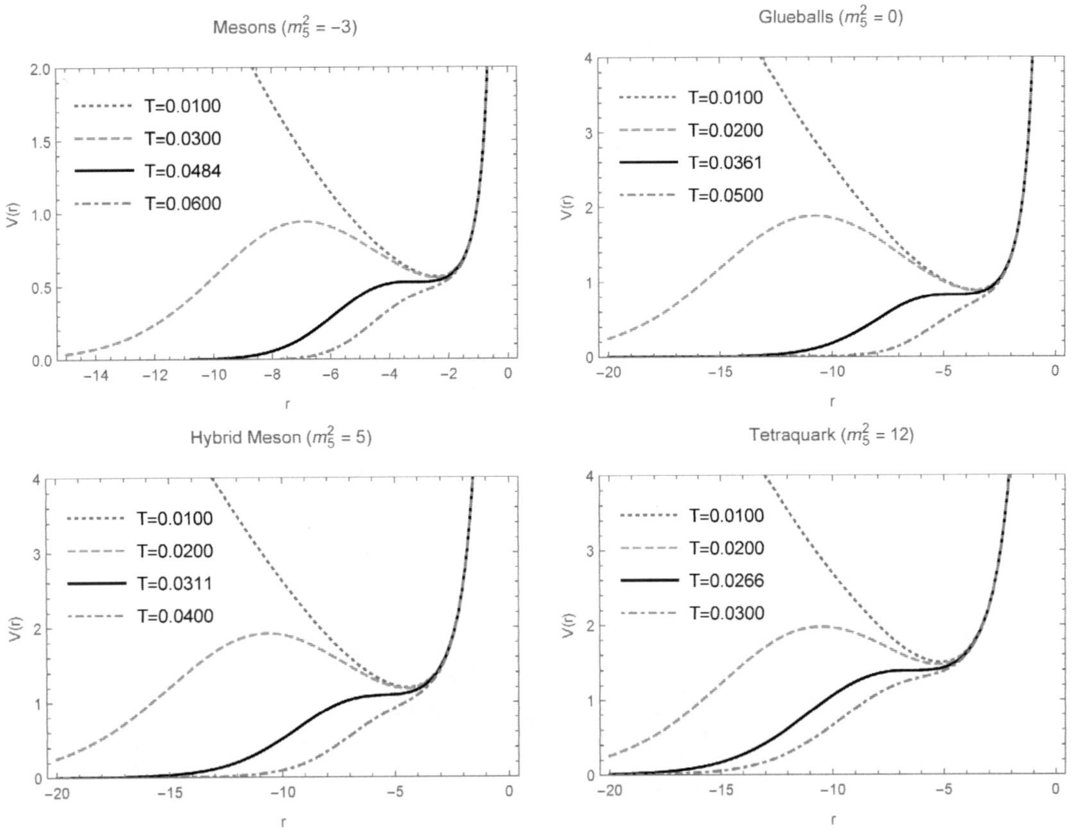

Figure 1. Plots shows the change in potential shapes for different temperatures for several scalar hadrons with a different number of constituents. Continuous lines correspond to cases considering the melting temperature for each hadron. Without lost of generality we use $R = 1$. All temperatures are in GeV.

$$\xi = \int_0^z \frac{1}{1 - \frac{t^4}{z_h^4}} dt = \frac{z_h}{2}\left(-\arctan\left(\frac{z}{z_h}\right) + \frac{1}{2}\log\left(\frac{z_h - z}{z_h + z}\right)\right) \tag{9}$$

where z_h is localization of event horizon, which is related with temperature by

$$z_h = \frac{1}{\pi T} \tag{10}$$

3. Results

At zero temperature we got the traditional soft wall model, and as it is possible to see in plots, increasing temperature the potential shape vary and at some characteristic temperature wells in potential disappear [9, 10, 16]. This is the melting temperature for each hadron in a thermal bath. In plots, continuous line shows potential at melting temperature, and as you can see, mesons are the more resistant hadrons to be melted in a thermal bath.

4. Conclusions

We consider an approach based on Gauge / Gravity ideas to calculate hadron temperature melting without to do calculations of spectral functions. Calculations consider in a simple way hadrons with arbitrary number of constituents (here we restrict us to mesons, glueballs, hybrid mesons and tetraquarks), and it shows that for scalars case, mesons are most resistant hadrons to be melted on a thermal bath.

This procedure can be applied to study another kind of hadrons (vectors, dirac, etc) and with small changes could be used to study properties of hadrons in dense media.

Appendix A. Liouville substitution

Lets consider the equation (e.g see [17])

$$\frac{d}{dz}\left[p(z)\frac{d}{dz}\psi(z)\right]+\left[\omega^2 g(z)-q(z)\right]\psi(z)=0. \tag{A.1}$$

The Liouville substitution

$$\psi(z)=v(\xi)[p(z)g(z)]^{-\frac{1}{4}}\xi=\int_0^z[\frac{g(t)}{p(t)}]^{\frac{1}{2}}dt \tag{A.2}$$

allows us to get a Schrödinger like equation as,

$$\frac{d^2v(\xi)}{d\xi^2}+\left[\omega^2-Q(\xi)\right]v(\xi)=0 \tag{A.3}$$

where

$$Q(\xi)=\frac{q(z(\xi))}{g(z(\xi))}+[p(z(\xi))g(z(\xi))]^{-\frac{1}{4}}\frac{d^2}{d\xi^2}(p(z(\xi))g(z(\xi)))^{\frac{1}{4}} \tag{A.4}$$

Acknowledgments

This work was supported by FONDECYT (Chile) under Grant No. 1141280.

[1] M. Gyulassy and L. McLerran, Nucl. Phys. A **750**, 30 (2005) doi:10.1016/j.nuclphysa.2004.10.034 [nucl-th/0405013].
[2] E. V. Shuryak, Nucl. Phys. A **750**, 64 (2005) doi:10.1016/j.nuclphysa.2004.10.022 [hep-ph/0405066].
[3] J. Erdmenger, N. Evans, I. Kirsch and E. Threlfall, Eur. Phys. J. A **35**, 81 (2008) doi:10.1140/epja/i2007-10540-1 [arXiv:0711.4467 [hep-th]].
[4] K. Kajantie, T. Tahkokallio and J. T. Yee, JHEP **0701**, 019 (2007) doi:10.1088/1126-6708/2007/01/019 [hep-ph/0609254].
[5] P. Colangelo, F. Giannuzzi, S. Nicotri and F. Zuo, Phys. Rev. D **88**, no. 11, 115011 (2013) doi:10.1103/PhysRevD.88.115011 [arXiv:1308.0489 [hep-ph]].
[6] P. Colangelo, F. Giannuzzi and S. Nicotri, JHEP **1205**, 076 (2012) doi:10.1007/JHEP05(2012)076 [arXiv:1201.1564 [hep-ph]].
[7] N. R. F. Braga, M. A. Martin Contreras and S. Diles, Eur. Phys. J. C **76**, no. 11, 598 (2016) doi:10.1140/epjc/s10052-016-4447-4 [arXiv:1604.08296 [hep-ph]].
[8] O. Aharony, S. S. Gubser, J. M. Maldacena, H. Ooguri and Y. Oz, Phys. Rept. **323**, 183 (2000) doi:10.1016/S0370-1573(99)00083-6 [hep-th/9905111].
[9] A. S. Miranda, C. A. Ballon Bayona, H. Boschi-Filho and N. R. F. Braga, Nucl. Phys. Proc. Suppl. **199**, 107 (2010) doi:10.1016/j.nuclphysbps.2010.02.013 [arXiv:0910.4319 [hep-th]].
[10] L. Bellantuono, P. Colangelo and F. Giannuzzi, Eur. Phys. J. C **74**, no. 4, 2830 (2014) doi:10.1140/epjc/s10052-014-2830-6 [arXiv:1402.5308 [hep-ph]].
[11] A. Vega and I. Schmidt, Phys. Rev. D **78**, 017703 (2008) doi:10.1103/PhysRevD.78.017703 [arXiv:0806.2267 [hep-ph]].
[12] T. Gutsche, V. E. Lyubovitskij, I. Schmidt and A. Vega, Phys. Rev. D **85**, 076003 (2012) doi:10.1103/PhysRevD.85.076003 [arXiv:1108.0346 [hep-ph]].

[13] A. Vega and P. Cabrera, Phys. Rev. D **93**, no. 11, 114026 (2016) doi:10.1103/PhysRevD.93.114026 [arXiv:1601.05999 [hep-ph]].
[14] P. Colangelo, F. De Fazio, F. Giannuzzi, F. Jugeau and S. Nicotri, Phys. Rev. D **78**, 055009 (2008) doi:10.1103/PhysRevD.78.055009 [arXiv:0807.1054 [hep-ph]].
[15] P. Colangelo, F. Giannuzzi and S. Nicotri, Phys. Rev. D **80**, 094019 (2009) doi:10.1103/PhysRevD.80.094019 [arXiv:0909.1534 [hep-ph]].
[16] M. Fujita, T. Kikuchi, K. Fukushima, T. Misumi and M. Murata, Phys. Rev. D **81**, 065024 (2010) doi:10.1103/PhysRevD.81.065024 [arXiv:0911.2298 [hep-ph]].
[17] G. Arfken, H. Weber and H. Weber, "Mathematical Methods for Physicist," 5th Edition, Academic Press. ISBN 10: 0120598256 / ISBN 13: 9780120598250

IOP Conf. Series: Journal of Physics: Conf. Series **1043** (2018) 012030 doi:10.1088/1742-6596/1043/1/012030

Family of dilatons for AdS/QCD models

Paulina Cabrera and Alfredo Vega

Instituto de Física y Astronomía, Universidad de Valparaíso,
A. Gran Bretaña 1111, Valparaíso, Chile.

E-mail: alfredo.vega@uv.cl

Abstract. We explore some possibilities for obtaining useful dilatons for AdS/QCD models. As a guideline, we consider dilatons that on the one hand reproduce the mesonic spectrum, and that on the other hand allow us a correct implementation of chiral symmetry breaking in AdS/QCD models. We discuss two procedures: one is based on supersymmetric quantum mechanics techniques and the other considers the interpolation between some limits on dilatons.

1. Introduction

The holographic approach to QCD has attracted a great deal of interest, primarily in two directions, i.e., top-down and bottom-up approaches [1]-[4]. Top-down approaches start from string theory and, after taking low-energy limits and compacting extra dimensions, attempt to obtain a theory close to QCD. In the bottom - up approach, well-known QCD properties are used as guidelines to build models in a gravity frame with asymptotically anti-de Sitter (AdS) spaces, thereby obtaining models denoted as AdS/QCD.

In the context of bottom-up models, several hadron properties can be easily calculated (see, e.g., [5]-[17]). The first version of an AdS/QCD model is known as hard wall (HW) [18, 19], where conformal invariance was broken by introducing a hard cut-off in a holographic coordinate; therefore, a field dual to hadrons cannot propagate in all space. The main problem with HW models is that, because the relationship between M^2 and J it is not linear, they do not reproduce Regge behavior. To improve this aspect, the soft wall (SW) model was proposed [20], where conformal invariance is broken by using a dilaton field. This scalar field allows us to introduce a soft cut-off along the holographic coordinate.

The most common dilaton used is quadratic in a holographic coordinate with AdS metric, and although in this case it is possible to obtain a spectrum with Regge behavior in the mesonic sector, this model is not free of problems. This situation has motivated several authors to search for improvements in this kind of model (see, e.g., [21, 22, 23] and references therein).

In AdS/QCD models the spectrum is calculated using a Schrödinger-type equation of motion; therefore, using techniques that can achieve strictly isospectral potentials could be very useful when exploring different alternatives for dilatons or metrics [24]. This offers us an opportunity to go beyond the typical choice for formulating AdS/QCD models. In this work we focus on different dilatons, but as was discused on [24] similar ideas can be used to study different metrics.

There are several methods for obtaining a family of strictly isospectral potentials for Schrödinger equations (see, e.g.,[25]-[29]). In this work we use one developed in context of supersymmetric quantum mechanics (SUSY QM) [29]. By using SUSY QM techniques we can explore dilatons (and/or metrics) that reproduce the same mesonic spectrum as in the usual models that consider quadratic dilaton and AdS metrics.

Applications of SUSY QM to AdS/QCD models have been considered previously [30]-[33], though in a different way; these authors consider transformations that produce potentials which are not strictly isospectrals, i.e., in one state the original potential and its supersymmetric partner differ. By using these transforms in they have shown that conformal invariance in the equation that describes bound states of quarks in light front holography produce interesting superconformal relationships that connect mesonic spectrum with a spectrum of baryons in light and heavy-light sectors.

The procedure propose by us in [24], considers first obtain a Schrödinger-type equation, and from this we derive a family of strictly isospectral potentials, and associated with these new potentials it is possible to extract a family of metrics and/or dilatons to AdS/QCD models (Here we focus on dilatons only). For the sake of simplicity, in this work we studied the family of dilatons related to Schrödinger equations that describe scalar glueballs in conventional SW models (i.e., AdS metric and quadratic dilaton) [34, 35, 17]. Because the mass of the gravity mode dual to scalar glueballs in AdS is zero, we obtained simplified equations.

Through the use of SUSY QM techniques, we suppose a family of strictly isospectral potential with a holographical coordinate z that has the same behaviour at small and high values. This means, for example, if we consider as original potential one associated with AdS metric with quadratic dilaton, each dilaton in the family with AdS metric will have $\phi(z \to 0) = \kappa^2 z^2$ and $\phi(z \to \infty) = \kappa^2 z^2$. By construction, each dilaton in this family produces the same spectrum for scalar glueballs, but unfortunately these dilatons have the same problem as quadratic ones at the moment of building AdS/QCD models with chiral symmetry breaking. The problem with chiral symmetry breaking in AdS/QCD models has motivated the study of some modifications to these models [36, 37, 38, 39, 40].

To solve the problem associated with the correct implementation of chiral symmetry breaking in SW models, some authors have considered dilatons that correspond to an interpolation between $\phi(z \to 0) = \kappa_0^2 z^2$ and $\phi(z \to \infty) = \kappa_\infty^2 z^2$, where $\kappa_0^2 \neq \kappa_\infty^2$ [37, 41, 42]; dilatons that interpolate between these limits can be obtained by introducing small changes in the isospectral potential calculated with SUSY QM transforms.

This article contains six sections. In Sec. II we briefly show how to obtain the spectrum for scalar hadrons with SW models that consider an symptotically AdS metric and an arbitrary z-dependent dilaton. In Sec. III we discuss the method we use to obtain a family of strictly isospectral potentials in the context of SUSY QM, and we explain how we used this to obtain a differential equation that allowed us to extract a family of dilatons. As an example, we present in detail the scalar glueball case. In Sec. IV we discuss the main problem in AdS/QCD models with chiral symmetry breaking. In Sec. V we present the possibility for using SUSY QM techniques to obtain families of dilatons consistent with the guidelines proposed; at the end of that section, we introduce a brief discussion about the interpolation functions for dilatons. Finally, in Sec. VI we present our conclusions and some future perspectives for additional work in this area.

2. Scalar fields

We consider a SW model, where action for scalar fields in spaces with 5D with dilaton is [36, 43, 44, 45]

$$S = -\int d^5x \sqrt{-g} e^{-\phi(z)} \frac{1}{2} [g^{MN} \partial_M \Phi \partial_N \Phi + m_5^2 \Phi^2], \tag{1}$$

with $M, N = 0, 1, 2, 3, z$, where z is a holographic coordinate, $\phi(z)$ is a scalar dilaton field (used to break the conformal invariance in SW models) and m_5 is the mass of AdS modes along bulk [which is related to the dimension of operators that creates hadrons in field theory according to $m_5^2 = \Delta(\Delta - 4)$].

The metric considered was

$$ds^2 = a^2(z)\eta_{MN}dx^M dx^N, \tag{2}$$

where $\eta_{MN} = diag(-1, +1, +1, +1, +1)$, and $a^2(z)$ is a warp factor; therefore, in general we can consider AdS spaces asymptotically.

We considered scalar modes according to

$$\Phi(x) = e^{-i\mathcal{P}\cdot\mathcal{X}} f(z), \tag{3}$$

where $\mathcal{P}$ and $\mathcal{X}$ correspond to momentum and position in boundary 4D and $\mathcal{P}^2 = M^2$, where M is the mass of hadron studied with this model.

Starting from equation of motion obtained from (1), and by using (3), it was possible to get an equation to $f(z)$

$$-f''(z)+\left(\phi'(z) - 3\frac{a'(z)}{a(z)}\right)f'(z) + a^2(z)m_5^2 f(z) = M^2 f(z), \tag{4}$$

which, by transforming

$$f(z) = \exp\left[\frac{1}{2}(\phi(z) - 3ln\ a(z))\right]\psi(z), \tag{5}$$

gives rise to a Schrödinger-type equation of the form

$$[-\partial_z^2 + V(z)]\psi(z) = M^2\psi(z), \tag{6}$$

where $V(z)$ is the potential

$$V(z) = m_5^2 a^2(z) + \frac{3}{4}\left(\frac{a'(z)}{a(z)}\right)^2 - \frac{3}{2}\left(\frac{a'(z)}{a(z)}\right)\phi'(z) + \frac{1}{4}\phi'^2(z) + \frac{3}{2}\frac{a''(z)}{a(z)} - \frac{1}{2}\phi''(z). \tag{7}$$

Through the use of $a(z) = R/z$ (where R is the AdS radius) and $\phi(z) = \kappa^2 z^2$, the conventional AdS/QCD model was obtained [36, 43, 44], which considers an AdS metric and a quadratic dilaton, and in this case, the potential is

$$V(z) = \frac{1}{z^2}\left(\frac{15}{4} + m_5^2 R^2\right) + z^2\kappa^4 + 2\kappa^2. \tag{8}$$

For this case, a solution to the eigenvalue problem (6) is

$$\psi(z) = N_n \exp^{-\frac{1}{2}z^2\kappa^2} z^{(\frac{1}{2}+\sqrt{4+m_5^2R^2})} L_n^{\sqrt{4+m_5^2R^2}}(z^2\kappa^2), \tag{9}$$

where N_n is a normalization constant, that for the special case $n = 0$ (that will be used later), is

$$N_0 = \frac{2\kappa^{2(1+\sqrt{4+m_5^2R^2})}}{\Gamma(1+\sqrt{4+m_5^2R^2})}. \tag{10}$$

The mass spectrum is given by

$$M_n^2 = 4\kappa^2\left(n + \frac{1}{2}\sqrt{4+m_5^2R^2} + 1\right). \tag{11}$$

As can be seen in the previous expression, SW models with an AdS metric and quadratic dilaton are in agreement with Regge trajectories because M_n^2 vs n is linear.

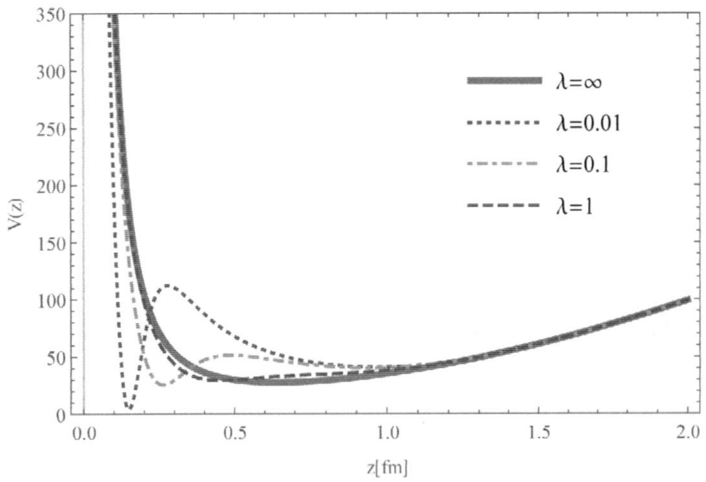

Figure 1. Some potentials of isospectral families with different values for λ. When $\lambda = \infty$ (8) it recovers the original potential (12). Values used are $m_5 = 0$ and $\kappa = 428\ MeV$.

3. Family of isospectral dilatons considerings SUSY QM techniques

As we saw in the case of SW models, if we start with an AdS metric and a quadratic dilaton it is possible to obtain a Schrödinger-type equation associated with different hadrons on the AdS side. Although not a common practice, using the reverse (i.e., starting from a given potential) it is possible to extract a metric and a dilaton that reproduce this potential. This is the base of procedure presented in [24], and discussed in this section.

For Schrödinger-type equations, there are several procedures that make it possible to obtain a family of strictly isospectral potentials associated with a given potential [25, 26, 27, 28, 29]. From among the different isospectral transformations, we used a procedure discussed in [29], that considers SUSY QM to obtain a family of strictly isospectral potentials depending on a parameter.

According to [29], two potentials V_1 and $\widehat{V}_1$ are strictly isospectral if

$$\widehat{V}_1(z) = V_1(z) - 2\frac{d^2}{dz^2}\ln[I(z) + \lambda]. \qquad (12)$$

In this case, $V_1(z)$ is the original potential (8), λ is and arbitrary parameter and the function $I(z)$, for coordinate between zero and infinity, is given by

$$I(z) = \int_0^z \psi_1^2(z')dz',$$

where ψ_1 corresponds to the ground state of V_1, in our case is given by (9) with $n = 0$; using $R = 1$ we obtained

$$I(z) = 1 - \frac{\Gamma(1+\sqrt{4+m_5^2}, z^2\kappa^2)}{\Gamma(1+\sqrt{4+m_5^2})}. \qquad (13)$$

Figure (1) shows several strictly isospectral potential that appeared from (12), starting with $V_1(z)$ given by (8) for scalar glueballs ($m_5 = 0$); it is possible to compare different potentials in this family in relation to the original one, and note also that changes in potentials are produced in a region close to the origin in z and depends on λ. At higher values of λ, changes in family

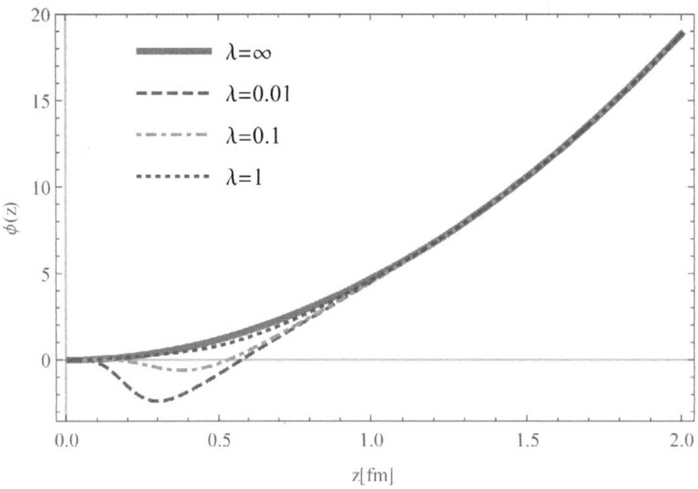

Figure 2. Some isospectrals dilaton associated with potentials shown in fig. (1), with $m_5 = 0$ and $\kappa = 428\ MeV$.

potentials are less, so when $\lambda \to \infty$ we recover the original potential. Again, all potentials in Fig. 1 produce the same spectrum for scalar glueballs, which in this case it is Regge-like.

Each member of this family was obtained by using (12) which could be considered as having been generated for a new metric and/or a new dilaton. For this reason we could match (12) and (7) and get

$$V_1(z) - 2\frac{d^2}{dz^2}\ln[I(z)+\lambda] = m_5^2 a^2(z) + \frac{3}{4}\left(\frac{a'(z)}{a(z)}\right)^2 - \frac{3}{2}\left(\frac{a'(z)}{a(z)}\right)\phi'(z) + \frac{1}{4}\phi'^2(z) + \frac{3}{2}\frac{a''(z)}{a(z)} - \frac{1}{2}\phi''(z). \tag{14}$$

Notice that if in the last expression we use a specified metric, we get a differential equation to the family of dilatons. In this case we considered an AdS metrics, and the equation for dilaton is

$$V_1(z) - 2\frac{d^2}{dz^2}\ln[I(z)+\lambda] = \frac{15}{4z^2} + \frac{m_5^2 R^2}{z^2} + \frac{3\phi'(z)}{2z} + \frac{1}{4}\phi'^2(z) - \frac{\phi''(z)}{2}. \tag{15}$$

Figure (2) shows several dilatons of family obtained solving (15).

It is important to verify that this procedure produces potentials with the same behaviour at low and high z as the original one. However, as we show in next section, this must be changed in order to obtain dilatons or metrics appropriated in AdS/QCD models. This must be considered when using boundary conditions to solve the differential equations suggested in this section, and it obligates us to introduce modifications on them.

4. Improvements in AdS/QCD models with chiral symmetry breaking

Although the spectrum in bottom-up models with an AdS metric and quadratic dilaton is well reproduced in the mesonic sector [30, 31, 32, 33, 46, 47], the incorporation of chiral symmetry breaking in this approach is far from QCD and needs improvements. This occurs because the scalar field in the bulk is dual to bilinear operator $q\bar{q}$, whose vacuum expectation value (VEV) —which is responsible for chiral symmetry breaking in this model— does not allow its explicit and spontaneous breaking in an independent and simultaneous way [36, 37, 38, 39, 40]. The origin of this problem can be found in the solution of the VEV equation, which produces a finite action at the limit $z \to \infty$, given by

$$v(z) \sim m_q z U(1/2, 0, z^2), \tag{16}$$

where $U(a, b, z)$ is a hypergeometric confluent function. According to the AdS/QCD dictionary established in [18], the VEV limit $z \to 0$ must be

$$v(z) \sim \alpha z + \beta z^3. \tag{17}$$

In these models, quark mass m_q and chiral condensate σ are related with a constant that appears in (17) through [18]

$$m_q = \frac{\alpha R}{\zeta} \quad and \quad \sigma = \beta R \zeta, \tag{18}$$

where ζ is a normalization parameter introduced in [48].

According to Gell-Mann–Oakes–Renner (GMOR) relationship, m_q and σ are related by $m_\pi^2 f_\pi^2 = 2 m_q \sigma$; if we expand (16) to small values of the holographic coordinate, we get $\alpha \sim m_q$ and $\beta \sim \sigma \sim m_q$. This means that when quark mass is zero in this model then the chiral condensate is also zero, and this is not in agreement with GMOR.

The literature has some alternatives to solve this problem. For example, a negative quadratic dilaton could be a possible alternative [49], but it predicts an extra massless mode that are not observed [50]. It was recently suggested [41] that an interpolation between two quadratic dilatons, one negative to small z and another positive to high z, could solve the problem without spurious modes that appear only with a negative dilaton. Another way to improve this model was studied in [37], where authors considered adding a quartic term in the scalar potential; in this case, they obtained

$$\phi' = \frac{1}{a^3 v'}\left[\partial_z(a^3 v') - a^5\left(m_X^2 v - \frac{\kappa}{2} v^3\right)\right] \tag{19}$$

which corresponds to an equation that relates the dilaton field $\phi(z)$, the warp factor in metric $a(z)$, VEV $v(z)$ and a constant κ introduced in quartic term in action.

We can properly incorporate chiral symmetry breaking in soft-wall models by using (19) with an AdS metric $[a(z) = R/z]$ and

$$v(z) = \frac{z}{R}(A + B \; tanh \; C z^2), \tag{20}$$

which is an interpolation between the limits to small and high values in z for VEV. In this case dilaton satisfies

$$\phi(z \to 0) = \kappa_0^2 z^2, \tag{21}$$

$$\phi(z \to \infty) = \kappa_\infty^2 z^2, \tag{22}$$

where $\kappa_0^2 = -\kappa_\infty^2$ in [41] and $\kappa_0^2 \neq \kappa_\infty^2$ in [37]. We notice that, without details, which dilatons corresponding to interpolations between limit (21) and (22) can, according to some authors, improve other facts in the SW model (e.g., [51]).

5. Family of dilatons for AdS/QCD models using SUSY QM techniques

Dilatons in agreement with (21) and (22) cannot be obtained from (15), because the original potential and the strictly isospectral potential has the same behaviour at low and high z values; i.e., dilatons in family found with (15) have $\phi(z \to 0) = \phi(z \to \infty) = \kappa^2 z^2$. However, it is possible to introduce small changes to (15) in order to obtain dilatons in agreement with (21)

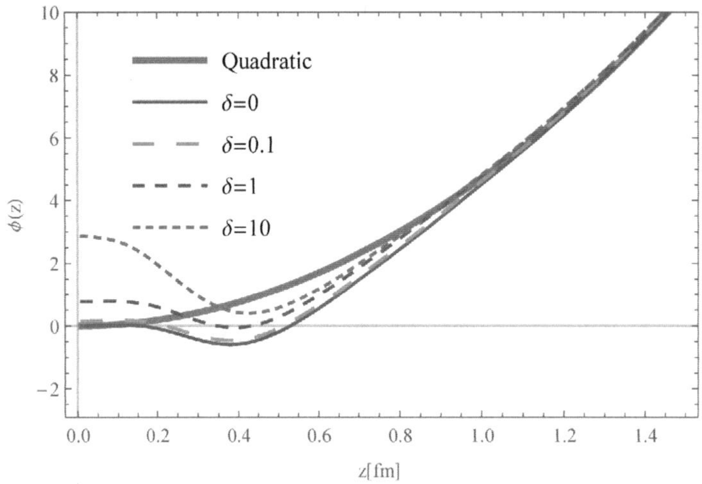

Figure 3. Dilatons for different values of δ, $p = even$, $\kappa = 428\ MeV$, $m_5 = 0$ and $\lambda = 0.1$

and (22). To achieve this, we added a new term in (15) that slightly modifies the behavior at small z.

We have two possible locations to introduce the change: on the right or left side in (15). Thus, instead of (15) we suggest:

$$\frac{1}{z^2}\left(\frac{15}{4}+m_5^2\right)e^{-\delta z^2}+z^2\kappa^4+2\kappa^2-2\frac{d^2}{dz^2}\ln[I(z)+\lambda] = \frac{1}{z^2}\left(\frac{15}{4}+m_5^2\right)+\frac{3\phi'(z)}{2z}+\frac{1}{4}\phi'^2(z)-\frac{\phi''(z)}{2} \quad (23)$$

or

$$\frac{1}{z^2}\left(\frac{15}{4}+m_5^2\right)+z^2\kappa^4+2\kappa^2-2\frac{d^2}{dz^2}\ln[I(z)+\lambda] = \frac{1}{z^2}\left(\frac{15}{4}+m_5^2\right)e^{-\delta z^2}+\frac{3\phi'(z)}{2z}+\frac{1}{4}\phi'^2(z)-\frac{\phi''(z)}{2}. \quad (24)$$

Notice that both equations can be summarized in

$$(-1)^p[e^{-\delta z^2}-1]\frac{1}{z^2}\left(\frac{15}{4}+m_5^2\right)+z^2\kappa^4+2\kappa^2-2\frac{d^2}{dz^2}\ln[I(z)+\lambda] = \frac{3\phi'(z)}{2z}+\frac{1}{4}\phi'^2(z)-\frac{\phi''(z)}{2}, \quad (25)$$

where δ and p are two parameters that modify the behavior of $\phi(z)$ at low z.

In Figs. 3 and 4 it is possible to see how δ modifies the magnitude of $\phi(z \to 0)$, and by using even or odd p values it was possible to choose the sign of $\phi(0)$ according to (23) or (24). Increasing δ produces differences between the spectrum calculated with quadratic dilaton and the one obtained from (23), (24) or (25). This constrains us to use small values for δ if we wish to obtain a dilaton that produces a spectrum very close to the original one. The behavior of the dilaton obtained in both extensions is $\phi \to C \pm \kappa_0^2 z^2$, with C a constant.

It is worth noting in relation to implementation of chiral symmetry breaking in AdS/QCD models that the constant term is not important, and it can be dropped to calculate other quantities. In addition, whether the change suggested was implemented according to (23) or (24) determines if is $\kappa_0 < \kappa_\infty$ or $\kappa_0 > \kappa_\infty$.

An alternative procedure to built dilatons or metrics could be to consider interpolation functions between known limits. For example, for dilatons in models with AdS metrics it is possible to consider (21) and (22) as known limits, and in this case a possible dilaton could be

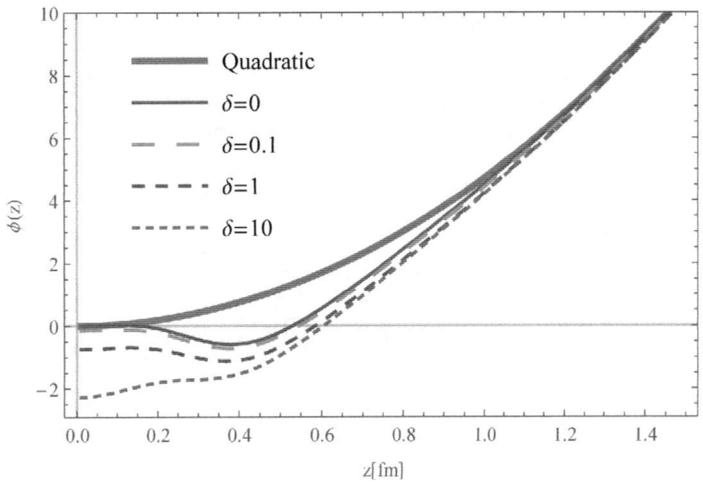

Figure 4. Dilatons for different values of δ, $p = odd$, $\kappa = 428\ MeV$, $m_5 = 0$ and $\lambda = 0.1$

$$\phi(z) = A_0 z^2 + (A_1 + B_1) z^2\ tanh(C\ z^2). \tag{26}$$

In this case, numerical calculation shows that this interpolation function produces a potential with Regge behaviour in a mass spectrum calculated with (6).

6. Summary and conclusions

In this work we discuss the ideas introduced in [24] to get families of dilatons useful for AdS / QCD models.

The main method discussed is based on SUSY QM techniques. This ensures that dilatons obtained with this procedure produce a proper mesonic spectrum. When we include as an additional condition the correct chiral symmetry breaking implementation, we found that in general this is not possible; however, introducing small changes in equations opens a possibility for obtaining dilatons that works in AdS spaces.

Another alternative studied considers dilatons that correspond to interpolating functions between two limits. In our case we consider limits when $z \to 0$ and $z \to \infty$, and following Ref. [37], the limits suggested in (21) and (22) for dilatons.

In our opinion, dilatons presented here, or other that could be obtained through the procedures discussed in this paper, could be interestingly used to calculate hadronic properties in other uses of AdS/QCD models. These possibilities will be explored in future work.

Acknowledgments

This work was supported by FONDECYT (Chile) under Grant No. 1141280 and by CONICYT (Chile) under Grant No. 7912010025.

[1] J. Erdmenger, N. Evans, I. Kirsch and E. Threlfall, Eur. Phys. J. A **35**, 81 (2008) doi:10.1140/epja/i2007-10540-1 [arXiv:0711.4467 [hep-th]].

[2] S. J. Brodsky, G. F. de Teramond, H. G. Dosch and J. Erlich, Phys. Rept. **584**, 1 (2015) doi:10.1016/j.physrep.2015.05.001 [arXiv:1407.8131 [hep-ph]].

[3] U. Gursoy and E. Kiritsis, JHEP **0802**, 032 (2008) doi:10.1088/1126-6708/2008/02/032 [arXiv:0707.1324 [hep-th]].

[4] U. Gursoy, E. Kiritsis and F. Nitti, JHEP **0802**, 019 (2008) doi:10.1088/1126-6708/2008/02/019 [arXiv:0707.1349 [hep-th]].

[5] T. Gutsche, V. E. Lyubovitskij, I. Schmidt and A. Vega, Phys. Rev. D **91**, no. 11, 114001 (2015) doi:10.1103/PhysRevD.91.114001 [arXiv:1501.02738 [hep-ph]].
[6] T. Gutsche, V. E. Lyubovitskij, I. Schmidt and A. Vega, Phys. Rev. D **86**, 036007 (2012) doi:10.1103/PhysRevD.86.036007 [arXiv:1204.6612 [hep-ph]].
[7] T. Gutsche, V. E. Lyubovitskij, I. Schmidt and A. Vega, Phys. Rev. D **87**, no. 1, 016017 (2013) doi:10.1103/PhysRevD.87.016017 [arXiv:1212.6252 [hep-ph]].
[8] A. Vega, I. Schmidt, T. Gutsche and V. E. Lyubovitskij, Phys. Rev. D **83**, 036001 (2011) doi:10.1103/PhysRevD.83.036001 [arXiv:1010.2815 [hep-ph]].
[9] A. Vega, I. Schmidt, T. Gutsche and V. E. Lyubovitskij, Phys. Rev. D **85**, 096004 (2012) doi:10.1103/PhysRevD.85.096004 [arXiv:1202.4806 [hep-ph]].
[10] S. J. Brodsky, F. G. Cao and G. F. de Teramond, Phys. Rev. D **84**, 075012 (2011) doi:10.1103/PhysRevD.84.075012 [arXiv:1105.3999 [hep-ph]].
[11] S. J. Brodsky and G. F. de Teramond, Phys. Rev. D **78**, 025032 (2008) doi:10.1103/PhysRevD.78.025032 [arXiv:0804.0452 [hep-ph]].
[12] S. J. Brodsky and G. F. de Teramond, Phys. Rev. D **77**, 056007 (2008) doi:10.1103/PhysRevD.77.056007 [arXiv:0707.3859 [hep-ph]].
[13] N. R. F. Braga and A. Vega, Eur. Phys. J. C **72**, 2236 (2012) doi:10.1140/epjc/s10052-012-2236-2 [arXiv:1110.2548 [hep-ph]].
[14] C. A. Ballon Bayona, H. Boschi-Filho and N. R. F. Braga, JHEP **0803**, 064 (2008) doi:10.1088/1126-6708/2008/03/064 [arXiv:0711.0221 [hep-th]].
[15] L. Bellantuono, P. Colangelo and F. Giannuzzi, JHEP **1510**, 137 (2015) doi:10.1007/JHEP10(2015)137 [arXiv:1507.07768 [hep-ph]].
[16] P. Colangelo, F. De Fazio, J. J. Sanz-Cillero, F. Giannuzzi and S. Nicotri, Phys. Rev. D **85**, 035013 (2012) doi:10.1103/PhysRevD.85.035013 [arXiv:1108.5945 [hep-ph]].
[17] H. Boschi-Filho, N. R. F. Braga, F. Jugeau and M. A. C. Torres, Eur. Phys. J. C **73**, 2540 (2013) doi:10.1140/epjc/s10052-013-2540-5 [arXiv:1208.2291 [hep-th]].
[18] J. Erlich, E. Katz, D. T. Son and M. A. Stephanov, Phys. Rev. Lett. **95**, 261602 (2005) doi:10.1103/PhysRevLett.95.261602 [hep-ph/0501128].
[19] L. Da Rold and A. Pomarol, Nucl. Phys. B **721**, 79 (2005) doi:10.1016/j.nuclphysb.2005.05.009 [hep-ph/0501218].
[20] A. Karch, E. Katz, D. T. Son and M. A. Stephanov, Phys. Rev. D **74**, 015005 (2006) doi:10.1103/PhysRevD.74.015005 [hep-ph/0602229].
[21] Z. Fang, D. Li and Y. L. Wu, Phys. Lett. B **754**, 343 (2016) doi:10.1016/j.physletb.2016.01.045 [arXiv:1602.00379 [hep-ph]].
[22] E. F. Capossoli and H. Boschi-Filho, Phys. Lett. B **753**, 419 (2016) doi:10.1016/j.physletb.2015.12.034 [arXiv:1510.03372 [hep-ph]].
[23] N. Evans, P. Jones and M. Scott, Phys. Rev. D **92**, no. 10, 106003 (2015) doi:10.1103/PhysRevD.92.106003 [arXiv:1508.06540 [hep-ph]].
[24] A. Vega and P. Cabrera, Phys. Rev. D **93**, no. 11, 114026 (2016) doi:10.1103/PhysRevD.93.114026 [arXiv:1601.05999 [hep-ph]].
[25] P. B. Abraham and H. E. Mosey, Phys. Rev. A **22**, 1333 (1980). doi:10.1103/PhysRevD.33.431
[26] M. Luban and D. L. Pursey, Phys. Rev. D **33**, 431 (1986). doi:10.1103/PhysRevD.33.431
[27] D. L. Pursey, Phys. Rev. D **33**, 1048 (1986). doi:10.1103/PhysRevD.33.1048
[28] A. Khare and U. P. Sukhatme, Phys. Rev. A **40**, 6185 (1989). doi:10.1103/PhysRevA.40.6185
[29] F. Cooper, A. Khare and U. Sukhatme, Phys. Rept. **251**, 267 (1995) doi:10.1016/0370-1573(94)00080-M [hep-th/9405029].
[30] G. F. de Teramond, H. G. Dosch and S. J. Brodsky, Phys. Rev. D **91**, no. 4, 045040 (2015) doi:10.1103/PhysRevD.91.045040 [arXiv:1411.5243 [hep-ph]].
[31] H. G. Dosch, G. F. de Teramond and S. J. Brodsky, Phys. Rev. D **91**, no. 8, 085016 (2015) doi:10.1103/PhysRevD.91.085016 [arXiv:1501.00959 [hep-th]].
[32] H. G. Dosch, G. F. de Teramond and S. J. Brodsky, Phys. Rev. D **92**, no. 7, 074010 (2015) doi:10.1103/PhysRevD.92.074010 [arXiv:1504.05112 [hep-ph]].
[33] S. J. Brodsky, A. Deur, G. F. de Tramond and H. G. Dosch, Int. J. Mod. Phys. Conf. Ser. **39**, 1560081 (2015) doi:10.1142/S2010194515600812 [arXiv:1510.01011 [hep-ph]].
[34] P. Colangelo, F. De Fazio, F. Jugeau and S. Nicotri, Phys. Lett. B **652**, 73 (2007) doi:10.1016/j.physletb.2007.06.072 [hep-ph/0703316].
[35] P. Colangelo, F. De Fazio, F. Jugeau and S. Nicotri, Int. J. Mod. Phys. A **24**, 4177 (2009) doi:10.1142/S0217751X09045443 [arXiv:0711.4747 [hep-ph]].
[36] P. Colangelo, F. De Fazio, F. Giannuzzi, F. Jugeau and S. Nicotri, Phys. Rev. D **78**, 055009 (2008)

doi:10.1103/PhysRevD.78.055009 [arXiv:0807.1054 [hep-ph]].

[37] T. Gherghetta, J. I. Kapusta and T. M. Kelley, Phys. Rev. D **79**, 076003 (2009) doi:10.1103/PhysRevD.79.076003 [arXiv:0902.1998 [hep-ph]].

[38] H. J. Kwee and R. F. Lebed, Phys. Rev. D **77**, 115007 (2008) doi:10.1103/PhysRevD.77.115007 [arXiv:0712.1811 [hep-ph]].

[39] A. Vega and I. Schmidt, Phys. Rev. D **82**, 115023 (2010) doi:10.1103/PhysRevD.82.115023 [arXiv:1005.3000 [hep-ph]].

[40] A. Vega and I. Schmidt, Phys. Rev. D **84**, 017701 (2011) doi:10.1103/PhysRevD.84.017701 [arXiv:1104.4365 [hep-ph]].

[41] K. Chelabi, Z. Fang, M. Huang, D. Li and Y. L. Wu, Phys. Rev. D **93**, no. 10, 101901 (2016) doi:10.1103/PhysRevD.93.101901 [arXiv:1511.02721 [hep-ph]].

[42] K. Chelabi, Z. Fang, M. Huang, D. Li and Y. L. Wu, JHEP **1604**, 036 (2016) doi:10.1007/JHEP04(2016)036 [arXiv:1512.06493 [hep-ph]].

[43] A. Vega and I. Schmidt, Phys. Rev. D **78**, 017703 (2008) doi:10.1103/PhysRevD.78.017703 [arXiv:0806.2267 [hep-ph]].

[44] T. Gutsche, V. E. Lyubovitskij, I. Schmidt and A. Vega, Phys. Rev. D **85**, 076003 (2012) doi:10.1103/PhysRevD.85.076003 [arXiv:1108.0346 [hep-ph]].

[45] W. de Paula and T. Frederico, Phys. Lett. B **693**, 287 (2010) doi:10.1016/j.physletb.2010.08.045 [arXiv:0908.4282 [hep-ph]].

[46] T. Branz, T. Gutsche, V. E. Lyubovitskij, I. Schmidt and A. Vega, Phys. Rev. D **82**, 074022 (2010) doi:10.1103/PhysRevD.82.074022 [arXiv:1008.0268 [hep-ph]].

[47] A. Vega and I. Schmidt, Phys. Rev. D **79**, 055003 (2009) doi:10.1103/PhysRevD.79.055003 [arXiv:0811.4638 [hep-ph]].

[48] A. Cherman, T. D. Cohen and E. S. Werbos, Phys. Rev. C **79**, 045203 (2009) doi:10.1103/PhysRevC.79.045203 [arXiv:0804.1096 [hep-ph]].

[49] F. Zuo, Phys. Rev. D **82**, 086011 (2010) doi:10.1103/PhysRevD.82.086011 [arXiv:0909.4240 [hep-ph]].

[50] A. Karch, E. Katz, D. T. Son and M. A. Stephanov, JHEP **1104**, 066 (2011) doi:10.1007/JHEP04(2011)066 [arXiv:1012.4813 [hep-ph]].

[51] D. Li and M. Huang, JHEP **1311**, 088 (2013) doi:10.1007/JHEP11(2013)088 [arXiv:1303.6929 [hep-ph]].

IOP Conf. Series: Journal of Physics: Conf. Series **1043** (2018) 012031 doi:10.1088/1742-6596/1043/1/012031

Majorana vs. Dirac sterile neutrinos at the LHC

Claudio O. Dib[1], **C.S. Kim**[2], **Kechen Wang**[3,4], **Jue Zhang**[4]

[1] Universidad Tcnica Federico Santa Mara, Valparaso, Chile,
[2] Yonsei University, Seoul, Korea,
[3] DESY, Hamburg, Germany,
[4] IHEP, Chinese Academy of Sciences, Beijing, China.

E-mail: claudio.dib@usm.cl

Abstract. We study leptonic decays $W^\pm \to e^\pm e^\pm \mu^\mp \nu$ and $W^\pm \to \mu^\pm \mu^\pm e^\mp \nu$ which would occur at the LHC if there exist sterile neutrinos with masses below M_W. We also study ways to discriminate their Majorana or Dirac character, a rather non trivial task, because lepton number conservation cannot be checked due to the missing neutrino in the final state. We find that it is indeed possible to discriminate between Majorana vs. Dirac sterile neutrinos by comparing the production of $e^\pm e^\pm \mu^\mp$ vs. $\mu^\pm \mu^\pm e^\mp$ if the N-e and N-μ mixings are sufficiently different. Alternatively, one could also distinguish the Majorana vs. Dirac character by studying the energy spectra of the opposite charge lepton, a method that works even for equal N-e and N-μ mixings.

1. Introduction

The origin of the small neutrino masses and their Dirac or Majorana character is a currently outstanding problem, since these are naturally zero in the Standard Model. Most explanations for the small neutrino masses are based on *seesaw* models [1]. These models predict additional heavy neutrinos, sterile in the Standard Model except for small mixings with the weak currents. The heavy neutrinos masses, m_N, in most scenarios should be of Majorana type and, depending on the model, they could lie anywhere from a few eV all the way to GUT scales. It is thus very important to explore the existence of these extra particles, and to resolve their mass character and scale, as they will shed light onto physics beyond the Standard Model.

Search for Majorana masses are usually done in *Neutrinoless Double Beta Decay* experiments [2]. Nevertheless, collider tests are also competitive for specific mass ranges. At the LHC, $W \to \ell^+\ell^+ jj$ is appropriate for $m_N > M_W$ [3, 4], while leptonic modes such as $W^\pm \to e^\pm e^\pm \mu^\mp \nu$ and $W^\pm \to \mu^\pm \mu^\pm e^\mp \nu$ are preferred for $m_N < M_W$ [5]. However, here the discrimination of Dirac vs. Majorana is a major challenge, because the final neutrino, which may or may not balance Lepton Number, escapes detection [6, 7].

Indeed, the mode e.g. $W^+ \to e^+e^+\mu^-\nu$ is actually a sum of two exclusive modes:

(i) $W^+ \to e^+e^+\mu^-\bar{\nu}_\mu$: a Lepton Number violating process (LNV),
(ii) $W^+ \to e^+e^+\mu^-\nu_e$: a Lepton Number conserving process (LNC).

These decays are forbidden in the Standard Model. They proceed only if there exist heavy sterile neutrinos, which would appear as intermediate states in these processes. These rates are very small, but they get resonantly enhanced if $m_N < M_W$ as N goes on its mass shell.

IOP Conf. Series: Journal of Physics: Conf. Series **1043** (2018) 012031 doi:10.1088/1742-6596/1043/1/012031

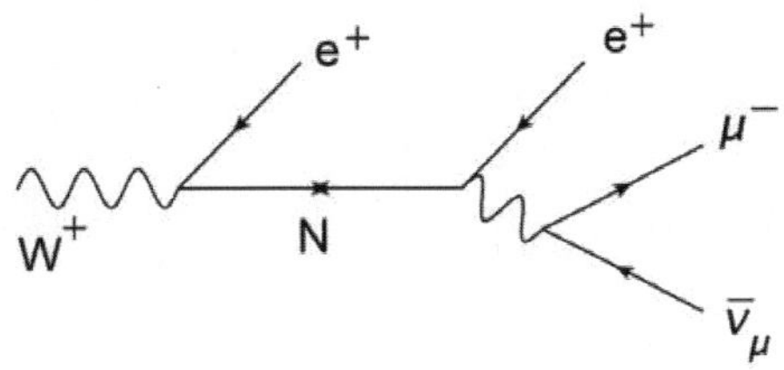

Figure 1. LNV process mediated by a sterile neutrino N (Majorana N only).

The amplitude for the LNV decay $W^+ \to e^+e^+\mu^-\bar{\nu}_\mu$ is depicted in the Feynman diagram of Fig. 1. Only a Majorana N can induce this process, since it violates lepton number. However, since the final neutrino $\bar{\nu}_\mu$ will escape detection in an experiment, it is not possible to directly verify this violation of lepton number.

The spectrum of the opposite charge lepton (i.e. the muon), in the N rest frame, is:

$$\Gamma(W^+ \to e^+e^+\mu^-\bar{\nu}_\mu) = \frac{G_F^3 M_W^3 m_N^5}{12\sqrt{2}\,\pi^4}\left(1-\frac{m_N^2}{M_W^2}\right)^2\left(1+\frac{m_N^2}{2M_W^2}\right) \times\frac{|U_{Ne}|^4}{\Gamma_N}\int_0^{1/2} d\epsilon_\mu\,(\epsilon_\mu^2 - 2\epsilon_\mu^3)\,. \quad (1)$$

where $\epsilon_\mu \equiv E_\mu/m_N$ is the normalized muon energy.

In turn, the integrated rate for this process is:

$$Br(W^+ \to e^+e^+\mu^-\bar{\nu}_\mu) \approx 4.8\times 10^{-3}\left(1-\frac{m_N^2}{M_W^2}\right)^2\left(1+\frac{m_N^2}{2M_W^2}\right)\frac{|U_{Ne}|^4}{\sum_{\ell=e,\mu,\tau}|U_{N\ell}|^2}. \quad (2)$$

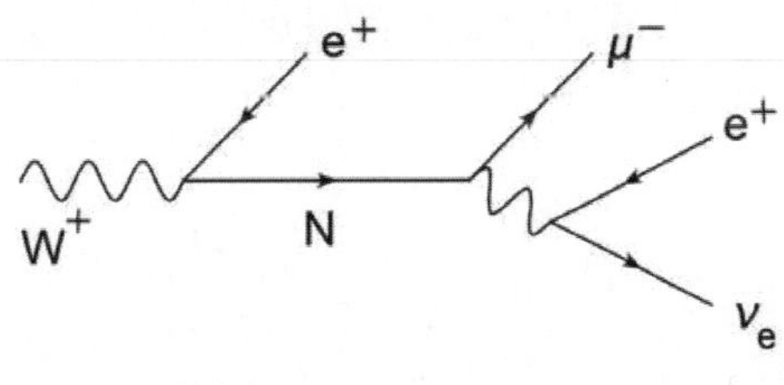

Figure 2. LNC process mediated by a sterile neutrino N (Dirac or Majorana N).

On the other hand, the LNC decay $W^+ \to e^+e^+\mu^-\nu_e$ is depicted in Fig. 2. This process conserves lepton number, and thus can be induced by a sterile neutrino of either Majorana or Dirac character. However, again such conservation cannot be directly verified in an experiment due to the neutrino ν_e in the final state, which would escape detection.

The spectrum of the opposite charge lepton in this process, measured in the N rest frame, is:

$$\Gamma(W^+ \to e^+e^+\mu^-\nu_e) = \frac{G_F^3 M_W^3 m_N}{12\sqrt{2}\,\pi^4}\left(1-\frac{m_N^2}{M_W^2}\right)^2\left(1+\frac{m_N^2}{2M_W^2}\right) \times\frac{|U_{Ne}\,U_{N\mu}|^2}{\Gamma_N}\int_0^{1/2} d\epsilon_\mu\left(\frac{1}{2}\epsilon_\mu^2 - \frac{2}{3}\epsilon_\mu^3\right), \quad (3)$$

IOP Conf. Series: Journal of Physics: Conf. Series **1043** (2018) 012031 doi:10.1088/1742-6596/1043/1/012031

where again $\epsilon_\mu \equiv E_\mu/m_N$ is the normalized muon energy. It is clear that this spectrum differs from that of the LNV process, given in Eq. (1). The two spectra are shown in Fig. 3 (left).

Even when the spectra are different, the integrated rate for the LNC and LNV processes are identical, except for the lepton mixing elements which appear as a global factor:

$$Br(W^+ \to e^+e^+\mu^-\nu_e) \approx 4.8 \times 10^{-3} \left(1 - \frac{m_N^2}{M_W^2}\right)^2 \left(1 + \frac{m_N^2}{2M_W^2}\right) \frac{|U_{Ne}U_{N\mu}|^2}{\sum_{\ell=e,\mu,\tau} |U_{N\ell}|^2}. \quad (4)$$

Not knowing a priori the values of these mixing elements, the two rates are, apparently, also undistinguishable in an experiment. Therefore some clever way must be devised to distinguish the Majorana vs. Dirac character of the intermediate neutrino, should these decay modes of W happen to be observed.

2. Analysis of the Spectra

The actual event rate of $e^+e^+\mu^-\nu$ at the LHC will depend on whether N is a Dirac or a Majorana neutrino. If N is a Dirac neutrino, it will be the LNC rate only, $\Gamma(e^+e^+\mu^-)_{LNC}$, while if N is a Majorana neutrino, it will be the sum of the LNC + LNV rates, $\Gamma(e^+e^+\mu^-)_{LNC} + \Gamma(e^+e^+\mu^-)_{LNV}$.

Our problem is then how to distinguish at the LHC a Majorana from a Dirac N (how to distinguish LNV from LNC), given that the two rates are the same, except for the mixing elements –so far unknown.

One way to discriminate between the Dirac and Majorana character of the sterile neutrino N is to analyze the spectrum of the opposite charge lepton, i.e. μ^- in the case of $(e^+e^+\mu^-)$, or e^- in the case of $(\mu^+\mu^+e^-)$ [1].

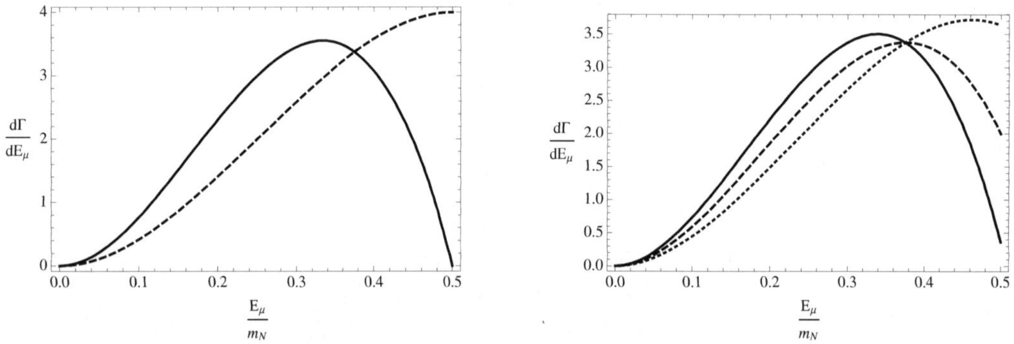

Figure 3. Muon energy spectra in the N rest frame, $(1/\Gamma)d\Gamma(e^+e^+\mu^-)/dE_\mu$ in the N. Left: Solid line: LNV (see Eq. 1); dashed line: LNC (see Eq. 3). Right: LNV+LNC for 3 mixing ratios $r_{\text{mix}} \equiv |U_{Ne}|^2/|U_{N\mu}|^2$: $r_{\text{mix}} = 10$ (solid), $r_{\text{mix}} = 1$ (dashed) and $r_{\text{mix}} = 1/10$ (dotted). For exchanged flavor $(\mu^+\mu^+e^-)$, the spectra correspond to the inverse of r_{mix}.

For a Dirac N, the spectrum will be that of the LNC process (fig. 3 left, dashed line), while for a Majorana N, the spectrum will be the sum of the LNC and LNV spectra, shown in fig. 3 right; the shape of this spectrum depends on the disparity ratio of mixings:

$$r_{\text{mix}} \equiv |U_{Ne}|^2/|U_{N\mu}|^2,$$

which defines whether the LNC or the LNV part dominates. For $r_{\text{mix}} \gtrsim 1$, the spectrum of $\Gamma(e^+e^+\mu^-)$ for a Majorana N vs. a Dirac N differ enough for the two cases to be discriminated.

For $r_{\text{mix}} \ll 1$ the spectrum of $\Gamma(e^+e^+\mu^-)$ for a Majorana N is indistinguishable from that of a Dirac N, in which case one should use the exchanged flavour mode $\Gamma(\mu^+\mu^+e^-)$, where the Majorana and Dirac spectra differ the most. Currently we are doing simulations for the reconstruction of these spectra at the LHC, a work that is still in progress.

Nevertheless, it is already clear that the discrimination of spectra in such rare decays is uncertain, as it will require a rather sizable sample of events.

3. Analysis of the Rates

Another way to discriminate between the Majorana vs. Dirac character if the sterile neutrino that induces the trilepton modes is to compare the rates with exchanged flavors, namely $\Gamma(e^+e^+\mu^-)$ and $\Gamma(\mu^+\mu^+e^-)$ [2]. From Eq. (4), one can see that the LNC rates are identical:

$$\Gamma(e^+e^+\mu^-)_{LNC} = \Gamma(\mu^+\mu^+e^-)_{LNC} \sim |U_{Ne}U_{N\mu}|^2, \tag{5}$$

while the LNV rates, Eq. (2), differ depending on the mixing –or more precisely on the disparity ratio r_{mix}:

$$\Gamma(e^+e^+\mu^-)_{LNV} \sim |U_{Ne}|^4, \quad \Gamma(\mu^+\mu^+e^-)_{LNV} \sim |U_{N\mu}|^4. \tag{6}$$

While the LNC rates are always equal, the LNV rates are equal only if $|U_{Ne}|^2 = |U_{N\mu}|^2$. With the fact that a Dirac sterile neutrino will produce the LNC process only, while a Majorana neutrino will produce both the LNC and LNV processes, one can compare the production of $e^+e^+\mu^-$ and $\mu^+\mu^+e^-$ (or their charge conjugates) induced by a Dirac or a Majorana sterile neutrino. Table 1 shows the comparison, where the rate of $W^+ \to e^+e^+\mu^-\nu$ in the Dirac case is chosen as the reference value, by which the rates of other cases are normalized: for a Dirac sterile neutrino the production rates of $e^+e^+\mu^-$ and $\mu^+\mu^+e^-$ should be equal, while for the Majorana case they will differ, depending r_{mix}.

For $r > 1$, the number of events $N(e^+e^+\mu^-)$ should be larger than $N(\mu^+\mu^+e^-)$ and, viceversa, if $r < 1$ then $N(\mu^+\mu^+e^-)$ will be more abundant than $N(e^+e^+\mu^-)$. Similar comparison also exists in the corresponding charge-conjugated processes.

	Dirac	Majorana
$e^+e^+\mu^-$	1	$1 + r_{\text{mix}}$
$\mu^+\mu^+e^-$	1	$1 + 1/r_{\text{mix}}$

Table 1. Relative factors in the branching ratios of $W^+ \to e^+e^+\mu^-\nu$ and $W^+ \to \mu^+\mu^+e^-\nu$ for either a Dirac or Majorana sterile neutrino scenario, where ν represents a standard neutrino or anti-neutrino. The same applies for the respective charge conjugate modes.

Consequently, if $r_{\text{mix}} = 1$, we will not be able to discriminate between Dirac or Majorana, because the number of $(e^+e^+\mu^-)$ and $(\mu^+\mu^+e^-)$ events will be the same in either case. However, for $r_{\text{mix}} \neq 1$ the discrimination will be possible. Our question is then how far from unity must r_{mix} be in order to discriminate Majorana from Dirac at the LHC.

To answer this question, we generate events in a Majorana neutrino scenario and test whether one can significantly discriminate the Majorana from a Dirac case. The analysis is done for several mixings, and two mass benchmarks: $m_N = 20$ GeV and $m_N = 50$ GeV.

In our simulation analysis, we first use FeynRules [8] to extend the SM, and then we generate simulated events with MadGraph 5 [9] using PYTHIA 6 [10] for the parton showers and DELPHES 3 [11] for detector simulations. The main backgrounds considered are (i) WZ production with $W \to$ *leptonic* and $Z \to \tau^+\tau^-$ and $\tau \to$ *leptonic*, and (ii) fake leptons from jets in processes γ^*/Z+jets or $t\bar{t}$.

IOP Conf. Series: Journal of Physics: Conf. Series **1043** (2018) 012031 doi:10.1088/1742-6596/1043/1/012031

We test for different cuts and find that the best cuts that reduce background are: (i) $p_T > 10$ GeV and $|\eta| < 2.5$ for leptons, (ii) $p_T > 20$ GeV and $|\eta| < 5.0$ for jets; (iii) $M_T(3\ell + \not{p}_T) < 90$ GeV; (iv) $\not{p}_T < 40$ GeV, no b-jets and $\sum p_T^{jet} < 50$ GeV.

The results are shown in Fig. 4. Here we show the plane s vs. r, where s is the average mixing factor that sets the magnitude of the rates (i.e. the average number of events), while $r = r_{\rm mix}$ is the mixing ratio. In this plane we show the contour lines for several confidence levels with which one can exclude the Dirac scenario from a Majorana case. As s gets larger, more events are produced and the exclusion becomes possible even for $r_{\rm mix}$ very close to unity. For both m_N benchmark scenarios, at least a 3σ exclusion level can be reached for mixing disparities $r_{\rm mix}$ as mild as e.g. $r_{\rm mix} \lesssim 0.7$ (or $1/r_{\rm mix} \lesssim 0.7$), provided the average mixing s is sufficiently large ($s \gtrsim 5$). For smaller s (smaller mixings), meaning fewer events, $r_{\rm mix}$ must be larger to reach the same level of discrimination; in the same way, as $r_{\rm mix}$ approaches 1, larger values of s are required as it becomes more and more difficult to exclude the Dirac case.

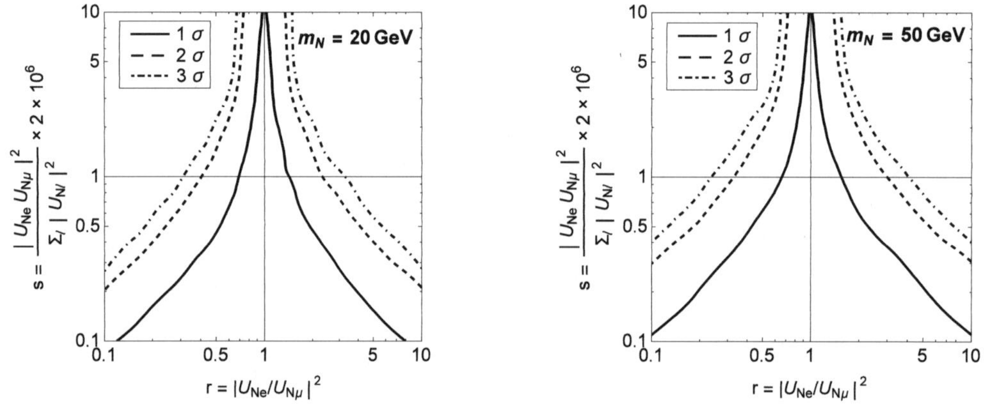

Figure 4. Confidence levels for excluding the Dirac case given a Majorana scenario, for two benchmarks: $m_N = 20$ GeV (left) and 50 GeV (right).

4. Conclusions

We studied the question of determining the Dirac vs. Majorana character of sterile neutrino with mass below M_W at the LHC. Because of such a low mass for the neutrino, in the conventional $\ell^+\ell^+jj$ search for Majorana sterile neutrinos at the LHC the jets are not energetic enough to pass the necessary detector cuts. Therefore, we focus on the alternative tri-lepton search channels. To avoid SM backgrounds we require no opposite-sign same-flavor lepton pairs in the final state. Although this tri-lepton search is ideal for such a low mass sterile neutrino search, the Dirac vs. Majorana discrimination becomes a non trivial task due to the missing neutrino (or antineutrino) in the final state.

We find two ways to distinguish between Dirac and Majorana in these trilepton events. Oe way is to analyse the energy spectrum of the opposite charge lepton. This is a robust method but has the difficulty of requiring a large sample of events, a rather difficult goal in such rare processes. Another way is to compare the rates of the two flavor configurations $e^\pm e^\pm \mu^\mp$ and $\mu^\pm \mu^\pm e^\mp$. Here the discrimination will be possible provided the mixing elements of the sterile neutrino N with electrons and muons are different enough. We find that one is able to exclude the Dirac sterile neutrino case by simply counting and then comparing the numbers of events in the $e^+e^+\mu^-$ and $\mu^+\mu^+e^-$ channels (or the corresponding charge-conjugated ones). We perform a careful collider simulation for this scenario. We have characterized the average size of the mixings by the parameter $s = 2 \cdot 10^6 \times |U_{Ne}U_{N\mu}|^2 / \left(|U_{Ne}|^2 + |U_{N\mu}|^2\right)$, and the disparity between the mixings by the parameter $r = |U_{Ne}/U_{N\mu}|^2$. We find that, in the 14 TeV pp collisions at the LHC with an integrated luminosity of 3000 fb^{-1}, at least a 3σ level of exclusion of the Dirac case with respect to the Majorana case can be achieved, depending on the size and disparity of the two relevant mixing parameters. For example, such level of discrimination is achieved for $s \gtrsim 5$ and $r \lesssim 0.7$ (or $1/r \lesssim 0.7$). For smaller values of s the number of events will be smaller as well, and therefore this level of discrimination can be reached only for larger values of the disparity ratio $r_{\rm mix}$.

Acknowledgments

Work supported in part by: Chile grants Fondecyt 1130617, Conicyt ACT 1406 and FB0821; the Korean gov. NRF grant of the MEST 2011-0017430 and 2011-0020333; the Innovation Program of the IHEP grant Y4515570U1, and the International Postdoctoral Exchange Fellowship Program 90, OCPC, 205, China.

References

[1] Valle J W F and Romao J C 2015 *Neutrinos in high energy and astroparticle physics* Physics textbook (Weinheim: Wiley-VCH) ISBN 978-3-527-41197-9, 978-3-527-67102-1 URL `http://eu.wiley.com/WileyCDA/WileyTitle/productCd-3527411976.html`
[2] Dell'Oro S, Marcocci S, Viel M and Vissani F 2016 *Adv. High Energy Phys.* **2016** 2162659 (*Preprint* `1601.07512`)
[3] Aad G *et al.* (ATLAS) 2015 *JHEP* **07** 162 (*Preprint* `1506.06020`)
[4] Khachatryan V *et al.* (CMS) 2016 *JHEP* **04** 169 (*Preprint* `1603.02248`)
[5] Izaguirre E and Shuve B 2015 *Phys. Rev.* **D91** 093010 (*Preprint* `1504.02470`)
[6] Dib C O and Kim C S 2015 *Phys. Rev.* **D92** 093009 (*Preprint* `1509.05981`)
[7] Dib C O, Kim C S, Wang K and Zhang J 2016 *Phys. Rev.* **D94** 013005 (*Preprint* `1605.01123`)
[8] Christensen N D and Duhr C 2009 *Comput. Phys. Commun.* **180** 1614–1641 (*Preprint* `0806.4194`)
[9] Alwall J, Frederix R, Frixione S, Hirschi V, Maltoni F, Mattelaer O, Shao H S, Stelzer T, Torrielli P and Zaro M 2014 *JHEP* **07** 079 (*Preprint* `1405.0301`)
[10] Sjostrand T, Mrenna S and Skands P Z 2006 *JHEP* **05** 026 (*Preprint* `hep-ph/0603175`)
[11] de Favereau J, Delaere C, Demin P, Giammanco A, Lemaitre V, Mertens A and Selvaggi M (DELPHES 3) 2014 *JHEP* **02** 057 (*Preprint* `1307.6346`)

IOP Conf. Series: Journal of Physics: Conf. Series **1043** (2018) 012032 doi:10.1088/1742-6596/1043/1/012032

On global aspects of duality invariant theories: M2-brane vs DFT

G. Abellán[2,a], **C. Las Heras**[1,b], **M.P. García del Moral**[1,c], **J.M. Peña**[2,d], **A. Restuccia**[1,3,e]

[1]Departamento de Física, Universidad de Antofagasta, Aptdo 02800, Chile.
[2] Departamento de Física, Facultad de Ciencias,
Universidad Central de Venezuela, A.P. 47270, Caracas 1041-A, Venezuela.
[3] Departamento de Física, Universidad Simón Bolívar,
Apartado 89000, Caracas 1080-A, Venezuela

E-mail: [a]gabriel.abellan@ciens.ucv.ve; [b]camilo.lasheras@uantof.cl;
[c]maria.garciadelmoral@uantof.cl; [d]jpena@fisica.ciens.ucv.ve;
[e]alvaro.restuccia@uantof.cl

Abstract. Supermembrane compactified on a $M_9 \times T^2$ target space is globally described by the inequivalent classes of torus bundles over torus. These torus bundles have monodromy in $SL(2,Z)$ when they correspond to the nontrivial central charge sector and they are trivial otherwise. The first ones contain eight inequivalent classes of M2-brane bundles which at low energies, are in correspondence with the eight type II gauged supergravities in $9D$. The relation among them is completely determined by the global action of T-duality which interchanges topological invariants of the two tori. The M2-brane torus bundles are invariant under $SL(2,Z) \times SL(2,Z) \times Z_2$. From the effective point of view, there is another dual invariant theory, called Double Field Theory which describe invariant actions under $O(D,D)$. Globally it is formulated in terms of doubled $2D$ torus fibrations over the spacetime with a monodromy given by $O(D,D,Z)$. In this note we discuss T-duality global aspects considered in both theories and we emphasize certain similarities between both approaches which could give some hints towards a deeper relationship between them.

1. Introduction

In the context of M-theory, String Theory or Effective Field Theories the search for T-dual/U-dual invariant theories is a relevant goal. In this note we want to discuss and briefly compare aspects of the bundle construction of two different duality invariant theories: Supermembrane theory, also called M2-brane theory and Double field theory (DFT). We consider of interest to understand whether or not there is any hint of a connection between them. In [1, 2] it was argued that a fundamental formulation of string/M-theory should exist in which T- and U-duality symmetries are manifest from the start. In particular, it was argued that many massive, gauged supergravities cannot be naturally embedded in string theory without such a framework [3–6].

Supermembranes are elements of M-theory, for that reason, dualities should appear as discrete symmetries of the theory. In this note we will restrict our analysis to the case of a supermembrane theory formulated on $M_9 \times T^2$. It contains two different topological sectors one with irreducible wrapping and a second one that is reducible. The theory corresponding to the irreducible wrapping sector called 'supermembrane with central charges' [7]. It is consistently defined at quantum level [8], in distinction with the reducible case [9,10]. At low energies it is described by the type II gauged supergravity theories [11,12]. The supermembrane without the central charge condition corresponds to the type II maximal supergravity in nine dimensions. The mass operator and the hamiltonian are U-dual invariant [11] and when supermembrane is dimensionally reduced to string theory T-duality for supermembranes agree with the standard one in string theory compactified on a circle [13]. Moreover, for the supermembrane with nontrivial central charge, it was found all the inequivalent classes of torus bundles with $SL(2,Z)$ monodromy that describes globally the theory, [12]. The M2-brane torus bundles are classified by their coinvariants of the base and the fiber and their associated monodromies. T-duality action determines the structure of the T-dual M2-brane bundle. M2-brane torus bundles with parabolic monodromy are the unique class of bundles invariant not only locally but also globally, i.e. preserving the coinvariants. This implies that at low energy the T-dual of a type IIB parabolic gauged supergravity is another type IIA parabolic gauged supergravity. The precise T-duality action on the classification of the M2-brane bundles: elliptic, parabolic, hyperbolic or with trombone monodromy is done in [12]. The construction is always globally well defined, hence geometric.

DFT is an effective field theory that describes sigma models on toroidal compactifications in which the number of spatial coordinates has been doubled to mimic the effect of winding modes for string theory. These coordinates form a doubled torus T^{2d} with d-coordinates conjugate to the momenta and the other d-coordinates conjugate to the winding modes [2]. In order to preserve the physical number of propagating degrees of freedom of the theory a constrain is imposed. The global formulation of the theory is done in terms of torus fibrations over a spacetime manifold such that the transition functions are evaluated not only under diffeomorphims and gauge shifts but also under the T-duality group $O(d,d,\mathbb{Z}) \sim SL(d,Z) \times SL(d,Z) \times Z_2 \times Z_2$ transformations [14]. The corresponding action is invariant under the duality transformations. String theory can be consistently defined in those backgrounds called non geometric backgrounds [1,2]. Such backgrounds can arise from compactifications with duality twists [15] or from acting on geometric backgrounds with fluxes with T-duality [1–3]. In special cases, the compactifications with duality twists are equivalent to asymmetric orbifolds which can give consistent string backgrounds [16–20]. Some global aspects of T-duality were analyzed in [3,14–19,21] . These torus bundles have an $O(d,d,Z)$ monodromy of the doubled torus over the base spacetime manifold. Examples of generalized T-folds [16,21] can be obtained by constructing torus fibrations over base manifolds with singularities or with non-contractible cycles, like for example over a torus. In section 2 we are going to review the supermembrane theory description in terms of bundles: we indicate the local (Hamiltonian and mass operator) and the T-duality global action on the Supermembrane theory. In section 3 we review very basic facts of DFT in particular focusing on its global description to introduce next section. In section 4 we compare both approaches indicating differences and similarities and we present our conclusions.

2. Supermembrane theory on a symplectic torus bundle

Let us consider a supermembrane theory with central charges formulated in the Light Cone Gauge (LCG) on a target space $M_9 \times T^2$ [7,22,23]. X^n are the embedding maps $\Sigma \to M_9$ where $n = 3,\ldots,9$ and $X = X^1 + iX^2$ from $\Sigma \to T^2$. They are scalars parametrizing the transverse coordinates of the supermembrane in the target space. The moduli of the target space 2-torus T^2 are the radius $\mathtt{R}$, and the complex Teichmüller parameter τ. The supermembrane has a nontrivial embedding over all the coordinates of the target. We define a matrix $\mathbb{W} = \begin{pmatrix} l_1 & l_2 \\ m_1 & m_2 \end{pmatrix}$ whose entires are the winding numbers of the embedded supermembrane. When the wrapping of the supermembrane is irreducible $det\mathbb{W} = \mathtt{n} \neq 0$ the theory has discrete spectrum [8]. For the case $det\mathbb{W} = 0$ the spectrum is continuous [9]. The topological condition $\mathtt{n} \neq 0$ defines a principal line bundle with first chern class $c_1 = n$ that algebraically implies the existence of a non-vanishing central charge in the supersymmetric algebra of the supermembrane. This sector of the wrapped supermembrane theory we denote as supermembrane with central charges. The Hamiltonian describing it is found in [11, 23, 24]:

$$\begin{aligned} H = &\int_\Sigma T_{\mathtt{M2}}^{2/3}\sqrt{\rho}\left[\frac{1}{2}(\frac{P_n}{\sqrt{\rho}})^2 + \frac{1}{2}\frac{P\overline{P}}{\sqrt{\rho}} + \frac{T_{\mathtt{M2}}^2}{4}\{X^m, X^n\}^2 + \frac{T_{\mathtt{M2}}^2}{2}(\mathcal{D}X^n)(\overline{\mathcal{D}}X^n)\right] + \\ + &\int_\Sigma T_{\mathtt{M2}}^{2/3}\sqrt{\rho}\left[+\frac{T_{\mathtt{M2}}^2}{4}(\mathcal{F}\overline{\mathcal{F}})\right] + (\mathrm{n}^2\, Area_{T^2}^2) + \\ + &\int_\Sigma T_{\mathtt{M2}}^{2/3}\sqrt{\rho}\left[-\overline{\Psi}\Gamma_-\Gamma_n\{X^n,\Psi\} - \frac{1}{2}\overline{\Psi}\Gamma_-\overline{\Gamma}\mathcal{D}\Psi\} - \frac{1}{2}\overline{\Psi}\Gamma_-\Gamma\overline{\mathcal{D}}\Psi\}\right] \\ + &\int_\Sigma \sqrt{\rho}\,L\left[\frac{1}{2}\overline{\mathcal{D}}(\frac{P}{\sqrt{\rho}}) + \frac{1}{2}\mathcal{D}(\frac{\overline{P}}{\sqrt{\rho}}) + \{X^n, \frac{P_n}{\sqrt{\rho}}\} - \{\overline{\Psi}\Gamma_-,\Psi\}\right], \end{aligned} \tag{1}$$

where L is a Lagrange multiplier, $T_{\mathtt{M2}}$ is the 11D tension of the supermembrane, ρ is the determinant of the spatial part of the non-flat worldvolume two torus of supermembrane Σ. The symplectic bracket is defined as $\{\mathsf{A},\mathsf{B}\} = \omega^{ab}\partial_a\mathsf{A}\partial_b\mathsf{B}$ with $\omega^{ab} = \frac{i\epsilon^{ab}}{2\sqrt{\rho}}$ and $a, b = z, \overline{z}$ the complex coordinates defined on Σ. The densities P_n are the canonical momenta associated to the X^n and P those of X. Ψ are scalars on the worldvolume but an $SO(7)$ spinor on the target space. Γ_n are seven Gamma matrices and $\Gamma = \Gamma_1 + i\Gamma_2$, denoting by $\overline{\Gamma}$ its complex conjugate. $\mathcal{F} = D\overline{A} - \overline{D}A + \{A,\overline{A}\}$. is a symplectic curvature defined on the base manifold with A a connection under the infinitesimal symplectomorphism transformation $\delta_\epsilon A = \mathcal{D}\epsilon$. See [22, 24] for a detailed analysis. The symplectic covariant derivative is defined as $\mathcal{D}\bullet = D\bullet + \{A,\bullet\}$ with $D\bullet = e_r^a\partial_a\bullet$ a rotated covariant derivative defined in terms of a zwei-bein e_r^a as, [11, 23] $e_r^a := -2\pi\mathrm{R}(l_r + m_r\tau)\Theta_{sr}\omega^{ba}\partial_b\widehat{X}^s$, with $r, s = 1, 2$ and being $d\widehat{X}$ are the harmonic one-form basis defined on Σ.

Symmetries of the Theory The Hamiltonian is invariant the residual symmetry under Area Preserving Diffeomorphisms (APD) connected and not connected to the identity. The theory is also invariant under two different $SL(2,Z)$ discrete symmetries: There is a $SL(2,Z)_\Sigma$ associated to the invariance under the change of the basis of the harmonic one forms defined on Σ [25] and the windings,

$$d\widehat{X} \to Sd\widehat{X}, \quad \mathbb{W} \to S^{-1}\mathbb{W}, \tag{2}$$

with $S \in SL(2,Z)$. This dependence is encoded in the symplectic covariant derivative of the Hamiltonian through the matrix $\Theta \in SL(2,Z)$ [23]. There is a second $SL(2,Z)_{T^2}$ global symmetry, the S-duality transformation, associated to an invariance of the mass operator related to the target 2-torus T^2,

$$\tau \to \frac{a\tau + b}{c\tau + d}, \quad \mathrm{R} \to \mathrm{R}|c\tau + d|, \quad A \to Ae^{i\varphi_\tau}, \mathbb{W} \to \begin{pmatrix} a & -b \\ -c & d \end{pmatrix}\mathbb{W}, \quad \mathbf{Q} \to \Lambda\mathbf{Q}, \tag{3}$$

where $\Lambda = \begin{pmatrix} a & b \\ c & d \end{pmatrix} \in SL(2,Z)$, $e^{-i\varphi_\tau} = \frac{c\tau+d}{|c\tau+d|}$ and $\mathbf{Q} = \begin{pmatrix} p \\ q \end{pmatrix}$ the Kaluza Klein (KK) charges of the supermembrane propagating on the target 2-torus T^2 considered.

Bundle description of the Supermembrane The embedding description did in [7, 22, 23] can be also understood in terms of a symplectic torus bundle over a torus with monodromy in $SL(2,Z)$. This global formulation show topological invariants that carry physical information. The total bundle space E is defined in terms of a fiber $F = M_9 \times T^2$ and Σ as the base manifold non trivially patched. The structure group $\mathtt{G}$ is the symplectomorphisms group leaving invariant the canonical symplectic structure in T^2. The action of $\mathtt{G}$ on F produces a $\pi_0(\mathtt{G})$-action on the homology and cohomology of F. In [12], the authors give a geometrical interpretation of the KK charges in terms of the $H_1(T^2)$ charges. The *monodromy* of the bundle $\mathcal{M}$ is defined as

$$\mathcal{M} : \pi_1(\Sigma) \to \pi_0(\mathtt{G}), \quad \text{with} \quad \mathtt{G} = Symp(T^2) \quad \text{and} \quad \pi_0(\mathtt{G}) = SL(2, Z)\,. \tag{4}$$

Consequently $\mathcal{M} = \Lambda^\gamma \in SL(2, Z)$ and it acts on the homology basis of the T^2 target torus with $\gamma = \gamma_1 + \gamma_2$ an integer defined in terms of the integer basis of the fundamental group $\pi_1(\Sigma)$. The global symmetries of the theory become restricted by the monodromy. The symplectic connection A defined on the base manifold transforms with the monodromy as $dA \to dAe^{i\varphi_{\mathcal{M}}}$ where $\varphi_{\mathcal{M}} = \frac{c\tau+d}{|c\tau+d|}$ is a discrete monodromy phase for a given modulus τ. In distinction the Hamiltonian is invariant. The torus bundles with a given monodromy $\mathcal{M}$ are classified according to the elements of the twisted second cohomology group $H^2(\Sigma, Z^2_{\mathcal{M}})$ of the base manifold Σ with coefficients on the module generated by the monodromy representation acting on the homology of the target torus [26]. There is a bijective relation between the elements of the coinvariant group $C_F = \{C_{\mathtt{a}}\}, \mathtt{a} = 1, \ldots, j$ associated with a particular monodromy group $\mathcal{M}_G$. A coinvariant class in the KK sector is given by

$$C_F = \{\mathbf{Q} + (\mathcal{M}_g - \mathbb{I})\widehat{\mathbf{Q}}\}\,, \tag{5}$$

for any element $\mathcal{M}_g \in \mathcal{M}$, and $\widehat{\mathbf{Q}}$ is any arbitrary element of the KK sector. There is also an induced action on the cohomology of the base manifold, which corresponds to a '*monodromy*' group of the winding sector $\mathcal{M}^*_G$. A coinvariant class in the winding sector is given by

$$C_B = \{\mathbf{W} + (\mathcal{M}^*_g - \mathbb{I})\widehat{\mathbf{W}}\}, \tag{6}$$

with $\mathcal{M}^*_g \in \mathcal{M}^* = \Omega\mathcal{M}\Omega^{-1}$ with $\Omega = \begin{pmatrix} -1 & 0 \\ 0 & 1 \end{pmatrix}$, and $\mathbf{W} = \begin{pmatrix} l \\ m \end{pmatrix} \in H^1(\Sigma)$. The $\mathcal{M}^*$ acts on the fields which define the Hamiltonian, through the matrix $\Theta = (V^{-1}\mathcal{M}^*V)^T$ that appears in the symplectic covariant derivatives $\mathcal{D}_r$ [23]. $\mathcal{M}$ and $\mathcal{M}^*$ groups lie in the same equivalence class but their respective coinvariants classifying bundles are not equivalent. Then in order to specify the physical content of the M2-brane on a symplectic torus bundles one also needs to determine (C_F, C_B, M_G). Each coinvariant class is invariant under the action of any element of the monodromy group $\mathcal{M}_g \in \mathcal{M}$. Among its elements we have the orbits of any element of the class. So the coinvariant class may be considered itself as a class of orbits under the action of $\mathcal{M}$. Given a symplectic torus bundle the Hamiltonian of the theory is defined for any orbit of the coinvariant class.

2.1. T-duality for supermembrane theory torus bundles

The T-duality transformation acts on the Hamiltonian H and the mass operator $\mathbf{M}^2$ and it has also a action on the structure of the bundle consequently on their topological invariants describing the M2-brane theory. Globally the T-duality transforms a bundle into a dual one, by interchanging the cohomological charges of the torus base manifold into the homological charges defined on the torus fiber with dual moduli. T-duality also interchanges the coinvariant class of the base and the fiber in the dual T-bundle, $(C_F, C_B) = (\widetilde{C}_B, \widetilde{C}_F)$, where we denote by tildes the quantities in the dual bundle. In general this transformation becomes non linear. At low energies this fact will be reflected in the change of the gauging group associated to the corresponding dual supergravity. In order to analyze it with more detail, we will consider separately the cases of trivial and non trivial monodromy:

- *Trivial monodromy*: In this case $\mathcal{M} = \mathbb{I}$, and the coinvariant classes, which classify the inequivalent torus bundles, have only one element $\mathbf{Q}$ in the KK sector and one element $\mathbf{W}$ in the winding sector. The duality transformation on the symplectic torus bundle has an action on the charges but also on the geometrical moduli. Following the notation of [12], we define dimensionless variables $\mathcal{Z} = (T_{\mathtt{M2}}\mathtt{A}Y)^{1/3}$ where $\mathtt{A} = (2\pi\mathtt{R})^2 Im\tau$ is the area of the target torus and $Y = \frac{\mathtt{R}Im\tau}{|q\tau-p|}$ is a variable proportional to the $\mathtt{R}$ radius of the complex torus. The T-duality transformation is given by:

$$\text{The moduli}: \quad \mathcal{Z}\widetilde{\mathcal{Z}} = 1, \quad \widetilde{\tau} = \frac{\alpha\tau + \beta}{\gamma\tau + \alpha}. \tag{7}$$

$$\text{The charges}: \quad \widetilde{\mathbf{Q}} = \mathcal{T}\mathbf{Q}, \quad \widetilde{\mathbf{W}} = \mathcal{T}^{-1}\mathbf{W}, \tag{8}$$

with $\mathcal{T} = \begin{pmatrix} \alpha & \beta \\ \gamma & \alpha \end{pmatrix} \in SL(2,Z)$. The symplectic torus bundles are classified in this case by two integers, the elements $Z \otimes Z$ and they are in one to one correspondence with the $U(1) \times U(1)$ principle bundle over the base manifold. Since the monodromy is trivial, the structure group may reduce to the group of symplectomorphism homotopic to the identity. The dual transformation is then completed by the transformation of the moduli as given in (7).

- *Non trivial monodromy:* In this case, the monodromy group $\mathcal{M}$ is non trivial and it is an abelian subgroup of $SL(2,Z)$. The T-dual transformation maps as before coinvariant classes on the KK sector onto coinvariant class in the winding sector and each class denoted as [ ] contains several elements. Hence the map between the coinvariant classes is only determined by $\mathcal{T}$ which is constructed from one element of each class $\mathbf{Q}$ and $\mathbf{W}$ respectively. The monodromy $\mathcal{M}$ can be parabolic, elliptic or hyperbolic, which can be linearly or non-linearly realized (trombone symmetries). The T-duality transformation is given by (7) and for the charges:

$$[\widetilde{\mathbf{W}}] = \mathcal{T}[\mathbf{Q}], \quad [\widetilde{\mathbf{Q}}] = \mathcal{T}^{-1}[\mathbf{W}], \quad \mathcal{M} \overset{\Omega}{\to} \mathcal{M}^*, \quad (C_F, C_B) \to (\widetilde{C}_B, \widetilde{C}_F). \tag{9}$$

The variables associated to the geometric moduli $\mathcal{Z}, Y$ and their duals are invariant on an orbit generated by $\mathcal{M}$ contained in the respective coinvariant class, provided that τ and $\mathbf{Q}$ transform as in (3). The symmetry of the Hamiltonian related to the basis of harmonic one-forms of Σ [25] allows to define the class of orbits associated to the winding matrices $[\mathbf{W}]$. T-duality defines a nonlinear transformation on the charges of the supermembrane since $\mathcal{T}$ is constructed from them, in distinction with the $SL(2,Z)_T$ action on the moduli which is a linear one. The condition $\mathcal{Z}\widetilde{\mathcal{Z}} = 1$ ensure that $(\text{T-duality})^2 = I$. This transformation becomes a symmetry for $\mathcal{Z} = \widetilde{\mathcal{Z}} = 1$ which imposes a relation between the tension, the moduli and the KK charges of the wrapped supermembrane, $T^0_{\mathtt{M2}} = \frac{|q\tau-p|}{\mathtt{R}^3(Im\tau)^2}$. Given the values of the moduli and the charges, the allowed tension is fixed $T^0_{\mathtt{M2}}$. For $\mathcal{Z} = 1$ the Hamiltonian and the mass operator of the supermembrane with central charges are invariant under T-duality:

$$\mathbf{M}^2 = (T^0_{\mathtt{M2}})^2\mathtt{n}^2\mathtt{A}^2 + \frac{k^2}{Y^2} + (T^0_{\mathtt{M2}})^{2/3}H = \frac{\mathtt{n}^2}{\widetilde{Y}^2} + (T^0_{\mathtt{M2}})^2 k^2\widetilde{\mathtt{A}}^2 + (T^0_{\mathtt{M2}})^{2/3}\widetilde{H}, \tag{10}$$

with $H = \widetilde{H}$. The mass operator of the supermembrane is U-dual invariant. Thought the mass operator is always invariant, under T-duality only the parabolic M2-brane bundles class is invariant since the action of T-duality is linear and the structure of the coinvariants is preserved, so the bundle. For the rest of the monodromies contained in $SL(2,Z)$ with linear or non linear realization the action of T-duality is non linear, the monodromy dual is always in the same monodromy class but the bundle is classified by different coinvariants [12].

2.2. Supermembrane theory and gauged supergravity in 9D

Scherk-Schwarz compactifications of supergravity may be expressed in terms of principal fiber bundles over circles with a twisting given by the monodromy [27,28]. The background possesses a group of global isometries G associated to the compactification manifold over which it is fibered. In the type II gauged supergravities in 9D, the monodromies are associated to the $GL(2,\mathbb{R}) = SL(2,\mathbb{R}) \times \mathbb{R}^+$ global symmetry group. In the $SL(2,\mathbb{R})$ sector, there are three inequivalent classes of theories, corresponding to the hyperbolic, elliptic and parabolic $SL(2,\mathbb{R})$ conjugacy classes, see [27]. The gauging of the R^+ scaling symmetry called trombone gives supergravities without lagrangians [29, 30]. At quantum level this last symmetry corresponds to an $SL(2,Z)$ non linearly realized and gives rise to a different symplectic torus bundle in comparison to the previous constructions in terms of linear representations [12,13]. Indeed, only when the monodromy is parabolic, the T-duality action is linear meanwhile for the elliptic and hyperbolic case, T-duality does not commute with the monodromy but forms a unique class of torus bundles in the type IIA side with an scaling symmetry $A(1)$. The T-duality of M2-brane trombone bundles maps also in the 'type IIA side' into two inequivalent classes of M2-brane dual trombone bundles. These classes are in perfect agreement with the eight type II gauged supergravities, four in the type IIB (elliptic, parabolic, hyperbolic and trombone) and other four in the type IIA (two trombone, one parabolic and one non abelian with group $A(1)$ [31, 32].

3. Global Aspects of T-duality in Double Field Theory

We review here very general facts of the global description of DFT. It is a reformulation of supergravity constructed in such a way that it is T-dual invariant and it makes explicit the duality group $O(D,D)$ of the theory [33]. So far, it has been studied extensively for the bosonic sector of the theory. T-duality is a transformation that changes moduli at the same time that KK charges are interchanged with the winding modes. Winding modes are a property of extended objects, hence to introduce T-duality invariance into an effective field theory, the spatial coordinates are duplicated $(x_D, \widetilde{x}^D)$ so that $\widetilde{x}^i, i = 1 \dots D$ are conjugated to the winding modes $\omega^i, i = 1 \dots D$ [34] and for consistency of the theory many algebraic structures become enlarged, see for a review, [35]. The string action on a toroidal general background is the following [36],

$$S = -\frac{1}{4\pi}\int_0^{2\pi} d\sigma \int d\tau \Big(\sqrt{\gamma}\gamma^{\alpha\beta}\partial_\alpha X^i \partial_\beta X^j G_{ij} + \epsilon^{\alpha\beta}\partial_\alpha X^i \partial_\beta X^j B_{ij}\Big), \tag{11}$$

where $\gamma_{\alpha\beta}$ is the induced worldsheet metric, B_{ij} is a constant target space 2-form and G_{ij} is the background metric. The associated Hamiltonian can be written in terms of a $2D \times 2D$ generalized metric $\mathcal{H}_G(E)$ constructed in terms of the background metric G_{ij} and the two-form B_2 in a non-trivial way [33]

$$\mathcal{H}_G(E) = \begin{bmatrix} G_{ij} - B_{ik}G^{kl}B_{lj} & B_{ik}G^{kj} \\ -G^{ik}B_{kj} & G^{ij} \end{bmatrix}, \tag{12}$$

such that worldsheet hamiltonian density H can be expressed as

$$H = \frac{1}{2}Z^T \mathcal{H}_G(E) Z + N + \bar{N}, \tag{13}$$

where N y $\bar{N}$ are the left and right number operators and $Z = \begin{pmatrix} \omega^i \\ p_i \end{pmatrix}$ is a generalized momentum that includes the KK momentum p_i and the winding modes ω^i, and a background matrix E_{ij} as

$$E_{ij} \equiv G_{ij} + B_{ij} = \begin{bmatrix} E_{mn} & 0 \\ 0 & g_{\mu\nu} \end{bmatrix}, \tag{14}$$

with $E_{mn} = G_{mn} + B_{mn}$, G_{mn} is the flat metric for the torus, B_{mn} the Kalb-Rammond 2-form components over the n-torus and $g_{\mu\nu}$ is background metric on a target space $M^{D-n-1,1}$. The hamiltonian density (13) is $O(D, D)$ invariant when it is induced a transformation law on the generalized metric $\mathcal{H}_G(E') = h\,\mathcal{H}_G(E)\,h^T$, where h is an element of $O(D, D, \mathbb{R})$ group and verifies $h\eta h^T = \eta$, with $\eta = \begin{pmatrix} 0 & \mathbb{I}_{D\times D} \\ \mathbb{I}_{D\times D} & 0 \end{pmatrix}$, the $O(D, D, \mathbb{R})$ invariant metric. Due to the charge quantization condition, ω^m and p_m are restricted to take discrete values. This implies that the symmetry group is restricted to be its arithmetic subgroup $O(n, n, \mathbb{Z})$, the T-duality group of string theory. An element of the T-duality group $h \in O(n, n, \mathbb{Z})$ can be expressed in terms of an $O(D, D)$ representation for that reason one can consider $O(D, D)$, as the global T-duality group.

The idea behind DFT is to find a T-duality invariant action effective field theory describing supergravity. To this end authors in [37, 38], construct an $O(D, D)$ tensor $\mathcal{H}_{MN}$, where $M, N = 1, \ldots, 2D$ are $O(D, D)$ curved indices, of the target space metric g_{ij}, the Kalb-Rammond 2-form b_{ij} and the dilaton ϕ, combined as in (12). The dilaton can be expressed as an $O(D, D)$ singlet $e^{-2d} = \sqrt{g}e^{-2\phi}$. A new set of coordinates $\tilde{x}_i$ conjugated to the winding modes ω^i are defined and with them some new generalized coordinates $X^M = (\tilde{x}^i, x_i)$ considering the original and the dual ones are introduced. These coordinates induce the generalized derivatives $\tilde{\partial}_M = \left(\frac{\partial}{\partial \tilde{x}_i}, \frac{\partial}{\partial x^i}\right)$. Additionally the fields depends on this generalized coordinates $\mathcal{H}_{MN}(X), d(X)$ and it is possible to construct the $O(D, D)$ invariant action as a generalization of Einstein Hilbert terms [39],

$$S = \int d^{2D} X e^{-2d} \mathcal{R}, \tag{15}$$

where the generalized curvature is given by

$$\begin{aligned} \mathcal{R} &= 4\mathcal{H}^{MN}\partial_M\partial_N d - \partial_M\partial_N\mathcal{H}^{MN} - 4\mathcal{H}^{MN}\partial_M d\partial_N d + 4\partial_M\mathcal{H}^{MN}\delta_N d, \\ &+ \frac{1}{8}\mathcal{H}^{MN}\partial_M\mathcal{H}^{KL}\partial_N\mathcal{H}_{KL} - \frac{1}{2}\mathcal{H}^{MN}\partial_M\mathcal{H}^{KL}\partial_K\mathcal{H}_{NL}. \end{aligned} \tag{16}$$

In the same way one generalizes the rest of supergravity terms and to do it is also necessary to modify several mathematical structures as the Lie derivative among others. Since the physical degrees of freedom have been doubled, a constraint is imposed to preserve the correct number of degrees of freedom. The strong constraint closes a generalized Lie derivative gauge algebra

$$\eta^{MN}\partial_M(A)\partial_N(B) = \partial^M(A)\partial_M(B) = \partial^M\partial_M(AB) = 0, \tag{17}$$

where η_{MN} is the $O(D, D)$ invariant metric, A and B represents any DFT field or generalized Lie Derivative Parameter $\xi^M = (\tilde{\lambda}_i, \lambda^i)$ defined by

$$\mathcal{L}_\xi A_M \equiv \xi^P\partial_P A_M + (\partial_M\xi^P - \partial^P\xi_M)A_P, \tag{18}$$
$$\mathcal{L}_\xi B^M \equiv \xi^P\partial_P A^M + (\partial^M\xi_P - \partial_P\xi^M)B^P. \tag{19}$$

After imposing the constraint, the field configuration depends only of the coordinates of the D-dimensional subspace, which in general is a linear combination of the original x^i and the dual coordinates $\tilde{x}_i$.

Global description String theory can be consistently defined in backgrounds in which the transition functions between the patches considers not only diffeomorphisms and gauge transformations but also T-duality transformations [2]. Those backgrounds that include T-duality transformations are called non geometrical. If one restricts to the case in which each transition function patches the charts with an element of $GL(n, Z) = SL(n, Z) \times Z_2 \subset O(n, n, Z)$, the large diffeomorphisms group acting on the fiber, then the background is geometrical. However, this is not the most general case since in string theory T-duality maps the transitions functions $S \in GL(n, \mathbb{Z})$ into $S' = gSg^{-1}$ with $g \in O(n, n, \mathbb{Z})$ and in general $S' \in O(n, n, \mathbb{Z})$ so the background becomes non-geometric. M-theory, as a theory of unification, contains the dualities as symmetries of the theory and DFT aims to be an effective description of M-theory, then it is natural to consider bundles whose transition functions are defined in $O(D, D, \mathbb{Z})$. In the DFT global formulation that we are considering, the target space M is locally a torus bundle with fiber T^D over the base manifold N, is extended by considering another T^D on the fiber, where both are subspaces of a doubled torus T^{2D} containing the original one and its dual, forming an extended target space $\tilde{M}$ that can be interpreted as a torus bundle with fiber T^{2D} over the base manifold N. Each of the two torus has an $SL(2, Z)$ invariance associated to its mapping class group. $\tilde{M}$ is a $2D$ torus bundle with a well defined monodromy in $O(D, D, \mathbb{Z}) \subset GL(2, \mathbb{Z})$, since it is contained on the large diffeomorphims group of the T^{2n}. If M is a geometric background and we have an $O(D, D, \mathbb{Z})$ monodromy on the fiber, then we have a fibration of $\tilde{M}$ over M [2]. However in general only locally one can define the T^D fibration with a G and B defined in the internal torus consequently there is no well defined global geometry on the physical space-time: it is locally the product of a D-torus embedded in T^{2D} and a sector of the base manifold N. This kind of construction which comes from torus fibrations over a torus base where the torus fiber undergoes a monodromy lying in the perturbative duality group $O(D, D, \mathbb{Z})$, were called monodrofolds [40] or T-folds [16, 21].

The polarization is the choice of the physical subspace forming a torus contained in the doubled T^{2D} for each point of the base manifold N. T-duality acts by changing the polarization, that is, by changing the physical subspace $T^n \subset T^{2n}$ and for each chart U of the manifold N, there is a local space-time chart $U \times T^n$ embedded in the doubled space $U \times T^{2n}$, while for local T-folds these do not form a spacetime manifold even though the entire doubled space forms a manifold which is a T^{2n} bundle over the base manifold N.

4. Discussion: A comparison of the global T-duality action in both approaches

In this section we would like to compare the bundle construction of Supermembrane theory with respect to the one associated to the Double Field Theory. For the DFT case the bundle construction has been mainly explored in the T-dual invariant string theory framework. It contains nongeometric vacua solutions [16, 20, 40]. Both theories are clearly different, to start with, DFT is an effective field theory while Supermembrane theory is part of M-theory. Supermembrane theory contains a topological sector (the one associated to the non trivial central charge) that can be consistently quantized. In spite of this, their bundle constructions have certain resemblances and differences that we would like to emphasize.

The Supermembrane bundle that we analyze in this note has the M2-brane worldvolume torus as its base and the fiber is the compactified target space $M_9 \times T^2$. The transition functions between the charts are given by the symplectomorphisms group preserving the symplectic 2-form on the base, and the bundle can be a trivial principal torus over a torus when the wrapped M2-brane has a vanishing central charge and a nontrivial symplectic torus bundles over a torus with monodromy in $SL(2, \mathbb{Z})$ linearly or non linearly realized when the wrapping is irreducible,i.e. has a nontrivial central charge. The M2-brane torus bundle is specified by their coinvariants: (C_B, C_F) which are defined in terms of two different monodromies (M_B, M_F) [12]. The hamiltonian is consistently defined on these M2-brane bundles. The global action of T-duality maps supermembrane class of bundles into dual ones, the homological charges of the torus of the fiber associated to the charges by the cohomological charges associated to the windings of the torus of the base. The tori geometry is not interchanged, the dual moduli of the Torus of the fiber is obtained through the T-duality rules. The mass operator and hamiltonian are U-dual invariant. The monodromy of the bundle is interchanged with the monodromy of the base [12]. The monodromy dual remains in the same conjugate class as the original one but in general the coinvariants that define the inequivalent classes of bundles, as we have discussed, are not preserved. The patching is always geometrical and consequently the global picture well defined. There is a $SL(2,\mathbb{Z})_\Sigma \times SL(2,\mathbb{Z})_{T^2}) \times Z_2$ invariance on the M2-brane bundle, where the Z_2 action is associated to nontrivial discrete interchange of global structures and topological invariants that we have already signaled.

On the other hand, in DFT global formulation, the fields are defined over the spacetime coordinates in distinction with the previous case. Its global description has been developed mainly inspired in the context of the corresponding duality invariant string worldsheet bundle. The bundle contains the spacetime manifold as a base and a doubled torus bundle as a fiber T^{2D}. The transition functions in general are not restricted to be just the diffeomorphisms and the gauge invariance but the charts are patched through elements of the T-duality group $O(D,D,\mathbb{Z})$. In the case in which duality group is included the torus bundle has a non trivial monodromy. The fibration of the dual torus over the original one is only possible when the monodromy of the fiber belongs to the duality group $O(D,D,\mathbb{Z})$. In order to make contact between the two approaches, let us particularize to the case in which the DFT describes a $T^4 = T^2 \times T^2$ bundle with $O(2,2,\mathbb{Z})$ monodromy. The group of duality invariance of the theory $O(2,2,\mathbb{Z})$ can be decomposed as $SL(2,Z)_\tau \times SL(2,Z)_\rho \times Z_2 \times Z_2$ where $SL(2,Z)_\tau$ is associated τ the Teichmüller parameter of the T^2 fiber, $SL(2,\mathbb{Z})_\rho$ with ρ the complexified Kähler parameter defined in terms of the constant element of the 2-form b and the area A , that is $\rho = b + iA$ parameterizing the dual $\tilde{T}^2$ and the Z_2 discrete symmetries acts on both parameters as follows [14],

$$\tau \Leftrightarrow \rho, \quad (\tau,\rho) \to (-\overline{\tau}, -\overline{\rho}). \tag{20}$$

In DFT theories, the T-duality action can be understood as a different choice of polarization which selects the physical subspace T^D inside the doubled tori T^{2D}. It mixes the spatial coordinates (X) with the doubled coordinates $(\widetilde{X})$. In those cases in which $T^{2D} = T^D \times \widetilde{T}^D$ and the polarization selects as the physical tori the one associated with the spatial coordinates, T-duality intertwine the two tori, interchanging the geometry and not only the moduli under T-duality. There are two monodromies defined one associated to the T^2_τ and another T^2_ρ on top of the $O(2,2,\mathbb{Z})$ monodromy associated to the T^4 bundle over the base manifold. These two monodromies become also interchanged under T-duality. It seems clear that are resemblances between both theories: Both theories realize similar symmetries in a different way, the worldvolume torus in the supermembrane and the dual 2-torus in the DFT are related to the winding charges. In both cases there are two tori construction one of which is related to the torus target and the other to the winding modes, under T-duality in DFT these tori are interchanged but in the supermemebrane case only the charges are interchanged but not the geometry. In both cases the two monodromies associated to each of the tori under T-duality are interchanged. Both theories share the same discrete invariance $O(2,2,\mathbb{Z})$ but differently realized since for example the Hamiltonian of the supermembrane involves the moduli in a nontrivial way. Finally, DFT is associated generically to non geometrical backgrounds while the supermembrane formulation -as it also happens for string theory- is always geometrical and globally well defined.

Since DFT has its roots in string theory and the five string theories have its origin in the supermembrane theory, a sector of M-theory, it is quite natural to expect -even when there are differences-, that DFT describing the factorized doubled torus with a its monodromy in $O(2,2,\mathbb{Z})$ could be related to an effective description of the M2-brane torus bundle with monodromy in

$SL(2, \mathbb{Z})$. Obviously this hint will need to be confirmed by a more exhaustive study which is outside of the scope of the present note.

5. Acknowledgements

MPGM is supported by Mecesup ANT1656, Universidad de Antofagasta, (Chile). A.R. and MPGM are partially supported by Projects Fondecyt 1161192 (Chile). J.M. Peña is grateful to Universidad de Antofagasta for its hospitality and its partial support during part of the realization of this work.

References

[1] Hull C *Duality and strings, space and time.* Contribution to Proceedings of "Mathematical Sciences beyond the Second Millenium" held at the Center for Advanced Mathematical Sciences, Beirut (*Preprint* hep-th/9911080)
[2] Hull C 2005 *JHEP* **0510** 065 (*Preprint* hep-th/0406102)
[3] Shelton J, Taylor W and Wecht B 2005 *JHEP* **0510** 085 (*Preprint* hep-th/0508133)
[4] Hull C and Reid-Edwards R 2009 *Fortsch.Phys.* **57** 862-894 (*Preprint* hep-th/0503114)
[5] Reid-Edwards R 2009 *JHEP* **0906** 085 (*Preprint* hep-th/0904.0380)
[6] Samtleben H 2008 *Class. Quant. Grav.* **25** 214002 (*Preprint* hep-th/0808.4076)
[7] Martin I, Ovalle J and Restuccia A 2001 *Phys. Rev.* D **64** 046001 (*Preprint* hep-th/0101236)
[8] Boulton L, Garcia del Moral M and Restuccia A 2003 *Nucl. Phys.* B **671** 343-358 (*Preprint* hep-th/0211047)
[9] de Wit B, Luscher M and Nicolai H 1989 *Nucl. Phys.* B **320** 135
[10] de Wit B, Peeters K and Plefka J 1997 *Phys. Lett.* B **409** 117-123 (*Preprint* hep-th/9705225)
[11] García del Moral M, Peña J, Restuccia A 2012 *JHEP* **1209** 063 (*Preprint* hep-th/1203.2767)
[12] García del Moral M, Peña J, Restuccia A 2016 *Preprint* hep-th/1604.02579
[13] García del Moral M, Peña J, Restuccia A 2013 *Int. J. Geom. Meth. Mod. Phys.* **10** 1360010 (*Preprint* hep-th/1211.2434)
[14] Lüst D, Massai S, Vall Camell V 2016 *JHEP* **1609** 127 (*Preprint* hep-th/1508.01193)
[15] Dabholkar A and Hull C 2003 *JHEP* **0309** 054 (*Preprint* hep-th/0210209)
[16] Flournoy A and Williams B 2006 *JHEP* **0601** 166 (*Preprint* hep-th/0511126)
[17] Kawai S and Sugawara Y 2008 *JHEP* **0802** 027 (*Preprint* hep-th/0709.0257)
[18] Cederwall M 2007 *M-branes on U-folds. Preprint* hep-th/0712.4287
[19] Dabholkar A and Hull C 2006 *JHEP* **0605** 009 (*Preprint* hep-th/0512005)
[20] Hull C 2005 *JHEP* **0510** 065 (*Preprint* hep-th/0406102)
[21] Hull C 2007 *JHEP* **0710** 057 (*Preprint* hep-th/0604178)
[22] Garcia del Moral M, Martin I, Peña J and Restuccia A 2011 *JHEP* **1109** 068 (*Preprint* hep-th/1105.3181)
[23] Garcia del Moral M 2011 A New Mechanism for Gauging a Theory *Pre-print* hep-th/11073255
[24] Martin I, Ovalle J and Restuccia A 2001 *Nucl. Phys. Proc. Suppl.* **102** 169-175.
[25] Garcia del Moral M, Martin I and Restuccia A (*Preprint* hep-th/0802.0573)
[26] Khan P 2005 *New York J. Math.* **11** 3555
[27] Hull C 1998 *JHEP* **9811** 027 (*Preprint* hep-th/9811021)
[28] Lavrinenko I, Lu H and Pope C 1998 *Class.Quant.Grav.* **15** 2239-2256 (*Preprint* hep-th/9710243)
Lavrinenko I, Lu H and Pope C 1997 *Nucl.Phys.* B **492** 278-300 (*Preprint* hep-th/9611134)
[29] Le Diffon A and Samtleben H 2009 *Nucl.Phys.* B **811** 1-35 (*Preprint* hep-th/0809.5180)
[30] Riccioni F 2010 *Class.Quant.Grav.* **27** 125009 (*Preprint* hepth-th/1001.1316)
[31] Bergshoeff E, de Wit T, Gran U, Linares R and Roest D 2002 JHEP **0210 061** (*Preprint* hep-th/0209205)
[32] Fernandez-Melgarejo J, Ortin T and Torrente-Lujan E 2011 *JHEP* **1110** 068 (*Preprint* hep-th/1106.1760)
[33] Hull C and Zwiebach B 2009 *JHEP* **09** 099 (*Preprint* hep-th/0904.4664)
[34] Tseytlin A 1990 *Phys. Lett.* B **242** 163-174.
[35] Aldazábal G, Marqués D and Nuñez C 2013 *Class. Quant. Grav.* **30** 163001 (*Preprint* hep-th/1305.1907v2)
[36] Kugo T and Zwiebach B , Prog. Theor. Phys. 87, 801 (1992)
[37] Giveon A, Rabinovici E and Veneziano G 1989 *Nucl. Phys.* B **322** 167.
[38] Maharana J and Schwarz J 1993 *Nucl. Phys.* B **390** 3 (*Preprint* hep-th/9207016)
[39] Hohm O, Hull C and Zweibach B 2010 *JHEP* **08** 008.
[40] Flournoy A, Wecht B and Williams B 2005 *Nucl. Phys.* B **706** 127149 (*Preprint* hep-th/0404217)

ERG and Reggeon Field Theory

Carlos Contreras H.

Departamento de Fisica,Universidad Tecnica Federico Santa Maria, Valparaiso, Chile

E-mail: carlos.contreras@usm.cl

Abstract. In this paper we present our recent non perturbative functional renormalization group analysis of Reggeon Field Theory to the interactions of Pomeron and Odderon fields. We establish the existence of a fixed point and its universal properties, which exhibits a novel symmetry structure in the space of Odderon-Pomeron interactions. We briefly discuss the implications of our findings for the existence of an Odderon in high energy scattering.

1. Introduction

In recents paper [1, 2] we have started an analysis of the flow equations of reggeon field theory (RFT) [3, 4, 5, 7, 6, 8], following the idea that RFT might provide a good description of strong interactions in the Regge limit and infrared region: rapidity $Y \to \infty$ and transverse distances (impact parameter) $|x_\perp| \to \infty$. Using the Wetterich formulation of the functional renormalization group equations [9, 10] we can study the problem in two transverse dimensions. As our main result [1] we have established the existence of a critical theory with one relevant direction: in the multidimensional space of the parameters of the effective potential, there exists one direction which is UV attractive (IR repulsive), whereas all other directions are IR attractive. We have verified the properties that RFT is in the same universal class with the simplest directed percolation model in statistical physics [11].

This investigation of RFT should be seen as a first step in searching for an effective theory which describes the high energy Regge limit of QCD. Regge theory is being used for analysing the nonperturbative $p\bar{p}$ scattering at the Tevatron (FermiLab), pp scattering at the ISR (CERN), RHIC (BNL) and at the LHC (CERN), and γp-scattering at HERA (DESY). On the perturbative side, high energy QCD has been analyzed using Regge theory (in particular, the BFKL Pomeron with various applications in e^+e^- scattering, forward jets in $\gamma^* p$ scattering, and Mueller-Navelet jets in pp or $p\bar{p}$). Whereas the first group of high energy scattering processes is characterized by transverse distances of hadronic sizes, the second one addresses scattering processes of small transverse diameters. This suggests to search, in the space of $2 + 1$ - dimensional reggeon field theories, for an interpolation between the two domains: for long transverse distances the Pomeron field has intercept very close to unity and a nonzero t-slope, for short transverse distances the BFKL intercept is significantly above one, and the slope is very small.

Within such a program in [1] we have restricted our analysis to one reggeon field, the Pomeron field. However, there exists one other Regge singularity the Odderon with intercept at or near one. In the nonperturbative region, the search for the Odderon has stimulated a long-lasting debate: the strongest evidence for its existence comes from the observed difference in the dip structure in the t-dependence of elastic cross section of pp or $p\bar{p}$ scattering. In contrast, in the perturbative region the existence of the Odderon is well-established [12]: in nonabelian $SU(3)$ gauge theory bound states of reggeized gluons exist: the BFKL Pomeron [13] and the Odderon. These two states represent the two equally important fundamental bound states of the $SU(3)$

IOP Conf. Series: Journal of Physics: Conf. Series **1043** (2018) 012033 doi:10.1088/1742-6596/1043/1/012033

gauge theory. Whereas the BFKL intercept is well above one, the odderon intercept has been found to be exactly at one [14, 15, 16, 17]. Self interactions of the Pomeron [18, 19, 20, 21] as well as interactions between Pomeron and Odderon naturally appear in perturbative QCD analysis [22, 23]. Analogous results are obtained also in the Color Glass Condensate, dipole and Wilson line approach [24, 25, 26].

The existence of the pertubative region motivates interest in the question whether the Odderon exists also in the nonperturbative region. To explore the connection between the UV region and the nonpertubative IR region we have to consider the Odderon, the IR fixed point structure should confirm whether the Odderon survives the flow from UV to IR. Also, such an analysis should provide information on the interaction between Odderon and Pomeron. In this paper we therefore extend our previous analysis to interactions of two fields, Pomeron and Odderon and it is organized as follows. In section 2 we describe our setup and in the next the calculations of the fixed point conditions (β-functions of the set of the parameters of the effective potential) is discussed. In the last section we present numerical results. In a concluding section we discuss first implication for real physics.

2. The setup

In the following we consider interactions between Pomeron and Odderon fields. As before, $\psi, \psi^\dagger$ denote the Pomeron field, and for the Odderon we introduce the field $\chi, \chi^\dagger$. The effective action $\Gamma_k[\psi^\dagger, \psi, \chi^\dagger, \chi]$ has the form:

$$\Gamma_k = \int d^2x d\tau \left(Z_P(\frac{1}{2}\psi^\dagger \overleftrightarrow{\partial}_\tau \psi - \alpha'_P \psi^\dagger \nabla^2 \psi) + Z_O(\frac{1}{2}\chi^\dagger \overleftrightarrow{\partial}_\tau \chi - \alpha'_O \chi^\dagger \nabla^2 \chi) + V_k[\psi, \psi^\dagger, \chi, \chi^\dagger] \right), \quad (1)$$

where D denotes the number of spatial dimension. $D = 2$ is the physical case of our interest, but for our analytic study we find it convenient to keep D as a continuous parameter.

For the lowest truncation the effective action takes the form:

$$V_3 = -\mu_P \psi^\dagger \psi + i\lambda \psi^\dagger(\psi + \psi^\dagger)\psi - \mu_O \chi^\dagger \chi + i\lambda_2 \chi^\dagger(\psi + \psi^\dagger)\chi + \lambda_3(\psi^\dagger \chi^2 + \chi^{\dagger 2}\psi). \quad (2)$$

but if we will study high order truncation we can consider the quartic truncation:

$$\begin{aligned} V_4 &= \lambda_{41}(\psi\psi^\dagger)^2 + \lambda_{42}\psi\psi^\dagger(\psi^2 + \psi^{\dagger 2}) + \lambda_{43}(\chi\chi^\dagger)^2 + i\lambda_{44}\chi\chi^\dagger(\chi^2 + \chi^{\dagger 2}) \\ &\quad + i\lambda_{45}\psi\psi^\dagger(\chi^2 + \chi^{\dagger 2}) + \lambda_{46}\psi\psi^\dagger\chi\chi^\dagger + \lambda_{47}\chi\chi^\dagger(\psi^2 + \psi^{\dagger 2}) \end{aligned} \quad (3)$$

and the quintic truncation has the following eleven terms:

$$\begin{aligned} V_5 &= i\left(\lambda_{51}(\psi\psi^\dagger)^2(\psi + \psi^\dagger) + \lambda_{52}\,\psi\psi^\dagger(\psi^3 + \psi^{\dagger 3}) + \lambda_{53}\,\chi\chi^\dagger(\psi^3 + \psi^{\dagger 3}) + \lambda_{54}\,\psi\psi^\dagger\chi\chi^\dagger(\psi + \psi^\dagger)\right) \\ &+ \lambda_{55}(\chi^2\psi^{\dagger 3} + \chi^{\dagger 2}\psi^3) + \lambda_{56}\,(\chi^2\psi^{\dagger 2}\psi + \chi^{\dagger 2}\psi^\dagger\psi^2) + \lambda_{57}\,(\chi^2\psi^\dagger\psi^2 + \chi^{\dagger 2}\psi^{\dagger 2}\psi) \\ &+ i\left(\lambda_{58}\,(\chi^4\psi^\dagger + \chi^{\dagger 4}\psi) + \lambda_{59}\,(\chi\chi^\dagger)^2(\psi + \psi^\dagger)\right) \\ &+ \lambda_{510}\,\chi\chi^\dagger(\chi^2\psi + \chi^{\dagger 2}\psi^\dagger) + \lambda_{511}\,\chi\chi^\dagger(\chi^2\psi^\dagger + \chi^{\dagger 2}\psi). \end{aligned} \quad (4)$$

Note the differences in the structure of the effective potential compared to the pure Pomeron case. As described in [1, 2], for the pure Pomeron case the couplings are real-valued for even powers of the Pomeron fields, whereas odd powers require imaginary couplings. This is a consequence of the even-signature of the Pomeron exchange. Since the Odderon has negative signature the situation is different. Considering signature conservation, t-channel states with an odd number of Odderons never mix with pure Pomeron channels and the transition $P \to OO$ is real valued:

the two-Odderon cut is positive (in contrast to the two Pomeron cut), and there is no need for an imaginary coupling. On the other hand, the transition $O \to OP$ has to be imaginary, since the Odderon-Pomeron cut carries a minus sign. As a result, all triple couplings are imaginary, except for the real-valued transition $P \to OO$.

In the sector of quartic couplings, all couplings involving Pomerons only are real-valued. Once the Odderon is included, again most quartic couplings remain real, but there are two exceptions: the transitions $O \to OOO$ and $P \to P + OO$ are imaginary. This can be easily understood considering a contribution to such quartic vertices coming by the composition of two triple ones. For the quintic part the 'exceptional' terms are in the second and fourth lines: in all these terms we either create or annihilate a pair of Odderons.

The signature-conservation rule, together with the structure of the quartic and quintic couplings of the effective potential suggests the following transition rules:
(i) states with even and odd numbers of Odderon never mix.
(ii) states will be labelled by the number of Odderon pairs, n. We assign a quantum number O_n. Signature rules imply that transitions changing n by odd numbers come with 'exceptional' couplings (e.g the transitions $P \to OO$, $O \to OOO$, or $P \to P + OO$), whereas transitions changing n by even numbers are 'normal' and have the same structure as pure Pomeron couplings (e.g., the transition: Pomeron *to* four Odderons is imaginary).

In the perturbative region, the transition $P \to OO$ has been computed [22, 23] and found to be nonzero. As one of our task is to verify this result.

Another important element is the introduction of dimesionless variables. The field variables are rescaled as: $\tilde{\psi} = Z_P^{1/2} k^{-D/2} \psi$, $\tilde{\chi} = Z_O^{1/2} k^{-D/2} \chi$ and for the potential we introduce $\tilde{V} = \frac{V}{\alpha'_P k^{D+2}}$. This choice implies that we introduce the dimensionless ratio $r = \frac{\alpha'_O}{\alpha'_P}$, and the Odderon slope α'_O will be written as

$$\alpha'_O = r\, \alpha'_P. \tag{5}$$

Finally, the couplings are rescaled in the following way:

$$\tilde{\mu}_P = \frac{\mu_P}{Z_P \alpha'_P k^2}, \quad \tilde{\mu}_O = \frac{\mu_O}{Z_O \alpha'_P k^2}, \ ; \ \tilde{\lambda} = \frac{\lambda}{Z_P^{3/2} \alpha'_P k^2} k^{D/2}, \quad \tilde{\lambda}_{2,3} = \frac{\lambda_{2,3}}{Z_O Z_P^{1/2} \alpha'_P k^2} k^{D/2}. \tag{6}$$

With these definitions the classical scaling (canonical) of the potential which would result by neglecting the quantum fluctuations in the flow equation becomes:

$$(-(D+2) + \zeta_P)\, \tilde{V} + \left(\frac{D}{2} + \frac{\eta_P}{2}\right) \left(\tilde{\psi} \frac{\partial \tilde{V}}{\partial \tilde{\psi}} + \tilde{\psi}^\dagger \frac{\partial \tilde{V}}{\partial \tilde{\psi}^\dagger}\right) + \left(\frac{D}{2} + \frac{\eta_O}{2}\right) \left(\tilde{\chi} \frac{\partial \tilde{V}}{\partial \tilde{\chi}} + \tilde{\chi}^\dagger \frac{\partial \tilde{V}}{\partial \tilde{\chi}^\dagger}\right). \tag{7}$$

The scale k dependent regulator functions are chosen as follows:

$$\begin{aligned} R_P(q^2) &= Z_P \alpha'_P (k^2 - q^2) \Theta(k^2 - q^2), \\ R_O(q^2) &= Z_O \alpha'_O (k^2 - q^2) \Theta(k^2 - q^2) = r Z_O \alpha'_P (k^2 - q^2) \Theta(k^2 - q^2). \end{aligned} \tag{8}$$

This optimized regulator [27] allows for a simple analytic integration in a closed form. Moreover, the anomalous dimensions are defined by:

$$\eta_P = -\frac{1}{Z_P} \partial_t Z_P, \quad \eta_O = -\frac{1}{Z_O} \partial_t Z_O \ ; \ \zeta_P = -\frac{1}{\alpha'_P} \partial_t \alpha'_P, \quad \zeta_O = -\frac{1}{\alpha'_O} \partial_t \alpha'_O. \tag{9}$$

3. RG flow

In order to find the flow equation of the potential we need to compute the rhs of the dimensionful flow equations which result from scale k controlled contributions from quantum fluctuations:

$$\partial_t \Gamma = \frac{1}{2} Tr[\Gamma^{(2)} + R]^{-1} \partial_t R. \tag{10}$$

The trace on the rhs extends over a $4x4$ matrix. The propagator matrix can be written the following form:

$$\Gamma^{(2)} + R = \begin{pmatrix} \Gamma_P^{(2)} & \Gamma_{PO} \\ \Gamma_{OP} & \Gamma_O^{(2)} \end{pmatrix}, \tag{11}$$

where the $2x2$ block matrices are:

$$\Gamma_P^{(2)} = \begin{pmatrix} V_{\psi\psi} & Z_P(-i\omega + \alpha'_P q^2) + R_P + V_{\psi\psi^\dagger} \\ Z_P(i\omega + \alpha'_P q^2) + R_P + V_{\psi^\dagger\psi} & V_{\psi^\dagger\psi^\dagger} \end{pmatrix}, \tag{12}$$

$$\Gamma_O^{(2)} = \begin{pmatrix} V_{\chi\chi} & Z_O(-i\omega + \alpha'_O q^2) + R_O + V_{\chi\chi^\dagger} \\ Z_O(i\omega + \alpha'_O q^2) + R_O + V_{\chi^\dagger\chi} & V_{\chi^\dagger\chi^\dagger} \end{pmatrix}, \tag{13}$$

and

$$\Gamma_{PO} = \begin{pmatrix} V_{\psi\chi} & V_{\psi\chi^\dagger} \\ V_{\psi^\dagger\chi} & V_{\psi^\dagger\chi^\dagger} \end{pmatrix} \quad \Gamma_{OP} = \begin{pmatrix} V_{\chi\psi} & V_{\chi\psi^\dagger} \\ V_{\chi^\dagger\psi} & V_{\psi\psi^\dagger} \end{pmatrix}. \tag{14}$$

The momentum integral contained in the trace can be done as described in [1]. The energy integral will be performed by complex integration. Unfortunately the analytic expression for the full flow of the potential is quite involved and difficult to handle. Since we are interested in an analysis based on polynomial expansions of the potential in terms of the Pomeron and Odderon fields, we find it more convenient to derive directly the flow equations for the polynomial coefficients (couplings).

In this work we shall limit ourself in analyzing the flow of the potential expanded around the origin (zero fields), i.e. we shall employ a weak field expansion. Therefore, for the derivation of the beta-functions we find it convenient to expand the inverse of (11) in the following way:

$$\begin{aligned} [\Gamma^{(2)} + R]^{-1} &= [\Gamma^{(2)}_{free} - V_{int}]^{-1} \\ &= G(\omega, q) + G(\omega, q) V_{int} G(\omega, q) + G(\omega, q) V_{int} G(\omega, q) V_{int} G(\omega, q) + \ldots \end{aligned} \tag{15}$$

Here we absorb the masses (intercepts minus one) into the free propagators:

$$G(\omega, q) = \begin{pmatrix} G_P(\omega, q) & 0 \\ 0 & G_O(\omega, q) \end{pmatrix}, \tag{16}$$

where

$$G_P(\omega, q) = \begin{pmatrix} 0 & (Z_P(-i\omega + \alpha'_P q^2) + R_P - \mu_P)^{-1} \\ (Z_P(i\omega + \alpha'_P q^2) + R_P - \mu_P)^{-1} & 0 \end{pmatrix} \tag{17}$$

and

$$G_O(\omega, q) = \begin{pmatrix} 0 & (Z_O(-i\omega + \alpha'_O q^2) + R_O - \mu_O)^{-1} \\ (Z_O(i\omega + \alpha'_O q^2) + R_O - \mu_O)^{-1} & 0 \end{pmatrix}. \tag{18}$$

The interaction matrix V_{int} is derived from the effective potential:

$$V_{int} = -\begin{pmatrix} V^r_{\psi\psi} & V^r_{\psi\psi^\dagger} & V^r_{\psi\chi} & V^r_{\psi\chi^\dagger} \\ V^r_{\psi^\dagger\psi} & V^r_{\psi^\dagger\psi^\dagger} & V^r_{\psi^\dagger\chi} & V^r_{\psi^\dagger\chi^\dagger} \\ V^r_{\chi\psi} & V^r_{\chi\psi^\dagger} & V^r_{\chi\chi} & V^r_{\chi\chi^\dagger} \\ V^r_{\chi^\dagger\psi} & V^r_{\chi^\dagger\psi^\dagger} & V^r_{\chi^\dagger\chi} & V^r_{\chi^\dagger\chi^\dagger} \end{pmatrix}, \tag{19}$$

where the upper script 'r' reminds that the reggeon masses have been removed.

Finally $\dot{R}$ consisting of two block matrices.

$$\dot{R} = \begin{pmatrix} \dot{R}_P & 0 \\ 0 & \dot{R}_O \end{pmatrix}, \tag{20}$$

where

$$\dot{R}_P = \partial_t R_P(q^2) O_+, \ \dot{R}_O = \partial_t R_O(q^2) O_+. \tag{21}$$

and

$$O_\pm = \begin{pmatrix} 0 & 1 \\ \pm 1 & 0 \end{pmatrix}. \tag{22}$$

After the momentum integrals and after the use of dimensionless variables, the factors of the block matrices can be replaced by

$$\dot{R}_P \to N_D A_D(\eta_P, \zeta_P) O_+, \ \dot{R}_O \to r N_D A_D(\eta_O, \zeta_O) O_+, \tag{23}$$

where the factors N_D and A_D are defined in [1], and η_P, ζ_P and η_O, ζ_O are the anomalous dimensions of the Pomeron and Odderon fields, respectively.

We are this left with the energy integrals in the expansion:

$$\int \frac{dz'}{2\pi} Tr\Big[\dot{R}\, G(z')\left(1 - V_{int}G(z') + V_{int}G(z')V_{int}G(z') - V_{int}G(z')V_{int}G(z')V_{int}G(z') + ...\right)\Big] \tag{24}$$

where $z' = i\omega'/(\alpha'_P k^2)$, and the free propagators in (16), as a result of the momentum integration and the use of dimensionless variables, have become:

$$G_P(\omega, q) \to G_P(z) = \begin{pmatrix} 0 & (-z + 1 - \tilde{\mu}_P)^{-1} \\ (z + 1 - \tilde{\mu}_P)^{-1} & 0 \end{pmatrix}, \tag{25}$$

$$G_O(\omega, q) \to G_O(z) = \begin{pmatrix} 0 & (-z + r - \tilde{\mu}_O)^{-1} \\ (z + r - \tilde{\mu}_O)^{-1} & 0 \end{pmatrix}. \tag{26}$$

For the derivation of the beta functions, we take derivatives of (24) with respect to the field variables and subsequently set the fields equal to zero.

3.1. β functions for cubic truncation

For this approximation of the effective potential, we keep on the rhs of (24) the terms with two and three V's. The z-integral is done by complex integration and we use the condition $r-\mu_O>0$ which is the physical relevant region. The flow equations we find:

$$\begin{aligned}
\dot{\mu}_P &= (-2+\eta_P+\zeta_P)\mu_P + 2A_P\frac{\lambda^2}{(1-\mu_P)^2} - 2A_O r\frac{\lambda_3^2}{(r-\mu_O)^2} \\
\dot{\mu}_O &= (-2+\eta_O+\zeta_P)\mu_O + 2(A_P+A_O r)\frac{\lambda_2^2}{(1+r-\mu_P-\mu_O)^2} \\
\dot{\lambda} &= (-2+D/2+\zeta_P+\frac{3}{2}\eta_P)\lambda + 8A_P\frac{\lambda^3}{(1-\mu_P)^3} - 4A_O r\frac{\lambda_2\lambda_3^2}{(r-\mu_O)^3} \\
\dot{\lambda}_2 &= (-2+D/2+\zeta_P+\frac{1}{2}\eta_P+\eta_O)\lambda_2 \\
&+\frac{2\lambda\lambda_2^2(6A_P+5A_O r)+4\lambda_2^3(A_P+A_O r)-4\lambda_2\lambda_3^2(A_P+2A_O r)}{(1+r-\mu_P-\mu_O)^3} \\
&+\frac{2A_P\lambda\lambda_2^2(r-\mu_O)^2}{(1-\mu_P)^2(1+r-\mu_P-\mu_O)^3} - \frac{4A_O r\lambda_2\lambda_3^2(1-\mu_P)^2}{(1-\mu_O)^2(1+r-\mu_P-\mu_O)^3} \\
&+\frac{2\lambda\lambda_2^2(3A_P+A_O r)(r-\mu_O)}{(1-\mu_P)(1+r-\mu_P-\mu_O)^3} - \frac{4\lambda_2\lambda_3^2(A_P+3A_O r)(1-\mu_P)}{(r-\mu_O)(1+r-\mu_P-\mu_O)^3} \\
\dot{\lambda}_3 &= (-2+D/2+\zeta_P+\frac{1}{2}\eta_P+\eta_O)\lambda_3 \\
&+\frac{2\lambda_2^2\lambda_3(A_P+2A_O r)}{(r-\mu_O)(1+r-\mu_P-\mu_O)^2} + \frac{4\lambda\lambda_2\lambda_3(2A_P+A_O r)}{(1-\mu_P)(1+r-\mu_P-\mu_O)^2} \\
&+\frac{2\lambda_2^2\lambda_3 A_O r(1-\mu_P)}{(r-\mu_O)^2(1+r-\mu_P-\mu_O)^2} + \frac{4\lambda\lambda_2\lambda_3 A_P(r-\mu_O)}{(1-\mu_P)^2(1+r-\mu_P-\mu_O)^2},
\end{aligned} \tag{27}$$

weher we have defined

$$A_P = N_D A_D(\eta_P,\zeta_P),\;\; A_O = N_D A_D(\eta_O,\zeta_O). \tag{28}$$

For the truncations of fourth order and beyond, one can use Mathematica and the results for the beta functions are already lengthy and we will not be listed here.

3.2. Anomalous dimensions

In order to study the critical properties, we need not only the beta functions for the coupling constants of the effective potential but also the anomalous dimensions. Using (9), the evolution equation for r then becomes:

$$\dot{r} = r\,(-\zeta_O+\zeta_P)\,, \tag{29}$$

which tells that at criticality the Pomeron and Odderon transverse space scaling laws do coincide.

To obtain the anomalous dimensions we define the two-point vertex functions:

$$\Gamma_P^{(1,1)}(\omega,q) = \frac{\delta^2\Gamma}{\delta\psi(\omega,q)\delta\psi^\dagger(\omega,q)}|_{\psi=\psi^\dagger=\chi=\chi^\dagger=0} \tag{30}$$

and

$$\Gamma_O^{(1,1)}(\omega,q) = \frac{\delta^2\Gamma}{\delta\chi(\omega,q)\delta\chi^\dagger(\omega,q)}|_{\psi=\psi^\dagger=\chi=\chi^\dagger=0}. \tag{31}$$

From the flow equations we obtain:

$$\dot{\Gamma}_P^{(1,1)}(\omega,q) = \alpha'_P \int \frac{dz' d^D q'}{(2\pi)^{D+1}} Tr\left[\dot{R}\, G(z',q') \frac{\delta V_{int}}{\delta \psi^\dagger} G(z+z',q+q') \frac{\delta V_{int}}{\delta \psi} G(z',q')\right]|_O + ... \tag{32}$$

$$\dot{\Gamma}_O^{(1,1)}(\omega,q) = \alpha'_P \int \frac{dz' d^D q'}{(2\pi)^{D+1}} Tr\left[\dot{R}\, G(z',q') \frac{\delta V_{int}}{\delta \chi^\dagger} G(z+z',q+q') \frac{\delta V_{int}}{\delta \chi} G(z',q')\right]|_O + ... \tag{33}$$

where the subscript 'O' indicates that, after differentiation, we have set all field variables inside the trace equal to zero and the dots indicate that there are more terms containing second derivatives of V_{int} with respect to the field variables which will not contribute. The anomalous dimensions are obtained by taking derivatives with respect to energy and momentum:

$$Z_P = \lim_{\omega\to 0, q\to 0} \frac{\partial}{\partial(i\omega)} \Gamma_P^{(1,1)}(\omega,q);\; Z_O = \lim_{\omega\to 0, q\to 0} \frac{\partial}{\partial(i\omega)} \Gamma_O^{(1,1)}(\omega,q) \tag{34}$$

and

$$Z_P\alpha'_P = \lim_{\omega\to 0, q\to 0} \frac{\partial}{\partial q^2} \Gamma_P^{(1,1)}(\omega,q);\; Z_O\alpha'_O = \lim_{\omega\to 0, q\to 0} \frac{\partial}{\partial q^2} \Gamma_O^{(1,1)}(\omega,q). \tag{35}$$

Introducing

$$\dot{\Gamma}_P^{(1,1)} = I_P^{(1,1)}(\omega,q),\; \dot{\Gamma}_O^{(1,1)} = I_O^{(1,1)}(\omega,q), \tag{36}$$

the anomalous dimensions are then given by:

$$-\eta_P = \frac{1}{Z_P} \lim_{\omega\to 0, q\to 0} \frac{\partial}{\partial(i\omega)} I_P^{(1,1)}(\omega,q);\; -\eta_O = \frac{1}{Z_O} \lim_{\omega\to 0, q\to 0} \frac{\partial}{\partial(i\omega)} I_O^{(1,1)}(\omega,q) \tag{37}$$

and

$$-\eta_P - \zeta_P = \frac{1}{Z_P\alpha'_P} \lim_{\omega\to 0, q\to 0} \frac{\partial}{\partial q^2} I_P^{(1,1)}(\omega,q), \quad -\eta_O - \zeta_O = \frac{1}{Z_O\alpha'_O} \lim_{\omega\to 0, q\to 0} \frac{\partial}{\partial q^2} I_O^{(1,1)}(\omega,q). \tag{38}$$

We obtain after the momentum integral and considering the expansion around zero fields:

$$\eta_P = -\frac{2A_P\lambda^2}{(1-\mu_P)^3} + \frac{2A_O r\lambda_3^2}{(r-\mu_O)^3};\quad \eta_O = -\frac{4(A_P + A_O\, r)\lambda_2^2}{(1+r-\mu_P-\mu_O)^3} \tag{39}$$

and

$$\eta_P + \zeta_P = -\frac{N_D\lambda^2}{D(1-\mu_P)^3} + \frac{N_D r^2\lambda_3^2}{D(r-\mu_O)^3} \tag{40}$$

$$\eta_O + \zeta_O = -\frac{4N_D\lambda_2^2}{D(1+r-\mu_P-\mu_O)^3}. \tag{41}$$

3.3. Analysis ϵ-expansion to the cubic truncation

We show the results of an analysis of the theory close to the critical dimension ($D = 4 - \epsilon$) at one loop. Such an analysis can help to identify a possible critical behavior of the system which may survive, at a qualitative level, down to $D = 2$.

Evaluating the Eq. (27), (28) and (39)-(41) in the ϵ-expansion we find only one fixed point:

$$\mu_P = \frac{\epsilon}{12},\quad \lambda^2 = \frac{8\pi^2}{3}\epsilon,\quad \eta_P = -\frac{\epsilon}{6},\quad \zeta_P = \zeta_O = \frac{\epsilon}{12}, \tag{42}$$

$$\mu_O = \frac{95+17\sqrt{33}}{2304}\epsilon,\quad \lambda_2^2 = \frac{23\sqrt{6}+11\sqrt{22}}{48}\epsilon,\quad \lambda_3 = 0,\quad \eta_O = -\frac{7+\sqrt{33}}{72}\epsilon,\quad r = \frac{3}{16}(\sqrt{33}-1).$$

Finally, the spectral analysis of the stability matrix show the other universal quantities of the system, apart from the anomalous dimensions. In particular we find two negative eigenvalues, associated to two relevant directions, and the corresponding critical exponents:

$$\lambda^{(1)} = -2 + \frac{\epsilon}{4} \quad \rightarrow \quad \nu_P = \frac{1}{2} + \frac{\epsilon}{16}; \quad \lambda^{(2)} = -2 + \frac{\epsilon}{12} \quad \rightarrow \quad \nu_O = \frac{1}{2} + \frac{\epsilon}{48}. \tag{43}$$

We have not found other solutions with all real couplings and $\lambda_3 \neq 0$ and we also note that the values of the couplings, the critical exponents and anomalous dimensions in the Pomeron sector are exactly the same as in the pure Pomeron case [1]. This seems to favour, at least in the vicinity of the critical dimension $D = 4$, the existence of just two non trivial fixed points, one with the pure pomeron content, and another one with both interacting fields, where the interaction responsible for creating the odderon fields is turned off. That is the scaling solution of Eq. (42) is a theory conserving the Odderon number.

4. Numerical results

We perform an analysis in the physical $D = 2$ transverse space searching the fixed point and the critical properties using polynomial truncation for the potential (LPA) with increasing order up to order 9.

Our analysis is essentially in the LPA approximation with the addition of an extra coupling r, depending on the anomalous dimensions ζ_P and ζ_O, which we have evaluated at the lowest order. In all the other beta functions the anomalous dimensions have been completely neglected. Such a strategy is based on our previous experience with the Pomeron theory. Indeed, the polynomial expansion around the origin without the inclusion of the anomalous dimensions was giving, a very good estimate of the critical exponent ν.

The fixed point for the cubic truncation is: $(\mu_P = 0.111111, \mu_O = 0.110753, \lambda = 1.05034, \lambda_2 = 1.44665, \lambda_3 = 0, r = 0.921810)$, which has three negative and three positive eigenvalues, i.e. there are three relevant directions. Two of the negative eigenvalues $\lambda_O = -1.9398$ and $\lambda_P = -1.8860$ are associated to the ν_P and ν_O critical exponents, respectively. The third negative eigenvalue $\lambda^{(3)} = -0.0916$ is close to zero and with an eigenvector mainly associated to the r coupling.

This solution is also the one found in the ϵ-expansion analysis. We observe a 'decoupling' of the two sectors: compared to the pure Pomeron case, the Pomeron is not affected by the presence of the Odderon, whereas the Odderon 'feels' the Pomeron. This decoupling is due to the vanishing of the triple coupling λ_3.

All these features also appear in the solution obtained in the quartic truncation: $(\mu_P = 0.274381, \mu_O = 0.26979, \lambda = 1.34738, \lambda_2 = 1.79401, \lambda_3 = 0., \lambda_{41} = -2.88712, \lambda_{42} = -1.27076, \lambda_{43} = -0.83228, \lambda_{44} = 0., \lambda_{45} = 0., \lambda_{46} = -5.2784, \lambda_{47} = -2.2078, r = 0.88018)$. The stability properties are the same as in the cubic case: three negative eigenvalues (-1.8159, -1.6751 and -0.20957). The Pomeron and Odderon sectors are decoupled, since the couplings $\lambda_3, \lambda_{44}, \lambda_{45}$ vanish.

We extend our analysis up to order 9 in the polynomial expansion. We collect the results found in Fig. 1 (the values for μ_P and μ_O and the the non zero couplings) and Fig. 2 (the critical exponents ν_P and ν_O and the third negative eigenvalue) in order to show the convergence with respect to the order of the truncation. We note that $\mu_P > \mu_O$ in all truncations and we see how at order 9 a good stability is reached.

Finally, we have the following estimates: for the critical exponents ($\nu_P \simeq 0.73$, $\nu_O \simeq 0.6$), for the anomalous dimensions ($\eta_P \simeq -0.33$, $\eta_O \simeq -0.35$) and $\zeta_P = \zeta_O \simeq +0.17$. Numerical studies of the flow of the dimensionless and dimensionful parameters, for the Pomeron-Odderon system remains a task for future analysis.

We observe that this special fixed point solution is associated to a critical theory conserving the Odderon number. We do not find any other physical critical solution with all couplings

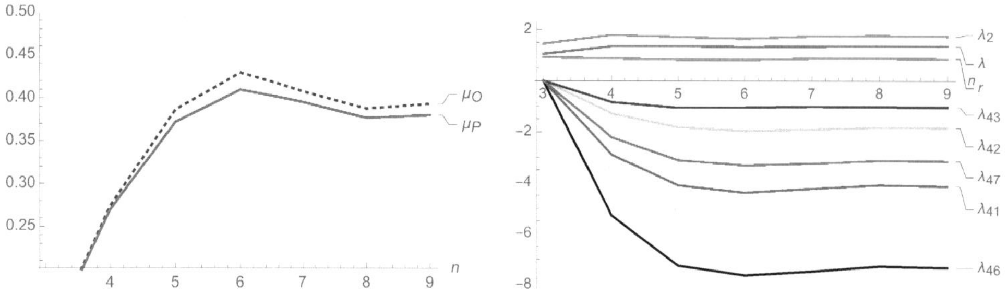

Figure 1. Values of the parameters of the fixed point solution of the LPA truncations for different orders n of the polynomial ($3 \leq n \leq 9$). The masses (which equal intercept minus one) μ_P (red curve) and μ_O (blue dotted curve) for the Pomeron and Odderon fields are in the left panel. The first non zero couplings $\lambda, \lambda_2, \lambda_{41}, \lambda_{42}, \lambda_{43}, \lambda_{46}, \lambda_{47}, r$ are reported on the right panel.

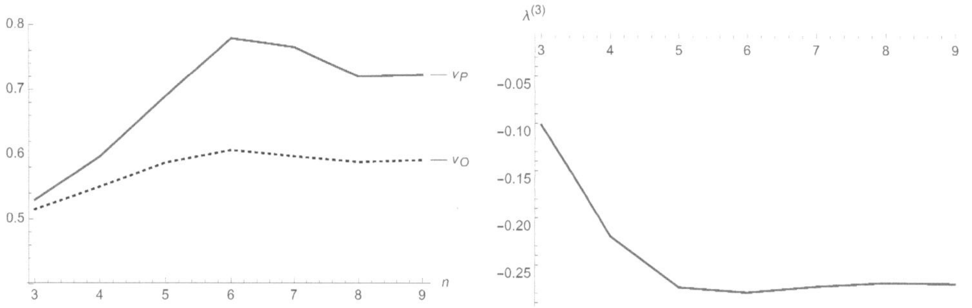

Figure 2. Values of the critical exponents of the fixed point solution of the LPA truncations for different orders n of the polynomial ($3 \leq n \leq 9$). The two negative leading eigenvalues define the two critical exponents ν_P (red curve) and ν_O (blue dotted curve) for the Pomeron and Odderon fields (left panel). We report also the value of a third negative eigenvalue found in our approximation (right panel).

being nonzero. For more details see ref. [2]

5. Discussion and outlook

In this paper we have extended our previous analysis of Pomeron RFT to a system of interacting Pomeron and Odderon fields in the infrared limit. Where the main motivation for including the Odderon comes from the observation that, in the UV region where perturbative QCD applies, there exist two fundamental composite states of reggeized gluons, the BFKL Pomeron with intercept well above one and a very small slope, and the Odderon with intercept at (or very close to) one and a small slope. This raises the question, when moving towards the nonperturbative IR region, to what extent the interactions between these fundamental fields lead to serious modifcations, e.g. a suppression of the Odderon exchange at high energies.

From (6) we see that near the fixed point both intercepts, $\alpha_P(0) - 1 = \mu_P/Z_P \sim k^{2-\zeta}\tilde{\mu}_P$ and $\alpha_O(0) - 1 = \mu_O/Z_O \sim k^{2-\zeta}\tilde{\mu}_O$ go to zero as k becomes small. Since the fixed point value of $\tilde{\mu}_P$ is slightly larger than $\tilde{\mu}_O$ we conclude that, for small but nonzero values of k, the Pomeron intercept is larger than the Odderon intercept. However, the most important conclusion to be drawn from this fixed point analysis is that the Odderon exists in the IR limit and does not die out with energy. Future study will be devoted to understand the QCD-RFT transition.

There are several questions to be addressed by future studies. Phenomenologically, not much is known about the Odderon slope [28, 29], and our result might be seen as an asymptotic prediction. Finally, the possibility that in the deep IR region the POO vertex is suppressed may also have phenomenological consequences. Processes involving a simple Odderon exchange, like hadron scattering pp - $p\bar{p}$ or meson photo-production [30] would be allowed in asymptotic IR [31].

Acknowledgements
This research was supported by the Fondecyt (Chile) grants 1140842 and by DGIP/USM.

References
[1] Bartels J, Contreras C and Vacca G P 2016 Could reggeon field theory be an effective theory for QCD in the Regge limit? *J. High Energy Phys.* JHEP **1603**(2016)201
[2] Bartels J, Contreras C and Vacca G P 2017 *Phys. Rev.* D **95** 014013
[3] Gribov V N and Migdal A A 1969 *Sov. Phys. JETP* **28** 014013 [Zh. Eksp. Teor. Fiz. **55** (1968) 1498]
[4] Gribov V N and Migdal A A 1969 *Sov. J. Nucl. Phys.* **8** 583 [Yad. Fiz. **8** (1968) 1002]
[5] Abarbanel H and Bronzan J 1974 *Phys. Rev.* D **90** 2397
[6] Sugar R L and White A R 1974 *Phys. Rev.* D **10** 4074.
[7] Migdal A, Polyakov A M and Ter-Martirosian K 1974 *Phys. Lett.* B **48** 239 [Pisma Zh. Eksp. Teor. Fiz. **68** (1975) 817].
[8] Moshe M 1978 *Phys. Rept.* **37** 255.
[9] Wetterich C 1993 *Phys. Lett.* B **301** 90.
[10] Morris T 1994 *Phys. Lett.* B **329** 241
[11] Cardy J L and Sugar R L 1980 Directed Percolation and Reggeon Field Theory *Journal Phys.* J. Phys. A **13**(1980)L423
[12] Bartels J, Lipatov L and Vacca G P 2000 *Phys. Lett.* B **477** 178
[13] Lipatov L 1976 *Sov. J. Nucl. Phys.* **23** 338; Kuraev E, Lipatov L and Fadin V 1975 *Phys. Lett.* B **60** 50, 1976 *Sov. Phys. JETP* **44** 443, 1977 *Sov. Phys. JETP* **45** 199; Balitsky I and Lipatov L 1978 *Sov. J. Nucl. Phys.* **28** 822.
[14] Bartels J and Vacca G P 2013 *Eur. Phys. J.* C **73** 2602
[15] Bartels J 1980 *Nucl. Phys.* B **175** 365
[16] Kwiecinski J and Praszalowicz M 1980 *Phys. Lett.* B **94** 413
[17] Bartels J, Fadin V, Lipatov L and Vacca G P 2013 *Nucl. Phys.* B **867** 827
[18] Bartels J and Wusthoff M 1995 *Z. Phys.* C **66** 157
[19] Braun M A and Vacca 1997 *Eur. Phys. J.* C **6** 147
[20] Bartels J, Ryskin M G and Vacca G P 2002 *Eur. Phys. J.* C **27** 101
[21] Bartels J, Lipatov L and Vacca G P 2004 *Nucl. Phys.* B **706** 391
[22] Bartels J and Ewerz C 1999 Unitarity corrections in high-energy QCD *J. High Energy Phys. JHEP* **9909**(1999)026
[23] Bartels J, Braun M and Vacca G P 2005 *Eur. Phys. J.* C **40** 419
[24] Kovchegov Y V, Szymanowski L and Wallon S 2004 *Phys. Lett.* B **586** 267
[25] Hatta Y, E. Iancu E, Itakura K and McLerran L 2005 *Nucl. Phys.* A **760** 172
[26] Kovner A and Lublinsky M 2007 Odderon and seven Pomerons: QCD Reggeon field theory from JIMWLK evolution *J. High Energy Phys.* JHEP **0702**(2007)058
[27] Litim D 2001 *Phys. Rev.* D **64** 105007
[28] Ewerz C 2003 The Odderon in quantum chromodynamics *Preprint* hep-ph/0306137
[29] Dosch H G, Ewerz C and Schatz V 2002 *Eur. Phys. J.* C **24** 561
[30] Bartels J, Braun M A, Colferai D and Vacca G P 2001 *Eur. Phys. J.* C **20** 323
[31] Bartels J, Braun M A and Vacca 2004 *Eur. Phys. J.* C **33** 511

IOP Conf. Series: Journal of Physics: Conf. Series **1043** (2018) 012034 doi:10.1088/1742-6596/1043/1/012034

The Ground State of the $D = 11$ Supermembrane: the External Dirichlet Problem

L. Boulton[1] María Pilar García del Moral[2], Alvaro Restuccia[2,3]

[1]Department of Mathematics and Maxwell Institute for the Mathematical Sciences, Heriot-Watt University, Edinburgh, EH14 4AS, United Kingdom.
[2]Departamento de Física, Universidad de Antofagasta, Aptdo 02800, Chile.
[3] Departamento de Física, Universidad Simón Bolívar, Apartado 89000, Caracas 1080-A, Venezuela

E-mail: [1]`l.boulton@hw.ac.uk`;[2]`maria.garciadelmoral@uantof.cl`; [3]`alvaro.restuccia@uantof.cl`

Abstract. We analyse the outer Dirichlet problem for the Hamiltonian of the $SU(N)$ regularized $D = 11$ Supermembrane. We show the existence and uniqueness of the solution for this problem. The solution belongs to a suitable Hilbert space, introduced in the present work, which is defined in terms of the supersymmetric charges. The existence and uniqueness of the ground state of the $D = 11$ supermembrane is directly related to the patching of solutions of the inner and outer Dirichlet problems.

1. Introduction

The quantization of the Supermembrane theory has been a challenge of fundamental importance since the introduction of the theory in [1] thirty years ago. Many aspects related to this quantization remain open. One in particular is the existence of the ground state, which to these days remains to be confirmed. If this ground state exists, it should be related to the $D = 11$ supergravity multiplet, which is the eigenfunction associated to zero eigenvalue for the Hamiltonian of the zero modes of the Supermembrane theory [5]. In order to confirm the existence of the ground state of the theory, one needs to show the existence and uniqueness of the ground state for the Hamiltonian of the non zero modes which should be invariant under $SO(9)$. This Hamiltonian was first derived from a $(0 + 1)$ reduction of the $D = 10$ Super Yang Mills Theory in a completely different context [2, 3, 4]. The $SU(N)$ regularized Hamiltonian for the nonzero modes of the Supermembrane [5] coincides with the Hamiltonian of the BFSS matrix model [6]. Hence the existence of the ground state for this matrix model, which is still undetermined, turns out to be exactly the missing step in the proof of existence of the ground state for the $D = 11$ Supermembrane.

Due to the complexity of this task, the problem of finding a nontrivial solution has been split into different regions of the the space. This has been considered asymptotically in several works [7, 8, 9, 10, 11, 14]. Under the presumption of existence, the structure of the possible ground state near the origin has been examined in [14]. In the works [15, 16, 17] we proposed a strategy to tackle the ground state problem involving the decomposition of $\mathbb{R}^D$ into an inner and outer region, and confirmed existence and uniqueness in the inner region. The present work is devoted to the complementary outer Dirichlet problem. Once the existence and uniqueness of this outer solution has been obtained, then it has to be patched in an appropriate sense with the inner solution. If the patch is smooth enough, then the overall state is indeed the ground state of the Hamiltonian of the non-zero modes of the $D = 11$ Supermembrane.

2. Existence of a solution for the outer region

Below we follow the notation set in [17]. Let Ω be a ball of radius R in $\mathbb{R}^D$ and let $\Omega_C = \mathbb{R}^D/\overline{\Omega}$. Let $\mathcal{S}_\mu$ denote the bilinear form

$$\mathcal{S}_\mu(u,v) := (Qu, Qv)_{L_2(\Omega_C)} + (Q^\dagger u, Q^\dagger v)_{L_2(\Omega_C)} + \mu(u,v)_{L_2(\Omega_C)} \tag{1}$$

where μ is a real, strictly positive parameter. This bilinear form gives rise to an inner product in the linear subspace $C_0^\infty(\Omega_C)$, the functions with compact support in Ω_C. That this is the case is a consequence of the condition (K) introduced in [17]. This condition is satisfied by the supersymmetric charges also in Ω_C. The completion of $C_0^\infty(\Omega_C)$ with respect to the metric associated to this inner product defines the Hilbert space, $S_\mu(\Omega_C)$. From the definition (1) it follows that $S_\mu(\Omega_C) \subset L^2(\Omega_C)$.

We introduce the operator $(-\Delta + V + \mu)$ with domain

$$\mathrm{Dom}(-\Delta + V + \mu) \subset S_\mu(\Omega_C) \cap H^2_{loc}(\Omega_C) \tag{2}$$

where V is the potential of the Supermembrane Hamiltonian. See [17] for details.

From the results in [17] it follows that given $f \in C_0^\infty(\Omega_C)$, there exists $\Psi \in \mathrm{Dom}(-\Delta+V+\mu)$ such that the outer boundary value problem:

$$\begin{cases} (-\Delta + V + \mu)\Psi = f & \text{in } \Omega_C \\ \Psi = 0 & \text{on } \partial\Omega \end{cases} \tag{3}$$

has a unique solution. Consequently the image of $(-\Delta + V + \mu)$ is dense in $L^2(\Omega_C)$. Also the operator $(-\Delta + V + \mu)$ is bounded from below, since $-\Delta + V$ is a positive operator and $\mu > 0$. Hence $(-\Delta + V + \mu)^{-1}$ is a bounded operator. From both these properties it follows that the domain of $(-\Delta + V + \mu)^{-1}$ is $L_2(\Omega_C)$ and the solution Ψ of the boundary value problem exists and it is unique for all $f \in L^2(\Omega_C)$.

The existence and uniqueness of Ψ also follows from the Riesz representation theorem. In fact, given $f \in L^2(\Omega_C)$, there always exists $\Psi \in S_\mu(\Omega_C)$ such that

$$\mathcal{S}_\mu(u, \Psi) = (u, f)_{L^2(\Omega_C)} \tag{4}$$

for all $u \in S_\mu(\Omega_C)$. This again implies that Ψ is the solution of the boundary value problem for any $L^2(\Omega_C)$. The regularity properties of the elliptic operator ensures that $\Psi \in L^2(\Omega_C) \cap H^2_{loc}(\Omega_C)$.

Now consider the bilinear form S defined on the subspace $C_0^\infty(\Omega_C)$:

$$\mathcal{S}(u,v) := (Qu, Qu)_{L_2(\Omega_C)} + (Q^\dagger u, Q^\dagger u)_{L_2(\Omega_C)}. \tag{5}$$

From property (K) in [17] which is also satisfied by te corresponding charges on Ω_C, this quadratic form also defines an inner product in C_0^∞. Its completion with respect to the associated norm defines the Hilbert space $S(\Omega_C)$.

Consider the operator $(-\Delta+V)$ with domain: $\mathrm{Dom}(-\Delta+V) \subset S(\Omega_C) \cap H^2_{loc}(\Omega_C)$. According to the Riesz representation theorem, for any $f \in C_0^\infty(\Omega_C)$ there always exists a unique solution in $\varphi \in S(\Omega_2) \cap H^2_{loc}(\Omega_C)$ such that

$$\begin{cases} (-\Delta + V)\varphi = f & \text{in } \Omega_C \\ \varphi = 0 & \text{on } \partial\Omega. \end{cases} \tag{6}$$

We notice that the solution φ has not been shown to lie in $L^2(\Omega_C)$. However, we can map it to the solution Ψ of problem (3) which indeed belongs to $L^2(\Omega_C)$. In fact,

$$\Psi = (-\Delta + V + \mu)^{-1}(-\Delta + V)\varphi = \varphi - \mu(\Delta + V + \mu)^{-1}\varphi \tag{7}$$

is well defined because $(-\Delta + V)\varphi \in C_0^\infty(\Omega_C)$, and it satisfies

$$\mathcal{S}_\mu(\Psi, u) = \mathcal{S}(\varphi, u) = (f, u)_{L^2(\Omega_C)} \quad \text{with} \quad \mu \in S_\mu(\Omega_C), \tag{8}$$

for any $f \in C_0^\infty(\Omega_C)$. Hence Ψ is the solution to problem (3), $\Psi \in S_\mu(\Omega_C) \cap H^2_{loc}(\Omega_C)$. if $\varphi \in L^2(\Omega_C)$, then $(-\delta + V + \mu)^{-1}\varphi \in S_\mu \cap H^2_{loc}(\Omega_C)$ and from (8) it follows that $\varphi \in S_\mu(\Omega_C) \cap H^2_{loc}(\Omega_C)$.

The outer Dirichlet problem for the Hamiltonian of the $D = 11$ Supermembrane may be formulated as follows: given $g \in \mathrm{Dom}(-\Delta + V)$, find $\Phi \in S_\mu(\Omega_C) \cap H^2_{loc}(\Omega_C)$ such that

$$\begin{cases} (-\Delta + V + \mu)\Phi = 0 & \text{in } \Omega_C \\ \Phi = g & \text{on } \partial\Omega \end{cases} \tag{9}$$

where Φ is restricted by the $SU(N)$ constraint. That is, the reduction under the $SU(N)$ regularization procedure of the area preserving constraint of the $D = 11$ Supermembrane.

Following the previous analysis we can prove the existence and uniqueness of $\Phi \in S(\Omega_C) \cap H^2_{loc}(\Omega_C)$ which solves the outer Dirichlet problem. It remains to show that Φ also belongs to $L^2(\Omega_C)$. This result surely depends on the dimension of the target space of the supermembrane. For the $D = 11$ Supermembrane we expect it to be true.

3. Conclusions

We argued on the existence and uniqueness of the outer Dirichlet problem associated to the Hamiltonian of the $SU(N)$ regularized $D = 11$ Supermembrane. The argument we have presented shows that the solution belongs to the Hilbert space $S(\Omega_C)$ and locally to $H^2_{loc}(\Omega_C)$. In order to construct the external ground state we require a in $L^2(\Omega_C)$. The missing step for the latter surely depends on the structure of the supersymmetric charges. However the improvement we have obtained compared to the results currently available in the literature is that we have shown the existence of a solution and based on it, the analysis of the asymptotic behavior is largely simplified.

4. Acknowledgements

MPGM is supported by Mecesup ANT1655, Universidad de Antofagasta, (Chile). A.R. and MPGM are partially supported by Projects Fondecyt 1161192 (Chile).

[1] E. Bergshoeff, E. Sezgin, P.K. Townsend,*Supermembranes and eleven-dimensional supergravity.* Phys.Lett. **B189**: 75-78, 1987.
[2] M. Claudson, M. B. Halpern, *Supersymmetric Ground State wavefunctions*, Nucl.Phys. **B250** (1985) no. 4, 689.
[3] R. Flume, *On Quantum Mechanics With Extended Supersymmetry and Nonabelian Gauge Constraints*, Ann. of Phys. **164** (1985) 189.
[4] M. Baake, M. Reinicke and V. Rittenberg, *Fierz Identities for Real Clifford Algebras and the Number of Supercharges* J. Math.**26** (1985) 1070.
[5] B. de Wit, J. Hoppe, H. Nicolai, *On the quantum mechanics of supermembranes.* Nucl. Phys. **B305**: 545,1988.
[6] T. Banks, W. Fischler, S.H. Shenker and L. Susskind, *M theory as a matrix model: A Conjecture* Phys.Rev. **D55** (1997) 5112-5128.
[7] J Hoppe, *Asymptotic zero energy states for SU(N) greater than or equal to 3*, hep-th/9912163.
[8] D. Hasler, J. Hoppe *Asymptotic factorization of the ground state for SU(N) invariant supersymmetric matrix models*, hep-th/0206043.

[9] J. Froehlich, J. Hoppe, *On Zero-Mass Ground States in Super-Membrane Matrix Models*, arXiv:hep-th/9701119.

[10] Y. Michishita, M. Trzetrzelewski, *Towards the Ground State of the Supermembrane*, Nucl.Phys.**B868** (2013) 539-553.

[11] J Hoppe, D Lundholm, *On the Construction of Zero Energy States in Supersymmetric Matrix Models IV*, arXiv:0706.0353.

[12] M.B. Halpern , C. Schwartz, *Asymptotic search for ground states of SU(2) matrix theory*, Int.J.Mod.Phys. **A13** (1998) 4367-4408.

[13] J. Frohlich, G.M. Graf, D. Hasler, J. Hoppe, S-T. Yau, *Asymptotic form of zero energy wavefunctions in supersymmetric matrix models*, Nucl.Phys. **B567** (2000) 231-248.

[14] J. Hoppe, D. Lundholm, M. Trzetrzelewski, *Construction of the Zero-Energy State of SU(2)-Matrix Theory: Near the Origin*, Nucl.Phys. **B817** (2009) 155-166.

[15] L. Boulton, M.P. Garcia del Moral, A. Restuccia, *On the ground state octonionic matrix models in a ball*, Phys. Lett. B **744** (2015) 260-262.

[16] L. Boulton, M.P. Garcia del Moral, A. Restuccia, *Massless ground state for a compact* $SU(2)$ *matrix model in 4D*, arXiv:1503.05462.

[17] L. Boulton, M.P. Garcia del Moral, Alvaro Restuccia, *The ground state of the* $D = 11$ *supermembrane and matrix models on compact regions*, Nucl.Phys. **B910** (2016) 665-684

IOP Conf. Series: Journal of Physics: Conf. Series **1043** (2018) 012035 doi:10.1088/1742-6596/1043/1/012035

The Gribov problem in presence of background field for $SU(2)$ Yang-Mills theory

Fabrizio Canfora[1], **Diego Hidalgo**[1,2], **Pablo Pais**[1,3]

[1] Centro de Estudios Científicos (CECS), Casilla 1469, Valdivia, Chile.
[2] Departamento de Física, Universidad de Concepción, Casilla 160, Concepción, Chile.
[3] Physique Théorique et Mathématique,Univérsite de Bruxelles and International Solvay Institutes, Campus Plaine C.P. 231, B-1050 Bruxelles, Belgium.

E-mail: [1] `canfora@cecs.cl`, [2] `dhidalgo@cecs.cl`, [3] `pais@cecs.cl`

Abstract. The Gribov problem in the presence of a background field is analyzed: in particular, we study the Gribov copies equation in the Landau-De Witt gauge as well as the semi-classical Gribov gap equation. As background field, we choose the simplest non-trivial one which corresponds to a constant gauge potential with non-vanishing component along the Euclidean time direction. We show that the Gribov copies equation is affected directly by the presence of the background field, constructing an explicit example. The analysis of the Gribov gap equation shows that the larger the background field, the smaller the Gribov mass parameter. These results strongly suggest that the relevance of the Gribov copies (from the path integral point of view) decreases as the size of the background field increases.

1. Introduction

The main tool to compute observable quantities in QFT is perturbation theory. In gauge theories, and in Yang-Mills (YM) theory in particular, a fundamental problem to solve in order to compute physical quantities is the over-counting of degrees of freedom related to gauge invariance (for a detailed analysis see [1]). The Faddeev-Popov (FP) gauge fixing procedure is the cornerstone which allows using the Feynman rules and Feynman diagrams in all applications of the standard model. The obvious fundamental hypothesis is that the gauge-fixing condition must intersect once and only once every gauge orbit. Locally, in the space of gauge fields, this hypothesis requires that the FP operator should not have zero modes so that the FP determinant is different from zero. The reason is that the existence of a proper gauge transformation preserving the gauge-fixing would spoil the whole quantization procedure since it would imply that the FP recipe does not completely eliminate the over-counting of degrees of freedom.

However, in [2], Gribov showed that in non-Abelian gauge theories (in flat, topologically trivial space-times) the FP procedure fails at non-perturbative level. The reason is that a proper gauge fixing is not possible due to the appearance of Gribov copies: namely, gauge equivalent configurations satisfying the Coulomb gauge. Later, Singer [3] showed that if Gribov ambiguities occur for the Coulomb gauge, they occur for all gauge fixing conditions involving derivatives of the gauge field.

Naively, one could expect to completely avoid the Gribov problem by simply choosing algebraic gauge fixings like the axial gauge or the temporal gauge, which are free of Gribov copies. However, these choices have their own, and even worse, problems (for a detailed reviews see [4] [5]).

On the other hand, the existence of Gribov copies is not just a problem since, as Gribov himself argued, the natural way to solve such a problem is able to shed considerable light on the infrared (IR) region of YM theory. Such solution is to restrict the path-integral only to a region Ω, which is called Gribov region, where FP operator is definite-positive [2, 6, 7, 8, 9] (detailed reviews are [10] and [5]) so that there are no Gribov copies connected to the identity. It is worth to mentioning that some Gribov copies are still left within the Gribov region [6]. In order to restrict the path integral to the Gribov region, one can use the Gribov-Zwanziger (GZ) approach [11, 12]. When the space-time geometry is flat and the topology trivial this method is able to reproduce the usual perturbation theory encoding, at the same time, the effects related to the elimination of the Gribov copies. For instance, it allows the computation of the glueball masses in excellent agreement with the lattice data [19, 20]. Within the same framework, it is also possible to solve the sign problem for the Casimir energy and force in the MIT-bag model [21].

Thus, it is natural to wonder whether or not this approach works so well also in the presence of a background gauge field. From the theoretical point of view, this analysis is very important as it discloses how strongly the presence of a background field can affect the Gribov region and the whole issue of Gribov copies. One of the most relevant applications of the background field method is the computation of the vacuum expectation value of the Polyakov loop [27] in which the presence of the Polyakov loop manifests itself as a constant background field with component along the Euclidean time. Formally, a constant background gauge field with only the timelike component non-vanishing is related to a bosonic chemical potential [29]. On the other hand, the physical interpretation of such Bosonic chemical potential is rather obscure in the case of non-perturbative gluons and so it will not be discussed in the present case. Another very important non-perturbative phenomenon in which the presence of a background gauge field plays a key role is the (both Abelian and non-Abelian) Schwinger effect [30, 31, 32]. Also in the case of the non-Abelian Schwinger effect, the relevant background gauge fields are constant A_μ which have components both along time and space directions. From the point of view of applications, such an analysis can also be quite relevant in relation with quark-gluon plasma [33, 34], color superconductivity in QCD [35], astrophysics [36, 34], and cosmology [37, 38].

The Background Field Method (BFM) [39, 40, 41] together with the techniques developed in [13, 14, 15, 16, 17] suggest that background fields can play a prominent role within the GZ approach to YM theory. Here we show that the Gribov copies equation and the gap equation are affected directly by the presence of a background field.

The paper is organized as follows. In the second section, the Gribov problem in the Landau-De Witt gauge is introduced. In the third section, explicit examples of Gribov copies in the Landau-De Witt gauge are studied. In the fourth section, the Gribov gap equation within a background field is analyzed. Some conclusions and discussions are drawn at the end.

2. A brief review of the Gribov-Zwanziger action

In this section we present an outlook of the GZ-approach without background consider background fields, which is the aim of Section 3. The Euclidean Yang-Mills action

$$S_{YM} = \frac{1}{4}\int d^4x F^a_{\mu\nu}F^{a\mu\nu}, \quad F^a_{\mu\nu} = \partial_\mu A^a_\nu - \partial_\nu A^a_\mu + g f^{abc} A^b_\mu A^c_\nu, \tag{1}$$

is invariant under the gauge transformation

$$A_\mu \to A_\mu = U^{-1} A_\mu U + U^{-1}\partial_\mu U, \tag{2}$$

with $U \in SU(N)$. In order to take into account the existence of Gribov copies due to the this gauge transformation, Gribov proposed [2] to restrict the domain of integration in the path integral to a region in functional space where the eigenvalues of the FP operator $\mathcal{M}^{ab}$ are strictly positive. This region is known as the Gribov region Ω, and is defined as

$$\Omega = \{A^a_\mu \mid \quad \partial_\mu A^a_\mu = 0; \quad \mathcal{M}^{ab} = -\partial_\mu(\partial^\mu \delta^{ab} - gf^{abc}A^c_\mu) = -\partial_\mu D^{ab}_\mu > 0\}. \tag{3}$$

where $D^{ab}_\mu = \partial^\mu \delta^{ab} - gf^{abc}A^c_\mu$ is the usual covariant derivative, which depends on A_μ. The boundary of this region is called the first Gribov horizon. Later on, Zwanziger [11] implemented the Gribov region Ω in Euclidean Yang-Mills theories resulting

$$S_{GZ} = S_{FP} + S_\gamma + S_0, \tag{4}$$

$$S_{FP} = \int d^4x \left(\frac{1}{4}F^a_{\rho\sigma}F^{\rho\sigma a} + ib^a\partial_\rho A^a_\rho + \bar{c}^a \mathcal{M}^{ad}c^d\right), \tag{5}$$

$$S_\gamma = \int d^4x \left(\gamma^2 g f^{abc} A^a_\rho(\varphi^{bc}_\rho - \bar{\varphi}^{bc}_\rho)\right) + \gamma^4 \int d^4x h(x), \tag{6}$$

$$S_0 = \int d^4x \left(-\bar{\varphi}^{ac}_\rho \mathcal{M}^{ab}\varphi^{bc}_\rho + \bar{\omega}^{ac}_\rho \mathcal{M}^{ab}\omega^{bc}_\rho + gf^{amb}(\partial_\rho\bar{\omega}^{ac}_\sigma)(D^{mp}_\rho c^p)\varphi^{bc}_\sigma\right), \tag{7}$$

where Greek indexes run from $\mu = 1 \ldots d$ and latin indexes from $a = 1 \ldots N^2 - 1$. The fields $(\bar{\varphi}^{ac}_\mu, \varphi^{ac}_\mu)$ are a pair of complex conjugate bosonic fields, while $(\bar{\omega}^{ac}_\mu, \omega^{ac}_\mu)$ are anti-commuting fields, and the Gribov parameter γ is determined by the following gap equation

$$\frac{\partial \varepsilon_{vac}}{\partial \gamma^2} = 0. \tag{8}$$

Therefore, equation (8) represents the horizon condition formula which will allow us to determine the Gribov parameter later on.

3. Gribov-Zwanziger action in a background field

As we present a brief introduction in Section 2 of the GZ-approach, now we analyze what happens if we take into account a background field. We consider the $SU(N)$ Yang-Mills theory in $d = 4$ Euclidean dimensions defined in Eq.(4). In the BFM (see [4, 42] for more details), one introduces a fixed background gauge field configuration B_μ through the splitting

$$A_\mu \rightarrow a_\mu \equiv A_\mu + B_\mu \ , \tag{9}$$

where A_μ and B_μ play completely different roles. On the one hand, A_μ represents the quantum fluctuations of the gauge field. On the other hand, the background field B_μ plays the role of a classical background, (this approach is quite relevant in the case of the Polyakov loop computation [27]). The gauge symmetry (2) changes with this background field as

$$A_\mu + B_\mu \rightarrow A'_\mu + B'_\mu = U^{-1}\partial_\mu U + U^{-1}(A_\mu + B_\mu)U \ . \tag{10}$$

Although it is not mandatory, in many applications (such as the already mentioned case of the Polyakov loop) it is convenient to demand that the background gauge field is fixed (namely, it does not transform under gauge transformations (2)). Consequently, it follows the natural requirement

$$\delta B^a_\mu = 0 \ .$$

In this case, the symmetry transformation in Eq. (10) can be written as

$$A_\mu \rightarrow A^U_\mu = U^{-1}\partial_\mu U + U^{-1}A_\mu U + \left(U^{-1}B_\mu U - B_\mu\right), \tag{11}$$

where it has been explicitly taken into account that B_μ is not affected by the gauge transformation. At the infinitesimal level, $U \approx \mathbb{I} + \omega^a \tau_a$, $\omega \ll 1$, one recovers the usual infinitesimal gauge transformations with a background gauge field [4, 42]

$$\delta A^a_\mu = f^{abc}\omega^b(A^c_\mu + B^c_\mu) + \frac{1}{g}\partial_\mu \omega^a, \tag{12}$$

$$\delta B^a_\mu = 0. \tag{13}$$

Correspondingly, the Landau gauge-fixing condition is also modified. In the presence of a background field, the most convenient gauge-fixing condition takes the form

$$\tilde{G}^a_\mu[B] \equiv \overline{D}^{ab}_\mu A^b_\mu = 0, \quad \overline{D}^{ab}_\mu := \partial_\mu \delta^{ab} + g f^{acb} B^c_\mu \,, \tag{14}$$

known as the Landau-DeWitt (LDW) gauge fixing condition. The GZ method can be applied to this situation by means a suitable choose of the background field B_μ in order to the new FP operator $\overline{\mathcal{M}}^{ac} \equiv -\overline{D}^{ab}_\mu(B) D^{bc}_\mu(a)$ is invertible inside the Gribov region Ω. Following the lines of [43] and [27], the GZ action under the LDW gauge acquires the form

$$S_{GZ} = \int d^4x \left(\frac{1}{4} F^a_{\mu\nu} F^{\mu\nu a} + \bar{c}^a \overline{D}_\mu(B) D_\mu(a) c^a - \frac{\left(\overline{D}_\mu(B) A_\mu\right)^2}{2\xi} + \bar{\varphi}^{ac}_\mu \overline{D}^{ab}_\sigma(B) D^{bd}_\sigma(a) \varphi^{dc}_\mu \right.$$
$$\left. - \bar{\omega}^{ac}_\mu \overline{D}^{ab}_\nu(B) D^{bd}_\nu(a) \omega^{dc}_\mu - g\gamma^2 f^{abc} A^a_\mu(\varphi^{bc}_\mu + \bar{\varphi}^{bc}_\mu) - \gamma^4 d(N^2-1) \right). \tag{15}$$

There are two key mathematical requirements (described in the references [2, 11]) necessary in order to write down a local GZ action in the presence of a background field. The first one corresponds to the condition that the FP operator at tree-level must be invertible. The second key requirement is the validity of the *Dell'Antonio-Zwanziger* theorem [9], which also could change in the presence of a background, but in the case of the LDW gauge provided the background field is constant and commutes with itself, and then the theorem remains valid.

4. The simplest non-trivial background field

In order to describe the effects of a background field, we will consider the simplest non-trivial background gauge field

$$B^a_\mu = \frac{r_0}{g}\delta^{a3}\delta_{\mu 0} \,. \tag{16}$$

Here we have chosen the above background gauge field with only Euclidean time component since it allows to construct explicitly analytic examples of Gribov copies as well as to solve the semi-classical Gribov gap equations (which, quite consistently, shows that the Gribov mass decreases with the increase of r_0).

In this case, the LDW gauge fixing reads

$$\tilde{G}^a_\mu[B] = 0 \,, \tag{17}$$

so that the Gribov copies equation becomes

$$\partial^\mu A^U_\mu + g[B^\mu, A^U_\mu] = 0 \,, \tag{18}$$

where A^U_μ is defined in Eq. (11). It is worth emphasizing that the background gauge field B^μ identically satisfies the LDW gauge-fixing: $\bar{D}^{ab}_\mu B^b_\mu = 0$.

IOP Conf. Series: Journal of Physics: Conf. Series **1043** (2018) 012035 doi:10.1088/1742-6596/1043/1/012035

The following standard parametrization of the $SU(2)$-valued functions $U(x^i)$ is useful

$$U = Y^0\mathbf{1} + Y^a\tau_a, \quad (Y^0)^2 + Y^aY_a = 1,$$

where Y^0 and Y^a are functions on the coordinates x^i, and the sum over repeated indices is understood also in the case of the group indices (in which case the indices are raised and lowered with the flat metric δ_{ab}). The $SU(2)$ generators τ^a satisfy $\tau_a\tau_b = -\delta_{ab}\mathbf{1} - \epsilon_{abc}\tau_c$, where $\mathbf{1}$ is the identity 2×2 matrix and ϵ_{abc} are the components of the totally antisymmetric Levi-Civita tensor with $\epsilon^{123} = \epsilon_{123} = 1$.

4.1. Gribov copies of the vacuum

In the present case, the gauge transformations of the vacuum have the expression (see Eq. (11))

$$0 \to U^{-1}\partial_\mu U + \left(U^{-1}B_\mu U - B_\mu\right) .$$

Correspondingly, the equation for the Gribov copies of the vacuum in the presence of a background field reads

$$\partial^\mu\left(U^{-1}\partial_\mu U + \left(U^{-1}B_\mu U - B_\mu\right)\right) + g[B^\mu, U^{-1}\partial_\mu U + \left(U^{-1}B_\mu U - B_\mu\right)] = 0 . \tag{19}$$

This, actually, is a system of coupled non-linear partial differential equations. In order to reduce it consistently to a single differential equation a particular hedgehog ansatz can be used [16] (see Appendix A for the details on the vacuum case). This corresponds to the following ansatz for the gauge copy

$$U = Y^0(x^i)\mathbf{1} + Y^a(x^i)\tau_a, \ , Y^0(x^i) = \cos\alpha(x^i), \quad Y^a(x^i) = \hat{n}^a\sin\alpha(x^i) \tag{20}$$

being $\hat{n}_a$ normalized with respect to the internal metric δ_{ab} as

$$\delta_{ab}\hat{n}^a\hat{n}^b = 1 . \tag{21}$$

4.2. Vacuum Gribov copies with T^3 topology

Let us analyze the Gribov copies equation in a flat spatial space with T^3-topology. Such choice of topology can be very useful in relation with lattice studies [44, 45]. We take the metric

$$ds^2 = \sum_{i=1}^{i=3}\lambda_i^2 d^2\phi_i, \tag{22}$$

where the $\lambda_i \in \mathbb{R}$ represents the length of the torus along the i-axis and the coordinates $\phi_i \in [0, 2\pi)$ corresponds to the $i-$th factor S^1 in T^3. In the T^3 case, the gauge transformation U is independent of the Euclidean temporal coordinate x^0 and is *proper* when [16]

$$U(\phi_i + 2m_i\pi) = U(\phi_i), \quad m_i \in \mathbb{Z}, \quad i = 1, 2, 3. \tag{23}$$

The generalized hedgehog ansatz adapted to this topology reads

$$\alpha = \alpha(\phi_1), \quad \hat{n}^1 = \cos(p\phi_2 + q\phi_3), \quad \hat{n}^2 = \sin(p\phi_2 + q\phi_3), \quad \hat{n}^3 = 0, \tag{24}$$

with p, q arbitrary integers. From this ansatz, the equation (A.1) is reduce to the following single scalar non-linear differential equation (see Appendix A for details),

$$\frac{d^2\alpha}{d\phi_1^2} = \xi\sin(2\alpha), \quad \xi = \frac{\lambda_1^2}{2}\left(\frac{p^2}{\lambda_2^2} + \frac{q^2}{\lambda_3^2} + 4\frac{r_0^2}{g}\right), \tag{25}$$

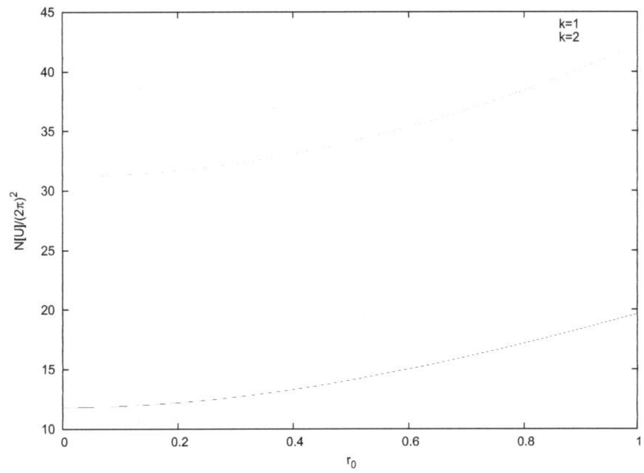

Figure 1. The norm of the copies $\frac{N[U]}{(2\pi)^2}$, according to (29), in the case $p = q = \lambda_i = 1$ versus the background r_0 for $k = 1$ (in red) and $k = 2$ (in green). The solutions of $\alpha(\phi_1)$ fulfil the condition (26).

and, according to (23), the condition

$$\alpha(\phi_1 + 2\pi) = \alpha(\phi_1) + 2\pi k, \quad k \in \mathbb{Z} \tag{26}$$

must be fulfilled. The equation (25) can be reduced to a first order conservation law

$$V = \frac{1}{2}\left[\left(\frac{d\alpha}{d\phi_1}\right)^2 + \xi\cos(2\alpha)\right] \Rightarrow \phi_1 - \phi_0 = \pm\int_{\alpha(\phi_0)}^{\alpha(\phi_1)} \frac{dy}{\sqrt{2V - \xi\cos(2y)}}, \tag{27}$$

where ϕ_0 and V are integration constants.

The most enlightening way to see the background's effect on the Gribov copies is by means its norm. In fact, as it is well known [46], the weight of a given copy U is related to its norm

$$N\left[U\right] = \int_{T^3} d^4x\sqrt{g}Tr\left[\left(U^{-1}\partial_\mu U + U^{-1}B_\mu U - B_\mu\right)^2\right], \tag{28}$$

where in this case g refers to the determinant of the metric associated to the line element (22), setting the coupling constant to be zero. In particular, the bigger is $N[U]$, the less relevant the copy is from the path integral point of view. As in this case there is a background potential, the integral (28) can be written as

$$N\left[U\right] = \frac{(2\pi)^2\lambda_2\lambda_3}{\lambda_1}\int_0^{2\pi} d\phi_1\left(2V + 3\xi\sin^2\alpha(\phi_1) - \xi\cos^2\alpha(\phi_1)\right), \tag{29}$$

where in the last equality we used the definition (27) of the constant V. In Figure 1, we show the norm $N[U]$ for $p = q = \lambda_i = 1$ increases when r_0 grows both for $k = 1$ and $k = 2$, at least in the range $r_0 \in (0.0, 1.0)$, for solutions $\alpha(\phi_1)$ such that fulfil the condition (26) and $\alpha(0) = 0$. Consequently, in this region, the bigger r_0 the smaller the importance of Gribov copies of the form considered here. It is necessary more computational power to see how is the behavior of the norm outside the region studied here (for instance, $|p| > 1$ and $|q| > 1|$).

5. Solving the GZ gap equation for $SU(2)$ with constant background field

In order to determine the gap equation, we will proceed first to show the effective potential to GZ action at one-loop approximation for the $SU(2)$ internal gauge group in the presence of a background potential discussed in Section 3. We will work at a small enough but non-zero temperature, taking into account the background field as

$$B^a_\rho = \frac{T}{g} r \delta^{a3} \delta_{\rho 0}, \tag{30}$$

where r is a dimensionless parameter related to the background field r_0 defined in (16) as $r = r_0/T$, with T the temperature. At a first glance, the above background field should have no physical at all as it is pure gauge. However, the gauge transformation which would remove it is not periodic in Euclidean time. Thus, such a gauge transformation is improper and it is not allowed. The present framework, despite its simplicity, it is still able to disclose in a very clean way the effects of the background gauge potential on the Gribov parameter. We will focus only on the dependence of the Gribov parameter with respect to the background parameter, keeping the temperature constant.

In order to obtain the vacuum energy at one loop, we consider only from (15) the quadratic terms in the fields which are functionally integrated (see (31) for higher order corrections). We find [27]

$$\varepsilon(r,\lambda^2) = -\frac{d(N^2-1)}{2Ng^2}\lambda^4 + \frac{T}{2V}(d-1)\mathrm{Tr}\ln\frac{\overline{D}^4+\lambda^4}{\Lambda^4} - d\frac{T}{2V}\mathrm{Tr}\ln\frac{-\overline{D}^2}{\Lambda^2}, \tag{31}$$

where V is the Euclidean space volume, $\lambda^4 = 2Ng^2\gamma^4$, being γ is the Gribov parameter, $\overline{D}$ is the covariant background derivative in the adjoint representation defined in (14), and Λ^2 is a scale parameter in order to regularize the result. We can rewrite (31), taking into account the Cartan subalgebra of $SU(2)$ is one-dimensional and the thermodynamic limit $V \to \infty$, as

$$\varepsilon(r,\lambda^2) = -\frac{d(N^2-1)}{2Ng^2}\lambda^4 + \frac{1}{2}(d-1)\sum_{s=-1}^{s=1}\left[I(sr,i\lambda^2)+I(sr,-i\lambda^2)\right] - \frac{d}{2}\sum_{s=-1}^{s=1} I(sr,0), \tag{32}$$

where s is the isospin $SU(2)$, and we defined the function

$$I(u,m^2) = \frac{T}{V}\mathrm{Tr}\ln\left(\frac{-\overline{D}^2+m^2}{\Lambda^2}\right) = T\sum_{n=-\infty}^{+\infty}\int\frac{d^3q}{(2\pi)^3}\ln\left[\frac{(2\pi nT+uT)^2+\vec{q}^2+m^2}{\Lambda^2}\right]. \tag{33}$$

In the last definition, we expanded in the Fourier space the zero-component momentum in the Matsubara bosonic frequencies $2\pi nT$ [47, 48], and $\vec{q}$ denotes the spatial momentum vector. We will compute first (33) using similar techniques which were already applied in GZ approach [27, 49], which lead us the result (see details in [27])

$$I(u,m^2) = \frac{m^4}{32\pi^2}\left[\ln\left(\frac{m^2}{\Lambda^2}\right)-\frac{3}{2}\right] - \frac{T^2m^2}{\pi^2}\sum_{n=1}^{+\infty}K_2\left(\frac{n\sqrt{m}}{T}\right)\frac{\cos(nu)}{n^2}. \tag{34}$$

where K_ν is the modified Bessel function of the second kind extended to the complex plane [50]. Inserting (34) into (32), we have

$$\begin{aligned}\varepsilon(r,\lambda^2) \;=\; & -\frac{d(N^2-1)\lambda^4}{2Ng^2} - 3\frac{(d-1)\lambda^4}{32\pi^2}\left[\ln\left(\frac{\lambda^2}{\Lambda^2}\right)-\frac{3}{2}\right] \\ & -\frac{i\lambda^2T^2}{2}(d-1)\sum_{s=-1}^{s=1}\sum_{n=1}^{+\infty}\left[K_2\left(\frac{n\sqrt{i\lambda^2}}{T}\right)-K_2\left(\frac{n\sqrt{-i\lambda^2}}{T}\right)\right]\frac{\cos(nrs)}{n^2} - \frac{d}{2}\sum_{s=-1}^{s=1}I(rs,0)\,.\end{aligned} \tag{35}$$

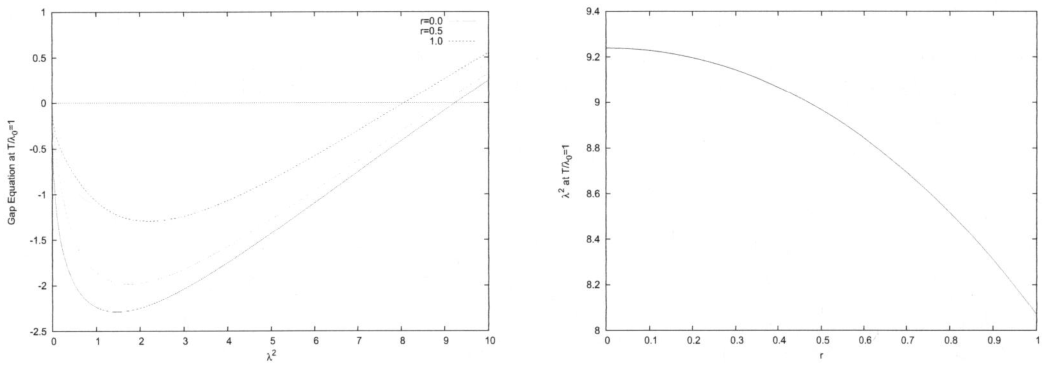

(a) (b)

Figure 2. (a) The gap equation (36) as a function of λ^2 for different values of r at $T/\lambda_0 = 1$. The value of λ which corresponds the curve intersects the y- axis is the solution of (36). (b) The zeros of the gap equation (36) as a function of the background field r at $T/\lambda_0 = 1$. We see clearly the Gribov mass parameter decrease when r in the range $[0, 1]$.

Because we are interested in solving the gap equation (8), we can neglect the last term of (35), as it does not depend on λ^2. In order to normalize the last equation, we shall choose Λ^2 such that for $T = 0$ the solution is $\lambda_0 = 1$. Now, one can write the gap equation (8) in the following way

$$\frac{\partial \varepsilon_{T\neq 0}(r,\lambda^2)}{\partial \lambda^2} - 3\frac{(d-1)\lambda^2}{16\pi^2}\ln\left(\frac{\lambda^2}{\lambda_0^2}\right) = 0\,, \qquad (36)$$

where

$$\varepsilon_{T\neq 0}(r,\lambda^2) = -\frac{i\lambda^2}{2}T^2(d-1)\sum_{n=1}^{+\infty}\frac{\left[K_2\left(\frac{n\sqrt{i\lambda^2}}{T}\right) - K_2\left(\frac{n\sqrt{-i\lambda^2}}{T}\right)\right]}{n^2}\left[1 + 2\cos(nr)\right]\,.$$

The gap equation (36) can be solved using numerical techniques. In Figure 2 (a), it is plotted the left hand side of the gap equation (36) as a function of λ^2 for different values of r. We see the intersection value of the curve (which is the solution for a given value of r) decrease when the background r grows, as it is shown more clearly in Figure 2 (b), where it is shown the parameters λ which are solution of gap equation at $T/\lambda_0 = 1$ versus r. We could interpret this as the theory becomes less confined as the Gribov parameter reduces (see Section 6) at least in the range $r \in [0, 1]$. The present expression of the Gribov parameter is only valid as long as $\frac{\partial\lambda^2}{\partial r} \neq 0$. The points where the latter derivative vanishes could signal a change on the phase diagram. Thus, when $\frac{\partial\lambda^2}{\partial r} \neq 0$, the present semi-classical approximation is not valid anymore. Therefore, we have included the plots in Figure 2 only the region in which our approximation can be trusted.

6. Conclusions and perspectives

In the present paper, it has been shown that the Gribov copies equation is affected directly by the presence of a background gauge field. In particular, explicit examples have been constructed

IOP Conf. Series: Journal of Physics: Conf. Series **1043** (2018) 012035 doi:10.1088/1742-6596/1043/1/012035

in which the norm of the Gribov copies satisfying the usual boundary conditions increases when the size of the background field is very large.

The analysis of the semi-classical Gribov gap equation in the chosen background gauge potential and of the dependence of the Gribov mass on the background potential itself, quite consistently, confirms the above results. Namely, we have shown that the larger is the size of the background gauge potential, the smaller is the corresponding Gribov mass.

Moreover, constant background gauge potentials are very important also in relation with the non-Abelian Schwinger effect [31, 32]. Although the constant gauge potentials considered in that references allow a complete analysis of the semi-classical Gribov gap equation, they make extremely difficult to construct explicit examples of Gribov copies.

Acknowledgements
We wish also thank David Dudal and David Vercauteren for pointing us out some mistakes in our previous version of this manuscript. This work is supported by Fondecyt Grant 1160137. P.P. was partially supported by Fondecyt Grant 1140155 and also thanks the Faculty of Mathematics and Physics of Charles University in Prague, Czech Republic, for the kind hospitality during part of the development of this work. The Centro de Estudios Científicos (CECs) is funded by the Chilean Government through the Centers of Excellence Base Financing Program of Conicyt.

Appendix A. Gribov copies with a constant background field

In this appendix we consider the derivations and properties of the equation of gauge-equivalent fields satisfying the LDW gauge in the presence of background field. Our aim is to calculate the condition for existence of Gribov copies in the vacuum. Thus, we must compute the following expression

$$
\begin{aligned}
U^{-1}\partial_\mu U + U^{-1}B_\mu U - B_\mu &= \left(Y^0\partial_\mu Y^c - Y^c\partial_\mu Y^0 + \epsilon_{abc}Y_a\partial_\mu Y_b\right)\tau_c \\
&\quad - \frac{2r_0}{g}\delta_{\mu 0}\left(\epsilon_{a3c}Y^aY^0 + Y^3Y^c\right)\tau_c + \frac{2r_0}{g}\delta_{\mu 0}Y^aY_a\tau_3 .
\end{aligned}
$$

The next step is to apply to this last expression the covariant background derivative and set it to be zero according to (18). This results in the following expression

$$
\begin{aligned}
&\left(-Y^0\Box Y^c - Y^c\Box Y^0\epsilon_{abc}Y^a\Box Y^b\right)\tau_c - \frac{2r_0}{g}\left(Y^c\dot{Y}^3 + \dot{Y}^cY^3 + \epsilon_{a3c}\left[Y^0\dot{Y}^a + \dot{Y}^0Y^a\right]\right)\tau_c \\
&\left[-2r_0\epsilon_{3bc}\left(Y^0\dot{Y}^b - \dot{Y}^0Y^b - \dot{Y}^cY^3 + Y^c\dot{Y}^3\right) - 4\frac{r_0^2}{g}\left(Y^0Y^c + \epsilon_{3bc}Y^3Y^b\right)\right]\tau_c + \frac{4r_0}{g}\dot{Y}^aY_a\tau_3 = 0 \qquad \text{(A.1)}
\end{aligned}
$$

where $\Box(\ldots) = \partial_\mu\partial^\mu(\ldots)$, and de dot represents the derivative with respect to the component which the background field belongs. In the particular case of flat spatial space T^3 (22) for the Y^μ prescription (20), and for the hedgehog ansatz (24) the set of equations (A.1) reduces to the following ordinary differential equation

$$
\frac{d^2\alpha}{d\phi_1^2} - \frac{\beta(p,q)}{2}\sin(2\alpha) - \frac{2r_0^2\lambda_1^2}{g}\sin(2\alpha) = 0, \quad \beta(p,q) = \lambda_1^2\left(\frac{p^2}{\lambda_2^2} + \frac{q^2}{\lambda_3^2}\right).
$$

References

[1] L. D. Faddeev, V. N. Popov, *Phys. Lett.* **B 25**, 29 (1967).
[2] V. N. Gribov, Nucl. Phys. B **139** (1978) 1.
[3] I. Singer, Comm. Math. Phys. 60 (1978), 7.
[4] B. DeWitt, "The global approach to Quantum Field Theory", Volume **I** and **II**, Oxford University Press (2003).
[5] G. Esposito, D. Pelliccia, F. Zaccaria, *Int. J. Geom. Meth. Mod. Phys.* **1** (2004) 423-441.

[6] P. van Baal, *Nucl. Phys.* **B 369** (1992), 259.
[7] M. Maggiore, M. Schaden, *Phys. Rev.* **D 50** (1994) 6616.
[8] J. A. Gracey, *JHEP* **0605** (2006) 052.
[9] G. F. Dell'Antonio, D. Zwanziger, *Nucl. Phys.* **B 326**, (1989) 333; *Comm. Math. Phys.* **138**, 291-299 (1991).
[10] R. F. Sobreiro, S. P. Sorella, hep-th/0504095.
[11] D. Zwanziger, *Nucl. Phys.* **B 209** (1982) 336; *Nucl. Phys.* **B 321**, (1989) 591; *Nucl. Phys.* **B 323** (1989), 513.
[12] D. Zwanziger, *Nucl. Phys.* **B 399** (1993), 477; *Nucl. Phys.* **B 518** (1998) 237; *Phys. Rev. Lett.* **90** (2003) 102001.
[13] F. Canfora, A. Giacomini and J. Oliva, *Phys. Rev.* **D 82** (2010) 045014.
[14] A. Anabalon, F. Canfora, A. Giacomini and J. Oliva, *Phys. Rev.* **D 83** (2011) 064023.
[15] F. Canfora, A. Giacomini and J. Oliva, *Phys. Rev.* **D 84** (2011) 105019.
[16] F. Canfora and P. Salgado-Rebolledo, *Phys. Rev.* **D 87** (2013) 4, 045023.
[17] M. de Cesare, G. Esposito and H. Ghorbani, *Phys. Rev.* **D 88** (2013) 087701.
[18] F. Canfora, A. Giacomini, A. Zerwekh, to appear on *Nucl. Phys.* **B**. [arXiv:1610.02886].
[19] D. Dudal, M.S. Guimaraes, S.P. Sorella, *Phys. Rev. Lett.* **106**, 062003 (2011).
[20] A. Cucchieri, A. Maas, T. Mendes, *Phys. Rev.* **D75**, 076003 (2007), hep-lat/0702022.
[21] F. Canfora, L. Rosa, *Phys. Rev.* **D 88** (2013) 045025.
[22] D. Zwanziger, *Phys. Rev.* **D 76**, (2007) 12504.
[23] F. Canfora, P. Pais and P. Salgado-Rebolledo, *Eur. Phys. J.* **C 74** (2014) 2855.
[24] P. Cooper, D. Zwanziger, *Phys. Rev.* **D 93** (2016) 105026.
[25] F. Canfora, A. Gomez, S. P. Sorella, D. Vercauteren, *Annals of Physics* **345**, (2014) 166.
[26] R. Benguria, M. Loewe, R. A. Santos, arXiv:0904.3364, 2009.
[27] F. Canfora, D. Dudal, I. Justo, P. Pais, L. Rosa, V. Vercauteren, *Eur. Phys. J. C.* **75**, 326 (2015).
[28] J. Fingber, U. M. Heller, F. Karsch, *Nucl. Phys.* B **392**, 493 (1993).
[29] J. Kapusta, *Phys. Rev. D* **24** (1981) 426.
[30] J. Schwinger, *Phys. Rev.* **82** (1951) 664.
[31] L. Brown, W. Weisberger, *Nucl. Phys. B* **157** (1979) 285.
[32] W. Weisberger, *Nucl. Phys. B* **161** (1979) 61.
[33] P. Arnold, *Int. J. Mod. Phys. E* **16** (2007) 2555 [arXiv:0708.0812 [hep-ph]]. U. W. Heinz, *Nucl. Phys. A* **721** (2003) 30 [arXiv:nucl-th/0212004]. L. McLerran, *Nucl. Phys. A* **787** (2007) 1.
[34] K. Yagi, T. Hatsuda, Y. Miake, *Quark-Gluon Plasma: From Big Bang to Little Bang* , Cambridge Monographs on Particle Physics, Nuclear Physics and Cosmology (No. 23), October, 2008.
[35] B. C. Barrois, *Nucl. Phys. B* **129** (1977) 390.
[36] F. Weber, Prog. Part. *Nucl. Phys.* **54** (2005) 193; D. H. Rischke, *Prog. Part. Nucl. Phys.* **52** (2004)197 (2004); M. Alford, *Prog. Theor. Phys. Suppl.* **153** (2004) 1; M. Buballa, *Phys.Rept.* **407** (2005) 205; I. Shovkovy, *Found. Phys.* **35** (2005) 1309.
[37] A. D. Linde, *Rept. Prog. Phys.* **42** (1979) 389.
[38] V.A. Matveev, V. A. Rubakov, A. N. Tavkhelidze, M. E. Shaposhnikov, *Sov. Phys. Usp.* **31** (1988), Usp. Fiz. Nauk, **156** (1988) 253; V. A. Rubakov, M. E. Shaposhnikov, *Phys. Usp.* **39** (1996) 461, Usp. Fiz. Nauk 166 (1996) 493.
[39] B. S. DeWitt, *Phys. Rev.* **162** (1967) 1195; *Dynamical Theory of Groups and Fields (Gordon and Breach, New York, 1965); in Quantum Gravity 2,* ed. C. J. Isham, R. Penrose and D. W. Sciama (Oxford University Press, New York, 1981), pp. 449;
[40] G. 't Hooft, Acta Universitatis Wratislavensis No 38 XII, Winter School of Theoretical Physics in Karpacz; *Functional and Probabilistic Methods in Quantum Field Theory Vol.* 1 (1975).
[41] L. F. Abbott, Acta Phys. Polon, *Introduction to the Background Field Method* **B** 13 (1982) 33. http://www.neurotheory.columbia.edu/Larry/AbbottActPhysPol82.pdf
[42] S. Weinberg: *The Quantum Theory of fields v.2*, (Cambridge University Press., 1996), pp. 95-109.
[43] V. Vandersickel, D. Zwanziger, *Phys. Rep.* **520** (2012) 175-251.
[44] M. Luscher, *Phys. Lett. l18* B (1982) **391**; Nucl. Phys. B **219** (1983) 233.
[45] J. B. Kogut, M. A. Stephanov, *The Phases of Quantum Chromodynamics*, Cambridge University Press (2004).
[46] M. Guimaraes, S. Sorella, *J. Math. Phys.* **52**, 092302 (2011).
[47] M. Le Bellac, "*Thermal Field Theory*", Cambridge University Press, Cambridge, 2000.
[48] J. I. Kapusta, C. Gale, "*Finite-Temperature Field Theory: Principles and Applications*", Cambridge University Press, Cambridge, 2006.
F
[49] F. Canfora, A. Giacomini, P. Pais, L. Rosa, A. Zerwekh, *Eur. Phys. J.* **C 76** (2016):443. [arXiv:1606:02271].
[50] L. V. Ahlfors, *Complex Analysis*, McGraw-Hill, Inc. (1979).

IOP Conf. Series: Journal of Physics: Conf. Series **1043** (2018) 012036 doi:10.1088/1742-6596/1043/1/012036

Optical and dielectric properties of β-Ta_2O_5

C. Valencia-Balvín[1], S. Orozco[2], J. M. Osorio-Guillén[3], S. Pérez-Walton[1]

[1] Facultad de Ingeniería, Instituto Tecnológico Metropolitano ITM, Calle 73 No 76A–354 Vía al Volador, Medellín, Colombia

[2] Facultad de Ciencias Exactas y Aplicadas, Instituto Tecnológico Metropolitano ITM, Calle 73 No 76A–354 Vía al Volador, Medellín, Colombia

[3] Instituto de Física, Universidad de Antioquia UdeA, Calle 70 No 52–21, Medellín, Colombia

E-mail: camilovalencia@itm.edu.co

Abstract. Tantalia (Ta_2O_5) is an interesting material which is used in several technological applications. In particular, this wide gap semiconductor is used as a new material for memristive devices, which could provide a paradigm shift on the memory devices fabrication. These new devices would be faster, denser and less power consuming than those available today. Because of this, we present a study of the optical and dielectric properties of β-Ta_2O_5 polymorph by means of first-principles Density Functional Theory calculations. Taking as a starting point the Nashed's model of the crystal structure of β-Ta_2O_5 and the generalized gradient approximation (GGA-PBEsol), we calculated the equation of state (EOS) of this polymorph. From EOS we identify the equilibrium volume and for this we calculated the frequency dependent dielectric tensor, refractive index ($n = 2.63$), absorption coefficient and transmittance. Also, we calculated the electronic ($\varepsilon_\infty = 6.98$) and ionic contribution ($\varepsilon_0 = 43.51$) to static dielectric constant and it was compared with some reported values in the literature.

1. Introduction

Ta_2O_5 is a wide gap semiconductor whose main characteristic is its high dielectric constant, which makes possible that this compound can be used in a variety of technological applications like memristor [1], anti-reflective films in solar cells [2] and water splitting processes [3]. For several years, numerous experimental studies have reported a set of polymorphs for Ta_2O_5, finding crystal structures at low and high temperature and high pressures. Mainly, at low temperature it has been reported an orthorhombic structure without oxygen vacancies, the β-phase, for which has been reported several experimental refinements [4]. Furthermore, for this structure has been proposed different theoretical models [5, 6, 7] to describe adequately the structural, electronic and optical properties of this phase. One such model is the one proposed by Nashed et al. [8], in which an orthorhombic structure described by the space group $Pmmm$ with two tantalum (Ta) and five oxygen (O) atoms in the unit cell is proposed, where local structures consist of chains of octahedral linked by the corners. In this work, we have taken this model as a reference to study the structural, electronic and optical properties of this compound in the β-phase and we had compared our results with existing experimental data and theoretical results obtained in other studies.

2. Simulation details

In this study, we have calculated the total energy as a function of the unit cell volume (E=E(Ω)), by fully relaxing the ionic positions and the shape of the unit cell for each calculated volume. Then, the obtained energies were fitted to a polynomial Eulerian strain Birch-Murnaghan equation of state (EOS) of order 4 [9, 10], where we extracted the equilibrium volume (Ω_O). All total energies, ionic forces and stress tensor components have been calculated via Density Functional Theory (DFT), where we have used for the exchange-correlation functional the generalized gradient approximation (GGA) with the PBEsol parameterization [11]. To solve the Kohn-Sham equations, we have used the projector augmented wave (PAW) method [12] as implemented in the VASP code [13, 14]. The PAW atomic reference configurations are: $5p^6\ 5d^4\ 6s^1$ for Ta and $2s^2\ 2p^4$ for O, where only electrons treated as valence electrons are explicitly enumerated. The energy cut-off in the planes waves expansion is 520 eV. Brillouin-zone integration has been performed on a Monkhorst-Pack 4×8×8 k-mesh with a Gaussian broadening of 0.01 eV for relaxation (ionic forces are converged to 0.1 meV/Å) and a Γ-centered 4×8×8 k-mesh using the tetrahedron method including Blöchl corrections [15] for the calculation of the total energy, the charge density and the optical properties. Furthermore, for electronic structure calculation we have also used the HSE06 hybrid functional [16, 17], which permitted us to obtain a better gap estimation. The calculation of the optical properties associated with the interband transitions are done by means of the imaginary part of the frequency dependent dielectric tensor ($\varepsilon^{(2)}(\vec{q},\omega)$) in the long-wavelength limit ($\vec{q} \to 0$), using the methodology developed for the PAW method [18]. The real part of the frequency dependent dielectric tensor is obtained from the imaginary part through the Kramers-Kröning relationship. From the knowledge of the dielectric tensor, we have calculated the refractive index $n(\omega)$, the extinction coefficient $\kappa(\omega)$, the absorption coefficient $\alpha(\omega)$ and the transmittance, using explicitly the equations listed in the reference [19]. Finally, we have also calculated the low-frequency dielectric tensor (ε_0) through density functional perturbation theory [20].

3. Results

3.1. Crystal structure

The crystal structure of β-Ta_2O_5 is formed from a set of TaO_6 octahedra where the Ta atom is in the octahedron center and O atoms are at the corners, where there is a re-orienting of the conventional structure [4] of β-Ta_2O_5 such that the b-vector becomes along the z-axis and the c-vector is in the $x-y$ plane [8], halving the number of atoms in the unit cell. In this new model the octahedral share corners as well (Figure 1).

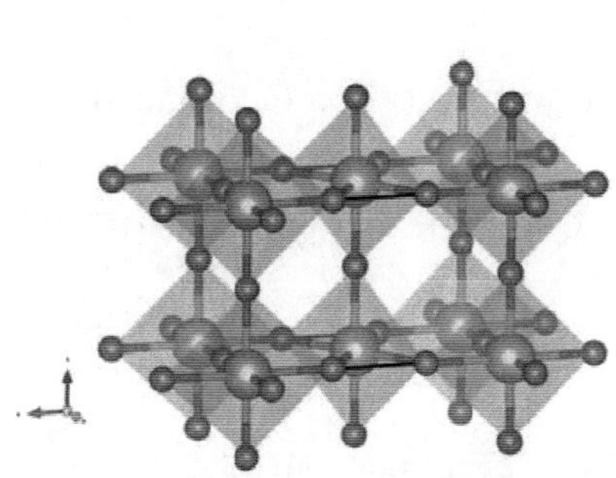

Figure 1. Calculated equilibrium crystal structure of β-Ta_2O_5. Ta and O atoms are represented by green and blue spheres, respectively.

In this model, β-Ta_2O_5 is an orthorhombic structure described by the space group $Pmmm$, with one formula unit per unit cell where two Ta ions are located at the Wyckoff positions: $1a$

(Ta_1) and $1f$ (Ta_2), while five O ions are distributed over the four Wyckoff positions: $1c$ (O_1), $2i$ (O_2), $1e$ (O_3) and $1h$ (O_4). In addition, we have found that the orthorhombic lattice has been conserved during the stress tensor optimization (unit cell vectors relaxation) for all the fixed volumes, there is no tilting on any direction ($\angle$ O-Ta-O $= 180^{\circ}$) and all Ta ions remain at the center of the octahedral.

3.2. Equation of state

The equation of state for β-Ta_2O_5 was obtained fitting the energies as a function to volume using the Birch-Murnaghan strain curve given by a fourth-order Eulerian polynomial, where the EOS is shown in the Figure 2. From EOS, we obtained the equilibrium geometry with a lattice parameters $a = 6.4489$ Å, $b = 3.6753$ Å and $c = 3.8670$ Å, values which are in good agreement with the parameters reported in the references [8] and they are comparable with values reported for low-temperature Ta_2O_5 structure [21], when the standard β-Ta_2O_5 is reconstructed from our calculated structure using a 1×2×1 supercell.

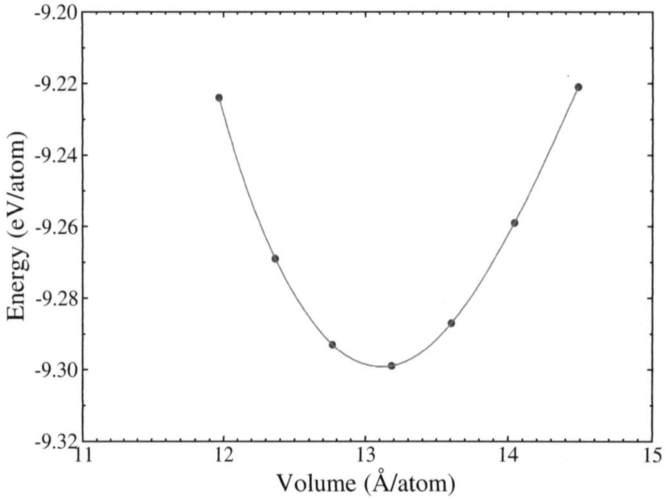

Figure 2. Calculated total energy per atom versus volume per atom for β-Ta_2O_5. GGA-PBEsol total energies (denoted by the filled symbols) are fitted using the methodology described in [9, 10].

For this structure, the cohesive energy is calculated using the equation:

$$E_c = \frac{(E_{\beta-Ta_2O_5})^{bulk}}{n} - \frac{(pE_{Ta}^{atom} - qE_{O}^{atom})}{n}, \tag{1}$$

where E_c is the cohesive energy, n is the total number of atoms, p and q are the number of Ta and O atoms, respectively. In this case, we obtained a value of -7.34 eV, which at being compared with the theoretical value reported by Wu et al. [22], we find that there is a relative error of 1%, evidencing that our estimate for the equilibrium geometry is good agreement with the Wu et al. calculation.

3.3. Electronic structure

The calculated dispersion relation for β-Ta_2O_5 is presented in Figure 3. The dispersion relation shows that β phase is an indirect wide gap semiconductor. The calculated energy gap with

PBEsol (HSE06) is 0.25 eV (1.54 eV), which we have compared with the experimental value of the gap for the orthorhombic low-temperature phase and others reported theoretical values in the Table 1. In this table, we can see that the calculated gap with HSE06 is in good agreement with calculated bandgap with other works for β-Ta_2O_5. Furthermore, we found that the valence band maximum (VBM) and the conduction band minimum (CBM) are located at R = (1/2, 1/2, 1/2) and Y = (0, 1/2, 0), respectively; as it can be observed in Figure 3.

Table 1. Calculated indirect and direct gaps for β-Ta_2O_5.

Functional	ΔE (eV)	Γ	X	R	T	Reference
PBEsol*	0.25	1.06	1.52	0.8	0.44	This work.
PBE*	1.45					[8]
PBE†	0.230					[23]
	0.2					[24]
HSE06*	1.54	2.61	2.94	2.36	1.96	This work.
HSE06†	0.9					[22]

* Nashed et al model
† Aleshina et al. structure

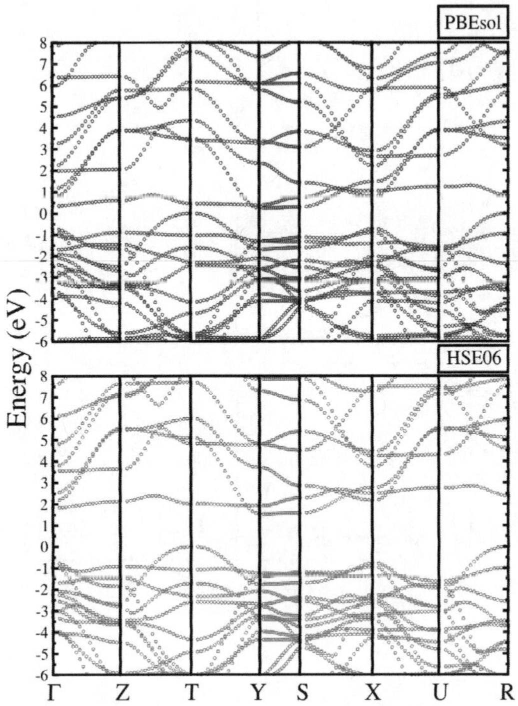

Figure 3. Band structure for β-Ta_2O_5 calculated with GGA-PBEsol and HSE06 functionals. The blue points represent GGA-PBEsol calculation and red points, represent HSE06 calculation.

A detailed analysis of the dispersion relation allows us to identify that the valence band is the result of the hybridization of O p-states and Ta d-states, and the conduction band is constructed mainly of Ta d-states and a small contribution of O p-states, where the character of the VBM is dominated by O p-states, whereas the CBM is dominate by Ta d-states. In the case of VBM, contributions are distributed as follows: O_1 p_y-states and O_3 p_z-states, while in the CBM the distribution of states is Ta_2 d_{xz}-states. Additionally, we can observe that HSE06 calculation produce a rigid shift of the conduction band (CB) to higher energies respect to VBM, where the topology of bands is basically the same.

3.4. Dielectric tensor

In the case of an orthorhombic crystal, the dielectric tensor has three components different from zero: the three components of the diagonal (ε_{xx}, ε_{yy} and ε_{zz}). In the Figure 4 (a-c), we show the calculated diagonal components for imaginary part of dielectric tensor $\varepsilon^{(2)}_{\alpha\beta}$ of β-Ta_2O_5 with PBEsol and HSE06 functionals. In this case, the dielectric tensor is anisotropic as it can be observed from the position of the main absorption peaks among its components, which are slightly located at different energies. The directional average of $\varepsilon^{(2)}_{\alpha\beta}$ with PBEsol (HSE06) shows the main peaks located at 4.1, 4.6, 6.5 and 9.4 eV (5.7, 6.1, 7.3 and 11.5 eV). These peaks are associated to interband transitions between O p-states in the valence band and Ta d-states in the conduction band.

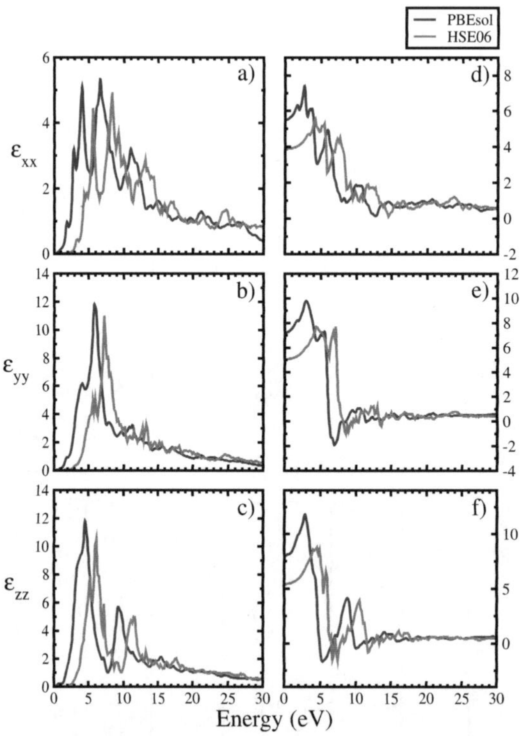

Figure 4. Imaginary (a), b) and c)) and real (d), e) and f)) parts of dielectric tensor for β-Ta_2O_5.

In Table 2, we show the calculated electronic (ε_∞) and ionic contribution (ε_0) to static dielectric tensor. Here, we can observe that ε_∞ presents some anisotropy, where ε_{xx} is lower than the other diagonal components ε_{yy} and ε_{zz}. On the other hand, ε_0 presents a high degree of anisotropy; the component ε_{zz} is much larger than ε_{xx} and ε_{yy}. Furthermore, ε_{yy} is approximately 1.8 times ε_{xx}. Finally, we have also calculated the directional average static dielectric constant ($\varepsilon(0) = \varepsilon_\infty + \varepsilon_0$), where ε_∞ and ε_0 are the directional averages of the electronic and ionic contribution to static dielectric tensor. These average values with PBEsol (HSE06) are $\varepsilon_\infty = 6.98$ (4.0) and $\varepsilon_0 = 43.51$, which permit us to obtain $\varepsilon(0) = 50.5$ with PBEsol functional. We have compared our results with the values reported in reference [25], finding relative errors around of 15% and 17%, respectively. These results show that our calculations are in agreement with the values reported in that study.

Table 2. Calculated optical-frequency and low-frequency dielectric tensor for β-Ta_2O_5 with GGA-PBEsol.

$$\varepsilon_\infty = \begin{pmatrix} 5.54 & 0.00 & 0.00 \\ 0.00 & 7.27 & 0.00 \\ 0.00 & 0.00 & 8.13 \end{pmatrix} \qquad \varepsilon_0 = \begin{pmatrix} 6.44 & 0.00 & 0.00 \\ 0.00 & 11.33 & 0.00 \\ 0.00 & 0.00 & 112.77 \end{pmatrix}$$

3.5. Optical properties
Once calculated the dielectric tensor along the three Cartesian directions, we have proceeded to calculate the generalized refractive index. The Figure 5-(a) correspond to the refractive index n, whereas the Figure 5-(b) correspond to the extinction coefficient κ. We can observe that the calculated PBEsol (HSE06) value of n at low-frequencies is 2.63 (2.2). For β-Ta_2O_5 we have not found in the literature a reported experimental value for n, but the reported value for an ultra-thin film of Ta_2O_5 is 2.15 [26], which compares well to our calculated values. In the Figure 5-(b), we can observe that the maximum absorption peaks are localized at 4.7, 6.2 and 9.9 eV (6.3, 7.2, 8.4 and 11.6 eV) with PBEsol (HSE06). These values are associated with inter-band transitions between O p-states in the valence band and Ta d-states in the conduction band.

After the calculation of the refractive index, also we have obtained the transmittance (Figure 6-(a)). We can observe that in the visible region (between 0.1 μm and 1 μm), the behavior of the transmittance is constant, with a value of 79% (85%) with PBEsol (HSE06), a result which indicates that in the visible region this material in β-phase is almost transparent. Finally, we calculated the absorption coefficient α which we show in the Figure 6-(b). This calculation shows that the absorption is nearly none in the visible region (around 2-3 eV), which confirms that this region β-Ta_2O_5 is transparent material and the first absorption peak is located in the middle ultraviolet region. Additionally, we observed the steepest increase with PBEsol (HSE06) of around 1.56 eV (2.83 eV), a value that is associated with the optical gap and it is consistent with our calculation.

4. Conclusions
In this paper we have studied the crystal and electronic structure and dielectric and optical properties of β-Ta_2O_5. We have identified that the valence band is a hybridization of O p and Ta d-states, while the conduction band is mainly form for Ta d-states. Further, this semiconductor in this phase presents an indirect gap, where the VBM and CBM is located in the points R = (1/2, 1/2, 1/2) and Y = (0, 1/2, 0), respectively. Also, we have calculated the static dielectric constant ($\varepsilon(0) = 50.5$), value which is in good agreement with values reported in others experimental and theoretical works. Finally, we have calculated the transmittance and absorption coefficient. Of this calculation we have identified that in the visible region tantalia in this phase is a transparent material.

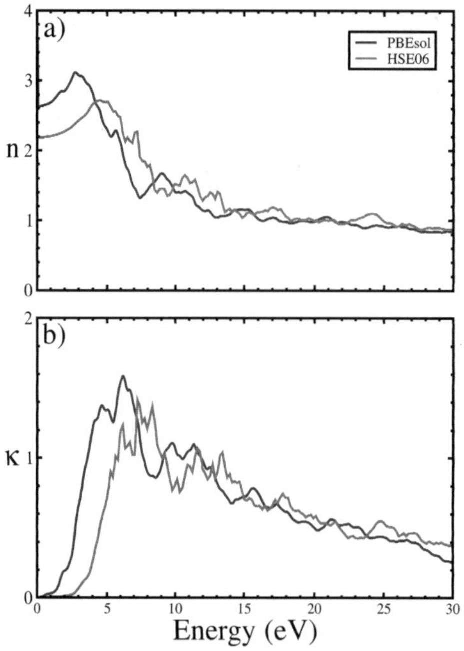

Figure 5. *a*) Refractive index n and *b*) extinction coefficient κ.

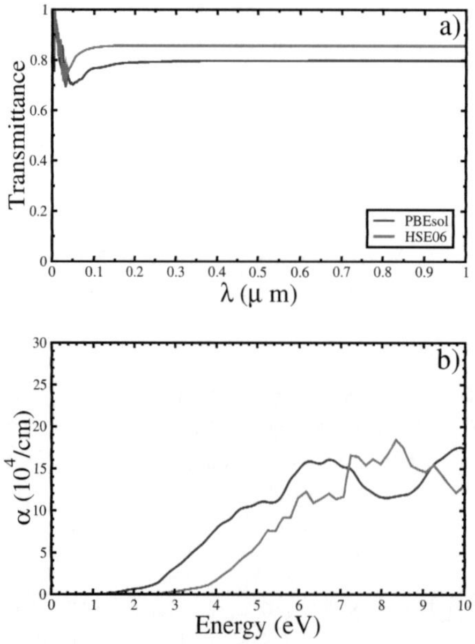

Figure 6. Transmittance and absorption coefficient (α).

Acknowledgments
This work was supported by Centro de Investigaciones–Instituto Tecnológico Metropolitano (Colombia) and CIEN–CODI–Vicerrectoría de Investigación–Universidad de Antioquia (Estrategia de Sostenibilidad 2016–2017).

References

[1] Lee M J, Lee C B, Lee D, Lee S R, Chang M, Hur J H, Kim Y B, Kim C J, Seo D H, Seo S, Chung U I, Yoo I K and Kim K 2011 *Nature Materials* **10** 625
[2] Kumar K J, Raju N and Subrahmanyam A 2011 *Surf. Coat. Technol.* **205** S261–S264
[3] Sato S, Morikawa T, Saeki S, Kajino T and Motohiro T 2010 *Angewandte Chemie International Edition* **49** 5101–5105
[4] Aleshina L and Loginova S V 2002 *Crystallogr. Rep.* **47** 415–419
[5] Ramprasad R 2003 *J. Appl. Phys.* **94** 5609–5612
[6] Lee S H, Kim J, Kim S J, Kim S and Park G S 2013 *Phys. Rev. Lett.* **110** 235502
[7] Andreoni W and Pignedoli C A 2010 *Appl. Phys. Lett.* **96** 062901
[8] Nashed R, Hassan W M I, Ismail Y and Allam N K 2013 *Phys. Chem. Chem. Phys.* **15** 1352–1357
[9] de-la Roza A O and Luana V 2011 *Comput. Phys. Commun.* **182** 1708–1720
[10] de-la Roza A O, Abbasi-Perez D and Luana V 2011 *Comput. Phys. Commun.* **182** 2232–2248
[11] Perdew J P, Ruzsinszky A, Csonka G I, Vydrov O A, Scuseria G E, Constantin L A, Zhou X and Burke K 2008 *Phys. Rev. Lett.* **100** 136406
[12] Blöchl P 1994 *Phys. Rev. B* **50** 17953–17979
[13] Kresse G and Furthmuller J 1996 *Phys. Rev. B* **54** 11169–11186
[14] Kresse G and Joubert D 1999 *Phys. Rev. B* **59** 1758–1775
[15] P E B, Jepsen O and Andersen O K 1994 *Phys. Rev. B* **49** 16223–16233
[16] Heyd J, Scuseria G E and Ernzerhof M 2003 *J. Chem. Phys.* **118** 8207–8215
[17] Heyd J, Scuseria G E and Ernzerhof M 2006 *J. Chem. Phys.* **124** 219906
[18] Gajdos M, Hummer K, Kresse G, Furthmuller J and Bechstedt F 2006 *Phys. Rev. B* **73** 045112
[19] Pérez-Walton S, Valencia-Balvín C, Dalpian G and Osorio-Guillén J 2013 *Phys. Status Solidi B* **250** 1644—1650
[20] Giannozzi P and Baroni S 2005 *Handbook of Materials Modeling* 189–208
[21] Fukumoto A and Miwa K 1997 *Phys. Rev. B* **55** 11155–11160
[22] Wu Y N, Li L and Cheng H P 2011 *Phys. Rev. B* **83** 144105
[23] Sahu B R and Kleinman L 2004 *Phys. Rev. B* **69** 165202
[24] Gu T, Wang Z, Tada T and Watanabe S 2009 *J. Appl. Phys.* **106** 103713
[25] Sergiu Clima G P, Elshocht S V, Gendt S D, Heyns M M, Wouters D J and Kittl J A 2009 *ECS Trans.* **19** 729–737
[26] Zhang D X, Zheng Y X, Cai Q Y, Lin W, Wu K N, Mao P H, Zhang R J, Zhao H B and Chen L Y 2012 *Appl. Phys. A* **108** 975–979

IOP Conf. Series: Journal of Physics: Conf. Series **1043** (2018) 012037 doi:10.1088/1742-6596/1043/1/012037

Analysis of optical and photoacoustic spectra quantum dots of CdS

E. Mera[1], **L. Gutierrez**[2] **and L. Da Silva**[2]

[1] Universidad Tecnológica Metropolitana, Physics Department, Box 9845, Santiago, Chile.
[2] Universidad Tecnica Federico Santa Maria, Box 110-V, Valparaiso, Chile.

E-mail: emera@utem.cl

Abstract. It was determined a relation between peaks of optical absorption and diameters of the quantum dots of CdS which permits to determine experimentally these diameters, and with this relation is used to determine the blueshift. Photoacoustic and optical spectra data of CdS quantum dots were analyzed. For this it was using the model of the Coordinates of Configuration. This model is applied under the condition of strongquantum confinement allowing to visualize the process of absorption and emission of photons and generation of phonons. It was determined the Huang-Rhys factor and the result indicates the type of coupling between the phonon and the quantum dot. Stokes shift was explained on the basis of the model of the Coordinates of Configuration. Recommendations are given for the resolution of photoacoustic spectra and requirements in optimization in measuring equipment for optical spectra.

1. Introduction

The semiconductor quantum dots, QD´s, in particular the CdS, have different physical properties of the semiconductors of the same bulk materials. In particular, these are presented in the optical properties of absorption and luminescence. The absorption and luminescence are produced by the generation of excitons in QD´s of CdS. In a semiconductor, the electron is in the conduction band and the hole in the valence band. The strong quantum confinement permits apply the model of Coordinates of Configuration, CC, and this confinement is responsible for the changes in the optical spectra, so that an analysis of these new optical properties is necessary to do with the purpose of display more carefully the physics of these QD's. In the publications related to this topic, spectra are shown in QD's semiconductors of CdS that are registered in absorption and luminescence spectra for QD´s of different diameters at room temperature [1][2][3][4]. These spectra show features which depend on the diameter of the QD so that while the diameter decreases the absorption as the luminescence spectra presented a blueshift. Beside absorption spectrum and its corresponding luminescent presented a Stokes shift. These transitions permit to analyze the non-radiative process. In addition, the non-radiative transitions can be analyzed using photoacoustic techniques and in this way it is desirable to do an analysis combining absorption, luminiscences and photoacoustic effect in the whole process in a QD. In the case of CdS QD's there are experimental results of these processes in literature [3][4]. In general, the theoretical models that attempt to explain the optical properties of these QD's differ from the experimental results by not explicitly consider the non-radiative transitions which leads to a slight discrepancy between the experimental results and these theoretical models [5][6][7].

2. Results

A brief analysis of experimental results of optical and photoacoustics spectral of CdS QD's is presented below with the purpose of showing the mechanism that link the absorption, and luminescence and photoacoustic spectra using the model CC. First, relations will be determined between the peaks of absorption and luminescence with the diameters of the QD's. In addition, it will use the CC model that applied under certain assumptions, including strong quantum confinement, will allow to observe the process as a whole.

There is a large number of publications related to optical spectra of QD's in CdS, however, it has opted to use the data from the publications of P. S. Nandakumar et al. [2] [3] and S. Abdallah et al. [4], because they have the most complete data for the purposes for this analysis. In this study are considered QD's whose diameters are between 1.2 nm and 2.0 nm, diameters to ensure strong quantum confinement. See Table 1.

Table 1. Diameters D(nm) and peaks of absorption and emission (eV)

D(nm)	Ea(eV)	Ee(eV)	Ea(eV)- Ee(eV)
1.2 [3]	2.88	2.88	0.00
1.4 [2]	2.77	2.77	0.00
1.6 [2]	2.75	2.71	0.04
1.8 [2]	2.69	2.66	0.03
2.0 [2]	2.66	2.59	0.07

Considering the diameters D (nm) and the peaks of absorption Ea (eV) in Table 1, was built the graph of Figure 1, ln Ea vs ln D.

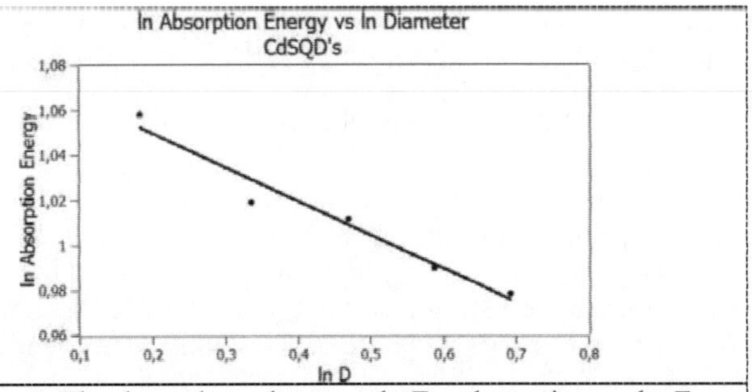

Figure 1. The dependence between ln Ea, absorption peaks Ea (eV) and ln D, QD diameters D (nm).

In this graph, there is a linear relation between these logarithmic quantities. From this relation it obtains the following expression for absorption is the Equation 1:

$$Ea * D^{0.1491} = 2.9281$$

Equation 1. Dependence between ln Ea, absorption peaks Ea (eV) and ln D, QD diameters D (nm).

Similarly, with the data from the peaks of luminescent energy Ee was constructed the graph of Figure 2.

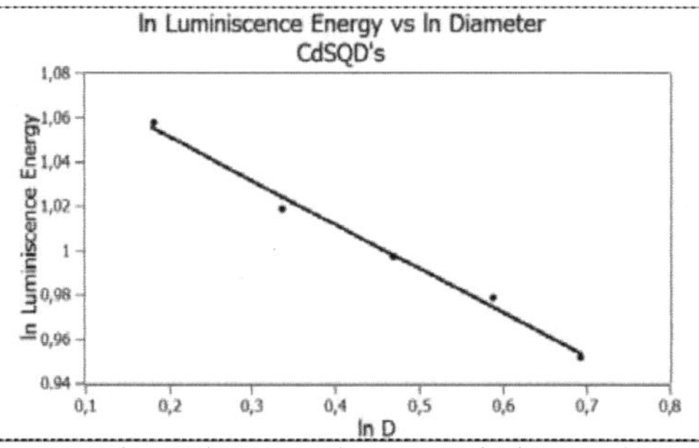

Figure 2. Dependence between ln Ee, emission peaks Ee (eV) and ln D, QD diameters D (nm), the linear relation in this case is better than graphs Figure 1

Where it gets the following Equation 2:

$$Ee * D^{0.1990} = 2.9774$$

Equation 2. Daily conversion factor to Tau Expression Angle

The equations (1) and (2) is achieved through the method of least squares. These relations are similar to the relation of Mollow-Ivey for F centers in alkali halides, where D is the distance between near neighbors and within that distance is hosted a vacancy of negative ion. From the equation (1) it is possible to obtain the diameter of the CdS QD's with an accuracy of 2 %.

The total energy of the non-radiative transitions can be determined by the difference between the energy of the photon absorbed Ea and the corresponding energy of the photon emitted Ee. This difference is shown in Table 1.

With regard to the data in Table 1, these differences correspond to the non-radiative energy transition of different QD diameters. In the QD's of 1.6 nm, 1.8 nm and 2.0 nm these differences suggest generation of phonons, but for QD's of 1.2 nm and 1.4 nm the difference is zero. However, there are photoacoustics recordings for these QD's as shown in reference [3]. The reason lies in the fact that when it takes the photoacoustic recording, sample is scanning by radiation from the 350 nm to 850 nm and at some moment the light source excites the upper levels of the QD. These states with higher levels are more efficient generating phonons. This will become evident when analyzing these transitions using the model of CC.

Due to the strong quantum confinement, when the electron absorbs the photon the electronic cloud expands compressing to their near neighbors generating a vibration in the QD. It is assumed that the electron is immersed in a harmonic potential generated by the near neighbors. Here whole process involves two types of energies: a quantum energy to absorb the incident photon by the part of the electron and other mechanical energy of vibration of the QD. This process is well represented in diagram of the CC in Figure 3.

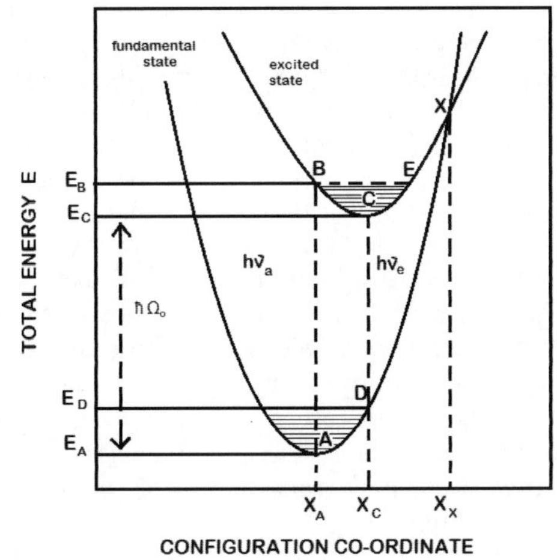

Figure 3. The absorption and emission mechanisms of a CdS QD are shown according to CC model. X point is the crossing of the two potentials, when B is below X then exist luminescence. Between B and C it has phonons similarly between D an A.

There it is indicated the step of the electron from its state A to state in B by absorbing the energy of the photon Ea = hγa . Then, the QD relaxes and the environment of the electron in its excited state changes to another harmonic potential. Within this new potential the electron passes from state B to state C creating phonons with a total energy Efl, then passes the state C the state D emitting a photon of energy Ee = hγe and returns to the final state, generating new energy phonons Ef2. By simple energy balance, it explains the Stokes shift hγa = hγe + Efl + Ef2, where Efl and Ef2 are the energies of transition between vibrational levels from the highest to the lowest level of the harmonic potential, γa and γe are the frequencies of optical absorption and luminescence. So it has the explanation of the of Stokes shift since $h\gamma a \geq h\gamma e$.

The CC model permits to determine, in addition, the Huang-Rhys factor So. This can be determined considering that frequencies ω differ slightly in the range from 1.2 nm to 2.0nm diameters of the QD´s and the factor of Huang-Ryhs is given by the relation: $So\hbar\omega + \hbar\Omega o = h\gamma a$, where $\hbar\Omega o = EC - EA$ according to diagram of Figure 3, doing the approximation that the energy of phonons in the excited state and ground state they are equal then the So Huang-Rhys factor [11] is given $So = (h\gamma a - \hbar\Omega o) / \hbar\omega \approx 1$

In this case the value of So for diameter of 2.0 nm of this QD and indicates a low generation of phonons.

To determine the equations of these harmonic potential, it knows that the differences between absorption and emission is 0.07 eV for D= 2.0 nm then we have the half of this energy for the ground state then the frequency ω = 217 8 (cm^{-1}) that is a value that corresponds to the range value given in publication of S. Biswas, et al. [8]. With these data is possible to write the equation of harmony potential: $E = \frac{1}{2} Kx^2$ where K = 29.6 (N/m). Note that the angular frequency ω depends on the diameter of the QD, so that if the diameter increases then ω grows [8]. Indicating that the harmonic potential in the excited state has a slope slightly greater than the potential of state of the state. Therefore in the excited state are generated more phonons than in the fundamental state. In addition, would explain why the electron to be excited with photons of energy corresponding to state at higher levels these generated the greatest number of phonons, contributing to the photoacoustic spectrum with new bands, which is why in the case of resolve with greater accuracy a photoacoustic spectrum is desirable to have records of optical absorption of high resolution that clearly distinguish the peaks of the different absorption bands and in this way properly attach the peaks in the spectra photoacoustics as starting point for solving these spectra. Taking into account, in addition, that the widths of the bands photoacoustics will be greater than the corresponding to the absorption optical experimental, fact already detected in other photoacoustics recordings [9].

Due to the volume expansion of QD every time that the electron is excited, it has been necessary to determine the increase in the diameter of the QD for the expansion of the QD. To that aim, it was considered very simple model that the electron is inside a box whose cube edges measure the diameter of the QD, then took the value of the energy of the photon with luminescent and it is again to recalculate the value of the edge [10]. In the case of the diameter of the QD of 2.0 nm, this was increased to 0.04 nm. This value must be taken into account in the graph of the harmonic potential in Figure 3. The minimum of the excited potential is moved in 0.02 nm respect to the minimum of the potential of the fundamental state. See Figure 3.

It is interesting to add it has been detected, in addition, shift toward the blue if the temperature of the sample decreases, This shift has an explanation according to this model, which in this case the diameter decreases by lowering the temperature of the QD [12]..

3. Discussion

It was obtained a relation that permits to determine the diameter of CdS QD's with an accuracy of 2% making use of the absorption spectrum of these QD's. It was explained why the photoacoustic spectra presented overlap of bands that must be adequately resolved by knowing the optical absorption spectra. This study focuses on cases of strong quantum confinement which allows to model these QD's with relatively simple models as is the case of the Coordinate of Configuration. The energy of a phonon is the order of 10 meV which leads to improve optical spectra to obtain a good resolution to the maximum of these spectra and hence to solve properly photoacoustics spectra. It is recalled that there were two absorption spectra with the same peaks giving a difference null and leaving without energy in order to be assigned to transitions non-radiative, however due to the low energy value of the phonons, the order of 10 meV, it may not have been able to detect by the spectroscopic equipment by low resolution. Beside it is necessary to take into account the presence of clusters of QD's that can generate photoacoustics bands shield against them and the registration of QD's isolated. In applying the model the CC it is possible to determine quantitatively the Stokes shift and the generation of the non-radiative transitions as vibrations of the QD and obtain the factor of Huang-Ryhs. These values obtained with this CC model can be compared with those obtained by other models. If the CC model fails to fit experimental data could be indicating the occurrence some new interesting phenomenon to analyze

References

[1] Yu-Shen Yuang, Yang-Fang Chen, Yang-Yao Lee and Li-Chi Liu 1994 ,*J. Appl. Phys.* **76** 3041-3044.

[2] Nandakumar P, Vijayan C and Murti Y. V. G. S 2002 *J. Appl. Phys.* **91** 1509-1514.

[3] Nandakumar P, Dobale A.R, Babu Y, Sastry M.D, Vijayan C, Murti Y. V. G. S, Dhanalakshmi K and Sundararajan G 1998 *Sol. St. Comm.* **106** 193-196.

[4] Abdallah S, Al-Hosiny N, and Badawi A 2012 *Journal of novel nanomaterials*, Article ID 498286, 6 pages.

[5] Efrons AI. L and Efrons A. L 1982 *Sov. Phys. Solid State* **16** 772-775.

[6] Brus L. E. 1984 *J. Chem. Phys.* **80** 4403-4409.

[7] Gangopadhyay S and Nag B. R 1997 *Nanotechnology* **8** 14-17.

[8] Biswas S, Dutta M, Snee P and Stroscio M.A 2001 *Chinese Journal of Physics* **49** 92-99.

[9] Da Silva L. F 2002 *Radiation Effects & Defects in Solids* **157** 621-627.

[10] Da Silva L. F 2005 *Radiation Effects& Defects in Soli*ds **160** 117-120.

[11] Bartram RH 1990 *J. Phys. Chem. Solids* **51** 641-651.

[12] Eychmuller A, Hasselbarth A, Katsicas L and Weller H 1991 *Phys Chem. Ber.Bunsen-Ges.Phys. Chem.* **95** 79.

Acknowledgments

The authors appreciates the support received by the Project Preis FONDECYT 1120639, 2013-2015 and the DGIP of the U.T.F.S.M and the Program of Environmental Applications of Nuclear Technology of the U.T.E.M. Also thank for the collaboration of Mr. Hector Da Silva P in the edition of this publication. One of the authors of the present investigation L. Da Silva dies in February 2017.

IOP Conf. Series: Journal of Physics: Conf. Series **1043** (2018) 012038 doi:10.1088/1742-6596/1043/1/012038

Melting curve of Si by means of the Z-method

Felipe González-Cataldo[1], Fernando Corvacho[2], Gonzalo Gutiérrez[2]

[1]Department of Earth and Planetary Science, University of California, Berkeley, California 94720, USA.

[2]Grupo de NanoMateriales, Departamento de Física, Facultad de Ciencias, Universidad de Chile, Casilla 653, Santiago, Chile

E-mail: f_gonzalez@berkeley.edu

Abstract. The melting curve of silicon is investigated through classical molecular dynamics simulations. We explore pressures from 0 to 20 GPa using the EDIP, Stillinger-Weber, and Tersoff interactomic potentials. Using the Z method, we demonstrate that the predicted melting temperature T_m can be significantly overestimated, depending on the potential chosen. Our results show that none of the potentials explored is able to reproduce the experimental melting curve of silicon by means of the Z-method. However, the EDIP potential does predict the change in the Clapeyron slope, associated with the diamond to β-tin phase transition.

1. Introducción

Solid silicon is a key material in the semiconductor industry, and has played a major role in the development of new technologies. At ambient pressure, Si crystallizes in the diamond structure, and transforms to the metallic β-tin structure around 12 GPa. Many aspects of the phase diagram of Si have been studied, such as the identification of different structures up to 250 GPa [1, 2], but an accurate determination of the melting curve still remains challenging, since high-temperature and high-pressure experiments on Si are difficult. Therefore, the available data need further theoretical confirmation.

The Z method is a novel technique that allows to determine the melting temperature of a material, based on a series of simulations at constant volume. The method has been applied to study wide variety of materials, such as Ar [3], Ta [4], Pt [5], SiO_2 [6], BeO [7], and many others, showing good agreement with other methods, and experiments. In this paper, we explore the predictions of the Z method for Si when different empirical potentials are used in classical molecular dynamics.

2. Methodology

2.1. Classical Molecular Dynamics

Since its development in 1960, Molecular Dynamics (MD) has been a powerful tool to study the properties of materials from their atoms interactions. Given an empirical interatomic potential, macroscopic properties of a material can be calculated by integrating the equations of motion of the atoms in the system and applying the theorems from Statistical Mechanics [8]. Nowadays, many softwares are available to perform MD, being LAMMPS [9] one of the most widely used. LAMMPS comes with several empirical potentials available, including the Tersoff [10, 11], EDIP [12], and Stillinger-Weber (SW) potentials [13], whose parameters have been adjusted from ab initio simulations and have shown to properly describe the mechanical properties of silicon.

The SW potential has been shown to reproduce the pair-correlation function $g(r)$ of the liquid, and was fit to reproduce the melting temperature T_m at ambient pressure to within a few hundred degrees of the experimental value of 1685 K [13]. The Tersoff potentials have been widely used to study silicon, but it is well known that it has problems describing the elastic constants and melting point. On the other hand, a modified version of Tersoff, so called MOD [14], shows good agreement with the experimental value of the melting temperature at ambient pressure. EDIP is an empirical potential that is also an improvement to the Tersoff potentials and, although it was developed earlier than MOD, it also reproduces the melting point of Si at ambient pressure, together with an accurate description of several structural properties. (Details about their functional form can be found in Appendix A). Thus, it is our purpose to benchmark the Z method for Si with all these potentials to show the differences in the prediction of the melting temperature for different pressures.

Using a cubic cell of 8000 atoms in the diamond structure, we carried out simulations of Si in the microcanonical ensemble for several densities, from 2.12 to 2.98 g/cm^3. We used a time step of $\Delta t = 1$ fs, and a simulation time of at least 50 ps to converge the thermodynamic properties.

2.2. Z Method

The Z method is a procedure that has been extensively used in multiple melting studies, even in materials with anomalous melting curves, such as Li [15] and H [16]. The idea is to perform ab initio or classical MD in the microcanonical ensemble (NVE) on a single solid system at different initial total energies. The total energy is controlled by setting a different initial temperature in each simulation. When the crystal is heated beyond its overheating limit, the temperature naturally drops to the melting temperature as the latent heat is removed from the kinetic energy (see Figure 1). The connected P–T points on the isochore form a Z-shaped curve. Several simulations for each isochore are needed in order to yield an accurate melting temperature.

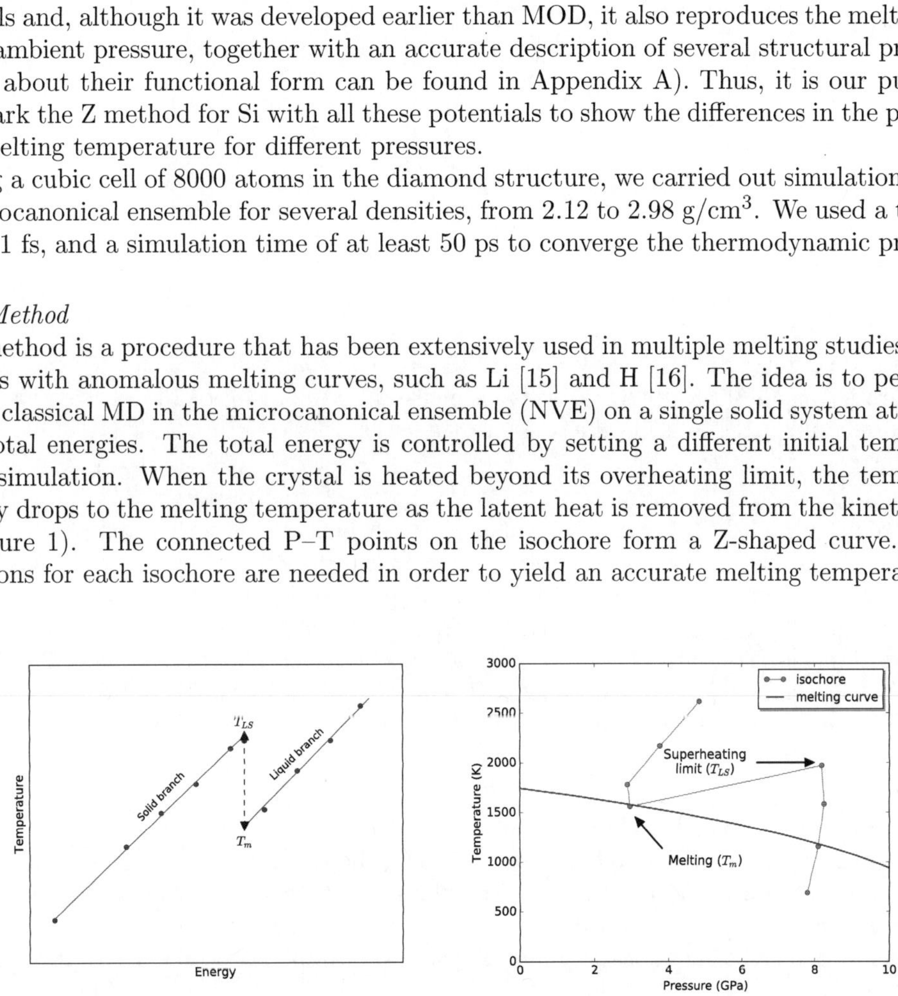

Figure 1. Schematic representation of the isochoric lines in Z-method simulations. T_{LS} is the maximum overheating temperature, and T_m the melting temperature.

2.3. Mean Squared Displacement

The Mean Squared Displacement (MSD) is a measure of the average displacement of atoms in a given sample, given by

$$\langle r^2(\tau) \rangle = \left\langle (\vec{r}_i(t_0 + \tau)^2 - \vec{r}_i(t_0)) \right\rangle_{t_0, i}, \tag{1}$$

where the average is taken over all particles i and time spans t_0. For the Z method, it allows to discriminate a solid from the liquid. When the atoms oscillate around the lattice positions, the slope of MSD is zero, meaning that the material is solid. If the MSD has a positive slope, the material is liquid. We calculated the MSD using the LPMD software [17], which allows to perform and analyze molecular dynamics simulations from different formats. In Figure 2 we show the MSD of Si for different temperatures along an isochore.

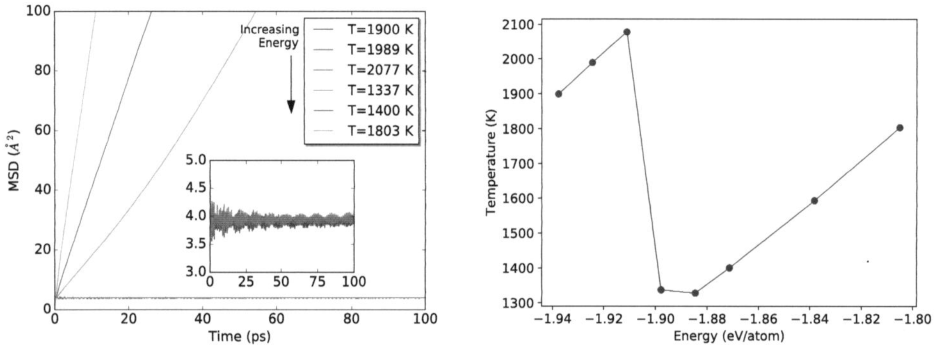

Figure 2. MSD for Si at a density $\rho = 2.515$ g/cm^3. The system melts when it is initial temperature is $T_0 = 4600$ K, which drops to a melting temperature of 1337 K for this density. The pressure equilibrates to 1.74 GPa after melting.

When we initialized the system at $T_0 = 4600$ K, it reached an equilibrium temperature of $T = 2077$ K still in solid state. When the energy was increased, the equilibrium temperature drops to a melting temperature of 1337 K. In the simulations that were close to the melting temperature, we observed that a long time was needed to completely melt the sample. Instabilities in the overheated diamond structure (Figure 3(a)) give rise to nucleation zones (Figure 3(b)), which grow until they completely melt the crystal (Figure 3(c) and 3(d)). We used OVITO [18] to identify the atoms that are coordinated in the diamond structure, and separate them from the liquid (red atoms in Figure 3).

3. Z method for different potentials

In Figures 4 and 5 we show different isochores Si using the EDIP and SW potentials. To reduce the size effects, for each isochore, we repeated all the simulations close to the melting point using 16000 atoms, and a relaxation time of 200 ps to ensure convergence.

We observe that, regardless of the potential used, the shape of the Z curve is inverted, as it has been observed in the case of lithium [15]. This is a clear signature of materials with negative Clausius-Clapeyron slope. We also notice that the variation in pressure between the overheated solid and the liquid is sometimes higher that 20 GPa, specially for Tersoff potentials. Each isochore varies significantly depending on the potential used. That has serious consequences on the prediction of the specific heat $C_v = (\partial U/\partial T)_V$, which is precisely the shape of the Z curve in the E-T phase space. On Figure 6 we show all the melting temperatures in the diagrams above, each one corresponding to a different isochore.

We observe that, consistent with previous results, the Tersoff potential T3 significantly overestimates the melting temperature of Si (around 20% higher than the melting temperatures reported by Deb). However, the MOD potential significantly underestimates the melting temperatures, despite the fact that it is an improvement to T3 and that it does reproduce

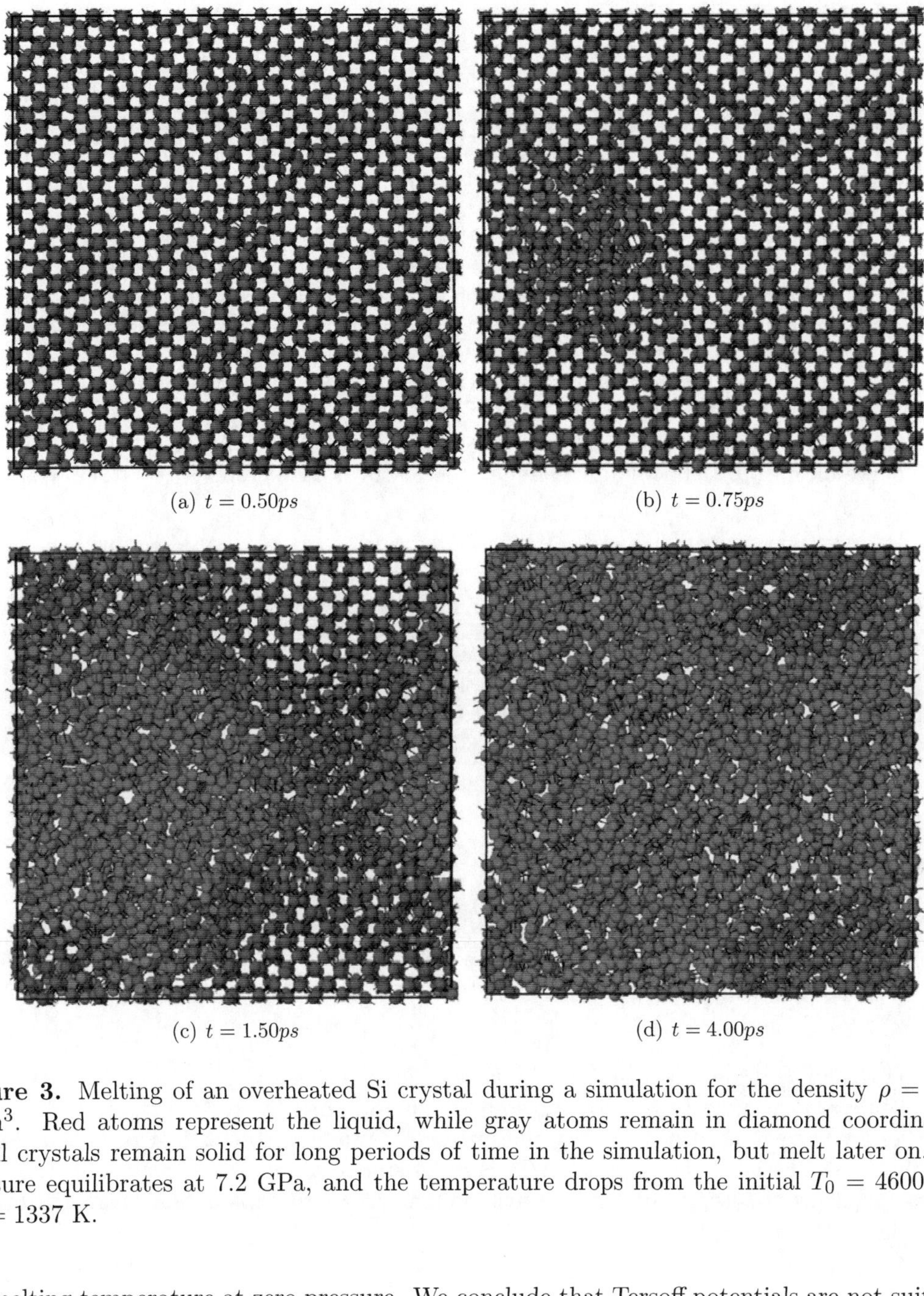

Figure 3. Melting of an overheated Si crystal during a simulation for the density $\rho = 2.515$ g/cm^3. Red atoms represent the liquid, while gray atoms remain in diamond coordination. Small crystals remain solid for long periods of time in the simulation, but melt later on. The pressure equilibrates at 7.2 GPa, and the temperature drops from the initial $T_0 = 4600$ K to $T_m = 1337$ K.

the melting temperature at zero pressure. We conclude that Tersoff potentials are not suited to simulate melting processes in any of its forms, using the Z method.

On the other hand, EDIP predicts a melting temperature of $T_m = 1500$ K at zero pressure that, although it is similar to the one predicted by MOD, is more than 100 K below the experimental melting temperature. It is interesting to notice that EDIP predicts a melting curve that changes slope around 7 GPa. Although it is far from the β-tin transition that occurs at 12 GPa, it is a signature of a solid-solid phase transition occurring below the melting curve. The SW potential predicts significantly high melting temperatures at high pressures. It also predicts a positive Clapeyron slope, regardless of the pressure range. However, it seems to agree

IOP Conf. Series: Journal of Physics: Conf. Series **1043** (2018) 012038 doi:10.1088/1742-6596/1043/1/012038

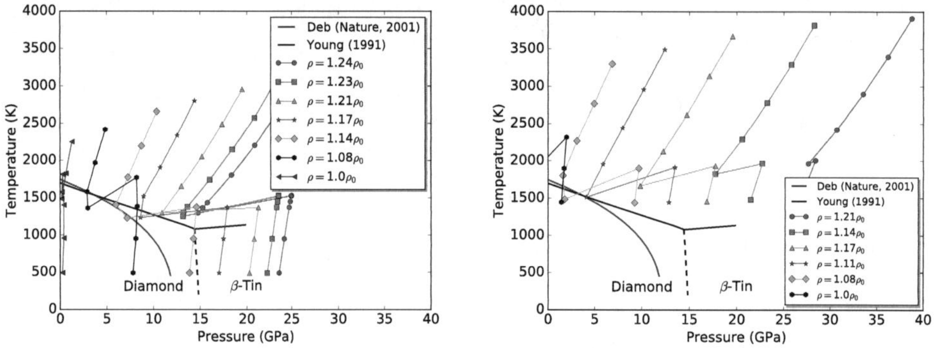

(a) EDIP (b) Stillinger-Weber

Figure 4. Silicon isochores obtained with the EDIP and SW potentials. The experimental melting curve of Si from Deb [19] and Young [20] are shown as reference. $\rho_0 = 2.33$ g/cm^3.

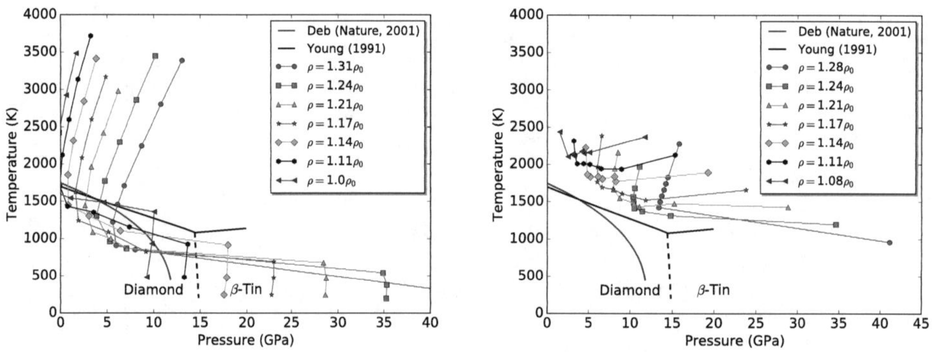

(a) Tersoff MOD (b) Tersoff T3

Figure 5. Silicon isochores obtained with the MOD and T3 potentials. The experimental melting curve of Si from Deb [19] and Young [20] are shown as reference. $\rho_0 = 2.33$ g/cm^3.

with the predictions from EDIP and MOD. Nevertheless, using the Z method, the SW is not capable of reproducing the 1685 K for the ambient pressure melting temperature.

Further research is needed regarding the use of Z method with classical potentials for Si. Improvements to the Z method, such as the inverse Z methodology [5], the two phase method [21], and first principles molecular dynamics must be considered to explain the mismatch on the predictions of the melting temperature of Si.

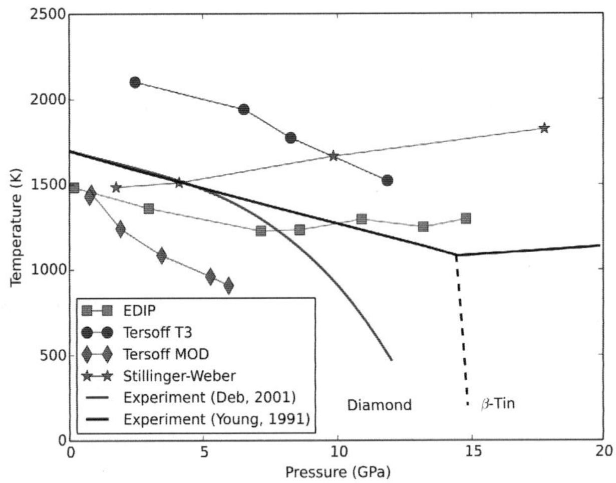

Figure 6. Melting points of Si obtained from the different interatomic potentials. The experimental melting curve of Si from Deb [19] and Young [20] are shown as reference.

Appendix A. Functional form of the potentials

The functional form of the EDIP potential is given by

$$E_i = \sum_{j\neq i} \phi_2(R_{ij}, Z_i) + \sum_{j\neq i} \sum_{k\neq i, k>j} \phi_3(R_{ij}, R_{ik}, Z_i) \tag{A.1}$$

where

$$\phi_2(r, Z) = A\left[\left(\frac{B}{r}\right)^{\rho} - e^{-\beta Z^2}\right] \exp\left(\frac{\sigma}{r-a}\right), \tag{A.2}$$

$$\phi_3(R_{ij}, R_{ik}, Z_i) = \exp\left(\frac{\gamma}{R_{ij}-a}\right) \exp\left(\frac{\gamma}{R_{ik}-a}\right) h\left(\cos(\theta_{ijk}), Z_i\right), \tag{A.3}$$

$$Z_i = \sum_{m\neq i} f(R_{im}) \tag{A.4}$$

$$f(r) = \begin{cases} 1 & r < c \\ \exp\left(\frac{\alpha}{1-\left(\frac{r-c}{a-c}\right)^{-3}}\right) & c < r < a \\ 0 & r > a \end{cases}, \tag{A.5}$$

$$h(l, Z) = \lambda[(1 - e^{-Q(Z)(l+\tau(Z))^2}) + \eta Q(Z)(l+\tau(Z))^2], \tag{A.6}$$

$$Q(Z) = Q_0 e^{-\mu Z} \tag{A.7}$$

$$\tau(Z) = u_1 + u_2(u_3 e^{-u_4 Z} - e^{-2u_4 Z}) \tag{A.8}$$

For a given configuration of atomic positions $\{R_i\}$, the energy over the atom i is given by E_i, where ϕ_2 represents the pair potential part, and ϕ_3 the three-body interactions. The effective coordination number, Z_i, is given as a sum of $f(R_{im})$, which represents the contribution of the m closest neighbors.

A = 7.9821730 [eV]	Q_o = 312.1341346
B = 1.5075463 [Å]	μ = 0.6966326
a = 3.1213820 [Å]	β = 0.0070975
c = 2.5609104 [Å]	α = 3.1083847
σ = 0.5774108 [Å]	u_1 = -0.165799
λ = 1.4533108 [eV]	u_2 = 32.557
γ = 1.1247945 [Å]	u_3 = 0.286198
ρ = 1.2085196	u_4 = 0.66
η = 0.2523244	

Table A1. Values of the parameters of the EDIP potential for the case of silicon, obtained from fits to ab initio calculations based on Density Functional Theory (DFT).

The values of the potential are given in the following table
For Stillinger-Weber, we have

$$E_i = \sum_{j>i} \phi_2(r_{ij}) + \sum_{j\neq i} \sum_{k>j} \phi_3(r_{ij}, r_{ik}, \theta_{ijk}) \tag{A.9}$$

where

$$\phi_2(r_{ij}) = A_{ij}\epsilon_{ij} \left[B_{ij} \left(\frac{\sigma_{ij}}{r_{ij}}\right)^{p_{ij}} - \left(\frac{\sigma_{ij}}{r_{ij}}\right)^{q_{ij}} \right] \exp\left(\frac{\sigma_{ij}}{r_{ij} - a_{ij}\sigma_{ij}}\right), \tag{A.10}$$

$$\phi_3(r_{ij}, r_{ik}, \theta_{ijk}) = \lambda_{ijk}\epsilon_{ijk} \left[\cos(\theta_{ijk}) - \cos(\theta_{0ijk})\right]^2 \exp\left(\frac{\gamma_{ij}\sigma_{ij}}{r_{ij} - a_{ij}\sigma_{ij}}\right) \exp\left(\frac{\gamma_{ik}\sigma_{ik}}{r_{ik} - a_{ik}\sigma_{ik}}\right) \tag{A.11}$$

The parameters used for silicon are

A = 7.049556277	λ = 21.0
B = 0.6022245584	γ = 1.20
a = 1.80	σ = 0.20951 nm
p = 4	ϵ = 3.4723x10^{-12} erg/at
q = 0	

For the Tersoff potential,

$$E_i = \frac{1}{2} \sum_i \sum_{j\neq i} V_{ij} \tag{A.12}$$

where

$$V_{ij} = f_C(r_{ij})[f_R(r_{ij}) + b_{ij} f_A(r_{ij})], \tag{A.13}$$

$$f_C(x) = \begin{cases} 1 & r < R - D \\ \frac{1}{2} - \frac{1}{2}\sin\left(\frac{\pi}{2}\frac{r-R}{D}\right) & R - D < r < R + D \\ 0 & r > R + D \end{cases}, \tag{A.14}$$

$$f_R(r) = A\exp(-\lambda_1 r), \tag{A.15}$$

$$f_A(r) = -B\exp(-\lambda_2 r), \tag{A.16}$$

$$b_{ij} = (1 + \beta^n \varsigma_{ij}^n)^{-\frac{1}{2n}}, \tag{A.17}$$

$$\varsigma_{ij} = \sum_{k \neq i,j} f_C(r_{ij}) g(\theta_{ijk}) \exp[\lambda_3^m (r_{ij} - r_{ik})^m] \tag{A.18}$$

$$g(\theta) = \gamma_{ijk} \left(1 + \frac{c^2}{d^2} - \frac{c^2}{[d^2 + (\cos(\theta) - \cos(\theta_0))^2]} \right) \tag{A.19}$$

Acknowledgements

G. Gutiérrez appreciates support from project PAIFAC-2016, Facultad de Ciencias, Universidad de Chile. F. González-Cataldo thanks support from CONICYT, postdoctoral scholarship Becas Chile 74160058.

References

[1] Hu J Z, Merkle L D, Menoni C S and Spain I L 1986 *Phys. Rev. B* **34** 4679–4684
[2] Mujica A, Rubio A, Muñoz A and Needs R J 2003 *Reviews of Modern Physics* **75** 863–912
[3] Belonoshko A, Skorodumova N V, Rosengren A and Johansson B 2006 *Physical Review B* **73** 012201
[4] Liu C M, Xu C, Cheng Y, Chen X R and Cai L C 2015 *Journal of Applied Physics* **118** 235901
[5] Burakovsky L, Chen S P, Preston D L and Sheppard D G 2014 *Journal of Physics: Conference Series* **500** 162001
[6] González-Cataldo F, Davis S and Gutiérrez G 2016 *Scientific Reports* **6** 26537
[7] Li D, Zhang P and Yan J 2014 *Scientific reports* **4** 4707
[8] Allen M P and Tildesley D J 1989 *Computer simulation of liquids* (Oxford university press)
[9] Plimpton S 1995 *Journal of computational physics* **117** 1–19
[10] Tersoff J 1989 *Phys. Rev. B* **39** 5566–5568
[11] Tersoff J 1988 *Phys. Rev. B* **38** 9902–9905
[12] Justo J a F, Bazant M Z, Kaxiras E, Bulatov V V and Yip S 1998 *Phys. Rev. B* **58** 2539–2550
[13] Stillinger F H and Weber T A 1985 *Phys. Rev. B* **31** 5262–5271
[14] Kumagai T, Izumi S, Hara S and Sakai S 2007 *Computational materials science* **39** 457–464
[15] Li D, Zhang P, Yan J and Liu H Y 2011 *EPL (Europhysics Letters)* **95** 56004
[16] Davis S, Belonoshko A, Johansson B, Skorodumova N V and van Duin A C T 2008 *The Journal of chemical physics* **129** 194508
[17] Davis S, Loyola C, González F and Peralta J 2010 *Computer Physics Communications* **181** 2126–2139
[18] Stukowski A 2009 *Modelling and Simulation in Materials Science and Engineering* **18** 015012
[19] Deb S K, Wilding M, Somayazulu M and McMillan P F 2001 *Nature* **414** 528–530
[20] Young D A 1991 *Phase Diagrams of the Elements* (University of California Press)
[21] Zhang W J, Liu Z L and Peng Y F 2014 *Physica B: Condensed Matter* **449** 144–149

IOP Conf. Series: Journal of Physics: Conf. Series **1043** (2018) 012039 doi:10.1088/1742-6596/1043/1/012039

Modeling field evaporation degradation of metallic surfaces by first principles calculations: A case study for Al, Au, Ag, and Pd

Teresita Carrasco[1], Joaquín Peralta[1], Claudia Loyola[1] and Scott R Broderick[2]

[1]Departamento de Física, Facultad de Ciencias Exactas, Universidad Andrés Bello, Sazié 2212, piso 7, Santiago, Chile.

[2]Dept. of Materials Designs and Innovation-University at Buffalo: State University of New York, USA.

E-mail: `joaquin.peralta@unab.cl`

Abstract. Under the effects of an extreme electric field, the atoms on a metallic surface evaporate by breaking their bonds with the surface. In this work, we present the effects of a high electric field, by the use of computational simulations, for different metallic surface chemistries: Al, Au, Ag, and Pd. To model this bond breaking procedrure (i.e. field evaporation), we use density functional theory through the Quantum-Espresso (QE) simulation package, which incorporates the electric fields by adding a saw-like funcion into the Hamiltonian. This approach, known as dipole correction, was applied to all simulations as is implemented in the QE package. In this work, we calculate the evaporation field (F_e) for all metallic species, which corresponds to the mean field at which atoms can break their bonds from the surface and evaporate. This result is compared with experimantal data from Atom Probe Tomography (APT) and computational data from prior simulations.

1. Introduction

The presence of high electric fields on the order of a hundred MeV/m is becoming increasingly common as science and engineering continues to push characterization boundaries. These kinds of electric fields are observed in particle accelerators, field emission microscopes, field ion microscopes, and atom probes, among others [1, 2, 3, 4]. The presence of these high electric fields on the surface induces the field evaporation phenomena on the surface. The field evaporation phenomena on a metal surface, in which atoms are evaporated under the effects of high electric fields, were first reported by Müller on a specimen tip in a Field Ion Microscope [2]. Nowadays, field evaporation plays a crucial role in Atom Probe Tomography experiments [5] and theory [6]. Also there is a growing relevance of this phenomena on computational simulations based on first-principles, and the link between simulated F_e values and experimental results [7, 8, 9, 10, 11], alongside the variety of atomic effects on the surface triggered by field evaporation that can be analyzed with this technique. The F_e plays an important role, particularly in the reconstruction process of the experimental data provided by APT experiments, and it is well established that preferential evaporation plays an important role in the quantitative interpretation of atom probe images [12, 13, 14, 15].

IOP Conf. Series: Journal of Physics: Conf. Series **1043** (2018) 012039 doi:10.1088/1742-6596/1043/1/012039

Field evaporation is usually considered as a thermally activated process [7, 9] in which the ion escapes over a potential barrier. The height of the activation energy (Q) for field evaporation is dependent on the electric field(F) and decreases with increasing field strength, as shown in Figure 1. As the external electric field increases from zero, the activation energy (Q) decreases and is accompanied by the formation of a hump. The electric field at which the hump just disappears is called the critical evaporation field (F_e) at 0K. Two principle models to describe the process have been deeply studied, (a) the classical image hump, and (b) the non-classical charge-draining hump [16, 17]. Multiple limitations are associated with these models. The classical image-hump model is not applicable to field evaporation [16], as the hump is almost always assumed to be physically a charge-draining hump. On the other hand, when the ad-atom is evaporated, the overall charge around the ad-atom decreases, and the atoms are expected to be positive ions; however, due to the nature of density functional theory (DFT) [18], the charge state of the evaporating atoms cannot be determined exactly by using the conventional formalism [9].

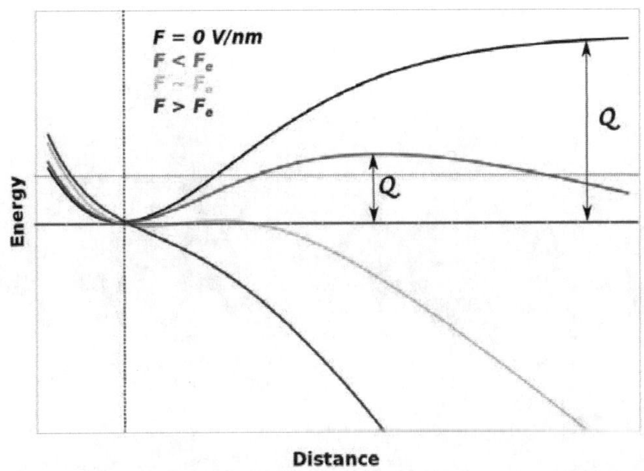

Figure 1. Schematic potential energy for pulling out an atom as a function of distance to the surface of the sample. The energy barrier Q decreases with increasing electric field. The black line represents the potential energy when there is no electric field. The red line represents the potential energy when electric field F is less than the critical evaporation field F_e, and the energy barrier decreases and is accompanied by the formation of a hump. The green line represents the potential energy when F is very close to F_e; this is the field at which the energy barrier (i.e., the hump) disappears. Finally, the blue line represents the potential energy when F greater than F_e is applied (please view color online). For all curves, zero energy is chosen at the equilibrium position of the ad-atom.

This work provides a value of the evaporation field F_e for four different metallic surfaces: Al, Au, Ag, and Pd. The values are obtained through DFT, which is one of the most relevant techniques used to understand phenomena at the atomic level because it is able to incoporate the atomic core, valence electrons, and relativistic effects in the process. Our results are compared with experimental data from APT and computational data from prior simulations [10, 7]. In our work, two different surface orientations for each surface, $< 111 >$ and $< 100 >$, have been considered.

In this paper the F_e obtained for different metallic surfaces is correlated to experimental and simulation results. We use DFT calculations to model the energetics of field evaporation

IOP Conf. Series: Journal of Physics: Conf. Series **1043** (2018) 012039 doi:10.1088/1742-6596/1043/1/012039

based on our previously developed methodology [10]. the rest of the document is organized as follows: The next section gives a description of the computational procedure used in the simulations; Section 3 shows the results for all of the metallic surfaces; and finally section 4 presents a discussion and conclusions of our work, along with future scopes.

2. Computational procedure

The surfaces of all different metallic compounds were built using 25 and 21 atoms for $< 111 >$ and $< 100 >$ directions, respectively. Figure 2 shows a representative structure for gold in the $< 111 >$ direction. All slabs were located in the middle of a supercell with a vacuum region of 15Å to the borders of the supercell in the z-axis. The structure sizes were considered smaller than other simulations, which have effects related to the periodicity of the cell and size. This allow us to simulate a large amount of data, as described in a previos work [10], and can take 8 to 10 times the regular time of a DFT simulation dye to the incorproation of a high electric field. In spite of this, the behavior of the field evaporation F_e for all the metallic compounds compares favorably with the reconstruction values.

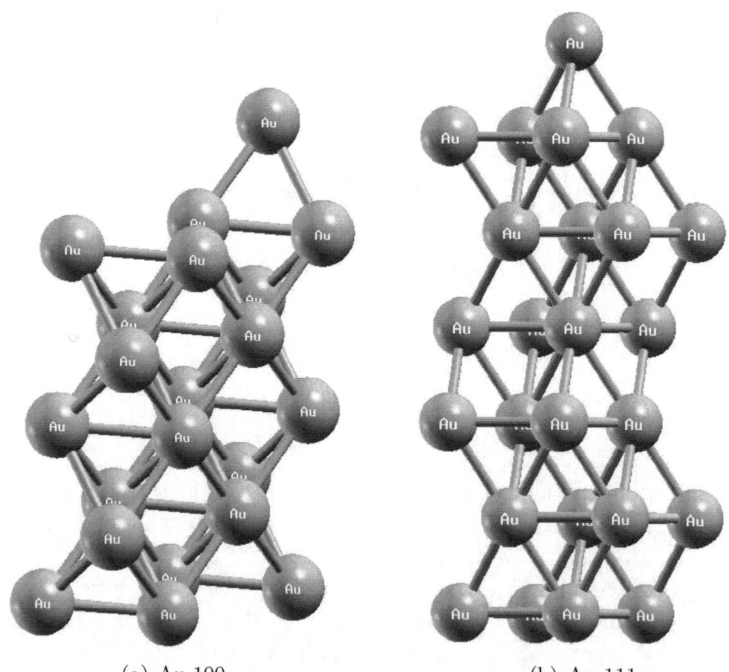

(a) Au 100 (b) Au 111

Figure 2. Representative (a) $< 100 >$/Au and (b) $< 111 >$/Au surface structures with an ad-atom on the surfaces in the $z-$direction.

The computational simulations were all performed using the Quantum-Espresso code [19] (QE) under the generalized gradient approximation (GGA). Metallic atoms were all represented by the Vanderbilt ultrasoft pseudopotential scheme. Convergence of energy and k-point mesh size were evaluated for each case. The energy cut-off used for each atomic species was 25, 40, 25, and 35 Rydberg for Ag, Al, Au, and Pd, respectively The k-point number used was 12x12x12 for the Au case and 8x8x8 for all of the rest. The Marzari-Vanderbilt [20] scheme was used for the simulations. The electric field was incorporated as an additional term in the Hamiltonian and was applied to all simulations using the dipole correction [21], implemented in the QE package.

In order to build the initial surfaces, we calculated the lattice parameters of all the metals by using the Bulk modulus. To proceed, a hydrostatic compression and expansion was used on a single lattice of each metal. With this information, we are able to determine the energy variation versus volume. With the results, we obtain the bulk modulus by fitting to the Murnaghan's equation of state [22]. The determined cell parameters and bulk modulus are presented in table 1.

Table 1. Calculation of bulk modulus B and lattice parameter a, as compared with data provided by other experimental and theoretical works.

Metal	B[GPa] [23]	B[GPa]This work	a[Å] [24]	a[Å] This work
Ag	0.69	0.61	4.085	4.149
Al	0.47	0.46	4.049	4.056
Au	0.99	0.86	4.078	4.174
Pd	1.25	1.13	3.890	3.964

The parameters are in good agreement with experimental results [25] and with previous first principles simulations [26, 10]. These results were used to build the initial structures. Four different slabs for < 111 > and < 100 > directions were built. They are composed of five layers, with an additional ad-atom on the surface. The direction perpendicular to the surface was chosen as the z direction. The slab was placed in the middle of a super-cell along the z direction and the vacuum spaces between the slab and the super-cell boundaries are equivalent to 15Å. All the structures (surfaces with ad-atoms) were relaxed with an ionic relaxation process, by the bfgs quasi-newton method without cell variation. The convergence criteria was set under the condition that the atomic forces on every atom are lower than 1.0×10^{-3} (in atomic units) for all directions.

As a function of the position of the ad-atom on the surface, the total energy of the system (surface and ad-atoms) was calculated under the influence of 12 different electric fields ranging from 0 to 49V/nm for Au and 0 to 27 V/nm for all other cases. The displacement of the ad-atom has been realized between -0.8Å and 6.0Å from the equilibrium position of the ad-atom. A seven-parameter function [7], at equation 1, was adopted here to fit the energy curve for the system formed by the evaporating atom and the slab under a particular electric field. In this equation, z is the position of the ad-atom, and a to h are parameters determined numerically.

$$E(z) = (a + bz + cz^2 + dz^3)\exp(-fz) + g + hz \tag{1}$$

F_e for each metallic surface has been calculated using a previously developed methodology [10]. We determined the variations of the hump-height in the presence of different electric fields (F). A sigmoid function, as shown in equation 2, has been used to fit the behavior of the hump-height under the effects of different electric fields, which have been shown to be a standard funtion in the phenomena [27, 10]. By numerical fit to the equation 2, we can evaluate the field evaporation value F_e for each type of metal surface.

$$H(F) = a + \frac{k - a}{(1 + q\exp(-b(F - m)))^{(1/n)}} \tag{2}$$

The following section shows the results for energies as a function of the ad-atom displacement and the evaporation field for each atomic configuration, and the conection of this with the F_e value.

IOP Conf. Series: Journal of Physics: Conf. Series **1043** (2018) 012039 doi:10.1088/1742-6596/1043/1/012039

3. Results

3.1. Evaporation Fields

A representative figure of the energy as a function of the displacement of the ad-atoms is presented. Figure 3 shows the energy required for evaporating an Au ad-atom from a $< 111 >$/Au surface as a function of the position of the ad-atom; wherein, each curve indicates a given electric field. The curves have been displaced in order to have better visual inspection of how the energy barrier decreases with increasing external electric field values.

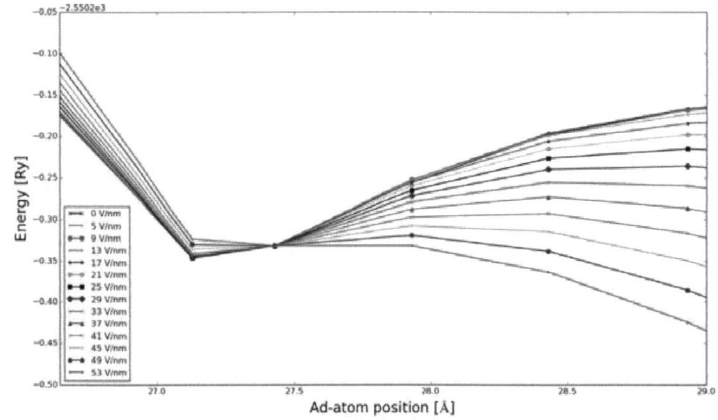

Figure 3. Energy as a function of the position of the ad-atom. Each curve represent a different intensity of the electric field. In order the electric field increases we observe the hump disappear, at the time hump completely disappear, then the ad-atom on the surface is evaporated.

A similar process is made for all different metallic surfaces in the direction $< 111 >$. For each curve, a fitting process of the data with equation 1 is used. A numerical analysis helps us to determine the the maximum and minimum of each curve in order to evaluate the hump in the sample. The results for all $< 111 >$ surfaces are presented in Figure 4.

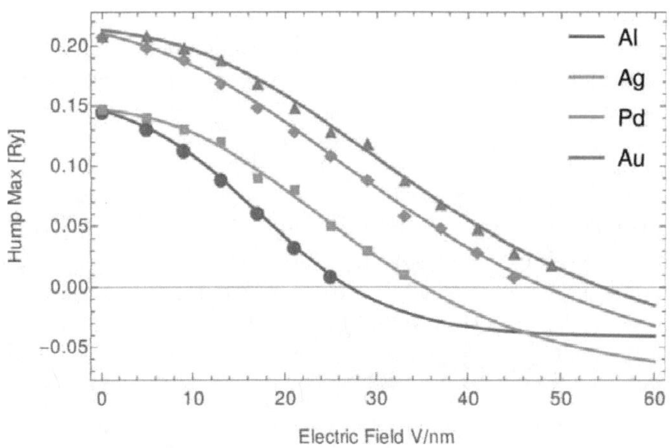

Figure 4. Energy as a function of the position of the ad-atom. Each curve represents a different intensity of the electric field. The electric field increases as we observe the hump disappear, and at the time the hump completely disappears then the ad-atom on the surface is evaporated.

The Figure 4 shows the variation of the hump in the ad-atom energy bond as the electric field is increased. The values of field evaporation for each case correspond to the zero hump for each curve for Al, Ag, Pd, and Au. We obtain F_e values of 27, 36, 48, and 54 V/nm respectively. In the particular case of Al, the lowest field evaporation observed is larger than expected. It has been shown in a previous work [10] that the largest slabs geometries result in lower field evaporation than the value obtained in the present work, and even lower than the regular value used for the reconstruction process (21 V/nm) in experiments. This means that the very local neighborhood of the ad-atom is responsible for the field evaporation necessary to remove the atom, and the effects of the periodicity based on the used methodology affects the field evaporation value, despite the distance between the atoms. It is also expected that vacancies, steps, or different ad-atom configurations modify this value [11].

For the case of $< 100 >$ surfaces, a similar procedure as previously described was made in order to obtain the variation of the hump for each metallic surface. Figure 5 presents the results of the variation of the hump for this case.

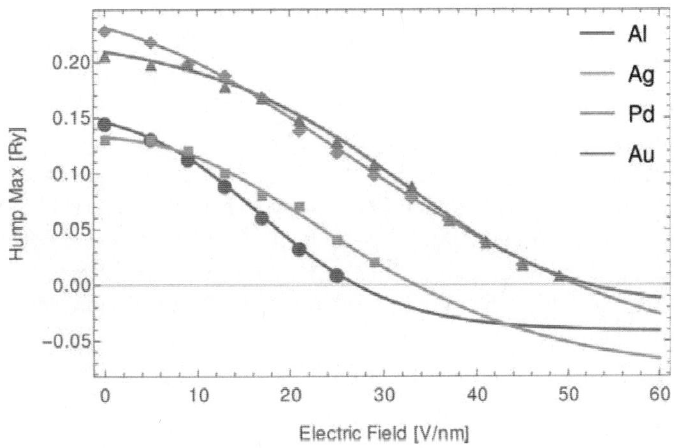

Figure 5. Energy as a function of the position of the ad-atom. Each curve represents a different intensity of the electric field. In order: the electric field increases, we observe the hump disappear, then the time hump completely disappears, and finally the ad-atom on the surface is evaporated.

As in the previous case for the $< 111 >$ direction, a clear order of the different metals is observed for the F_e: Au $>$ Pd $>$ Ag $>$ Al. The table 2 shows the F_e for the different surfaces and the values used as a reference in the reconstruction process in the experimental procedures.

We observe a clear difference established between the determined values provided from the DFT procedure and the reference values for the reconstruction. However, the procedure for small size surfaces is still a good representation of the magnitude of order for all of the materials. This ordering of the evaporation fields has been attempted previously [27]; in that case, the values were almost correct with the exception of the ordering of Au and Pd. In our case, with the methodology described, we are able to reproduce the order as expected. It also has been well demostrated that the values obtained by this technique typically overestimates the values of F_e which are evaporated mostly from pristine surfaces [7, 10, 11] and do not consider the presence of imperfections at the surface like vacancies, steps, and grain boundaries, among others. For all cases, no difference larger than 3 V/nm is observed, which is larger for the Au and Pd cases.

Table 2. F_e values obtained for two different surfaces and four different metallic species. Despite the similar values for the two directions, there are still important differences in the cases of Pd. Comparison shows a clear difference with references taken from the values used in the reconstruction process.

Metal	F_e for $< 111 >$	F_e for $< 100 >$	Ref F_e
Ag	35	33	24
Al	27	27	19
Au	54	52	53
Pd	48	51	37

4. Conclusions

A well established process is used to estimate the critical evaporation field of a (111) and (100) oriented surface for four different metallic specimen: Al, Ag, Pd, and Au. A critical evaporation field (F_e) is obtained using the proposed approach. The estimated evaporation field shows differences with previos simulations based on the size of the surfaces used. Additionally, it is well-known that differences are expected because of the use of pristine surfaces which omit the presence of imperfections at the surface. Despite the differences in the comparison of each F_e, which is motivated by simulation times using the incorporation of high electric fields, a behavior congruent with all metals is observed and easily related to the reconstruction field parameter (F) with all metals. We observe also how this methodology reproduces the expected order for each metal in terms of F_e, and how the changes for different directions do not strongly affect the values necessary to evaporate the atoms.

Acknowledgements

This work was supported by the Proyecto FONDECYT Iniciación 11130501. JP Also acknowledges partial support from Proyecto FONDECYT Regular 1140514 and Proyecto UAB-775. CL acknowledges support from Proyecto FONDECYT Iniciación 11150279, Proyecto PAI-79140025, and Proyecto DI-1350-16/R.

References

[1] Grudiev A, Calatroni S and Wuensch W 2009 *Phys. Rev. ST Accel. Beams* **12**(10) 102001 URL `http://link.aps.org/doi/10.1103/PhysRevSTAB.12.102001`
[2] Müller E W and Bahadur K 1956 *Phys. Rev.* **102**(3) 624–631 URL `http://link.aps.org/doi/10.1103/PhysRev.102.624`
[3] Mller E W 1966 *The Physics Teacher* **4** 53–56 (*Preprint* `http://dx.doi.org/10.1119/1.2350901`) URL `http://dx.doi.org/10.1119/1.2350901`
[4] Kelly T F and Larson D J 2012 *Annual Review of MAterials Research* **42** 1–31
[5] Kelly T F, Vella A, Bunton J H, Houard J, Silaeva E P, Bogdanowicz J and Vandervorst W 2014 *Current Opinion in Solid State and Materials Science* **18** 81 – 89 ISSN 1359-0286 URL `http://www.sciencedirect.com/science/article/pii/S1359028613000983`
[6] Silaeva E P, Karahka M and Kreuzer H 2013 *Current Opinion in Solid State and Materials Science* **17** 211 – 216 ISSN 1359-0286 atom Probe Tomography URL `http://www.sciencedirect.com/science/article/pii/S1359028613000685`
[7] Sánchez C G, Lozovoi A Y and Alavi A 2004 *Molecular Physics* **102** 1045–1055
[8] Ono T and Hirose K 2004 *Journal Applied Physics* **95** 1568–1571'
[9] Ono T, Sasaki T, Otsuka J and Hirose K 2005 *Surface Science* **577** 42–46
[10] Peralta J, Broderick S and Rajan K Mapping energetics of atom probe evaporation events through first principles calculations. submitted to Ultramicroscopy.

[11] Loyola C, Peralta J, Broderick S R and Rajan K 2016 *Journal of Vacuum Science & Technology A: Vacuum, Surfaces, and Films* **34** 061404 (*Preprint* `http://dx.doi.org/10.1116/1.4964833`) URL `http://dx.doi.org/10.1116/1.4964833`
[12] Marquis E A and Vurpillot F 2008 *Microscopy and Microanalysis* **14** 561–570
[13] Miller M K 1987 *Journal de Physique Colloques* **48** 565–570
[14] Philippe T, Gruber M, Vurpillot F and Blavette D 2010 *Microscopy and Microanalysis* **16** 643–648
[15] Vurpillot F, Bostel A and Blavette D 2000 *Applied Physics Letters* **76** 3127–3129
[16] Gomer R 1994 *Surface Science* **1** 129–152
[17] Forbes R G 1995 *Applied Surface Science* **87** 1–11
[18] Hohenberg P and Khon W 1964 *Physical Review* **136** B864–B871
[19] Giannozzi P, Baroni S, Bonini N, Calandra M, Car R, Cavazzoni C, Ceresoli D, Chiarotti G L, Cococcioni M, Dabo I, Corso A D, Fabris S, Fratesi G, de Gironcoli S, Gebauer R, Gerstmann U, Gougoussis C, Kokalj A, Lazzeri M, Martin-Samos L, Marzari N, Mauri F, Mazzarello R, Paolini S, Pasquarello A, Paulatto L, Sbraccia C, Scandolo S, Sclauzero G, Seitsonen A P, Smogunov A, Umari P and Wentzcovitch R M 2009 *Journal of Physics: Condensed Matter* **21** 395502 (19pp)
[20] Marzari N, Vanderbilt D and Payne M 1997 *Physical Review Letters* **79** 1337–1340
[21] Bengsston L 1999 *Physical Review B* **59** 12301–12304
[22] Murnaghan F D 1944 *Proceedings of the National Academy of Sciences* **30** 244–247 URL `http://dx.doi.org/10.1073/pnas.30.9.244`
[23] Sutton A P and Chen J 1990 *Philosophical Magazine Letters* **61** 139–146 (*Preprint* `http://dx.doi.org/10.1080/09500839008206493`) URL `http://dx.doi.org/10.1080/09500839008206493`
[24] each reference visit the website F 2016 We the references provided for each element in http://www.webelements.com
[25] Harada Y and Hunand D C 2002 *Materials Science and engineering A-Structural materials properties microstructure and processing* **329** 686–695
[26] Wang J Y and Zhou Y C 2004 *Physical Review B* **69** 144108–1 144108–13
[27] Sánchez C G and Lozovoi A Y Field evaporation of adatoms from metal surfaces - personal communication. personal communication.

IOP Conf. Series: Journal of Physics: Conf. Series **1043** (2018) 012040 doi:10.1088/1742-6596/1043/1/012040

Strain rate and temperature effect on $Zr_{50}Cu_{50}$ metallic glass under pure shear

M. Sepulveda–Macias[1], G. Gutierrez[1] and F. Lund[2]

[1] Grupo de NanoMateriales, Departamento de Física, Facultad de Ciencias, Universidad de Chile, casilla 653, Santiago, Chile.

[2] Departamento de Física, Facultad de Ciencias Físicas y Matemáticas, Universidad de Chile, Santiago, Chile.

E-mail: msepmacias@gmail.com

Abstract. We present a molecular dynamics study of the mechanical properties of $Zr_{50}Cu_{50}$ glass under pure shear. The samples have 145.200 atoms with dimensions of $40 \times 20 \times 2$ nm in the x, y and z directions, respectively. After quenching, we obtain 3 different systems at temperatures of 10, 50, 300 K. All systems were then subjected to dynamic pure shear at constant shear rates of 2.5×10^8 s^{-1}, 5×10^8 s^{-1} and 2.5×10^9 s^{-1}. We investigate the changes of mechanical properties as a function of temperature and shear rate, and find that the shear modulus is not sensitive to the changes of the shear rate, but decreases as the temperature rises. Ultimate strength is enhanced at higher shear rates and decreases with temperature. A study of the atomic structure reveals that the low values of ultimate strength are associated with high values of local atomic shear strain. These results are reminiscent of what happens at the jamming transition.

1. Introduction

Metallic glasses (MGs) are noncrystalline solids obtained by continuous cooling from the liquid state. They are much stronger than most metals and alloys, and have high hardness and elastic limit [1–3]. These properties make MGs attractive for structural applications [4]. However, they have limited ductility and suffer catastrophic failure without noticeable plastic deformation at room temperature. It has been hypothesized that this behavior can be explained by means of localization of structural and dynamical heterogeneities called shear transformation zones (STZs) [5]. They lead to the generation of free volume and the nucleation of shear bands (SBs), leading the MG to a fragile fracture. Despite this explanation, the microscopic understanding of the transition from elastic to plastic behavior, and the formation and evolution of shear bands submitted to external forces is an open research area.

To explore this area, molecular dynamics simulations are an excellent tool widely used in recent years. This technique allows us to track the position of atoms and calculate their mechanical properties avoiding the experimental limitations. Salient are the works presented by Dasgupta *et al* . [6] and Albaret *et al* . [7] who have studied the role played by the non-affine atomic displacement field, particularly "plastic events", as described by Eshelby inclusions [8, 9] on Lennard-Jones glasses [6] and amorphous silicon [7]. The present work shows results of a preliminary study on the relation between microscopic structure and macroscopic plastic behavior for $Zr_{50}Cu_{50}$ metallic glass.

We study the evolution of the atomic structure of a $Cu_{50}Zr_{50}$ metallic glass system using molecular dynamics simulation at three temperatures according to the applied shear via a dynamical shearing test, where the local structure of atoms is analysed using the local atomic shear strain.

2. Methodology

In our molecular dynamics (MD) simulations, we adopt the embedded–atom method (EAM) potential proposed by Mendelev *et al* . for ZrCu systems [10] as implemented in LAMMPS code [11]. The EAM gives the total energy of an atomic system in the form

$$E_i = F_\alpha\left(\sum_{j\neq i}\rho_\beta(r_{ij})\right) + \frac{1}{2}\sum_{j\neq i}\phi_{\alpha\beta}(r_{ij})\,, \tag{1}$$

where F is the embedding energy which is a function of the atomic electron density ρ. In the second term, ϕ is a pair potential interaction and α and β are the element types of atoms i and j.

The $Zr_{50}Cu_{50}$ metallic glasses are prepared as follows: First, a crystalline Cu system is created and then 50% of Cu atoms are randomly chosen and replaced by zirconium, thus obtaining a system composed by 145.200 atoms with dimensions of $40 \times 20 \times 2$ nm in the x, y and z directions, respectively. Free boundary conditions are applied along the x and y direction and periodic boundary conditions along z. As an example we describe in detail the procedure to obtain the system at 10 K. In this case the temperature and pressure are set at 2200 K and 0 GPa respectively, allowing the system to relax for 200 ps using NPT ensemble to ensure a well–equilibrated sample. The integration timestep is set at 1 fs. Then, the system is cooled down to the final temperature during 1370 ps, using the Berendsen thermostat [12]. This new configuration is allowed to relax during 100 ps, at 10 K, to ensure a well–equilibrated metallic glass. After quenching, we obtain 3 different systems at temperatures of 10, 50, 300 K, each one with a cooling rate of 2.19, 3 and 1 K/ps, respectively. All systems were then subjected to dynamic pure shear in the axial direction at constant shear rates of 2.5×10^9 s^{-1}, 5×10^8 s^{-1} and 2.5×10^8 s^{-1} in the NVE ensemble. For that purpose we fix the boundaries of the material, and assign to them a fixed velocity in the axial direction to obtain the pure shear. All the MD simulations are carried out using the classical molecular dynamics code LAMMPS [11].

We have investigated the change of the mechanical properties as a function of temperature and shear rate. To this end, stress–strain curves have been measured, using the (xy) component of the stress and strain tensors. To quantify the microstructural changes of the sample we use the local atomic shear strain η^{Mises} [13]. This parameter requires two atomic configurations, the current and reference one, defined as

$$\eta_i^{\mathrm{Mises}} = \sqrt{\eta_{yz}^2 + \eta_{xz}^2 + \eta_{xy}^2 + \frac{(\eta_{yy}-\eta_{zz})^2 + (\eta_{xx}-\eta_{zz})^2 + (\eta_{xx}-\eta_{yy})^2}{6}}\,, \tag{2}$$

where $\eta_{\alpha\beta}$ are the components of the strain tensor of atom i. We visualize η_i^{Mises} using the software OVITO [14].

3. Results and Discussion

The stress–strain curves for the three temperatures and strain rates under study are presented in Figure 1. As a function of strain γ, all curves present an elastic behavior up to a strain of 4%, reach a maximum between 4% and 6%, and slowly but steadily decrease beyond 6% up to the maximum strain used in the simulation, 25%. The shear modulus G and ultimate shear stress σ_{u} (obtained as the maximum stress value at the stress–strain curve) of each curve are

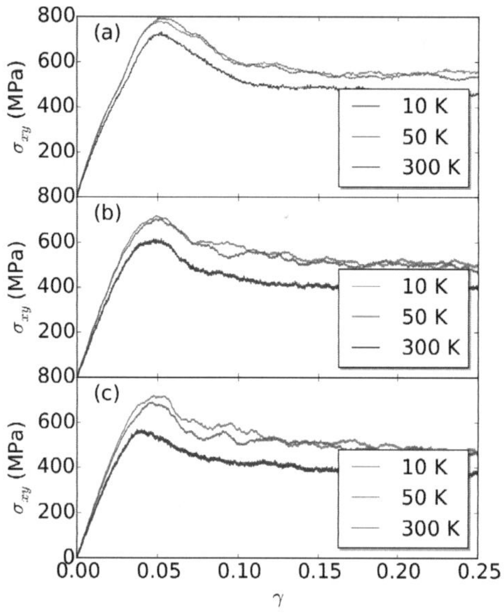

Figure 1. Calculated stress–strain curves for quasi–twodimensional $Zr_{50}Cu_{50}$ metallic glasses under the effect of pure shear at different shear rates. In (a) we present the case of $2.5 \times 10^9\,s^{-1}$, (b) $5 \times 10^8\,s^{-1}$ and (c) for $2.5 \times 10^8\,s^{-1}$.

summarized in Table 1. From Table 1 we observe that for a given temperature, T, there is no significant change in G as a function of the shear rate. For a given shear rate however, G is a decreasing function of T, as expected. Concerning σ_u, at fixed temperature it increases by 15–30% with increasing shear rate, with higher increases at higher T.

Table 1. Shear modulus G and ultimate shear stress σ_u, of the $Cu_{50}Zr_{50}$ metallic glasses under the effect of pure shear at three different temperatures and shear rates.

		$2.5 \times 10^9\,s^{-1}$	$5 \times 10^8\,s^{-1}$	$2.5 \times 10^8\,s^{-1}$
Property	Temperature (K)	(GPa)	(GPa)	(GPa)
G				
	10	24.51	24.80	24.01
	50	20.96	21.05	20.88
	300	17.19	17.33	19.24
σ_u				
	10	8.07	7.38	7.22
	50	7.95	7.04	6.93
	300	7.35	6.12	5.60

In order to shed light on the behavior of σ_u as a function of shear rate, we have computed the local atomic shear strain η_i^{Mises} at a strain of 6% namely, immediately after the ultimate shear stress σ_u is reached, for the three temperatures and three shear rates studied. The result is shown in Figure 2. The reason to choose a strain of 6% to sample the microstructure is that this point is well within the region of plastic behavior, and σ_u is still a useful tool for qualitative analysis. At higher strains, certainly beyond 8%, the different regions become blurred due to the formation of shear bands. To calculate the η^{Mises} showed in Figure 2 we use, as a reference configuration, the initial condition. The current configuration is the one at 6% of strain. The salient feature that is observed is that the lower the shear rate and, hence, the lower the ultimate

IOP Conf. Series: Journal of Physics: Conf. Series **1043** (2018) 012040 doi:10.1088/1742-6596/1043/1/012040

shear stress, the greater the local atomic shear strain. This relation between a macroscopic mechanical property with the microscopic structure of the metallic glass is reminiscent of what happens near the jamming transition [15] where, at a slow shear rate, the system has enough time to explore the phase space, and facilitate the movement of the particles, resulting in a larger plastic deformation. In contrast, for the case of a rapid shear rate, the atoms have no time to explore their vicinity, get stuck, and a higher stress is needed to unlock them.

Another feature that can be gleaned from Figure 2 is that atoms with smaller local atomic shear strain (dark blue) get clustered in regions of a few nm, and are surrounded by atoms with larger local shear strain (light blue) that are spatially connected. This feature becomes more marked at higher temperatures and lower shear rates. A similar behavior has been observed by Feng *et al* . [16] in a compression test, using molecular dynamics simulation, of $Cu_{64}Zr_{36}$.

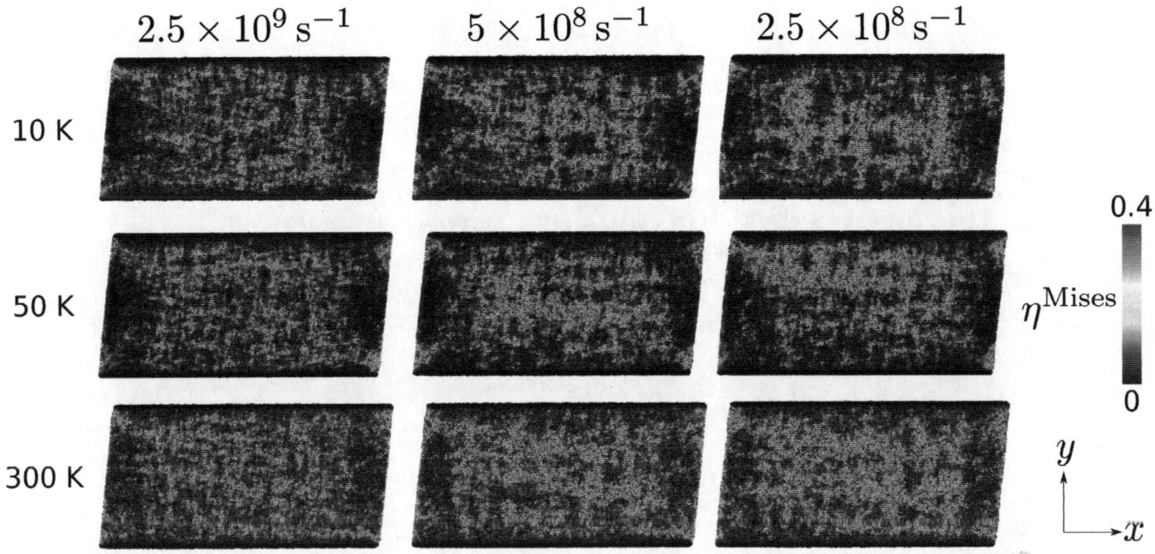

Figure 2. Image of the structure of $Zr_{50}Cu_{50}$ at $\gamma = 0.06$ and for three shear rates under study. Here we present the situation for (a) 10 K, (b) 50 K and (c) 300 K of temperature . Each sample contains 145.200 atoms and are colored according to their local atomic shear strain.

4. Concluding remarks

The changes in mechanical properties and atomic structure for $Zr_{50}Cu_{50}$ metallic glass under the effect of pure shear have been calculated. From the explored range of parameters, preliminary conclusions are that the shear modulus is not sensitive to changes in the shear rate, but is sensitive to changes of temperature. The ultimate strength increases at higher shear rates and decreases at higher temperature. A measurement of the local atomic shear strain reveals that it has higher values at low shear rates, with a geometric distribution similar to the one reported by Feng *et al* . [16] The high values of local atomic shear strain associated with low strain rates are similar to what is observed in the jamming transition [15], where rapid strain rates do not allow atoms to move far apart, thus preventing the appearance of high local shear strain.

Acknowledgments

M.S-M. thanks to CONICyT PhD. fellowship No. 21140904. G.G. thanks PAIFAC-2016, Facultad de Ciencias, Universidad de Chile. F.L. acknowledges to Fondecyt grant 1160823.

References

[1] Telford M 2004 *Materials Today* **7** 36 – 43 ISSN 1369-7021
[2] Wang W, Dong C and Shek C 2004 *Materials Science and Engineering: R: Reports* **44** 45 – 89 ISSN 0927-796X
[3] Ashby M and Greer A 2006 *Scripta Materialia* **54** 321 – 326 ISSN 1359-6462
[4] 2015 *Nat. Mater.* **14** 541–541 ISSN 1476-1122
[5] Falk M L and Langer J S 1998 *Phys. Rev. E* **57**(6) 7192–7205
[6] Dasgupta R, Hentschel H G E and Procaccia I 2012 *Physical Review Letters* **109** 1–4 ISSN 00319007 (*Preprint* `1207.3591`)
[7] Albaret T, Tanguy A, Boioli F and Rodney D 2016 *Phys. Rev. E* **93**(5) 053002 URL `https://link.aps.org/doi/10.1103/PhysRevE.93.053002`
[8] Eshelby J D 1957 *Proceedings of the Royal Society of London A: Mathematical, Physical and Engineering Sciences* **241** 376–396 ISSN 0080-4630
[9] Eshelby J D 1959 *Proceedings of the Royal Society of London A: Mathematical, Physical and Engineering Sciences* **252** 561–569 ISSN 0080-4630
[10] Mendelev M I, Kramer M J, Ott R T, Sordelet D J, Yagodin D and Popel P 2009 *Philosophical Magazine* **89** 967–987
[11] Plimpton S 1995 *Journal of Computational Physics* **117** 1 – 19 ISSN 0021-9991
[12] Berendsen H J C, Postma J P M, van Gunsteren W F, DiNola A and Haak J R 1984 *The Journal of Chemical Physics* **81** 3684–3690
[13] Shimizu F, Ogata S and Li J 2007 *Materials Transactions* **48** 2923 – 2927
[14] Stukowski A 2010 *Modelling and Simulation in Materials Science and Engineering* **18** 015012
[15] Biroli G 2007 *Nat Phys* **3** 222–223 ISSN 1745-2473 URL `http://dx.doi.org/10.1038/nphys580`
[16] Feng S, Qi L, Wang L, Yu P, Zhang S, Ma M, Zhang X, Jing Q, Ngai K, Greer A, Li G and Liu R 2016 *Scripta Materialia* **115** 57 – 61 ISSN 1359-6462

IOP Conf. Series: Journal of Physics: Conf. Series **1043** (2018) 012041 doi:10.1088/1742-6596/1043/1/012041

Normal Modes in Graphene for Different Geometries

Gabriel Cáceres[1], **Felipe González-Cataldo**[2] **and Gonzalo Gutiérrez**[1]

[1]Departamento de Física, Facultad de Ciencias, Universidad de Chile, Casilla 653, Santiago, Chile

[2]Department of Earth and Planetary Science, University of California Berkeley, Berkeley, CA 94720, USA

E-mail: gabriel.caceres@ug.uchile.cl

Abstract. Using classical molecular dynamics, we study the time evolution of out-of-plane standing waves on a square and circular single-layer graphene membrane. We explore the first six normal modes, obtaining the oscillation frequencies from the atomic motion. The modes show long-time stability in the harmonic regime, with no decoupling on multiple frequencies. Using the frequencies of oscillation, we calculated the transverse speed of sound in graphene and the tension on the membrane.

1. Introduction

The high flexibility, remarkable electronic properties and exceptional mechanical strength of graphene make it a useful material in technological applications [1–3]. Another fascinating property of this material is its high Q-factor [4] which makes graphene an interesting material to study the oscillation of normal modes, as the rate of energy loss is very low relative to the stored energy. Recent advances in nanotechnology allow for the production of nanoscale mechanical resonators [3,5], a field where graphene is considered promising for many applications. Therefore, a further characterization of different vibration modes of graphene for different geometries is needed, both experimental and theoretical [6].

In this work, we use classical molecular dynamics simulations with interactions modelled by the an empirical potential to study the flexural normal modes, also called lowest frequency acoustic branch ZA in phonon analysis, of single-layer graphene membranes of square and circular shapes. The primary aim is to compare the simulation with the theoretical results, obtaining a criteria for harmonic oscillation in the membrane. In addition, we estimate the speed of transverse waves in graphene.

2. Theory

The wave equation with Dirichlet boundary conditions governs the dynamics of a membrane of any shape, whose borders are fixed. That is, if $z(x, y, t)$ is the out-of-plane deviation of the membrane at time t in the position (x, y), then its evolution is controlled by the equation

$$\nabla^2 z(x,y,t) = \frac{\rho}{T}\frac{\partial^2 z(x,y,t)}{\partial t^2} \tag{1}$$

with the boundary condition

$$z(x,y,t)|_{\partial\Omega} = 0 \ , \tag{2}$$

where $\partial\Omega$ is the surface boundary, ρ the mass per area unit, and T is the tension per length unit. The ratio of the last two quantities corresponds to the square of the speed of sound in the media, given by

$$c = \sqrt{\frac{T}{\rho}} . \tag{3}$$

For a rectangular membrane, the solution is given by

$$z_{n_x,n_y}(x, y; t) = z_0 \sin\left(n_x \frac{x\pi}{L_x}\right) \sin\left(n_y \frac{y\pi}{L_y}\right) \sin\left(\phi + \omega_{n_x,n_y} t\right) , \tag{4}$$

where L_x and L_y are the dimensions of the rectangle, n_x and n_y integers numbers, and

$$\omega_{n_x,n_y} = \pi c \sqrt{\frac{n_x^2}{L_x^2} + \frac{n_y^2}{L_y^2}} \tag{5}$$

is the oscillation frequency. The phase ϕ and the height z_0 in Eq. 4 are determined by the initial conditions. If the membrane is square-shaped, then $L_x = L_y = L$, and

$$\omega_{n_x,n_y} = \pi \frac{c}{L} \sqrt{n_x^2 + n_y^2} . \tag{6}$$

For a circular membrane of radius a, the solution takes the form

$$z_{m,n}(r, \theta; t) = z_0 \, J_m\left(\frac{r}{a}\alpha_{m,n}\right) \sin(m\theta + \psi) \sin(\omega_{m,n} t + \phi) , \tag{7}$$

with $m = 0, 1, 2, \ldots$, $n = 1, 2, 3, \ldots$, where J_m is m-th Bessel function of the first kind, $\alpha_{m,n}$ the n-th positive root of J_m, and ϕ, ψ and z_0 are constants to be determined by the initial conditions. In this case the oscillation frequencies are given by

$$\omega_{m,n} = \frac{\alpha_{m,n} c}{a} , \tag{8}$$

where c is the speed of transverse wave in the media.

3. Methodology

The classical molecular dynamics simulations were performed using the Tersoff potential [7,8] for the interatomic interactions. The lattice constant a for this potential was found by minimizing the energy respect to the area. We found the energy is minimized for $a = 2.46$ Å, in agreement with the value accepted in the literature [6, 9]. Then, the graphene monolayer is strained uniformly in every direction and fixed at its edges to generate tension in the membrane, such that carbon atoms are separated 1.56 Å from each other. We considered a nearly square membrane of side 224 Å, fixing the position of all the atoms that were 10 Å from the border of the membrane, so that nearly 14500 atoms are allowed to move. For the circular membrane, we use the same strained atom lattice, but we only allow the movement of atoms that are at a distance $d = 107.47$ Å from the center. The atoms outside this region are interacting with the moving atoms, but they are not allowed to move. In this case, the dynamic region is composed by nearly 11500 atoms. For the initial conditions, we chose an initial height $z_0 = 0$ at time $t = 0$ for all atoms, and a velocity profile corresponding to the shape of a given eigenfunction, that is,

$$v_{n_x,n_y}(x, y; t = 0) = v_0 \sin\left(n_x \frac{x\pi}{L}\right) \sin\left(n_y \frac{y\pi}{L}\right) \tag{9}$$

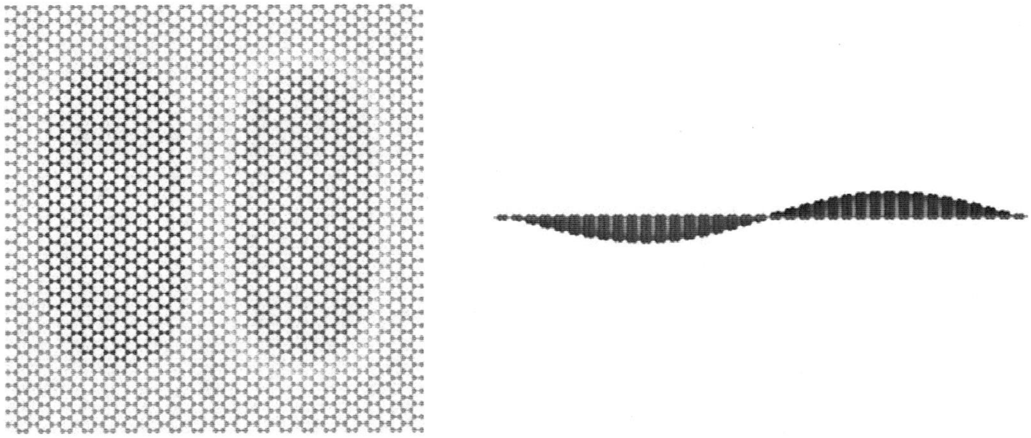

Figure 1. Oscillation of the second normal mode in the square-shaped graphene membrane. Left panel: top view; right panel: lateral view.

for the square-shaped membrane, and

$$v_{m,n}(r,\theta;t=0) = v_0\, J_m\left(\frac{r}{a}\alpha_{m,n}\right)\sin(m\theta) \tag{10}$$

for the circular membrane. The amplitude of these velocity profiles determines the amplitude of the oscillation in the out-of-plane direction, and also controls harmonicity of the motion. An initial velocity that is too high induces large vibration amplitudes that can introduce anharmonic contributions that decouple the normal mode in a combination of normal modes. To ensure harmonicity, we tried different values for the amplitude of the velocity profile in Eqs. (9) and (10). For each value, we took the Fourier transform of the kinetic energy of the membrane, which is periodic in the harmonic regime, to guarantee that no other modes appear in the simulation. The simulations were performed in the microcanonical ensemble (NVE), using the LAMMPS [10] software package, with a timestep of 1 fs, and at least 100 ps were used to monitor stability of the modes.

4. Results

Following the described protocol, we performed molecular dynamics simulation for several normal modes. For instance, in Figure 1 it can be seen the form of the actual oscillation corresponding to the second normal mode for the square-shaped graphene sheet. The frecuency of each mode is obtained from the atomic motion. The Fourier analysis performed over each simulation showed that there was only a single frequency in each mode generated. The modes show long-term stability, and the dependence of the frequency on the oscillation amplitude, which is controlled by the initial velocity v_0, has not been observed on our simulations.

In Figures 2 and 3 we show the measured frequencies in both membranes. From a linear fit to the data in Figure 2, we can obtain a slope of 0.156 ± 0.002 ps^{-1} that, according to Eq. (6), corresponds to the speed of sound of $c = 1.060 \pm 0.010$ km/s. The linear fit to the frequencies of the circular membrane in Figure 3 gives a slope 0.102 ± 0.001 ps^{-1}. Using Eq (8), we deduce a speed of transverse waves of $c = 1.096 \pm 0.007$ km/s. These two independent measures give an average value of $c = 1.078 \pm 0.006$ km/s which, according to Eq. 3, corresponds to a tension of $T = c^2\rho = 0.756$ N/m.

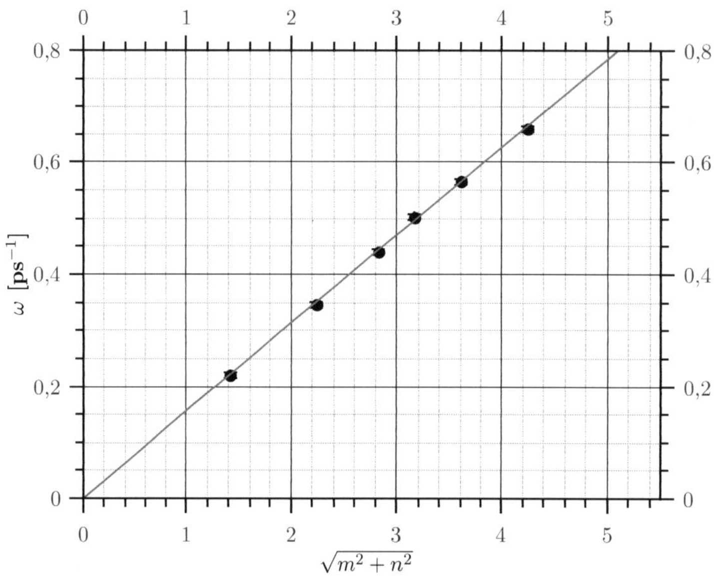

Figure 2. Frequencies of the square-shaped graphene membrane as a function of m and n, the integers that characterize the mode in Eq. (4). A linear fit (solid, red line) gives a slope of 0.156 ± 0.002 ps^{-1}.

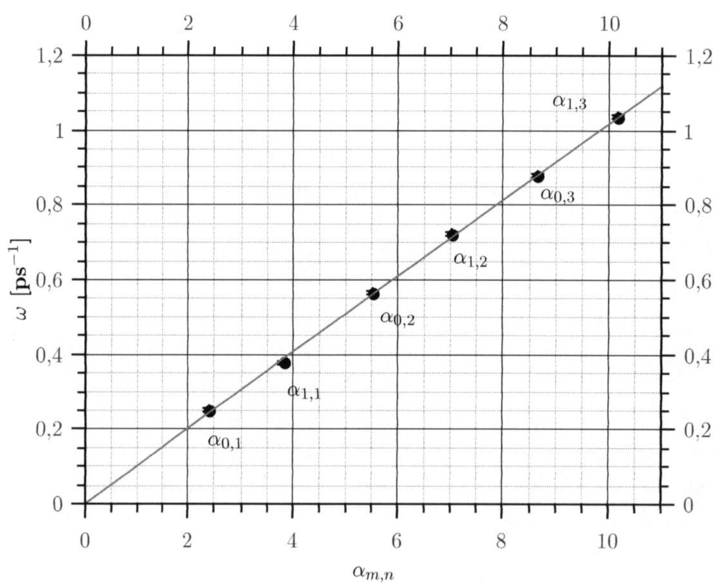

Figure 3. Frequencies of the circular-shaped graphene membrane as a function of α_{mn}, the n-th zero of the Bessel function J_m. A linear fit, in red, gives a slope of 0.102 ± 0.001 ps^{-1}.

5. Conclusion

We have measured the transverse speed of sound of strained graphene, associated with the lowest frequency acoustic branch *ZA*, using the classical molecular dynamics simulations. The normal modes generated in both shapes, square and circular, show a long-time stability that gives a speed of the transverse waves of $c = 1.078 \pm 0.006$ km/s for a graphene membrane that has been strained 10% from the original lattice constant. This velocity is lower than in-plane sound velocities associated with the *TA* and *LA* phonon branches of unstrained graphene [11]. We conclude that a ten percent of strain in graphene corresponds to a tension of $T = 0.756$ N/m. The velocities inferred from both geometries only differs by 3%, which means that this result is independent of the size or shape of the membrane.

The stability of the modes shows that it is indeed possible to generate single-layer graphene membranes that can be used as nanoresonators. Although other factors, such as temperature and impurities, have not been included in this work, they would be valuable in future works, as well as the study of isospectral membranes.

Acknowledgments

G. Gutiérrez appreciates support from project PAIFAC-2016, Facultad de Ciencias, Universidad de Chile. F. González-Cataldo thanks support from CONICYT, postdoctoral scholarship Becas Chile 74160058.

References

[1] Novoselov K S, Geim A K, Morozov S V, Jiang D, Zhang Y, Dubonos S V, Grigorieva I V and Firsov A A 2004 *Science* **306** 666–669
[2] Castro Neto A H, Guinea F, Peres N M R, Novoselov K S and Geim A K 2009 *Rev. Mod. Phys.* **81** 109–162
[3] Bunch J S, van der Zande A M, Verbridge S S, Frank I W, Tanenbaum D M, Parpia J M, Craighead H G and McEuen P L 2007 *Science* **315** 490–493
[4] Barton R A, Ilic B, van der Zande A M, Whitney W S, McEuen P L, Parpia J M and Craighead H G 2011 *Nano Letters* **11** 1232–1236
[5] Chen C, Rosenblatt S, Bolotin K I, Kalb W, Kim P, Kymissis I, Stormer H L, Heinz T F and Hone J 2009 *Nature Nanotechnology* **4** 861–867
[6] Jiang J W, Wang B S, Wang J S and Park H S 2015 *Journal of Physics: Condensed Matter* **27** 083001
[7] Tersoff J 1989 *Phys. Rev. B* **39** 5566–5568
[8] Tersoff J 1990 *Phys. Rev. B* **41** 3248 3248
[9] Chung D D L 2002 *Journal of Materials Science* **37** 1475–1489
[10] Plimpton S 1995 *J Comp Phys* **117** 1–19
[11] Dmitriev S V, Baimova J A, Savin A V and Kivshar Y S 2012 *Computational Materials Science* **53** 194–203

Description of the surface contour and dispersion relation of subharmonical waves in Q-2D granular media.

I. Cáceres[1], D. Risso[1]

1 Department of Physics, Physics Faculty, University of Bio-Bio, Avda Collao 1202, Chile

E-mail: icaceres@ubiobio.cl

Abstract. When a granular media having a subharmonic pattern is simulated, the first challenge is to define the surface contour of the granular media. This work proposes a method to define this contour. After define this contour, we propose an automated method to detect the presence or absence of the subharmonic oscillation pattern. Finally, with this information, we used the results to study the dispersion relation of the subharmonic pattern and its dependence with the depth of the granular media, resulting in a coincidence with the dispersion relation for shallow water.

1. Introduction

In the last decades several works have been carried out to study the behavior of the granular media [1, 2, 3, 4]. A granular media is composed of particles interacting with each other by energy dissipating interactions, if no energy is injected into the granular media, then the granular media rapidly decants and is immobilized. One of the mechanisms for injecting energy into the granular media is to vibrate the vessel in which the granular media is located and fluidize it [5]. This method, under certain circumstances, causes the granular media to oscillate subharmonically with respect to base osclation [6]. In this work, a granular media is computationally simulated, we propose a method to define the contour of a granular media, we propose a method to automatically discriminate the presence of the subharmonic pattern and finally the dispersion relation of the subharmonic wave pattern and its dependence with the depth of the granular media.

2. Characterization of the surface contour of a granular media

If we have a distribution of two-dimensional grains inside a rectangular box in which the grains have been compacted around one of the walls of the box, which we will call "background", it may be convenient for our investigation to define the surface contour of the granular media. This compaction could be due to the fact that the granular media has been compacted in the lower part of the box by the effect of an external gravitational force or other external force. If the information we have is the position of the geometric center of each grain, then we can define the surface contour as follows:

Once all the X and Y positions of all the particles inside the box are known, we separate the arrangement into N_c horizontal boxes of D_x width. For each of the boxes we consider only the particles whose geometric center is inside the box and in each box we calculate the average vertical position and corresponding standard deviation of the set of particles inside of the box. Finally we define the upper surface of the granular media within the box as the sum of the average vertical position plus the corresponding standard deviation of the particles within the box by a factor of $\sqrt{3}$. This surface will be associated with the horizontal position of the center of the box x_n. By repeating this process in all the boxes we can obtain a discretized function $C_s(x)$ to we define the upper contour of the granular media as shown in the figure 1. Likewise, if for some reason the granular media has been separated from the bottom, we can also define the lower bound by subtracting from the average the standard deviation by the factor of $\sqrt{3}$. If the granular media is in evolution, repeating this method for different instants of time we can define a surface contour $C_s(x,t)$ dependent on both the horizontal position and time. With this tool we can study the spatial and temporal evolution of the surface of a granular media.

If the granular media is subject to some fluidization mechanism, it is very likely that some particles of the granular media will be distanced from the total grain set. In this case, the distant particles will increase the standard deviation distorting the surface contour, reason why we must use a criterion to ignore the particles that are far removed from the others.

Before calculating the upper contour as already described, the distance between each of the particles of the system is evaluated. If a particle is less than a certain distance of discrimination d_0 from another particle, we will say that the other particle is its neighbor. If a particle does not have at least two neighbors, then that particle is not taken into account for the calculation of the contour. In this way, the particles that are sufficiently close together are part of the granular media and the particles that move away from the whole are excluded except in the rare case where three isolated particles are very close. The distance d_0 that we recommend is of two particle diameters, in this way, two particles are considered to be neighboring if it is impossible for another particle to pass between them.

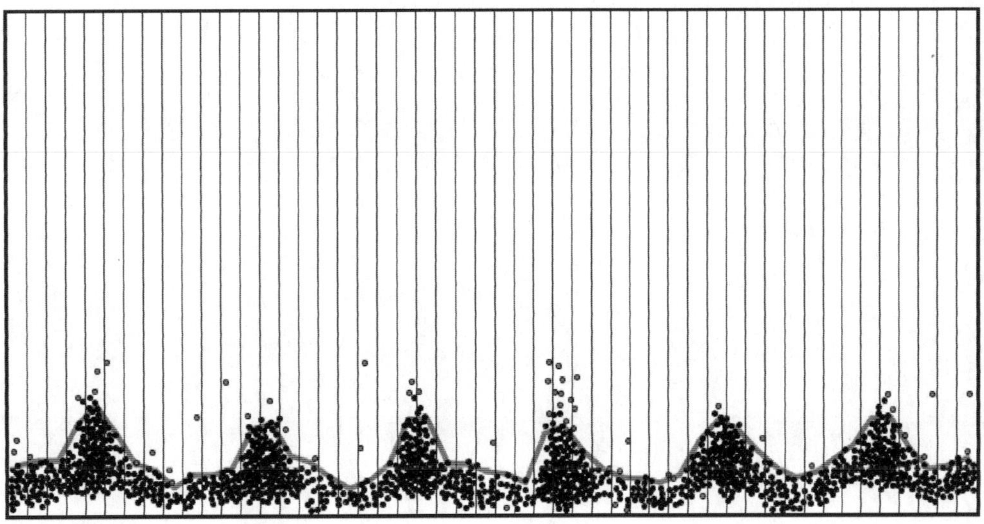

Figure 1. Example of discretization of the granular media and creation of the upper contour. The red line represents the surface contour and the red spheres are the particles that are considered outside the granular media because they do not have enough neighbors.

3. Automatic identification of surface pattern

A useful method for finding patterns in some feature of a system is by studying the autocorrelations of the fluctuations of that feature. If the system to be studied repeats a

IOP Conf. Series: Journal of Physics: Conf. Series **1043** (2018) 012042 doi:10.1088/1742-6596/1043/1/012042

sinusoidal characteristic, the autocorrelation of the fluctuations will also be a sinusoidal function. On the other hand, if the characteristic does not present a defined pattern and any fluctuations are a product of system noise, then the autocorrelation will tend to be a delta centered at 0. For the case that we are studying first we need to extract surface fluctuations. If the granular media has decanted becose gravity or another homogeneous external field, the equilibrium surface without fluctuations should be a line $C_p(x) = mx + n$ perpendicular to the external field. This straight line is easily obtained from $C_s(x)$ using a basic linear fit. This straight line is not necessarily parallel to the base of the box, so if the box is inclined or subject to some asymmetry along it, it is expected that the slope m is different from 0. Then, the function of spatial fluctuations with respect to equilibrium will be the function $C_f(x) = C_s(x) - C_p(x)$. Once we have our contour of fluctuations, we compute our autocorrelation function (ref ecu: autocorrela), using the values x_n corresponding to the position of the centers of the boxes used in calculating the surface contour.

$$A(d_n) = \sum_{i=1}^{N-n} \frac{C_f(x_i)C_f(x_{i+n})}{N-n} \qquad n = 0, ..., N \qquad d_{n+1} - d_n = x_{n+1} - x_n \qquad d_0 = 0 \quad (1)$$

Once we have calculated the autocorrelation function $A(d)$, we proceed to extract significant values from it. One of them is that if we assume that the surface contour has the sinusoidal shape, then the average amplitude a will be defined by $a = \sqrt{2A(0)}$. Another value to extract is the dominant wavelength λ_1, this wavelength is the distance for the autocorrelation function in which the first local maximum other than $A(0)$ is found. In addition, we define the discrimination factor f_{di} as the ratio between the value of the autocorrelation function for the dominant wavelength and the value of the autocorrelation function at distance zero $f_{di} = A(\lambda_1)/A(0)$, so if the fluctuations have perfectly sinusoidal behavior, then $f_{di} = 1$ whereas if the fluctuations only respond to a noisy behavior, then $f_{di} = 0$, an example of this criterion can be observed in the figure 2.

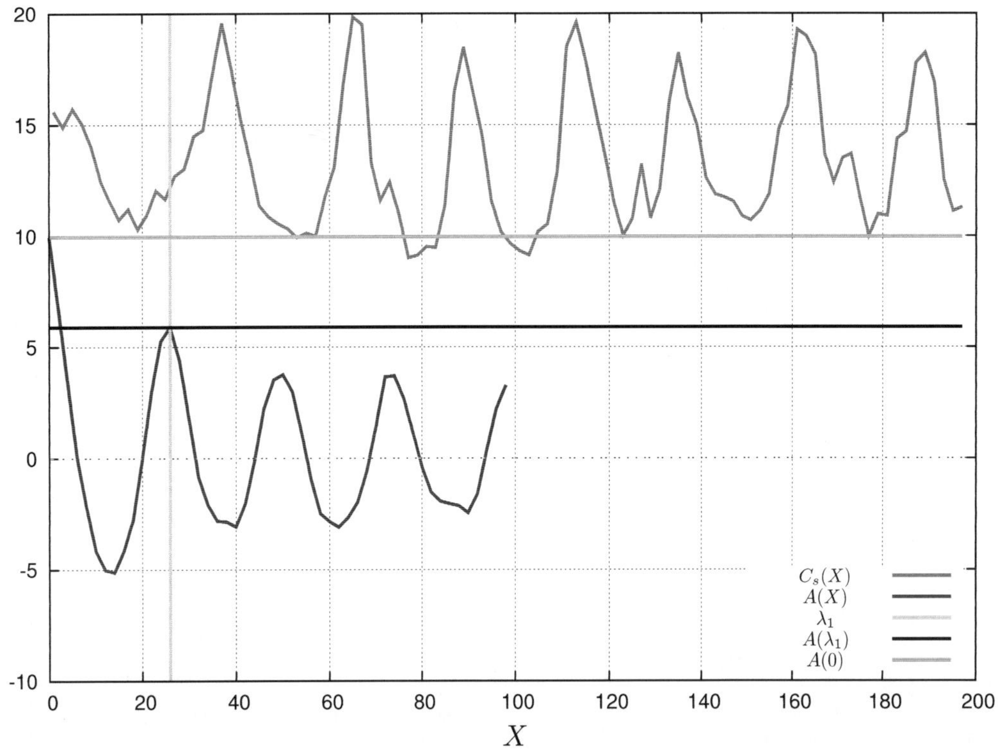

Figure 2. Example diagram of upper contour and autocorrelation of fluctuations, where $C_s(X)$ is the upper contour, $A(X)$ the autocorrelation of the fluctuations of $C_s(X)$ and λ_1 is the dominant wavelength.

Using the factor f_{di} we can begin to discriminate the existence or non existence of a periodic pattern in the contour. For this study we will use the criterion that if $f_{di} > e^{-1}$ will assume that there is a pattern, and otherwise there is no pattern. In the case of a subharmonic oscillation pattern, along the oscillation cycle this method detects that the pattern appears and disappears two times per period. In such a case, we can keep track of the amplitude a over time, detect the time instant in which a is a maximum for the first time in each cycle, and thus do a stroboscopic study of the granular media having as reference the phase of the cycle in which the pattern reaches its maximum amplitude.

4. Dispersion relation and its dependence on depth

In this study, multiple simulations were executed in a vertically vibrated granular media, using an event driven molecular dynamics algorithm [7, 8]. The simulations were carried out within a box of width $L_X = 200$ and height $L_Y = 100$ containing a set of particles of diameter d and coefficient of restitution 0.6. To fluidize the granular media, the box is vertically vibrated with an angular frequency ω_B and an amplitude A_B. In each of the simulations the non-dimensional acceleration $\Gamma = A_B\omega_B^2/g$, the dimensionless velocity $\zeta = A_B\omega_B/(\sqrt{dg})$ was varied, and the amount of particles , Which becomes a change in the mean depth of the granular media $h/d = \frac{Nd}{L_x}$. For all these simulations the previous criterion was applied to discriminate the existence of the subharmonic pattern to preserve only the results in which the pattern exists, also considering that the pattern generated is subharmonic, then the oscillation frequency of the granular media will be $\omega = \omega_B/2$. In studying this dataset and testing different settings, we find a good fit for $\omega_f^2(k, h)$ (2):

$$\omega_f^2 = 1.1 \pm 0.1\, g\, k\, \tanh(0.8 \pm 0.2\, h\, k) \tag{2}$$

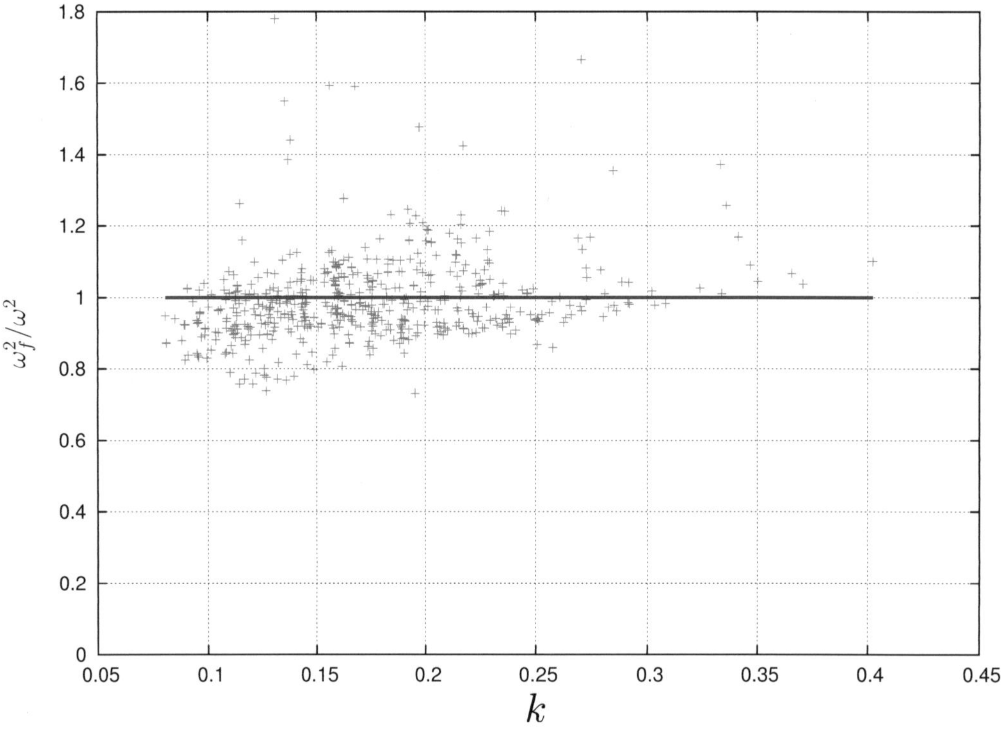

Figure 3. Quotient ω_f^2/ω^2 versus wave number k for different values of h.

This dispersion relation and its dependence on depth is very similar to the dispersion relation for shallow water. In the manner in which h is defined, this value should be 1 for a crystallized compaction ordered as squares and 0.81 for a triangular compaction. Our value is closer to 0.81, which coincides with the fact that, upon striking the base, the granular media is compacted triangularly. This result is quite interesting, since considering that the dispersion relation is comparable to the dispersion relation for shallow water implies that a large part of the causes that generate this dispersion ratio occur when the base is impacted, since the dispersion ratio for shallow water requires that the liquid media always remain in contact with the bottom of the vessel, which in this medium only occurs when the granular media impacts the bottom of the box.

5. Conclusion and comments

We have proposed a method to define the surface contour of a granular media from the individual position of each grain. Using this method, we were able to study the dispersion relation for a vibrated granular media in which there is a subharmonic vibration which gives us the same dispersion relation for shallow water, which leads us to suppose that the guiding mechanism for the subharmonic waves occur when the granular media strikes the vibrating base. However, in the case of the surface waves of a liquid, its dispersion relation requires that the medium has a surface tension, which makes us assume that for the case of the granular media there could be some phenomenon that plays the role of the surface tension.

5.1. Acknowledgments
Authors wish to acknowledge partially supported financial by Fondecyt grant 1120775, and Universidad del Bio-Bio.

References
[1] I. Ortega, M. G. Clerc, C. Falcón, and N. Mujica, "Subharmonic wave transition in a quasi-one-dimensional noisy fluidized shallow granular bed". Phys. Rev. E **81**, 046208 (2010).
[2] J. E. Macías, M. G. Clerc, C. Falcón, and M. A. García-Ñustes, "Spatially modulated kinks in shallow granular layers". Phys. Rev. E **88**, 020201(R) (2013).
[3] F. Melo, P. B. Umbanhowar and H. L. Swinney, "Hexagons, Kinks, and Disorder in Oscillated Granular Layers". Phys. Rev. Lett. **75**, 3838 (1995).
[4] J. Garay, I. Ortega, M. G. Clerc, and C. Falcón, "Symmetry-induced pinning-depinning transition of a subharmonic wave pattern". Phys. Rev. E **85**, 035201(R) (2012).
[5] C. Bizon, M. D. Shattuck, J. B. Swift, W. D. McCormick, and H. L. Swinney, "Patterns in 3D Vertically Oscillated Granular Layers: Simulation and Experiment". Phys. Rev. Lett. **80**, 57 (1998).
[6] K. M. Aoki and T. Akiyama, "Spontaneous Wave Pattern Formation in Vibrated Granular Materials". Phys. Rev. Lett. **77**, 4166 (1996).
[7] D. C. Rapaport. "The Art of Molecular Dynamics Simulation" (Cambridge University Press, 1996).
[8] S. Gonzalez, D. Risso, and R. Soto, "Extended event driven molecular dynamics for simulating dense granular matter", Eur. Phys. J. Special Topics 179, 3341 (2009).

IOP Conf. Series: Journal of Physics: Conf. Series **1043** (2018) 012043 doi:10.1088/1742-6596/1043/1/012043

Density functional theory study of defects in cadmium telluride: a PBC and QM/MM comparison

M Casanova-Páez[1,2] **and E Menéndez-Proupin**[1]

[1] Departamento de Física, Facultad de Ciencias, Universidad de Chile, Casilla 653, Santiago, Chile

[2] Departamento de Física, Facultad de Ciencias Físicas y Matemáticas (FCFM), Universidad de Chile, Casilla 487-3, Santiago, Chile

E-mail: `marcos.casanova@ug.uchile.cl`

Abstract. Ab initio calculations, in the framework of density functional theory (DFT), are employed to determine the formation energies of tellurium vacancy (V_{Te}) for different ionized states. We performed a periodic boundary conditions (PBC) calculation, applying quasiparticle corrections and establishing the thermodynamic transition level. In addition, we have also used the quantum mechanics/molecular mechanics (QM/MM) approach in order to check its capability predicting formation energies, finding similar results compared to the previous supercell method.

1. Introduction

It is important for photovoltaic devices to achieve the control over the charge carriers and the management of impurities and native defects. Cadmium telluride (CdTe) is an iconic example of II-VI compound and one of its main uses is as photovoltaic material, gaining fast acceptance due its competitiveness to respect the traditional silicon solar cells, becoming a cheap alternative. CdTe is also widely used as radiation detectors [1]. One advantage of CdTe as photovoltaic material is its direct band gap (1.5 eV at room temperature) which positions it in the range of maximum spectral irradiance and reaching the maximum efficiency for a single absorber solar cell [2]. Defects in CdTe are widely studied through DFT, since they have been subject of controversy for a long time, based on local or semi-local density approximation (LDA or GGA). Furthermore, most native defects in CdTe occur in both their neutral and one or more ionized states [3]. There is no direct evidence of tellurium from experiments, but theoretical calculations predict that this is one of the most important intrinsic donors.[1; 4–6]

Point defects formation energies are calculated by taking the difference in total energies of a system with and without the defect of interest. One complication that arises is that isolated point defects energies are desired in most cases, making necessary the use of a big supercell in order to reduce the size effects as much as possible. The formation energy of V_{Te} of a crystal containing n CdTe units is computed as

$$\Delta H_f(V_{\mathrm{Te}}^q) = E(\mathrm{Cd}_n\mathrm{Te}_{n-1}) + E(\mathrm{Te}) - E(\mathrm{Cd}_n\mathrm{Te}_n) + \Delta\mu_{\mathrm{Te}} + qE_{\mathrm{F}} + \Delta E_{\mathrm{size}} + \Delta E_{\mathrm{q.p.}}, \quad (1)$$

where $E(\mathrm{Te})$ is the energy per atom of bulk Te. $E(\mathrm{Te}) + \Delta\mu_{\mathrm{Te}}$ is the chemical potential of the corresponding species, and it depends on the thermodynamic equilibrium conditions. In Te-rich conditions is set to 0 eV but in Te-poor conditions it equals the CdTe heat of formation [7]. Also, the analogue of the chemical potential for charge is given by the chemical potential of electrons, i.e. , the Fermi energy E_{F}. Finally, ΔE_{size} and $\Delta E_{\mathrm{q.p}}$ are correction terms that accounts for finite k-point sampling, interaction between supercells, quasiparticle and band-edge corrections. More detailed information about these corrections can be found in Ref. [7]

Most calculations of CdTe defects have been studied using the 64-atoms supercell (SC64) of the perfect crystal [1; 4–6; 8–10]. This supercell is $2 \times 2 \times 2$ times the 8-atoms cubic unit cell. A previous work [7] showed that the SC64 supercell is too small to obtain reliable results for the V_{Te} band structure, they overcame this problem using a 216-atoms supercell (SC216). In this work we have used a 512-atoms supercell (SC512), which is a $4 \times 4 \times 4$ times the 8-atoms cubic unit cell with only one k-point. On the other hand, we have also performed a calculation using the QM/MM approach. For an embedded cluster calculation, a part of the crystal structure containing the defect and surrounding atoms is chosen. This set of atoms are treated fully quantum-mechanically (QM region). The cluster is embedded in a set of point charges (MM region) to account for long-range electrostatic interactions due to atoms of the crystal lattice outside the cluster region. In addition to point charges, the embedding needs an intermediate region of pseudopotentials which are placed at the first shell of embedding cadmium atoms of a CdTe cluster to minimize non-physical polarization. Within this model, an ion in the intermediate region is treated as a point core and a shell connected by a spring, simulating its dipole polarizability. The positions of the cores and shells in the intermediate region can be optimized with respect to the total energy of the system, in accordance with the relaxed geometric and electronic structure of the inner cluster. Despite the fact that CdTe is not an ionic material, we are assuming that the ions are in the formal oxidation state, i.e. the crystal consists of Te^{-2} and Cd^{+2} ions.

2. Computational details

2.1. Supercell approach

Supercell calculations were performed using the all-electron, full-potential electronic structure code FHI-aims [11]. Our calculations were performed using the generalized gradient approximation (GGA) for the exchange-correlation functional proposed by Perdew, Burke and Ernzerhof (PBE) [12]. We have used the tight basis set for the wavefunctions (which are radial functions proposed by FHI-aims for well converged energies). The lattice constant was obtained through the Birch-Murnaghan equation of state, differing by 2.2 % from its experimental value of 6.48 Å. The SC512 supercell was constructed using 4 times the converged theoretical lattice constant of the 8-atoms cubic unit cell calculated before (6.625 Å). For all the structures, forces were minimized until residuals fell below 10^{-3} eVÅ^{-1} per atom.

The heat of formation of CdTe is found to be -0.93 eV, which is in good agreement with respect to its experimental value of -0.96 eV [3]. The energies of the bulk Cd and Te were calculated by relaxing the 2-atoms and 3-atoms unit cells of the experimental structures, respectively. The Brillouin zone was sampled by a $22 \times 22 \times 10$ and $9 \times 9 \times 6$ Monkhorst-Pack k-points grid sampling for Cd and Te, respectively.

2.2. QM/MM appoach

The embedded cluster calculation was performed by the computational chemistry code Chemshell [13], which acts as a linker between FHI-aims and GULP [14], an interatomic potential code used in the classical region. FHI-aims performs all the quantum calculations obtaining the interacting forces which goes directly to GULP in order to move the atoms, point charges, etc. Our QM region setup is a Cd_{32} Te_{43} cluster embbeded into a MM region of about 5000 point charges and 200 cadmium pseudopotentials. The forcefield for classical interactions include the Buckingham and Lennard-Jones potentials and a harmonic-type term for the polarizability. The basis used in the QM region are tight (as explained before), and the lattice parameter used to construct the cluster is the one calculated by the Birch-Murnaghan equation of state (6.625 Å). The norm-conserving pseudopotential was constructed by the FHI98PP pseudopotential program [15] using the procedure of Troullier and Martins.

3. Results

3.1. Tellurium vacancy in SC512 supercell: Geometric and electronic properties

Geometric relaxations are important as they play an important role in formation energies, specially for ionized states. When a Te atom in removed from the CdTe lattice, the arrangement of atoms in the neighborhood of the defect will adjust to lower the energy of the system.

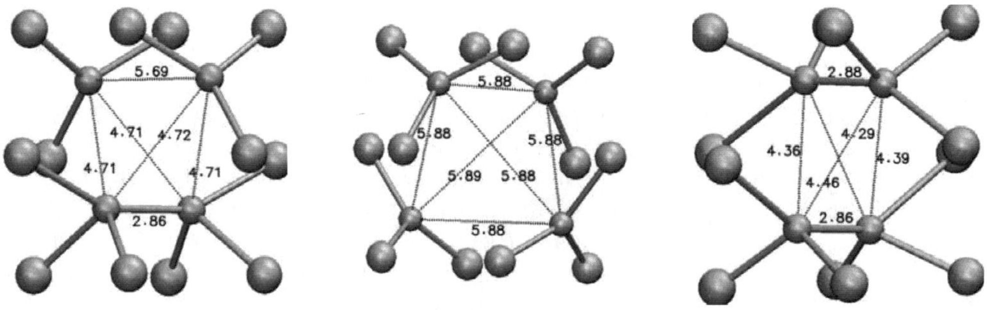

Figure 1. C_{2v} symmetry (left) for neutral and negative charge states, T_d^{out} symmetry (middle) for positive charge states and D_{2d} symmetry (right) for 2- charge state in the QM/MM approach.

The T_d symmetry of the atomic arrangment around the vacancy is conserved in the relaxation for both the 1+ and 2+ ionized states. The neighboring cadmium atoms relaxed outward, keeping the symmetry, locating approximately at the center of the triangles formed by the neighboring tellurium atoms. Let us denote this configuration as T_d^{out} in order to differenciate it from the configuration with inward relaxation T_d^{in} and from the absence of relaxation T_d^{ideal}. The Cd-Cd distance in this configuration is 5.88 Å. On the other hand, for the neutral and negative states, two of the four Cd atoms get closer inducing a C_{2v} distortion. This distortion leads to formation of a Cd_2 dimer with a bond length between 2.86 Å and 2.88 Å, depending of the charge state. It is important to mention that our calculations show that C_{2v} symmetry is the most stable structure for the 2− charge state, at difference with previous reports. For example, some groups have found a D_{2d} symmetry for this charge state [4; 7], which has a higher energy with respect to our structure by 0.4 eV, approximately. We found this symmetry as well in the QM/MM approach and it will be explained in the corresponding section 3.2.

Some advantages of using a 512-atoms supercell is that the band-filling correction is not necessary, since we are using only one k-point in the calculation. On the other hand, the Coulomb interaction between charged defects is reduced as much as possible, reaching its maximum value of 0.17 eV for 2± charge state. For all the ionized states considered here, the effects of the vacancy are small band shifts and splits.

3.1.1. Defect energy states The defect energy states were found in both the neutral and 1+ charge states. In the neutral state (with C_{2v} symmetry) one level is located in the band gap and the other level is in the conduction band (Fig. 2(a) 2(b)). On the other hand, the 1+ state (with T_d^{out} symmetry) both levels are located in the conduction band (Fig. 2(c) 2(d)). This can be confirmed through the total density of states (DOS) of the pristine crystal calculation and

IOP Conf. Series: Journal of Physics: Conf. Series **1043** (2018) 012043 doi:10.1088/1742-6596/1043/1/012043

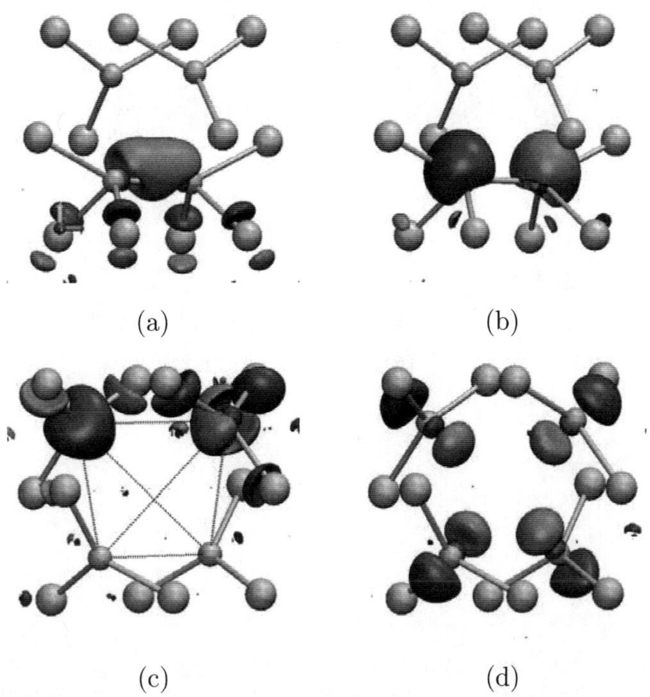

Figure 2. Wavefunction isosurface for neutral charge state (top panels) and 1+ state (bottom panels). (a) level in the band gap (b) level in the conduction band (c) and (d) levels in the conduction band, respectively.

the charged states. In addition, we also plotted the wavefunction isosurface onto those states and we checked their energy location.

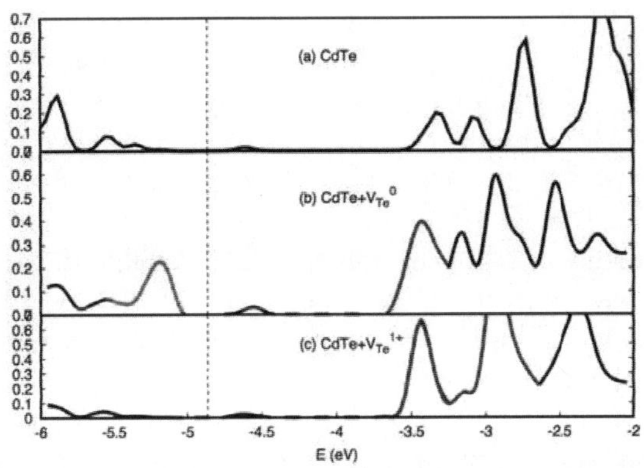

Figure 3. PBE total DOS (a) in ideal CdTe (b) tellurium vacancy with neutral charge state and (c) tellurium vacancy with single positive charge state through the PBC approach. The vertical dashed line indicates the Fermi level of the ideal CdTe and also the defect levels are shown in red and blue colors.

The total density of states shown in Fig. 3 correspond to the supercell calculation using only the Γ-point and, for this reason, the Van Hove singularity can not be seen. Nevertheless, using a denser K-point mesh will fix this problem.

The value of the wavefunction isosurface was set to ±0.044 a.u., in both charge states. The location of these defect energy levels are: For neutral charge state, one level is located in the

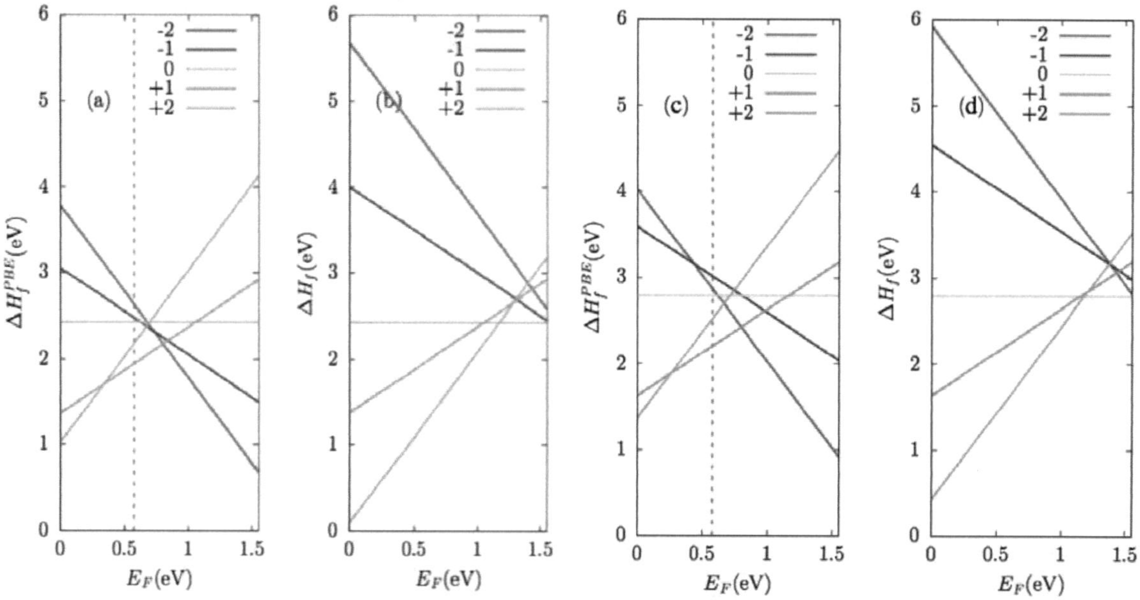

Figure 4. Formation energies as a function of the Fermi energy for the Te vacancy in Te-rich conditions. (a) and (c) are PBE formation energies. The vertical dashed line indicates the gap at PBE level. (b) and (d) are GW corrected formation energies. The left pair shows the formation energies in the PBC approach and the right pair for the QM/MM approach. The value E_F=0 eV represents the VBM and E_F=1.55 eV the CBM.

band gap at -5.15 eV and the other level is in the conduction band at -3.42 eV (red and blue lines in Fig. 3(b)). In the 1+ charge state, these levels are both located in the conduction band at -3.43 eV and -2.94 eV (blue lines in Fig.3(c)).

3.1.2. Formation energy It is well known that GGA underestimate the band gap of materials, causing errors in the formation energy. To obtain the band edges using semilocal exchange-correlation functionals one needs a subsequent beyond-DFT calculation, like the GW approximation. The corrections to the band edges were reported in Ref. [7].

Fig. 4 shows the formation energies with and without corrections. From here we have that 0 and 2+ states are the only stable states. The quasiparticle corrections have been applied, shifting the VBM by -0.470 eV and the CBM by +0.481 eV [7]. The 1+ state is insensitive to the quasiparticle corrections due a correction cancellation between the modification of the Fermi level and the donor correction. These corrections are not required for the neutral state. As this method gives a band gap of 1.55 eV, this is the maximum value of the Fermi level. The calculated thermodynamic transition level is $\varepsilon(2+/0)$=1.17 eV.

3.2. Tellurium vacancy in the QM/MM approach: Geometric and electronic properties
The geometric relaxation was computed by the Chemshell code and the structure was fully relaxed until residual forces fell below 10^{-3} eV/Å. In order to take account for the polarization response, we have calculated the static dielectric constant (ε_0=11.972) and the high-frequency dielectric constant (ε_∞=8.707).

All the strucures were constrained to relax only in the inner QM region (i.e. the QM boundary layer was fixed during the relaxation) in order to avoid asymmetric movement of ions near the coupling region. The relaxed structures correspond to the same structure in the supercell approach, except for the 2− ionized state, which presents a D_{2d} distortion (Fig.1), leading to the formation of two perpendicular Cd dimers with bond length of 2.88 Å. This well known size problem is described in Ref. [7]. Bond lengths are almost kept for both approaches, varying between 0.04 Å to 0.09 Å in some cases.

3.2.1. Electronic properties In the QM/MM approach it is important to recover the electronic properties of the extended periodic material. For this reason, we have calculated the total density of states (DOS) and we compared it with respect to a 64-atoms supercell calculation (Fig.5).

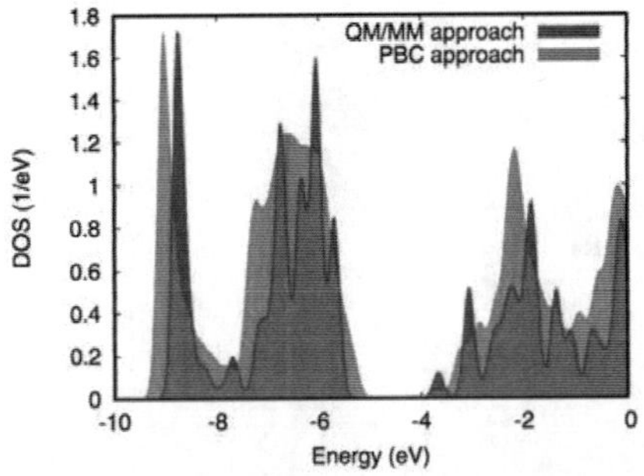

Figure 5. Comparison of the total density of states (DOS) using pseudopotentials, point charges and polarization response. Even a relatively small embedded $Cd_{32}Te_{43}$ cluster recovers important features from the reference bulk calculation (green color).

The HOMO-LUMO gap is overestimated and it is more pronounced in small clusters. As reference, the periodic boundary conditions calculation leads to a gap of 0.57 eV while in the QM/MM approach, using a $Cd_{32}Te_{43}$ cluster, the gap reaches up to 2.03 eV. This is known as quantum confinement effect and is qualitatively described in chapter 1 in ref.[16].

3.2.2. Defect energy states The defect energy states were found for the same charge states that were studied with periodic boundary conditions in section 3.1.1. The neutral charge state (with C_{2v} symmetry) the defect states are located in the band gap and conduction band (Fig. 7(a) and (b)). Meanwhile in 1+ charge state (with T_d^{out} symmetry) these levels are both located in the conduction band (Fig. 7(c) and (d)). We confirmed this through the density of states of both charge states and the pristine calculation, as well as plotting the wavefunction isosurface onto

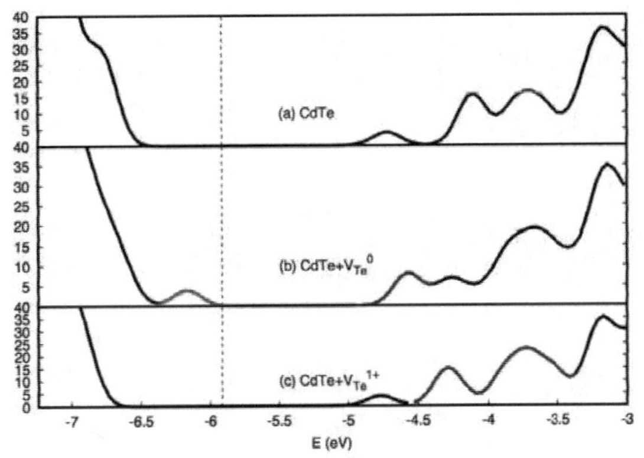

Figure 6. PBE total DOS (a) in ideal CdTe (b) tellurium vacancy with neutral charge state and (c) tellurium vacancy with single positive charge state through the QM/MM approach. The vertical dashed line indicates the Fermi level of the ideal CdTe and also the defect levels are shown in red and blue colors.

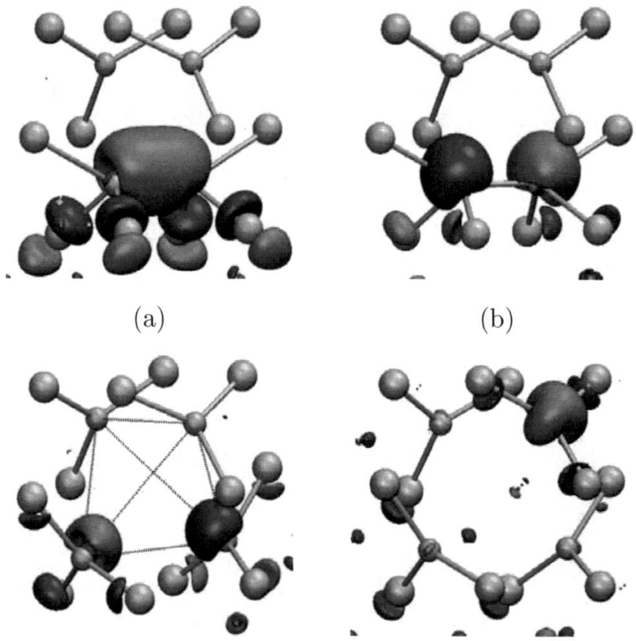

(a) (b)

(c) (d)

Figure 7. Wavefunction isosurfaces in the QM/MM approach for neutral charge state (top panels) and 1+ state (bottom panels). (a) level in the band gap (b) level in the conduction band (c) and (d) levels in the conduction band, respectively.

those states. The value of the wavefunction was set to ±0.085 a.u. The location of these states are: For neutral, the one in the band gap is located at -6.17 eV and the other is at -4.56 eV, inside the conduction band. In the 1+ charge state, these levels are located in the conduction band at -4.33 eV and -3.76 eV (Fig. 6).

3.2.3. Formation energies Potential alignment was carried out comparing by the density of states (DOS) for all the charge states. They were aligned with respect to deep energy levels (~ -17 eV) of the pristine cyrstal calculation. Fig. 8 shows this alignment in a simple way.

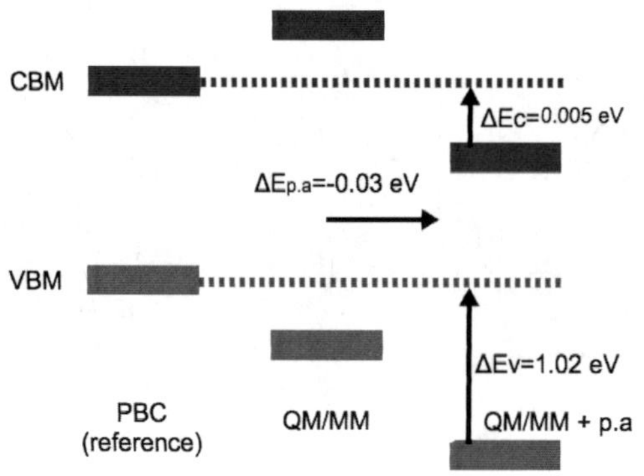

Figure 8. Diagram of the potential alignment for neutral charge state from QM/MM to PBC approach. The red dashed line is the VBM reference and the blue dashed line is the CBM reference from the PBC approach.

Table 1. Potential alignment for all the charge states (in eV). ΔE_v and ΔE_c are the energy differences with respect to the periodic VBM and CBM, respectively.

Ionized state	2-	1-	0	1+	2+
E_v	-3.59	-5.14	-6.17	-7.95	-8.99
E_c	-2.45	-3.57	-4.58	-5.82	-6.82
ΔV	-2.14	-1.05	-0.03	1.05	2.14
$E_v^{p.a}$	-5.73	-6.19	-6.20	-6.90	-6.85
$E_c^{p.a}$	-4.59	-4.62	-4.61	-4.77	-4.68
ΔE_v	0.55	1.01	1.02	1.72	1.67
ΔE_c	-0.02	0.02	0.01	0.17	0.08

Once again, in Fig. 4 we observe that the 0 and 2+ states are the only stable states found. Compared to the previous periodic boundary approach, all the energies are shifted upwards. However, despite of this shifting, the tendency is kept in both approaches. The calculated thermodynamic transition level in this approach is found to be $\varepsilon(2+/0)$=1.18 eV.

4. Conclusions

In the supercell approach the relaxed structures found have C_{2v} symmetry for the neutral and negatives charge states and T_d^{out} symmetry for the positives states. We observe that the first symmetry has a localized state above the VBM that is shifted upward when the symmetry is changed. In addition, our calculations predict that in the 2- charge state, V_{Te}^{2-} takes the C_{2v} symmetry, which was not found previously. The thermodynamic transition level found is $\varepsilon(2+/0)$=1.17 eV. Nevertheless, one calculation has predicted a negative formation energy for V_{Te}^{2+} in Te-poor conditions (-0.84 eV). Negative formation energies indicate a trend to high defect concentration and this scheme needs to be studied by a different methodology that correct this result [17].

On the other hand, the relaxed structures found in the QM/MM approach are: D_{2d} for the 2− state, C_{2v} for the 1− and neutral state and T_d^{out} for positive states. All the ionized states are the same in both approaches but for the 2− state, which presents a different structure due size effects. One crucial step was the potential alignment that was performed by comparing the total density of states of the pristine crystal with respect to the charged ones. Finally, the formation energies are slightly shifted upward compared with the supercell calculations, but the tendency was kept in this approach. Once again, only two states seems to be stables and the thermodynamic transition level found is $\varepsilon(2+/0)$=1.18 eV, just one hundredth of eV above the supercell result.

Acknowledgments
We would like to thank to FONDECYT project 1130437 for the funding, to Volker Blum (Duke University), Daniel Berger (TÜM) and Andy Logsdail (UCL) for all their help and support through this work and finally to the NLHCP for the computational resources offered.
Powered@NLHPC: This research was partially supported by the supercomputing infrastructure of the NLHPC (ECM-02).

References

[1] Biswas K and Du M H 2012 *New. J. Phys.* **14** 063020
[2] Shockley W and Queisser H J 1961 *J. Appl. Phys.* **32** 510

IOP Conf. Series: Journal of Physics: Conf. Series **1043** (2018) 012043 doi:10.1088/1742-6596/1043/1/012043

[3] Triboulet R and Siffert P 2010 *CdTe and Related Compounds; Physics, Defects, Hetero- and Nano-structures, Crystal Growth, Surfaces and Applications* (Radarweg 29, PO Box 211, 1000 AE Amsterdam, The Netherlands: Elsevier)

[4] Du M H, Takenaka H and Singh D J 2008 *Phys. Rev.* B **77** 094122

[5] Ma J, Kuciauskas D, Albin D, Bhattacharya R, Reese M, Barnes T, Li J V, Gessert T and Wei S H 2013 *Phys. Rev. Lett.* **111** 067402

[6] Yang J H, Park J S, Kang J, Metzger W, Barnes T and Wei S H 2014 *Phys. Rev.* B **90** 245202

[7] Menéndez-Proupin E and Orellana W 2015 *Physica Status Solidi* B **252** 2649

[8] Run X, Xu H T, Tang M Y and Wang L J 2014 *Chin. Phys.* B **23** 077103

[9] Wei S H and Zhang S B 2002 *Phys. Rev.* B **66** 155211

[10] Menéndez-Proupin E, Amézaga A and Cruz Hernández N 2014 *Physica* B **452** 119

[11] Blum V, Gehrke R, Hanke F, Havu P, Havu V, Ren X, Reuter K and Scheffler M 2009 *Computer Physics Communications* **180** 2175

[12] Perdew J P, Burke K and Ernzerhof M 1996 *Phys. Rev. Lett.* **77** 3865

[13] Chemshell, a computational chemistry shell `www.chemshell.org`

[14] Gale J D 1997 *J. Chem. Soc., Faraday Trans.* **93**

[15] Fuchs M and Scheffler M 1999 *Comput. Phys. Commun.* **119** 67

[16] Borrelli N 2003 *Semiconductor Nanocrystals: From Basic Principles to Applications* Nanostructure Science and Technology ed Efros A L, Lockwood D J and Tsybeskov L (New York: Kluwer Academic / Plenum Publishers)

[17] Richter N A 2013 *Phys. Rev. Lett.* **111** 045502

IOP Conf. Series: Journal of Physics: Conf. Series **1043** (2018) 012044 doi:10.1088/1742-6596/1043/1/012044

Theoretical study of bismuth-doped CdTe

J A Ríos-González[1], **E Menéndez-Proupin**[2] **and J L Peña**[1]

[1] Departamento de Física Aplicada, CINVESTAV Unidad Mérida, C.P. 97310 Mérida, Yucatán, México

[2] Departamento de Física, Facultad de Ciencias, Universidad de Chile, Las Palmeras 3425, 780-0003 Ñuñoa, Santiago, Chile

E-mail: emenendez@uchile.cl

Abstract. The electronic structure and thermodynamic properties of bismuth impurities in cadmium telluride (CdTe) are studied by means density functional theory calculations. The exchange and correlation are treated using the hybrid functional HSE, where the range-separation parameter is tuned to fit the CdTe band gap. The spin-orbit coupling is included in the calculations, showing a strong effect on the Bi-related energy levels. The band diagram and the formation energies are obtained for substitutional and interstitial configurations. It is shown that simple substitutional impurities can generate in-gap bands, which can act as intermediate levels in two-photon absorption and can be appropriate for intermediate-band solar cells.

1. Introduction

Cadmium telluride (CdTe) is a semiconductor compound employed as photovoltaic material which is growing rapidly in acceptance, becoming a cheap alternative to the silicon solar cells. CdTe is also widely used for radiation detectors [1]. A recent study [2] has shown a significant increase of the short-circuit current in solar cells doped with Bi. This increase has been attributed to the presence of an intermediate band (IB) that allows the promotion of electrons from the valence band (VB) to the conduction band (CB) by two-step transitions VB to IB to CB, through the absorption of sub-bandgap photons. Previous results had also suggested the existence of an intermediate band in Bi-doped CdTe [3–5]. Seminovski et al [6] have proposed several complex defects, i.e., $2Bi_{Cd}$, Bi_i–V_{Cd}, Bi_{Cd}–O_{Te} that can create an IB, but did not find IB associated to simple substitutional defects. Here, it is shown that the simple Bi_{Cd} and Bi_{Te} can create IB. The crucial ingredient is the spin-orbit coupling (SOC).

2. Methodology

2.1. Electronic structure calculations

Quantum calculations if Bi impurities in supercell models have been performed using density functional theory (DFT). DFT calculations have been done using a plane-wave projector augmented wave [7, 8] scheme, as implemented in the Vienna Ab Initio Simulation Package (VASP) [9]. The band diagrams and total energies have been obtained using the the hybrid functional proposed by Heyd, Scuseria and Ernzerhof (HSE06) [10] with the range separation parameter modified to reproduce the band gap of 1.55 eV given by G_0W_0@HSE06 calculation. SOC is included in this framework. This functional will be called HSE(ω)+SOC, with the fitted value $\omega = 0.0811$ Å^{-1}. The wavefunctions have been expanded in plane waves with cutoff energy

IOP Conf. Series: Journal of Physics: Conf. Series **1043** (2018) 012044 doi:10.1088/1742-6596/1043/1/012044

of 265 eV. This is sufficient to obtain converged forces and total energies using the library of PAW potentials for GW calculations. This PAW library is generally useful for DFT calculations, a fact that we have verified computing the band diagram, lattice parameter and bulk modulus (6.54 Å and 41.05 GPa, respectively). The latter have been obtained by fitting the Birch-Murnaghan equation of state.

In our computational model, the Bi impurities substitute one Cd or Te atom in a 64-atoms supercell (SC64). To sample the reciprocal space, a Γ-centered $2 \times 2 \times 2$ k-point grids has been used. The atomic positions were relaxed using the generalized gradient approximation (GGA) for the exchange-correlation functional as proposed by Perdew, Burke, and Ernzerhof (PBE) [11].

The calculated CdTe heat of formation, $\Delta H(\mathrm{CdTe})$ is found to be of -0.78 eV with PBE and -1.19 eV with HSE(ω)+SOC. The latter is in good agreement with the updated experimental value -1.04 eV [12]. The energy of bulk Cd has been obtained by relaxing the 2-atoms crystallographic unit cell. The Brillouin zone was sampled by a $22 \times 22 \times 10$ k-point grid. The energy of bulk Te has been obtained relaxing the crystallographic 3-atoms unit cell, while the tellurium Brillouin zone was sampled by a $9 \times 9 \times 6$ k-point grid grid.

The formation energy of substitutional bismuth impurities, $\mathrm{Bi_X}$ (X=Cd, Te), has been obtained computing the difference between the total energies of a crystal supercell containing n CdTe units, $E(\mathrm{Cd}_n\mathrm{Te}_n)$, and the energy of the same supercell with a substitution. Therefore, the formation energy of $\mathrm{Bi_{Cd}}$ is

$$\begin{aligned} \Delta H_f(\mathrm{Bi}^q_{\mathrm{Cd}}) &= E(\mathrm{Cd}_{n-1}\mathrm{Te}_n\mathrm{Bi}) - E(\mathrm{Cd}_n\mathrm{Te}_n) + [E(\mathrm{Cd}) + \Delta\mu_{\mathrm{Cd}}] \\ &\quad -[E(\mathrm{Bi}) + \Delta\mu_{\mathrm{Bi}}] + q[E_V + E_F] + \Delta E_{size}. \end{aligned} \tag{1}$$

To obtain the formation energy of $\mathrm{Bi_{Te}}$, it suffices to exchange Cd with Te in Eq.(1). Here, $E(\mathrm{Cd})$ is the energy of Cd in a reference state, in this case it is the energy per atom of bulk Cd. $\Delta\mu_{\mathrm{Cd}}$ is the chemical potential relative to the reference state, and it depends on the thermodynamic equilibrium conditions. In Cd-rich conditions $\Delta\mu_{\mathrm{Cd}} = 0$, while in Cd-poor conditions $\Delta\mu_{\mathrm{Cd}}$ equals the CdTe heat of formation $\Delta H(\mathrm{CdTe})$ [13]. For a charged vacancy, q is the number of electrons donated to the environment, while $E_V + E_F$ is the electron chemical potential expressed as the sum of the valence band maximum (VBM) E_V and the Fermi level E_F relative to the VBM. ΔE_{size} includes several size corrections that are explained in [14]. To avoid precipitation of pure Bi or $\mathrm{Bi_2Te_3}$, the chemical potentials must satisfy

$$\Delta\mu_{\mathrm{Bi}} \leq 0, \text{ and } \quad 2\Delta\mu_{\mathrm{Bi}} + 3\Delta\mu_{Te} \leq \Delta H(\mathrm{Bi_2Te_3}), \tag{2}$$

where $\Delta H(\mathrm{Bi_2Te_3}) = -0.90$ eV is the calculated $\mathrm{Bi_2Te_3}$ heat of formation. The above inequalities imply that in Cd-rich conditions $\Delta\mu_{\mathrm{Bi}} \leq 0$, while in Te-rich condition, $\Delta\mu_{\mathrm{Bi}} \leq -0.45$ eV.

3. Electronic structures

In this section we analyze the band structure of substitutional defects $\mathrm{Bi_{Cd}}$ and $\mathrm{Bi_{Te}}$. The band diagrams and density of states (DOS) have been calculated for SC64 models.

3.1. Bi_{Cd}

$\mathrm{Bi_{Cd}}$ adopts a configuration with T_d symmetry, where the neighboring Te are 3.00 Å far from Bi. Figure 1 shows the band structure calculated for perfect CdTe and for $\mathrm{Bi_{Cd}}$ in the SC64 model. It can be seen that the impurity introduce a band below the CBM. This band is localized, as suggested by its flatness, as well as by comparing the PDOS on the first- and second-neighbors

IOP Conf. Series: Journal of Physics: Conf. Series **1043** (2018) 012044 doi:10.1088/1742-6596/1043/1/012044

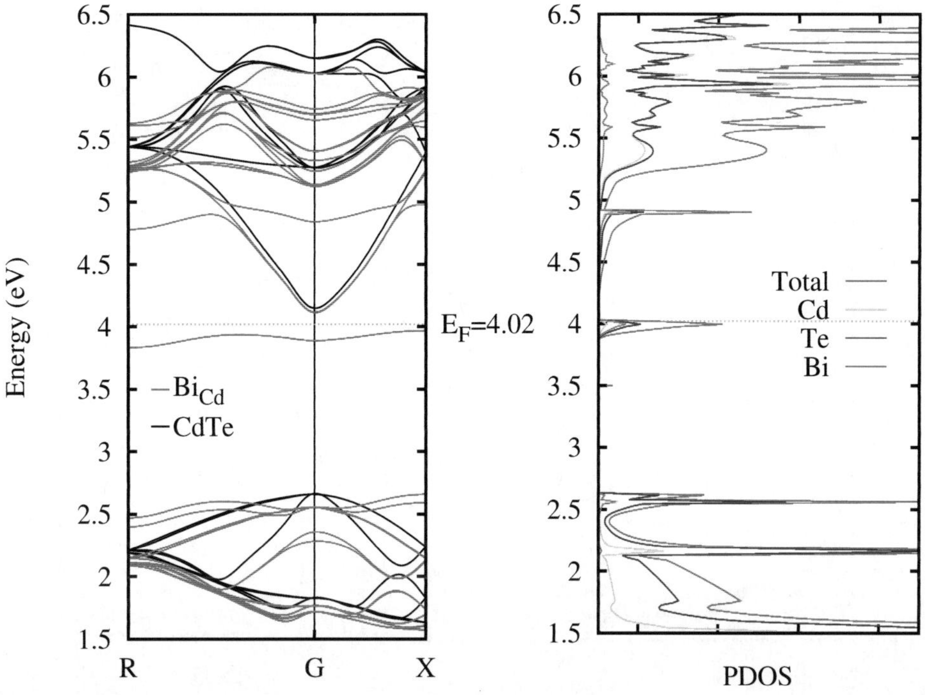

Figure 1. (Color online) Band structure of Bi_{Cd} in CdTe performed with a 64-atom supercell. Black lines are the bands for the perfect crystal. Red lines are the bands of the defective supercell in neutral charge state. Dashed lines indicate the Fermi level (E_F). The PDOS is shown at the right.

(not shown). This band is non degenerate and it is occupied in the neutral state of the impurity. The center of the IB is 0.105 eV below the CBM. The charge state 1+ is also possible, making Bi_{Cd} a donor impurity. A half-filled IB, as wanted for IB solar cells [15], can be obtained in combination with acceptor impurities or intrinsic defects, e.g., cadmium vacancies.

3.2. Bi_{Te}

Figure 2 shows the band structure calculations of the perfect CdTe and Bi_{Te} in the SC64 supercell. Bi_{Te} is a deep acceptor with charge states 0 and 1−. In sufficient concentration, as in the SC64 model, Bi_{Te} can create a non-degenerate IB. The center of the IB is 0.80 eV below the CBM. The PDOS shows that the IB has significant contributions from the Te atoms in the supercell. This contribution comes mainly from the second neighbors of Bi (recall that the first neighbors are Cd). The top of the valence band is modified by the presence of Te. The upper part of the valence band in pure CdTe is mainly due to the Te 5p orbitals [16]. Here, Bi_{Te} splits the VBM and causes a complex band structure with some dispersion, formed by Bi and Te orbitals, as it can be appreciated in the DOS. This upper band is occupied, and may act as a hole killer in case of low Bi concentrations.

3.3. Formation energies

Figure 3 shows the calculated formation energies using the SC64 model. For Bi_{Cd}, only the neutral and 1+ charge states are stable for Fermi level within the CdTe band gap. The transition

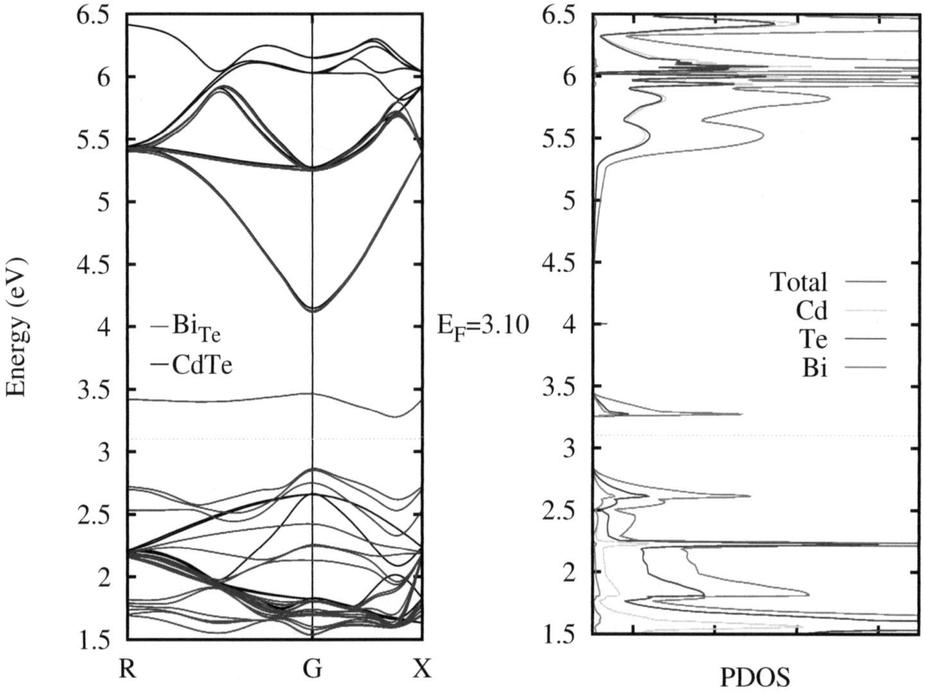

Figure 2. (Color online) Band structure of Bi_{Te} in CdTe performed with a 64-atom supercell. Black lines are the bands for the perfect crystal. Blue lines are the bands of the defective supercell in neutral charge state. Dashed lines indicate the Fermi level (E_F). The PDOS is shown at the right.

level is $\varepsilon(1+/0) = 1.29$ eV. For Bi_{Te}, the 1+, 0 and 1− charge states are stable, with transition levels $\varepsilon(1+/0) = 0.05$ eV and $\varepsilon(0/1-) = 0.84$ eV.

The relative stability of Bi_{Cd} vs Bi_{Te} depends strongly on the relative abundance of Cd and Te. In Cd-rich condition Bi_{Te} has lower formation energy than Bi_{Cd}, being most probable in thermodynamic equilibrium. This order is reversed in Te-rich condition. Therefore, the predominant type of substitution can be tuned if the Cd and Te chemical potentials can be controlled. In Te-rich conditions, Bi_{Cd} is favored. Bi_{Cd}^{1+} acts as a deep donor, compensating the Cd vacancies. Bi_{Te} is favored in Cd-rich condition, acting as a deep acceptor. It can also donate electrons for Fermi levels close to the VBM, acting as compensator.

4. Conclusions

A modified hybrid functional of the HSE family has been proposed to obtain accurate band structures of CdTe and use it to study defects that have active levels in the gap. This framework has been applied to the study of Bi impurities of CdTe. It is shown that Bi substituting both Cd and Te cause deep levels in the gap and, in sufficient concentrations, can induce an IB. The positioning of the IB in the band gap is accomplished only when the SOC is included in the calculations. These results provide a model that may explain the observed increase of the short-circuit current in Bi-doped CdTe solar cells [2].

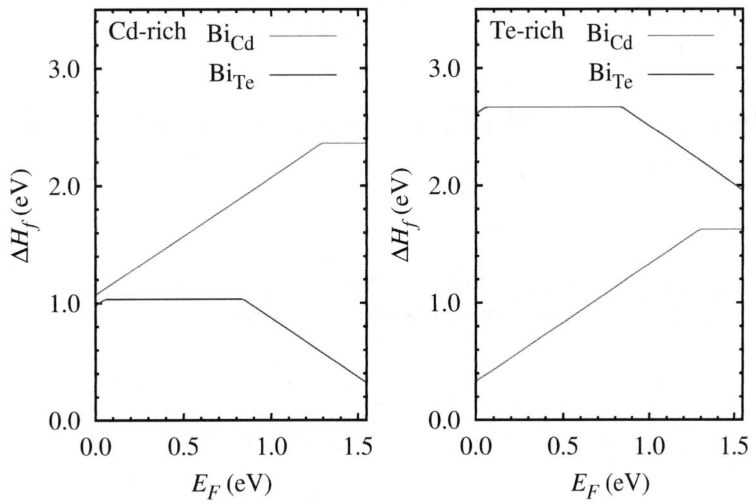

Figure 3. (Color online) Formation energies as a function of the Fermi energy for the Bi_{Cd} and Bi_{Te}, computed with a 64-atoms supercell (SC64).

Acknowledgments

This work was supported by the FONDECYT Grant No. 1130437, CONACYT-SENER SUSTENTABILIDAD ENERGETICA (México) under project CeMIE-Sol PY-207450/25. J.A.R.-G. thanks CONACYT fellowships (for PhD@México and for visit to Chile). Powered@NLHPC: This research was partially supported by the supercomputing infrastructure of the NLHPC (ECM-02). P. Palacios, Y. Seminovski, and P. Wahnón are acknowledged for introducing the topic. M. Flores is acknowledged for stimulating discussions.

References

[1] Biswas K and Du M H 2012 *New J. Phys.* **14** 063020
[2] Vigil-Galán O, Courel M, Cruz-Gandarilla F and Seuret-Jiménez D 2016 *J. Mater. Sci.: Mater. Electron.* **27** 6088–6095
[3] Saucedo E, Ruiz C M, Bermudez V, Dieguez E, Gombia E, Zappettini A, Baraldi A and Sochinskii N 2006 *J. Appl. Phys.* **100** 104901
[4] Saucedo E, Franc J, Elhadidy H, Horodysky P, Ruiz C M, Bermudez V and Sochinskii N 2008 *J. Appl. Phys.* **103** 094901
[5] Kolosov A, Krivobok V S, Klevkov Y V and Adiyatullin A F 2013 *Semiconductors* **47** 561–568
[6] Seminovski Y, Palacios P and Wahnón P 2013 *Sol. Energy Mater. Sol. Cells* **114** 99–103
[7] Blöchl P E 1994 *Phys. Rev. B* **50** 17953
[8] Kresse G and Joubert D 1999 *Phys. Rev. B* **59** 1758
[9] Kresse G and Furthmüller J 1996 *Phys. Rev. B* **54** 11169–11186
[10] Heyd J, Scuseria G E and Ernzerhof M 2006 *J. Chem. Phys.* **124** 219906
[11] Perdew J P, Burke K and Ernzerhof M 1996 *Phys. Rev. Lett.* **77** 3865–3868
[12] Brebrick R F 2010 *J. Phase Equilib. Diffus.* **31** 260–269
[13] Wei S H and Zhang S B 2002 *Phys. Rev. B* **66** 155211
[14] Menéndez-Proupin E and Orellana W 2015 *Phys. Status Solidi B* **252** 2649–2656
[15] Luque A and Martí A 1997 *Phys. Rev. Lett.* **78** 5014
[16] Menéndez-Proupin E, Gutiérrez G, Palmero E and Peña J L 2004 *Phys. Rev. B* **70** 035112

IOP Conf. Series: Journal of Physics: Conf. Series **1043** (2018) 012045 doi:10.1088/1742-6596/1043/1/012045

Radiographic imaging of insects using a copper x-pinch plasma backlighting source

Milenko Vescovi[1], **Felipe Veloso**[1,2,3], **Gonzalo Muñoz-Cordovez**[1], **Vicente Valenzuela-Villaseca**[1], **Mario Favre**[1,2] **and Edmund Wyndham**[1]

[1] Instituto de Física, Pontificia Universidad Católica de Chile, Santiago, Chile.
[2] Centro de Investigación en Nanotecnología y Materiales Avanzados "CIEN-UC", Pontificia Universidad Católica de Chile, Santiago, Chile

E-mail: fveloso@fis.puc.cl

Abstract. The use of copper x-pinch plasmas as backlighting source for radiographing insects is described. The experiments use 4x 25μm Cu wires driven by a ~400kA, ~350ns current pulse to produce a localized source of soft x-rays in the micrometer range. Both source size and energy range of the x-rays photons produced by the x-pinch enable the imaging of low contrast objects such as the rich inner structures of insects with excellent spatial resolution. The technique is used in three different insects: a moth, a fly and a beetle. These radiographies allow clear observation of the details of outer (e.g., wings) and internal systems of the insects. A description of the method and ways to improve the quality of the radiographs are shown and discussed.

1. Introduction

Micron-scale imaging of small insects typically uses optical microscopy techniques and further details (i.e., in the nanometer scale) can be obtained using electron microscopy techniques. In both cases, the inner structures of the insects cannot be observed and the use of x-rays is required. However, most insects are either transparent or unresolvable to standard medical radiographies given the photon energies (>50keV) and source size (~100s of μm) they usually produce. Therefore, in order to have well-defined imaging of the rich internal structures of insects, smaller sources of lower energy x-ray photons ($E \lesssim 30$keV) are required. In this matter, synchrotron x-ray sources have appear as ideal backlighting sources [1,2], but the accessibility to this kind of equipment is very limited. On the other hand, plasma backlighting sources are able to produce small size sources of soft x-ray photons (<10keV) at a relatively low cost. Such is the case of the x-pinch plasma which produces a localized, micron-sized bright spot of soft x-rays in the nanosecond temporal scale [3,4]. These x-pinch plasmas are produced after the vaporization of thin metallic wires by a fast (<μs), high current pulse (>100kA) provided by a pulsed power device. These wires are mounted as the load of the pulsed power driver connecting cathode and anode in a crossed-configuration forming the shape of the letter "X". When the high current passes through the wires, they vaporize and become a high-temperature plasma which

[3] To whom any correspondence should be addressed.

is confined by the resulting Lorentz force on the configuration. Several experiments have used x-pinches as backlighting source for probing high energy density plasmas (mostly, wire array z-pinches) and biological objects [5,6]. This paper presents radiographic imaging of small insects using copper x-pinches as backlighting source driven at current levels of ~1kA/ns.

2. Experimental setup

The experiments were conducted at the Llampudken pulsed power generator [7], which delivers ~400kA in ~350ns. The x-pinch load is composed by four copper wires (purity > 99.9%), 25μm diameter each crossing on a single centred point in the inter-electrode region. Since more than two wires are used in our configuration, magnified images of the load are taken before each shot in order to ensure that wires are touching on a single point instead of multiple crossing. Experiments were performed at a background pressure lower than 5 10^{-4} mbar. In order to determinate the size of the x-ray source, a slit-wire camera system is used, which basically consists on analysing the shadows produced by thin wires onto an x-ray film. Using different wire diameters, x-ray filters and geometrical optics considerations, this slit-wire system provides the source size for a given energy range [8]. This parameter is particularly important for the resolution of the soft x-ray imaging system, since higher resolution images require smaller backlighting sources. The insects to be radiographed were located inside of the vacuum chamber attaching them to a holder using either entomological pins or double sided tape. Carestream D-Speed x-ray films filtered with 20μm Al foil were used for imaging and placed behind the insect, thus producing point-projection images of the insect with a magnification of ~1. Figure 1 shows an schematic diagram of the experiment, together with images of the x-pinch load. In addition, a frequency doubled Nd:YAG laser (532nm, 4ns FWHM) and filtered pinhole imaging are used to obtain time-resolved and time-integrated images of the plasma respectively. Since the dynamics of the x-pinch plasma as point-like source of x-rays has been reported in the literature [3,4], this kind of images are good for monitoring the plasma conditions in our radiography experiments. Regarding time-resolved images, properly filtered open-shutter CMOS detectors are used to record laser shadowgraphy imaging. Time-integrated images are taken using the same 20μm Al filtered x-ray films used for radiography images.

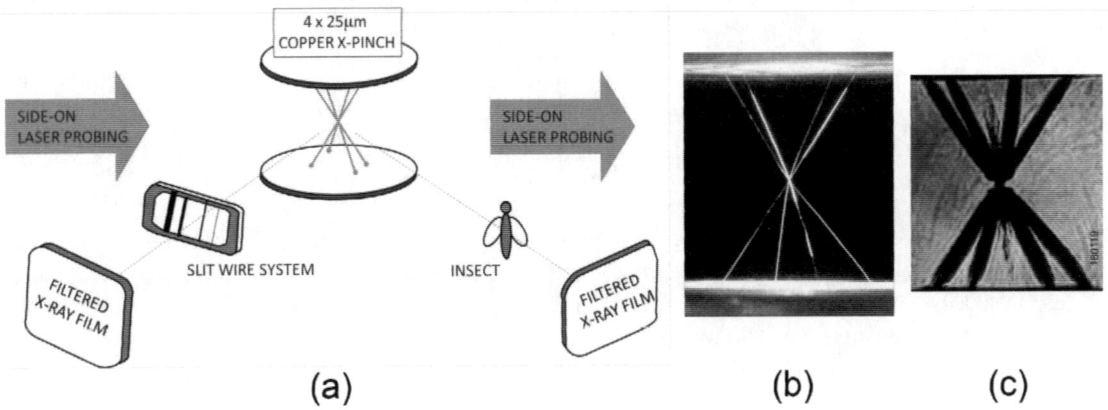

Figure 1. (a) Schematic view of the x-pinch load and diagnostics, (b) Photograph of a 4x 25μm copper x-pinch before discharge. (b) Laser shadowgraphy of the x-pinch plasma. It can be seen the formation of the diode gap at the wires crossing point which serves as x-ray source.

3. Experimental results and discussion

In X-pinch experiments, the quality of the backlighting source in terms of spatial localization and brightness is strongly dependant of the formation of a dense hot-spot at the crossing point of the wires. Since four wires are used in our experiments it is possible to obtain several crossings producing spatially separated hot-spots and therefore, a lower quality backlighting source for radiography purposes. This effect is qualitatively investigated by intentionally mounting the copper wires load using either a single or multiple crossing points, as shown in Figure 2. In single crossing point experiments, the time-integrated images clearly show half of the x-pinch plasma with a strong emission at the crossing. This strong and localized emission at the crossing is the source used for backlighting in the radiographic images of insects shown in Figure 3. The observation of half x-pinch plasma has been ascribed in the literature as the acceleration of electron beams from the crossing through the plasma of the wires after the formation of the gap that separates the top and bottom sides (known as diode-gap) towards the anode surface [3]. However, when multiple crossings are used the quality of the backlighting source is significantly reduced. In order to obtain some observable contrast in the self-emission images, more processing time of the x-ray film in both the developer and fixer solution is required. Regarding radiographic images, when these multiple crossings are observed the images become either blurred or no radiographic image at all.

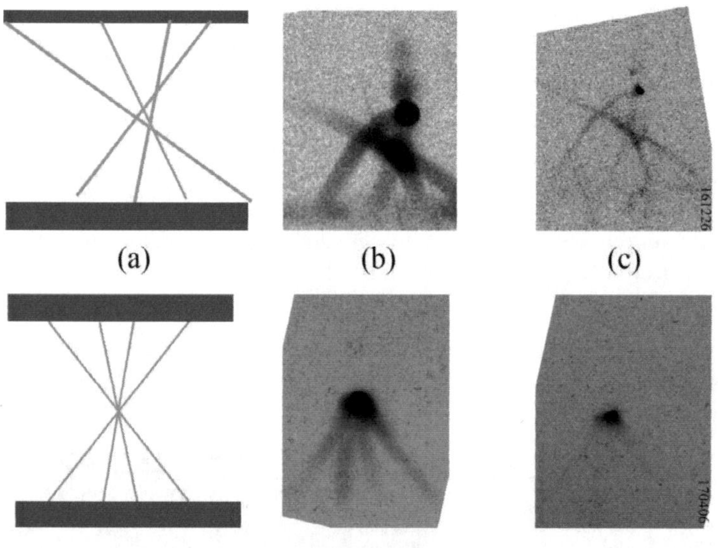

Figure 2: Pinhole images of different quality 4-wires x-pinch loads. Top row corresponds to multiple crossing points where no radiographies were obtained. Bottom row shows a single crossing with good quality radiographies. (a) Schematic diagram of the loads. (b) 500μm pinhole diameter, and (c) 150μm pinhole diameter

In order to characterize the backlighting source size, a typical slit-wire image is shown in Figure 4. In this case, the system is filtered with a 50μm Al filter and a set of 13μm, 25μm, 50μm 75μm diameter wires is used. From this kind of images, it is determined that the source size of soft x-ray photon in our present experiments must be $\lesssim$30μm [9]. Hence, the use of this backlighting source allows us to observe structures within the insects of the order of a few tens of microns. In order to test the capability to obtain radiographies of insects, Figure 3 shows three different insects: a moth, a housefly and a beetle where their rich structure can be observed. In moth and housefly radiographs, their inner structures can be easily observed, together with a detail view of their wings and legs. Nonetheless, the exoskeleton of beetles has considerable more absorption to the soft x-rays. This situation could be overcome by using fewer x-ray filtering the film, but this can also be detrimental to the observation of the areas not covered the exoskeleton (such as the abdomen). If both areas are needed to be observed simultaneously, special careful on the x-ray filtering must be paid.

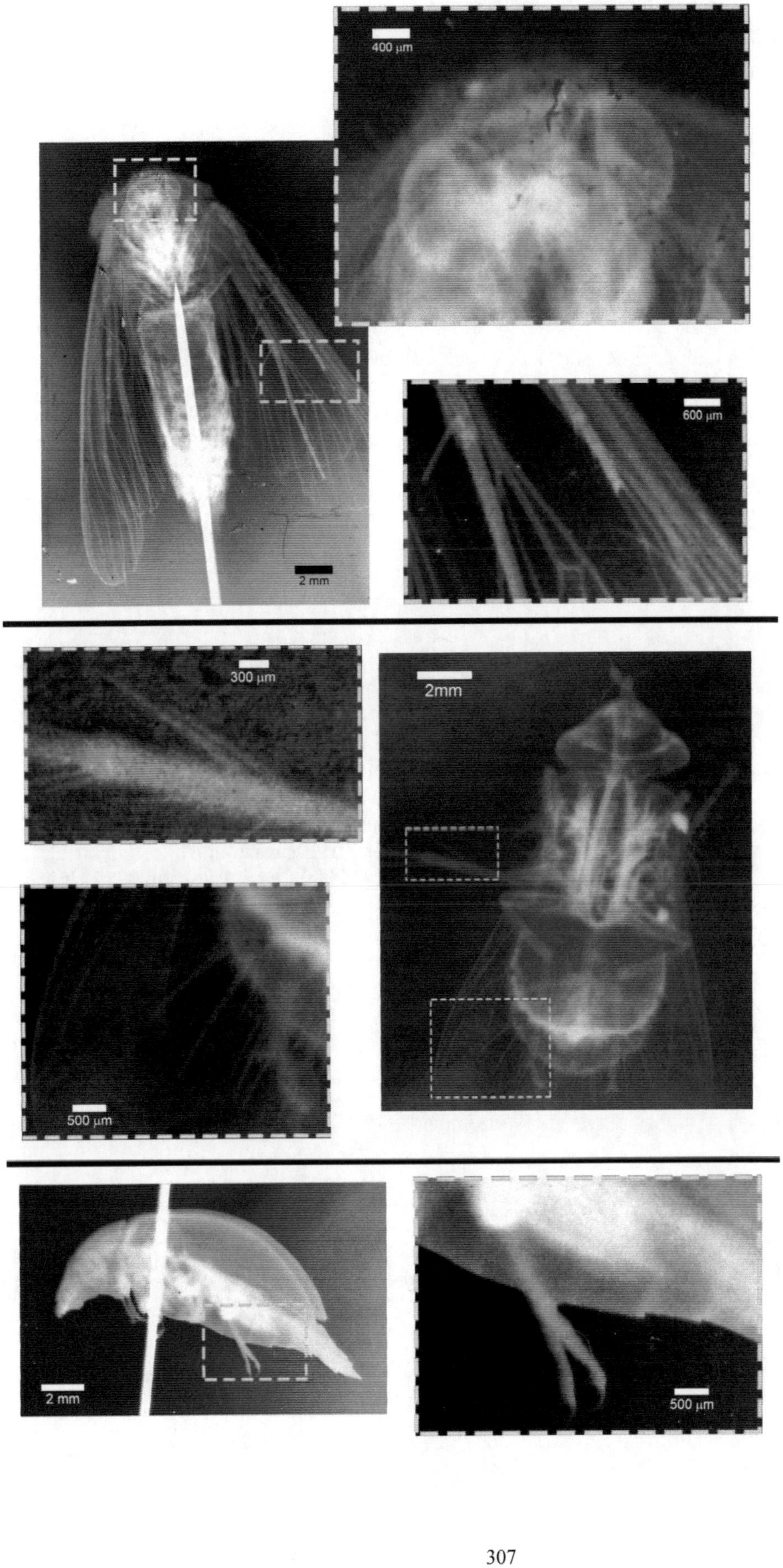

Figure 3. Radiography of different insectarium insects. Top: A moth (noctuidae) Center: A housefly Bottom: A beetle. Both moth and beetle are positioned in a 0.5mm diameter entomological pin. Image contrast in zoomed regions has been digitally adjusted.

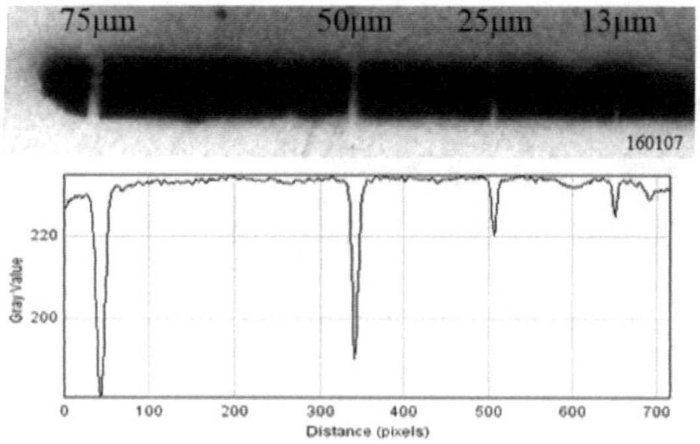

Figure 4: Slit-wire image of a typical shot. Wire diameters are indicated, together with a lineout of the image.

The capability of obtaining radiographic images of insects using 4x25μm Cu wires in the Llampudken generator has been demonstrated. Nevertheless, there are some aspects that must be considered if higher resolution imaging is required. Firstly, similar experiments have reported that source size in x-pinches can be reduced to spatial scales of the order of one micron or less [3]. This can be achieved by properly optimizing the x-pinch load to the available pulsed power driver characteristics. In addition, higher magnification images facilitate the detection of wave properties of the soft x-rays (such as refraction and diffraction) produced by the edges on the object being radiographed which also increases the resolution [10]. Future experiments will consider the improvement on the resolution of our radiographic images using these features.

4. References

[1] Socha J J, Westneat M W, Harrison J F, Waters J S and Lee W 2007 BMC Biology 5, 6
[2] Socha J J, Forster T D and Greenlee K J 2010 Resp Physiol Neurobio 1735, 565-573
[3] Pikuz S A, Shelkovenko T A and Hammer D A 2015 Plasma Phys Rep 41, 291
[4] Pikuz S A, Shelkovenko T A and Hammer D A 2015 Plasma Phys Rep 41, 445
[5] Lebedev S V et al 2001 Rev Sci Instrum 72, 671
[6] Shelkovenko T A, Pikuz S A and Hammer D A 2016 Plasma Phys Rep 42, 226
[7] Chuaqui H, Wyndham E, Friedli C and Favre M 1997 Laser Part Beams 15, 241
[8] Choi P, Dumitrescu C, Wyndham E, Favre M and Chuaqui H 2002 Rev Sci Instrum 73, 2276
[9] Valdivia M P et al 2016 Rev Sci Instrum 87, 11D501
[10] Pikuz S A, Shelkovenko T A, Sinars D B, Chandler K M and Hammer D A 2001 Proc SPIE 4504, 234

IOP Conf. Series: Journal of Physics: Conf. Series **1043** (2018) 012046 doi:10.1088/1742-6596/1043/1/012046

Dynamics of laser produced annular plasmas

V. Valenzuela-Villaseca[1], F. Veloso[1], G. Muñoz-Cordovez[1], M. Favre[1], H. M. Ruiz[2], and E. Wyndham[1]

[1] Instituto de Física, Pontificia Universidad Católica de Chile, Av. Vicuña Mackenna 4860, Santiago, Chile

[2] Departamento de Física, Universidad Técnica Federico Santa María, Av. España 1680, Valparaíso, Chile

E-mail: mfavre@fis.puc.cl

Abstract. We present preliminary experimental observations of the plasma dynamics of laser produced annular plasmas, expanding freely in a 80 mTorr Argon background. A Nd:YAG laser, 340 mJ, 3.5 ns, at 1.06 μm, operating at 10 Hz, is used in the experiments. To produce the annular plasma the laser pulse is focused on a titanium target using a combination of axicon and convergent lens. The dynamics of the laser plasma is characterized by 50 ns time resolved plasma imaging of the visible plasma emission. The inwards radial expansion of the initial plasma ring result in on-axis stagnation, giving rise to an expanding axial laser plasma plume. A quantitative description of the laser plasma dynamics is presented.

1. Introduction

An annular plasma can be produced by focusing a collimated laser beam over a flat solid target, using a combination of axicon prism and convergent lens [1]. These ring-like plasmas have been used to initiate an annular gas-embedded z-pinch [2] or to enhance plasma emission in laser induced breakdown spectroscopy (LIBS) [3,4], among other applications. Here we present preliminary results on the study of the plasma dynamics, both radial and axial, of a laser produced annular plasma.

2. Experimental set-up

In the experiments we use a Nd:YAG laser, 340 mJ, 3.5 ns, at 1.06 μm, operating at 10 Hz. The laser beam is focused over a flat Titanium target, using a combination of axicon and focusing lens, which produces an initial plasma ring of 5 mm radius and 200 μm thickness. Laser power density on target was $1.6 \cdot 10^9$ W/cm^2. In order to be able to compare with point focus laser produced plasmas studied before [5], the experiments were conducted in Argon background, at 80 mTorr. The laser plasma is characterized by 50 ns time resolved plasma imaging of the visible plasma emission using an Andor ICCD camera. The laser produced plasma expands either freely. A schematic of the experimental set-up is presented in figure 1.

3. Results and discussion

Figure 2-a shows images of the initial stages of the laser annular plasma. At $t = 34$ ns the plasma is localized close to the target surface, on the left hand side of the image. The shape corresponds to the emission that of a rather uniform annular plasma, of $\sim$ 5 mm radius. The onset of an

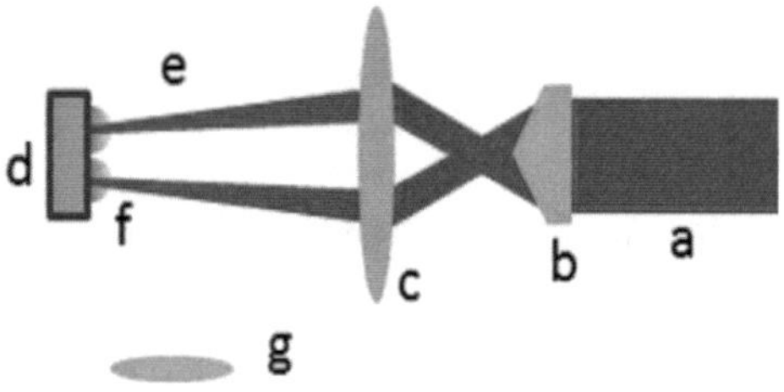

Figure 1. Experimental set-up: a) Laser beam, b) Axicon prism, c) Focusing lens, d) Titanium target, e) Annular focused laser beam, f) Laser produced annular plasma, g) Light collecting assembly for plasma imaging.

axially expanding laser plasma plume is also observed, which is due to stagnation on axis of the initial annular laser produced plasma [1]. At $t = 74$ ns the plasma is already expanding axially, being the dominant feature the expanding plasma plume resulting from on-axis stagnation. This is observed clearly in the radial profile of the light intensity at 2.5 mm from the target surface, shown in figure 2-b. Here it can also be appreciated that at this time the radial profile of light intensity is not symmetrical, which can be explained due to small misalignments in the optical system and/or inhomogeneities of the laser beam.

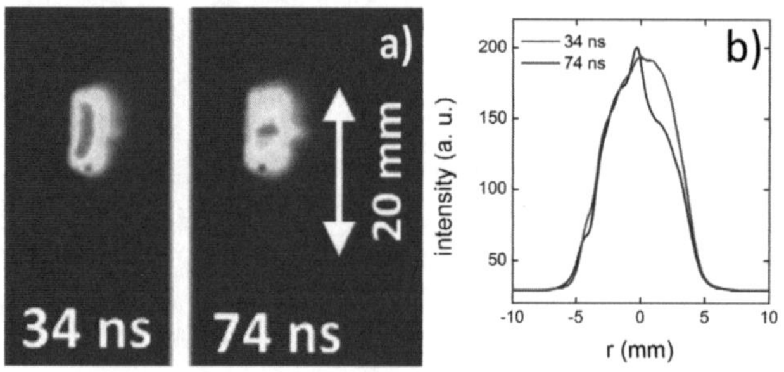

Figure 2. a) Plasma images showing the initial stages of the annular plasma evolution, and b) Radial profile of the light intensity in the same images, at 2.5 mm from the target surface.

Figure 3 shows the time evolution of the laser produced plasma. For times up to 340 ns, the distinctive feature of the laser produced plasma is that of an axially expanding plasma jet, with a characteristic bow-like, wider plasma front at the leading edge. This can be explained in terms of a combination of a tight axial expanding plume resulting from on axis stagnation of the initial annular plasma ring, and the axial displacement of the disk-shaped initial plasma. At later times the axial plasma jet is seen to decay, being the dynamics dominated by the expansion, both axial and radial, of the bow-like initial plasma front. In particular, the plasma is seen to decay by expansion and cooling in a time scale of $\sim 4\ \mu$s.

In order to characterize the plasma dynamics we plotted the radial and axial positions of the leading plasma front, defined by the light emitting edge of the plasma, as a function of time. This is shown in figures4-a and 4-b, respectively. The radial expansion was measured at 2.5 mm from the target surface. The initial annular plasma is seen to expand radially with constant velocity $1.2 \cdot 10^4$ m/s. This is shown by the linear fitting in figure 4-a. In contrast with the radial dynamics, the axial expansion of the plasma does not shown a constant velocity feature. This can be described quantitatively by the the *drag model* [6–8]. During the expansion process the plasma is not subject to energy input, so the later dynamics is determined by the energy released on target at plasma creation. At later times, and in the presence of a neutral background gas, the expanding laser plasma drags background gas, which results in and increasing mass load at the expanding plasma boundary, leading to an eventual stop. In this case the expansion

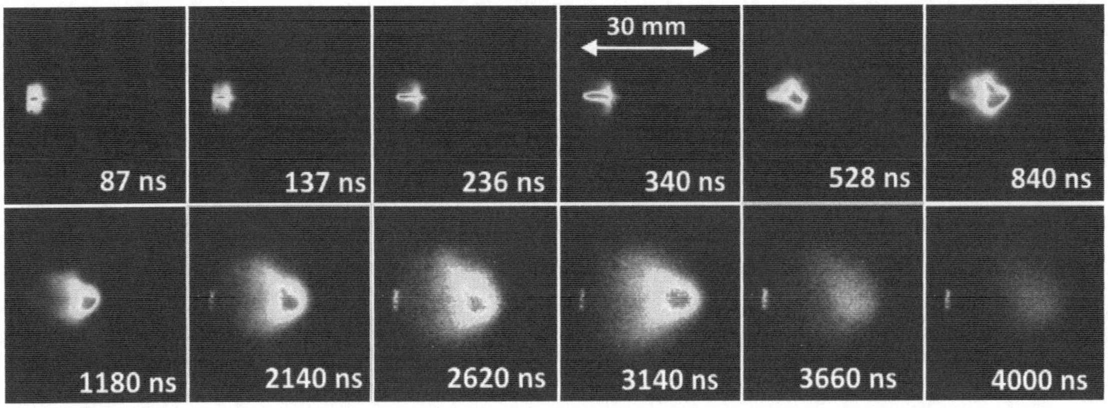

Figure 3. Plasma images showing the time evolution of the initial annular plasma.

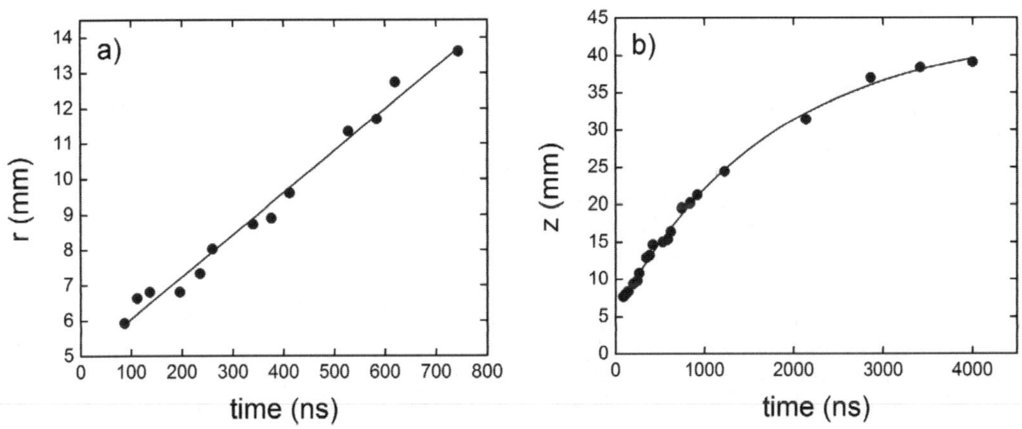

Figure 4. Expansion dynamics of the laser plasma. a) Radial expansion at 2.5 mm from the target, and b) Axial expansion along the symmetry axis.

dynamics as a function of time is given by

$$z(t) = z_s\left(1 - e^{-\beta t}\right) - z_0 \tag{1}$$

where z_s is the stopping distance, β is the dragging coefficient, and z_0 is an initial condition that allows a delayed onset of the plasma plume emission, as compared with the time of initial laser plasma formation. From Eq. 1, the initial velocity of the plasma front is $v_0 = \beta z_s$. In this case, the application of the drag model leads to the solid fitted lines in figure 4-b. It can be inferred that the drag model describes well the axial dynamics of the plasma. Details of the drag model fitted parameters are presented in table 1.

In general, the stopping distance z_s is determined mainly by the presence of the background gas. The adjustment parameter z_0 is mainly determined by the time interval required for light emission to take place, following the creation of excited state in the laser plasma at $z = 0$ [7]. This time interval depends on the laser power density and target material. The initial axial

	z_s (mm)	β (s^{-1})	z_0 (mm)	v_0 (m/s)
Fitting parameters	37.8	$5.67 \cdot 10^{-4}$	-5.7	$2.15 \cdot 10^{4}$

Table 1. Drag model adjusting parameters

velocity, v_0, results to be almost twice that of the constant radial velocity. This is due to the fact that the initial axial velocity is determined by the explosive behavior of the initial laser produced plasma, whereas the radial velocity is determined mainly by the thermal expansion of the plasma in a low pressure background, which does not experience dragging effects on the spatial scale involved in the experiments. In general, the plasma images in figure 3, at later times, show a distinctive behavior, qualitatively similar to that of the single focal point laser produced plasma, expanding in a neutral gas background [5–8].

4. Conclusions

We have performed preliminary experiments to characterize the plasma dynamics of laser produced annular plasmas. A point to highlight is the observation of on-axis stagnation, which leads to the formation of a jet-like plasma. Further experiments, with additional diagnostics and different parameter regimes for the formation of the laser produced annular plasma are being performed, which will be reported elsewhere.

Acknowledgments

This work has been funded by project FONDECYT 1141119.

References

[1] Veloso F, Chuaqui H, Aliaga-Rossel R, Favre M, Mitchell I H, and Wyndham E 2006 *Rev. Sci. Instrum.* **77**, 063506
[2] Veloso F, Chuaqui H, Correa N, Favre M, and Wyndham E 2009 *Plasma Sources Sci. Technol.* **18**, 045013
[3] Cabalín L M and Laserna J J 2004 *J. Anal. At. Spectrom.* **19**, 445
[4] Malevich P N and Chumakov A N 2012 *J. Appl. Spectroscopy* **79**, 838
[5] Ruiz H M, Guzmán F, Favre M, Bhuyan H, Chuaqui H and Wyndham E S 2012 *Plasma Sources Sci. Technol.* **21** 034014
[6] Geohegan D B 1992 *Appl. Phys. Lett.* **60** 2732
[7] Gonzalo J, Afonso C N, and Madariaga I 1997 *J. Appl. Phys.* **81**, 951
[8] Harilal S S, Bindhu C V, Tillack M S, Najmabadi F, and Gaeris A C 2003 *J. Appl. Phys.* **93**, 2380

IOP Conf. Series: Journal of Physics: Conf. Series **1043** (2018) 012047 doi:10.1088/1742-6596/1043/1/012047

In vitro irradiation of colorectal cancer cells by pulsed radiation emitted from a hundred joules plasma focus device and its comparison with continuous irradiation.

J. Jain[1,2,3], J. Moreno[2,3,4], R. Andaur[5], R. Armisen[6,7], G. Avaria[2,3,4], B. Bora[2,3,4], S. Davis[2,3,4], C. Pavez[2,3,4], K. Marcelain[5], and L. Soto[2,3,4*]

[1]Comisión Chilena de Energía Nuclear, Casilla 188-D, Santiago Chile
[2]Center for Research and Applications in Plasma Physics and Pulsed Power, P4
[3]Universidad de Talca, Chile
[4]Universidad Andres Bello, Departamento de Ciencias Fisica, Republica 220, Santiago, Chile
[5]Deparatamento de Oncología Básico-clínica, Facultad de Medicina, Universidad de Chile, Independencia 1027, Independencia, Santiago, Chile
[6]Centro de Investigación y Tratamiento del Cáncer, Facultad de Medicina, Universidad de Chile, Independencia 1027, Independencia, Santiago, Chile
[7]Current affiliation: Center for Excellence in Precision Medicine, Pfizer Chile, Santiago, Chile

**E-mail: jalajjain83@gmail.com, lsoto@cchen.cl*

Abstract. In the last years, pulsed reduced low dose radiation has been proposed as an alternative for treatment of recurrent cancer. Nonetheless, distinction between the effects of low dose pulsed and continuous radiation is barely known at cellular level. In order to study the effects of low dose pulsed radiation at cellular level, *in vitro* experiments are important to further advance the basic understanding in this area. In the present work we demonstrate the usefulness of a low-energy plasma focus device PF-400J as a potential source of low-dose pulsed radiation for *in vitro* cancer cell experiments. Colorectal cancer cell line, DLD-1, were irradiated by pulsed x-rays. Fifty pulses of x-rays provide ~0.12 Gy dosis, which were measured using thermoluminescence detectors (TLD-100 dosimeters). Irradiation-induced DNA damage was assessed at different time points after irradiation. A statistically significant double strand break (DSB) DNA damage was observed at 30 minutes after irradiation. A comparison of DSB induced by continuous source in the same type cancer cells and pulsed irradiation is made at 30 minutes post-irradiation. In the case of pulsed irradiation, DSB per unit dose found higher. Our findings suggest that low-energy plasma focus devices could have potential application as pulsed radiation source in the area of *in vitro* cancer cell experiments.

1. Introduction

Dense plasma focus (DPF) devices produce short-lived plasma in various phases. This short-lived plasma produces various kinds of radiations such as pulsed x-rays [1-5], charged particles [6-10] and neutrons if working gas is deuterium [11-14, 22].

Normally a DPF consists in a central electrode (usually anode) that is partially covered by an insulator. Cathode bars symmetrically surround the anode. Plasma in DPF is categorized in various phases that are described in the following. On the application of pulsed-high voltage, at first, the discharge takes place over the insulator and form a plasma current sheet (PCS). This stage is known as breakdown phase. Later, due to Lorentz force the PCS starts to expand and moves along the length of the anode. This phase is called rundown phase. At the top of the anode, the PCS starts to move in radial inward direction and compress the neutral gas there. This is known as compression phase. The fast compression of neutral gas produce radial shock waves that ionize neutrals at the top of the anode. These shock waves travel to the axis of the anode and reflect back. At the time, when reflection and compression balance, a plasma column forms. This is known as pinch phase. Various physical phenomena take place during compression and pinch phases and cause emission of electromagnetic and particle radiation. Indeed, induction of electromagnetic fields cause acceleration of charged particles. Electrons accelerate toward the anode and produce x-rays via *bremsstrahlung* upon impinging the anode. On the other hand, ions accelerate away from the anode and produce neutrons via *beam-target* fusion mechanism, if the discharges are produced with deuterium. Disruption of the pinch phase provides axial shock and plasma jet [15-16]. These later phases have been recently characterized and found applications in plasma wall interaction in fusion reactor and astrophysical jet phenomena. Electromagnetic radiation, emitted from DPF, covers a wide range of its spectrum from visible to hard x-rays that can be used for *in vitro* experiments on cancer cells.

Radiation therapy is often used for cancer treatment, which includes the exposure of cancerous tumor sites to high dosis of radiation. However, high dosis are fractionated over a long period. Still, the fractionated dosis have certain side effects. Due to this reason, studies in the area of low dose effects on cancer cells have been carried out [17-18].

Pulsed reduced low dose radiation (PLDR) is useful in the treatment of recurrent cancerous tumor [19]. In order to study the effects of low dose pulsed radiation effects at cellular level (*in vitro*), DPF will be a good choice. DPF emits neutrons as well as x-rays. X-rays are known to provide low linear energy transfer (LET). On the other hand, particles (neutrons, heavy ions) are high LET. The peculiarity of DPF that it emits pulsed x-rays and neutrons; can be explored further in order to study the effects of pulsed radiation (x-rays and neutron) on cancer cell lines in *in vitro* experiments that is the focus of present paper.

In section 2, experimental setup and cancer cell culture preparation is presented. The results and discussion are presented in section 3. The work is concluded in section 4.

2. Experimental setup

PF-400J (880 nF, 38 nH, 20-30 kV, 176-539J, ~300 ns time to achieve peak current) [1, 6, 11, 15] consists in a stainless steel (SS) hollow anode of diameter 12 mm, symmetrically surrounded by eight stainless steel cylindrical cathode bars. The effective length, was 7.0 mm in this case. A schematic of PF-400J with electrical signals and various phases is shown in figure 1. During pinch phase, the impedance, mainly the inductance, increases and current decreases. This signature of pinching action (that is important to have radiation emission from DPF), can be seen in current derivative and voltage signals (dI/dt), please see figure 1. In order to irradiate cancer cells it is mandatory to characterize radiation with respect to dosis. Thermoluminiscence detectors (TLD-100) have been used in order to measure the dosis for x-rays. TLD-100 can be used to measure the dosis for wide range of x-rays (5 keV – tens of MeV). In addition, TLD-100 are tissue equivalent, therefore, are suitable for present study. Here it is worth to mention that PF-400J emits x-rays in low (SXR) and high-energy (HXR) range. HXR are the one that are detected outside the plasma focus vacuum chamber and SXR are detected inside and cannot penetrate the vacuum chamber. In order to confirm the detection and distinction, P-I-N diode

(BPX65, 0.5 < E < 30 keV) was mounted inside and photomultiplier tube with scintillator (BC408, 0.1 < E < 5 MeV) was mounted outside the vacuum chamber along the same line of sight. It was observed that when there was signal in P-I-N diode no signal was observed in photomultiplier tube. With this observation, dosis outside and inside the vacuum chamber were measured simultaneously. The outside dosimeter array that was placed over the aluminum vacuum window of width 1mm provides hundred times lower dosis that were obtained inside the vacuum chamber. It was realized that most of the low-energy x-rays would have been attenuated while passing through the aluminum window. With this observation, the aluminum vacuum window was replaced by plastic vacuum window (polyethylene) of width ~ 1 mm, and with this arrangement the dosis outside the vacuum chamber were about 0.12 Gy for fifty pulses of x-rays, while placing a lead piece inside the hollow anode.

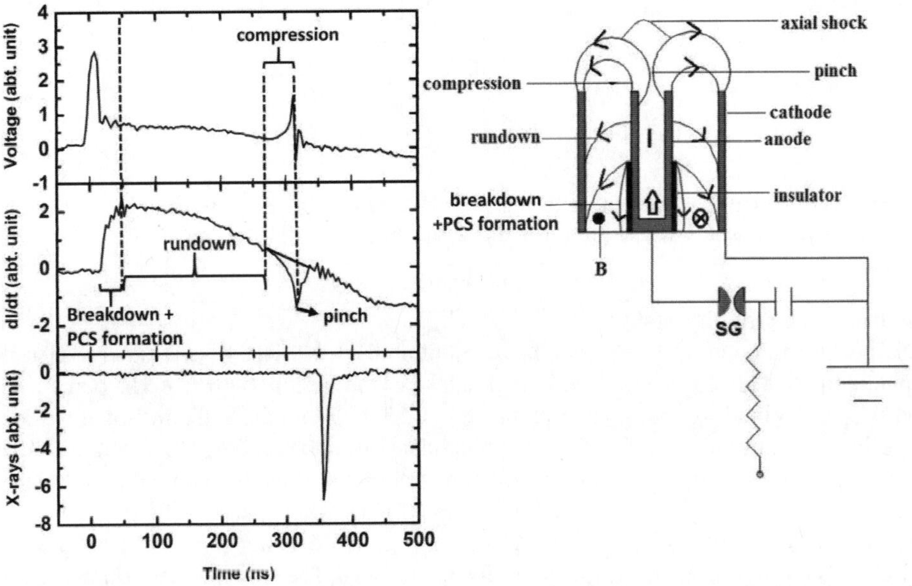

Figure.1 Various phases of the PCS dynamics are visible in electrical signals and a schematic of DPF.

Based on the abovementioned observations cell cultures were irradiated by fifty pulses of x-rays, outside the PF-400J vacuum chamber by keeping the cell culture over vacuum window that was made of plastic material, along the PF axis. Petri dishes of diameter ~ 3.5 cm with four partitions were used to culture cancer cells and dosimeter measurements. Three partitions were used to culture cells and one to keep dosimeters. In this way, it was possible to monitor dosis provided by pulsed x-rays in real time experiment. The distance between the top of the anode and cell culture petri dish was ~ 7 cm. In addition, a photomultiplier tube was kept at ~ 80 cm from top of the anode along the PF-axis. In this study, photomultiplier was used as a referential device in order to count x-ray pulses. Hydrogen gas at nine mbar was used to produce discharges. Figure 2 shows the experimental setup for cell culture irradiation by pulsed x-rays.

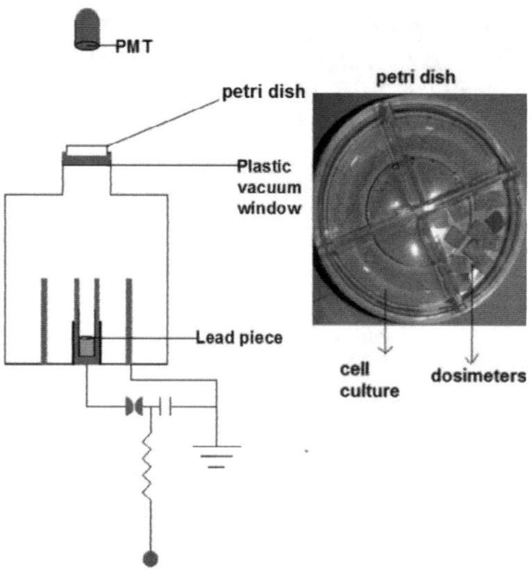

Figure 2 Experimental arrangement for cancer cell irradiation and dose measurement in the case of pulsed x-rays. PF-400J is used for this experiment

2.1 Cell culture and viability assays

DLD-1 human colorectal cancer cells were cultured in RPMI1640 (Mediatech, Herndon, VA, USA) supplemented with 10% heat-inactivated fetal bovine serum (Mediatech), penicillin G (100 U/ml), and streptomycin (100 µg/ml) and incubated in a 5% CO_2 humidified atmosphere at 37ºC. Cells were seeded in 4-chamber 35 mm dish (In vitro Scientific, Sunnyvale, CA, USA) at a density of 9.6 x 10^5 cells/well a day before radiation. After irradiation, viability was evaluated by trypan blue exclusion assay. Briefly, a volume of represented cells were mixed with and equal volume of 0.4 % trypan blue solution (Logos Biosystems, Gyunggi-Do, Korea) and counted using a LUNA™ Automated Cell Counter (Logos Biosystem). All assays were performed at least three independent times.

2.2 Immunofluorescence Assay

For immunofluorescence assays, 2.4 x 10^5 cells/well were grown on glass coverslips in 4-chamber 35 mm dish (In vitro Scientific, Sunnyvale, CA, USA). At 30, 60 and 120 minute after radiation, cells were fixed using 4% p-formaldehyde for 10 minutes, permeabilized with 0.25% Triton X-100 at room temperature and blocked with 3% BSA/PBS for 45 minutes. Cells were incubated overnight with an anti-γ-H2AX antibody (1:1000, Millipore, Temecula, CA, 05-636) diluted in 0.05% Triton X-100 and 1% BSA/PBS. After washing steps, cells were incubated with an Alexa Fluor-488 secondary antibody (1:500, Molecular Probes, A-21042). Slides were mounted with ProLong Gold Antifade Reagent with DAPI (Life Technologies, NY). Cells were observed under a fluorescence microscope (BX53, Olympus). For γ-H2AX quantification, foci were counted using a Find Maxima plugin and normalized by nuclei numbers using ImageJ software (Rasband, National Institutes of Health, USA). Cells were counted in 5 different fields and at least 100 cells were evaluated per sample.

3. Results and discussion

3.1 X-rays dose measurement

Pulsed x-ray dosis were characterized using TLD-100 dosimeters. PMT was used as referential device. Average dose in 50 x-ray pulses were found ~ 0.12 Gy with the insertion of lead piece inside the hollow anode.

3.2 Double strand break (DSB)

DLD-1 cells cultured in a petri dish, were irradiated with fifty pulses of x-rays. DNA damage was assessed after irradiation. A mock condition (cell culture that was not irradiated but follow the same path as the irradiated cell culture) was used as control. DNA damage was evaluated at 30, 60, and 120 minutes after irradiation by detecting the phosphorylation on Ser-139 in histone H2AX (named γ-H2AX). Whenever there is induction of double strand break (DSB), a focus of γ-H2AX protein is detected in the vicinity of DSB. Hence, DSB can be quantified by counting γ-H2AX foci. Figure 3 shows the results of DSB quantification. An increase in DSB induction is evident at 30 minutes after pulsed x-rays irradiation.

Three independent experiments (n=3) for DNA damage study were performed. Increase in the γ-H2AX foci was significant only at 30 minute post irradiation. At later time points, statistical test showed no significance difference between pulsed x-rays irradiation and mock control. It is highly possible that DNA damage is been repaired later. Moreover, if DNA damage cannot be repaired cell death process take place [20]. However, in these cells, pulsed x-rays irradiation with 0.12 Gy, did not induce cell death, as analyzed using trypan blue dye assay. Since we did not find any positive results of the cell death, in the case of 50 pulses of x-rays, cell death measurement was limited to n=1.

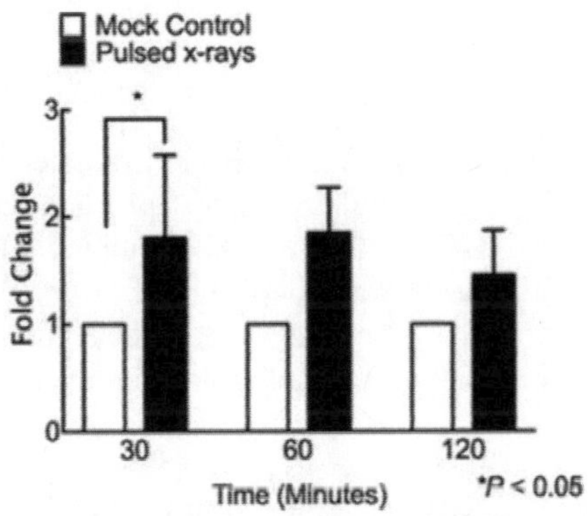

Figure 3 DNA damage was evaluated by detection of γ-H2AX foci in mock and irradiated cells at 30, 60, 120 minutes post irradiation quantification of γ-H2AX foci fold change. * Difference was statistically significant (* P<0.05); Student's T-test.

3.3 Comparison of DSB induced by pulsed and continuous irradiation

Figure 4 shows a comparison between DSB induced per unit dose (DSB/dose) by continuous high (12 Gy) and low dose (0.6 Gy) irradiation [21] and pulsed low dose irradiation (~0.12 Gy, emitted from PF-400J). It can be seen from figure 4 that DSB per unit dose is higher in the case of pulsed radiation. It can be say that pulsed radiation is more effective in DSB induction than continuous radiation. The number of DSB in the case of continuous low dose (0.6 Gy) x-ray irradiation was found about 3.7-fold at 30 minutes [21]. In the case of pulsed x-rays, this data is ~ 1.7-fold. The ratio of dosis ~ 0.6/0.12 = 5 and the ratio of the fold change of the DSB will be 3.7/1.7 = 2.18 at 30 minute. With these estimations, it can be concluded that the effect of DSB induction in the case of low dose pulsed x-rays irradiation cannot be linearly extrapolated from low dose continuous x-rays DSB induction. In addition, a similar comparison between high dose (12 Gy), obtained from continuous x-rays [21] and low dose pulsed x-rays (~0.12 Gy) irradiation is made. The ratio of dosis is 12/0.12 = 100 and the ratio of fold change in DSB is 8/1.7 = 4.7, at 30 minutes. This comparison reveals that doses are 100 times higher in the case of continuous x-rays irradiation but the induced fold change in DSB ~ 4 – 5 times.

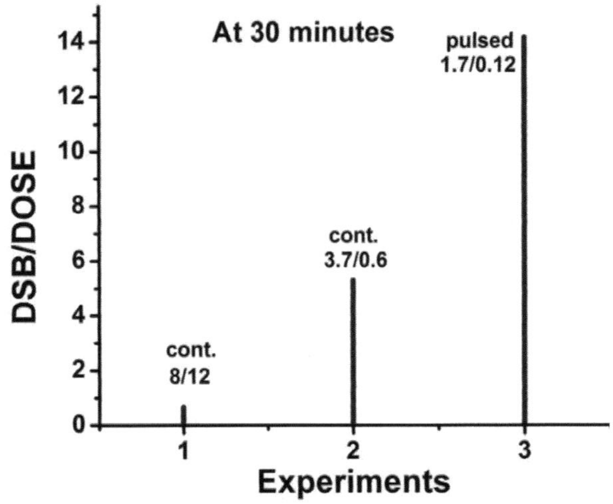

Figure 4 Variation in DSB/dose with continuous high/low dose and pulsed low dose irradiation. Pulsed low dose irradiation shows higher DSB/dose that indicates that pulsed radiation is more effective in DSB induction than continuous radiation.

4. Conclusion

Radiation emitted from DPF device PF-400J was applied to irradiate colorectal cancer cell lines DLD-1 in *in vitro* experiments. Dosis for 50 x-ray pulses were ~ 0.12 Gy, while placing a lead piece inside the hollow anode. Double strand break (DSB) in DNA was found at 30 minutes post-irradiation with statistical significance. In this case, cell death was not observed. A comparison between high and low dose irradiation, provided by continuous x-ray source and low dose pulsed radiation, provided by PF-400J DPF device was made. It was found that pulsed radiation is more effective in inducing DSB in DNA than continuous radiation.

Acknowledgement
The work is supported by grant ACT-1115, CONICYT, Chile.

References

[1] Zambra M, Silva P, Moreno M, Pavez C and Soto L 2009 *Plasma Phys. Control. Fusion* **51** 125003
[2] Barbaglia M, Bruzzone H, Acuña H, Soto L and Clausse 2009 *Plasma Phys. Control. Fusion* **51** 045001
[3] Barbaglia M, Soto L and Clausse A 2012 *Journal of Fusion Energy* **31** 105
[4] Pavez C, Pedreros J, Zambra M, Veloso F, Moreno J, Tarifeño-Saldivia A and Soto L 2012 *Plasma Phys. Control. Fusion* **54** 105018
[5] Neog N K, Mohanty S R and Borthakur T K 2008 *Physics Letters A* **372** 2294
[6] Jain J, Moreno J, Pavez C, Bora B, Inestrosa-Izurieta M J, Avaria G and Soto L 2016 J. Phys.: Conf. Ser. 720 012042
[7] Bhuyan H, Chuaqui H, Favre M, Mitchell I and Wyndham E 2005 *J. Phys. D: Appl. Phys.* **38** 1164
[8] Mohanty S R, Bhuyan H, Neog N K, Rout R K and Hotta E 2005 *Jpn. J. Appl. Phys.* **44** 5199
[9] Jakubowski L, Sadowski M and Zebrowski J 2001 *Nucl. Fusion* **41** 755
[10] Neog N K and Mohanty S R 2007 *Physics Letter A* **361** 377
[11] Silva P, Moreno J, Soto L, Birstein L, Mayer R and Kies W 2003 *Appl. Phys. Lett.* **83** 3269
[12] Verma R, Rawat R S, Lee P, Krishnan M, Springham S V and Tan T L 2009 *Plasma Phys.*

Control. Fusion **51** 075008
[13] Soto L, Silva P, Moreno J, Zambra M, Kies W, Mayer R E, Clausse A, Altamirano L, Pavez C and Huerta L 2008 *J. Phys. D: Appl. Phys.* **41** 205215
[14] Ellsworth J L, Falabella S, Tang V, Schmidt A, Guethlein G, Hawkins S and Rusnak B 2014 *Rev. Sci. Instrum.* **85** 013504
[15] Soto L, Pavez C, Moreno J, Inestrosa-Izurieta M J, Veloso F, Gutierrez G, Vergara J, Castillo F, Clausse A, Bruzzone H and Delgado-Aparicio L 2014 *Phys. Plasmas* **21** 122703
[16] Pavez C, Pedreros J, Tarifeño-Saldivia A and Soto L 2015 *Phys. Plasmas* **22** 040705
[17] Cuttler J M and Pollycove M 2003 *J. American Physicians and Surgeons* **8** 108
[18] Litvyakov N V, Rostov V V, Bol'shakov M A, Buldakov M A, Afanas'ev K V, Astapenko A N, Kutenkov O P and Cherdyntseva N V 2005 **50** S51-S58
[19] Charlie Ma C M, Luxton G and Orton C G 2011 *Med. Phys.* **38** 4909
[20] Roos W P, and Kaina B 2006 *Trends in Molecular Medicine* **12**, 440
[21] Private communication with Faculty of Medicine, University of Chile group
[22] Soto L, Pavez C, Moreno J, Altamirano L, Huerta L, Barbaglia M, Clausse A, and Mayer R E 2017 Physics of Plasmas **24** 082703

IOP Conf. Series: Journal of Physics: Conf. Series **1043** (2018) 012048 doi:10.1088/1742-6596/1043/1/012048

Pulsed x-rays dose measurements from a hundred joules plasma focus device.

J. Jain[1,2,3,*], J. Moreno[1,2,4], R. E. Avila[2], G. Avaria[1,2,4], C. Pavez[1,2,4], B. Bora[1,2,4], S. Davis[1,2,4] and L. Soto[1,2,4]
[1]Comisión Chilena de Energía Nuclear, Casilla 188-D, Santiago Chile
[2]Center for Research and Applications in Plasma Physics and Pulsed Power, P4
[3]Universidad de Talca, Chile
[4]Universidad Andres Bello, Departamento de Ciencias Físicas, República 220, Santiago, Chile

**E-mail:* jalajjain83@gmail.com, lsoto@cchen.cl

Abstract. Present work is aimed to perform dosimetric measurements to characterize dosis obtained from pulsed x-rays emitted from a hundred joules plasma focus device PF-400J using thermoluminescent dosimeters (TLD-100). Two dosimeter arrays (containing 21 dosimeters in each) were used. One of the arrays was kept inside the PF-400J vacuum chamber and other outside the vacuum chamber, simultaneously. It was found that dosis obtained from the inside array (~200.7 mGy) were hundred times larger than the outside array (~1.1 mGy) for hundred pulses of x-rays. Later, the vacuum window of PF-400J, which was made of 1 mm aluminum, was replaced by a plastic window and a similar dosimeter array was kept outside the chamber over the plastic window. With this arrangement, the obtained doses (100 pulses of x-rays) were of the same order of magnitude (~106 mGy) as it was inside the vacuum chamber. Later, a lead piece was inserted inside the hollow anode of PF-400J, which increased dose (~250 mGy) per hundred pulses of x-ray outside the vacuum chamber using plastic vacuum window. Our results suggest that PF-400J could be a useful device to study low dose pulsed radiation effects on cancer cell lines in *in vitro* experiments.

1. Introduction

Plasma focus (PF) devices were built, mainly, to study fusion reactions and neutron emission [1-6]. The idea was to use PF devices as an efficient nuclear energy source. Besides the fact that PF devices have high neutron yield, it was found that ignition conditions were not fulfilled, which are mandatory to convert fusion devices in an efficient nuclear energy source. Due to this reason, PF devices left abandoned. Recently, studies on PF devices got a thrust due to their use in various applications [7-10] and a new generations of table top plasma focus devices begun [26-29]. PF devices are rich in various physical phenomena and provide a platform to study the fundamental aspects of physics. Emission of various kinds of radiation from PF devices make them useful in biological [11-15] and material sciences [16-19]. PF devices consist in a co-axial electrode assembly in which cathode bars surround the central electrode, anode, symmetrically. Partial length of the anode is covered by an insulator. The uncovered length of the anode is knows as effective length. After anode connects to high voltage, at first discharge takes place over the insulator and forms a plasma current sheet (PCS), which develops a current density between anode and cathode. Later, due to self-generated magnetic field the PCS expands and runs over the effective length of the anode under the action of Lorentz force. At the open end of the anode, the PCS compresses neutral gas and forms a plasma column, typically known as pinch. During compression and pinch phases, electromagnetic forces are induced that accelerate charged particles. Electrons impinge the anode and produce x-rays via *bremsstrahlung*. Plasma dynamics and radiation emission under the framework of PF devices can be found in [20].

PF devices emit radiation in pulsed form, which can be used to study the effects of pulsed radiation on cancer cells in *in vitro* experiments, which has importance to study post-irradiation effects on cells at cellular level. Indeed, pulsed low-dose-rate radiation therapy is used to treat the recurrent cancer [21]. Cancer research has advanced a lot from technology point of view but at cellular level, it requires more studies. For instance, studies about the distinct effects of pulsed and continuous radiation on cells. In order to use pulsed radiation (x-rays) emitted from PF devices for cancer cells irradiation, it is mandatory to characterize doses. S. Zapryanov *et al* [15] used a 3 kJ PF device in order to irradiate live microorganisms by soft x-rays. Dose measurement was performed using thermoluminescent detectors (TLD). Inside the PF chamber at 15 cm from the plasma column, dosis were ~11 mSv for 4 shots using 20 μm Al foil and ~ 65 mSv for 14 shots using 100 μm Al foil. In addition, it was mentioned that the total energy released in hard x-rays region is lower than the total energy released in soft x-rays region.

Pavez *et al* [22] used a hundred joules PF device, PF-400J, in order to study x-rays emission and equivalent dosis. Dose measurements were performed using TLDs (TLD-100). Based on the observations, dose per shot of the order of 17 μSv around the symmetry axis was reported. In the present work, TLD-100 dosimeters are used in order to measure the doses inside and outside the PF-400J [3, 22-25] simultaneously. These measurements were performed keeping in mind the use of pulsed x-rays to irradiate cancer cells in *in vitro* experiments.

In section 2 experimental setup for dose measurements is presented. The results are presented and discussed in section 3. Work is concluded in section 4.

2. Experimental setup

PF-400J consists in a stainless steel (SS) hollow anode, symmetrically surrounded by eight stainless steel cylindrical cathode bars. The effective length, was 7.0 mm in this case. A schematic of experimental arrangement used for dose measurements inside and outside PF-400J vacuum chamber simultaneously is shown in figure 1(a). The chamber's vacuum window, which was made of 1 mm Al sheet, was at a distance ~ 7cm from the top of the anode. Two dosimeter arrays, each containing 21 TLD-100 dosimeters, were placed inside and outside the PF-400J chamber, simultaneously, along the PF axis as shown in figure 1(a). Plastic petri dishes (used for cell culture preparation) of diameter ~ 3.5 cm were used to make the dosimeter array. The array located inside the chamber was kept at ~ 4 cm from the top of the anode. The dosimeter array outside the chamber was kept at ~ 7 cm from the top of the anode just over the chamber's vacuum window. The inside array was covered by a ~ 15 μm aluminum foil in order to prevent charged particles and plasma interaction with the dosimeter array. In addition, a photomultiplier tube (PMT) was kept at ~ 80 cm from the top of the anode. In this study, PMT was used as a referential device in order to count x-ray pulses. Hydrogen gas at 9 mbar was used to produce discharges in order to get 100 pulses of x-rays. With this experiment, let us say first experiment, we learnt that most of the x-rays were attenuated during their interaction with 1 mm Al window. Keeping in mind this observation, a window was prepared of the same material as petri dish, that is plastic (~1.2 mm) to replace with 1 mm Al window in order to measure dosis outside the chamber, let us say the second experiment. During the second experiment, the dosimeter array was arranged in the same box and placed over the vacuum window that was made of petri dish material. This time, only one dosimeter array outside the chamber was placed, as shown in figure 1(b).

IOP Conf. Series: Journal of Physics: Conf. Series **1043** (2018) 012048 doi:10.1088/1742-6596/1043/1/012048

In all the experiments, annealed dosimeters were used prior to use them for pulsed x-rays dose measurement emitted from PF-400J. Annealing is useful to re-configure the internal structure of the dosimeters. Sensitivity of TLD-100 dosimeters used in these experiments showed a 5% standard deviation. To reduce this effect, 100 dosimeters were uniformly irradiated by a calibrated x-ray source, and 60 of them were selected, with deviation less than 3% from the common average. Further, the response from that irradiation was used for the individual calibration of the selected dosimeters, so, later, their signals from exposure to PF-400J radiation was scaled with that calibration factor.

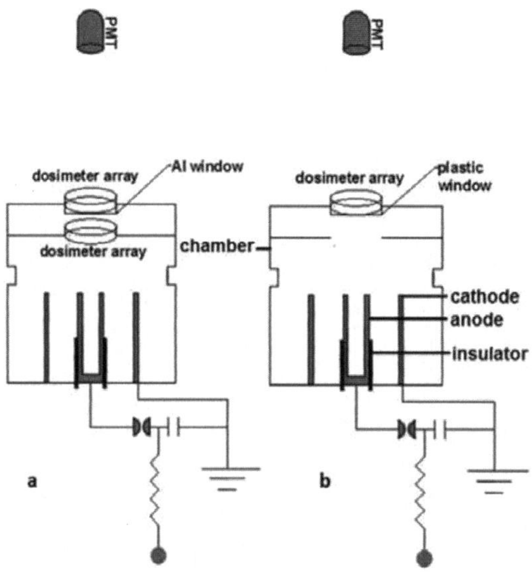

Figure 1 Experimental arrangement for dose measurement using TLD-100 dosimeters (a) inside and outside the PF-400J vacuum chamber simultaneously (b) outside the vacuum chamber only, with the use of plastic material as vacuum window. PMT was used as a referential device to count x-ray pulses.

3. Results and discussion

Figures 2(a), 2(b), 2(c), and 2(d) show schematic of the dosimeter arrays and accumulated dosis obtained from 100 pulses of x-rays inside (figures (2a), (2c)) and outside (figures (2b), 2(d)) of PF-400J vacuum chamber, respectively. The average dose in 100 x-ray pulses was found ~ 200.7 $\pm$ 72.69 mGy and 1.1 $\pm$ 0.23 mGy, inside and outside the vacuum chamber respectively. In addition, relatively higher dose zones inside the chamber were identified. In figures (2a) and (2c) it can be seen that the central dosimeters (F5, H5, J8, G5, C7, I4, B9, C4, and G4), shown by italic font style, acquire relatively higher doses.

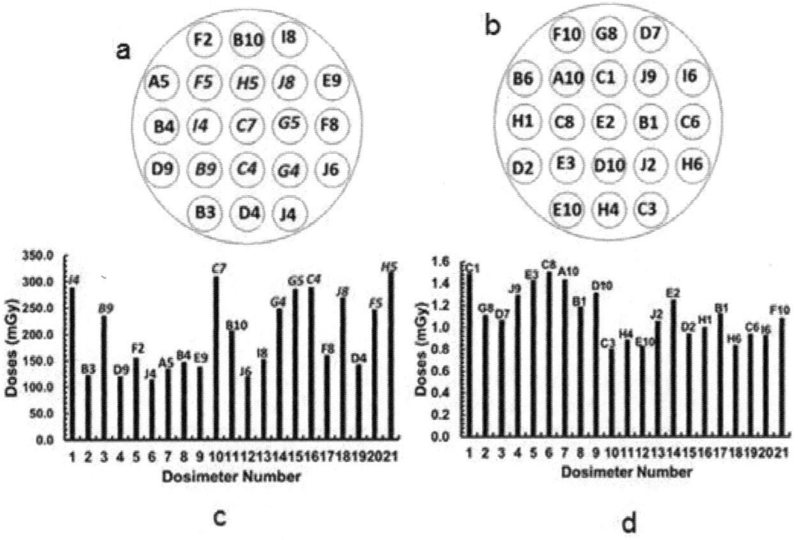

Figure 2 Schematic arrangement of dosimeters (TLD-100) in (a) inside and (b) outside arrays and obtained dosis (c) inside and (d) outside the PF-400J vacuum chamber, respectively. Inside the vacuum chamber dosis are hundred time larger than outside the chamber.

These results suggest that most of the x-rays would have been attenuated during their passage through aluminum window. Keeping in mind this observation, a vacuum window made of petri dish material, that is plastic of width 1.2 mm was used, please see figure 1(b). A dosimeter array (as shown in figure 2) was kept over this window. Figures 3(a) and 3(b) show the schematic arrangement of dosimeter arrays and obtained results, respectively with this arrangement. In this case, average dose (100 pulses of x-rays) outside the chamber was 106 ± 31.18 mGy, which has same order of magnitude as was inside the chamber. From figures 2(a) and 2(b) it can be seen that the central dosimeters (A10, B6, J4, J6, G8, D10, H4, and D7), shown by italic font, acquire relatively higher dosis. This pattern is similar as found inside the chamber, as mentioned earlier.

A lead piece was inserted inside the hollow anode of PF-400J to measure the dosis outside the vacuum chamber. In this case, 100 pulses of the x-rays provide dosis ~ 250 mGy. Please note the plastic window was used for this experiment. Insertion of lead increases dose per hundred x-ray pulses that are almost two times than without insertion of lead.

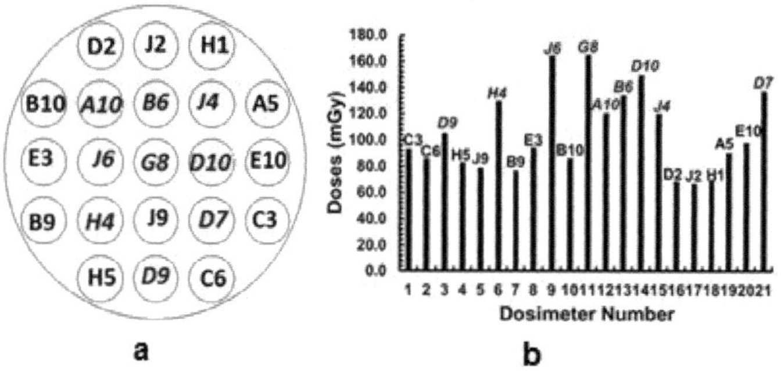

Figure 3 Dosis obtained with the use plastic vacuum window. Dosimeter array was kept outside the vacuum chamber over the plastic window.

In the first experiment, dosis outside the vacuum chamber were found hundred times smaller than inside the chamber. Please note, the dosimeter array that was kept outside had various obstacles in the path of x-rays to reach it. X-rays first interact with 15 μm aluminum foil then plastic of petri dish (~1.2 mm) which contains the dosimeter array inside the chamber. Later, x-rays will interact with the upper cap of petri dish, made of plastic that was kept inside the chamber and after with 1 mm Al window, over which the dosimeter array was kept outside the chamber. It has been reported that the emission of low energy x-rays is higher than the higher energy x-ray in PF devices [13]. Since transmission percentage of x-rays below 15 keV through 1 mm Al window is less than 10%, therefore low energy x-rays (say below 15 keV) are strongly attenuated by those various materials. Hence, much higher dosis inside the chamber, two order of magnitude higher than outside the chamber. These observations are consistent with the concept that low energy x-ray emission dominates in PF discharges [13].

It was observed that the use of plastic vacuum window provides similar dosis outside the vacuum chamber of PF-400J as was inside, see figure 3. Nonetheless, the dosis in this configuration were almost half the dosis inside the chamber. The decrement in dosis with the use of plastic window might be due the fact that x-rays have to pass through 1.2 mm window material and later through the box in which the dosimeter array was arranged. X-rays will be attenuated while passing through these obstacles.

Increment in dose per hundred x-ray pulses was observed with the insertion of a lead piece inside the hollow anode. Prior to discuss the effects of lead insertion inside the hollow anode on x-rays emission, let us first discuss how x-rays produce in PF devices. During compression and pinch phases of PF devices, there is generation of various kinds of instabilities and electromagnetic fields. Charged particles accelerate in the presence of the induced electromagnetic fields in various directions. Electrons accelerate toward the central anode and produce x-rays via *bremsstrahlung* upon impinging the anode. Insertion of lead may have following cause that could affect the doses. Interaction of electron beams with lead target can sputter lead material and introduce impurities in pinch phase. These impurities increase the resistivity of the pinch that further allow the induced electromagnetic fields to penetrate inside the pinch. This penetration of electric fields will allow relatively large number of electrons within the pinch to accelerate and produce x-rays via *bremsstrahlung*. In addition, these impurities themselves may work as target. Nonetheless, in the present work only an experimental observation that insertion of lead inside the hollow anode increases dose per hundred pulses of x-ray is presented. More computational and experimental efforts are needed in order to explain x-rays emission from PF-400J.

4. Conclusion

Two experiments were carried out in order to measure the dosis inside and outside the vacuum chamber of plasma focus device PF-400J, simultaneously (first experiment). In this case, average dosis were ~ 200.7 mGy and 1.1 mGy inside and outside the vacuum chamber respectively. Note that, in the first experiment the vacuum window over which the dosimeter array was kept outside the chamber was made of aluminum of thickness 1mm. Based on these observations, it was realized that most of the x-rays would have been attenuated while passing through aluminum window. In the second experiment, a vacuum window that was made of plastic was used in order to have similar dosis outside the vacuum chamber as was inside. In this case, the dosis outside the chamber were ~106 mGy. Later, insertion of a lead piece inside the hollow anode increased dose per hundred x-ray pulses that were ~ 250 mGy, outside the PF-400J vacuum chamber while using plastic vacuum window. These findings suggests that PF-400J could be a useful device for *in vitro* experiments to study the effects of low dose pulsed radiation on cancer cells.

Acknowledgement
The work is supported by grant ACT-1115, CONICYT, Chile.

References

[1] Castillo F, Milanese M, Moroso R and Pouzo J 2000 *Journal of Physics D: Applied Physics* **30** 1499-1997
[2] Michel L, Schonbach K H and Fischer H 1974 *Applied Physics Letter* ***24*** *57*
[3] Silva P, Moreno J, Soto L, Birtein L, Mayer R, Mayer E and Kies W 2003 *Applied Physics Letter* **83** 3269
[4] Wang X, Han M, Wang Z and Kun L 1999 *China Technological Sciences* **42** 83-87
[5] Verma R, Rawat S R, Paul L; Augustine T L T, Shariff H, Ying G J, Springham S V, Talebitaher A, Ilyas U and Shyam A 2012 *IEEE Transactions on Plasma Science* **40** 3280 – 3289
[6] Castillo F, Herrera J J E, Rangel J, Milanese M, Moroso R, Pouzo J, Golzarri J I and Espinosa G 2003 *Plasma Physics and Control Fusion* **45** 289
[7] Zambra M, Moreno J, Silva P, Soto L, Sylvester G and Pavez C 2008 *Journal of Physics: Conference Series* **134** 012047
[8] Kato Y and Be S H 1986 *Applied Physics Letter* **48** 686
[9] Beg F N, Ross I, Lorenz A, Worley J F, Dangor A E and Haines M G 2000 *Journal of Applied Physics* **88** 3225
[10] Moreno C, Vénere M, Barbuzza R, Del Fresno M, Ramos R, Bruzzone H, González Florido P J and Clausse A 2002 *Brazilian Journal of Physics* **32** 20-25
[11] Hussain S, Ahmad S, Khan M Z, Zakullah M and Waheed A, 2003 *Journal of Fusion Energy* **22** 195-200
[12] Gemishev O, Zapryanov S, Blagoev A, Markov M and Savov V 2014 *Biotechnology & Biotechnological Equipment* **28** 850-854
[13] Dubrovsky V, Gazaryan I, Gribkov G V, Ivanove A, Yu P, Kost O A, Orlova M A and Troshina N N 2003 *Journal of Russian Laser Research* **24** 289-300
[14] Virelli A, Zironi I, Pasi F, Ceccolini E, Nano R, Facoetti A, Gavoc E, Fiore M R, Rocchie, F, Mostacci D, Cucchi G, Castellani G, Sumini M and Orecchia R 2015 *Radiation Protection Dosimetry* **166** 1–5
[15] Zapryanov S, Goltsev V, Galutsov B, Gelev M and Blagoev A 2012 *European Physical Journal of Applied Physics* **58** 11201
[16] Lee S, Lee P, Zhang G, Feng X, Gribkov V A, Liu M, Serban A and Wong T K S 1998 *IEEE Transactions on Plasma Science* **26** 1119-1126
[17] Gribkov V A, Srivastava A, Keat P L C, Kudrryashov V and Lee S 2002 *IEEE Transactions on Plasma Science* **30** 1331-1338
[18] Inestrosa-Izurieta M J, Ramos-Moore E and Soto L 2015 *Nuclear Fusion* **55** 093011
[19] Bernard A, Bruzzone H, Choi P, Chuaqui H, Gribkov V, Herrera J, Hirano K, Krejci A, Lee S, Luo C, Mezzetti F, Sadowski M, Schmidt H, Ware K, Wong C S and Zoita V 1998 *J. Moscow Phys. Soc.* **8** 93 – 170
[20] Kang S, Lang J, Wang P, Li J, Lin M, Chen X, Guo M, Chen F, Chen L and Ming Ma C 2014 *Journal of Applied Clinical Medical Physics* **15** 102 – 113
[21] Pavez C, Pedreros J, Zambra M, Veloso F, Moreno J, Tarifeño-Saldivia A and Soto L 2012 *Plasma*

Physics and Control Fusion **54** 105018 (9pp)
[22] Silva P, Moreno J, Pavez C, Soto L and Arancibia J 2006 *AIP Conf. Proc* **875** 442
[23] Jain J, Moreno J, Pavez C, Bora B, Inestrosa-Izurieta M J, Avaria G and Soto L 2016 *Journal of Physics: Conference Series* **720** 012042
[24] Jain J, Moreno J, Avaria G, Pavez C, Bora B, Inestrosa-Izurieta M J, Diez D, Alvarez O, Tapia J, Marcelain K, Armisen R and Soto L 2016 *Journal of Physics: Conference Series* **720** 012043
[25] Soto L, Pavez C, Moreno, J, Inestrosa-Izurieta, M J, Veloso F, Gutierrez G, Vergara J, Clausse A, Bruzzone H, Castillo F and F Delgado-Aparicio L 2014 *Physics of Plasmas* **21** 122703
[26] Tarifeño A, Pavez C, Moreno J and Soto L 2011 *IEEE Trans. Plasma Science* **39** 756
[27] P. Silva, J. Moreno, L. Soto, L. Birstein, R. Mayer, W. Kies 2003 *Applied Physics Letters* **83** 3269
[28] Soto L, Silva P, Moreno J, Zambra M, Kies W, Mayer R E, Clausse A, Altamirano L, Pavez C, and Huerta L 2008 J. Phys. D: App. Phys. **41** 205215
[29] Soto L, Pavez C, Moreno J, Altamirano L, Huerta L, Barbaglia M, Clausse A, and Mayer R E 2017 *Physics of Plasmas* **24** 082703

IOP Conf. Series: Journal of Physics: Conf. Series **1043** (2018) 012049 doi:10.1088/1742-6596/1043/1/012049

Analysis of Deterioration in a Plasma Focus Device

Daniel Zanelli[1,*], Enrique López[1], Cristian Pavez[2,3,4], José Pedreros[3], Jalaj Jain[3,5], Gonzalo Avaria[2,3,4], José Moreno[2,3,4], Biswajit Bora[2,3,4], Sergio Davis[2,3,4] and Leopoldo Soto[2,3,4,*]

1 Facultad de Ciencias Físicas y Matemáticas, Universidad de Chile

2 Comisión Chilena de Energía Nuclear, Casilla 188-D, Santiago Chile

3 Center for Research and Applications in Plasma Physics and Pulsed Power P4, Chile

4 Universidad Andres Bello, Departamento de Ciencias Físicas, República 220, Santiago, Chile

5 Universidad de Talca, Chile

∗ email: *zanelli.93@gmail.com, lsoto@cchen.cl*

Abstract. The Plasma Focus (PF) is a kind of dense transient plasmas in with high-pulsed voltage. To produce devices for eld application it is necessary to obtain PF equipment able to operate for a long period of time. Thus, a reliability analysis is indispensable. In this work a reliability analysis program for plasma focus devices is presented. The program considers a criticality analysis using Failure Modes and Effects Criticality Analysis (FMECA) to identify the most important failure modes of the system. Said failure modes are studied operating the Plasma Focus for many cycles, obtaining from them the characteristic curves of $V(t)$ and $\dot{I}(t)$. Feature Extraction (FE) techniques are applied to obtain a list of parameters that correlate to the degrading process. Furthermore, Machine Learning tools are used to learn from the obtained data, linking the changes in these parameters during its life cycle to the decay of the system in hope for future implementation of a predictive maintenance system and a reference for data analysis and prediction in PFs. The study was applied to a portable plasma focus device operated at 2 joules of stored energy.

1. Introduction

In plasma focus (PF) devices high voltage is applied to a low pressure gas between coaxial cylindrical electrodes separated by an insulator. Due to its capacity to produce hot/warm and dense plasmas [1, 2]; intense beams of charged particles, neutrons [3-5], x-ray emission [6-8], plasma shocks [9], filaments [10] and jets are generated [11]. Thus, it has become a laboratory for fundamental and applied research related to fusion, neutron production [3-5], hard x-ray [6-8], high brightness soft x-ray production, material for fusion reactors [9, 12], astrophysical phenomena [10, 11] and biology [13]. Also, the PF has attracted the attention of the plasma research community as non-radioactive radiation source for industrial and field applications. In particular, at the Chilean Nuclear Energy Commission (CCHEN) the scale rules to develop miniaturized plasma focus devices has been determined [14]. In the last year several table top and portable plasma focus devices have been designed and constructed at CCHEN [3-5, 15]. In this study a deterioration model of the system as the life cycle progresses is developed. In which, a Failure Modes and Eects Criticality Analysis (FMECA) is done to a small tabletop plasma focus operating at few joules (PF-2J). Following an analysis of the characteristic electrical signals of plasma focus devices is implemented to correlate an advance in its wear. This is done in the

Figure 1. Photograph of PF-2J

most critical failure mode using Feature Extraction techniques to compare a linear model and a nonlinear Neural Network (NN) model.

1.1. The PF-2J Device

A portable plasma focus operating at few joules for field applications was designed and constructed at CCHEN. The device, PF-2J, including the capacitor, spark-gap, discharge chamber and power supply is portable in a hand luggage with a total weight of less than $1.5kg$. The capacitor has $120nF$. The measured total external inductance is $40nH$. The total impedance of the generator is in the order of 0.58Ω. The device operates from 6 to $8kV$ charging voltage (i.e. $\sim 2-4J$), achieving 9 to $13k$A in $120ns$. Figure 1 shows a photograph of the PF-2J.

2. Failure Modes and Effects Criticality Analysis

When performing the analysis, a series of key elements were considered, with their respective failure modes: the severity that this failure mode entails, the detectability of the failure and its probability of occurrence.

Each of these elements are assigned a value between 0 and 10 to describe the failure mode. Later a criticality factor is determined by the multiplication of these values. The results of this preliminary analysis is shown in table 1.

From this table a tendency of certain failure modes to be much more critical than others can clearly be seen. Specifically for the Insulator, the deposition of a conductive film around it has become the most critical failure mode, as well as the most frequent cause of failure for most PF equipments.

This is why this study will be focused on the detection of wear associated to the thin layer of material that degrades the insulator from its dielectric properties.

Table 1. FMECA Results for PF-2J

Elements	Failure Modes	Sev.	Prob.	Det.	Criticality
Capacitor	Burning	6	5	4	120
	Short Circuit	5	6	5	150
Spark Gap	Corrosion Electrodes	5	3	4	60
Anode	Corrosion	5	3	3	45
Insulator	Deposition of Conductive Film	6	7	5	210
Cathodes	Corrosion	5	3	3	45
Discharge Chamber	Gas Filtrations	3	4	5	60
Spark Gap Chamber	Gas Filtrations	3	4	5	60
Gas Connectors	Gas Filtrations	3	4	5	60
Viewers	Reduced Visibility	2	3	1	6
	Breaking	4	1	1	4
Electrical Connectors	Burning	6	5	3	90
	Short Circuit	5	5	3	75
Measurers	External Failure	2	4	5	40
Power Supply	Burning	5	2	5	50
Vacuum Pump	External Failure	2	2	5	20
Gas Tank	External Failure	2	2	3	12

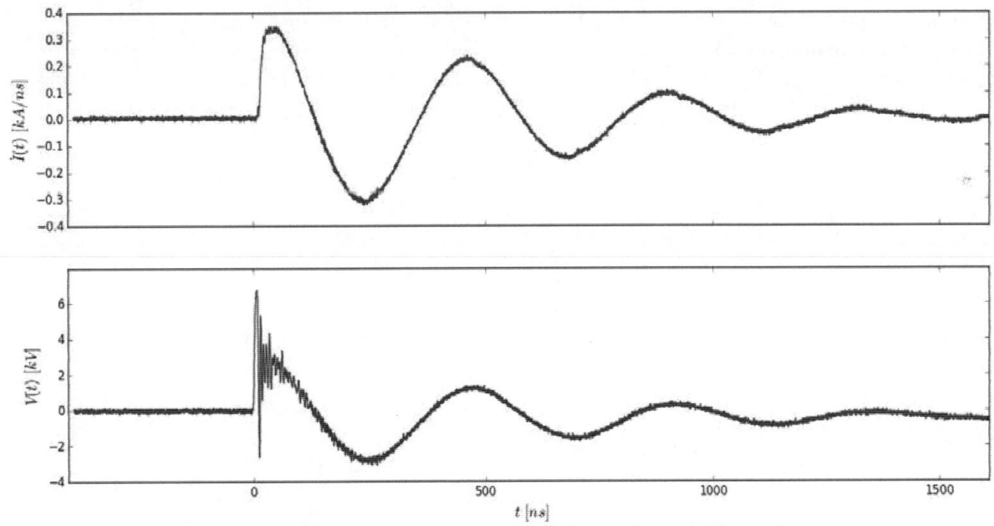

Figure 2. Caracteristic Signals of PF-2J in a shot with absence of z-pinch

3. Caracteristic Signals and Feature Extraction

The characteristic electrical signals of the PF-2J are shown in figure 2 without dense plasma pinching and in figure 3 with pinching $95ns$ after the discharge. It should be noted that these curves correspond to a damped harmonic oscillation coming from the RLC circuit that is formed in the equipment. The following characteristic signals were generated by charging the PF-2J at $6kV$ with H_2 gas filled in the discharge chamber at $8mbar$.

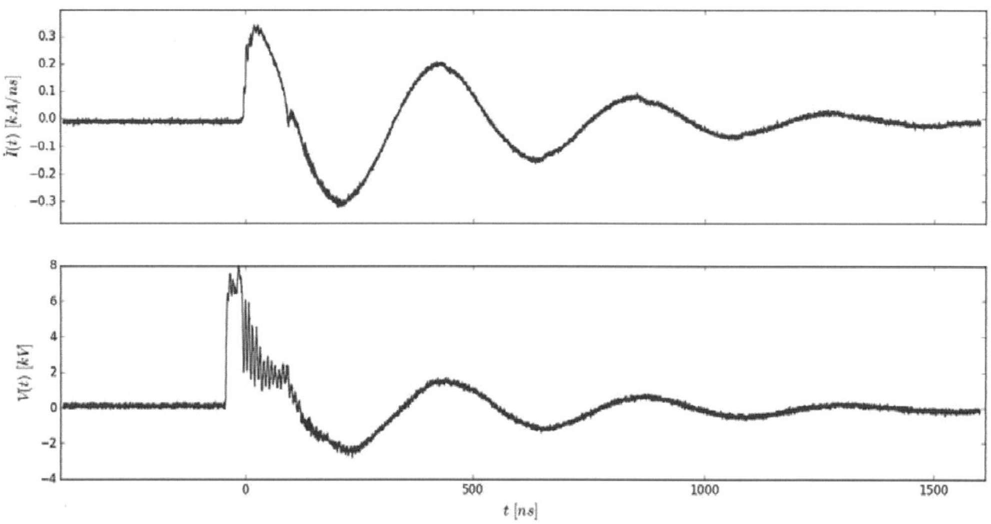

Figure 3. Caracteristic Signals of PF-2J in a shot with z-pinch at $95[ns]$

For this reason, another program has been developed in *python* that analyzes the signals and extracts parameters that describe it in several ways. Here comes the *Feature Extraction* concept, as it is presented in [16], where signal data analysis is performed. In this case, the chosen features are described in equations 1 to 4, where n_j is the j element of the function, u is the mean, σ is the standard deviation and N is the number of elements in the function.

$$Kurtosis = \frac{\sum_{j=1}^{N}(n_j - u)^4}{(N-1)\sigma^4} \tag{1}$$

$$Clearance\ Indicator = \frac{max|n|}{\sqrt{\frac{1}{N}\sum_{j=1}^{N}(n_j)^2}} \tag{2}$$

$$Shape\ Indicator = \frac{\sqrt{\frac{1}{N}\sum_{j=1}^{N}(n_j)^2}}{\frac{1}{N}\sum_{j=1}^{N}|n_j|} \tag{3}$$

$$ClearanceIndicator = \frac{max|n|}{(\frac{1}{N}\sum_{j=1}^{N}\sqrt{|n_j|})^2} \tag{4}$$

The results of the evolution of the mentioned features as a function of the number of discharges are shown below. In the figure 4 the values of Kurtosis, Crest Indicator, Clearance Indicator and Shape Indicator for $\dot{I}(t)$ are shown, in the figure 5 the same features for the $V(t)$ signal are shown.

Considering that the experiment presented instances of Pinch from the vicinity of $N = 1000$, this moment (after 1000 discharges) is considered as the reference point of the beginning of the degradation, since the values before this correspond to the conditioning of the insulator prior to the life cycle. Taking this into account, it is convenient to find the proportional change to the reference point, where F is a feature of the system, the percentage change is defined as in the equation 5.

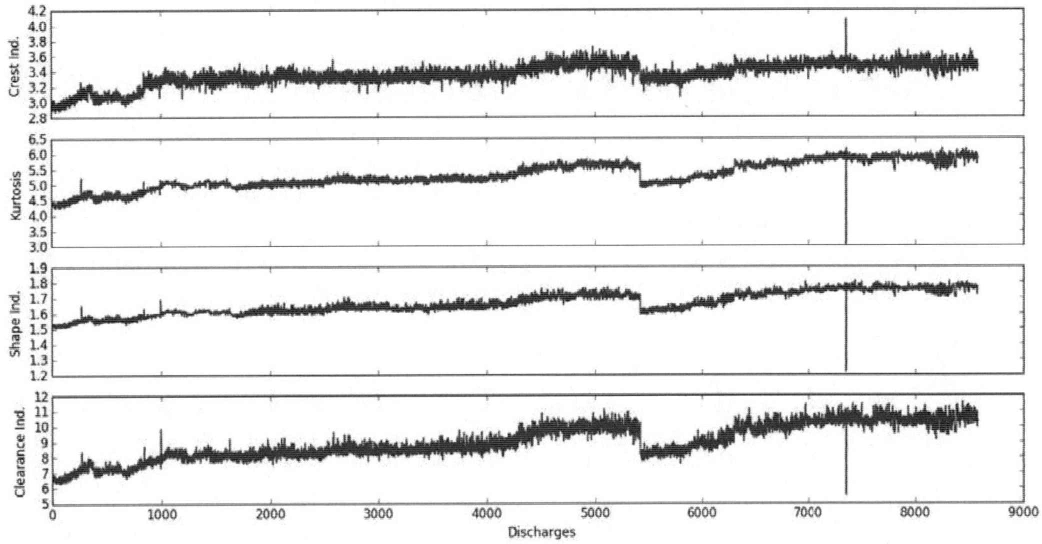

Figure 4. Feature Extraction of $\dot{I}(t)$

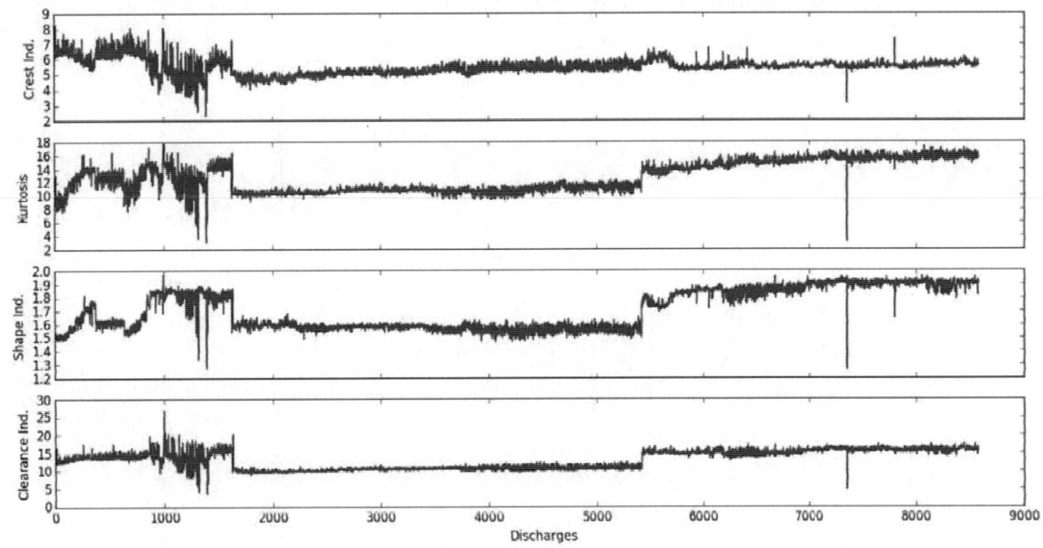

Figure 5. Feature Extraction of $V(t)$

$$\Delta F_x = \frac{F(N_x) - F(N = 1000)}{F(N = 1000)} \times 100 \tag{5}$$

4. Degradation Models

According to the data that arise from the experiment, relationships can be established between the evolutions of the different characteristic features over time with the advance in the life cycle of the equipment.

To approach the model that allows to correlate these states, an analytical relationship can be established between a multidimensional space, the features, towards a one-dimensional space, the wear and tear. Hence it can be denoted $\overrightarrow{F}$ to the vector that contains all features after normalization. It is crucial to normalize the features to a point of reference so that the model can be applied to multiple experiments.

If a relationship exists between the change in features and the wear of the equipment, it can be assumed that there is a function Y such that:

$$Y(\cdot) : \Re^n \rightarrow \Re \mid Y(\overrightarrow{F}) = N \quad (6)$$

Where N is the amount of discharges that the insulator of the equipment has endured until an arbitrary time.

4.1. Estimating N With a Linear Combination

The first approach to estimate this function could be to opt for linear behavior. Although there is a loss of information given by the approximation of the curves 4 and 5 to a linear behavior, the fact that the phenomenon is seen in several dimensions allows for eliminating sources of noise, while maintaining a simple and direct approach.

For this, the noise of the curves $F_i(N)$ is reduced by using the *Signal* library of python using the function $signal.lfilter()$ and then a linear interpolation in order to obtain an analytical resolution for the estimation of N, for each of the F_i. That is to say:

$$\overrightarrow{N}_{estimated} = \begin{bmatrix} N_1 \\ N_2 \\ N_3 \\ \vdots \\ N_n \end{bmatrix} = \begin{bmatrix} y_1(F_n) \\ y_2(F_n) \\ y_3(F_n) \\ \vdots \\ y_n(F_n) \end{bmatrix} \quad (7)$$

Where $y_i = m_i \; cdotF_i + n_i$ represents the equation of the interpolated line for the inverse graphs of $F_i(N)$.

This way, a linear estimation of the number of discharges made for each parameter can be obtained. However, now there are many estimates for the same result. Considering that it is possible to make a global estimate of N as a linear combination of the estimates for each feature:

$$\begin{gathered} N_{est} = C_1 \cdot N_1^{est} + C_2 \cdot N_2^{est} + C_3 \cdot N_3^{est} + ... + C_n \cdot N_n^{est} \\ N_{est} = \sum_i C_i \cdot N_i^{est} \end{gathered} \quad (8)$$

In this equation all the N_i^{est} are known for each of the n features and N discharges, so the coefficients that best represent reality can be found, contrasting them with the real number of shots corresponding to the feature vector. To find the coefficients C_i of each of these features, a crucial problem occurs in the change of dimensionality, because n corresponds to the total number of features and N to the total number of discharges , there are $n \sim 10$ variables and $N \simeq 8000$ equations. So it is proposed to solve the problem as if there were just n equations, being these the first n, then solving it again with the equations n to $2n$, and so on until solving all the N equations using N/n linear combinations:

$$\begin{aligned}
1 &= C_1 \cdot N_1^{est}(1) + C_2 \cdot N_2^{est}(1) + ... + C_n \cdot N_n^{est}(1) \\
2 &= C_1 \cdot N_1^{est}(2) + C_2 \cdot N_2^{est}(2) + ... + C_n \cdot N_n^{est}(2) \\
3 &= C_1 \cdot N_1^{est}(3) + C_2 \cdot N_2^{est}(3) + ... + C_n \cdot N_n^{est}(3) \\
&\cdots \\
n &= C_1 \cdot N_1^{est}(n) + C_2 \cdot N_2^{est}(n) + ... + C_n \cdot N_n^{est}(n)
\end{aligned} \tag{9}$$

And the same for the next n equations:

$$\begin{aligned}
n+1 &= C_1 \cdot N_1^{est}(n+1) + C_2 \cdot N_2^{est}(n+1) + ... + C_n \cdot N_n^{est}(n+1) \\
n+2 &= C_1 \cdot N_1^{est}(n+2) + C_2 \cdot N_2^{est}(n+2) + ... + C_n \cdot N_n^{est}(n+2) \\
n+3 &= C_1 \cdot N_1^{est}(n+3) + C_2 \cdot N_2^{est}(n+3) + ... + C_n \cdot N_n^{est}(n+3) \\
&\cdots \\
2n &= C_1 \cdot N_1^{est}(2n) + C_2 \cdot N_2^{est}(2n) + ... + C_n \cdot N_n^{est}(2n)
\end{aligned} \tag{10}$$

This way N/n linear systems can be solved to obtain the n coefficients C_i, N/n times. These systems can be expressed in a matrix form for simplicity:

$$\begin{bmatrix} 1 \\ 2 \\ 3 \\ \vdots \\ n \end{bmatrix} = \begin{bmatrix} N_1^{est}(1) \; N_2^{est}(1) \; ... \; N_n^{est}(1) \\ N_1^{est}(2) \; N_2^{est}(2) \; ... \; N_n^{est}(2) \\ N_1^{est}(3) \; N_2^{est}(3) \; ... \; N_n^{est}(3) \\ \vdots \\ N_1^{est}(n) \; N_2^{est}(n) \; ... \; N_n^{est}(n) \end{bmatrix} \cdot \begin{bmatrix} C_1 \\ C_2 \\ C_3 \\ \vdots \\ C_n \end{bmatrix} \tag{11}$$

$$\overrightarrow{N}_{real} = \overrightarrow{N}_{est} \; \overrightarrow{C} \tag{12}$$

This way the number of discharges made by the equipments can be estimated through a linear combination of features.

$$N_{est} = \sum_i C_i \cdot F_i \tag{13}$$

Using the experimental data, the predictions of N_{est} can be found by solving the linear system for the features presented in 4 and 5, obtaining an estimate of N for each set of features as shown in figure 6.

4.2. Estimating N With a Neural Network

Considering that a linear analysis is simple, fast and of good results, it is an analysis that can be too basic for such a complex problem. We are not only extrapolating the data getting away from the real spectrum once, but twice, thanks to the estimates of N_i^{est} and C_i.

Which is why a model of neural networks is proposed. It allows to iteratively relate the collected data, in order to approximate the ideal function $Y(\overrightarrow{F})$. In this study a neural network of 4 layers is used, 2 visible and 2 hidden, as shown in the figure 7.

The weights of the neural network are the state variables of the system, they describe the function $Y(\overrightarrow{F})$. These are contained in matrices W_0, W_1 and W_2, which determine the status of the network's computation after prolonged training.

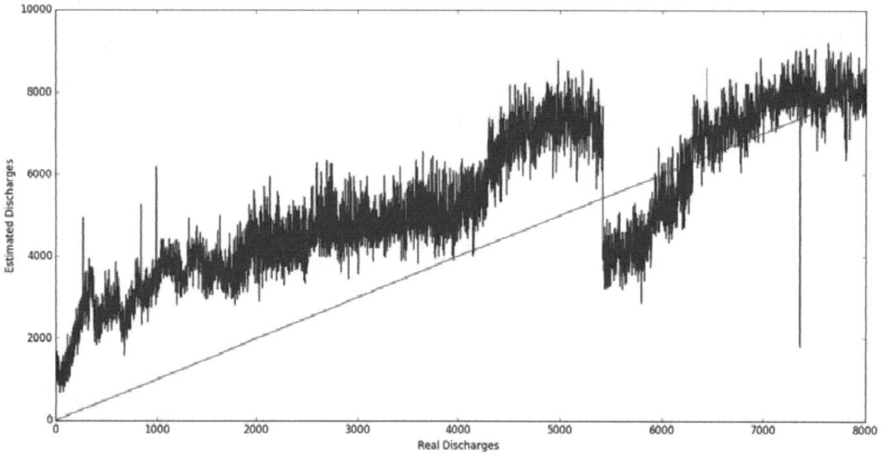

Figure 6. Estimates of N with Linear Combination. Note that the discontinuity at $N \approx 5500$ present in all features (figures 4 and 5) is due to the disassembly of the spark-gap, nevetheless it returns to the trend.

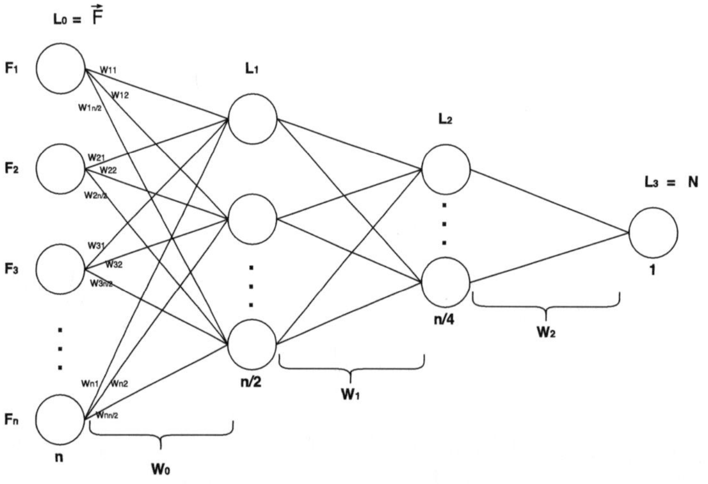

Figure 7. Diagram of Neural Network Used

For simplicity, the number of nodes in layers 1 and 2 was chosen as half of nodes in the previous layer, thus ensuring a smooth transition in dimensional reduction. Therefore the dimensions of the matrices W_0, W_1 y W_2 are $(n \times \frac{n}{2})$, $(\frac{n}{2} \times \frac{n}{4})$ and $(\frac{n}{4} \times 1)$, respectively.

This neural network consists of a basic *Perceptron* of two hidden layers, where each neuron of the hidden layers acts according to the sigmoid function of the linear combination of each F_i weighted by each of its w_{ji}. Meaning that each node i in any layer has a value given by the previous layer, as described in equations 14-16.

$$L_1[i] = \sigma\left(\sum_j w_{ji} \cdot L_0[j]\right) = \sigma\left(\sum_j w_{ji} \cdot F_j\right) \tag{14}$$

$$L_2[i] = \sigma\left(\sum_j w'_{ji} \cdot L_1[j]\right) \tag{15}$$

$$N = L_3 = \sigma\left(\sum_j w''_j \cdot L_2[j]\right) \tag{16}$$

With the sigmoid function defined as:

$$\sigma(x) = \frac{1}{1+e^{-x}} \tag{17}$$

The neural networks initially operate using matrices of weights generated at random, then the error is calculated according to the prediction made in contrast to the actual value provided by the training data. This error is propagated from the last layer to each of the nodes, allowing to know what the impact of each weight to the final error using the *backpropagation* technique. This technique along with many neural network operations are extensively detailed in [17].

With this neural network an analysis can be made to the data to obtain an estimate of N, as in the linear combination analysis. Using the same parameters studied in the previous model, the neural network is trained by associating the series of features with a value of N.

It should be noted that the data was delivered to the neural network in a random order, without losing the correspondence between a series of values $\overrightarrow{F}$ and the N associated to each. This ensures that learning is not distorted by the chronological order of the data, resulting in the graph shown in figure 8

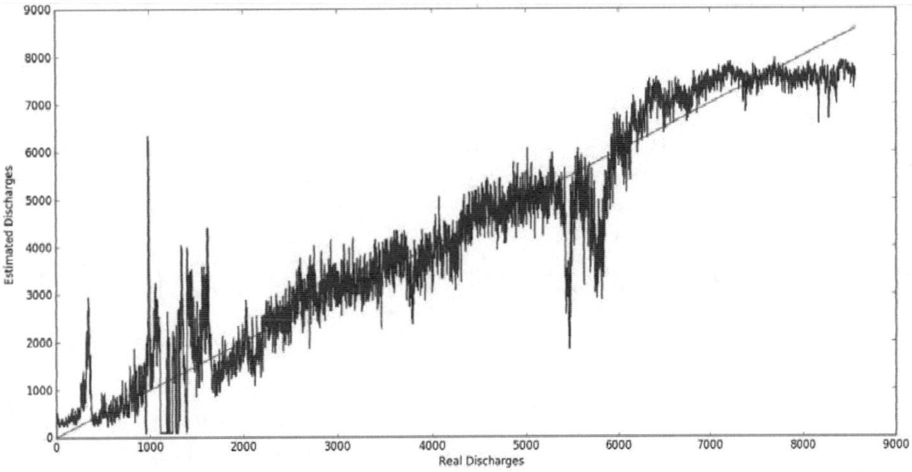

Figure 8. Estimates of N With Neural Networks. Note that the discontinuity at $N \approx 5500$ is much less severe using this method than with lineal combination

5. Conclusion

From this analysis it can be said that by observing the characteristic signals emitted during a regular discharge of a PF equipment, certain descriptive features can be obtained and aid in determining the current state of the equipment.

Furthermore, it can be concluded that rough predictions can be made about the discharges made by observing certain key parameters. In particular, Kurtosis, Shape Indicator, Clearance Indicator and Crest Indicator of the characteristic signals $V(t)$ and $\dot{I}(t)$. Furthermore the actual state of degradation of the PF equipment can be estimated through linear combination and neural network approaches.

Considering that the shape indicator, clearance indicator, kurtosis and crest indicator increase as the experiment evolves, this can be characterized as one phenomenon: the damping of the oscilation is greater. This explains the evolution of these parameters simultaneously. On the other hand, it is a good option to rely on a range of parameters that make evidence of the same phenomenon since they ensure that it is a global trend and not a corruption of the data.

In such a problem it can be seen from the figures 6 and 8 that a nonlinear neural network approach is much more suitable for determining a function that reduces the dimensionality spectrum, as is $Y(\overrightarrow{F})$.

Describing the overall performance and deterioration is crucial to arrive at better designs and best practices in plasma focus devices. Although this represents an initial model for understanding the degradation of the PF-2J, future research is important to understand the wear phenomenon in plasma focus devices and similar equipment.

Acknowledgement

The work is supported by IAEA CRP contract 20370 and by grant ACT-1115, CONICYT, Chile.

References

[1] Leopoldo Soto, Plasma Physics and Controlled Fusion 47, A361 (2005) *New Trends and Future Perspectives on Plasma Focus Research*

[2] Ariel Tarifeño, Cristian Pavez, José Moreno and Leopoldo Soto, IEEE Trans. Plasma Science, 39, 756 (2011) *Dynamics and Density Measurements in a Small Plasma Focus of Tens of Joules Emitting Neutrons*

[3] P. Silva, J. Moreno, L. Soto, L. Birstein, R. Mayer, W. Kies, Applied Physics Letters 83, 3269 (2003) *Neutron Emission from a Fast Plasma Focus of 400 Joules*

[4] Leopoldo Soto, Patricio Silva, José Moreno Moreno, Marcelo Zambra, Walter Kies, Roberto E. Mayer, Alejandro Clausse, Luis Altamirano, Cristian Pavez, and Luis Huerta J. Phys. D: App. Phys. 41, 205215 (2008). *Demonstration of neutron production in a table top pinch plasma focus device operated at only tens of joules.*

[5] Leopoldo Soto, Cristian Pavéz, José Moreno, Luis Altamirano, Luis Huerta, Mario Barbaglia, Alejandro Clausse, and Roberto E. Mayer, Physics of Plasmas 24, 082703 (2017) *Evidence of nuclear fusion neutrons in an extremely small plasma focus device operating at 0.1 Joules*

[6] Mario Barbaglia, Horacio Bruzzone, H. Acuña, Leopoldo Soto and Alejandro Clausse, Plasma Physics and Controlled Fusion 51, 045001 (2009) *Experimental study of the hard x-ray emissions in a Plasma Focus of hundreds of Joules*

[7] Cristian Pavez and Leopoldo Soto, IEEE Trans. Plasma Science, 38, 1132 (2010) *Demonstration of x-ray Emission from an ultraminiature pinch plasma focus discharge operating at 0.1 J. Nanofocus*

[8] C Pavez, M Zambra, F Veloso, J Moreno and L Soto, Journal of Physics: Conference Series 511, 012028 (2014) *Potentiality of a table top plasma focus as X-ray source: Radiographic applications*

[9] L. Soto, C. Pavez, J. Moreno, M. J. Inestrosa-Izurieta, F. Veloso, G. Gutiérrez, J. Vergara, A. Clausse, H. Bruzzone, F. Castillo, and L. F. Delgado-Aparicio, Physics of Plasma 21, 122703 (2014) *Characterization of the axial plasma shock in a table top plasma focus after the pinch and its possible application to testing materials for fusion reactors*

[10] Leopoldo Soto, Cristian Pavez, Fermin Castillo, Felipe Veloso, José Moreno, and S. K. H. Auluck, Physics of Plasmas 21, 072702 (2014) *Filamentary structures in dense plasma focus: current filaments or vortex filaments*

[11] C. Pavez, J. Pedreros, A. Tarifeño Saldivia and L. Soto, Physics of Plasmas 22, 040705 (2015) *Observations of plasma jets in a table top plasma focus discharge*

[12] M.J. Inestrosa-Izurieta, E. Ramos-Moore and L. Soto, Nuclear Fusion 55, 093011 (2015) *Morphological and structural effects on tungsten targets produced by fusion plasma pulses from a table top plasma focus*

[13] J. Jain, J. Moreno, R. Andaur, R. Armisen, D. Morales, K. Marcelain, G. Avaria, B. Bora, S. Davis, C. Pavez, and L. Soto, AIP Advances 7, 085121 (2017) *Hundred joules plasma focus device as a potential pulsed source for in vitro cancer cell irradiation*

[14] Leopoldo Soto, Cristian Pavez, Ariel Tarifeño, José Moreno and Felipe Veloso, Plasma Sources Sci. and Technol. 19, 055017 (2010) *Studies on scalability and scaling laws for the plasma focus: similarities and differences in devices from 1MJ to 0.1J*

[15] L. Soto, C. Pavez, J. Moreno, J. Pedreros and L. Altamirano, Journal of Physics: Conference Series 511, 012032 (2014) *Non-radioactive Source for Field Applications Based in a Plasma Focus of 2J: Pinch evidence*

[16] Kwok L. Tsui, Nan Chen, Qiang Zhou, Yizhen Hai, Wenbin Wang. 2015. Mathematical Problems in Engineering Volume 2015, Article ID 7931612015. *Prognostics and Health Management: A Review on Data Driven Approaches*

[17] Ryszard Tadeusiewcz. 1992 *Sieci Neuronowe, Akademicka Oficyna Wydawnicza*

IOP Conf. Series: Journal of Physics: Conf. Series **1043** (2018) 012050 doi:10.1088/1742-6596/1043/1/012050

Didactic Proposal to include Nanoscience and Nanotechnology at high School curriculum linking Physics, Chemistry and Biology

Nayareth Quirola[1], Valentina Marquez[1], Silvia Tecpan[1], Samuel E. Baltazar[1,2]

[1] Physics Department, Universidad de Santiago, Chile
[2] Center for the Development of Nanoscience and Nanotechnology, Universidad de Santiago, Chile

Abstract. Science and technology at the nanoscale is an important field, not only in modern science, but also an economic activity of vertiginous growth, and therefore, nanoscience should be included as a solid part of the educational curriculum. An interdisciplinary didactic proposal is presented to incorporate nanoscience and nanotechnology (N&N) at the current Chilean high school curriculum. Throughout the didactic sequence, it is expected a better understanding of the science courses followed by the students and, at the same time, of the fundamental concepts of N&N, identifying their benefits and risks to the society with a critical view. This article presents the steps followed in the design of the didactic proposal and characteristics of the learning activities, as well as the methodology foreseen for its validation. Possible risks for the implementation of the proposal are discussed.

1. Introduction

Nanoscience and nanotechnology is one of the fastest growing areas of scientific knowledge, leading to a major impact in the global economy. The presence of more and more products within reach of any citizen, make it necessary to promote literacy in N&N that has been called nanoliteracy [1]. Additionally, the current educational materials of Nanotechnology are mainly available from introductory to advanced levels at the scientific community [2]. However, the access of these contents to the rest of the community is very limited and mainly focused to outreach activities [3].

There are several authors arguing that it is not enough for a nation to direct its efforts to the development of research and applications of N&N. It is also necessary to instruct its citizens to make well-informed decisions, regarding the consumption of nanotechnology products, support of new research projects and participation in public debates on these issues [1, 4-6]. According to scientists, politicians and business leaders, nanotechnology is conceived as the beginning of the next great industrial revolution [1]. Therefore, it is necessary for the citizens to develop a critical vision of the potential benefits and risks that this technology offers, since ignorance unnecessarily increases the perception of risk [4-7].

The scientists involved with nanoscience and nanotechnology agree that it is necessary to include these contents within the curriculum of secondary and elementary education [5, 8, 9], not only as dissemination activities but as an intended formal education in N&N. It implies that educational policy

IOP Conf. Series: Journal of Physics: Conf. Series **1043** (2018) 012050 doi:10.1088/1742-6596/1043/1/012050

makers must establish a systematic and deliberate effort to promote changes to the official curriculum. This process could be long, because it requires the consensus of different members in the educational system [10]. Therefore, an option to immediately integrate N&N contents is to incorporate them into the current school science lessons as a contemporary and motivating theme for the new generations of students who have already heard concepts of nanotechnology at informal educational environments [6, 9].

The emergence of nanotechnology has implied that dissimilar disciplines are merging and converging at the nanoscale to develop innovative applications. However, even within the academic community, knowledge of N&N is often fragmented. It is necessary that the didactic proposals reflect the trans-, inter-, and multidisciplinary nature of this field of knowledge, showing the convergence, not only between the exact sciences, but also with the social sciences through the Science, Technology and Society (STS) framework [1].

One of the main advantages of including N&N to the current curricula of secondary school is to give an interconnected view of Physics, Chemistry and Biology courses throughout the different concepts and applications of N&N, providing a more comprehensive view of these courses to the students.

2. Teaching proposals in nanoscience education

Nanotechnology educational programs are recently designed [8, 9], and theoretical framework is still an ongoing work. These features are important because of the necessity to explain the methodological and content decisions to be included. The advances in science education are helpful to introduce N&N and some agreements have been reached [6].

There is a consensus about the need to develop teaching proposals on nanoscience with a solid foundation in the basic knowledge of school science, mainly in physics, chemistry, biology and mathematics [1, 3, 10]. It also coincides with the need to incorporate elements of applied sciences such as technology of materials, environment and medicine, among others [11]. However, it is less common to find proposals that incorporate the view from the social sciences and the humanities [1, 9].

The amount of didactic proposals increases as the school level increases, so there are few proposals for elementary school, more of them in high school and still more at higher levels [9, 10]. This might be due to the level of abstraction necessary to understand the laws present at nanometric scales [11]. However, experts also agree on the need to begin with nanoscience education from elementary education [10].

Although there are many nanoscience learning resources available on the Internet with interesting demonstrations or experiments, these are not part of a didactic sequence. Additionally, many of them require purchasing special equipment or material, so it is difficult to adapt some of them to different contexts [6].

2.1. Didactic proposals in nanoscience and nanotechnology in Latin America

In Latin America, there are fewer nanoscience training initiatives than in English-speaking countries. Particularly, in Chile, until 2011 there were only two training initiatives in nanoscience for elementary and secondary education, however, they were not supported by textbooks or manuals for their teaching, because at that time, there was no record of any corresponding literature on the subject [10].

Regarding the training of high school science teachers, only two of the Chilean universities have incorporated nanoscience topics, suggesting that only a few high school science teachers are trained to teach N&N. The lack of initiatives for the training of science teachers in nanoscience by universities might occur because currently nanoscience education in high school is partially included only in the academic curriculum [12] and within an elective course (Chemistry) [13], implying that few teachers

IOP Conf. Series: Journal of Physics: Conf. Series **1043** (2018) 012050 doi:10.1088/1742-6596/1043/1/012050

are required with this preparation. As a consequence, the students choosing educational programs have marginal options to get nanoscience education.

At the Latin America region, five didactic proposals on nanoscience training were found at secondary education level. It was found that four of them are not linked to their current scholar science curriculum, and only one establishes the need to link nanoscience contents with curriculum, but does not indicate how to do it. From this we can infer that, at regional level, there is no general criterion of didactic orientations about how to introduce contents of N&N in the courses of sciences, specifically at high school [10].

One of the contributions of this proposal is that it is based on the current Chilean high school science curriculum. After an analysis process, the precise curricular units were determined to incorporate the contents of N&N in the three science subjects (Physics, Chemistry, Biology) in a sequential and progressive way, linking the three scholar science courses, and favoring that students recognize the interdisciplinary nature of N&N [1].

3. Background

Currently, there is not a Chilean state policy to incorporate N&N education in a coherent and homogeneous way, despite the economic importance that this area of knowledge represents worldwide. This implies that there is not a national strategy to organize and integrate the different initiatives that have emerged from public and private research centers to systematize N&N education [10]. On the other hand, the curriculum of high school is undergoing a process of revision and updating, particularly, the science curriculum of the last two years of high school is being analyzed by the different members of the educational system involved in the scientific education. In this context, the need arises to provide a didactic proposal for N&N education that can be implemented immediately in the current curriculum of high school.

It is expected that in the short term, the proposal can be implemented to get evidence of its usefulness in basic N&N education (This process is explained in the section four). With these results, it will be possible to propose their formal incorporation into the national science curriculum.

3.1 Importance of nanoscience and nanotechnology in Chile

N&N have entered the Chilean market through different companies, however, there is currently no long-term government policy for the development of N&N. It is necessary that the government gives priority to the promotion of local nanotechnology and, therefore, the resources and guarantees necessary for a greater production of Research and Development (R&D) done in Chile. This would mean for Chilean science and technology a significant impact towards the technological transfer of innovation processes of the productive sector in general. In a country with a rich diversity of natural resources such as Chile, the N&N would be able to raise quality standards to become a technologically independent country with added value, since nanotechnology can be applied in a myriad of materials, improving results of use and application [11].

3.2 N&N in the high school sciences curriculum Chilean

In Chile, at high school, students can choose between vocational education and academic education. In the year 2010, approximately 54% of students in the last two years of high school selected academic education [12]. Currently, N&N education in high school is limited to those students who choose the curriculum of Chemistry, which is only available for academic education. This limitation implies that there is a small proportion of students who have access to N&N education. Furthermore, N&N is included as a content in the Chemistry curriculum, but there is no learning objective for that content. This could happen because of time constraints for an extensive curriculum [13].

The scientific curriculum for the last two years of high school is undergoing a process of revision and updating. This current curriculum, which includes N&N content for a small proportion of students, will remain in place until 2020 [14], making it necessary to start with N&N education in high school as soon as possible. Waiting for the new national science curriculum implies to stop training the current generations of high school in the big ideas of N&N [5].

4 Characteristics of the didactic proposal

The design of the didactic proposal presented in this paper was carried out following the stages of the process presented in Figure 1. The goal of this proposal is to increase the understanding of the current contents of the three school sciences, and at the same time, of the big ideas of N&N [5]. In this context, the first stage was the analysis of the current science curriculum looking for contents related with N&N. In the next stage, the theoretical advances of N&N education were analyzed [15]. Right away, on the next stage, different educational resources of N&N were analyzed evaluating their pertinence and usefulness. In the fourth stage, the design of educational resources was carried out. These educational resources are in context with the Chilean reality. In the next stage was designed the didactic sequence, in order to advance progressively in the basic N&N knowledge. The level of difficulty of the proposed activities advances progressively along with the proposal. Finally, in the last stage, a teacher's handbook was developed with the basic instructions to understand and carry out the learning activities included into the didactic sequence, even when the teachers are not training in N&N education. At this moment, the design process is finished.

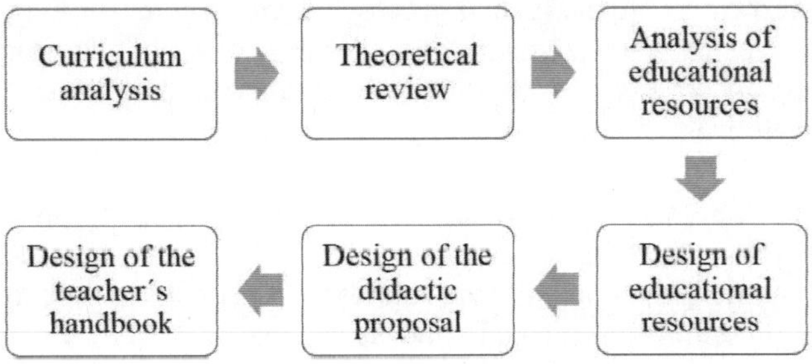

Figure 1. Design process of the didactic proposal to incorporate N&N in high school.

Currently, the proposal is in a process of validation through an expert judgment. The validation results will be used to make the relevant adjustments on the didactic proposal [4]. Subsequently, a training workshop will be held for teachers interested in the implementation of this proposal. Once the stages are completed, the implementation will be carried out (see Figure 2). It is hoped to obtain evidence of the achievement of the learning objectives for both, science subjects and N&N big ideas [5, 15].

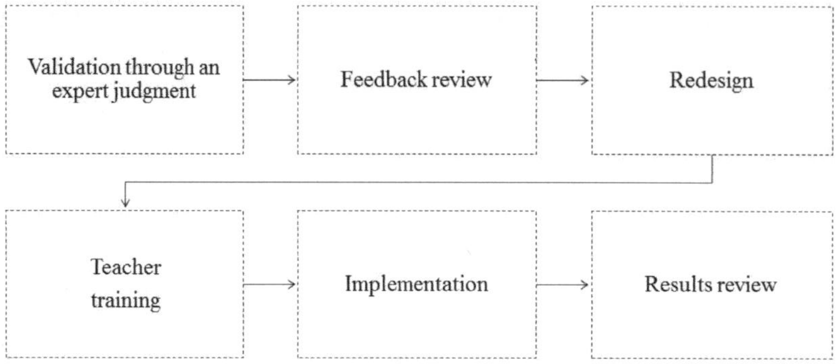

Figure 2. Next steps after validation of the didactic proposal to incorporate N&N in high school through an expert judgment

4.1 Relationship between the didactic proposal and the current high school science curriculum.

After a process of analysis, the precise curricular unit was determined to incorporate the contents of N&N (see Figure 1). The third unit of the last year, the science curriculum of high school allows to incorporate in a sequential and progressive way the contents of N&N. In the three subjects of school science there are contents that are related to N&N. In Physics, we have the curriculum unit "The atom and its nucleus", while in Chemistry, the curricular unit is called "Polymers" and in Biology, the curricular unit is "Immune system". The proposal to incorporate N&N advances progressively starting in Physics, then continuing in Chemistry and finally concluding in Biology. The N&N contents developed in Physics will be used in Chemistry and later in Biology. The experts on N&N education agree that it is important to develop the abilities to communicate and to work across and between conventional science disciplines [2].

There are many applications of nanotechnology, for example in the production of active food packaging, or the use of nanomaterials for the treatment of various diseases. In this proposal, Technology, Medicine and the Environment were selected as unifying topics between the subjects (see Figure 3). Additionally, these topics incorporate the STS framework to make evident the relationship between scientific and technological advances with society [7, 11]. It is important that students understand the interdisciplinary nature of N&N. The sequence of activities is designed to make evident the relationship between disciplines [1, 2].

On the other hand, the proposal also intends to promote collaboration between teachers. It is necessary that the Physics, Chemistry and Biology professors organize the previous activities of each course to execute the didactic sequence in the proposed order. The science curriculum of the three courses is extensive. Based on this, the proposal has only 90 minutes of face-to-face work assigned to learn the N&N concepts at each subject. They can be two sessions of 45 minutes each or a single session of 90 minutes

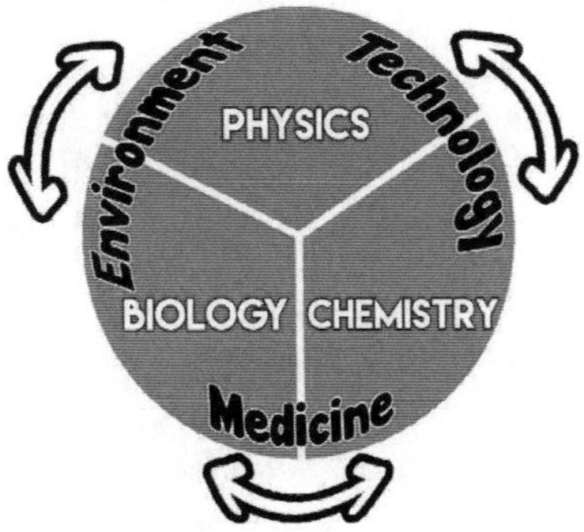

Figure 3. Linking among disciplines in the didactic proposal

4.2 Themes of N&N included, teaching methods selected and expected learning outcomes.

In this section, we present the different activities developed for each one of the Physics, Chemistry and Biology courses. The activities will be driven to introduce and develop some well-known key concepts of N&N [8]. We have used different educational methods and approaches in the activities because it is necessary to incorporate the different contents of the proposal to the curriculum. These methods are necessary because each unit has specific goals and cover different aspects of N&N. Particularly, we do not need to give an introduction of the fundamental concepts of nanoscience and nanotechnology at each course, prolonging unnecessarily our proposal. Instead of this, we emphasize the collaboration between the courses and cover the main aspects such as Nanoscale (Physics), synthesis and functionalization (Chemistry) and application (Biology) [15]. Additionally, the inclusion of several methods to implement the activities relies in the requirements of various student groups with different needs and realities [8]. The set of proposed activities allows incorporating the STS framework, so the relation with de social sciences becomes evident [7, 11].

First, we present the proposal of Physics, where the fundamental concepts such as nanoscale and surface-volume ratio are given as well as the study of the properties of the materials at different scales. These contents are incorporated in the third unit of the curriculum at the last year of high school. Then, we introduce the main concepts of synthesis and functionalization of nanomaterials in the unit three of the Chemistry program. Finally, a discussion and debate activities are prepared at the end of the unit three of Biology to include the consequences of the application of nanotechnology.

4.2.1 Activities in Physics

In the Physics course, we have prepared a suitable audiovisual material to introduce the basic concepts of N&N such as nanometer, scale, self-assembly or tunneling effect, top-down and bottom-up views and their consequences. The use of multimedia materials has already been proven to be suitable in the introduction of Nanotechnology to the scholar community [8]. One of the first challenges to introduce nanoscience is related to the concept of scale [6]. We have used a classification activity to associate objects to different world scales, helping the students to explore the perception of the size of different objects. Therefore, the first task of the teachers is to help to the students to identify and classify the different worlds across to the size of the objects. These objectives follow the expected learning of the Physics curriculum "To describe the atomic nucleus and some of its properties". In this context, we have defined a learning result "To recognize the Nano-world and its dimensions, comparing objects, properties and laws of it with those belonging to other worlds". The corresponding activity consists in to classify objects inside of a scale: Macro-world, Micro-world, Nano-world and subatomic scale.

The next topic is related to the properties of the materials at nanoscale. Here the first activity is to identify the constituents of the atom, and the evolution of the atomic models. The second activity is related to the surface-volume ratio concept [15]. We propose a hands-on math activity, where the students calculate geometrical features of materials, such as surface area, at different scales [6, 9]. Here the students will explore how the differences in the scale leads to differences in materials properties such as reactivity. This is achieved evaluating the surface area of a cube being cut in halves and associate the increased surface area to the reactivity of the material.

Finally, with the help of a short video, we can introduce the tunneling effect, which is a concept of quantum mechanics and very useful in the characterization of nanostructures [15] (see STM in Chemistry section). We summarize this section with a mind map activity, where the main concepts and vocabulary are included. This visual aid will be helpful for the students in the following courses.

4.2.2 Activities in Chemistry

The proposal is incorporated in the Chemistry curriculum at the end of the unit three “Polymers”, where the expected learning “To present remarkable polymers at industrial processes such as clothes production and plastics identifying their importance in the society”. From the expected results, we have defined “To identify the use of N&N in the production of plastics, declaring their benefits and risks to the society”.

In this section, we must consider the previous knowledge obtained from the introduction of N&N in physics section (Nanoscale, surface-volume ratio), as well as the contents of the Chemistry curriculum dedicated to polymers. The first part of this module is focused to the fabrication and functionalization of nanopolymers. The first activity considers the introduction of carbon nanotubes from the geometrical construction and their mechanical, electronic and chemical properties [8, 9, 15]. We propose the application of a hands-on activity, where the students build a representation of nanotubes from a sheet with a hexagonal pattern. This activity will be complemented with a description of the properties of nanotubes as a function of their geometric construction [8].

The second part is devoted to the characterization of the nanopolymers, emphasizing the development of tools to “see” and manipulate nanostructures. The first activity shows an analogy of how the scanning tunneling microscope (STM) works, through the *sensing* concept applied to different properties [6, 9, 15].

The next activity is related to the functionalization of nanotubes with polymers by an interactive representation of the creation of active sites and formation of different bonds. We propose a role-playing activity, where the students will represent the functionalization of nanotubes with different polymers. This resource has already been implemented in undergraduate biology classroom [16], as well as physics teachers training [17], and helps improving the explanation of several processes to the students, focusing in their memory-oriented learning abilities through the visual representation of the functionalization processes.

The following activity is intended to identify the usefulness of nanostructured polymers in society according with the STS framework [7]. The students will extract information from a newspaper column related with the application of nanopolymers in the national food industry. The last activity for this module focuses on the identification of nanomaterials of different commercial products. The students should relate the properties and applications of these nanomaterials with the characteristics of the product.

4.2.3 Activities in Biology

This proposal is focused in the application of nanotechnology to biological systems and will be incorporated at the end of the third unit in the Biology curriculum "Immune system: diseases and treatments". The previous knowledge requires the concepts already covered in the physics and chemistry sections (nanoscale, functionalization, etc.), along with the formal content of the original curriculum (immune system). As a learning result of this section, the students should be able "To analyze the risks and benefits of N&N to the society, particularly, in the application to medicine and treatments autoimmune diseases".

We are considering two activities. The first one corresponds to the search and critical discussion of several information sources related to three topics: "Use of vaccines with nanotechnology in the treatment of human immunodeficiency virus (HIV)", "Nanotechnology in the cancer treatment" and "Transplant treatment with Nanotechnology". The students will prepare a report based on the provided literature, identifying nanomaterials, benefits and risks of these technologies, as well as the reliability, objectivity and relevance of the sources. Through this activity, the students will investigate the benefits and risks involved in the application of nanotechnology at health area.

Finally, at the biology section, we have included a debate activity in classroom, based on the benefits and risks of nanotechnology applied to diseases treatments [1, 7]. This method is focused in the discussion and identification of arguments proposed by different students groups [18, 19]. The class will be divided in three groups: Moderator (teacher), jury and experts (in favor and against the proposed topics). At the end of the activity, the teacher will give closing statements to emphasize the different aspects of nanoscience (scale, interdisciplinary, etc.) and their effects in the society, considering positives and negative aspects, as well as the necessity to acquire the basic knowledge on N&N to make well-informed decisions.

4.3 Teacher's handbook

One of the main difficulties to implement this proposal is related to the lack of scientific knowledge in nanoscience from the teachers, because most of them were trained before the inclusion of N&N in the science educational programs [20]. This is especially important in Latin America, where the inclusion of nanoscience to the educational program of science teachers has been only recently incorporated at specific curricula [13]. As part of the proposal, a manual was developed for the teacher with all the instructions to effectively implement the didactic proposal. Each learning activity is described in detail and additional information is provided, so even teachers with no formal education in N&N can guide learning activities. Our proposal includes not only the worksheets with completed learning activities for each unit, but also it has additional comments to implement the activities, suggested answers and expected conclusions given to the teachers. In this context, we have included a suggested assessment for the learning activities. This is helpful to identify the learning level reached by the students, as well as the achievement degree of the expected goals for each activity.

4.4 Validation Strategy

We choose expert judgment as validation strategy [4]. To validate the scientific contents as well as the educational goals of the proposal, we have considered the evaluation of our proposal from two groups: scientific researchers of Physics, Chemistry and Biology, and high school teachers of the corresponding areas. In the case of scientific evaluation, the researchers are also involved in different topics of nanoscience and nanotechnology, and their feedback will be important in order to get current definitions, concepts and state-of-art applications for each course. From the teachers, we will get the

IOP Conf. Series: Journal of Physics: Conf. Series **1043** (2018) 012050 doi:10.1088/1742-6596/1043/1/012050

useful comments related not only to the implementation of the activities and the evaluation of the level reached by the students, but also the feedback of the appropriate inclusion of the proposal at each class to follow and complete the objectives of the Chilean curricula.

5 Discussion and Conclusions

This proposal is characterized inside the Chilean curricular discussion process at the scientific basis of education. One of the main strengths of this proposal is its adequate application into the educational curricula, due to its abbreviated extension, not only keeping its contents, but also supporting the current curriculum of each science course. The inclusion of N&N into the secondary curricula is an important aspect that has been previously considered at N&N [9], but considering only two specific activities in Physics and Chemistry. In our case, the interdisciplinary characteristics of N&N make it suitable to incorporate into the sciences courses of secondary school, *e.g.*, the nanoparticles are defined in Physics and later applied in Biology, or properties changes proposed in Physics are later incorporated in carbon nanotubes discussed in Chemistry. Because of these connected topics, a more comprehensive view of the science course is expected, and additionally the Science-Technology-Society (STS) aspects are included because of the contents in Chemistry and Biology related to environment and medicine.

In Physics, we propose, based on the theoretical framework of nanoscience education [6, 15], as a learning result: "Recognizes the nanoworld and its dimensions, comparing the objects, properties and laws that belong to it and those belonging to other worlds". This could be achieved through the proposed learning activities. At the same way, in Chemistry we propose as a learning result: "Identifies the use of nanoscience and nanotechnology in the production of plastics, declaring its benefits and risks for society". Finally, in Biology the proposed learning result is: "Analyzes the risks and benefits of nanoscience and nanotechnology in society by discussing their use in medicine for the treatment of autoimmune diseases".

The use of different educational methods is an important aspect in order to capture the attention and to stimulate several groups of students with different interests. These methods have been previously applied at high school level and were discussed using a Delphi study [15].

The proposal is easy to apply from the point of view of its extension since it requires only 90 minutes for each science subject. In addition, it supports and complements the current contents established in the curriculum. However, there are some risks to its implementation. These are: a) lack of education in N&N by teachers and b) difficulties with teacher's coordination resulting from collaborative work between subjects. The didactic proposal includes possible actions to minimize these risks.

Finally, the revision of our proposal is currently underway by science teachers and scientific experts. Once we have collected and revised the feedback from the reviewers, we will follow the implementation process (see Figure 2). This will consider a workshop for teachers focused in the implementation of this proposal in classroom, and finally the analysis of the implementation results. We agree on the need to improve N&N training programs for high school science teachers [1, 3, 10], making this didactic proposal a useful tool to achieve this goal.

The authors thank the support of the Basal funding for Scientific and Technological Centers under Project FB0807. Also, S. E. B. would like to thank the support of DICYT under the grant 041631BR and the PID-USACH project 051-2016.

References

[1] Yawson RM, An epistemological framework for nanoscience and nanotechnology literacy 2012 *Int. J. Technol. Des. Ed.* **22** 297-310

[2] Jackman JA, Cho DJ, Lee J, Chen JM, Besenbacher F, Bonnell DA, Hersam MC, Weiss PS and Cho NJ, Nanotechnology Education for the Global World: Training the Leaders of Tomorrow 2016 *ACS Nano* **10**(6) 5595-9
[3] Sullivan TS, Geiger MS, Keller JS, Klopic JT, Peiris FC, Schumacher BW, Spater JS and Turner PC, Innovations in Nanoscience Education at Kenyon College 2008 *IEEE T. Educ.* **51**(2) 234-41
[4] Lin SF, Lin HS and Wu YY, Validation and Exploration of Instruments for Assessing Public Knowledge of and Attitudes toward Nanotechnology 2013 *J. Sci. Educ. Technol.* **22**(4) 548-59
[5] Tretter TR, Instructional impact on high school physics students' nanoscience conceptions 2015 *Nanotechnology Reviews* **4**(1) 33-50
[6] Planinsic G, Lindell A and Remskar M, Themes of nanoscience for the introductory physics course 2009 *Eur. J. Phys.* **30**(4) S17-S31
[7] Gardner G, Jones G, Taylor A, Forrester J and Robertson L, Students' Risk Perceptions of Nanotechnology Applications: Implications for science education 2010 *Int. J. Sci. Educ.* **32**(14) 1951-69
[8] Blonder R and Sakhnini S, Teaching two basic nanotechnology concepts in secondary school by using a variety of teaching methods 2012 *Chem. Educ. Res. Pract.* **13**(4) 500-16
[9] Greenberg A, Integrating Nanoscience into the Classroom: Perspectives on Nanoscience Education Projects 2009 *ACS Nano* **3** 762-9
[10] Tutor-Sanchez JD, Formación en nanociencia y nanotecnologia: un reto en iberoamerica 2013 *Momento* **46** 42-53
[11] Sakhnini S and Blonder R, Nanotechnology applications as a context for teaching the essential concepts of NST 2016 *Int. J. Sci. Educ.* **38**(3) 521-38
[12] Farias M and Carrasco R, Diferencias en Resultados Académicos entre Educación Técnico-Profesional y Humanista-Científica en Chile, *Consejo Nacional de Educación* 2012
[13] Química, Programa de Estudio 4to medio, Ministerio de Educación de Chile 2009
[14] Nuevas Bases Curriculares y Programas de Estudio, *Ministerio de Educación de Chile* 2015
[15] Sakhnini S and Blonder R, Essential Concepts of Nanoscale Science and Technology for High School Students Based on a Delphi Study by the Expert Community 2015 *Int. J. Technol. Des. Ed.* **37**(11) 1699-738
[16] Stevens SY, Sutherland LM and Krajcik JS, The Big Ideas of Nanoscale Science and Engineering: A Guidebook for Secondary Teachers *NSTA Press* 2009
[17] Scherr RE, Close HG, Close EW and Vokos S, Representing energy. II. Energy tracking representations 2012 Phys. Rev. St Phys. Educ. Res. **8**(2) 020115
[18] Kolsto SD, Patterns in Students' Argumentation Confronted with a Risk- focused Socio-scientific Issue 2006 *Int. J. Sci. Educ.* **28**(14) 1689-716
[19] Erduran S and Jiménez-Aleixandre MP, Argumentation in Science Education *Springer* 2007
[20] Jones MG, Blonder R, Gardner GE, Albe V, Falvo M and Chevrier J, Nanotechnology and Nanoscale Science: Educational challenges 2013 *Int. J. Sci. Educ.* **35**(9) 1490-512

Incidence of Lamy's Theorem in static learning: Balance of forces

N J Sánchez Salcán[1], F P Londo Yachambay[2]
[1]Universidad Nacional de Chimborazo, Avda. Antonio José de Sucre, Km 1.5 Vía a Guano Riobamba-Ecuador
[2]Escuela Superior Politécnica de Chimborazo, Panamericana Sur km 1 ½, Riobamba-Ecuador

Email: nsanchez@unach.edu.ec

Abstract. This research is within the thematic area education, especially within Didactics of physics, which is based on the existing problems in the teaching and learning of physics both at the level of secondary education and higher education. This research is based on meaningful learning theory raised by Ausubel, which provides valuable elements for the teaching of physics, with higher levels of understanding of the information and resistant to oblivion. The objective was to determine the incidence of the theorem of Lamy in the learning of the equilibrium of three forces, applied to the students of exact sciences of the Universidad National de Chimborazo. It is considered a quantitative correlational research with a quasi-experimental design. The method used was the hypothetical deductive because it allowed to observe the phenomenon, consider a hypothesis for the explanation of the same. By the test of normality of the data is used evidence not parametric and, considering the statistical test of Wilcoxon resulted in a significant difference, between the application of the theorem of Lamy and the analytical method of decomposition of forces in problem solving on equilibrium of three forces.

1. Introduction

Studies show that in addition to the intrinsic complexity of the mathematics and physics, one of the main causes of apathy that have students towards Physics is the usual training system that usually lead teachers to practice, resulting in many cases of desertion or the choice of other races. The attitude of the students within the classroom is memorization, leaving aside the reasoning and provoking anxiety called the number, creating a false impression of having understood the contents proposed for the study of physics, which generates disadvantages in the ongoing development of learning during the following academic content. This, in addition to also exist in the classroom students who are an eminence solving theoretical problems of physics and any other subject, but in reality they do not know what real meaning has studied. On the other hand, according to the results obtained from the Secretariat of the career of exact sciences of the Universidad Nacional de Chimborazo, students obtained low scores of the test that was applied to all students who belong to the network of mathematics and physics, it is evident that students do not possess minimal knowledge both in mathematical physics at the level of basic education High school and higher.

In the teaching of physics and in particular the static, are commonly taught its foundations and fundamental concepts for understanding analysis. It is very common to confuse the physics as an

application of the mathematical equations and not as a science that deals with observation and knowledge of nature.
As researcher and also teacher of the third semester of exact sciences have seen the little domain presenting the students with regard to basic concepts, difficulty understanding trigonometric functions, application of science, field of study, in addition, lacks domain geometric causing difficulty in understanding it, as Physics makes use of geometric shapes for the representation of problem situations. The above difficulties are modelled in this study, in order to find alternative solutions that allow students significantly improve the problem mentioned above, using strategies appropriate for the resolution of problems and give you treatment that facilitates the mathematical approach that physics is often prioritized both the mathematical calculation as the analysis and interpretation of the results obtained during each problem solved.
On the other hand according to data provided by the National Institute of educational evaluation (Ineval) knowledge test disciplinary master 2016 - Ecuador, 84770 applicants who aspire to enter to the public teachers were evaluated, obtaining a basic score that falls on the proposed scale greater or equal to 600 points and less than 800 points evidenciandose unable to reach excellence [1].
Worth mentioning as we are training future teachers of the Ecuadorian education, they need both cognitive and methodological tools enabling it to communicate successfully in the world of physics and solve the problems that warrant their use but understanding it.
This research focused on the application of the theorem of Lamy to facilitate the study of static in the resolution of environmental problems, it was intended to awaken the interest of students to address issues that are closely linked with mathematical demonstrations. As well as providing future teachers of the career of Sciences of a methodological strategy for the learning of static with a view to the development of competencies in students; in where strategies become tools to break the pattern of one-way knowledge-centered learning, for flesh give step to a participatory approach, based on the model competency-based, and the exchange of experiences, where student expose their experiences, lessons learned, following a critical posture, support and encouragement to creativity, to meet the different situations that arise in the educational context from a perspective of psychology in order to develop a competent teaching, and students prepared for today's world.
Finally this research generates good academic performance of the student, and thus set aside the traditional education, which is not involved in their learning, is simply a being that it is limited to watching and listening; and consequently the learning of physics is not only to approve it, but to be applied in everyday life and allow learners actively develop in the process teaching and learning.
The question that arises then is focused to conduct research so it will be present in each one of the processes it. To what extent Lamy's theorem contributes to the improvement of the learning of the students for the resolution of environmental problems related to the static?

2. Theoretical framework

To carry out the investigation was necessary to perform a literature review about the proposed theme, which allowed to take into account the investigations or work late to it, here is a summary of the research. At the international level the thesis "study of the static of solids applying methodological strategies that facilitate understanding in problem solving in students of 10th grade C of the Institute Licenciado Miguel Larreynaga San Juan de Río Coco, matrix during the second half of 2014" [2], focused on the implementation of methodological strategies to facilitate the study of the static of solids in the resolution of environmental problems It was intended to awaken the interest of students to address themes that are closely linked with mathematical demonstrations and that therefore they are apathetic to face these situations. Based on his work the researcher managed to develop three methodological strategies taking into account the context, the characteristics of the selected group, the subject of study, the level of depth that was intended to achieve, selecting activities that facilitated the coordination of elaborate organizing properly every teaching moment methodological strategies for students.

Research "Proposal of teaching of balance of rigid body by the action of forces coplanar with the use of the plane geometry in static, of the University course Pascual Bravo" [3]. The methodology consisted of knowing first know the student, emphasizing the concepts of physics you must strengthen before passing to static, link what you know with the new concepts and through class work to build the concept of what you should learn. With these experiences are new skills that will be carried to new contexts and situations. The investigator concludes that the exploration of knowledge tests, construction of free-body diagrams is achieved see difficulties in the management of interpret the laws of equilibrium in physics. To sense that a single force exercised balance, shows that they make no difference between the first and the second law of Newton.
At the local level there are no researches related to the proposed research topic.
The theorem of Lamy, was enunciated by the French religious Bernard Lamy (1645-1716), which States the following:
When a rigid body in balance is subject to the action of three concurrent forces, each module is directly proportional to the sine of its respective opposite angle.

$$\frac{F_1}{\text{sen}\alpha} = \frac{F_2}{\text{sen}\beta} = \frac{F_3}{\text{sen}\gamma} \qquad [4]$$

This theorem is a consequence of the law of Sines, applied after forming the triangle of forces.
Learning should develop mainly in productive and meaningful ways that revitalize the study methodology, in order to reach Metacognition, i.e. to apply knowledge to solve problems for life. The skill is the expression of know-how in students, which characterizes the domain of action. The knowledge, skills and attitudes, constitute the main guide of the educational process. Within the teaching process learning is important to develop critical logical thinking of the students that are evident in the approach to skills and knowledge to interpret and solve problems of everyday life.
This research is based on the Polya model, for the resolution of problems, namely: identifying relevant concepts, the problem, run the solution, and evaluate the response.

3. Methodological framework

He was a descriptive-comparative study with a quasi-experimental design, where the same Group learned how to solve problems of balance of forces using the analytical or method known as vector decomposition, which is that usually apply the teachers in the classroom and the other learned to solve problems by using the theorem of Lamy.
The method used was hypothetical-deductive, as the imagination of the researcher rather than the inductivismo works with the experimentation to find empirical conditions suitable for the validation of the hypothesis.
Population was constitutida by 70 students of the career of Sciences exact of the Faculty of Sciences of the education, human and National University technologies of Chimborazo, which through sampling non-probability of intentional type, will chose the representative sample for research being 14 students of the 3rd semester of exact sciences. The criterion for the selection of the sample was held because the subject to be treated is found in the silabo this semester.
The implementation of the proposal was carried out in 8 sessions with a duration of two hours, at the seventh session applied testing related specifically to the theme of balance of forces.
The following describes the procedure followed by the students for the solution of problems related to the balance of forces using vector decomposition method and the theorem of Lamy and shown in detail because solving a problem following protocol.
For example, consider the wooden packing 75 kg shown in the spatial diagram of the figure. It rested between two buildings and is now raised to the platform of a truck that will remove it from there. The package is supported by a vertical cable attached in A two strings that pass over pulleys fixed to buildings in B and C. You want to determine the voltage at each of the strings AB and AC. [5].

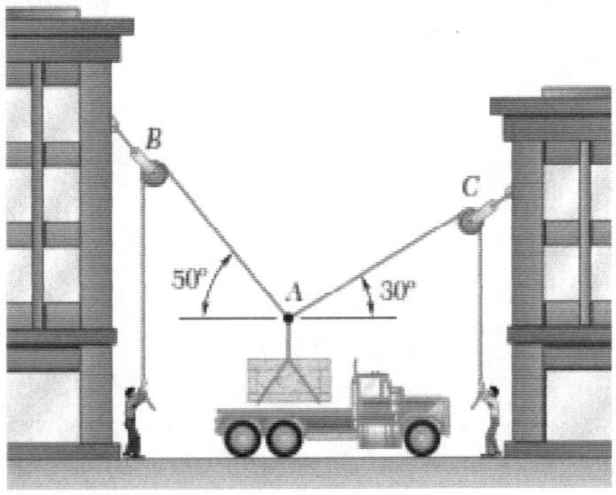

Figure 1. Space diagram of forces
Problem of mechanical engineering, which is derived from a real physical situation, the methods of analysis are applied to a system of forces acting on a particle. [5]

Based on the Polya model for problem solving, they proceeded in the following manner:
Identify the relevant concepts. This problem concerns the conditions of equilibrium of a particle, this is must know the conditions that should be considered that a body is in balance. That is, when a particle is in equilibrium, the resultant of all forces acting on the particle must be equal to zero.
Pose the problem. To solve the problem should draw a diagram of free body showing particle in equilibrium. Since discussed tensions in the strings, the free body diagram must include at least one of these tensions and if possible to both.

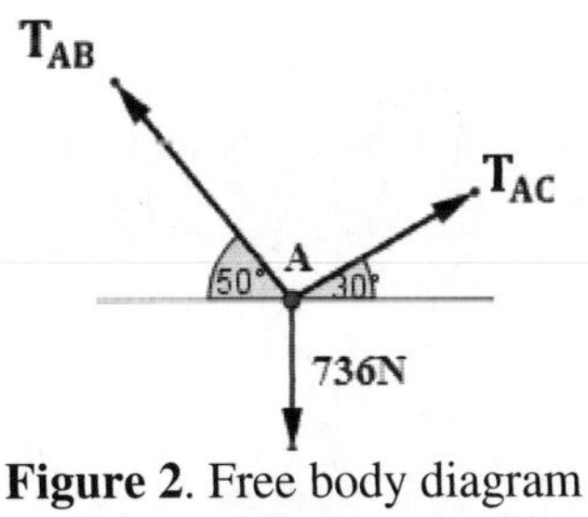

Figure 2. Free body diagram
The point A is in equilibrium, the forces exerted on A by the vertical cable and the two strings are evident

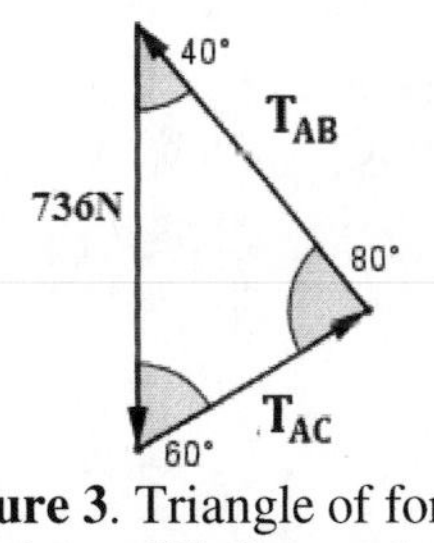

Figure 3. Triangle of forces
The particle is in equilibrium under three forces, the problem can be solved by drawing a triangle of forces.

Run the solution: Lamy theorem and the analytical solution by decomposition of forces was used for the solution to this problem. The forces exerted by the two strings are not known, but are equal in magnitude to the tension in the rope AB and AC cord.

Method N ° 1. Application of theorem of Lamy

$$\frac{T_{AB}}{\text{sen}120^0}=\frac{T_{AC}}{\text{sen}140^0}=\frac{736\text{N}}{\text{sen}100^0}$$

$$T_{AB}=647\text{N} \qquad T_{AC}=480\text{N}$$

Method N ° 2. Analytical solution or vectorial decomposition. The equations of equilibrium should be considered

$$\sum F_x=0$$

$$T_{AC}\cos 30^0-T_{AB}\cos 50^0=0 \qquad (1)$$

$$\sum F_y=0$$

$$T_{AC}\mathrm{sen}30^0+T_{AB}\mathrm{sen}50^0-736N=0 \qquad (2)$$

We solve T_{AC} from equation 1

$$T_{AC}=\frac{T_{AB}\cos 50^0}{\cos 30^0} \quad (3)$$

Replacing (3) into equation (2) is obtained:

$$\frac{T_{AB}\cos 50^0\mathrm{sen}30^0}{\cos 30^0}+T_{AB}\mathrm{sen}50^0-736=0$$

$$T_{AB}\mathrm{tg}30^0\cos 50^0+T_{AB}\mathrm{sen}50^0=736N$$

$$T_{AB}\left(\mathrm{tg}30^0\cos 50^0+\mathrm{sen}50^0\right)=736N$$

$$T_{AB}=\frac{736N}{\mathrm{tg}30^0\cos 50^0+\mathrm{sen}50^0}$$

$$T_{AB}=647\ N \qquad (4)$$

Replace (4) into equation (3) and get

$$T_{AC}=\frac{647\ N\cos 50^0}{\cos 30^0}$$

$$T_{AC}=480N$$

Evaluate the response: it corroborates what manifested itself by identifying the problem, that is, a particle is in equilibrium, the resultant of all forces acting on the particle must be equal to zero.

$$\sum F_x=0$$

$$\left(415{,}69\vec{i}+240\vec{j}\right)+\left(-415{,}88\vec{i}+495{,}63\vec{j}\right)+\left(0\vec{i}-736\vec{j}\right)=0$$

$$0=0$$

To this end, the techniques and instruments used to obtain data were:
The written test that was applied to the 14 students of the 3rd semester of the Race of Exact Sciences, the same that consisted of 10 items related to the subject of balance of force, the test was scored on 10 points.
In addition, an observation sheet was applied in order to corroborate that with the application of Lamy's Theorem in solving problems related to the equilibrium of three concurrent forces, the student was able to pass from the simple to the complex.
It should be mentioned that before applying the instruments these were validated as well as the degree of reliability was known using the Cronbach alpha coefficient.
For the verification of the hypothesis: The Lamy theorem significantly influences the learning of static: Balance of forces in the 3rd semester students of the Exact Sciences career of the National University of Chimborazo, worked with a confidence level of 95% and a significance level of 5%; We verified the normality of the data with the shapiro Wilk test because it was applicable to samples smaller than 50, obtaining a value that denotes the use of non-parametric tests, for that reason the Wilcoxon distribution

was selected, taking into consideration the criterion: if the Probability obtained P-value $\leq \alpha$, reject Ho, otherwise Ha is accepted, the SPSS software was used for the analysis of the data.

4. Analysis and discussion of results

As a general criterion, George and Mallery state that the Cronbach alpha coefficient calculated to be acceptable must be greater than 0.90 [6]; The alpha obtained was 0.894, so the instrument applied had a high degree of reliability. (See table No. 1)

Table 1. Reliability of the data collection instrument

Cronbach's alpha	Cronbach's Alpha based on the elements tipicados	Strength n
0,90	0,894	10

The results tab of the observation (see table N ° 2) applied 14 students of the career of Sciences shows the influence of the application of the theorem of Lamy in the resolution of problems related to the balance of force, of which of deduces:

- ✓ Better understanding in the resolution of problems, related to the balance of forces
- ✓ Appropriate application of the steps for solving problems linked to the balance of forces according to indicators of achievement established for study.
- ✓ Correct graphic and geometric representation of problem situations proposed during class.
- ✓ Reduction of the time in the resolution of problems, with a good quality of answers to problems.

94% of students were able to move from the simple to the complex in the resolution of problems, using the theorem of Lamy.
80% of students using the theorem of Lamy, resolve problems in less than the set time.

Table N ° 2. Tabulation of observation guide

	METHOD 1. Vector decomposition		METHOD 2: Lamy's theorem	
	YES	NO	YES	NO
1. Identifies the physical magnitudes existing in the problem	80%	20%	85%	15%
2. Makes the graphic representation of the problem	91%	9%	93%	7%
3. Determines if the data are insufficient to solve the problem	81%	19%	90%	10%
4. Identifies the formulas useful for solving the problem	75%	25%	91%	9%
5. Run in detail each operation	70%	30%	95%	5%
6. Shows that the response corresponds to what is asked in the problem	78%	22%	92%	8%
7. He was able to move from the simple to the complex, using each method	65%	35%	94%	6%
8. The time to resolve the problem was less than the stipulated time	63%	37%	80%	20%

IOP Conf. Series: Journal of Physics: Conf. Series **1043** (2018) 012051 doi:10.1088/1742-6596/1043/1/012051

As evidenced in table N° 3, significant is the difference between the averages of the methods applied for the resolution of problems of equilibrium of three forces, the average obtained using analytic or known as vector decomposition method was of 5,821 and using the theorem of Lamy Gets an average of 8.15.

Table N ° 3. Descriptive statistics

	N	Minimum	Maximum	Media	Desv. tìp
Method 1: decomposition of forces	14	5	7,50	5,821	0,77
Method 2: Lamy theorem	14	7	9	8,150	0,75

The hypothesis testing was carried out in SPSS software. (See table N ° 4). Taking into account the statistical decision described above for the acceptance or rejection of the hypothesis, it found that the p-value retrieved 0.001 is less that the 0.05 significance level is therefore rejected the null hypothesis and accepted the alternative hypothesis, this is: Lamy theorem significantly influences learning balance of forces in the students of 3rd semester of exact sciences of the Universidad National de Chimborazo. This verifies the hypothesis. Resulting analysis to the application of the theorem of Lamy in trouble help effectively to the understanding of the static

Tabla N°4. Contraste[b] Statistics

	Method 1 – Method 2
Z	-3,296 [a]
Sig. Asintòt (bilateral)	0,001

5. Conclusions

With respect to the analysis of the results achieved demonstrate that for solving problems where three forces are involved, method 2 referred to the theorem of Lamy is achieved best quality learning compared to the analytical method that is the breakdown of forces.
On the other hand has been observed that students show greater interest and motivation for learning in statics, using the theorem of Lamy in the resolution of problems related to the context.

6. References

[1] National Institute of Educational Evaluation. (2016). Teaching evaluation Ecuadorian. Retrieved from http://sure.evaluacion.gob.ec/ineval-dagi-vree-web-1.0-SNAPSHOT/publico/vree.jsf

[2] Espinoza, N. (2015). Study of the static of solids applying methodological strategies that Facilitate comprehension in the resolution of problems in students of 10th grade C of the Miguel Larreynaga Institute San Juan de Rio Coco, Madriz during the second semester of 2014 (undergraduate thesis). National University Autonoma of Nicaragua. Estelí.

[3] Giraldo J (2014) Proposal of teaching of balance of rigid body by the action of forces Coplanar with the use of flat geometry in the course of Statics, of the University Institution Pascual Bravo (undergraduate thesis). National university of Colombia. Medellín, pp.3

[4] Perez W. 2000. Theoretical physics. 1st ed. Perù: San Marcos. p. 51-53

[5] Beer F Jhonston, E., Mazurek, D., Eisenberg, E. (2010) .Mechanical vector for static engineers. 9ª ed. Mexico: McGraw-Hill. P. 37.

[6] George, D., & Mallery, P. (2003). SPSS for Windows step by step: A simple guide and reference. 11.0 update (4th ed.). Boston: Allyn & Bacon. p. 231

IOP Conf. Series: Journal of Physics: Conf. Series **1043** (2018) 012052 doi:10.1088/1742-6596/1043/1/012052

Learning and Sustainability: an approach from the interaction

> The attitude that this new man maintains facing nature will be radically different from the attitude he maintained in previous times.
>
> Werner Karl Heisenberg

Patricio Pacheco, Cecília Rios
J.P.Alesandri 1242, Ñuñoa,Santiago
Universidad Tecnológica Metropolitana, Santiago de Chile

patricio.pacheco@utem.cl

Abstract. We present a study of behaviors, belonging to the domain of emotions, inquiry, persuasion and the use of information and that are deployed in time by teams of students immersed in learning processes. These are transformed into time series on which non-linear dynamics techniques are applied, obtaining the connectivity for each team. From this value and the quotient Positivity / Negativity (referred to the emotions) can be characterized learning dynamics and its performance. The Lyapunov coefficients, calculated to the time series, make it possible to obtain the Kolmogorov entropy. From this value, associated to the experimental time series, the complexity of the system and the loss of information are deduced. This allows us to construct, for the learning process, a quantitative estimate of sustainability.
Keywords: time series, connectivity, Lyapunov coefficients, entropy, sustainability

1. Introduction

Several authors ([1, 2, 3, 4, 5, 6] point out that emotion fulfills an important cognitive role: knowledge of life and the universe is not only intellectual, since the subtle nuances of it brings us the emotiveness (emotivity). Emotions are the other way of knowing about the world and about themselves [4]. In learning, value should be given to facts, people and situations (which form the initial contextualized condition), according to their influence on the emotionality of individuals and teams. This value assignment manifests, neurobiologically, in attendance and perceptive selection. In the selection of what is remembered by the long-term memory, and in the perception that dispositions and attitudes are "felt" as more appropriate [4, p.180, 7]. This makes learning sustainable over time. Thus the question which gives the title to this work arises:

Can appropriate mathematical indicators be built to be measured that provide information about sustainability in times of a learning process? [8, 9, 10].

In order to elaborate a quantitative answer to this question, observational follow-up is carried out, during a semester with a group of students whose major is engineering in a technological university.

They constitute a segment of students who enter with their completed secondary studies, without a university selection test, and a significant proportion, form a first-generation human group to access to higher education. In total, five Physical Sciences courses are monitored. The reasons to choose this subject point, essentially, to the emotional commitment to it, from which everything becomes: interest or lack of interest, absenteeism, desertion, etc. Which is supported each semester in the percentages of approved and failed. Each course consists of an average of 32 students (approximate) who are divided, for the specific sessions of follow-up, into two of 16 students forming 4 work teams of 4 students each. The "typical" performance matrix per student team is divided into small time intervals to complete the 90-minute class [5, 11, 12].
It is important to observe the behaviors deployed by the students in the process, in which 6 "macro" variables are identified: IND (IND), Persuasion (PER), Positivity (POS), Negativity (NEG), Internal Information External (IE), constructing the normalized proportions: X = IND / PER, Y = POS / NEG, Z = II / IE. Both numerator and denominator of each fraction contain a subset of 13 behaviors, which are assigned a numerical range from -6 to +6, as shown in the example # 1 schema for Positivity and Inquiry [12, 13].

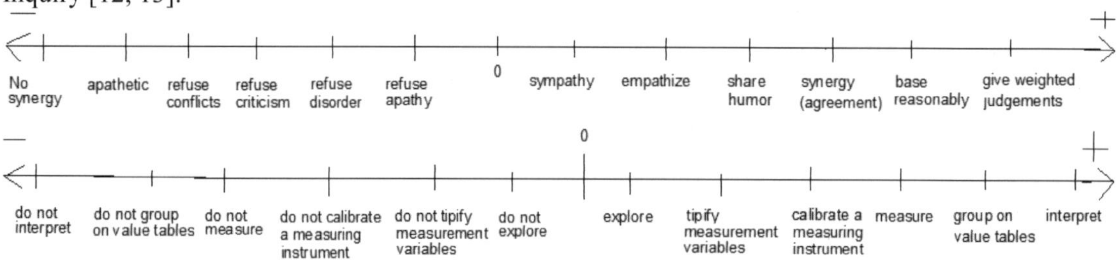

Scheme No. 1. It indicates the dual polarity domain and the coding scale, in its double polarity, of Positivity and Inquiry, with its thirteen associated behaviors.

This hierarchy determines a total of 78 behaviors to observe during the 90-minute session.
Observational monitoring is a complex process that can be addressed in two ways:
On the one hand, through videos of the class sessions from which the information is extracted and on the other hand, is to train observers to record the behaviors deployed by the teams for each variable "macro" (13), PER (13), POS (13), NEG (13), II (13), and IE (13).
With the purpose of exploring collaborative work and self-organization, it is suggested to form four-student teams (2 men and 2 women, etc.). These teams are called Experimental Group. An initial condition is applied in the form of an experimental or theoretical activity, with an emphasis on positive emotionality (solidarity, founded criticism, respect, shared humor, etc.) allowing to each team to deploy their own emotionality interactively. This same initial condition gives the purpose of the activity through questions or challenges. The team must be self-sufficient regarding the proposed question or challenge.
In so far as, the process progresses in time, the respective data and tables of values are generated, a "type" table is shown below:

	codification
6	interpret
5	group data on value tables
4	measure
3	calibrate a measuring instrument
2	tipify measurement variables
1	explore
0	neutro
-1	do not explore
-2	do not tipify measurement variables
-3	do not calibrate a measuring instrument
-4	do not measure
-5	do not group on value tables
-6	do not interpret
total:13	

t (min)	Conduct of inquiry
0	
9	
18	
27	
36	
45	
54	
63	
72	
81	
90	

Table 1. Presents the behavior to be followed in time and its coding taxonomy.

With the purpose of studying validity and reliability criteria of the instruments of measurement of the research, control courses, called Control Groups, without initial condition and without emphasis on the emotional interaction and with the classic format of development of the academic activity, followed by the same variables [6, 13 a]. It is convenient that the constructed intervals of time are very fine, of the order of 1 second for each record of observation. Some of the graphs in the time, for the case with initial condition with emphasis on the emotions are presented below:

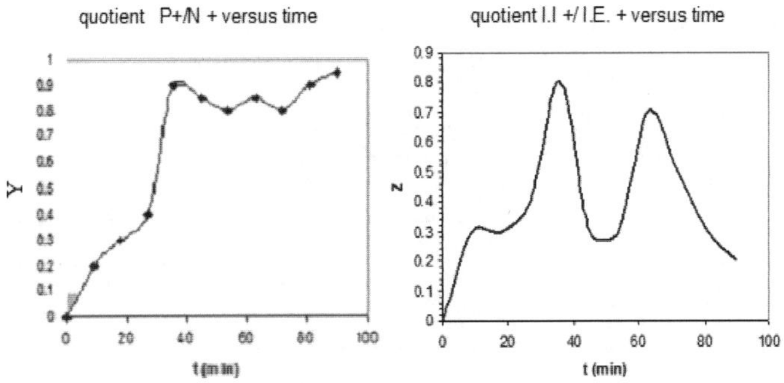

Figure1.
It represents the variation in the time of the Time Series of the Positivity / Negativity (= Y) and the Internal Information / External Information (= Z).

These graphs are the Time Series. These can be represented by Fourier Series and so they are usually called Fourier Time Series [14, 15].
Finally, from the Time Series of X (t), Y (t) and Z (t) we construct the discretized column vectors (although vectors with a minimum of 1000 elements allow to make good estimates, ideally they contain over the 5000 components of the analysis of Lyapunov coefficients). Their graphics in the Space of Phases, characteristically acquire forms such as:

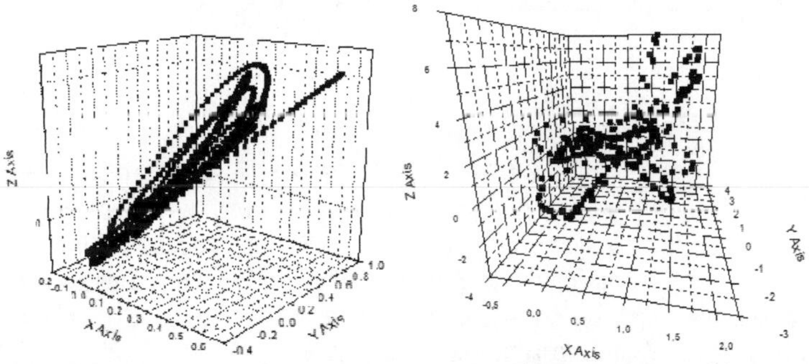

Figure 2. The graphs show the time evolution of the Time Series X, Y, Z - each one of 1000 data - generated from the time tracking of the different teams of students that satisfy fixed (left) and attractor Chaotic (right).

The graphs allow to classify the dynamics, according to the performance of the teams that make up the Experimental Group, in: Low, Medium and High. It contrasts with the performance of Control Groups, traditional courses with no initial condition and emotional deployment in line with a traditional approach. The cross-correlations by group according to the influence exerted by the variety of emotions Y (= Positivity / Negativity) on the variable X (= Inquiry / Persuasion) [5, 12]:

Group	Control	Experimental	Comparison: Experimental / Control
Cross correlation	0.3	0.5	1.7

That is, the experimental group, that is treated with contextualized initial conditions and climate that leads to high emotional connectivity within each team, shows that the balanced presence of positivity / negativity in their relational ties exerts a $1.7 \approx 2$ influence, two times approximately, superior on the variable X, that is the inquiry / persuasion (the most rational part of the work of the team), leading the team in a more effective and secure towards the achievement of high performance. This influence is translated into connectivity and emotional field evolution, which is reflected in the value that the students give to learning and in the achievements of the same that covers from the experience of

collaborative work, each component is determined in the learning process to formal assessment procedures.
These corollaries allow to concentrate in the courses with an initial condition and influence of the emotional fields, Experimental Group. It is possible to apply modeling to the study of Time Series through the MatLab Software, either by the Lorenz equations [15, 16], Neuronal Networks [17] and Cellular Automata [18, 19]. For this study the Lorenz equations [15] are applied through the fourth order Runge - Kutta numerical method [14] to the discretized column vectors of X, Y and Z [5]. These equations have a parameter, called control parameter or connectivity [43, 44, 45], showing that it is directly related to the performance and the Pos / Neg quotient of the emotions of each team [6] Table 2:

	CONNECTIVITY CALCULATION ACCORDING:				
PERFORMANCE	THEORETICAL	CENTROID	CORRELATIONAL	AVERAGE	POS / NEG
Low(weak attractor)	16.5	20.15	16	17.5	0.375
Medium(medium attractor)	19.5	22.1	20	20.5	1.95
High(strange or chaotic attractor)	31	26	29	28.7	5.25*

Table 2. It shows the performance according to connectivity and the Positivity / Negativity (POS / NEG). (*: POS / NEG = 5 / 1, [11, 46])

Teams with connectivity and high Pos / Neg (emotional flowering [11]) are sustained over time and achieve the objectives of the activity. Looking at Table 2, we can see an increase in connectivity as we approach to chaotic or complex dynamics, as shown in Graphic 3:

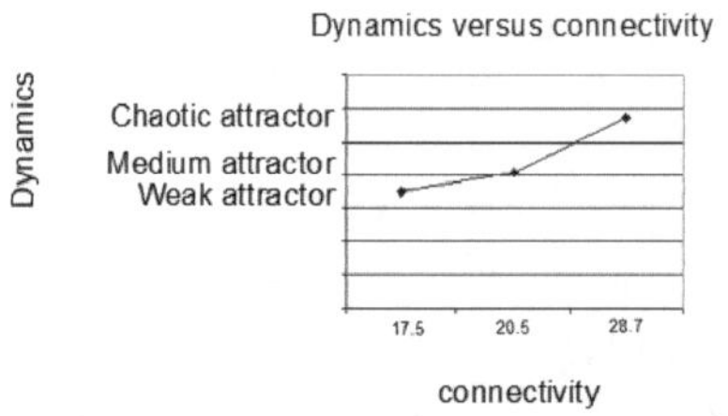

GraphicN°3. Learning dynamics versus connectivity

What does this entropic connectivity behavior (called entropic connectivity) mean for learning? Is it possible to calculate it? And how does it relate to the complexity of the learning process under study?
To answer these questions, there are several numerical procedures applied to the Time Series (Sprott, 2006), which allow the determination of Lyapunov coefficients [20], Kolmogorov (SK) entropy [21, 22, 23], complexity [22] and finally uncertainty in information [23].

2. Theoretical approach [22]

From a point of view of a physical-mathematical theory of collaborative learning, it is possible from the free energy of Helmholtz [24] to obtain a definition characterization of the sustainability of a learning process.
Thus, for every open thermodynamic system, the total entropy, S, can be written as the sum of the internal entropy of the system, S_{INTERN}, and the external entropy, S_{EXTERN}, to it:

$$S = S_{INTERN} + S_{EXTERN} = S_I + S_E$$

The production of entropy expressed as a variation over time.

$$\frac{dS}{dt} = \frac{dS_I}{dt} + \frac{S_E}{dt}$$

The second law of thermodynamics states that the internal production of entropy due to irreversible processes occurring within the system: $\frac{dS_I}{dt} > 0$

For the special case where constrictions are constant, irreversible classical thermodynamic theory states that eventually the system will reach a steady state thermodynamic state in which all microscopic variables, including total entropy, are stationary in time, so:

$$\frac{dS}{dt} = 0 \text{ , then: } \frac{dS_I}{dt} = -\frac{dS_E}{dt} \Rightarrow \frac{dS_E}{dt} < 0$$

A system of such characteristics, to be maintained in a stable stationary thermodynamic state, requires a continuous negative flow of entropy into the system. This can be visualized from the perspective of game theory and thinking of two teams A and B and assuming that some players of A assume that one strategy, say A, is better than the other, no matter how the other team, B, plays. Clearly the population of A who assumed this position will decrease over time since their adaptability is lower than the average and will eventually disappear. Then the whole population will be with strategy B, so the adaptability of B and the average of adaptability are equal and the production of entropy will be zero reaching the steady state. From the social point of view, this process requires a continuous negative flow of entropy that is not sustainable (Think of the countries that have opted for a bad development strategy and another country for a good one, sooner or later everyone will want to be with the "best" strategy in such a way that the negative flow of entropy derived from the decline of those who opted for the "bad" strategy would prove to be ultimately unsustainably planetary, as we see it today).
Sustainability, understood as the ability of a system to reach states of greater longevity, can be obtained by minimizing Helmholtz free energy [24]:

$$F = U - TS$$

where T is the system temperature associated with internal randomness and U is the internal energy associated with the energy due to the interactions.
The free energy of Helmholtz can be minimized in two ways. One is to minimize U and the other is to maximize S. Most of the time, the internal energy U is a conserved quantity or cannot be controlled externally, it remains as the only alternative to maximize the entropy.
If the variation in the time of the Entropy is greater than zero:

$$\frac{dS}{dt} > 0$$

indicates that the entropy grows and the system is in a more sustainable configuration if the variation in time of the Entropy is less than zero:

$$\frac{dS}{dt} < 0$$

means that the entropy decreases and the system is in a configuration far from sustainability.
A positive entropy production is reached when the average adaptability is greater than the local one and this corresponds to cooperative games (or processes), thus a quantitative indicator for the measurement of the sustainability of a system is obtained.
In the real world interactions are an integral part. As the entropy of the system increases, it becomes more stable over time, more sustainable. This allows to make explicit the relationship between information and entropy since the complexity is measured from the information once all the computational calculations of the system have been realized.
Adaptability is a measure of the Helmholtz free energy minimization capacity

3. Entropy and complexity

Let the equivalence be between Kolmogorov's entropy and entropy: $S_K \Leftrightarrow \frac{dS}{dt}$

as indicated before, the production of positive entropy is achieved if the average adaptability is greater than the local one. This observation is also consistent with connectivity for the chaotic case, which is on the order of 30 [12], which would indicate that a high connectivity between components of a computer is a condition of high entropy production and vice versa: means that a j-nth component of the equipment is increasingly establishing connections with the other components and, in turn, these with each other and with it.

Real-world observations show that interactions are a characteristic feature of their evolutionary relational dynamics. This allows us to visualize, as it happens in nature, that the open systems "propitiate" an increasing order in connectivity, being dynamically sustainable, increasing its complexity in time.

If a sequence of characters, X, composed of values (time series) or symbols, which follow a probability distribution P (x), is constructed for information, I [25]:

$$I = -\sum P(x) \log P(x)_i = S$$

This expression has the same functional form of the entropy of the Physical Sciences.

According to Gershenson [26, 27] the complexity can be defined:

$$C = aI_{out}(1 - I_{out})$$

a is a normalization constant that makes it possible for quantity C to be a percentage value without units.

Using the entropy, S [22]:

$$C = aS_{obs}(1 - S_{obs}) = aS(1 - S)$$

If S_{MAX} is the maximum entropy and accounts for the total information capacity of the system, then S_{OBS} is the observed entropy and corresponding to the present information content. The constructed formal structure leads "naturally" to a definition of the complexity of a system, which can be estimated through its entropy. Since Kolmogorov's entropy is time-dependent, by its relation to the Lyapunov coefficients of the time series, the complexity remains:

$$C(t) = aS_{K_{obs}}(1 - S_{K_{obs}})$$

4. Experimental results

From the point of view of the Lyapunov coefficients, the Kolmogorov entropy, the complexity and the sustainability for two learning dynamics, the three (weak attractor, medium attractor, strange or chaotic attractor) were studied. Teams of students in learning processes: medium attractor and strange or chaotic attractor.

Medium or fixed attractor Case

The Lyapunov coefficients λx, λy and λz are associated to the time series of the Inquiry / Persuasion (X (t)), Positivity / Negativity (Y (t)) and Internal Information / External Information (Z (t))

The numerical calculation of the Lyapunov Coefficients [43, 44] for the series yields the values:

$$(\lambda_x, \lambda_y, \lambda_z) < 0$$

The average loss of information, over time, is equal to the mean sum of positive Lyapunov exponents [28], the Kolmogorov entropy:

$$S_K = \sum_{\lambda_i > 0} \lambda_i = \lambda_x + \lambda_y + \lambda_z = 0$$

As it was mentioned before, $S_K \Leftrightarrow \frac{dS}{dt}$, then $S_K \Leftrightarrow \frac{dS}{dt} = 0$

There is no disorder increase, which is consistent with the value of the control or connectivity parametric, low or medium if compared to that of the chaotic dynamics, according to calculations by centroid, cross correlation [12], by the theoretical recursive method according to MatLab [29] of the applied nonlinear model [12, 15, 30, 31, 32], shown in Table 2.

Strange or chaotic attraction case:
The numerical calculation of the Lyapunov coefficients [43, 44] for the time series X (t), Y (t), Z (t) [12], yields the values:

$$\lambda_x \approx 0.004,\ \lambda_y \approx 0.001\ ,\lambda_z \approx 0.004$$

And the average loss of information over time [28] according to the Kolmogorov entropy:

$$S_K = \sum_{\lambda_i > 0} \lambda_i = \lambda_x + \lambda_y + \lambda_z \approx 0.009 \approx 0.01 > 0$$

Following the reasoning of the previous case $S_K \Leftrightarrow \frac{dS}{dt}$, then: $S_K \Leftrightarrow \frac{dS}{dt} > 0$

5. Calculation of complexity according to fixed and strange attractors

From the previous results it is clear that the production of entropy grows, allowing the system to enter a more sustainable configuration, confirming that the average adaptability is greater than the local one since the character of the activity developed by the teams of students corresponds to processes of collaborative or cooperative learning, generating a quantitative indicator for the measurement of the sustainability of the system.
The value of the entropy production is directly related to the control parameter, r, called connectivity, of the chaotic case with high performance computers, with a numerical value of 30 [5]. This shows that high connectivity between components of a computer produces high entropy and vice versa (an i - th component of the equipment during the learning process leading to high performance increasingly establishes, in the context of the Positivity / Negativity ratio, increasing connections with other components and thus for the remaining n-1 components). In everyday life in the real world connectivity [45] is an integral part. As noted, as the number and order of interactions of the system increases, it is found to be more stable over time, in essence more sustainable.
Using the equivalence between Kolmogorov's entropy and entropy, one can write:
Fixed attractor case:

$$C(t) = aS_{K_{obs}}(1 - S_{K_{obs}}) = a\,0.0\,(1 - 0.0) \approx 0.0.$$

The complexity of the system is zero.
Strange or chaotic attraction case:

$$C(t) = aS_{K_{obs}}(1 - S_{K_{obs}}) = a\,0.01(1 - 0.01) \approx 0.0099a \approx 0.01a > 0$$

That is, the complexity of the system grows.
The numbers show that the entropic element of the system for the learning process and that leads to high performance is the connectivity (r, control parameter) of averaged values:

$$\left.r_{WEAK\ ATTRACTOR} = 18;\ r_{MEDIUM\ ATTRACTOR} = 21\right\} \Rightarrow C(t) = 0\ ;\ \left. r_{CHAOTIC\ ATTRACTOR} = 29\right\} \Rightarrow C(t) > 0$$

By increasing connectivity, the system will become characteristically complex and sustainable. Thus, the test based on the observational monitoring of the behaviors deployed by the teams during the process confirms, in a first approximation, that it becomes sustainable as connectivity grows (for this reason it is the entropic element). At lower values of connectivity, the system is "dumped", turning it rigid, weak or adaptive, which
This result confirms the absolutely relational nature of all the elements that make up the learning process. So what is done to the supposedly most insignificant component of the network that sustains life ends up affecting each other.

6. Study of the horizon of projection

Chaotic Dynamic Case: High Performance

What is the maximum time that the process of Learning based on increasing entropic connectivity can be projected, regarding a given activity?
The learning process in chaotic dynamics, within the framework of this research, is considered to be collaborative, which favors the connectivity network of a team of 4 people, regarding an activity set out as a challenge or entrepreneurship for an academic period which is equivalent to 90 minutes.
The Kolmogorov entropy [28]: $S_K = - p_i \log 2\ p_i = 0.01$ bits / sec for the academic period, with the λ_i positive maximum values of Lyapunov coefficients of the Time Series.
This numeric value indicates that 0.01 information bits per second are lost in the 90 minutes. Although there is a loss this is relatively low, which indirectly indicates that there is a high percentage of retention during the session.
The projection time for the loss of the system information is:

$$\text{Lyapunov time} = 1 / S_K = 1 / 0.01 = 100 \text{ sec} = 1.666 \text{ min} \approx 2.0 \text{ min}.$$

According to this result, in order to maintain an optimal ratio of meaningful learning (with high connectivity generating high performance), for each activity deployed by the student teams, the "ideal" current information should be in a time of the order of 2.0 minutes. Above this limit it is not possible to make predictions, which can be trusted, about the process characterized by contextual initial conditions and proposed activity as a challenge or undertaking, so any strategy that is applied will be useless.
This result also advises that before making an accurate projection in the long term, it is indispensable to calculate the possibility of doing it in any other type of scenario. It is notable that this time horizon does not depend on the historical past of the process on the basis of which the projection has been made. In chaotic systems, it is more favorable to look at the future, to increase the level of reliability of a projection, than to look back.
Assuming that for every second of the 90 minutes an "average information loss rate" of 0.01 bits / sec is generated, then by dividing the 90 minutes of the academic class by the average information loss rate, we get an approximate time of 6.25 days that could be interpreted as a maximum "optimistic" projection time of prediction.

7. Determination of loss information

Considering the relationship between Kolmogorov's entropy and the characteristic parameter of chaos, the Lyapunov exponent, λ, it follows that this is proportional to the loss of information, if $\lambda > 0$ the movement is chaotic, information is lost, since the temporal prediction is less accurate and, therefore, since the loss of information is greater, no prediction can be made outside of that time horizon [23].
Then, the relationship between Kolmogorov's entropy and the characteristic parameter of chaos is shown, the Lyapunov exponent, λ, is proportional to the loss of information [23]:

$$\langle \Delta I \rangle_i \log 2 = -\lambda(i_0(t)) \Rightarrow \langle \Delta I \rangle_i = \frac{-\lambda(i_0(t))}{\log 2}, \quad i = X,.Y,Z$$

Fixed attractor case, $\lambda < 0$:

$$\{\langle \Delta I \rangle_X ; \langle \Delta I \rangle_Y ; \langle \Delta I \rangle_Z\} = \{\, 0.01827 \;\; > 0; 0.003654 \;\; > 0; 0.0003321 > 0 \;\; \}$$

A system with negative Lyapunov exponent does not progressively lose the information that had initially and is, from a certain time, predictable (it is the matrix idea of the current educational paradigm).

Strange or chaotic attractor case, $\lambda > 0$:

$$\{\langle \Delta I \rangle_X ; \langle \Delta I \rangle_Y ; \langle \Delta I \rangle_Z\} = \{\, -0.014616 \;\; < 0; \; -0.0043185 \;\; < 0; -0.0132877 < 0 \;\; \}$$

There is loss of information in the time series for X (t), Y (t) and Z (t). For X (t), which corresponds to the time series for the Inquiry / Persuasion ratio, we obtain since it indicates there is information loss.

how do we understand this result? And what do you suggest? It can be interpreted that the algorithmic routines, deployed by the teams of high performance students, lead to certain results (positive feedback process) where there is discovery that makes meaningful learning, but with great efficiency to dispense with superfluous information, which, definitely, decreases the information towards what it is actually necessary or relevant.
Regarding Y (t), associated with the Positivity / Negativity Emotion Ratio, the question is: Is it desirable that a system in the process of learning to behave in this way? It seems prudent, assuming there is more openness in teams working collaboratively, in looking forward than still loaded with stereotypes and emotional fixations (fixed point attractors). Emotions exhibit complications in their classification because they are very remote and complex neurobiological mechanisms (their beginnings can be investigated in unicellular ones attracted by variations of light or oxygen, that are able to "feel", pheromones that incite behaviors in social insects, in the Interactions between predator-prey ...) [4,7]. Thus the chaotic dynamics copes with an acceptable break with respect to behavior patterns that favor certain repetitive behaviors [33] that restrain the creativity demanded in contexts of high connectivity - between people, people with the ecosystem and the ecosystem with himself - characteristic complexity of the world we live in, which reminds us that "a butterfly beats its wings in Bibury and a Tsunami affects the coast of Constitución" [34].
In the quotient between internal and external information the ratio Z (t), the calculations show loss of information, how can this corollary be explained? The ratio manifests the relationship between the behaviors exhibited by the team participants in their process of searching for internal and external information to the team, and which of them prevails. In innovative processes (rupture of archetypes), it seems to be more what is discarded than what it is preserved (try new strategies which involves, relatedly, a disposition or a different emotional tone). Thus, it is aligned with the observed.
A process characterized this way shows how they are influenced, affecting the final result, the three-time series. Connectivity, 30, associated with chaotic dynamics and leading to high performance is not understood without the participation and nonlinear interference between emotions, inquiry, persuasion and information, which also marks distance with the reductionist models.

8. Conclusions

The numerical ranks, in line with the theoretical models analyzed [22], indicate the characteristics that make the learning process sustainable [35]. It is shown, in the first approximation, that these depend on: *increasing connectivity, quotient of emotions Positivity / Negativity (P / N), Information and uncertainty* [21, 25, 36] *and Entropy.* Parameters that arise from the experimental observation and that allow to characterize the complexity of the process. Then and only then, a "healthy system of learning for human species" will lead to: performance, self - organization, diversity and sustainability. A system with such characteristics will go into deep tension with everything that surrounds it [47]. In general, chaotic systems are characterized by consuming considerable energy and information to maintain and increase their level of complexity, while being very sensitive to initial conditions (which may be environmental fluctuations).
From this different approach, it is possible to reflect with a deeper and more specific learning, on our position and activity on this planet which is, without minimizing our very long-term efforts for interplanetary travel, the only safe place we have to live on. [37, 38, 39, 40, 41, 42].

References

[1] Maturana, H. (2001). *Emociones y Lenguaje en Educación y Política.* Santiago de Chile: Dolmen
[2] Ibáñez, N. (2002). Las emociones en el Aula. *Estudios. Pedagógicos*, nº 28, 31 – 45. dói: http://dx.doi.org/10.4067/S0718-07052002000100002.
[3] Goleman, D. (2005). *La Inteligencia Emocional.* Barcelona: Kairós.
[4] Lavados, J. (2013). *El cerebro y la educación.* Santiago de Chile : Taurus
[5] Pacheco, P. (2015). *Aprendizajes, una aproximación desde el caos.* Madrid: EAE.
[6] Pacheco, P & Correa, R. (2016). "Study of connectivity in student teams by observation of their

learning processes". *Journal of Physics: Conference Series.* doi:10.1088/1742-6596/720/1/012 056

[7] Gracia-Bafalluy, M & Escolano, E. (2014). "Aportaciones de la neurociencia al aprendizaje de las habilidades numéricas". *Revista de Neurología,* 58(2), 69 – 76. PMID: 24399623

[8] Wackernagel, M, Schulz, N B, Deumling, D, Linares, A C, Jenkins,M, Kapos, V, Monfreda, C, Loh, J, Myers, N, Norgaard, R & Randers, J. (2002). "Tracking the ecological overshoot of the human economy". *Proc.Natl. Acad.Sci.* U.S.A., 99 (14), 9266 – 9271. doi: 10.1073/pnas. 142033699

[9] Hui, C. (2006). "Carrying capacity, population equilibrium, and envrionment's maximal load". *Ecological Modelling*, 192, 317-320. http://dx.doi.org/10.1016/j.ecolmodel.2005. 07.001

[10] Sayre, N F. (2008). "The Genesis, History, and Limits of Carrying Capacity", *Annals of the Assoc. of American Geo.*, 98(1), 122. http://dx.doi.org/10.1080/00045600701734356.

[11] Bakeman, R, Gottman, J.M, (1989). Observación de la interacción: introducción al análisis secuencial. Madrid:Morata.

[12] Pacheco, P, Villagran, S & Quiroz, E. (2013). "Dinámica no lineal y rendimiento académico: verificación experimental e interpretación". *Revista Internacional de Educación y Aprendizaje*, Vol 1, Issue 1, nº1, 49 – 73.

[13] Pacheco, P. (2015 a) "Construcción y validación de los instrumentos para la medición de las influencia de los campos emocionales en los aprendizajes significativos". *Revista Internacional de Educación y Aprendizaje*, Vol. 3, Issue 1, nº1, 1-24.

[14] Zill, D.(1988). Ecuaciones Diferenciales con Aplicaciones. Mexico, D.F.: Grupo Editorial IberoAmérica S.A. de C.V.

[15] Lorenz, E.N. (1960). "Maximun simplification of the dynamics equations". *Tellus*, **12**, 243 – 254

[16] Lorenz, E. (1963). Deterministic nonperiodic flow. *Journal of atmospheric Sciences* 20, 130–41.1. doi: 10.1177/0309133308091948

[17] Frank, R J, Davey, N, & Hunt, S P.(2001). "Time Series Prediction and Neural Networks" *Journal of Intelligent and Robotic Systems*, Vol. 31, 91–103. doi: 10.1023/A:10120 74215150

[18] Wolfram, S. (1984). "Cellular automata as models of complexity". *Nature*, Vol. 311, No. 5985, 419 – 424. doi:10.1038/311419a0

[19] Steeb, W H. (2005). *The Nonlinear Workbook.* London: 3rd edn, World Scientific.

[20] Wolf, A, Swift, J B, Swinney, H L & Vastano, J A. (1985).Determining Lyapunov exponents from a time series. *Physica 16D*, 285 – 317. North-Holland, Amsterdam. doi:10.1016/0167-2789(85)90011-9

[21] Kolmogorov, A N. (1983). "Combinatorial foundations of information theory and the calculus of probabilities". *Russian Mathematical Surveys*, 38:4, 29 – 40.

[22] Lopez-Corona, O, Padilla, P, Huerta, A, Mustri – Trejo, D, Perez, K, Ruiz, A, Valdés, O, & Zamudio, F. (19 de febrero del 2015). "Measuring social complexity and the emergence of cooperation from entropic principles".*Physics-soc-ph.*arXiv:1502.05741v2[physics Soc-ph]

[23] Martínez, J A y Vinagre, F A. (2016). La entropía de Kolmogorov; su sentido físico y su aplicación al estudio de lechos fluidizados 2D. Departamento de Química Analítica e Ingeniería Química, Universidad de Alcalá, Alcalá de Henares. Madrid: Academia. Recuperado el 2 de Noviembre de 2016 de:www.academia.edu/2479372/La_Entropia_de_Kolmogorov_su_Sentido_Fisico_y_su_ Aplicación_al_Estudio_de_Lechos_Fluidizados 2D

[24] Wheeler, L P. (1998). *Josiah Willard Gibbs, The History of a Great Mind.* Woodbridge, CT: Ox Bow Press.

[25] Shannon, C. (1948). "A mathematical theory of communication". *Bell System Technical Journal* 27: 379-423 y 623 – 656. doi:10.1002/j.1538-7305. 1948.tb01338. x.

[26] Gershenson, C, Fernandez, N. (10 de agosto del 2012). "Complexity and Information: Measuring Emergente, Self – organization, and Homeostasis at Multiple Scales". *arXiv*:1205

-2026, v 2[cs.TI]. doi: 10.1002/cplx.21424

[27] Gershenson, C. (2013). "The Implications of Interactions for Science and Philosophy". *Foundations of Science* 18(4):781-790. doi: 10.1007/s10699-012- 9305-8.

[28] Pesin, Y B. (1977). "Characteristic Lyapunov Exponents and Smooth Ergodic Theory". *Russian Mathematical Surveys*, Volume 32, Number 4, 55 – 144. http://dx.doi.org/10.1070/

[29] Nakamura, S. (1997). *Análisis Numérico y Visualización Gráfica con Matlab.* México D.F.: Pearson

[30] Grassberger, P & Procaccia, I. (1983 a). Characterization of Strange Attractors. *Physical Review Letters* 50, 346–349. doi: https://doi.org/10.1103/PhysRevLett.50.346

[31] Grassberger, P. & Procaccia, I. (1983 b). Measuring the Strangeness of Strange Attractors. *Physica D* **9**, 189 - 208. doi: 10.1016/0167-2789(83)90298-1

[32] Kaspar, F & Schuster, H G. (1986). Easily calculable measure for the complexity of spatiotemporal patterns. *Physical Review,* A 36, 842 - 848. doi: https://doi.org/10.1103/ PhysRevA.36.842

[33] Bijleveld, E, Scheepers, D & Ellemers, N. (2012). "The cortisol response to anticipated intergroup interactions predict self-reported prejudice". *PLoS One*, *7*(3), e33681. PMID 22442709. doi: 10.1371/journal.pone.0033681

[34] Brian, A, Durlauf, S N, Lane, D A. (1997) *Introduction: Process and Emergence in the economy. The Economy as an Evolving Complex System II.* Massachusets: Addison -Wesley

[35] Pacheco, P. (2016). Caos: conexión de aprendizajes, entropía, sustentabilidad y complejidad. Madrid: EAE.

[36] Chaitin, G J. (1982). "Gödel's theorem and information". *International Journal of Theoretical Physics*, Volume 21, Issue 12, 941-954. doi: 10.1007/BF02084159.

[37] Zimmerer, K S. (1994). "Human Geography and the "New Ecology": The Prospect of Promise and Integration", *Annals of the Assoc. of American Geo.*, 84(1), 108-125. Recuperado el 18 de Noviembre de: dusk.geo.orst.edu/prosem/PDFs/human_geog.pdf

[38] Lutz, W, Sanderson, W & Scherbov, S. (2001). "The End of World Population Growth", Nature, n. º 412, Issue: 6846, 543-545. doi:10.1038/35087589

[39] Lutz, W, Sanderson, W & Scherbov, S. (2004). The End of the World Population Growth in the 21st Century: New Challenges for Human Capital Formation and Sustainable Development. London: Earthscan Publications Ltd.

[40] Delp, M D, Charvat, J M, Limoli, C L, Globus, R K & Ghosh, P 2016 "Apollo Lunar Astronauts Show Higher Cardiovascular Disease Mortality: Possible Deep Space Radiation Effects on the Vascular Endothelium". *Scientific Reports*. 6, 29901. doi: 10.1038/ srep29901.

[41] Schuschny, A. (2001). Auto organización en sistemas económicos, Buenos Aires: Facultad de Ciencias Económicas. Recuperado el 22 de Diciembre de: bibliotecadigital.econ.uba.ar/ download/tesis/1501-1161 SchuschnyAR.pdf

[42] Pacheco, P, Villagrán, S & Guzmán, C. (2015b)."Estudio del campo emocional en el aula y simulación de su evolución durante un proceso de enseñanza-aprendizaje para cursos de ciencias". *Estudios Pedagógicos,* XLI, N° 1: 199-217. http://dx.doi.org/10.4067/S07 18- 070 52015000100012.

[43] Sprott, J C. (1995). Software CDA, Chaos Data Analyzer Programs.

[44] Sprott, J C. (2006). *Chaos and Time – Series Analysis.* New York: Oxford University Press.

[45] Green, D G. (1994). "Connectivity and the evolution of Biological System". *Journal of Biological Systems* 2, 91 – 103.

[46] Gottman, J M, Rose, F, Mettetal, G. (1982). "Time Series analysis of social interaction data". En T. Field y A. Fogel (Eds), *Emotion and interactions* (pp. 261 – 289). Hillsdale, N.J: Lawrence Erlbaum

[47].Bar-On, Y M., Phillips, R. & Milo, R. (May 21, 2018). The biomass distribution on Earth. PNAS https://doi.org/10.1073/pnas.1711842115.

IOP Conf. Series: Journal of Physics: Conf. Series **1043** (2018) 012053 doi:10.1088/1742-6596/1043/1/012053

Radiation leading to long-lived current states in a superconducting circuit

Gilberto A Urzúa and Alvaro Restuccia

Abstract. In this work we show that radiation can have interesting effects on a superconducting inductive circuit. Consider that for some time a battery supplies a constant voltage to a ring made of a superconductor wire above its critical temperature. At some instant the battery is disconnected and the ring is immersed in a low temperature bath, its resistance becoming zero. We study the system right after it undergoes the phase transition and enters the superconducting state. The system is modeled as an ideal selfinductance, and the radiation emitted is modeled as magnetic dipole radiation. We analyze qualitatively the differential equation that models the electrical current evolution to obtain some of its main characteristics. We find that, depending upon its initial conditions when it enters the superconducting state, the electrical current can evolve in one out of three very well differentiated sets of states. The first set consists of just one state where the current decays exponentially, in a similar fashion as it would do if the phase transition would have not taken place. The second set is comprised of short-lived states, where the current decays faster than the exponential state, and dies suddenly. The third set is comprised of long-lived states, where the current decays slower than the exponential state and reaches smoothly a constant value different from zero at a finite time. It stays in that state and actually lives in it forever.

Departamento de Física, Universidad de Antofagasta, Casilla 170, Antofagasta, Chile

E-mail: gilberto.urzua@uantof.cl, alvaro.restuccia@uantof.cl

1. Introduction

Consider a physical system consisting of a ring made of a superconducting wire, at the beginning above its critical temperature, that is connected to a battery supplying a constant voltage V_0. The battery provides energy to the system and a positive defined electrical current $i > 0$ starts circulating. As the current i grows, some fraction of the provided energy is stored as magnetic energy around the ring, some fraction is dissipated as Joule heat in the wire, and the minor complementary fraction is radiated to infinity. After a while the current asymptotically reaches a steady-state value that maintains a constant magnetic field. In this regimen there is no radiation to infinity [1], and the energy being delivered by the battery is being dissipated as Joule heat.

The battery can be disconnected at any time during this process, ceasing to deliver power. When the battery is disconnected, the current does not drop to zero at once, but decreases smoothly, since the magnetic field around the ring now gives power back to the wire. While the current decays, some fraction of the delivered back energy is dissipated as Joule heat and the complementary fraction is radiated to infinity. Being the resistance finite, the final current is zero and in the end no energy remains stored as magnetic energy around the ring. Usually in this kind of physical system the energy radiated to infinity is ignored, and the system is modeled as an RL circuit.

In this paper we are interested in studying what happens to the electrical current after the ring has been cooled through its transition temperature for superconductivity. We assume that once the battery is disconnected, the superconductor ring is immersed in a low temperature bath below its transition temperature, its resistance R dropping essentially to zero in a transition time ε. The fact that now the resistance is zero, implies that energy is not dissipated as Joule heat anymore, and we are compelled to include radiation in our analysis.

Steady electrical currents without attenuation have been observed in superconducting solenoids [2]. The question arises, does radiation plays some role in the birth of such constant current states? We study the system after the phase transition has taken place. The physical system is modeled by a circuit that have an ideal selfinductance, and the radiation emitted is modeled as magnetic dipole radiation. After imposing energy balance, we obtain the differential equation that models the evolution of the current. Analyzing it we find that if, after the system undergoes the phase transition, the current enters the superconducting state satisfying certain conditions, after a finite time it will reach a state of constant current different from zero that stays forever.

2. Background: RL model

In a normal non-superconducting system, usually the energy radiated to infinity is ignored and the system is studied within the RL model: The physical system is modeled by a circuit that have an ideal resistance R and an ideal selfinductance L. The model does not incorporate radiation to infinity. Within this model, when the battery is connected the electrical current increases as

$$i = \frac{V_0}{R}\left[1 - \exp\left(-(R/L)t\right)\right]. \tag{1}$$

When enough time goes by, the current reaches asymptotically the value $i = V_0/R$. If the battery is disconnected, it decays exponentially as

$$i = \frac{V_0}{R}\exp\left(-(R/L)t\right). \tag{2}$$

Being the resistance finite, the final current is zero and in the end no energy remains stored around the ring. Any energy radiated to infinity that might have occurred, in this model is counted as energy dissipated as Joule heat, which becomes overestimated.

If for describing the system after the disconnection of the battery one improved the model incorporating the radiation to infinity, one would expect that just some fraction of the stored energy in the magnetic field would be dissipated as Joule heat, while the remaining fraction would be radiated to infinity. Again, being the resistance finite, the final current would be zero.

3. Model with selfinductance L plus radiation for the superconducting system

We will set $t = 0$ right after the system leaves the phase transition and enters into the superconducting state, denoting the current by $I(t)$. The current at $t = 0$ is positive, $I|_{t=0} \equiv I_0 > 0$, and we assume that its derivative at $t = 0$ is $dI/dt|_{t=0} \equiv dI/dt|_0 < 0$. The superconducting system is modeled as an ideal selfinductance L and an ideal resistance $R = 0$. Now we include the radiation emitted by the ring to infinity, modeling it as magnetic dipole radiation. After imposing balance of energy, the power equation gives the second order non-linear differential equation

$$K\left(\frac{d^2I}{dt^2}\right)^2 + LI\frac{dI}{dt} = 0, \tag{3}$$

for the current $I(t)$, where $K > 0$ is the radiation constant [3]. Being a second order equation, its initial conditions are given by two independents constants: One for the current and another one for its derivative. For the initial condition $I|_0 > 0$ of this problem, Eq. (3) allows as initial conditions $dI/dt|_0 \leq 0$.

For $dI/dt|_0 = 0$, its second derivative must be $d^2I/dt^2|_0 = 0$, what leads to a constant current equal to its initial value I_0, which does not dissipate energy by Joule effect, nor by radiation to infinity. It just keep a constant magnetic field around the loop.

For $dI/dt|_0 < 0$ its second derivative can be either positive or negative, and to solve Eq. (3) we can proceed as follows:

While $I(t) \neq 0$, we can use the logarithmic derivative of the current I as a variable x,

$$x = \frac{1}{I}\frac{dI}{dt}. \tag{4}$$

Taking its time derivative, after rearranging terms we obtain

$$\frac{d^2I}{dt^2} = \left(\frac{dx}{dt} + x^2\right) I. \tag{5}$$

Plugging Eqs. (5) and (4) back into Eq. (3), and dividing by I^2 we get

$$K\left(\frac{dx}{dt} + x^2\right)^2 + Lx = 0. \tag{6}$$

Being K and L positives, Eq. (6) implies that x must be nonpositive. Equation (6) can be rewritten as

$$\frac{dx}{dt} + x^2 = \pm\left(\frac{K}{L}\right)^{-1/2}\sqrt{-x} \qquad x \leq 0. \tag{7}$$

From here on we focus the discussion of Eq. (7) on the *plus* sign. From Eq. (5) this implies that we will be discussing solutions with $d^2I/dt^2 \geq 0$, i.e. convex solutions for I. Among these ones, those solutions which approach to zero behave more smoothly than the concave ones.

Making the change of variable

$$w = \left(\frac{K}{L}\right)^{1/6}\sqrt{-x} \qquad w \geq 0, \tag{8}$$

the Eq. (7) with the *plus* sign, becomes

$$-2\left(\frac{K}{L}\right)^{1/3} w\frac{dw}{dt} + w^4 = w. \tag{9}$$

We notice that if $w = 0$ at some time T, then $w = 0$ for all $t \geq T$ is a solution of Eq. (9). Since from Eq. (4),

$$w = \left(\frac{K}{L}\right)^{1/6}\sqrt{-\frac{1}{I}\frac{dI}{dt}}, \tag{10}$$

it follows that, for $w = 0$, $dI/dt = 0$, i.e. $I(t) = I|_T$ for all $t \geq T$.

For $w \neq 0$, Eq. (9) can be divided by w to get

$$2\left(\frac{K}{L}\right)^{1/3}\frac{dw}{dt} = w^3 - 1, \qquad w > 0. \tag{11}$$

In the following we study the system of Eqs. (10)-(11), that is equivalent to the original power equation given in Eq. (3).

4. Exponential solution

From Eq. (11), we see that the static solution for w is the constant $w = 1$. Squaring Eq. (10) for $w = 1$ one obtains

$$\frac{1}{I}\frac{dI}{dt} = -\left(\frac{K}{L}\right)^{-1/3}, \tag{12}$$

whose solution is the exponential function

$$I(t) = I_0 \exp(-\,(K/L)^{-1/3}\, t), \tag{13}$$

where $I_0 \equiv I|_{t=0} > 0$.

5. Sudden death solution

We now study the case for $w_0 > 1$, where $w_0 \equiv w|_{t=0}$. Evaluating Eq. (10) on $t = 0$ and squaring it, we obtain the initial condition

$$\left(\frac{1}{I}\frac{dI}{dt}\right)\bigg|_0 < -\left(\frac{K}{L}\right)^{-1/3}. \tag{14}$$

After separating variables in Eq. (11) and integrating, we obtain

$$\int_{w_0}^{w_t} \frac{dw}{w^3 - 1} = \frac{1}{2}\left(\frac{K}{L}\right)^{-1/3} t, \tag{15}$$

where $w_t \equiv w|_t$. When t increases, the right side increases and therefore the integral increases. Since in the neighborhood of $w_0 > 1$ the argument of the integral is positive, it must hold that $w_t > w_0$. For this integral, as $w_t \to \infty$, the area under the curve remains finite, i.e.

$$\int_{w_0}^{\infty} \frac{dw}{w^3 - 1} < \infty, \tag{16}$$

and therefore, when $w_t \to \infty$, $t \to T_1$ finite. We can analyze the implications of this for the current I as $t \to T_1$. From Eq. (5) and Eq. (7) with the *plus* sign we have that

$$\frac{1}{I}\frac{d^2I}{dt^2} = \left(\frac{K}{L}\right)^{-1/2} \sqrt{-x} > 0. \tag{17}$$

Since $I > 0$, then $d^2I/dt^2 > 0$. The fact that $d^2I/dt^2 > 0$ implies that the initial negative value $dI/dt|_0$ is a lower bound of the nonpositive quantity dI/dt, for all t. Moreover from Eq. (10),

$$w_t = \left(\frac{K}{L}\right)^{1/6} \sqrt{-\frac{1}{I}\frac{dI}{dt}}. \tag{18}$$

Since dI/dt is bounded below, $-dI/dt$ is bounded above. Therefore if $w_t \to \infty$ then necessarily $I \to 0$. Therefore

$$\text{as } t \to T_1, \quad I(t) \to 0. \tag{19}$$

The previous analysis is valid up to $t \to T_1$. As the solution is continuous when $I \to 0$, then

$$\text{for} \quad t = T_1, \quad I(T_1) = 0. \tag{20}$$

Now we need to analyze the solution for $t > T_1$. From the power equation (3)

$$K\left(\frac{d^2I}{dt^2}\right)^2 + \frac{1}{2}L\frac{d(I^2)}{dt} = 0, \tag{21}$$

integrating we get

$$\frac{1}{2}LI^2\bigg|_t + \int_{T_1}^{t} K\left(\frac{d^2I}{dt^2}\right)^2 = \frac{1}{2}LI^2\bigg|_{T_1}. \tag{22}$$

If at the instant T_1 the current is zero $I^2\big|_{T_1} = 0 \Rightarrow I = 0$ para todo $t \geq T_1$: The solution $I(t)$ decreases to $I = 0$ in a finite time T_1 and it stays $I = 0$ for all $t > T_1$. We have found similar behavior in capacitive system when radiation is considered [4].

6. Long-lived solution

Finally we study the case with $w_0 < 1$. Let us consider again the integral

$$\int_{w_0}^{w_t} \frac{dw}{w^3 - 1} = \frac{1}{2}\left(\frac{K}{L}\right)^{-1/3} t. \tag{23}$$

When t increases, the right side increases and therefore the integral increases. Since in the neighborhood of $w_0 < 1$ the argument of the integral is negative, it must hold that $w_t < w_0$. For this integral, as $w_t \to 0$, the area under the curve remains finite, i.e.

$$\int_{w_0}^{0} \frac{dw}{w^3 - 1} < \infty, \tag{24}$$

and, therefore as $w_t \to 0$, $t \to T_2$ finite. From Eq. (18) if $w_t \to 0$ then $x \to 0$. From the definition

$$x = \frac{1}{I}\frac{dI}{dt}, \tag{25}$$

as I is bounded above by its initial value $I|_0$, if $x \to 0$ necessarily $dI/dt \to 0$. Therefore, there exists a value I_f such that

$$\text{as } t \to T_2, \quad I \to I_f, \tag{26}$$

and $dI/dt \to 0$. By continuity

$$I(T_2) = I_f. \tag{27}$$

The current I_f is different from zero because for $t > 0$ the solution with the initial condition $I|_0 = I_0$ and $w|_0 < 1$ never crosses the curve $I(t) = I_0 \exp(-(K/L)^{-1/3}t)$, that corresponds to $w = 1$. In fact, during the evolution $w < 1$ for all $t > 0$ and initially the slope of $I(t)$ is lower than the slope of the exponential.

Let us analyze what does occur for $t \geq T_2$. We have to find the solution of the power equation (3)

$$K\left(\frac{d^2I}{dt^2}\right)^2 + LI\frac{dI}{dt} = 0, \tag{28}$$

with the initial conditions at $t = T_2$: $I(T_2) = I_f$ and $dI/dt|_{T_2} = 0$. It is straightforward to notice that the solution is

$$I(t) = I_f \neq 0 \text{ for } t \geq T_2. \tag{29}$$

We notice that at $t = T_2$, $w = 0$. We call this the long-lived solution.

7. Conclusions

We have studied an inductive physical system after it has undergone a phase transition to a superconducting state, its resistance being zero. We have modeled the system as a circuit with an ideal selfinductance L. We have included the radiation emitted to infinity modeling it as magnetic dipole radiation. After imposing energy balance, we have obtained the differential equation that models the system after the phase transition. Having analyzed it qualitatively we have found that, among the convex electrical current states available, the current can live in one of three classes of well differentiated states, depending upon its initial conditions when it enters the superconducting state after the phase transition. The relevant parameter is the value of its logarithmic derivative. If the electrical current enters the superconducting state decaying with a negative logarithmic derivative equal to $-\,(K/L)^{-1/3}$, then it will keep decaying exponentially. If it enters decaying with a logarithmic derivative more negative than $-\,(K/L)^{-1/3}$ it will keep decaying faster than the exponential, and will have a short life and a sudden death at a finite time. If the electrical current enter the superconducting state decaying with a logarithmic derivative less negative than $-\,(K/L)^{-1/3}$ it will keep decaying slower than the exponential, but after a finite time it will reach smoothly a longed-lived state of constant current different from zero. Actually it would live in that state forever, in agreement with the experimental results shown by File and Mills [2].

References

[1] Landau L D and Lifshitz E M 1987 *Classical Theory of Fields* (Butterworth-Heinemann, Oxford)
[2] File J and Mills R G 1963 *Phys. Rev. Lett.* **10**(93)
[3] Boykin T B, Hite D and Singh N 2002 *Am. J. Phys.* **70**(4) 415–420
[4] Urzúa G A, Jiménez O, Maass F and Restuccia A *Journal of Phys. Conference Series* **720**

A cognitive approach to the museography of an interactive science museum: a worked example

Luis Huerta[1,2]

[1] Facultad de Ingeniería, Universidad de Talca, 3340000 Curicó, Chile.

[2] P4-Center for Research and Applications in Plasma Physics and Pulsed Power Technology, 7600713 Santiago, Chile.

Glauco Cohen-Pantoja[1,3]

[1] Facultad de Ingeniería, Universidad de Talca, 3340000 Curicó, Chile.

[3] Universidade Federal do Oeste do Pará, CEP 68040-070, Santarém, Brazil.

E-mail: `lhuerta@utalca.cl, glaucopantoja@hotmail.com`

Abstract. Exhibits in interactive science museums (science centers) achieve their learning objectives as the interaction of visitors produces the inferences that are expected. We present here a methodology to assess communicational effectiveness of interactive exhibits. The methodology is developed by following Vergnaud's conceptual fields theory and permits us to determine the cognitive processes that exhibits demand from the visitors. We applied this methodology to a particular exhibit on electromagnetic induction at the Museo Interactivo Mirador (MIM), in Santiago, Chile. The necessary data were obtained from the audiovisually recorded visit of a group of secondary students, whose knowledge on the subject was previously analyzed and contrasted with their behavior during the visit.

Keywords: Informal learning, science centers, interactive exhibits, museography

1. Introduction

Evidence shows that interactive museums (science centers, in Anglo-Saxon countries) are an important resource for complementing formal education in science. A global survey conducted in 2013 by the Association of Scientific and Technological Centers (ASTC) shows that no less than 17% of visitors come from scheduled school visits [1]. This number is larger in Chile, where the Mirador Interactive Museum (MIM) reports that school visits account for a third of the total [2]. Evidence also demonstrates that in general terms most informal learning in science topics occurs in science museums [3, 4]. This highlights the importance of research in the area of informal education, particularly in the effectiveness of interactive exhibits and museography of science centers to produce the expected learning [5]–[10].

The design of interactive exhibits with the purpose of producing learning of ideas and concepts only through the visitors' autonomous interaction poses a significant challenge [11]. Without the orientation or sistematicity provided by a formal environment, visitors may interpret exhibits differently than expected. In addition, the visitor's prior knowledge can only be vaguely assumed and may exist at a wide variety of levels. In order to get visitors to infer the concepts and relationships relevant to the exhibits' goals, it is crucial to ensure the communicational effectiveness of the exhibits and the entire museography. In prior studies regarding this topic, we find some approaches that have been used to study effective communication for learning in an informal environment [12]-[15].

When the visitors' experience results in an increase in their knowledge, regardless of the level of that knowledge, it is an outcome of the cognitive processes they develop when engaging with the interactive exhibits. Consequently, every museological strategy should consider these processes. However, previous studies regarding these subjects are scarce and most do not incorporate a cognitive approach [16, 17]. In fact, exhibit design is primarily guided by intuition and classroom/laboratory experience or based on certain technical guidance that is learned when exhibit developers share their experiences [18, 19]. Notwithstanding, if the objective is to enhance visitors' learning, exhibit design could greatly benefit from a better understanding of how cognitive processes are developed.

In this article, we address the communicational ability of exhibits to transmit their contents by introducing a methodology that is based on determining the cognitive processes that are displayed. This report is based on an investigation that we conducted in the electromagnetism area of the Mirador Interactive Museum (MIM) in Santiago, Chile. To date, no systematic studies have been developed at MIM regarding the impact of a visit on the visitors' knowledge and understanding of the corresponding scientific concepts. We observed a group of 15 high school students during their visit to the Electromagnetism area of the museum. The main goal of our research was to examine how students' thinking evolves by analyzing the cognitive processes that they developed as a result of their interaction with the different exhibits on electromagnetism. Two full reports on those primary goals will be published elsewhere [20, 21]. However, a necessary and complementary byproduct of the prior study is the observation of elements to ensure the exhibits achieve their purposes. Another study with a similar purpose was published [22]; however, that analysis did not include cognitive processes.

The communicational abilities of exhibits are also strongly limited by visitors' prior knowledge. Exhibits are effective if the pertinent concepts-in-action and theorems-in action are recognized by visitors. Therefore, an in-depth analysis of the importance of prior knowledge and experiences must precede the museography and exhibit design. In our approach we focus on cognitive processes underlying visitors' interaction with exhibits. Following Vergnaud, the cornerstone of cognition is conceptualization, in whose construction participate implicit ideas and categories which Vergnaud describes under the concepts of concepts-in-action and theorems-in-action: a close relationship between reality –situations– and its representation –operational invariants and linguistic-symbolic codes. In order to elucidate the cognitive processes involved, we present a methodology based on the identification of the operational invariants put in action by the visitor and, from them, we derive his/her evolving thinking as a result of experience. As a corollary, we derive a methodology useful for designing exhibits and museography.

2. Theoretical foundations

Learning is a transformational process, as purported by Piaget, who highlighted the adaptive function of cognition [23, 24]. Piaget defined the building blocks of cognition as schemas, or to clarify, "a cohesive, repeatable action sequence possessing component actions that are tightly interconnected and governed by a core meaning" ([23], p. 7). In alignment with the same constructivist approach, Vygostky emphasizes the importance of social and cultural elements in learning [25, 26]. Vygostky introduces the concept of the Zone of Proximal Development (ZPD) as "the distance between the actual developmental level as determined by independent problem solving and the level of potential development as determined through problem solving under adult guidance, or in collaboration with more capable peers" (in [25], p. 86). Vygotsky has inspired designers of interactive experiences because the concept of the ZPD suggests that designers provide "scaffolding" to enable learners to participate in a more complex discourse than they could experience on their own [27].

A visitor's experience when interacting with an exhibit cannot be described thoroughly in terms of the intended ultimate goal. In general, one exhibit poses multiple problems to visitors. Then, a variety of schemas are necessary for visitors to display. A more elaborate theory is needed to analyze visitor-exhibit interactions in a science museum. These interactions are not solely a psychological issue but also epistemological. The contents of the subject to be learned through the interaction with an exhibit play a fundamental role. G. Vergnaud, a psychologist and Piaget's disciple, devoted much of his work to understanding how mathematics and science are learned [28, 29]. In Vergnaud's theory of conceptual fields, he recognizes that cognition must be referred to the specific content and complexities related to the topic that is being learned [30]–[32]). He also agrees with Vygotsky that "constructivism concerns not only the development of individuals, but also of culture, especially of scientific culture" (free translation of the original text in [33]).

For Vergnaud [28, 29], the cornerstone of cognition is conceptualization, which refers to "the identification of objects, of their properties and relationships" [34]. Vergnaud asserts that our thinking includes implicit ideas and categories by referring to the concept of knowledge-in-action or the operational invariant. These concepts are the building blocks of conceptualization and have a close relationship between reality (situations) and representation (operational invariants

IOP Conf. Series: Journal of Physics: Conf. Series **1043** (2018) 012054 doi:10.1088/1742-6596/1043/1/012054

and linguistic-symbolic representations). In Vergnaud's theory of conceptual fields, a different non-classical perspective is assumed and it is established that a concept cannot be described without a reference to a specific situation (we follow hereafter Ref. [35]). Therefore, the relevance of situations is important in a psychological context through the difficulties that students must manage. It is also epistemologically relevant because the problem that situations involve is a source of knowledge. In this manner, Vergnaud developed a concept $\mathcal{C}$ as the triplet $\mathcal{C} = (\mathcal{S}, \mathcal{I}, \mathcal{R})$ where $\mathcal{S}$, the reference of the concept, is the set of situations that ensures that the concept is significant and useful, $\mathcal{I}$ represents the meaning of the concept and is the set of operational invariants that assign the meaning and contents to make the concept operate, and $\mathcal{R}$ represents the signifier of the concept and is the set of linguistic and symbolic representations used to display the invariants and situations. To study the development of a concept during the learning process or to analyze how it is used, it is necessary to consider the triplet as a whole.

Because situations give meaning to a concept, a situation makes sense because of the schemas. A schema is the invariant organization of behavior for a certain class of situations. Vergnaud developed a different conception of schema than Piaget and established a subject-in-situation relationship rather than using the subject-object interaction of Piaget. The primary components of a schema are the operational invariants, which represent the knowledge contained in the schemes. These invariants constitute the implicit or explicit basis for obtaining the relevant information for the objective to be achieved and to develop appropriate standards for action. The operational invariants may be referred to either as concepts-in-action, which are concepts that may or may not be relevant, or theorems-in-action, which are propositions related to concepts that may be either true or false. Theorems-in-action involve concepts-in-action. The concepts-in-action relate to each other through theorems-in-action.

From the theoretical perspective, the concept of schema provides a link between behavior and representation. The relationship between situations and schemas is the primary source of representation and therefore conceptualization. Conversely, operational invariants ensure the essential articulation between theory and practice because perception, search and the selection of information are based entirely on the system of concepts-in-action that is available to an individual (objects, attributes, relationships, and conditions) and the theorems-in-action that underlie his/her behavior.

For every situation, the data to be analyzed and the inferences that result depend on theorems-in-action and the identification of different types of relevant elements. Most concepts and theorems-in-action remain entirely implicit, but they can also become explicit. From this perspective, concepts-in-action and action-theorems can progressively become true scientific concepts and theorems. The goal to be achieved by a set of exhibits regarding a specific topic is to help the visitor to make the corresponding concepts and theorems explicit and scientifically accepted from the implicit knowledge the visitor displays. Although the entire process can take longer than one visit to the museum, the goal for every museum experience is that there is sufficient correlation between the visitor's prior knowledge, displayed as concepts and theorems-in action, and the interactions that are made available by the exhibit.

Vergnaud's approach provides us with adequate tools to analyze visitor-exhibit interactions. First, the various ways that an exhibit can be manipulated must be considered. They include different situations where students can display their knowledge-in-action (operational invariants). In this manner, prior knowledge is more pertinent to the situation and can be revealed as implicit concepts and relationships. Because we focus on determining the operational invariants and their representations, we are able to examine whether the expected inferences are possible.

Learning is based on prior knowledge. To reliably evaluate the success of the museography and exhibits, the theoretical framework of our methodology focuses on clarifying the prior knowledge of visitors in regard to the specific topic of the exhibits. The goal is to determine how the problems that are posed to the visitor by the exhibits fit that prior knowledge.

3. The conceptual field

Vergnaud purported that a conceptual field is a set of problems and situations, and the solution of those problems requires concepts, procedures and representations of different but closely related types [28]. In the research we partially report here, we consider museum exhibits on electromagnetism. An example we will develop below is focused on electromagnetic induction. A variety of concepts are involved, from electric charges to currents, from currents to magnetism, etc. A thorough review of the concepts that are involved in the exhibits and the museography of the area is necessary to approach the complexity of the topic. In particular, regarding the subject of electromagnetic induction, almost every concept in electromagnetism must be considered.

Among other general physics concepts, magnetic field, electric current, conductivity, magnetic force, magnetization, voltage, energy are relevant in the subject fo electromagnetic induction. In addition, more elaborate concepts that are derived from electromagnetic theory, such as magnetic flux and electromotive force (emf), could appear. But also mathematical concepts such as the rate of change in time and vectors are essential. Experiences that demonstrate how a current is produced, when a material becomes magnetized and attracted by a magnet, or how an electromagnet works could be crucial to understand the concepts involved in electromagnetic induction phenomena.

Historically, electromagnetic induction is an outcome of nineteenth-century science which, along with deep theoretical consequences, has been the scientific basic for electric power production technologies. The phenomenon is based on the generation of an electric current that is caused by the action of a time-varying magnetic field over a conductor. This variable magnetic field can be generated in various manners that include the relative motion of a magnetic field source (for example, a magnet) and a conductor. The conductor is typically a closed electrical circuit where the current is induced (induced electrical current). The effect is not perceptible by human senses but can be detected using instruments, such as an ammeter. The exposure to a phenomenon that cannot be seen "by naked eye" is clearly a challenge for exhibit designers.

The standard instruments that are used to detect the effect must also be understood through a complementary conceptualization of a related subject.

Classical electromagnetism is widely understood through Maxwell's theory, and its concepts and relationships can be described accurately through mathematical relationships. However, museography regarding this topic cannot be based on the mathematical Maxwell's theory if it is adressed to general public, or even secondary school students. Instead, approaches recurring to partial descriptions, such as Faraday's Law, could be in principle more accessible. However, as Zuza et al. stated, "interpreting Faraday's law has been a challenge in the physics community and most students experience learning difficulties when attempting to understand electromagnetic induction phenomena and Faraday's law" [36]. Therefore, it becomes apparent that the topic of electromagnetism poses a significant challenge for museographers and exhibit designers.

Following Vergnaud, conceptualization is achieved through the situations the learner faces. These situations are provided by the exhibit, in a variety that depends on the exhibit's design. In the following section, we present an interesting example of an exhibit on electromagnetic induction. We also describe the different, possible situations this exhibit could pose to the visitor.

4. An exhibit on electromagnetic induction phenomena

In this section, we illustrate with an example the methodology we exposed above. "The Electromagnetic Pendulum" is an exhibit of MIM's electromagnetism area, which we describe in detail below.

This exhibit features a metallic swinging bar, that includes a permanent magnet in its moving end, that oscillates when released from a higher position. Under the trajectory of this pendulum, a cart that carries a plate in a vertical position and is composed of various materials can move on a semicircular track. The horseshoe-shaped magnet produces a magnetic field that is perpendicular to the plane of oscillation, and when the pendulum meets the cart, the plate passes through the gap in the magnet. A photograph of the exhibit is provided in Fig. 1, with details about the cart where the plates are mounted. Depending on the plate material and geometry, there will be an interaction between the magnet and the plate when both are in relative motion.

Six plates are available: three copper plates (one full solid, one with a circular hole in its center, and one comb-shaped with slits), one solid aluminum plate, one solid acrylic plate, and one acrylic plate with a solenoid embedded and including two LEDs in the circuit of the solenoid). Fig. 2 provides a photograph of the plates.

None of the plates are ferromagnetic, and the plates are only attracted by the magnet because of the electrical current that is induced by the magnet. Once the cart is placed with the plate in the bottom of the rail at rest and the pendulum swings, a force can be produced between the plate and the magnet of the pendulum. This force is due to the action of the magnetic field on the electric current that is induced on the plate. This process causes the cart to move by the rail, following the pendulum.

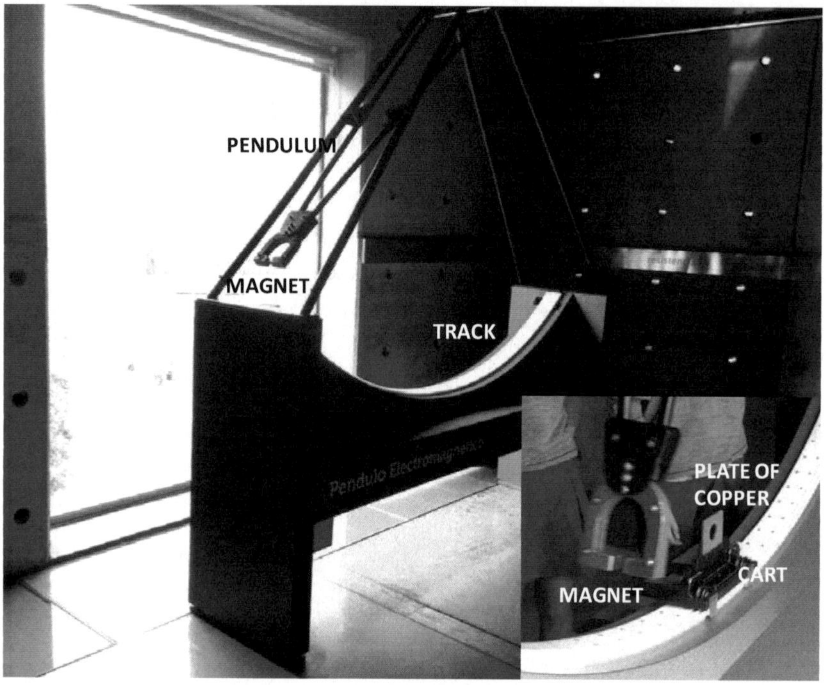

Figure 1: The exhibit "Electromagnetic Pendulum". In the lower right corner of the figure a detail of the cart where the plates are mounted is shown.

Figure 2: The plates of the exhibit. The plates at the top of the figure are solid and constructed of copper, aluminum and acrylic. At the bottom, the first two plates are constructed of copper. The latter has a solenoid embedded with a pair of LEDs on its upper edge (not seen in the picture), which are connected to the solenoid.

The goal of the exhibit is to ensure that the visitor notices the distinctive behavior of the cart for the various mounted plates and to deduce certain hypotheses to explain what is observed. Although the exhibition was designed for autonomous operation by the visitor, for safety reasons, the MIM has provided a guide for conducting the experience.

5. The exhibit as a set of situations

Every interactive exhibit poses a problem to visitors, requiring actions and then conceptualization. The exhibit becomes a situation that encourages the development of concepts. Therefore, it is necessary to accurately describe the exhibition as a situation and, in cognitive terms, beyond the specific aspects that arise from a scientific perspective or the exhibit instructions. In alignment with Vergnaud, it is not sufficient to conceive situations solely in terms of the operations that are demanded of visitors (sensorimotor, cognitive) because these are concatenated with more generic thinking [28]. To better analyze the museum exhibits, we define three fundamental elements to describe situations: objects, variables and unknowns [21].

Objects include the interacting elements and building blocks that are included in the situations. Objects are characterized by consistent properties that are independent from the context. In the case analyzed in this study, objects include magnets, solenoids, conductors, circuits, ammeters, and voltmeters. A magnet is recognized according to the concept of magnetism, and a circuit is related to the concept of electric current.

Variables are descriptions that relate objects to their mutual interactions and define the dynamic context of the problem. In this study, the magnets and conductors are in relative motion, and LED bulbs light up when an electric current is induced. In addition, variables define the degree of complexity; using cognitive processing to solve a problem depends on dynamics that are exposed by those variables. An example of this process is when electromagnetic induction results from a time-varying magnetic field that acts on a conductive material. This process poses various levels of difficulty for conceptualization if the variation of the magnetic field is produced by a magnet?s movement relative to the conductor or when an AC voltage source powers a solenoid.

Unknowns define the framework for responding to the problem and correspond to the domain of searching for the visitor's explanations. Unknowns can be quantitative or qualitative. For this study, an unknown could be, for example, why the ammeter detects an electrical current in a closed circuit when a magnet is moving near it.

The three components described above permit us to classify situations in different domains that are specific to the contents of the exhibits and also to account for the different degrees of freedom that the situations represent for the visitors. For example, because a magnetic field can be generated by a magnet as a solenoid with an electric current (electromagnet), the respective effect on each object will have a different epistemological characteristic if the field comes from one source or another. To clarify, the concept of magnetic field cannot be considered devoid of the object that constitutes its source. Although in scientific terms the source of the magnetic field is indistinguishable in the phenomenon of induction, this will not be the case for every visitor.

We use this methodology to characterize the exhibit we described previously. The various possible situations that could result from operating this exhibit are described in Fig. 3.

Objects	Pendulum, magnet, cart, plates (solid copper, copper with hole, copper with slits, solid aluminum, solid acrylic, and solenoid with LEDs), and a semicircular rail for the cart.
Variables	The pendulum with the magnet oscillates; the cart with the plate (of copper, aluminum, acrylic, etc.) moves along the track; the cart is initially located at the lower section of the track; the magnet and the plate interact; the pendulum and the cart oscillate synchronously; the cart with the aluminum plate moves while interacting with the magnet of the pendulum; the pendulum oscillates and the LEDs light up; the cart is released from the higher section of the track; etc.
Unknowns	Why does the car move when the pendulum oscillates although they are not in contact? Why does the cart move sometimes and other times it does not move? Why are the aluminum or copper plates attracted to the magnet although they do not become magnetized? Why does the plate with a solenoid not move when the pendulum swings, but rather, the LEDs light up? Etc.

Figure 3: The situations the exhibit "Electromagnetic Pendulum" provides are composed of the objects, variables and unknowns listed in the table.

6. Discussion

The exhibit described in this report was designed to provide a number of situations that permit, in principle, the conceptualization of electromagnetic induction phenomena. The primary concepts involved include magnetic field, relative motion, magnetization properties of materials, conductivity, induced electric current, magnetic force, and energy transfer. Depending on the specific situation (objects, variables, unknowns), it is expected that visitors put into action certain related concepts and manage the corresponding theorems-in-action, that is, the corresponding relationships between the concepts.

This article describes various situations that the exhibit "Electromagnetic Pendulum" provides for visitors. Our analysis allows us to determine if the concepts and relationships the students use allow them to explain what they observe and ultimately complement what they know about electromagnetic induction. In this study we applied a methodology to evaluate the exhibit's effectiveness.

An examination of the dialogues that were held between the students indicates the operational invariants, that is, the concepts-in-action and theorems-in-action for each situation. Generally, we observed that students were influenced by the exhibit to deduce relevant features that characterize electromagnetic induction. Two primary conclusions that the students inferred are that non-ferromagnetic metals, such as aluminum and copper plates, are attracted by the magnet because of their conductive nature and the induced currents in those metallic plates are due to the relative motion between the magnet and the cart with the plate. Although these two conclusions, which appear as strong signals of what is occurring, are not necessarily understood as they are described in formal electromagnetic theory.

The exhibit was designed to establish certain crucial facts about electromagnetic induction. One fact is not to use ferromagnetic metals to avoid the well-known magnet-iron interaction. Another fact involves a coil circuit model for the induced current inside the metal. In this model, a copper plate with a large central hole provides an approach that is similar to the model. Furthermore, this plate is similar to the solenoid, which was previously described (see Fig. 2). In addition, the slotted plate provides a situation that emphasizes the coil model for induced current, since for that plate, no significant interaction is observed when the magnet moves relative to the cart. In spite of all that, It is remarkable that the students in this study were unable to infer any conclusion that reliably supported the model. The mechanism that was proposed by students to explain the absence of a force between the magnet and the slotted plate or even why the LEDs in the solenoid light up differed from the intentions of the exhibit designers. Additional exhibits exploring these aspects could be useful if the model, which is important to be able to understand the phenomenon occurring in the solenoid, is a relevant issue.

The exhibit is effective in revealing certain primary aspects of the phenomenon, such as the relevance of the conductive nature of the material that interacts with the magnet and the need for relative motion between the conducting plate and the magnet (incorporated in the pendulum). These are important aspects to understand what produces the basic phenomenon; however, it is not enough to develop adequate explanations and infer other forms to produce the phenomenon. The ultimate reasons of why the plates are attracted by the magnet and why the magnetic force affects the current remain hidden. This latter issue is important if we pursue a simple explanation of what the visitors observe in the exhibit operation beyond the physics of electromagnetic induction. The exhibit is not successful in this aspect, and the reasons for this lack of success may be found in other elements of the entire museography. By examining the other exhibits in the area, we do not find an exhibit showing explicitly the action of a magnetic field on an electric current, in the same terms of the force that acts on a permanent magnet.

Conversely, to infer a modern scientific explanation of the induction phenomenon, it is

IOP Conf. Series: Journal of Physics: Conf. Series **1043** (2018) 012054 doi:10.1088/1742-6596/1043/1/012054

essential to understand the concept of a varying magnetic field. Although the mechanistic approach that results from highlighting the relative motion between the parts could be considered relevant, the success of Maxwell's electromagnetic theory is derived from focusing on fields. This must not be considered an advanced topic, because a comprehensive understanding of this concept is the cornerstone of modern physics. In principle, the exhibit provides examples of this and allows a comparison of the results that are obtained by first oscillating the pendulum or initiating the experience by releasing the cart from its highest position in the track with the pendulum at rest. Why does this exhibit not work in the aspect we discuss? One specific exhibit cannot provide all the elements. It is a matter that concerns the entire museography. Unless the guide helped visitors understand that from the perspective of the cart, we have a varying magnetic field (in both situations that are mentioned above), it is unlikely for visitors to infer this aspect of the exhibit.

Two primary questions arise here. One question is whether the students have learned more about electromagnetic induction because of the visit to the museum, which is a matter for a different report [21]. The other question is whether the visitors that we considered in this research are representative of the public that visits the museum. The answer to this question is no. The students are part of a privileged group in the sense that they have previously studied the subject as a part of their formal education. However, their experiences in the museum are representative of how effective the exhibit is in conveying ideas and facts to produce scientific inferences regarding electromagnetic induction. The discussion above demonstrates that the exhibit partially accomplishes this task. However, we believe that this is a matter related to the entire museography, and we do not place any blame on one exhibit.

7. Conclusions

The issue we analyze in this article is how an interactive exhibits can lead visitors to infer the relationships that are involved in its contents. These inferences refer to the understanding of facts and ideas through the organization, comparison and interpretation based on the visitors' prior knowledge. We developed a methodology that permits us to determine the cognitive processes through the identification of the operational invariants that are put in action during the experience. We are aware that we cannot completely describe the relevant cognitive operations because the invariants are possible representations of those processes. However, the determination of the operational invariants provides more information regarding the most basic elements of visitors' thoughts that are related to what we seek to understand, or rather, the evolution of the visitor's thinking as a result of the experience.

A proper conceptualization requires a variety of situations, and this affects the design of the museography. In informal contexts, distributed learning becomes relevant, particularly in relation to educational communication [37]. In regard to the design of an exhibition on electromagnetic induction, a set of exhibits will provide different situations that allow visitors to experience and learn all the concepts that are needed for the expected inferences. The exhibit we discuss here provides, in principle, a diversity of situations. However, the learning process varies for different people. Museum exhibition planning provides the possibility of considering this issue by introducing parallel experiences for visitors' interactions.

A practical corollary is relevant here in relation to the objective of complementing the methodologies involved in the design of interactive exhibits. The method of analysis we utilized in this article is suitable for use in the conceptual design of both museography and exhibits. Fig. 4 illustrates the steps.

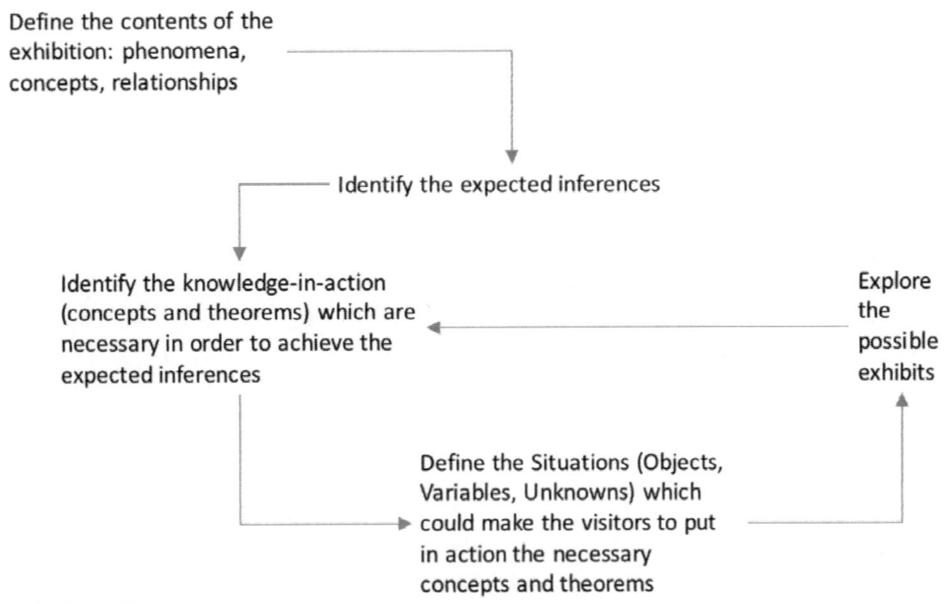

Figure 4: Scheme of a methodology for the conceptual design of museography and interactive exhibits.

The primary goal of a science museum is to inform the public of the phenomena, concepts, methods and ideas of science. In this manner, learning goals must be understood according to different levels. Although this study focused on examining the understanding of concepts and relationships at a level compared to what is expected in formal education, we consider that the methodology and conclusions are also valid for lower levels of learning. In addition, we emphasize that interactive science museums have become an important resource for science teachers, whose expectations certainly highlight the interactive experiences as an important complement to formal classes. We believe that the conclusions of this article could lead to improvements that would not be difficult to achieve in the design of museography and exhibits.

Acknowledgments

We are very grateful to Valeria Vera, Head of Education at MIM, for providing access to the museum and providing everything that we needed to conduct this investigation. We thank the guides of the Electromagnetism section for their assistance in collecting data. In addition, we are grateful to Patricio Torres, Leonardo Herrera, and Rocío Parrao y Maximiliano Montoya for producing the audiovisual recording of the students' visit to MIM. We acknowledge and thank Prof. Rodrigo Pérez, Director of the Centro Educacional Vygostky, and the students who participated in the investigation. G.P. acknowledges the hospitality of the Universidad de Talca during his postdoctoral stay and the scholarship that was provided by Banco Santander of Brasil and UFOPA. The conclusions of this study do not represent in any manner a judgment on MIM's educational proposal.

References

[1] Association for Science–Technology Centers (ASTC) (2013). "2013 Science Center and Museum Statistics". http://www.astc.org/wp-content/uploads/2014/10/2013-Science-Center-Statistics.pdf. Retrieved January 13, 2017.

[2] Memoria Anual MIM (2013), http://www.mim.cl/index.php/quienes-somos/memorias.

[3] Bell, P. et al. (Eds.) (2009), "Learning Science in Informal Environments". The National Academies Press: Washington DC.

[4] McPherson, E. (2014), "Informal Learning in Science, Math, and Engineering Majors for African American Female Undergraduates"

[5] Screven C. G. (1993), "Museums and Informal Education". CMS Bulletin Vol. 1 No. 1.

[6] Anderson, D. (1998), "An analysis of the importance of informal and formal science learning contexts to each other: An overview perspective. Learning Science in Informal Contexts", Questacon, Canberra, Australia.

[7] Anderson, D. (1999), "The development of science concepts emergent from science museum and post-visit activity experiences: Studentsćonstruction of knowledge". Brisbane, Australia, Queensland University of Technology.

[8] Anderson, D.; Lucas, K.B.; Ginns, I. B. and Dierking, L. (2000), "Development of Knowledge about Electricity and Magnetism during a Visit to a Science Museum and Related Post-Visit Activities". Science Education, 84, p. 658-679.

[9] Tran, L. U. (2006), "Teaching science in museums: The pedagogy and goals of museum educators". Science Education 91(2), 278?297.

[10] Shin, H., Park, E. J., and Kim, C. (2014). "Learning affordances: Understanding visitors' learning in science museum environment". In C. Bruguire, A. Tiberghien, and P. Clément (Eds.), Topics and Trends in Current Science Education (pp. 307-320). New York: Springer.

[11] Allen, S. and Gutwill, J. (2004), "Designing with multiple interactives: Five common pitfalls". Curator 47(2): 199- 212.

[12] Silverman, L.H. (1999). "Meaning making matters: communication, consequences and exhibit design", in Exhibitionist, Fall issue, AAM-NAME (National Association for Museum Exhibition).

[13] NRC (National Research Council) 2009. "Learning science in informal environments: People, places, and pursuits", edited by P. Bell, B. Lewenstein, A.W. Shouse, and M.A. Feder. Washington, DC: The National Academies Press.

[14] Ded, M. "Rigour in Communicating Maths: A Mathematical Feature or an Unnecessary Pedantry?" in Raising Public Awareness of Mathematics, Eds. Behrends, E. , Crato, N., Rodrigues, J. F., Springer (2012).

[15] Conference on "Discursive Space: breaking barriers to effective spatial communication in museums" (2013), by Ryerson University and the Art Gallery of Ontario, June 20-23, Toronto, Canada. Retrieved January 11, 2017, from http://www. discursivespace.com/

[16] by McKenna-Cress, P. and Rice, K. (2014) "Teaching Exhibition Development and Design", in Exhibitionist, Fall issue, AAM-NAME (National Association for Museum Exhibition).

[17] Bedford, L. (2014). "The art of museum exhibitions: How story and imagination create aesthetic experiences". Walnut Creek, CA: Left Coast Press.

[18] McLean, K. (1993). "Planning for People in Museum Exhibitions". Association for Science–Technology Centers Publications.

[19] Humphrey, T., P. Gutwill, J. P. (Eds.) (2005) "Fostering Active Prolonged Engagement: The Art of Creating APE Exhibits". Exploratorium Museum Professional Series. San Francisco : Exploratorium.

[20] Huerta, L. and Cohen-Pantoja, G. (2017), "The communicational effectiveness of interactive exhibits at a science center: a new methodology of analysis applied to an exhibit on electromagnetism", submitted to Phys. Rev. PER.

[21] Cohen-Pantoja, G. y Huerta, L. (2017), "Análisis cognitivo de las concepciones de estudiantes de Enseñanza Secundaria en una visita a la sala de Electromagnetismo de un Centro de Ciencias: un estudio exploratorio de casos en el Museo Interactivo Mirador en Chile", to be submitted.

[22] Mortensen, M. F. (2011), "Environment: Visitors' experience with and understanding of an immersion exhibit". Int. J. Sci. Educ. 33, 517-545 http://dx.doi.org/10.1080/09500691003754589.

[23] Piaget, J. (1952). Translated by Cook, M. T. "The origins of intelligence in children". New York, NY: International University Press.

[24] Piaget, J. (1971). Biology and Knowledge. Chicago: University of Chicago Press.

[25] Vygotsky, L. S. (1978). "Mind in society: The development of higher psychological processes" (Trans. Cole, M. et al), Cambridge, MA: Harvard University Press.

[26] Vygotsky, L. (1986). "Thought and Language". Cambridge, MA: MIT Press.

[27] Roschelle, J. (1995). "Learning in interactive environments: Prior knowledge and new experience". In J. H. Falk and L. D. Dierking (Eds.), Public institutions for personal learning: Establishing a research agenda, pp. 37-51. Washington, DC: American Association of Museums.

[28] Vergnaud, G. (1983). "Multiplicative structures". In Lesh, R. and Landau, M. (Eds.) "Acquisition of Mathematics Concepts and Processes". New York: Academic Press Inc. pp. 127-174.

[29] Vergnaud, G. (1988). "Multiplicative structures". In Hiebert, H. and Behr, M. (Eds.). "Research Agenda in Mathematics Education. Number Concepts and Operations in the Middle Grades". Hillsdale, N.J.: Lawrence Erlbaum. pp. 141-161.

[30] Vergnaud, G. (2009). "The Theory of Conceptual Fields", Human Development 52, pp.83?94 DOI: 10.1159/000202727.

[31] Vergnaud, G. (1993). "Teoria dos campos conceituais". In Nasser, L. (Ed.) Anais do 1o Seminário Internacional de Educação Matemática do Rio de Janeiro. p. 1-26.

[32] Vergnaud, G. (1994). "Multiplicative conceptual field: what and why?" In Guershon, H. and Confrey, J. (1994). (Eds.) The development of multiplicative reasoning in the learning of mathematics. Albany, N.Y.: State University of New York Press. pp. 41- 59.

[33] Vergnaud, G. (2007) "In what sense the conceptual fields theory might help us to facilitate meaningful learning?" (My translation), Investigações em Ensino de Ciências ? V12(2), pp.285-302.

[34] Vergnaud, G. (2013), "Conceptual development and learning". Revista Qurriculum, 26, 39-59; ISSN: 1130-

5371, p. 41.

[35] Moreira (2002), "A teoria dos campos conceituais de Vergnaud, o ensino de ciências e a pesquisa nesta área". Investigações em Ensino de Ciências 7, 7.

[36] Zuza, K., Almud, J.-M., Leniz, A., and Guisasola, J. (2014), "Addressing students' difficulties with Faraday's law: A guided problem solving approach", Phys. Educ. Res. **10**, 010122.

[37] Fossard, E. D. (2005), "Communication for Behavior Change, Volume I: Writing and Producing Radio Dramas". Sage Publications, New Delhi.

IOP Conf. Series: Journal of Physics: Conf. Series **1043** (2018) 012055 doi:10.1088/1742-6596/1043/1/012055

Mechanical mass separator: an integrating approach in teaching Mechanics for Engineering and Physics majors

Américo Cuchillo Flórez[1,2]

[1]Departamento de Física, Universidad de Atacama (UDA), Copiapó, Chile
[2]Centro para el Desarrollo Educacional Yachachiy (CEDEYA), Cusco, Perú

e-mail: americo.cuchillo@uda.cl

Abstract. An Integrating Approach (IA) in the teaching and learning of a first formal course of Mechanics for future engineers and physicists is presented. In this approach, the course is introduced as a practical and technological problem to resolve that delineates a single route during its development. The need to solve the different aspects of the problem leads progressively to establish the contents and concepts of the subject; i.e., these arise as tools to understand and resolve the aforementioned problem. The approach allows to elucidate and assess the approximations, limitations and corrections that must be performed to an own model of a scientific discipline such as physics. The approach potentiality relies in allowing the student, through a concrete and simple problem, to articulate Kinematics, Newton's laws, and Energy, appreciating the subject in a holistic and non-fragmentary fashion. Moreover, it allows the students value Physics and to estimate its usefulness, because they find it useful and necessary for understand the basics behind all technological development, even although it seems very simple. To this purpose, is proposed use a simple device that shoots different masses, through a spring placed on a horizontal surface at a certain height, and measure the distance to which they fall to the floor. This device facilitates the explanation of the physical concepts that emerge in a first analysis of the problem. This approach, as well, facilitates to present in an elegant way the influence of the mass on motion as well as the distinction between constant and variable forces. The experimental results show mass effects according with theoretical predictions, allowing going beyond the usual kinematics taught at this level.

1. Introduction

Physics teaching seeks the abstraction of reality by translating it into simple and universal models that allow describing and predicting the behavior of the natural phenomena that these models pretend to represent [1]. However, this same abstraction may interfere with students' understanding of Physics, especially if the students are facing for the first time a formal course of Mechanics at university level. Concepts such as "point" object or particle, ideal string, ideal pulley, to name a few may induce to think they are aspects disconnected from reality and with little or none practical application. This may generate lack of interest for Physics and a decreasing motivation for subject learning. Textbooks, as professors in general, try to neutralize this by relating these abstract models with practical and real applications. This is mainly done through examples and exercises where concepts and mathematical relations are used in a wide range of common contexts and even interdisciplinary ones. While these practical applications allow students to value its usefulness, the mere fact the used model came from the above-mentioned simplifications may generate suspicion and confusion on the student about if these practical examples are directly related to reality or they are just "practical situations in paper". In other words, it is necessary to apply the model to a concrete and real situation during the teaching of Mechanics and in general Physics.

In general, lab activities account for the latter work and can be carried out prior to, during or after theory teaching. They allow the student to link different activities to the concepts and models they have previously studied.

Both applications of the model through exercises and lab activities are oriented towards diversifying its use and applications. This is necessary since we are dealing with a basic science with an interdisciplinary reach. However classroom exercises are not always articulated within the same topic, least with exercises related to other topics within the same subject, save for concepts and mathematical relations. The same situation occurs with lab activities for a variety of reasons, such as time, equipment access at institutions with less money or others.

When there is no clear articulation between the abstraction activities (modelling), application of a given model (exercises) and lab (experiments), these are seen as fragments of something called Mechanics, which is good and useful, but lacking something concrete which can articulate the different activities carried out during Mechanics teaching.

On the other hand, even if the student performs well, not knowing when and why a given activity is carried out may lead him/her to gradually lose both direction and initial motivation. Lacking a define direction and something concrete that articulate the subject as a whole the student sees himself/herself in a sea of contents and activities, achieving learning without a clear purpose other than passing the subject and the hope of using this knowledge in the future in his/her own field of expertise. A few of above-mentioned problems has been extensively studied in the past by Physics teaching researchers, by means of robust tools, with findings such as that described here [2-6].

This paper sets a direction by using a concrete and real problem for the student to solve while, along the subject, a mechanical system (simple device) to solve the problem is designed by him/her. In this way the modelling and abstraction of a real physical system may be made easier and more understandable, presenting models as useful and predictive tools, with validity limited by the imposed approximations and, therefore, open to future corrections. Using this approach the future engineer learns why he/she needs Physics while the future physicist learns how a Physics theory is built.

The use of a concrete problem, being it real and practical, allows to organize the subject highlighting that solution requires understanding and interconnection of concepts being developed with already studied ones. This allows to achieve an integral articulation of different topics of the Mechanics through a unique and practical problem.

Finally this approach suggests to the Physics teacher he/she can always set the direction of the subject through a problem adequate to the educational context.

2. The Problem

The problem is to separate solid spheres of equal radius according to their different masses. This situation may occurs, for example, when separating beads with bimodal or even multimodal mass distribution within defined mass ranges. For the given case, it is considered a system composed by spheres of equal volume with different masses m and M. As a device to achieve the stated purpose, a "mass launcher" consisting of a tube with a spring inside it is proposed. The masses are shot from the end of the device, located at a height H of the ground, and collected at some point at ground level, located a horizontal distance D from the launch point as shown in Figure 1. In principle, this device allows to give a complete solution to the proposed problem and sets the direction of the subject. It is also the concrete element of application and articulation.

During the development of the subject the student will design and build, at least partially, the device. The most important thing is that the student will learn in the subject the concepts of mechanics through the construction and operation of the apparatus, being able to perform this in a group or individual way.

At the end of the course (Mechanics), the student must know in detail both the operation of the device and the physical principles that underlie it.

IOP Conf. Series: Journal of Physics: Conf. Series **1043** (2018) 012055 doi:10.1088/1742-6596/1043/1/012055

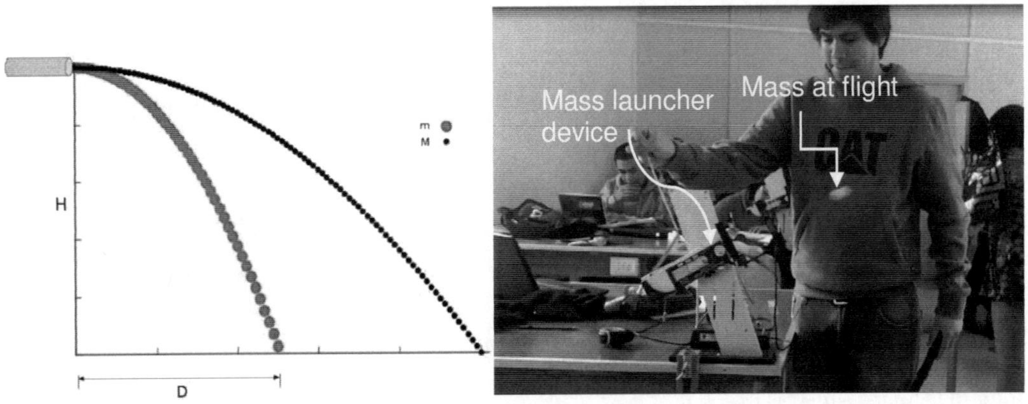

Figure 1. Right: Mechanical mass separator. For simplicity, the masses will always be launched under the same geometric conditions: height and angle (horizontal throw for the case). Right: Shooting done by one undergraduate student. It was done at angle different to zero by picture facilities during the flight.

The problem and the proposed solution do not limit, nor much less replace, the varied applications that should be taken both in theoretical classes and laboratory activities. However, the subject is oriented to achieve the final task that has been proposed, articulating implicitly the intermediate actions.

3. Subject Development

When observing the device and the mass selection mechanism the first problem that arises is the need to describe the movement of a body, in this case the sphere. It is necessary to introduce the physical concepts of: observer, reference system, position, speed and acceleration, among others; that is, the kinematics of a particle.

The application of the theoretical model of the movement of a projectile allows to predict that the horizontal distance to which the sphere falls from the launch point is given by:

$$D = v_0 \cdot \sqrt{\frac{2H}{g}} \qquad (1)$$

Where H, v_0 and g are the height of the launch, the initial velocity of the sphere, and the acceleration of gravity, respectively.

With respect to this result, the only reason that two different mass spheres launched from the same point can impact at two different positions is because of the difference between their initial velocities. Therefore, the different speeds that they acquire only depend on their masses. Then, the initial velocity of the sphere is a function of its mass: $v_0 = v_0(m)$.

At this stage the urgent need of introducing the concept of mass appears. It should be noted that a mass cannot reach, on its own, to a greater or lesser speed. The action of one or more external agents is required to cause any change in velocity. This, then, leads to the introduction of the concept of force and the laws of Newton, i.e., the study of dynamics. However, it should be remembered that in the above case the mass only affects the initial velocity of the projectile but not its subsequent movement; this will be discussed later.

At this stage, it is important to introduce the mass launcher mechanism. This consists of a smooth tube in which there is a spring with an elastic constant *k* whose natural length reaches the point where the sphere leaves the mass launcher. The spring is compressed into a deformation $\Delta x = L$ and when released moves the mass *m* a length *L* from A to B, as shown in Figure 2.

IOP Conf. Series: Journal of Physics: Conf. Series **1043** (2018) 012055 doi:10.1088/1742-6596/1043/1/012055

The net force on the sphere is due only to the restoring force of the spring which is a variable force. However, in a first tentative approach to solve the problem, the force will be considered constant and equal to its initial value at A (see footnote) [1].

With this assumption, the acceleration is given by:

$$a = \frac{k\,\Delta x}{m} \tag{2}$$

and applying the equations of constant acceleration, the speed with which the sphere comes out of the tube is:

$$v_{salida} = \sqrt{\frac{2k\,\Delta x}{m}L} = v_0(m) \tag{3}$$

Recalling that $\Delta x = L$ results in:

$$v_0(m) = \sqrt{\frac{2k}{m}}L \tag{4}$$

with which the dependence of the initial velocity, as a function of mass, and hence the expression for the horizontal range D, is determined by the equation (1). By substituting (4) into (1) we can see the kinematics and dynamics articulated since D(k,m,L,H,g) contains elements of both but only roughly.

However, it is clear that there is an erroneous consideration in the previous application, which facilitates discussion, reinforces and makes note of the differences between a variable force, such as that manifested in an ideal spring, and the constant forces usually studied up until this stage.

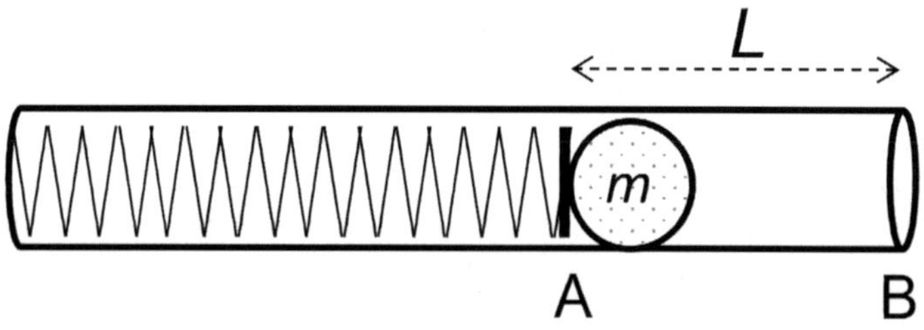

Figure 2. Mass launcher schematic representation. The points A and B correspond to the points of maximum compression and null elongation of the spring respectively.

This difficulty, when working with variable forces, gives reason for noting the need to reformulate Newton's second law, which is discussed later, introducing the physical concepts of work and mechanical energy, without which it is very difficult to solve the problem. Once work, mechanical energy and its conservation have been studied, it is possible to apply the theoretical model and perform the correct calculation of the exit speed of the sphere.

Applying the conservation of energy, with the approximation of disregarding friction, results in the corrected initial velocity of the sphere:

$$v'_{salida} = v'_0(m) = \sqrt{\frac{k}{m}} \cdot L \tag{5}$$

[1] This consideration is a fairly common mistake for a student who has only worked with constant forces; however, it can also be induced by the mathematical limitations of their level of training.

which differs from the result previously obtained. Finally, by substituting (5) into (1) we can see the kinematics, dynamics and energy articulated, now allowing for an exact solution to the problem.

From the correction made, it can be argued and emphasized that the variable forces are not applied as the constant forces and when not properly considered, the output velocity is overestimated by a factor of $\sqrt{2}$. This demonstrates the need and utility of formulating new models with different potentialities such as energy.

In this way, it is possible to present the contents and the physical principles underlying the operation of the proposed device, by means of an approach which helps develop an integrated learning focused on a single problem in all its aspects indicated and discussed above.

4. Discussions

The mass of the particle in the movement of a projectile or kinematics is usually ignored. However for the inquisitive student it is conflicting to accept that the mass of the object does not affect the movement of the body, since the experience of throwing an object suggests that with a larger mass a smaller reach is achieved. Kinematics does not manage to explain this to the student, but the texts at this level do not clarify it either. However, through the approach presented this behavior emerges clearly and at the same time naturally through equations (5) and (1). Although the mass does not affect the movement during its flight, it only affects the initial velocity and therefore the horizontal range. In the studied system the launch velocity is dependent on the mass of the particle and its effect is seen in the achieved reach. The approach considered does not avoid this reality, since it makes imperative to consider the mass, making it clear that it only intervenes in a stage of the movement, which does not generate any conflict with everyday life.

It is also immediately appreciated that the same force (mass launcher) produces different effects (v_0) on different masses. This forces the immediate joint (articulation) between the kinematic and Newton's laws.

In Figure 3 are shown results from experimental measurements under the same conditions done by some students, as a way to illustrate what would be expected from measurements made by all physics students. For this, a set of spheres with masses concentrated in two regions was considered. As can be seen, those with the largest mass obtain lower (initial) output velocity as opposed to the smaller ones. As a consequence of this, the larger masses achieve a lower (horizontal) range, as can be seen in the upper box. On the other hand the experimental range is compared with the theoretically expected value in the lower box and the adjusted straight line shown has slope 0.94. In an ideal situation (hypothetical) the line should have slope 1.00, however, the obtained result is quite clear in showing that there is an expected correspondence between the theoretical and experimental values. This should be sufficient to illustrate the usefulness (validity) of the theory and the performance of the apparatus as a mass separator.

On the other hand, the approach given to the problem allows to clarify a very common problem in variable forces that is the confusion between position, displacement and deformation, noting that this can lead to erroneous conclusions similar to the one observed in the relations (2) and (3).

Finally, it should be noted that in the usual transition to the concepts of work and energy, these usually arise as new topics unrelated to Newton's laws except to define the concepts of kinetic and potential energy (work-energy theorem). However, the beauty of the approach allows them to be introduced in a natural, useful and necessary way to overcome the difficulty imposed by the variable forces, thus articulating the dynamics with energy and work, allowing a different view of the problem through tools with other potentialities and limitations with respect to Newton's laws.

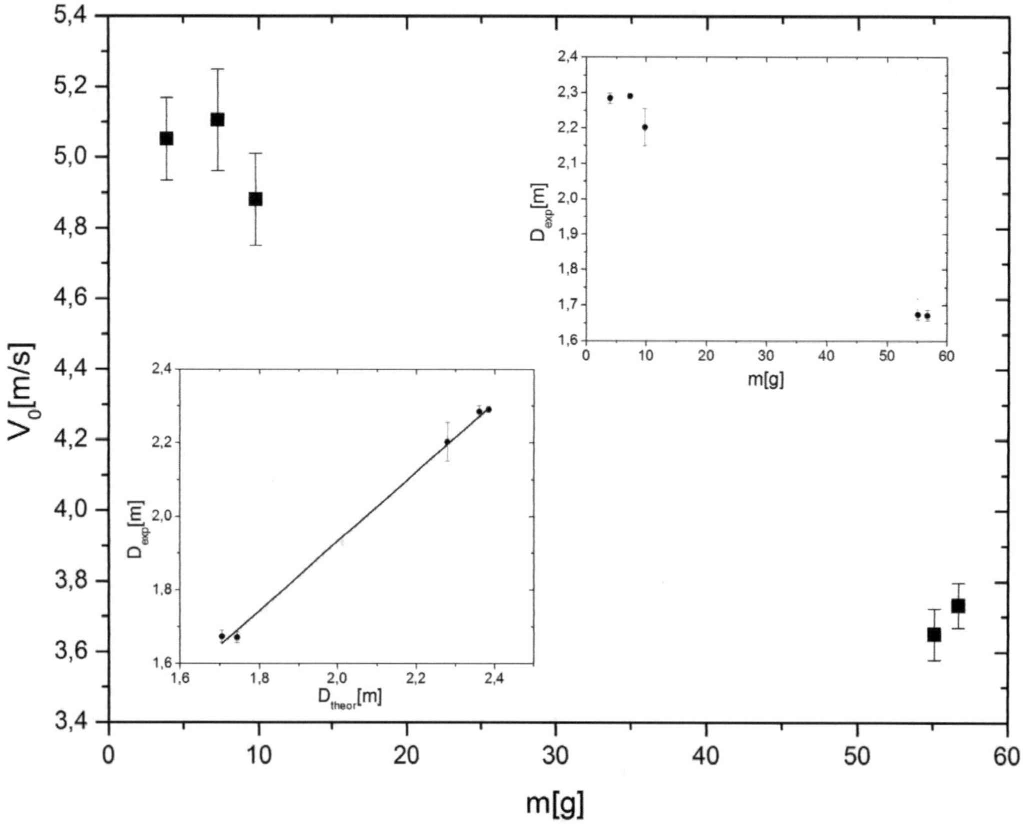

Figure 3. Output velocity as a function of the particle mass. Upper inset: Reach achieved as function on mass and, lower inset: Comparison between experimental and theoretically predicted range (only item usually studied).

5. Conclusions

Through a visual problem, real, simple, but also challenging for the student, it is possible on the one hand to articulate the contents and activities of the subject; and on the other hand establishes a heading, central course, from beginning to end, during the development of the subject. This allows us to conceive an integral vision of Mechanics and specially avoid getting lost in the contents and the Mathematics.

The presented approach allows to reflect and to go beyond the second law of Newton with constant force, articulating it with the concepts of energy as an alternative formulation for the study of the movement. This is a valuable aspect since a progressive transition is made from a concrete (problem) object to one of the most successful abstractions of physics and at the same time one of the most difficult concepts to understand, that is energy.

Finally, the approach facilitates the student, to follow and evaluate the different stages and tools that are required when constructing a model or a physical theory of something concrete, and at the same time allows him to confirm that a physical model, in spite of the simplifications, is useful in a real problem and transcends its limitations (especially crucial in a first course of abstraction and/or physical modeling).

Regarding the impact of the Integrating Approach (IA) on student learning, research will be carried out and reported in a timely manner.

Acknowledgments
The author gives special thanks to Professors Gonzalo Fuster Roa, Nicolás Porras Rojas and Roberto Rojas Vega, collaborators with a long history in teaching subjects from the Department of Physics of the Universidad Técnica Federico Santa María, for: a) to have corrected and suggested important modifications presented in this article; b) to be great managers of the initiation and conclusion of the author in the investigation in this matter; and c) to have generously provided material requirements and their knowledge. My students have been, over the years, my best systems of study, learning as well as inspiration, for which I thank you for their willingness. The work was partially funded through the INCUBA project of the Faculty of Natural Sciences UDA.

References
[1] Young H D, Freedman R A and Ford A L 2008 *Sears and Zemansky's University Physics with Modern Physics* 12th ed, (San Francisco: Pearson Education, Inc.)
[2] Hammer D 1994 Epistemological beliefs in introductory physics *Cogn. Instruct.* **12** 151-83
[3] Redish E F, Steinberg R N and Saul J M 1997 The Distribution and Change of Student Expectations in Introductory Physics (in The Changing Role of Physics Departments in Modern Universities) *AIP Conf. Proc.* **399** 689-98
[4] Redish E F, Steinberg R N and Saul J M 1998 Student Expectations in an Introductory Physics *Am. J. Phys.* **66** 212-24
[5] Adams W K, Perkins K K, Podolefsky N S, Dubson M, Finkelstein N D and Wieman C E 2006 A new instrument for measuring student beliefs about physics and learning physics: The Colorado Learning Attitudes about Science Survey *Phys. Rev. ST Phys. Educ. Res.* **2** 010101
[6] Sahin M 2009 Exploring university students' expectations about physics and physics learning in a problem-based learning context *Eurasia J. Math. Sci. Technol. Educ.* **5**(4) 321–33

Correct answers with wrong justifications? Analysis of explanations in classical mechanics with FCI test

Carla Hernández[1] and Silvia Tecpan[1]
[1] Physics Department, Universidad de Santiago, Chile

carla.hernandez.s@usach.cl

Abstract. This study presents the results of an analysis of the explanations constructed by future physics teachers about their chosen answers in a conceptual test in classic mechanics. Additionally, to selecting a multiple-choice answer in the Force Conceptual Inventory test [1] the teachers were required to give an explanation for their answers in order to determine the relationship between their level of conceptual knowledge and the quality of the explanations they gave using this knowledge. The results suggest that the 28 participants have difficulties in constructing adequate physical explanations. They suggest too that they have deficiencies in the development of their scientific thinking skills and in the use of their prior knowledge. We propose indicators for the construction of explanations in physics and we discuss the implications for the initial formation of physics teachers.

1. Introduction

Over the past two decades, various standardized tests have been developed with the intention of quantitatively measuring the level of disciplinary knowledge acquired by university and school students in a Physics course. Using a methodological design of an experimental type, several studies have compared the knowledge acquired by students after instruction, which is usually based on some methodology of the active learning, i.e., non-expository [2,3] establishing normalized gain percentages to determine the results [4].

In the case of classical mechanics content, one of the most used test has been the Force Conceptual Inventory (FCI) [1]. Although this test have served as a learning indicator in basic physics courses for engineering students [5,6], there have been few investigations into the in service teachers' content learning and even fewer into that of pre-service physics teachers [7].

On the other hand, although the quantitative result is important to consider as evidence of the impact that the new teaching methodologies have on the learning of physics, this article focuses on the justifications that the students gave to their selected answers in the test. Below are shown the theoretical foundations that justify this study, followed by an explanation of the methodology of analysis performed, and finally the results and pedagogical implications are described.

2. Usefulness of FCI test

Internationally there is a consensus on the usefulness of the Force Concept Inventory as a tool to measure conceptual knowledge in classical mechanics [4,8]. And it is that the content associated with the concepts of force and movement is often difficult to learn for students from the rigorous point of view of science, since the interaction of each subject with the world around him generates a series of preconceived ideas on the movement of objects that are difficult to change with formal instruction.

Student's ideas about force and movement have been extensively studied [9] and are known to condition the way they generate explanations for the phenomena of nature.

The inventory includes the most common misconceptions that students present in different contexts, which have been determined through research processes related to: kinematics, impetus, active force, action/reaction pairs, concatenation of influences, other influences on motion like resistance and gravity. Several misconceptions present in the inventory are based on poorly differentiated concepts and function properly as distractors that discriminate the level of knowledge among students who have succeeded in overcoming them from those who do not.

However, there is debate on the implications of obtaining a high score in the FCI [10,11]. Studies have reported the discrepancies between the quantitative results obtained with the known Conceptual Physical Inventories, and explanations given by students to answer questions raised them [12,13].

In this way, a study by [13], where engineering students had to justify the answers chosen in the test, determined that a low percentage of students could develop the so-called "Newtonian thinking", having difficulties applying concepts consistently in their explanations. In the case of teachers, these difficulties are also present and affect their educational practices [14].

3. Importance of explanations in science

In the last two decades, it has been recognized in the sciences teaching the importance of considering the argumentation and explanations building as a central scientific activity that students must learn [15,16]. Consistently, there are numerous researches about the characteristics of student's scientific explanations [15,17,18].

In words of [16], *"constructing explanations requires both a conceptual understanding of relevant theories and their application to a specific problem and the epistemic understanding of the criteria for a good explanation"* (pp.32).

For its part, [19] highlight three important elements within an explanation:

a) Claim. Is a statement that responds directly to the question posed.
b) Evidence. Explaining scientific phenomena, students should be able to collect, select and use the data as evidence to support their claims.
c) Reasoning. It is the logic that allows to link the evidence with the scientific principles involved, to support their claims. Usually, students have more difficulty with the reasoning component of scientific explanations than with gathering evidence[19].

Although scientific explanations are important for science, they are often omitted in practice in the classroom [20, 21]. Moreover, when students must formulate an explanation or construct an argument, often they struggle to articulate and justify their claims [22].

However, several studies on the subject have suggested that this ability is not natural for most individuals, but is learned through practice [23]. To support students in building explanations and arguments have been used Toulmin model of argumentation (1958) [16,23,24].

Even so, student's understanding of content may also influence their ability to effectively use the evidence in their explanations. Nevertheless, understanding the scientific content itself is not a sufficient condition for properly reasoning a scientific explanation [22].

4. Methodology

A qualitative study was carried out with exploratory scope and survey design. The instrument used was the Force Conceptual Inventory test [1] and applied to 28 pre-service physics teachers who had already approved the course of classical mechanics in a Chilean state university. The revised version of the FCI of 30 items was used [10] and translated into Spanish by [25], where for each item response it was requested to write an explanation regarding the situation of the question to justify and verify whether the choice of each option was accompanied by an appropriate reasoning. This is to ensure that the student performed an exercise of analysis of the situation presented when answering each question.

Explanations built by students were analyzed by content [26] which allowed, on one side, to identify which are the concepts most used to explain the different situations, and other, how they relate these concepts to existing evidence to generate explanations consistent with theory and/or corresponding laws [27]. Given the characteristics of the study, should be mentioned that the students were not previously instructed on how to construct an explanation or in its characteristics according to the literature.

Based on the proposal of [28] the following analysis criteria were established to determine whether an explanation was considered correct or not:

1) Use the context of the question to extract the evidence and data that, linked to theory, support a claim
2) Establishes logical relationships between concepts and phenomena involved and the evidence, consistent with theories and/or laws of physics
3) Incorporates all the theoretical concepts involved accurately
4) Considers and establishes the limitations of validity of the theory involved and/or law of physics involved
5) Uses adequate and rigorous scientific language

However, it must be considered that the students do not have additional information or evidence than those that can be deduced from the situation presented. Furthermore, the choice item response in the FCI corresponds to what [19] consider the statement or conclusion directly answer the question.

Based on the proposed criteria that were validated by construct and content [29] with a group of university physics teachers, the explanations were classified into three categories:

- Correct: Those that meet all the proposed criteria.
- Incomplete: When you meet certain criteria, regardless of how many and which are. In this case, it is considered that the explanations are not necessarily incorrect, but could be improved upon the criteria already fulfilled.
- Wrong: In case you do not meet all the criteria suggested.

5. Results

Statistical analysis of student's performance was performed in the complete test, on a scale of 1 to 100 (Table 1).

Table 1. Descriptive statistics of the performance in the test

Statistic	Percentage (%)
Average	73
Median	73
Standard deviation	16
Mode	67
Range	70
Minimum	30
Maximum	100

Analysis of the frequency of correct responses for each item were also made. Graphic 1 shows the results obtained in the application of FCI.

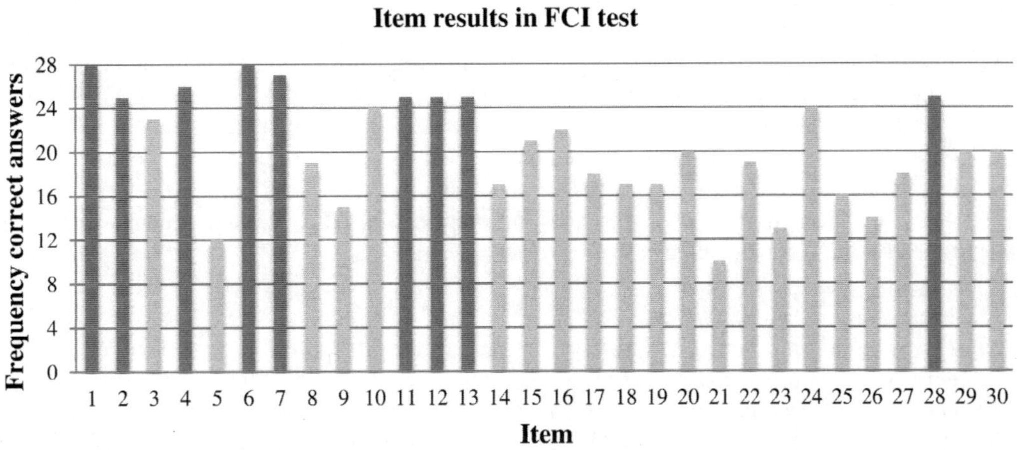

Figure 1. Graph representing the results of students in the FCI by item. The bars represent student performance for each of the 30 questions. Dark bars are those with 25 or more correct answers.

Of total items was established as a cut-off criterion to analyze the explanations, those where at least 25 students selected the correct option in each item, representing an 83% efficiency in answering each. This decision is based on the hypothesis that a greater number of correct answers to the item could imply a greater number of correct explanations. Applying the defined categories to the students' explanations, we identified differences between the number of correct answers obtained with the test [quantitative result] and the number of explanations considered correct (qualitative results). Table 2 shows the contrast of these results.

Table 2. Classification of the explanations for each item that exceeds the cut-off criterion for the analysis.

	Item								
	Q1	Q2	Q4	Q6	Q7	Q11	Q12	Q13	Q28
Frequency of correct answers to the item [n]	28	25	26	28	27	25	25	25	25
Correct explanation	0	0	0	1	0	0	0	1	0
Incomplete explanation	1	3	2	7	6	5	3	2	4
Wrong explanation	27	22	23	20	19	20	22	22	21

6. Analysis and discussion

Details of the analysis performed on the explanations literals constructed by the students for two questions of the FCI are presented below, by way of example. On one side, the question 1 [Q1] where all students (S) selected the right choice in the test and yet none of the explanations satisfies all of the proposed criteria.

Q1. *Two metal balls are the same size, but one weighs twice as much as the other. The balls are dropped from top of a two-story building at the same instant of time. The time it takes the balls to reach the ground below will be: [...]*

In this case, the only explanation classified as incomplete was:

IOP Conf. Series: Journal of Physics: Conf. Series **1043** (2018) 012056 doi:10.1088/1742-6596/1043/1/012056

S3: "free-fall time is not dependent on the mass when located in the same space [gravity same]"

As explanation, according to the criteria, consideration of two physical concepts mainly "time" and "mass" is appreciated, establishing a relationship of independence between them that referred to the use of a theory and an adequate reasoning. In addition, a limitation of the theory is included to mention that this is valid if it occurs "in the same space".

However, it does not meet the criteria of using appropriate scientific language, since it refers to the concept of "gravity" and not "gravitational acceleration". Moreover, no reference to the context of the question, given by the fall of two metal spheres, in particular.

The other 27 explanations given for this question were classified as wrong, because they do not meet any of the criteria. Here are some cases:

S1: "The mass is negligible"

S10: "Because they have the same size"

S11: Bodies subject to the same acceleration, Galileo movement studies

S14: The mass does not depend on the time

S27: Sum of the forces on the Y axis equals zero

In general, conceptual difficulties related to the presence of misconceptions, consistent with Section II of this article is appreciated. In the case of S14, a difficulty related to the distinction between dependent and independent variables is observed, which is repeated in other students.

Regarding the criteria, explanations do not consider the evidence given by the context of the question nor the limitations of the theory involved. Nor are relations established between concepts such as "mass", "free-fall time objects" and "gravitational acceleration".

Table 3 shows the diversity of concepts used by students in their explanations where "mass" is distinguished as the most used concept. Table 4 shows the relations established between some of them.

Table 3. Frequency [*f*] of most used concepts in the explanations for Q1

Concepts	*f*
Mass	19
Time	5
Acceleration	6
Gravity	3
Movement	2
Size	1
Friction	1
Gravitational field	1
Weight	1
Gravitational acceleration	1
Velocity	1

Table 4. Frequency [*f*] of most used relationship between concepts in the explanations for Q1

Relationship between concepts		***f***
Mass	- Time	5
	- Acceleration/gravitational acceleration	4
	- Movement	1
	- Velocity	1
Gravity – Friction		1

The diversity of concepts used suggests confusion about the content since for the same question (Q1), students build various explanations. Besides, lack of rigor is distinguished in the use of vocabulary, for example, only one student mentioned "gravitational acceleration" while the rest only refers to the concept "acceleration". Similarly, the use of the concept "gravity" to refer to the "force of gravity" or to "gravitational acceleration" is confusing.

On the other hand, there is the case of Q13 where 25 students marked the correct answer option, but only one (S17) could to construct an adequate explanation for it according to the established criteria.

Q13: A boy throws a steel ball straight up. Disregarding any effects of air resistance, the force (s) acting on the ball until it returns to the ground is [are]...

The correct answer to this question, and therefore, the claim of the explanation is … *a constant downward force of gravity only.* When requesting an explanation for this answer, the only one classified as correct was as following:

S19: "There is no other force acting on the ball since there is no friction and the ball is within the gravitational field of the Earth that is uniform"

In this case, the explanation satisfies the criteria set because it includes the necessary concepts, uses extracted evidence from the context of the question, also considers the limitations of physical theory indicating that the phenomenon is valid just near Earth and considering the uniformity of the gravitational field. From the point of view of theory, the explanation is adequate and indicates a right use of the language when using the corresponding terms without confusion.

On the other hand, the explanations classified as incomplete were the following:

S10: "Only the gravitational field interacts upon the object"

S17: *"The child gives an impulse to the ball and then only affects the weight during the fall as it is on Earth"*

In the case of S10, although it alludes to the gravitational field, the language used is not appropriate since it mentions that it acts "upon" the object, which is incorrect from a theoretical point of view. Moreover, it does not refer to the context since it mentions any object and does not establish limitations of validity of the theory that supports the explanation.

Among the explanations considered incorrect for Q13 are:

S2: "Gravity only depends on the falling objects"

S3: "Only gravity acts at that moment"

S14: "Action and reaction force"

S18: "The hand strength is not maintained"

These responses generally do not mention limits of validity to the theory, not use evidence or explanation to contextualize the case of question where the object was a steel ball. Particularly in the case of S3, the use of the term "gravity" is inappropriate and should refer to the weight. In this sense, the language is not precise and to understand the meaning of the answer requires an interpretation that is in the opinion of the reader.

On the other hand, in the S2 response, a confusion in the determination of dependent and independent variables is again evident.

For S14, the use of Newton's third law, known as "action and reaction" would be inappropriate because the laws are descriptive but explanatory, therefore, not enough to sustain an explanation [27]. Finally, in the case of S18, also to not meeting the criteria is using the concept of "hand strength" is incorrect.

Table 5 shows the concepts most used by students to explain the answers and the relationship between concepts for Q13.

Table 5. Frequency [*f*] of concepts most used in the explanations for Q13

Concepts	*f*
Force	6
Weight	6
Gravity	5
Movement	3
Gravitational Field	2
Gravitational Acceleration	2
Gravitation	1
Action and reaction force	1
Impulse	1
Hand strength	1
Inertia	1

Table 6. Frequency [*f*] of most used relationship between concepts in the explanations for Q13

Relationship between concepts		*f*
Weigth	- Force	1
	- Impulse	1
	- Movement	1
Movement - Gravitational acceleration - inertia		1
Force – Gravity		1

As in Q1, and all the questions whose answers were analyzed, the various relationships established by students between the concepts, offer diverse ways of relating the evidence provided by the question and previously learned theory. This is indicative that they have made distinct types of logical reasoning to construct their explanations suggesting the existence or persistence of misconceptions in future teachers.

In general, the results show an inconsistency between the quantitative results obtained with the test, which indicates a level of knowledge consistent with a Newtonian thinking in students [13], and the result of the qualitative analysis carried out on the explanations which suggests important conceptual confusion.

Moreover, regarding the proposed criteria for the analysis, the lack of consideration to the context of the questions and the absence of limitations for theories and/or laws used to explain, were the most frequent absences built explanations for all questions discussed.

7. Conclusions

In this paper, results of an application of the FCI test [1] with two variants have been presented. On one side, it was applied to pre-service physics teachers in the context of a teaching course, where the objective was to learn to teach classical mechanics, and other, was requested to explain each of the answers selected in the test.

Regarding the hypothesis, it is rejected because a high score in the test has not involved, in this case, a high number of correct explanations under the proposed criteria. The results suggest that the ability to explain is incipiently developed, which can happen because few spaces in the classroom are provided for pre-service teachers learn to build their own explanations. This is indicative of weaknesses in the level of reasoning and low understanding of the content by the participants, in coherence with other studies where these factors have been positively correlated [19].

In this line, it is proposed to use conceptual inventories as tools to promote the construction of scientific explanations, requesting to justify each response to the item and later to establish criteria for analysis of explanations such as those proposed in this study. The analysis of results in this type of interventions requires mixed methods to deepen qualitative results complementary to the quantitative ones.

Finally, the implemented strategy allows orienting training actions focused on the development of skills related to reasoning and the construction of scientific explanations, which are exercised infrequently in classes. Given the specific characteristics of the sample, it is not possible to generalize the results. It is suggested to replicate the methodology with a probabilistic sample, as well as to extend this type of studies to other areas of teacher education in science to strengthen the criteria for analysis of explanations.

Acknowledgment

The authors acknowledge financing from the Universidad de Santiago de Chile, through the projects DICYT 041531HS and Basal USA 1555.

References

[1] Hestenes D, Wells M and Swackhamer G 1992 Force concept inventory *Phys Teach* **30**141

[2] Meltzer DE and Thornton RK 2012 Resource Letter ALIP–1: Active-Learning Instruction in Physics *Am J Phys* **80** 478.

[3] Redish EF 2003 *Teaching Physics with the Physics Suite*. (Somerset, NJ: John Wiley & Sons)

[4] Hake RR 1998 Interactive-engagement versus traditional methods: A six-thousand-student survey of mechanics test data for introductory physics courses *Am J Phys* **66** 64

[5] Martin-Blas T, Seidel L, Serrano-Fernandez A 2010 Enhancing Force Concept Inventory diagnostics to identify dominant misconceptions in first-year engineering physics *Eur J Eng Educ*. **35** 597–606.

[6] McKagan SB, Perkins KK, Wieman CE. Reforming a large lecture modern physics course for engineering majors using a PER-based design. *AIP Conf. Proc.* 2007 ed L McCullough, L Hsu and P Heron, 34–7
[7] Zavala G, Alarcón H and Benegas J 2010 A professional development course with an introduction of models and modeling science eds R Lesh, PL Galbraith, CR Haines and A Hurford *Modeling Students' Mathematical Modeling Competencies* (NY: Springer) pp 491–500.
[8] Fagen AP, Crouch CH and Mazur E 2002 Peer instruction: Results from a range of classrooms *Phys Teach* **40** 206–9
[9] McDermott LC, Shaffer PS and Somers MD 1994 Research as a guide for teaching introductory mechanics: An illustration in the context of the Atwood's machine *Am J Phys.* **62** 46–55
[10] Hestenes D and Halloun I 1995 Interpreting the force concept inventory: A response to March 1995 critique by Huffman and Heller *Phys Teach.* **33** 502.
[11] Huffman D and Heller P 1995 What does the force concept inventory actually measure? *Phys Teach.* **33** 138–43.
[12] Wood AK, Galloway RK, Donnelly R and Hardy J 2016 Characterizing interactive engagement activities in a flipped introductory physics class. *Phys Rev Phys Educ Res.***12** 10140
[13] Corona A 2010 Opciones Newtonianas de estudiantes no-Newtonianos, análisis de alumnos Universitarios: FCI. *Latin-American J Phys Educ.*; **4** 422–8.
[14] Fernández M, Perez R, Peña S and Mercado S 2008 Concepciones de maestros de primaria sobre el planeta Tierra y gravedad. Implicaciones en la enseñanza de la ciencia. *Rev Electrónica Investig Educ.***10** 1–25.
[15] Driver R, Newton P, and Osborne J 2000 Establishing the norms of scientific argumentation in classrooms. *Sci Educ* **84** 287–312
[16] Sandoval WA and Millwood KA 2005 The quality of students use of evidence in written scientific explanations. *Cogn Instr.* **23** 23–55.
[17] Kelly GJ and Takao A 2002 Epistemic levels in argument: An analysis of university oceanography students' use of evidence in writing. *Sci Educ.* **86** 314–42
[18] Toth EE, Suthers DD and Lesgold AM 2002 Mapping to know: The effects of representational guidance and reflective assessment on scientific inquiry. *Sci Educ.* **86** 264–86.
[19] McNeill KL, Lizotte DJ, Krajcik J, and Marx RW 2006 Supporting students' construction of scientific explanations by fading scaffolds in instructional materials. *J Learn Sci.*; **15** 153–91.
[20] Kuhn D 1993 Science as argument: Implications for teaching and learning scientific thinking. *Sci Educ.***77** 319–37.
[21] Newton P, Driver R and Osborne J 1999 The place of argumentation in the pedagogy of school science. *Int J Sci Educ.* **21** 553–76.
[22] Sadler TD 2004 Informal reasoning regarding socioscientific issues: A critical review of research. *J Res Sci Teach.* **41** 513–36
[23] Osborne J, Erduran S, and Simon S 2004 Enhancing the quality of argumentation in school science. *J Res Sci Teach.* **41** 994–1020
[24] Jimenez-Aleixandre MP, Rodriguez AB and Duschl RA 2000 Doing the lesson "or" doing science": Argument in high school genetics. *Sci Educ.* **84** 757–92.
[25] Macia-Barber E, Hernández M V and Menéndez J 1995 Cuestionario sobre el concepto de fuerza.
[26] Merriam SB 2009 *Qualitative Research. A guide to design and implementation* (San Francisco, CA: Jossey Bass)
[27] Abd-El-Khalick F 2012 Examining the sources for our understandings about science: Enduring conflations and critical issues in research on nature of science in science education. *Int J Sci Educ.***34** 353–74.
[28] McNeill KL and Krajcik J 2009 Synergy between teacher practices and curricular scaffolds to support students in using domain-specific and domain-general knowledge in writing arguments to explain phenomena. *J Learn Sci.* **18** 416–60

[29] Wiersma W 1986 *Research Methods in Education: an Introduction.* (Boston, MA, EE.UU: Allyn and Bacon)

IOP Conf. Series: Journal of Physics: Conf. Series **1043** (2018) 012057 doi:10.1088/1742-6596/1043/1/012057

Aerodynamic test with self-propelled scale model

Freddy Faúndez

Academia de Ciencias Aeronáuticas. Universidad Técnica Federico Santa María.
Av. Santa María 6400. Santiago, Chile
freddy.faundez@usm.cl; freddy.faundez@venti.cl

Abstract. This research presents a new experimental technique for the testing prototypes of aerodynamic devices of heavy freight vehicles. This technique allows the evaluation and recording of aerodynamic variables, being an alternative to the experimental limitations present in a wind tunnel, or the complexity and costs associated to the real-scale experimental test. For this new methodology a radio-controlled reduced-scale model is used, equipped with different instruments for recording aerodynamic variables. The experiment was performed at the runway in the Chilean Air Force Base in Santiago, at a speed of 90 [km/hr]. The results demonstrate that this technique allows recreating more realistic conditions of the aerodynamic phenomena, obtaining reliable values for the aerodynamic drag, at a low cost.

1. Introduction

Today, the effects of global warming have driven the development and introduction of energy efficiency policies aimed at reducing pollutant emissions in the industry, whether by ecological incentives or by multinational agreements that seek to stop global warming. This, coupled with the variability of fossil fuel prices, pushes freight transport to develop new ways to reduce fuel consumption. Of the technologies available to maximize the use of diesel in transportation, the use of aerodynamics stands out for its simplicity, quick installation and low impact in its operations.

Research carried out in Chile [1] shows that the incorporation of aerodynamic technologies allows a reduction in fuel use of up to 15.34%, making it one of the most promising ways to reduce its energy consumption in this industrial sector. However, the use of aerodynamics devices requires a thorough knowledge and understanding of the fluid movement of air around a moving vehicle, so the design of these aerodynamic devices is closely linked to the applications of fluid equations and their turbulent phenomena, or the development of meticulous experimental trials.

For the design of an aerodynamic device there are three possible options: the use of numerical models, the scale model test or the real-scale evaluations. Each of them has strengths and weaknesses that can limit their effectiveness. In the case of scale trials the main problem is the size of the model with which it will be worked, which is closely linked to the size of the wind tunnel to be used.

Considering the wind tunnel tests performed in different international laboratories [2], it is seen that in order to achieve a dynamic similarity in which the three main parameters: the Prandtl, Reynolds and Mach numbers, But for these parameters to be in optimal ranges, a larger laboratory is needed than those available in Chile.

On the other hand, the use of Computational Fluid Dynamics (CFD) allows the modeling of a large number of different geometries and configurations, but the resolution and accuracy of its results is conditioned by the ability of the computer to use (FLOPS), and same time to apply a reductionist criterion of variables in order to guarantee a convergent result, which eliminates certain turbulent

phenomena that are highly complex to simulate, such as the rotating movement of the wheels of the vehicle and the displacement of the floor. Finally, the last type of test is the real-scale tests, which because of their high costs allows the evaluation of a very small number of prototypes.
Due to the limitations of each type of test, a new experimental aerodynamic evaluation technique is proposed, using a model of a transport vehicle, self-propelled and 1:6 scale of the original called GRSV (Ground Research Scale Vehicle), thus combining the different advantages of a full-scale test and the use of scaled-down models to obtain an appropriate phenomenological similarity to evaluate different aerodynamic devices.

2. Objetives

The objectives of the present research work are to evaluate the effectiveness of this methodology, considering the size and the speed of the GRSV. The specific objectives are:

- Implement and evaluate the technical feasibility for performing aerodynamic scale tests with self-propelled vehicles.

- To determine the effectiveness of the new experimental procedure, comparing the results with the real scale test, wind tunnel experiment and CFD analysis.

3. Metodology

The calculation methodology to be used to obtain the aerodynamic drag of the GRSV will be the Energy Conservation Method [3]. In addition, two control points will be used; comparing the static pressure measurements with the results obtained using CFD.
The first stage of the project was the realization of several computational simulations, in order to evaluate the sensitivity of the instruments to be built and installed on the GRSV and the magnitude of the aerodynamic forces involved in this experiment, considering the Reynolds number in each case.

Table 1. Reynolds number in each study.

Type of analysis	Re. Number
CFD	805.000
GRSV 25[m/s]	805.000
GRSV 30[m/s]	1.006.000
Real Truck Scale Model	4.360.000

3.1. CFD Simulation

Several fluid dynamics simulations were performed with the digital model of the GRSV test vehicle, using ANSYS software with different air flow velocities.
In order to optimize computational capacities, we used the mean symmetry configuration of the three-dimensional model, with a total of 6,486,766 tetrahedral elements in the mesh, taking care of the size relation between elements, its geometric distortion and distribution around the study model and domain.
The simulation was performed in a steady formulation, with gravitational effects and first-class Dirichlet-type boundary conditions, without Poiseuille profile in the velocity function [5].
The upper and lateral limits of the domain were considered without friction, whereas for the lower limit it was considered as a real rough wall with displacement of the same speed and direction as the air flow. Furthermore, a standard isotropic turbulence of 0.1% was established for the entire domain [6].

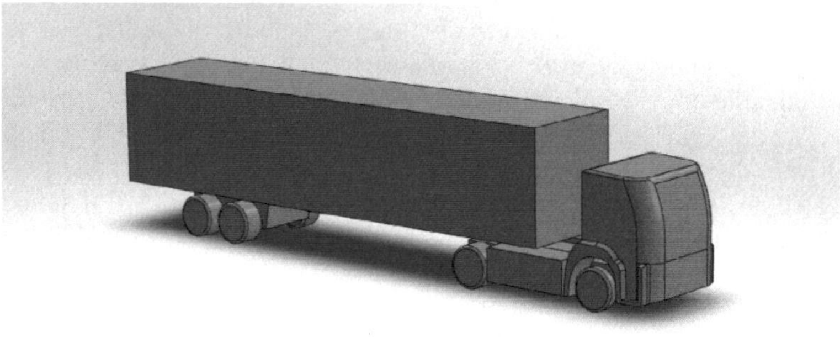

Figure 1. Three dimensional model of GRSV.

The air was considered as ideal gas without temperature variation and with one atmosphere of absolute pressure, while for the exit of the domain was configured with barometric pressure equal to zero.
The mathematical model used to calculate the turbulence was the two equations k-e model, with standard values for turbulence constants and kinetic energy [4].

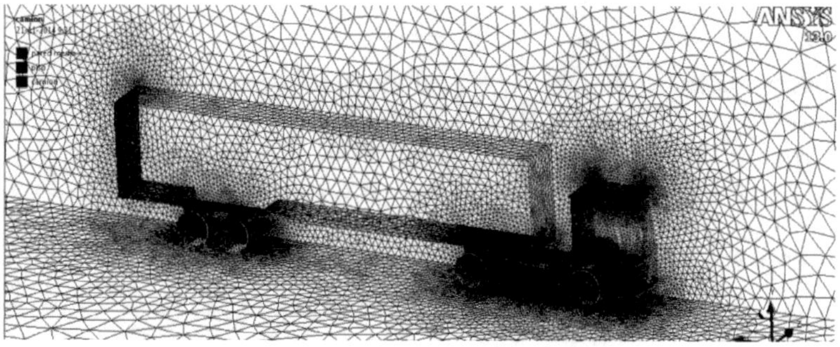

Figure 2. Mesh symmetric configuration.

For an interpolation that ensures numerical stability, second order models were implemented for all convergence criteria, while the algorithm to solve the iterative process was the SIMPLE method (Semi Implicit Method for Pressure Linked Equations). The results were verified by numerical convergence and the mass balance between the inflow and outflow zone of the domain.

Figure 3. Pressure distribution on GRSV.

Figure 4. Distribution of turbulent kinetic energy around GRSV.

At this stage, the two control points were chosen for the reading and comparison of the static pressure between the simulation and the scale test. These two points are located in two areas where robust pressures are measured: In the center of the front face (called P1) and the other in the back face (P2), in order to capture the influence of the formation of the boundary layer [7] between the floor and the bottom of the GRSV.

Figure 5. Control points for static pressure measurement: P1 (left) in the front and P2 (right) in rear zone.

The results obtained at this stage indicate that the aerodynamic drag force is approximately 190 [N], a magnitude that can be detected by decelerating GRSV at a test speed of more than 15 [m/s].

4. Experimental Development

The scale model of the truck is used to design and build the GRSV is an average geometric design characteristic of a configuration of tract and semitrailer, with a short cabin and two tractor axles. For the trailer, an 18-foot format was chosen with two rear axles, while all components and parts of the trailer remained the same 1:6 scale. The cabin was built with composite materials, while for the trailer was used high strength aluminum and wood. The constructive design of the GRSV allows to install numerous instruments and sensors within it, up to a maximum of 20 kg of equipment in a safe way.

Figure 6. Final aspect of GRSV in runway.

In order to achieve the objectives proposed in this research, the vehicle was equipped with a long-range radio-control system and a 10 Hp electric motor with high-performance LiPo batteries, and to guarantee the uniformity of its displacement, it was equipped with a inertial electronic system to maintain the vehicle's steering at high speeds. Additionally, a camera was installed in the upper part of the cabin that transmits in real time to the operator, facilitating its safe driving over long distances. Within the GRSV were installed three accelerometers, a GPS and 30 static pressure sensors with digital data logger, with sampling rates of 20 readings per second. These pressure sensors were built using three ATMega2560 microcontrollers and BMP180 digital sensors. A Pitot tube was also installed on the front of the GRSV and several temperature sensors, both for recording environmental conditions and for monitoring their systems interns. For the environmental variables, a portable weather station was used. Optionally the GRSV was equipped with two Pitot rake arrangements, constructed by 3D printing that allows recording the profile of the boundary layer in areas of interest. Finally, the GRSV with batteries and instruments on board has a total weight of 41 [kg], with 2.95 [m] long, 0.47 [m] wide and 0.67 [m] high. Its maximum speed is 110 [km/hr] using 8-cell LiPo.

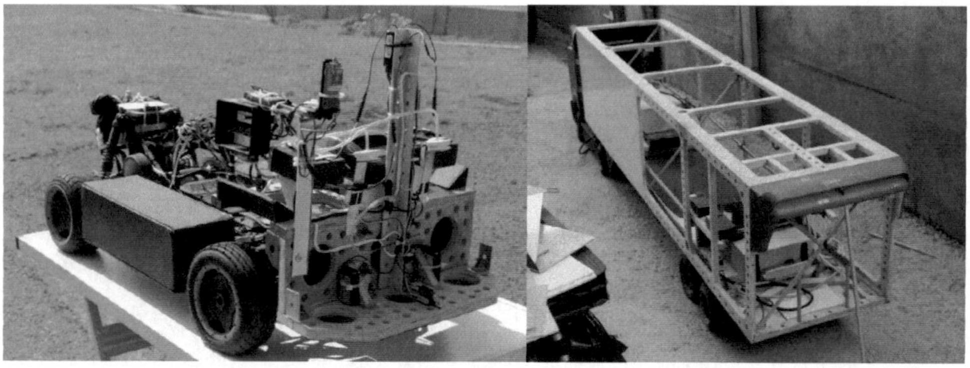

Figure 7 and 8. Interior view of GRSV: Truck (left) and trailer (right).

5. Experimental tests

The calibration tests were carried out at the Aeródromo de Vitacura in Santiago, while for the experimental trials the main runway in the Base Aérea El Bosque of the Chilean Air Force in Santiago was used.

Two-way displacements were performed on the 1.5 km long runway to eliminate environmental effects, at a maximum speed of 70 [km/h]. A group of displacements were made to record the static pressure in the control points at constant speed, while the other group of displacements were used exclusively to apply the Coast Down method, where the speed decay was recorded in between the 70 [km/hr] and 40 [km/hr], because in this interval the braking effects due to the aerodynamic drag is greater than the mechanical friction [2].

Figure 9. GRSV in runway of the El Bosque Chilean Air Force Base.

6. Results obtained

Approximately 80,000 data are obtained for each displacement, which helps to discriminate and compare different displacements to obtain significant values and valid tests, discarding those where the velocity was not stable or wind was detected in the test zone.

At the truck control points, mean static pressure values of 120 [Pa] for the front zone and -15 [Pa] on the rear face, which are compared with those obtained with the CFD simulations, presenting a difference greater than 10% in both cases, which could be explained by the type of turbulence model used in numerical modeling.

For the values of the aerodynamic drag, the GPS readings a continuous decay of the speed (Figure 10), and using the energy conservation equation of the Coast Down tests in the different speed sections, is obtained the values of the coefficient of drag according to each Reynolds number. These values are compared with those in the literature to similar Reynolds numbers [2] and those obtained with CFD at different speeds (Figure 11).

Table 2. Static pressure in control points at 25 [m/s] with different techniques.

	GRSV	CFD	Error [%]
P1	120	140	15
P2	-15	-12	20

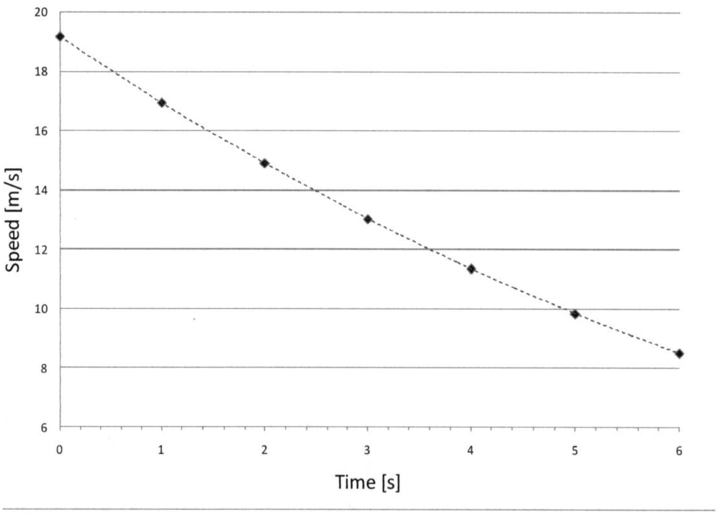

Figure 10. Continuous decay of the speed of GRSV during coast down test.

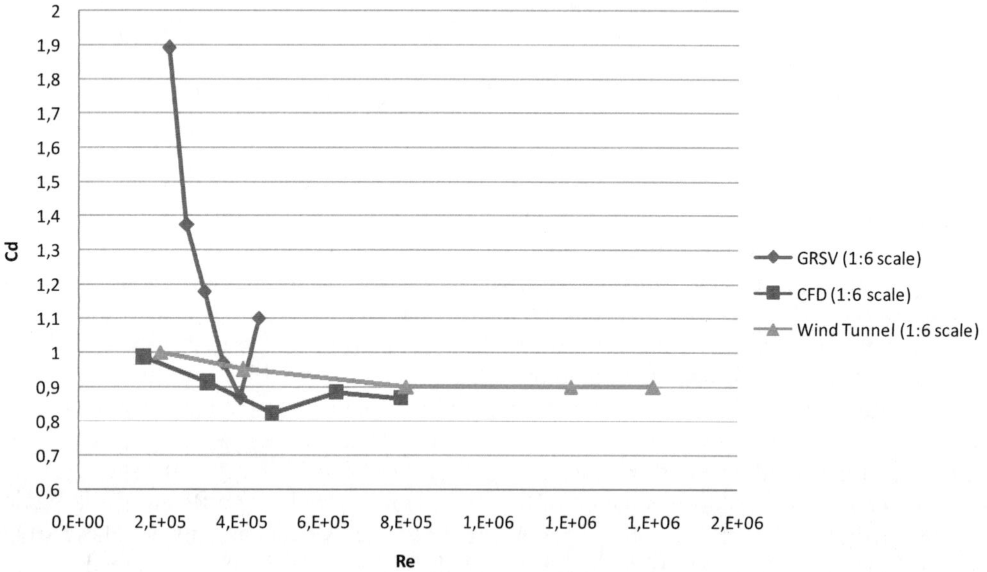

Figure 11. Variation of drag to different Reynolds number and scale model

For the case of the maximum test speed, the drag coefficient values are closer to those expected, as seen in Figure 12, when compared to those expected for scale trials [2] and in the real-size tests [3].

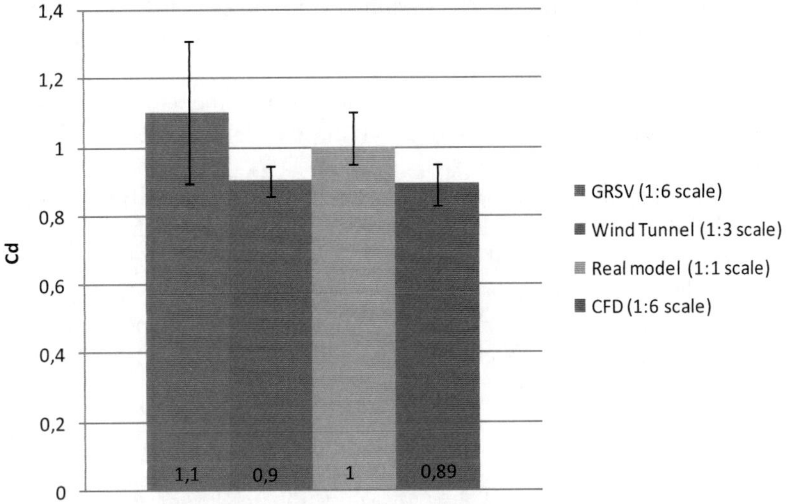

Figure 12. Drag coefficient according to experimental technique.

If we compare the aerodynamic drag values obtained by the Coast Down method using the GRSV, with those resulting from the CFD calculation for the different values of the Reynolds number and decomposing between the effects of mechanical friction and the air flow, we obtain a coherent quadratic relation between the values of the total drag force and the speed of GRSV (Figure 13).

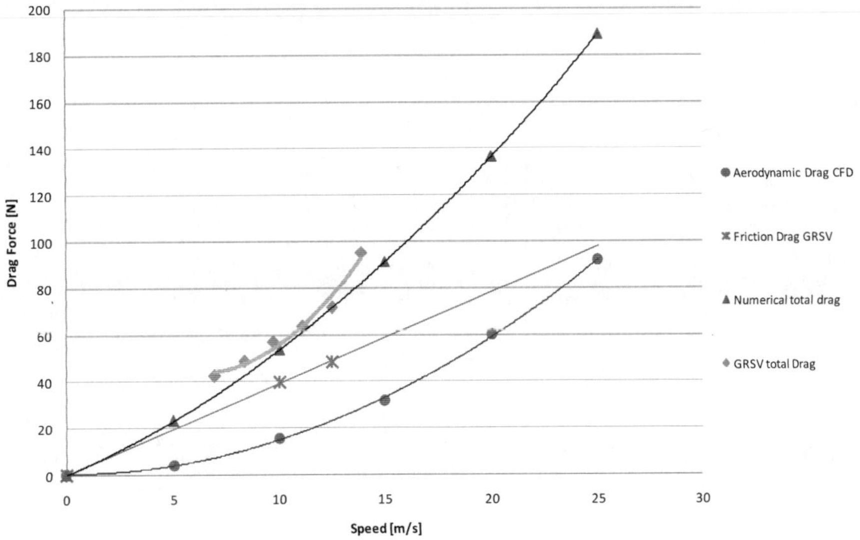

Figure 13. Forces acting in the GRSV for each test speed.

IOP Conf. Series: Journal of Physics: Conf. Series **1043** (2018) 012057 doi:10.1088/1742-6596/1043/1/012057

7. Conclusions

From the results obtained with the tests using the GRSV, it is possible to conclude that:

a) This test method is a viable, inexpensive and rapid experimental technique, compared with wind tunnel trials. It also incorporates the effect of ground movement and wheel rotation, making this method a more realistic technique because reproducing complex turbulent phenomena, while facilitating and betterment the testing of multiple aerodynamic devices, compared to CFD simulations, thanks on-board instrumentation and current 3D printing.

Also, it is verified the stability in the taking of data in the displacements at high speed, which will allow carrying out specific studies of the behavior of the air flows under the GRSV and in the turbulent wake of the same.

b) It is a reliable procedure at speeds above 65 [km/h] due to the influence of the effect of mechanical friction and rolling drag, which increase the value of the total drag. This negative effect cannot be easily corrected, because due to the high test speeds, the vehicle needs a mass and inertia that allows it to maintain the control and stability in the displacements.

c) The Energy Conservation Method to determine the drag coefficient presents an important dispersion at low velocities, which is consistent with the recommendations related to the work scale [2 & 3], which forces the GRSV to perform tests at its maximum speed.

Reference

[1] Agencia Chilena de Eficiencia Energética 2015 *Catálogo Tecno Eficiencia para el Transporte de Carga Pesado* (Santiago: Ministerio de Energía Gobierno de Chile)

[2] Rae & Pop 1984 *Low-Speed Wind Tunnel Testing* (New York: Wiley Publications)

[3] Corey D 2002 *A Ground-Based Research Vehicle for Base Drag Studies at Subsonic Speeds* (California: NASA Dryden Flight Research Center)

[4] Thwaites B 1960 *Incompressible Aerodynamic* (New York: Dover Publications)

[5] Shames I 1995 *Mecánica de Fluidos* (Santa Fé: McGraw-Hill Interamericana)

[6] Anderson J 1991 *Fundamentals of Aerodynamics* (New York: McGraw-Hill)

[7] Schlichting H 1972 *Teoría de la Capa Límite* (Bilbao: Ediciones Urmo)

IOP Conf. Series: Journal of Physics: Conf. Series **1043** (2018) 012058 doi:10.1088/1742-6596/1043/1/012058

Front Microrheology of Biological Fluids

C.A. Trejo-Soto[1,2,3], E. Costa-Miracle[3], A.I. Rodríguez-Villarreal[3], J.Cid[4], M. Castro[5],T. Alarcón[3,6,7,8] and A. Hernández-Machado[1,3,7].

[1] Departament de Física de la Matèria Condensada, Facultat de Física, Universitat de Barcelona, Av. Diagonal 645, E-08028 Barcelona, Spain.
[2] Departamento de Ingeniería en Construcción, Facultat de Arquitectura y Construcción, Universidad Autónoma de Chile, Av. Alemania 1090, Temuco, Chile.
[3] Centre de Recerca Matemàtica, Edifici C, Campus de Bellaterra, 08193 Bellaterra (Barcelona), Spain.
[4] Servicio de Hemoterapia y Hemostasia, Hospital Clinic de Barcelona, Barcelona, Spain.
[5] GISC and Grupo de Dinámica No Lineal (DNL), Escuela Técnica Superior de Ingeniería (ICAI), Universidad Pontificia Comillas, E-28015 Madrid, Spain.
[6] Departament de Matemàtiques, Universitat Autònoma de Barcelona, 08193 Bellaterra (Barcelona), Spain.
[7] Barcelona Graduate School of Mathematics (BGSMath), Barcelona, Spain.
[8] ICREA, Pg. Lluís Companys 23, 08010 Barcelona, Spain.

E-mail: `claudia.trejo@uautonoma.cl`

Abstract. We present a study of front microrheology through the development of a microfluidic device and method that describes accurately the non-linear rheology of blood, by means of a simple optical detection method based on tracking the fluid-air interface moving inside a microchannel. We study the behavior of Newtonian fluids of different viscosities and densities, as well, we performed measures for blood at different red blood cells concentration and at different days from its extraction. We have developed a scaling method which allows us to determine a relation between the red blood cell properties at different days from its extraction, according to the agreggation properties of red blood cell. Our results have been compared with theoretical and bibliographical results, which shows realiable results with an error around 6%. In general, our device and method is usefull as a viscometer and rheometer, as well as, it enables to establish a relation between blood viscosity and its red blood cells characteristics.

1. Introduction

Since Poiseuille times, several techniques have been developed to measure the viscosity of blood. During the 60s and 70s, with the appearance of the first rheomethers, the rheological properties of blood were accurately measured and it was observed that blood had shear thinning properties, meaning, that if the velocity of the blood flow increases, the viscosity of blood decreases. From a medical point of view, blood and its blood plasma are the most effective fluids to detect global pathologies in human and animals. These pathologies may be related to their viscosities, their plasma proteins, or the properties of its red blood cells, as their aggregation, deformability or the elastic properties of its cellular membrane. With the birth of microfluidics at the beginnings of the 90s, new techniques for the diagnostic of diseases have been developed. The avantages in the use of microfluidic devices for diagnostics are: the low amount of sample required to perform a measure, their portability, that they are easy to use and the low cost of its fabrication. We have developed a microfluidic device and method that describes accurately the non-linear rheology of blood, by means of a simple optical detection method based on tracking the fluid-air interface moving inside a microchannel. We present the results for two Newtonian fluids water and blood plasma. As well, we presents viscosity results for blood at different red blood cells concentration and at different days from its extraction.

2. Experimental Setup

The microsystem is a rectangular channel, of heights $b = 150$ and $b = 300\,\mu m$, width $w = 1\,mm$ and length $l = 4\,cm$, molded in a biocompatible hydrophobic silicone, PDM (polydimethylsiloxane), on a glass substrate according to Replica Moulding and Soft Lithography microfabrication methods [1, 2]. The bottom surface of the microchannel is made of glass and the top and lateral surfaces are made of PDMS. The inlet and the outlet are perpendicular holes set at each extremity of the channel. The observation of the flow inside de microchannel is made using and inverted microscope Optika XDS-3 with a 4× objective and a high speed camera, Photrom Fastcam Viewer 3, recording between 1000 and 125 frames per second.

We perfomed measurements of the velocity of the fluid interface across the length of the channel at different injected pressures. Pressures range from $\Delta P = 4116$ to $490\,Pa$. The mean velocity of the fluid, is obtained by tracking the fluid front mean position, h, as a function of time between several contigous images and averaging its values thought the channel length. Pressure is controlled through a fluid column inside a reservoir set at heights from $H = 0.40$ to $0.050\,m$ and conected to a biocompatible tube of uniform internal circular cross-section of diameter $d = 0.254\,mm$ and length L = 0.430 m, see Figure 1.

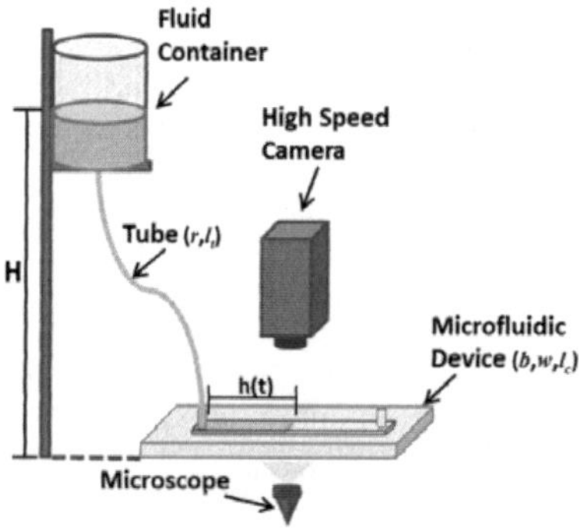

Figure 1. Schematic representation of the experimental setup [3].

3. Results

3.1. Viscosity of Newtonian Fluids

The viscosity of the newtonian fluids is calculated considering the coupled system, reservoir-tube-microchannel, by means of the relation between the shear rate, $\dot{\gamma} = \dot{h}/b$, and a rescaled stress defined as:

$$\sigma_{rs} = \frac{\pi r^4}{8wb^2 l_t}\left(\rho g H - P_L\right), \tag{1}$$

where P_L is the Laplace pressure due to the curvature of the fluid front. The viscosity value is then obtained as a function of the velocity, $\dot{h}$, by means of:

$$\eta = \frac{\sigma_{rs}}{\dot{\gamma}} = \frac{\pi r^4}{8wbl_t}\frac{(\rho g H - P_L)}{\dot{h}} \tag{2}$$

The viscosity results for Newtonian fluids are shown in Figure 2, for two different channel heights.

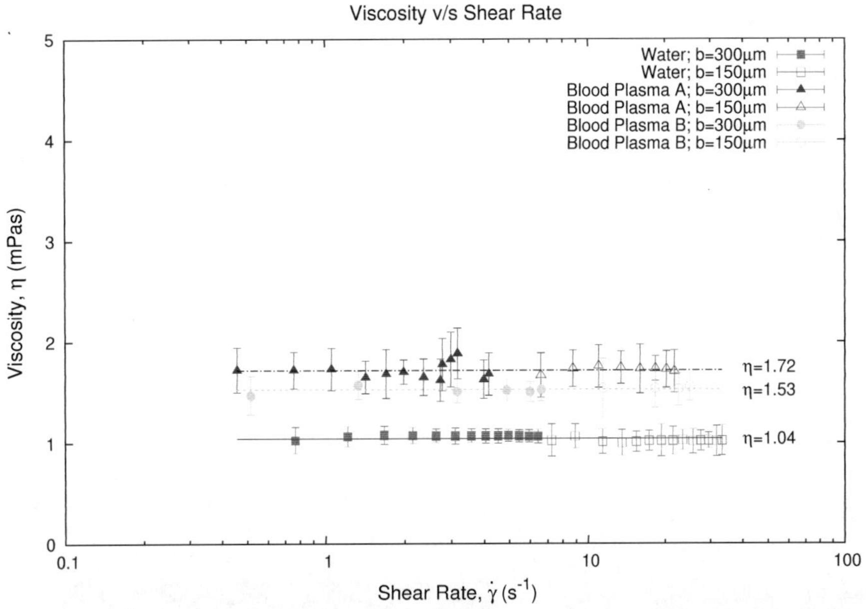

Figure 2. The figure shows the viscosity of two different blood plasma samples: sample A (blue) and sample B (cyan) compared with water (red). The viscosity values are obtained by mean of Eq. (2) as a function of the shear rate for two different channels heights. The viscosity values for each samples are $\eta_{pA} = 1.72 \pm 0.01\, mPas$ and $\eta_{pB} = 1.53 \pm 0.02\, mPas$.

We observe from Figure 2 that combining the results for different channel heights we are able to extend the viscosity estimation for almost 2 decades [4].

3.2. Viscosity of Blood

In the case of blood, which is non-Newtonian, we obtained the viscosity as a function of the shear rate using a Power Law Model [5]:

$$\eta = m\dot{\gamma}^{n-1}, \tag{3}$$

where n is the viscosity exponent, m is the base viscosity value and the shear rate is defined as:

$$\dot{\gamma}_F = \frac{b^2 w}{\pi r^3}\left(3 + \frac{1}{n}\right)\frac{\dot{\gamma}}{b} \tag{4}$$

according to the Weissenberg-Rabinovich-Mooney correction for non-Newtonian rheometry.

Studying a 48% hematocrit blood sample, measured at different days from extraction, we observe a shear thinning behavior for blood viscosity, see Figure 3. Furthermore, the viscosity values of the fifth day sample are higher than the viscosity values of the fresh sample.

3.3. Aggregation of Red blood cells

Comparing images of the sample at diferents days from its extraction we observed an increased aggregation of the red blood cells, see figures 4 and 5.

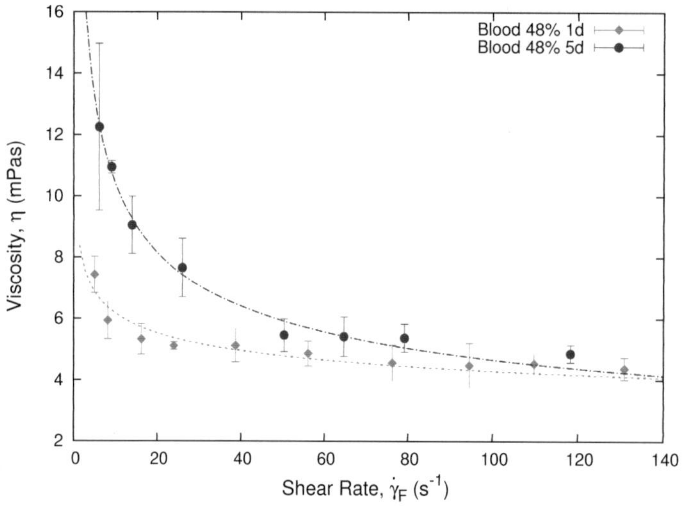

Figure 3. The figure shows the viscosity of a blood sample at a 48% hematocrit, at different days from its extraction, in magenta for a fresh blood sample and in blue for a 5 days blood.

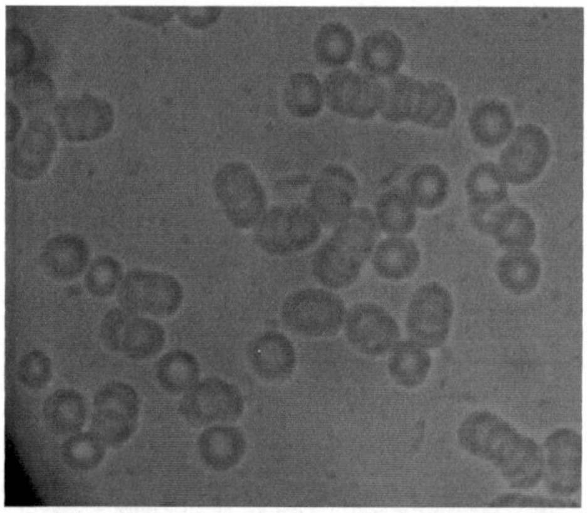

Figure 4. Image of the red blood cells at the first day from its extraction. The image was taken using a x100 objective.

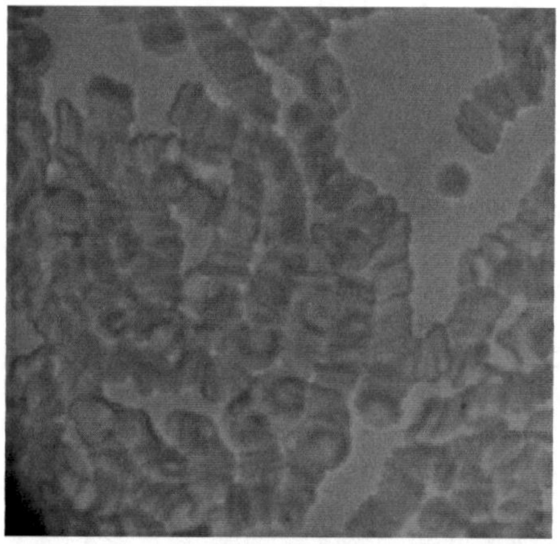

Figure 5. Image of the red blood cells at the fifth day from its extraction. The image was taken using a x100 objective.

To analyze the effects of aging, we fix the value of the hematocrit and study the behavior of the blood sample as it ages. We introduce a new, non-linear scaling parameter, the adhesion scaling number, A, which quantifies the effects of aging on RBC's aggregation [6]. This quantity is defined as:

$$A = \frac{\eta_0 \dot{\gamma}_F d^3}{E(\dot{\gamma}_F, a, E_0)} \tag{5}$$

where η_0 is the viscosity of plasma, $d \approx 7.8\,\mu m$ is the average diameter of a red blood cell, E is the energy scale associated to the aggregation forces between RBCs, where $E = (\dot{\gamma}_F, a, E_0) = k_a(\dot{\gamma}_F, a)E_0$. The scaling factor k_a is a function of the front shear rate $\dot{\gamma}$ and a the age of the sample and $E_0 \approx 500 k_B T$ is a reference value of RBC's aggregation [7].

The scaling factor is obtained by collapsing the curves for the variation of the viscosity as the shear rate changes. In order to quantify how aggregation induces this viscosity increment we establish a relation between the sample viscosities by means of the aging factor, $k_a(\dot{\gamma}_F, a)$, defined as:

$$k_a(\dot{\gamma}_F, a) = \left(\frac{\eta_{n5d}}{\eta_{n1d}}\right)^2. \tag{6}$$

The last step of the scaling, consist of writing the normalized viscosity of blood as a function of the adhesion scaling number, A, which depends on the aggregation energy correspondent to each sample. For an unaggregated sample, $E_0 \approx 2 \times 10^{20}\,J$ which using Eq. 6, is increased by $k_a = 2.79$ for the non-Newtonian regime and $k_a = 1.22$ for the Newtonian regime, for the 5 days aggregated sample. In this case, the viscosity of both regimes show changes which are assumed a consequence of the aggregation of RBCs, due to the aging of the sample.

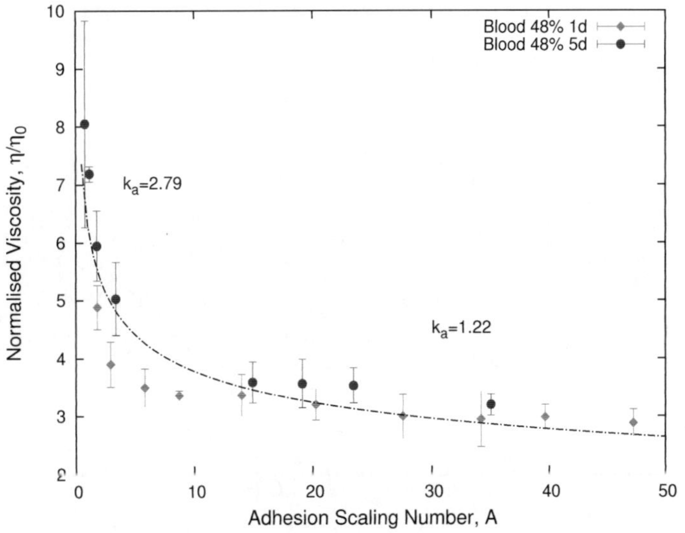

Figure 6. The figure shows the normalized viscosity of the blood samples against the Adhesion Number, A, correspondent to the non-Newtonian $k_a(\dot{\gamma}_F, a) = 2.79$ an $k_a(\dot{\gamma}_F, a) = 1.22$ for the Newtonian regime. We appreciate the collapse of the viscosity curves as a result of our scaling method.

4. Discussion

We have shown that variations in the non-linear viscosity of blood associated to aging can be explained by means of a scaling theory based on the introduction of an adhesion scaling number, 5. This quantity is defined as the ratio between the viscous energy and the aggregation energy. Our scaling theory allows us to quantify the increase in aggregation energy by aging with respect to its reference characteristic scale, E_0. The increase in the characteristic aggregation energy is given by the aging factor, $k_a(\dot{\gamma}_F, a)$, which is observed to increase as the sample ages. A strength of our approach comes from the introduction of a novel front microrheometer that allows to extract bulk non-linear viscosities from the simple observation of a moving fluid-air interface.

Acknowledgments
C.T-S and A.H-M. gratefully acknowledge partial financialsupport from MINECO (Spain) under project FIS2013-47949-C01-01-P and AGAUR project 2014SGR-878. E.C-M, I.R-V, and T.A. gratefully acknowledge the spanish Ministry for Science and Innovation (MICINN) for funding under grants MTM2015-71509-C2-1-R and Generalitat de Catalunya for funding under grant 2014SGR1307. C.T-S acknowledges CONICYT and the Chilean Ministry of Education for funding PhD fellowship, Becas Chile.

References

[1] Y. McDonald and G. Whitesides, Accounts of chemical research 35(7):491-499 (2002).
[2] JN. Vulto et al., Lab on a Chip 5(2):158-162 (2005).
[3] C.A. Trejo Soto, Ph.D. Thesis Dissertation (2016).
[4] C. Trejo-Soto, E. Costa-Miracle, A.I. Rodríguez-Villarreal, J. Cid, T. Alarcón and A. Hernández-Machado, PLoS One 11(4)e:0153559 (2016).
[5] F.A. Morrison (2001), *Understanding rheology*, (Oxford University Press).
[6] C. Trejo-Soto, E. Costa-Miracle, A.I. Rodríguez-Villarreal, J. Cid, M. Castro, T. Alarcón and A. Hernández-Machado, Soft Matter, 13(16), 3042-3047 (2017).
[7] D. Fedosov et al., PNAS 108(29):11772-11777 (2011).

IOP Conf. Series: Journal of Physics: Conf. Series **1043** (2018) 012059 doi:10.1088/1742-6596/1043/1/012059

Secondary targeting EDXRF system optimized for detection of gold, silver and gadolinium nanoparticles

B Casanelli[1], R Saavedra[1], M Vásquez[1] y M Santibáñez[1,2]

[1] Departamento de Ciencias Físicas, Universidad de la Frontera, Temuco, Chile.
[2] Centro de Física e Ingeniería en Medicina (CFIM), Facultad de Ingeniería y Ciencias, Universidad de La Frontera, Temuco, Chile.

E-mail: b.casanelli01@ufromail.cl

Abstract. In this work we seek to increase the amount of X-rays photons capable of excite nanoparticles of Au, Ag and Gd from a common emission spectrum of a mini-tube of X-rays, with the purpose of optimize its detection. For this, we used the secondary target method and we investigated three targets for each element, in order to find which is the one that improves better the detection limit. A 3D model is designed to mount the system in a Cartesian geometry. It was used a CAD (computer-aided design) software to model parts and components which are necessary to locate the mini-tube and the detector. Also, Monte Carlo code (through XMIM-SIM software) was used to simulate the interactions and study the spectrum of each target and sample, and a numerical computer software to correct and interpret the data. Finally we obtained that the best targets for Au and Gd are Y and Zn, respectively, achieving improvements up to 140% in the detection limit. For Ag nanoparticles the results did not show improvements regarding the case without target.

1. Introduction

Energy Dispersive X-ray Fluorescence (EDXRF) is an effective, fast and non-destructive technique to determine trace elements in different types of samples [1–3] based on the stimulated photoelectric absorption (predominant effect for photons with energy lower than 100 keV [4]) and the spontaneous emission, that allow to identify the element through the analysis of the photons emitted when the atom returns to its ground state.

The secondary target method was described and designed in the late last century [5–9]. In this technique the primary radiation from the X-ray tube excite a target which emits a secondary radiation that is used as a source to excite the sample [10]. The intensity of the primary beam is strongly reduced by the process, but the new quasi-monochromatic spectrum can produce a great improvement of the detection limit, as demonstrated by Lindgren and Selin [11], getting an increase in the fluorescent signal and a decrease in the produced background.

Recently, different nanoparticles have been functionalized to act as tumor markers and simultaneously be applied on highly localized treatments. Gold, silver and gadolinium nanoparticles have been the focus on several studies, since they are enhancers of local dose delivery for orthovoltage radiotherapy [12–16] as well as a contrast agent with a better attenuation than the traditional compounds used [12,17]. Several EDXRF systems have been implemented to identify and quantify nanoparticles *in vivo*, reaching different detection limits [18,19]. The most common

problems of studying these nanoparticles agents *in vivo* are: how to increase the sensitivity of the systems in order to reduce measurement times required, and how to reduce the dose of radiation given to the patient, allowing the study of different nanoparticles of interest. To increase the sensitivity in EDXRF systems is necessary increase the production of fluorescence photons exciting the sample with energies near to the absorption edge of the nanoparticles.
In this work is shown the form of optimize the primary spectrum through secondary targets, and the criteria that allow to choose the best target to excite nanopaticles of Au, Ag and Gd, reducing the background and increasing the characteristic peak of interest.

2. Method

2.1. Selection of elements and targets

The composition of the targets is directly related with the element under study, since its $K_{\alpha 1}$ transition should have the right amount of energy to excite mainly the atom edges of the nanorpaticles before the rest of the sample [10]. The emission $K_{\alpha 1}$ had been used as a criterion for choosing the target since is the one with the highest probability to occur.
To select the targets a database was set up and programmed to compare and select the best alternatives for each element and absorption edge. The data was extracted from the National Institute of Standards and Technology (NIST) [20].
After using the algorithm, the results are studied with the purpose of selecting targets acquirable and manageable at normal conditions and discarding, for example, targets made of Xe to observe the transition $K_{\alpha 1}$ of Ag, because its ordinary state is gaseous.
The finally selected targets are presented in table 1.

Table 1. Elements, targets and transitions of interest.

Element	Transition	Targets
Gd	$L_{\alpha 1}$ and $L_{\beta 1}$	Ni, Cu and Zn
Au	$L_{\alpha 1}$ and $L_{\beta 1}$	Sr, Y and Zr
Ag	$K_{\alpha 1}$ and $K_{\beta 1}$	Sb, I and Ba

2.2. 3D Design

The design and construction of a 3D structure that allows mounting the tube of X-rays, the targets and the detector are necessary for a second stage of the project, in which, already selected the best targets through computer simulations, real tests are made.
The montage, shown in the figure 1-a-b, consist in a fixed part where is possible to put a mini-tube of X-rays (model M237), a changeable part in where is the carousel of targets, and the detector (model XR100SDD). In the design it is satisfied a Cartesian geometry, since this form the X-rays from the target and tube have a low probability of scattering from the sample into the detector, as proven by Standzenieks and Selin [21].
Alternatively, the changeable carousel can be replaced by a specially mount for the mini-tube (figure 1-c), and then realize EDXRF without targets, which serves to compare results.

2.3. First spectrum simulation

Next stage is know the spectrum emitted from the targets when the primary radiation interact with it. To predict it, Monte Carlo code is used through the software XMI-MSIM [22].

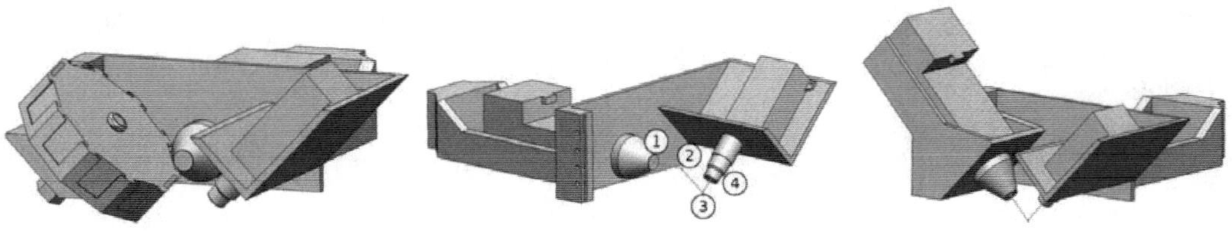

(a) (b) (c)

Figure 1. a) 3D design with carousel system. b) Travel of the beam. b-1) Primary radiation is emitted from the mini-tube. b-2) This radiation interacts with the target, and a secondary quasi-monochromatic radiation is scattered. b-3) The secondary spectrum interacts with the sample, and a third beam is produced, in which the background noise is very low and the characteristics X-rays of the element under study can be measured clearly. b-4) The final beam arrives to the detector. c) Alternative montage in which the carousel has been replaced with the mini-tube, and the EDXRF is without targets.

For the simulation was used a tube of 35 keV whit filament of W and window of Be; samples (targets) with thickness of 0.2 cm with purity of 100% and 3 cm of air on it; 3 cm sample-source distance, 3 cm sample-detector distance, 45° between sample and beam, 90° between source and detector; and Solid-State Drive (SDD) detector of Si with thickness of 0.045 cm.
After simulate the nine targets, is used numerical computer software to correct the deconvoluted spectra with the SDD Efficiency and adjust it with the Lambert-Beer equation and the mass attenuation coefficient of dry air (obtained from NIST) with the purpose of find the real spectrum that comes from the target.

2.4. Second spectrum simulation

Now the spectrum of each target is loaded as primary radiation and shot on the corresponding samples. The configuration is similar to the previous one, except for the composition of the samples, which are composed of 2.5 cm of water (that simulates the human tissues), 1 cm of 1% nanoparticles solution in water (as reference layer), and 3 cm of dry air.
In the same way is simulated a spectrum without the 1% of nanoparticles in the reference layer, to get the background noise only.

3. Results

After simulating the same process described in the 2.3 and 2.4 subsections for all targets, are compared the intensity, background noise (BN) and the limit of detection (LOD) of the three targets for each element in energy intervals corresponding to transitions under study, obtaining the results shown in the tables 2, 3, and 4. Additionally, the process was simulated without targets, obtaining the results shown in the table 5. LOD improvements are shown in table 6.
For the calculus of LOD, is used the equation

$$LOD = 3\frac{\sqrt{B}}{H} \tag{1}$$

where B is the number of counts in background, and H is the sensitivity [10, 23]. To determine **H** was used the background subtraction method, in which is subtracted the simulation without nanoparticles (background) to the simulation with nanoparticles [14, 24], as shown in figure 2.

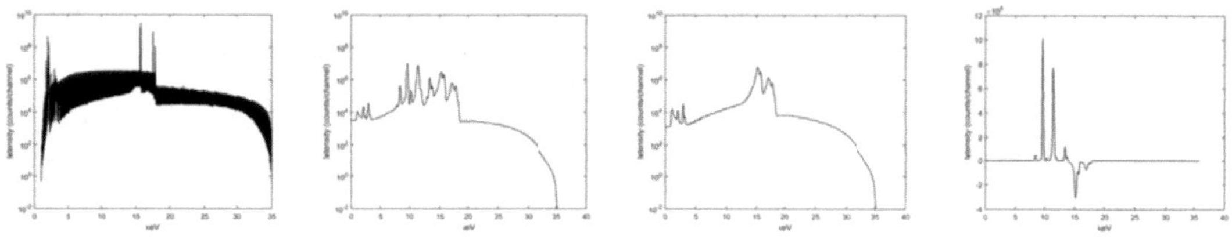

(a) (b) (c) (d)

Figure 2. Spectrum of Zr for analysis of Au (note that graph (a), (b) and (c) has a logarithmic scale on the y-axis). a) The corrected unconvoluted spectrum that is emitted from the target of Zr. b) The obtained spectrum from the sample using the target. c) The obtained spectrum from the sample using the target but without nanoparticles (only background noise). d) The spectrum without background noise ((b) less (c)), in which can be see the transition $L_{\alpha 1}$ (first left large peak) and the transition $L_{\beta 1}$ (second left large peak) of the Au.

Table 2. Results for Gd.

Target	Intensity (counts/channel)					
	$L_{\alpha 1}$	BN $L_{\alpha 1}$	LOD $L_{\alpha 1}$	$L_{\beta 1}$	BN $L_{\beta 1}$	LOD $L_{\beta 1}$
Ni	2.9308E7	1.6560E5	4.1655E-5	2.9256E6	5.4347E5	7.5595E-4
Cu	**2.9337E7**	1.9433E5	4.5080E-5	2.0713E7	**2.2897E5**	6.9304E-5
Zn	2.8220E7	**1.2636E5**	**3.7789E-5**	**2.5901E7**	2.5158E5	**5.8095E-5**

Table 3. Results for Au.

Target	Intensity (counts/channel)					
	$L_{\alpha 1}$	BN $L_{\alpha 1}$	LOD $L_{\alpha 1}$	$L_{\beta 1}$	BN $L_{\beta 1}$	LOD $L_{\beta 1}$
Sr	6.8937E7	2.6084E5	2.2226E-5	7.1574E7	7.8513E5	3.7139E-5
Y	**7.1377E7**	**2.0207E5**	**1.8894E-5**	**7.6535E7**	6.1379E5	**3.0709E-5**
Zr	5.9723E7	2.2063E5	2.3594E-5	6.5606E7	**5.2656E5**	3.3182E-5

Table 4. Results for Ag.

Target	Intensity (counts/channel)					
	$K_{\alpha 1}$	BN $K_{\alpha 1}$	LOD $K_{\alpha 1}$	$K_{\beta 1}$	BN $K_{\beta 1}$	LOD $K_{\beta 1}$
Sb	**1.2916E7**	1.2696E6	**2.6171E-4**	**8.8583E5**	2.0841E6	4.8891E-3
I	3.3858E6	6.6232E5	7.2110E-4	3.0458E5	2.4137E5	**4.8391E-3**
Ba	1.4271E6	**5.7659E5**	1.5962E-3	1.2019E5	**1.5080E5**	9.6929E-3

Table 5. Results without targets.

Element	Intensity (counts/channel)					
	Transition	BN	LOD	Transition	BN	LOD
Au	$L_{\alpha 1}$: 4.09E8	3.99E7	4.63E-5	$L_{\beta 1}$: 3.65E8	7.10E7	6.94E-5
Ag	$K_{\alpha 1}$: 1.50E8	9.66E6	6.22E-5	$K_{\beta 1}$: 1.28E7	1.59E7	9.36E-4
Gd	$L_{\alpha 1}$: 1.38E8	1.79E7	9.16E-5	$L_{\beta 1}$: 1.20E8	2.06E7	1.14E-4

Table 6. Detection limit improvements.

Element	Transition	Target	Intensity (counts/channel)		Improvement (%)
			LOD without target	LOD with target	
Au	$L_{\alpha 1}$	Y	4.63E-5	1.89E-5	144.98
	$L_{\beta 1}$		6.94E-5	3.07E-5	126.06
Ag	$K_{\alpha 1}$	Sb	1.67E-4	2.62E-4	-36.25
	$K_{\beta 1}$		9.36E-4	4.89E-3	-80.86
Gd	$L_{\alpha 1}$	Zn	9.16E-5	3.78E-5	142.33
	$L_{\beta 1}$		1.14E-4	5.81E-5	96.21

As expected, the intensity of the primary beam was reduced, as shown in figure 3.

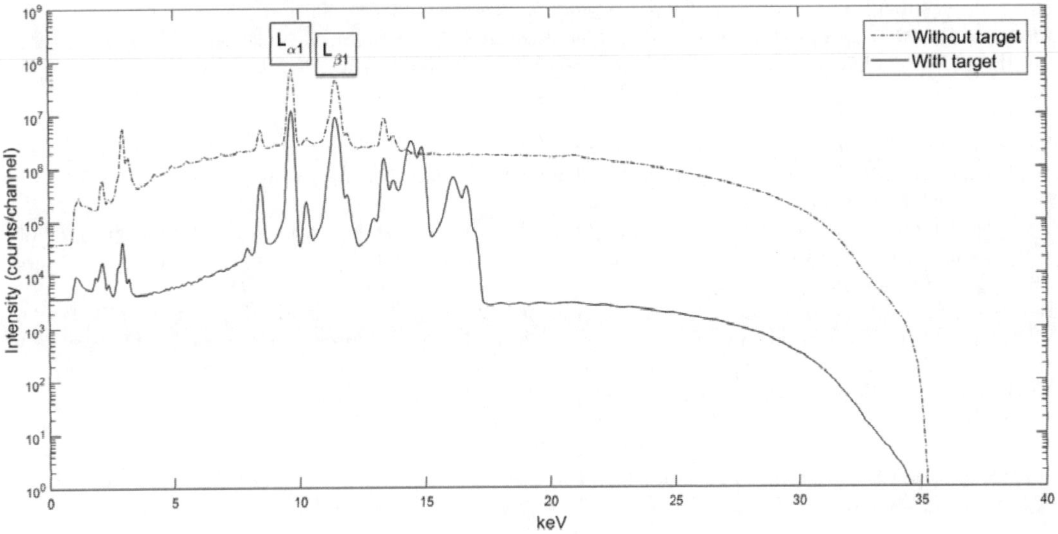

Figure 3. Spectra for Au with and without Y target (log scale on the y-axis).

4. Conclusions

It was possible to design an EDXRF system consisting of a X-rays mini-tube, an SDD detector and a carousel of secondary targets, in a Cartesian geometry which optimizes the signal-to-noise ratio. Likewise, a secondary targeting method was optimized, re-obtaining results shown in the literature about the background reduction and the LOD improvement. Also, we detected some uncertainty in the simulated background, which can introduce errors in the background subtraction method, as suggested by Wu et. al. [24]. Finally, and based on the results exposed in tables 2-6, we conclude that the best targets for Gd and Au are Zn and Y, respectively, in which the LOD can be improved up to 140%, while for Ag the LOD has been worsened approximately between a 40% and 80%. Therefore, in this work, no target for Ag present improvements regarding the case without targets. However, future studies incorporating administered dose criteria could improve these results.

5. Acknowledgments

This project is financed by CONICYT through the project FONDECYT N°11150673 and by the Direction of Investigation of the Universidad de La Frontera DI 16-2016.

6. References

[1] Geraki K, Farquharson M J and Bradley D A 2002 *Physics in Medicine and Biology* **47** 2327
[2] Longoni A, Fiorini C, Leutenegger P, Sciuti S, Fronterotta G, Strüder L and Lechner P 1998 *Nuclear Instruments and Methods in Physics Research Section A: Accelerators, Spectrometers, Detectors and Associated Equipment* **409** 407 – 409
[3] Khuder A, Bakir M, Karjou J and Sawan M 2007 *Journal of Radioanalytical and Nuclear Chemistry* **273** 435–442
[4] Sethi A 2006 *X-Rays: Interaction with Matter* (John Wiley & Sons, Inc.)
[5] Standzenieks P and Selin E 1979 *Nuclear Instruments and Methods* **165** 63 – 65
[6] Spatz R and Lieser K H 1979 *X-Ray Spectrometry* **8** 110–113
[7] Cristensen L H, Rasmussen S E, Pind N and Henriksen K 1980 *Analytica Chimica Acta* **116** 7–17
[8] Porter D E 1973 *X-Ray Spectrometry* **2** 85–89
[9] Rao D V, Cesareo R and Gigante G E 1993 *X-Ray Spectrometry* **22** 406–409
[10] Lindgren E S 2006 *Energy Dispersive, X-Ray Fluorescence Analysis* (John Wiley & Sons, Ltd)
[11] Ida H and Kawai J 2005 *Spectrochimica Acta Part B: Atomic Spectroscopy* **60** 89 – 93
[12] Hainfeld J F, Smilowitz H M, O'Connor M J, Dilmanian F A and Slatkin D N 2013 *Nanomedicine* **8** 1601–1609
[13] Hainfeld J F, Slatkin D N and Smilowitz H M 2004 *Physics in Medicine and Biology* **49** N309–N315
[14] Mattea F, Vedelago J, Malano F, Gomez C, Strumia M C and Valente M 2017 *Radiation Physics and Chemistry* **130** 442–450
[15] Ngwa W, Kumar R, Sridhar S, Korideck H, Zygmanski P, Cormack R A, Berbeco R and Makrigiorgos G M 2014 *Nanomedicine* **9** 1063–1082
[16] Zhang D G, Feygelman V, Moros E G, Latifi K and Zhang G G 2014 *PLoS ONE* **9** e109389
[17] Hainfeld J F, Slatkin D N, Focella T M and Smilowitz H M 2006 *The British Journal of Radiology* **79** 248–253
[18] Figueroa R, Santibañez M, Malano F and Valente M 2015 *Radiation Physics and Chemistry* **117** 198–202
[19] Santibáñez M, Vásquez M, Figueroa R and Valente M 2016 *Radiation Physics and Chemistry* **122** 28–34
[20] NIST National institute of standards and technology URL `https://www.nist.gov`
[21] Standzenieks P and Selin E 1979 *Nuclear Instruments and Methods* **165** 63–65
[22] Schoonjans T Xmi-msim URL `https://github.com/tschoonj/xmimsim`
[23] Gilfrich J V and Birks L S 1984 *Analytical Chemistry* **56** 77–79
[24] Wu D, Li Y, Wong M D and Liu H 2013 *Medical Physics* **40** 051901

A framework for e^+-e^- annihilation detection using nano- particles for tumour targeting in radiotherapy

F. Geser[1,2], A. Leiva Genre[2], R. Figueroa[3,4], M. Vásquez[3], F. Malano[1,2], F. Mattea[2,3], M. Santibáñez[4,5], J. Velásquez[6], M. Valente[1,2,4,5]

[1]Instituto de Física E. Gaviola – CONICET, Argentina.
[2]Laboratorio de Investigaciones e Instrumentación en Física Aplicada a la Medicina e Imágenes por Rayos X – LIIFAMIR[X], FaMAF, Universidad Nacional de Córdoba, Argentina.
[3]Departamento de Química Orgánica, Facultad de Ciencias Químicas, Universidad Nacional de Córdoba, CONICET, Córdoba, Argentina.
[4]Centro de Física e Ingeniería en Medicina, Universidad de La Frontera, Chile.
[5]Departamento de Ciencias Físicas, Universidad de La Frontera, Chile.
[6]ICOS Inmunomédica, Temuco, Chile.

E-mail: fgeser@famaf.unc.edu.ar valente@famaf.unc.edu.ar

Abstract. The quality of modern patient radiotherapy treatment is strongly dependent on reliable and accurate dosimetry techniques. Although desirable, *in situ* and *in vivo* dosimetry is usually hard to implement in routine radiotherapy procedures. The potential use of nanoparticles has been recently considered for biomedical applications achieving promising performance for diagnosis as well as therapeutic practices. The present work reports about a novel proposal based on the use of high atomic number nanoparticles for online estimation of absorbed dose during conventional radiotherapy treatment. In this first phase, the investigation handles with experimental and theoretical tasks regarding the potential use of the variations in the annihilation peak and its correlation with the presence of high atomic number nanoparticles acting as signal enhancers. Dedicated Monte Carlo simulations are also presented in order to complement experimental results and theoretical models. Although the correlation between annihilation peak with the presence of nanoparticles was successfully confirmed, further investigations are still necessary in order to estimate feasible absorbed dose from determinations of annihilation peaks.

1. Introduction

Imaging and therapeutic techniques aimed to deal with neoplasic diseases incorporate the use of ionizing radiation as part of the medical procedure. In this context, it is crucial to achieve reliable and accurate dosimetry. This task can be addressed by experimental methods for in-vivo or in-phantom dosimetry, as well as calculations based on models for radiation transport in biological systems. It may be desirable to develop methods capable of in situ and online dosimetry. However, it is difficult to find worldwide established commercial techniques available for this purpose.Significant efforts are currently dedicated to the study of novel methods with such capabilities. During the last years, the potential use of nano-particles (NPs) in radiation oncology applications was largely investigated focusing the attention of radiology and treatment improvement by radio-sensitivity enhancement[1,2]. High atomic number (Z) NPs are becoming one of the most promising contrast agents for computed tomography (CT) due to their remarkable properties including high X-ray absorption coefficient, tailored surface chemistry, and excellent biocompatibility[3]. Thus, the increased photoelectric absorption probability produces an increased absorbed dose close around the target site minimizing harmful radiation to normal tissues. Although iodine was the first element investigated to enhance local tumor control by using high Z materials[4], gold NPs appear nowadays probably as the most promising technique to this aim[5,6]. Gold NPs have also the due to strong NIR light absorption, the majority of interest in potential for photo-thermal cancer therapy due to the large amount of heat generated by near infrared (NIR) irradiation[6]. On the other hand, the principle of contrast enhancement by high Z agents is based on the fact that high Z materials present, normally, higher physical density than most of the atoms in biological tissues.Current radiology techniques are mainly based on absorption contrast imaging, therefore the more efficient X-ray absorption of high Z materials leads to the improvements in radiology image contrast[2,4].

It is well known that when using X-rays with energy around 100 keV, the presence of high Z NPs produces the emission of characteristic fluorescent lines than can be accurately detected[7,8]. However, during typical radiotherapy treatments, the irradiation beams have significant higher energies according to accelerating voltages around 6-18 MV. Clinical implementation of radiology techniques based on NPs detection, and associated potentiality for tumor localization and radio-sensitivity enhancement,requires specific investigations of such processes for megavoltage photon beams. Therefore, the annihilation peak, that is dependent on irradiated material properties, might appear as a suitable parameter to be investigated to this aim.

This work reports on preliminary results regarding the design and experimental setup to determine the feasibility and reliability of using prompt gamma emission from positron-electron (e^+-e^-) annihilation as qualitative/quantitative correlated parameter for detecting the presence and localization of high Z NPs within biological tissues and, eventually, to establish a correlation with local absorbed dose. The first step consisted of modeling the effect on the radiation-matter interaction properties by means of calculating effective macroscopic cross sections according to the actual concentration of NPs within the water-equivalent phantoms that reasonably mimic low Z biological tissues for radiology purposes.Special interest was dedicated to the theoretical calculation of e^+-e^-pair production-annihilation. Additionally, Monte Carlo (MC) subroutines were developed in order to perform simulations to investigate both, radiation production and further detection processes with the aim of seeking for correlations between e^+-e^- annihilation photon emissions with the type, spatial distribution and concentration of the NPs.

2. Materials and Methods

This section reports about theoretical approaches, as well as experimental and virtual (Monte Carlo)methodologies for radiation-matter interactions within systems consisting of small fractions of high Z NPs diluted in low Z and quite water equivalent media.

2.1. Models for pair (e^--e^+)production and annihilation for systems of NPs in biological tissues

The pair production-annihilation processes are already well described in literature[9, 10], therefore only a brief summary is presented in this Section providing the main issues about the underlying physics for e^--e^+ production and associated annihilation photons.

The e^--e^+ pair production concerns a transition between an initial state comprehended by two photons of different origin, and a final state corresponding to a positron and an electron. One of the photons is generated in the laboratory frame, possibly by Bremsstrahlung, and must have a total energy E_γ satisfying $E_\gamma \geq 2m_0c^2 \approx 1.022\ keV$, where m_0 is the electron rest mass and c is the speed of light in vacuum. The other photon comes from the Coulomb field of another particle such as an atomic electron or the nucleus. For the purposes of the present work, the theoretical approach is based on the fact that the pair production is due to the nuclear Coulomb field, as sketched in the laboratory reference system in figure 1.

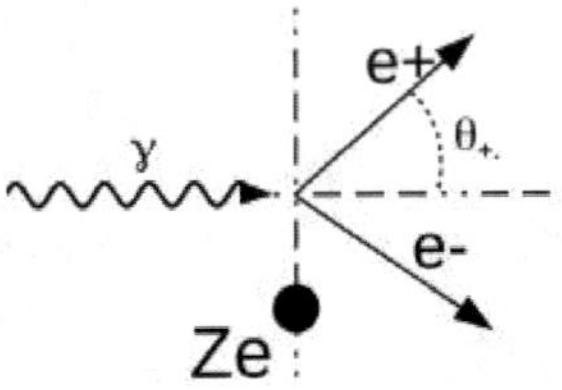

Figure1: Sketch of e^--e^+ pair production in the laboratory inertial system due to a Z atomic number nucleus.

Determination of the microscopic cross section for e^--e^+ pair production in nuclear Coulomb field is a canonical problem that can be solved in the first Born approximation as proposed by Bethe and Heitler[11]. The result is stated as a three differential cross section on positron energy E_+ and momentum angles $\theta_\pm$. Then, introducing the unit system $\hbar = c = 1$, it results for electron and positron momenta:

$$d\sigma^3 = \frac{Z^2\alpha^2}{(2\pi)^2}\frac{1}{|\vec{q}|^4}\frac{dE_+\,d\Omega_+\,d\Omega_-}{\omega^3}\Theta(E_\gamma - E_+ - m_0)\ F(p_+^\mu, p_-^\mu; k^\mu)\,, \quad (1)$$

where α is the fine structure constant, the F function depends on the 4-momentum of the three particles[12] and $|\vec{q}|$ is the magnitude of the 3-momentum transferred to the nucleus, while the Heaviside step function Θ ensures a zero value for the cross section if $E_\gamma - K_+ < 2m_0c^2$, *i.e.* the threshold for pair production. Although this approach for the cross section does not account for the electron cloud screening or the internal structure of the nucleus, it may be useful to integrate expression (1) with respect to angular variables ($\Omega_\pm$) in order to assess an approximation for the dependence on positron energy E_+, as reported in figure 2.

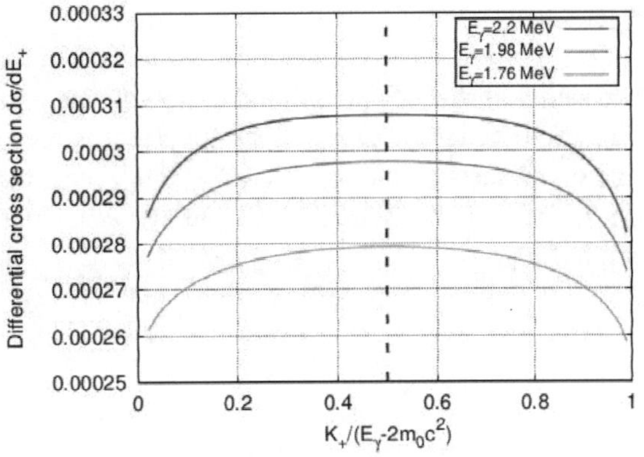

Figure 2: Theoretical approach for the microscopic cross section for e^--e^+ pair production in Au (Z=79) as a function of positron energy E_+ for different E_γ photon energies.

IOP Conf. Series: Journal of Physics: Conf. Series **1043** (2018) 012060 doi:10.1088/1742-6596/1043/1/012060

Similarly, fixing the positron energy E_+ to the most probable case, *i.e.* half the photon available energy, formula (1) can be plotted as a function of both, positron and electron angles (θ_+ and θ_+, respectively) as shown in Figure 3.

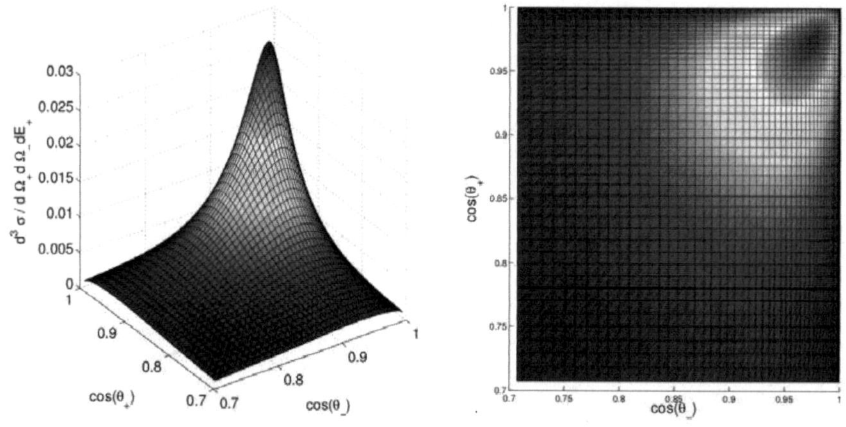

Figure 3: e^--e^+ pair production microscopic cross section as a function of the scattering angles θ_+ and θ_- considering symmetric distribution of incident photon energy.

This result is useful to predict the most probable directions for the creation of the e^--e^+pair, which correspond to the forward direction according to the incident γ photon. As expected, the angle corresponding to the maximum in this surface ($\theta_\pm$) is symmetrical for the e^--e^+pair having the value $\theta_\pm \approx 12.6°$ (0.2199 rad). As for the energy, the positron angle should be slightly greater because of the Coulomb repulsion with the nucleus, but the effect was neglected in this approach.

The proposed approach to investigate the macroscopic cross sections for systems consisting of high Z NPs in biological soft tissues is based on some assumptions: the mixture can be modeled, for radiology purposes, as being composed of a low percentage (weight in weight, w/w) of high Z NPs diluted in water equivalent media, therefore constituting a homogeneous soup of atoms, as usually treated by Monte Carlo techniques. However, for other physicochemical purposes, it may be necessary to take into account that NPs are in a solid phase suspended in liquid buffer.

According to the aim of the present work, it is necessary to estimate the effective macroscopic cross section for systems consisting of high Z NPs diluted in liquid water, which are irradiated by high energy (6 MV) photon beams used for radiotherapy. Specifically, it is required to know the influence on the effective macroscopic cross section due to NPs type and concentration. The spectrum of the incident photon beam corresponds to the Varian 610C linear accelerator (linac) is available in literature [14] and it is shown in Figure 4.

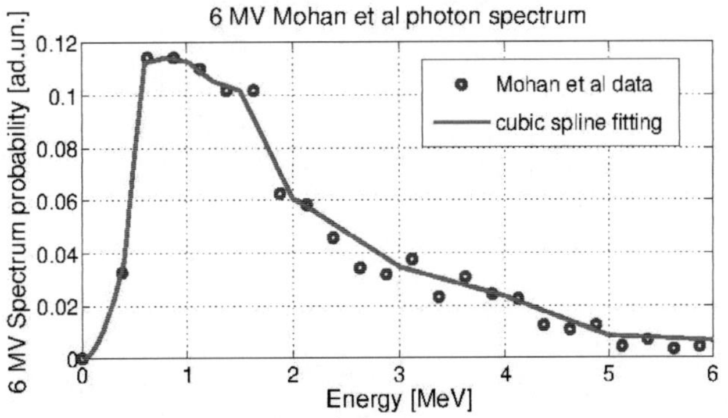

Figure 4: Photon spectrum from Varian 610 C linac reported by Mohan et al. [14] (blue circles) along with corresponding cubic spline by Hermite polynomials (green solid line).

The effective macroscopic cross sections, weighted over photon spectrum, were calculated based on data provided by NIST database about cross sections for pure elements and some specific compounds, like liquid water. For the purposes of the present work, regarding radiology properties of high Z NPs, the proposed approach consists on estimating the compound cross section in terms of pure element cross sections according to Bragg's additivity rule[15, 16]. If σ_i is the cross section of *i*-th element, and N_i is the number of atoms of that element present in the compound, then the total cross section σ for the compound can be approached as:

$$\sigma \approx \sum_i N_i \sigma_i \,. \qquad (2)$$

Therefore, in the framework of the proposed approach, the effective cross section σ_{eff} can be estimated for the corresponding photon beam by means of a proper weighted average over photon spectrum ($\varphi(E)$) and NPs concentration (x % w/w):

$$\sigma_{eff} \approx \frac{\int dE\ \sigma(x\%\tfrac{w}{w})\ \varphi(E)}{\int dE\ \varphi(E)} \,. \qquad (3)$$

This formalism was applied to estimated effective cross sections (pair production and total) for systems containing different concentrations (w/w) of NPs of Ag, Gd and Au, as summarized in figures 5 and 6.

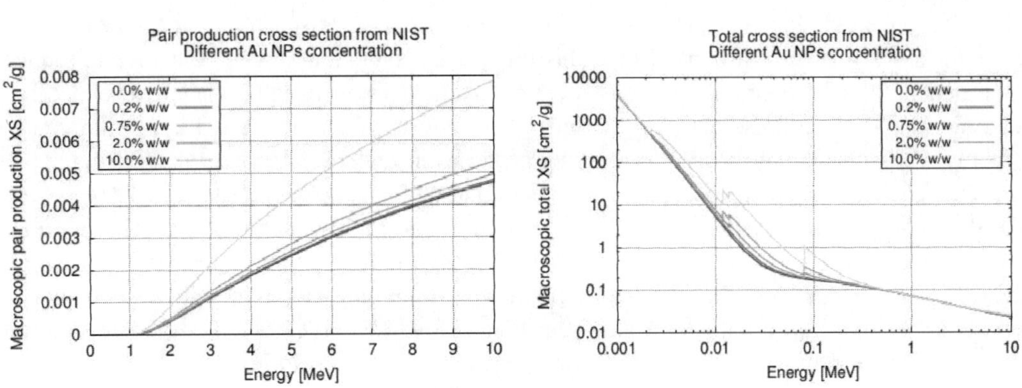

Figure 5: Pair production (left) and total (right) macroscopic cross sections for different concentrations of Au NPs.

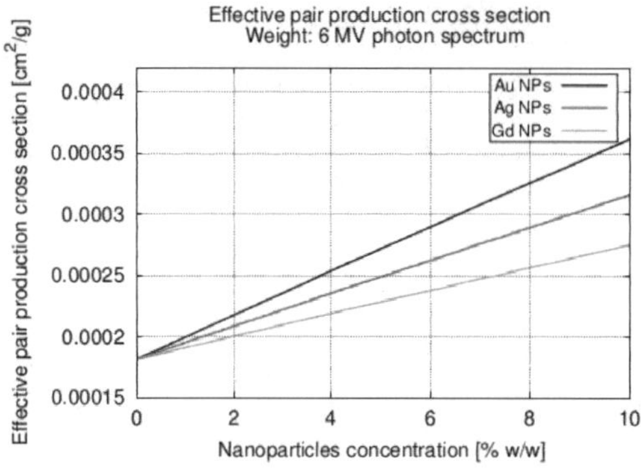

Figure 6: Pair production effective cross section for different NPs types.

As reported in Figures 5 and 6, it is verified that the effective pair production cross section increases according to NPs concentration, and this effect is greater for higher Z NPs. Hence, the probability of producing a pair and associated annihilation photons is increased in terms of NPs atomic number and concentration. This characteristic stands as a key issue and primary motivation for the present work.

2.2. Experimental setup for NPs detection during radiotherapy irradiation

Experimental setup was designed and implemented in bunker facility with a Varian Clinac 610C linac dedicated to routine clinical use for patient treatment. As mentioned, the linac works delivering a 6 MV photon beam that can be adjusted to different dose rates according to the required monitor units (MU) value determined in monitor ionization chamber. Field size, source-to-surface distance, incident direction and other irradiation parameters were configured according to the purposes of the different objectives of the present work. The detection system consists of a Canberra high energy germanium detector model GL1010 based on semiconductor diodes having a P-I-N structure with intrinsic region (I) sensitive to ionizing radiation, particularly X- and γ-rays. The operating principle of this kind of detectors are widely described in literature, then only a brief description about its main issues is presented. Incident photons interact with the material within the depleted volume of a detector producing charge carriers (holes and electrons) that are swept by the electric field to the P and N electrodes. This charge, which is in proportion to the energy deposited in the detector by the incoming photon, is converted into a voltage pulse by an integral charge-sensitive preamplifier. Because germanium has a relatively low band gap, this type of detectors requires to be cooled to reduce the thermal generation of charge carriers (thus reverse leakage current) to an acceptable level. Otherwise, leakage current induced noise destroys the energy resolution of the detector. Liquid nitrogen is the commonly recommended to be used for cooling purposes in such detectors, as followed in the present work. The GL1010 detector operates with the digital pulse processor (DPP) AMPTEK® model HPGe-PX5.The PX5 replaces both the shaping amplifier and multi-channel analyzer(MCA) found in analogue systems and it is capable of controlling the preamplifier output, applying real-time digital processing to the signal, detecting the peak amplitude, and binning this in its histogram memory. The spectrum is then transmitted to the user's computer. Actually, PX5-HPGe is a modified version of PX5 signal processor and power supply. A custom power supply board is installed in a standard PX5, providing the higher bias voltages needed by a germanium detector along with the higher pre-amplified power voltages and currents required by most preamplifiers. This DPP includes dedicated software (DPPMCA) capable ofchanging acquisition parameters such as coarse and fine gain, low and high thresholds of the respective slow and fast channels of pulse processing, number of channels in the spectrum, energy range calibration, among others.

Different phantoms were used for experimental measurements in form of cylinders made of cellulose ($H_{10}C_5O_6$) and having (3.2±0.1) cm and (2.6±0.1) cm in diameter (*d*) and height (*h*), respectively. The mass was determined to be m = (20.23 ± 0.01) g and the mass density was estimated as ρ =(0.97±0.7) gcm^{-3}.One of the phantoms has a small cavity were plastic vials with NPs-buffer are located, whereas the second one is absolutely solid.

The gold NPs used in the present work were provided by AuroVist™ certifying 15 nm as dimension distribution of the Gold BloodPool X-ray Contrast Agent distributed in nano-probes. Silver nano-particles are made at home, following the protocol described elsewhere[8] with dimensions characterized by size distributions with 15 nm as mean diameter. A vial was prepared with this Au NPs diluted in a buffer consisting of 20 mM of sodium phosphate and 150 mM of sodium chloride (NaCl). The sodium phosphate is a mixture of monobasic sodium phosphate (Na_2HPO_4) and dibasic sodium phosphate (NaH_2PO_4).

The setup for the experiments was previously investigated by analytic and Monte Carlo approaches. A sketch of the experimental disposition of irradiation source, phantoms, and collimation and detection systems is presented in figure 7.

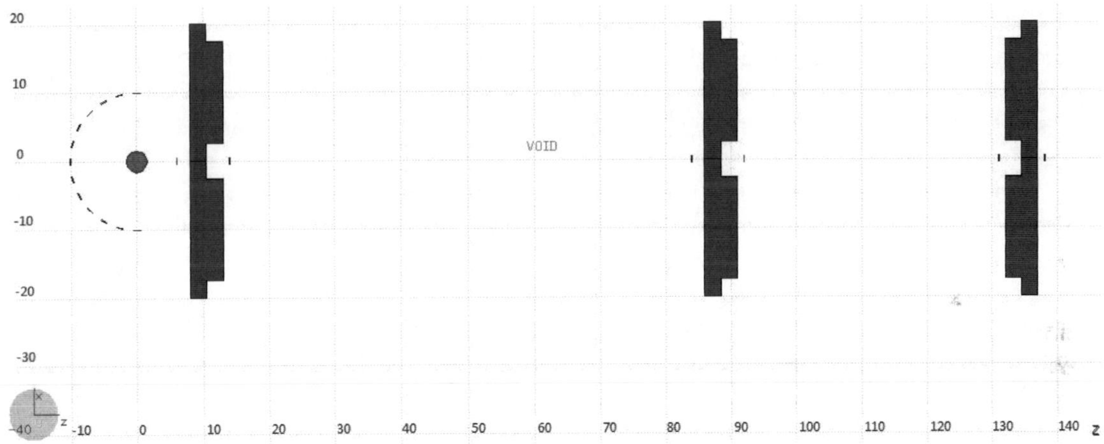

Figure 7: Experimental setup including phantom with NPs (red cylinder) and collimation system (3 sets of lead shutters). Ge detector (not shown) was placed at all positions immediately before and after shutters, as indicated in black lines.

The incident photon beam was set to 3.5×3.5 cm^2 field size placing the center of the phantom at the linac isocenter with the gantry at 180°. The collimation aperture of each shutter was carefully investigated in order to establish the optimal configuration providing good enough signal-to-noise ratio for the annihilation peak but avoiding saturation and maintaining appropriate detection dead times (lower than 5%). Nevertheless, an extra 1mm diameter Pb collimator is attached at the Ge entrance window in order to avoid lateral scattering contributions. In this framework, detection was realized at 90° respect the incidence photon beam performing, in all cases, two separate set of measurements: homogeneous cellulose phantom and cellulose phantom containing vial with NPs.

Finally, once photon spectra are recorded by Ge detector, the proposed procedure consists of performing offset correction and channel-energy calibration, and after that, for comparison purposes, the corresponding spectra processed according to the following formalism: The recorded and corrected signal Φ is then normalized by means of:

IOP Conf. Series: Journal of Physics: Conf. Series **1043** (2018) 012060 doi:10.1088/1742-6596/1043/1/012060

$$\Phi_{Norm}(E) = \frac{\Phi(E)}{\int dE\ \Phi(E)} \quad (4)$$

2.3. Monte Carlo simulations for NPs detection and dosimetry purposes

Monte Carlo simulations are nowadays one of the most useful and frequently used techniques for particle interaction and radiation transport modelling. Applications of Monte Carlo methods for medical physics purposes have grown systematically during the last decades mainly due to advances in computation technology. FLUKA[17] and MCNP6[18] main codes were used for the purposes of the present work. FLUKA main code is a general purpose Monte Carlo simulation package that fully integrates particle physics being capable of dealing not only with electron and photon transport, but also with complete models along with crosstalk for all radiation components with particular interest on heavy ions.FLUKA uses a multiple scattering approach for charged particle transport implementing a dedicated algorithm based on Moliere's theory. The lowest transport limit is 0.1 keV for photons and 1 keV for electrons. Multiple scattering is implemented by Moliere models providing also single scattering options that may be useful for low energy ranges.MCNP6 is a general purpose, continuous-energy, generalized-geometry and time-dependent Monte Carlo radiation transport code, designed to track particles of many types and broad ranges of energies. Contrary to previous versions, MCNP6 has been expanded to handle a multitude of particles and physical models improving transport and tracking issues. According to the aim of the present work, both Monte Carlo codes are capable of satisfactory handle incoherent and coherent scattering, fluorescent emission after photoelectric absorption, Bremsstrahlung and absorption in pair production with local emission of annihilation radiation.

Simulation setup for FLUKA and MCNP codes were prepared exactly according to disposition shown in figure 7. The inputs for the simulation subroutines define the incident photon beam 3.5×3.5 cm^2 with spectrum according to Figure 6.Tallies were introduced to compute photon fluence at different positions where the Ge detector was placed in experimental measurements, as shown in figure 7. Simulations were performed for 5×10^9 primary particles with absorption energy thresholds set to 50 keV for both photons and electrons, thus requiring 72 and 120 hours for FLUKA and MCNP6, respectively.

In addition to the calculations dedicated to simulate experimental setup including the NPs, other Monte Carlo simulations were prepared aimed to validate the theoretical approach proposed to model cross sections of systems consisting low percentages (weight in weight) of high Z NPs diluted in low Z buffers. In these cases, the geometrical setup consists on a thin layer irradiated by a mono-directional beam and transmitted/scattered radiation is recorded by impact detectors at different positions as shown in figure 8.

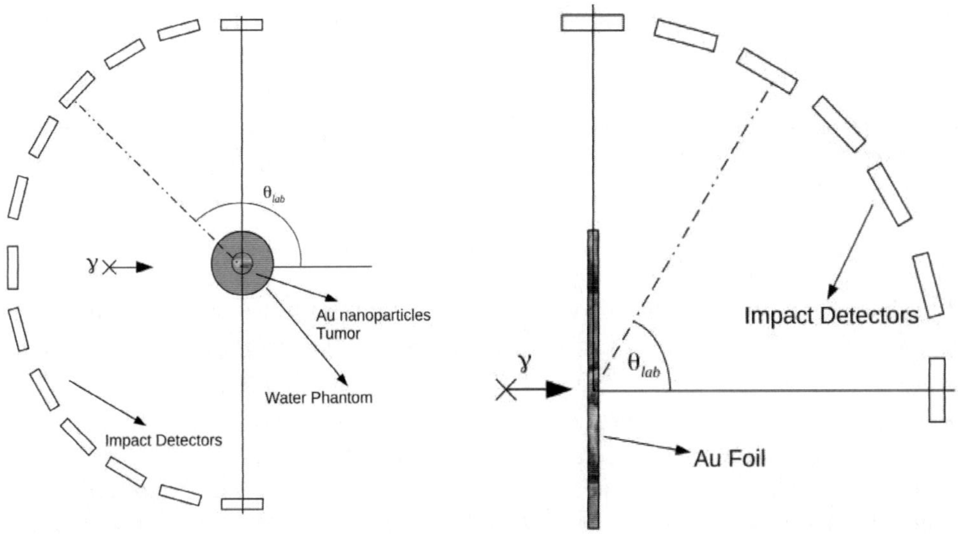

Figure 8: Monte Carlo simulations setups for: study of annihilation peak using spherical phantom containing a NPs irradiated by 2.2 MeV photons along with detectors at different angles (left) and study of positron angular distribution (right).

3. Results and Discussion

The first step consisted on investigating the performance of the proposed theoretical approach to model radiation-matter interaction in systems composed by low percentage (w/w) of high Z NPs diluted in water-equivalent media. This goal was accomplished by means of evaluation and comparison of Monte Carlo simulations carried out by different and independent validated Monte Carlo codes, FLUKA and MCNP6. As mentioned above, expression (1) can be integrated using the angles in order to get an approximate estimation for the cross section differential in positron energy, as reported in figure 2 for the theoretical approach. The corresponding results for Monte Carlo simulations are shown in figure 9.

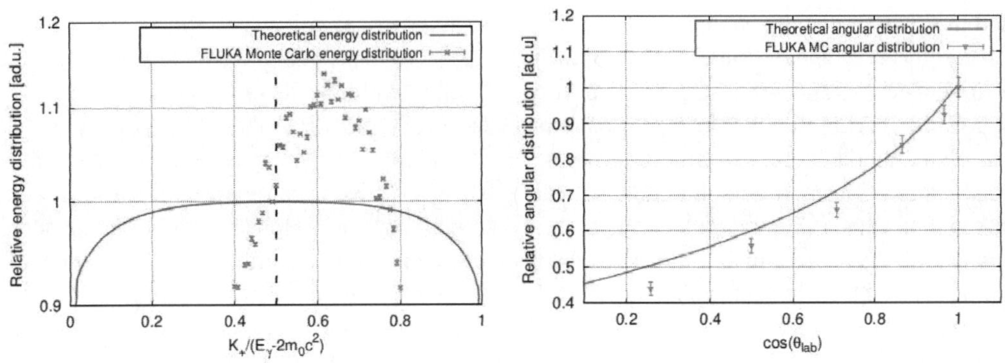

Figure 9: Comparisons between theoretical approach and FLUKA for: cross section for pair production in Au for 2.2 MeV incident photon beam (left) and positron angular distribution (right).

In order to contrast the theoretical predictions on pair production, simulation of an Au foil of $\Delta z = 0.01$ cm of thickness was performed the FLUKA code using a photon pencil beam with energy equivalent to the effective energy [19]of the linac spectrum of Figure 6 (2.2 MeV), so that the energy expected for the positrons is in the range $0 \leq K_{+} \leq 2.2\ MeV - 2m_0 c^2 = 1.18\ MeV$.

As can be seen by inspection of Figures 2 and 11, neglecting nuclear Coulomb field yields slightly more energy to the positron due to repulsion, thus making the cross section symmetrical respect to $\frac{K_{+}}{E_{\gamma}-2m_0 c^2} = \frac{1}{2}$. The difference of energy delivered to the photon taking into account the Coulomb repulsion is always less than $\delta(E) = 0.0075 \cdot Z\ MeV$ [20]. As obtained in figures 2 and 9, both the extreme cases when the positron takes all available energy from the photon, or it takes any, have zero probability of happening.Complete positron current leaving the foil was calculated.These relative energy distributions were normalized to the maximum value of the theoretical distribution at $\frac{K_{+}}{E_{\gamma}-2m_0 c^2} = \frac{1}{2}$ that explains why the simulated energy distribution is asymmetrical respect to this point, showing that FLUKA code includes actually an adequate treatment of the nuclear Coulomb repulsion effect over created positrons. Additionally, as the theoretical distribution was calculated in the Born approximation, it tends to fail when photon energy is close to the pair production threshold,or when electron or positron takes all the available energy. This issue corresponds to a failure in the assumptions of point nucleus without Coulomb screening effect due to the electronic cloud for the theoretical calculations. Nevertheless, the energy difference between the maximum of both distributions is $\delta(E) = 0.13\ MeV$ satisfying the condition: $\delta(E) = 0.13\ MeV < 0.0075.Z\ MeV = 0.59\ MeV$ for Au (*Z=79*).

The angular distribution of positrons calculated through Monte Carlo simulation, reported in Figure 11, was obtained with the half-crown geometry of impact detectors, as shown in figure 8 locating the impact detectors (cylinder 2 cm radius and 1 cm height) were located at 25 cm from the center of the foil. According to the obtained results reported in figure 9, it is possible to observe that: sticking to the most probable cases (available energy uniformly distributed and directions close to the that of the initial photon), FLUKA code makes a good approximation of the pair production phenomena, giving correct energy distributions with fine details as the Coulomb repulsion and electron cloud screening effect. Furthermore, the angular distribution of positrons is shown to have approximately the same behavior predicted theoretically for MC simulations, allowing this tool to model pair production in an accurate way. Finally, and according to the purposes of the present work, it should be mentioned that the ~ $O(Z^2)$ behavior of expression (1) adequately describes the relevant behavior for the effective cross section of systems composed by low percentage (w/w) of high Z NPs, such as Gd, Ag, Au, diluted in water-equivalent media.

Regarding the e^{-}-e^{+} annihilation, it was confirmed by FLUKA and MCNP6 Monte Carlo simulations with the concentric spheres that maximum of cross section corresponds to directions close to that of the pair in the CMS reference frame. Simulations were performed using the internal sphere ($r_{in} = 0.5$ cm) as emulatingthe tumor containing 0.1% of Au NPs dilute in water-equivalent tissue; whereas the external sphere ($r_{out} = 1.0$ cm) emulating patient body was made of water-equivalent tissue. Therefore, the tumor is located 0.5 cm depth in water-equivalent material. As sketched in figure 8, the source is positioned in the region inside the half-crown of detectors 14 cm shifted from the center of the phantom. The detected spectra obtained with FLUKA and MCNP6 by means of the quarter-crown detection system are reported in figure 10.

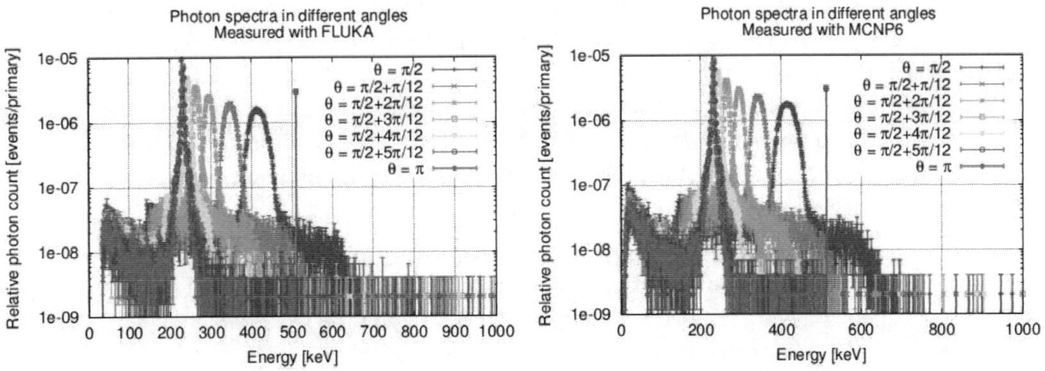

Figure 10: Photon spectra obtained with FLUKA (left) and MCNP6 (right) recorded by the quarter-crown detection system for a 2.2 MeV incident photon beam irradiating the two spheres system.

As shown in the results summarized in figure 10, the angular behavior is similar in both Monte Carlo codesand the peak probabilities are almost the same, as can be seen in figure 11.

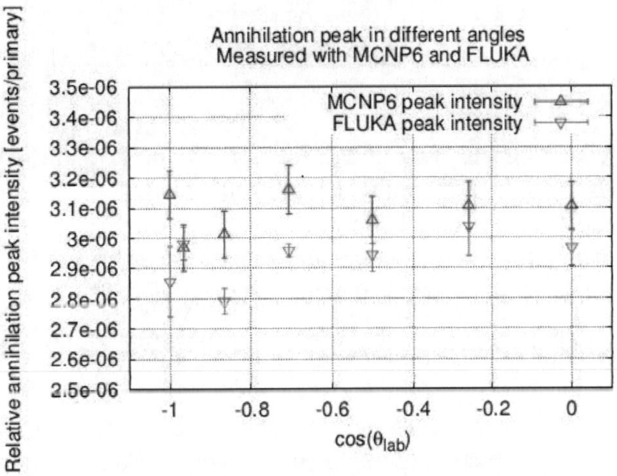

Figure 11: Photon spectra obtained with FLUKA (left) and MCNP6 (right) recorded by the quarter-crown detection system for a 2.2 MeV incident photon beam irradiating the two spheres system.

According to the results reported in Figure 11, angular dependencies show similar trends for FLUKA and MCNP6.Considering that integral of peakcounts is approximately constant for different directions, it might be acceptable to place a detector 90° respect initial photons incidence direction in order to achieve the good signal-to-noise ratio (SNR) but making it easier to design and implement the experimental setup.

3.1. Detection of NPs during radiotherapy irradiation by Monte Carlo simulations

The difference of the annihilation peak intensity with and without NPs in the anthropomorphic phantom was calculated using Monte Carlo code FLUKA. The geometry is depicted in figure 7. The input file included an external subroutine defining the source as a 3.5×3.5 *cm*2 extended field with energy distributed according to Mohan et al spectrum. The materials for cellulose and buffer plus NPs were defined using the chemical composition already explained. The spectra were measured in the different impact detectors shown in figure 7, but those after the first lead shutter were not able to measure anything even for 10^9 primary photons. The impact detector that simulates the Ge detector shows zero signal, and thus the first impact detector in the direction of the Ge detector was used for comparison with experimental results. The spectra measured in this case, shows a Dirac's delta annihilation peak, as can be seen in figure 12.

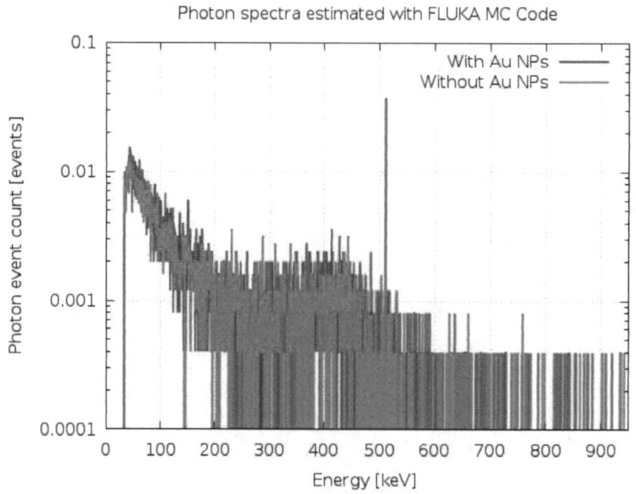

Figure 12: Photon spectra estimated with FLUKA MC Code for the anthropomorphic phantom with and without Au NPs

In this case, using equation (4) with the values at the peak, the intensity of the annihilation signal with NPs is of the order of 2.1% more intense than the peak without NPs.

3.2. Experimental detection of NPs during radiotherapy treatment

As described above, the experimental setup, preliminary validated by Monte Carlo simulations, was designed and configured exactly according to the simulations, as sketched in figure 7 using the linac couch was used as bench top for the positioning of the different accessories: phantoms, collimation systems and detectors. Figure 13 shows a typical example of the spectra recorded by the Ge detector for Au NPs. Contrary to Monte Carlo simulation, a spread annihilation peak was obtained in every measurement due to the intrinsic limitation of experimental detector to discriminate photons with similar energies, thus assigning the count to adjacent channels in the spectrum.

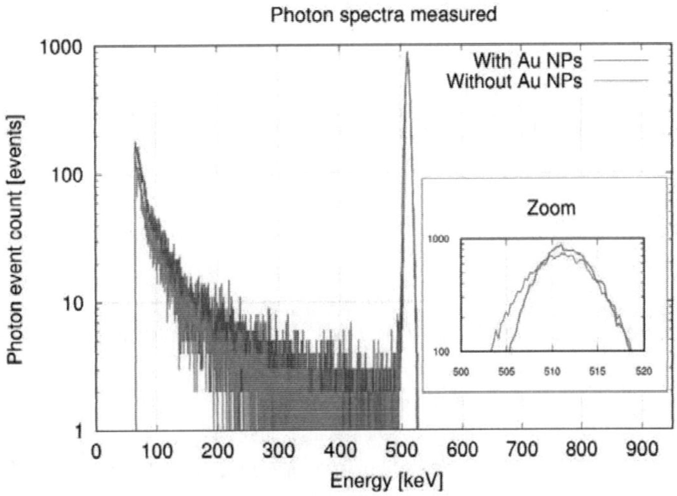

Figure 13: Experimental photon spectra obtained with the Canberra Ge detector for the cellulose phantoms with (red) and without (green) Au NPs.

The annihilation peak intensity in the case of cellulose phantom doped with Au NPs is, in average, 1.73% more intense than the case without NPs. Even though this tendency is marked, this difference is smaller than the percentage experimental error at the annihilation peak, that is 3.5%, thus being indistinguishable.

4. Conclusions

A theoretical approach was proposed and implemented to model and preliminary estimate cross sections of systems composed by water with low quantity of high atomic number nanoparticles diluted within it. Dedicated Monte Carlo subroutines were adapted to investigated basic radiation-interaction process for this kind of systems obtaining very good agreements with theoretical predictions.

Moreover, experimental setups were carefully designed and investigated in the virtual environment of Monte Carlo simulations using to independent Monte Carlo codes. Finally, it was possible to perform complex and high demanding experimental measurements that confirmed variations in annihilation peak due to the presence of nanoparticles during standard irradiation with photon beam from linear accelerator for patient treatment.

Although the obtained results appear as promising for establishing a preliminary framework for theoretical, simulation and experimental approaches to study these processes, further investigations are still necessary in order to achieve definitive and conclusive evidence and correlation between annihilation peak variations, nanoparticle presence/concentration and, potentially, absorbed dose during patient radiotherapy treatments.

5. References

[1]Pottier A., Borghi E., Levy L. *The future of nanosized radiation enhancers.* British Journal of Radiology **88**, 2015.

[2]Hainfeld J., Slatkin D., Smilowitz H. *The use of gold nanoparticles to enhance radiotherapy in mice*. Phys Med Biol. **49**, 2004.

[3]Butterworth K., Coulter J., Jain S., Forker J., McMahon S., Schettino G., Prise K., Currell F., Hirst D.*Evaluation of cytotoxicity and radiation enhancement using 1.9 nm gold particles: potential application for cancer therapy*. Nanotechnology. **21**, 2010.

[4]Matsudaira H., Ueno A., Furuno I. *Iodine contrast medium sensitizes cultured mammalian cells to X rays but not to γ rays*. Radiat Res. **84**, 1984.

[5]Chithrani D., Jelveh S., Jalali F.,van Prooijen M., Allen C., Bristow R., Hill R., Jaffray D.*Gold nanoparticles as radiation sensitizers in cancer therapy*. Radiat Res. **173**, 2010.

[6]Chithrani D., Jelveh S., Jalali F.,van Prooijen M., Allen C., Bristow R., Hill R., Jaffray D.*Recent Progress in Cancer Thermal Therapy Using Gold Nanoparticles*. J. Phys. Chem. C**120**, 2016.

[7]Alivov, Y., Baturin, P., Le, H., Ducote, J., Molloi, S. *Optimization of K-edgeimaging for vulnerable plaques using gold nanoparticles and energy resolved photoncounting detectors: a simulation study*. Phys. Med. Biol. **59**, 2014.

[8]Mattea F., Vedelago J., Malano F., Gomez C., Strumia M., Valente M. *Silver nanoparticles in x-ray biomedical applications*. Radiation Physicsand Chemistry **130**, 2017.

[9]Mandelstam S. *Determination of the pion-nucleon scattering amplitude from dispersion relations and unitarity. General theory*. Phys. Rev. **112**, 1958.

[10]Nagirner D. *Two-photon processes of electron-positron pair production and annihilation: kinematics and cross sections*. Astrophysics **42**, 1999.

[11]Bethe H. and Heitler W. *On the stopping of fast particles and on the creation of positive electrons*. Proc. of the Royal Society of London Series A Containing Papers of a Mathematical and Physical Character (1905-1934) **146**, 1934.

[12]Greiner W. and Reinhardt J. *Field Quantization*. Ed. Springer-Verlag, ISBN 978-3-642-61485-9, 1996.

[13]Peskin M. and Schroeder D. *An Introduction to Quantum Field Theory*. Ed. Addison-Wesley, ISBN 9780201503975, 1995.

[14]Mohan R., Chui C., Lidofsky L. *Energy and angular distributions of photons from medical linear accelerators*. Med. Phys. **12**, 1985.

[15]Bragg W. and Kleeman R. *On the α particles of radium, and their loss of range in passing through various atoms and molecules*. Trans. Royal Soc. South Australia**29**, 1905.

[16]Thwaites D. *Bragg's rule of stopping power additivity: A compilation and summary of results.* Radiation Research **95**, 1983.

[17]Battistoni, G., Cerutti, F., Fassò, A., Ferrari, A., Muraro, S., Ranft, J., Roesler, S., Sala, P. 2007. *The FLUKA code: Description and benchmarking*. AIP Conf. Proc. **896**, 2007.

[18]Pelowitz, D., Durkee, J. Elson, J., Fensin, M., James, M., Johns, R., Mckinney, G.,Mashnik, S., Waters, L., Wilcox, T., Hendricks, J., Verbeke, J. *MCNPX 2.7.0 Extensions*, 2011.

[19]Khan F. *The physics of radiation therapy*. Lippincott, Williams and Wilkins, 4th Ed., 2009.

[20]Attix, F. *Introduction to radiological physics and radiation dosimetry*. Wiley-VCH, 2004.

IOP Conf. Series: Journal of Physics: Conf. Series **1043** (2018) 012061 doi:10.1088/1742-6596/1043/1/012061

Detection Limits of a TXRF Analysis System in Double Reflection Mode

V H Poblete[1*], L Canales[2]

[1]Universidad Tecnológica Metropolitana, Dieciocho 390, Santiago-Chile.

[2] Master's Program in Nuclear Technology, Universidad Tecnológica Metropolitana

Corresponding author: victor.poblete@utem.cl

Abstract. Total reflection X-ray fluorescence (TXRF) analysis system is described with regard to develop a new prototype in double reflection mode. We discuss the basic principles of the technique to design and set-up a new TXRF module, coupled to an old X-ray generator. Different geometries are analyzed and the system is calibrated with standard solutions of several elements and the detection limits are established in the range of 4×10^{-11} to 13×10^{-11} g. Finally, a certified standard sample is measured and the results are compared for all studied geometries. This research is focused on the analysis of micro traces, in scientific and technological applications.

1. Introduction

Total reflection X-Ray Fluorescence (TXRF) technique Is a powerful analytical tool for micro and trace multi-elemental analysis and has the ability to simultaneously detect a wide range of elements (B-U) in a few minutes [1,2].

TXRF analysis is also of simple quantification, easy sample preparation and small sample amount is required, usually in the order of the micro liters. These characteristics make it a valuable tool in many fields of research, with varied applications to the environment and monitoring [3,4], mineralogy [5], oceanography [6], biology [7], forensic [8], industry and medicine. Liquid samples can be analyzed by TXRF directly. Soil, sediments, airborne particulate and solid samples in general, also can be analyzed after some pre-treatment like separation of suspended matters and acidic digestion. In another cases TXRF is highly suitable for nondestructive and micro analytical capabilities. In another cases TXRF is highly suitable for nondestructive micro analytical applications like as Biological tissue thin samples in medical and forensic applications [9, 10].

In this work we discuss the basic principles of the technique and the design and the development of a module in double reflection. An old Siemens Kristalloflex 710 / 710H X-ray generator is used, with Mo target and Zr filter, which is operated at 40kV / 20mA and the differents TXRF geometries are coupled for comparatives measurements. The systems are calibrated with standard solutions of several elements and the limits of detection are established, comparing them with those obtained by the different TXRF geometries. Significant increase in signal-to-noise ratio is expected to establish detection limits in the order of picograms.

2. Basic principles

TXRF technique is a variation of energy dispersive X-ray fluorescence (EDXRF) technique because a significant difference in the excitation and detection geometry is observed. In the EDXRF technique (figure 1), the primary radiation excites the sample at an angle of about 45°, wwhile TXRF is characterized by two fundamentally attractive factors: first, the primary beam interacts with the specimen at a critical angle less than to the critical angle at which the total reflection of the X-rays occurs (usually less than 0.1°) and second, the primary beam interacts on a plane with polished surface, e.g. quartz plate, containing the sample (figure 2).

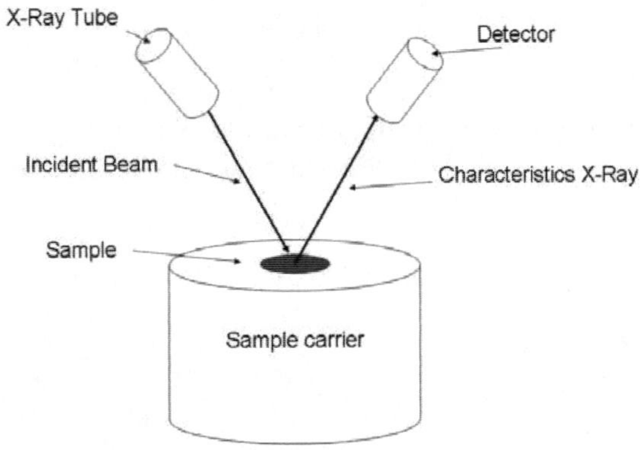

Figure 1: Instrumental arrangement for conventional EDXRF.

The totally reflected beam having nearly the same intensity as the primary beam, is generated at the glancing angle smaller than the critical angle, when the primary beam is totally reflected due the grazing incidence.

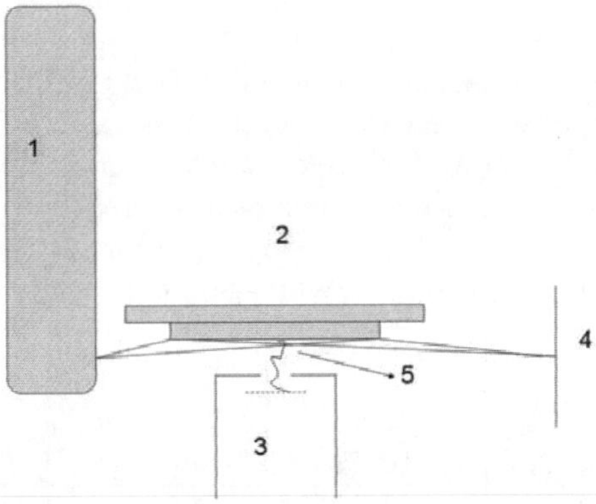

Figure 2: Instrumental arrangement for a simple TXRF. 1: X-Ray Generator; 2: quartz plate; 3: detector; 4: X-Ray stopper and 5: Characteristics X-Ray Fluorescence from the sample.

The instruments that use the TRXRF phenomenon have been designed and operated based on the classical dispersion theory of light and the Fresnel equations [11-13].

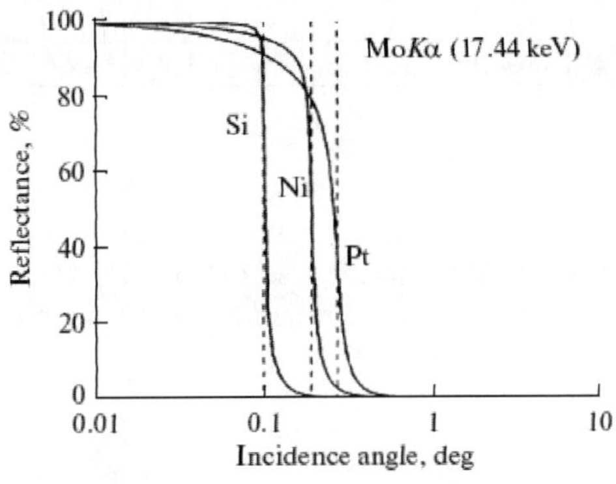

Figure 3: Angular dependence of the Si, Ni and Pt surface reflectance under Mo*K*α X-ray irradiation.

In this technique, the most important parameters studied are: the critical angle $\phi_C = \frac{99.1}{E}\sqrt{Z\rho / A}$, the penetration depth *Zp*, and the reflectivity *R*. The angular dependence of the reflectivity under MoKα X-ray irradiation, and the penetration depth of MoKα X-ray radiation versus incidence angle, for different reflectance surfaces are showing in figure 3 and figure 4, respectively [14].

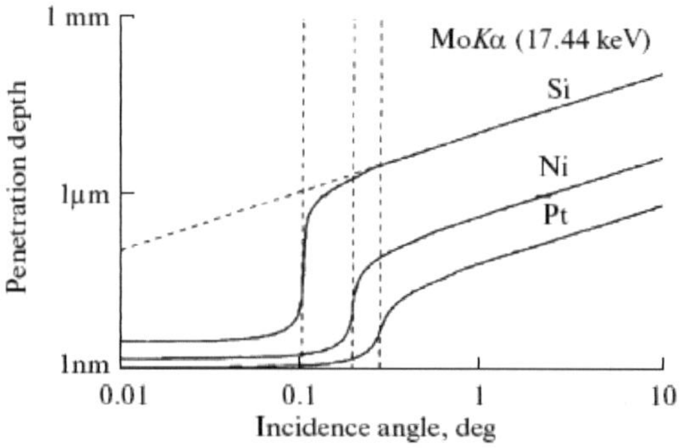

Figure 4: Penetration depth of Mo*K*α X_ray radiation for Si, Ni, and Pt versus incidence angle.

Penetration depth, for very small incidence angles, is independent of the energy of the primary beam. This gives an idea about the important role played by Zp in the superficial analysis, when the TXRF analysis technique is used.. Table I shows the critical angles, penetration depths and reflectivity for different materials. Two energies for the X-ray beam are considered.

Table 1. Critical angle (ϕ_C), Reflectivity (R) and Penetration Depth (Z_p), for different materials

Material	ρ g/cm^3	ϕ_C (min) E_1/E_2	Z_p (nm) $\phi = 3min$ E_1/E_2	Z_p (nm) $\phi = \phi_C/2$ E_1/E_2	Z_p (nm) $\phi = \phi_C$ E_1/E_2	R $\phi = \phi_C/2$ E_1/E_2
Plexiglass	1.20	9.9/4.6	4.5/5.7	4.9/4.9	127.2/247	0.997/0.999
Quartz	2.20	12.9/5.9	3.4/3.8	3.8/3.8	40.1/84.3	0.985/0.997
Silicon	2.32	13.2/6.1	3.3/3.7	3.7/3.7	31.1/62.0	0.973/0.994
Aluminum	2.69	14.0/6.4	3.1/3.4	3.5/3.5	30.5/64.4	0.978/0.995
Titanium	4.54	17.8/8.2	2.4/2.6	2.7/2.7	11.6/26.2	0.907/0.976
Germanium	5.31	18.9/8.7	2.3/2.4	2.6/2.6	18.3/13.3	0.966/0.937
Steel	7.80	23.5/10.8	1.8/1.9	2.1/2.1	7.5/14.2	0.875/0.963
Tantalum	16.60	31.9/14.6	1.3/1.4	1.5/1.5	6.8/6.2	0.915/0.900
Platinum	21.41	36.0/16.6	1.2/1.2	1.4/1.4	5.4/4.9	0.897/0.979

2.1 *Instrument design*

In the design conditions of a TRXRF instrument, the energy levels that generate background noise must be minimized, but it is more important to maximize the intensity of the analyte signal. The operation of the X-ray tubes is limited to the loss of solid angle. This problem can be optimized or improved by reducing the distances between the tube anode and the position of the sample. Also the optical focus of the X-Ray can be improved.

2.1.1 TRXRF in Simple Reflection

The simplest arrangement used for surface analysis is shown in figure 5. It is noted that the equipment settings for performing a scanning displacement for large samples is simple.

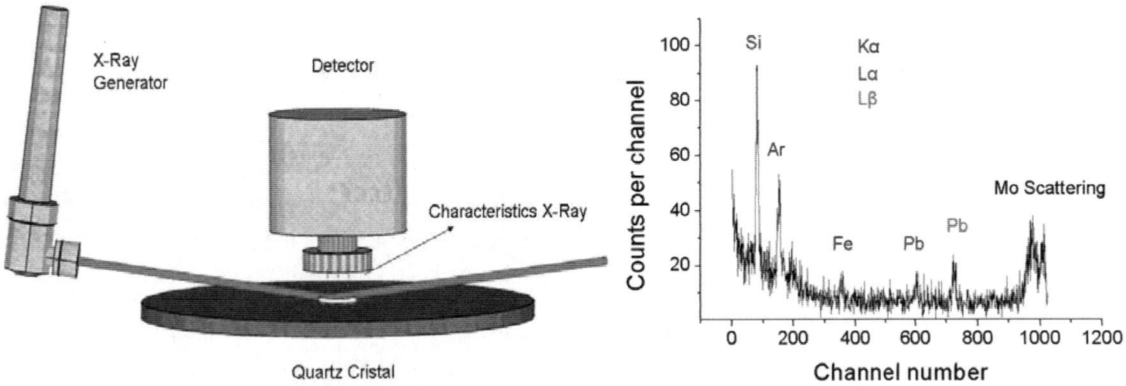

Figure 5: Left side: Simple TRXRF Arrangement used for quarz crystal analysis. Right side: Spectrum of the quartz crystal; counting time 200 sec.

In this experiment, the quartz crystal is excited under its critical angle, with an energy of 40 KeV. Here in the spectrum one clearly observes the characteristic peak of the Mo tube, and the "backscattering Compton" produced by high energy photons. Naturally in this spectrum the background is very high and the limits of detection are drastically dependent on the signal-to-noise ratio. Here we can appreciate the characteristic peaks of the elements Si, Ar, Fe and Pb, corresponding to the quartz crystal.

2.1.2 TRXRF in Standard Reflection

A more general arrangement is shown in figure 6. This facility was acquired through the International Atomic Energy Agency [15] for trace element analytical applications. In this geometry the primary radiation Attacks under the critical angle on a first reflecting crystal, with an energy of 40 KeV, which in turn reflects the X-rays towards de sample position (quarz crystal).

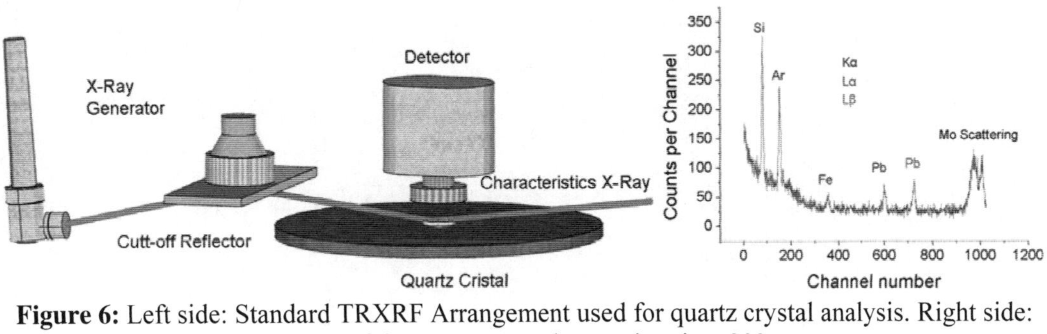

Figure 6: Left side: Standard TRXRF Arrangement used for quartz crystal analysis. Right side: Spectrum of the quartz crystal; counting time 200 sec.

Figure 6 shows a significant decrease in background noise. For the elements Si, Ar, Fe and Pb, the signal/noise ratio increases considerably, with respect to the arrangement in simple reflection. This geometry allows to reduce the elastic dispersion which is absorbed mainly in the first reflecting crystal, due the penetration depth of the primary beam.

The reflecting crystal interposed in the X-ray path serves as an effective filter, cutting all X-rays with energies greater than adjustable edge energy. On the other hand, the effect of filter on X-ray reflection allows the operation of the X-ray tube more effectively at maximum voltage, because the "Bremsstrahlung" normally present, is cut by the insertion of this lens ("cut-off").

2.2 TRXRF Module in Double Reflection

The new design described in this work, uses as an excitation system an old Siemens Kristalloflex 710 / 710H X-ray generator, with Mo target and Zr filter. The excitation system is operated at *40kV / 20mA*. The X-ray beam passes through a lead collimator that has an aperture of *0.1 mm x 10 mm*. Then two crystals of *100 mm* long, *20 mm* wide and *7 mm* thick, are attached to the collimator and the crystals are separated from each other at a distance of 45 μm. These crystals have adequate optical characteristics that allow the production of the physical phenomenon of total reflection: perfectly flat, highly polished surfaces with an average surface roughness of 1 *nm* and optical quality ($\lambda/20$).

The expected effect is to produce the collimation of the primary beam, by using mechanical devices that allow an attack angle less than or equal to the critical angle (6.4 arc minutes for the Mo anode). Later, an Amptek detector, model XR-100CR, with Si-PIN X Ray Detection, active area of 7 mm^2, silicon thickness of 300 μm, with a beryllium window of 25 μm thickness, at a distance of 5 *mm* between the beryllium window and the glass surface is coupled to the double reflection module. Subsequently the system is connected to an associated electronics, for signal processing. More background information can be obtained directly from the corresponding author of this article.

Figure 7 shows the new prototype developed as an original contribution for the use of the TRXRF technique. This device allows a significant reduction of background noise with respect to the other systems described above. Figure 8 shows the comparative spectra of three TRXRF systems, corresponding to the surface analysis of a quartz crystal, measured in the same experimental conditions. Clearly, the new prototype exceeds the detection limits obtained by the other two TRXRF systems.

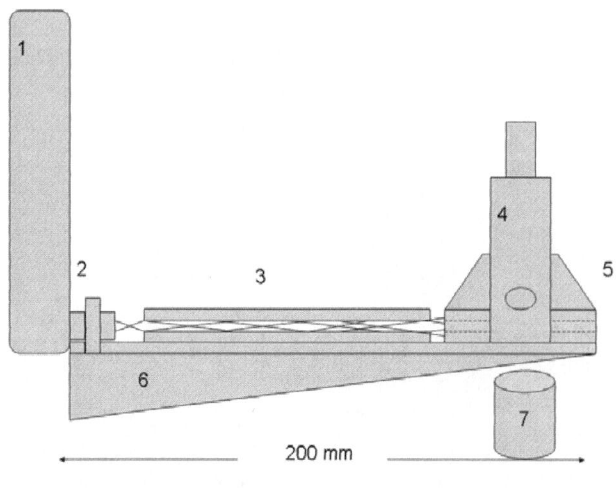

Figure 7: TRXRF Module in Double Reflection: 1: Siemens, Kristalloflex X-Ray generator. 2: Collimator and Zr filter. 3: Parallel quartz crystals. 4: X-Y geometry. 5: quartz simple holder. 6: System support. 7 Amptek Detector, model XR-100CR, Si-PIN X-Ray Detection.

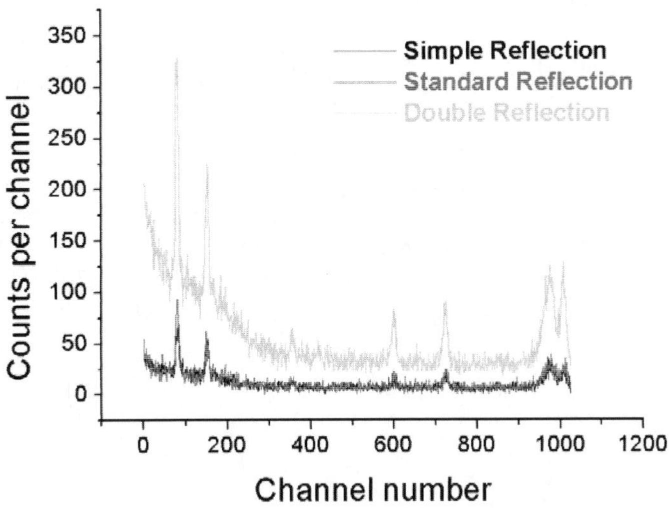

Figure 8: Comparative X-Ray Spectrum of the 3 TRXRF systems

2.2.1 Calibration of the TRXRF Module in Double Reflection

The total reflection module was calibrated to obtain the best geometric configuration, so that the excitation radiation illuminates on the sample holder under full reflection conditions. The critical angle ϕ_{crit} *(min)* can be calculated by Eq. (7), where E represents the energy of incident radiation (in keV), and ρ, Z, A, correspond to the density (g/cm^3), Z at the electron number in an atom or molecule and A is the atomic mass (g / mol) of the sample carrier, respectively.

$$\phi_{crit} - \frac{99}{E}\sqrt{\frac{\rho Z}{A}} \tag{1}$$

For the proposed system a quartz crystal (SiO2) is used as the sample carrier. The parameters are: $\rho = 2{,}5\ g/cm^3$; $Z = 30$ *electrons/molecule*, $A = 60{,}0863\ g/mol$ and an energy $E = 17{,}44\ keV$ of the $K\alpha$ X-ray from the incident radiation. Finally a critical angle of 6.4 min is obtained. This angle is adjusted using the micrometer screws of the reflector module.

The detection limit is obtained by:

$$L.D. = \frac{Concentration}{Peak\ Area} 3\sqrt{Background} \tag{2}$$

2.2.2 Quantitative analysis

The results are obtained by evaluating the measured spectra of both samples and calibration standards using only the net intensity or area under the curve, according to the following expression:

$$I_i = G\varepsilon_i(E_i)K_i\rho_i \frac{(1-e^{-X_i \rho_M D})}{X_i \rho_M} \tag{3}$$

Where $S_i = G\varepsilon_i(E_i)K_i$ is called the sensitivity of the element i and ρ_i/ρ_M is the weight fraction of the element in the sample. In this way the sensitivity of each element is expressed as:

IOP Conf. Series: Journal of Physics: Conf. Series **1043** (2018) 012061 doi:10.1088/1742-6596/1043/1/012061

$$S_i = \frac{I_i X_{M,i}}{W_i(1 - e^{-X_{M,i}\rho_M D})} \tag{4}$$

Here, X_i Is the absorption coefficient of the element i in the sample and D represents the thickness of the sample.

The used methodology, corresponds to the elemental sensitivities method applied to the multi-elemental analysis, considering the thin film approximation [16]. In order to establish the concentration of the elements present in the sample, these data are interpolated in the calibration curve (Figure 9 and Figure 10), considering the following expression:

$$S_i = a_0 + a_1 Z_1 + a_2 Z_2^2 + ... \tag{5}$$

Where a_i are numerical factors and Z_i, corresponds to the atomic number of the analyte.

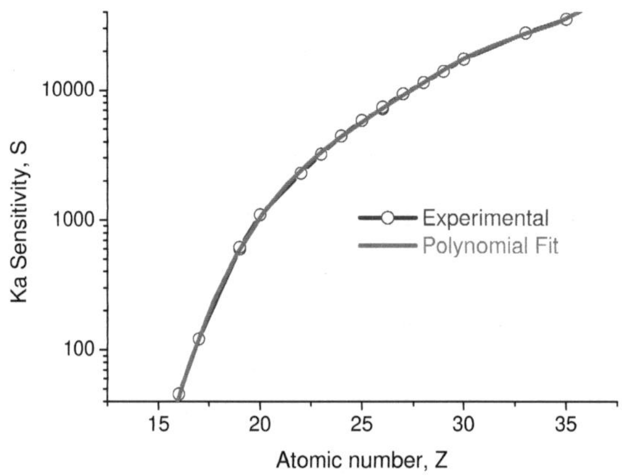

Figure 9: Experimental and fit Sensitivities of the $K\alpha$ lines (polinomial of order 6) as a function of atomic number of the elements.

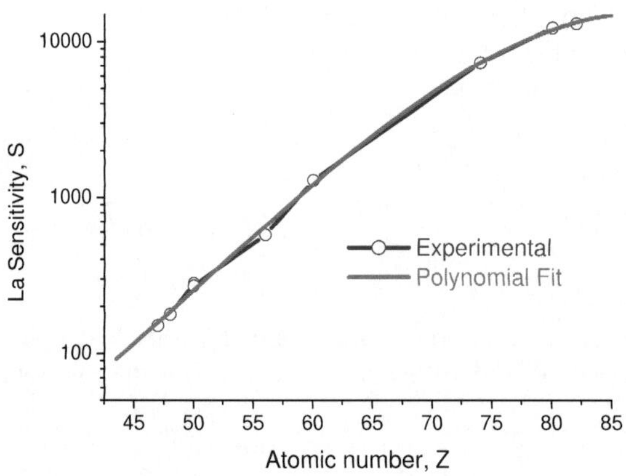

Figure 10: Experimental and fit Sensitivities of the $L\alpha$ lines (polinomial of order 4) as a function of atomic number of the elements.

This calibration curve is constructed by measuring standard samples containing elements of high purity, with different and known concentrations. Later, using the AXIL software [17], the sensitivities of all the elements within that range of energies are calculated. Figure 11 shows the X-ray spectrum of two multielemental standards of $2x10^{-10}$ g and 10^{-9} g, respectively for each one of the elements.

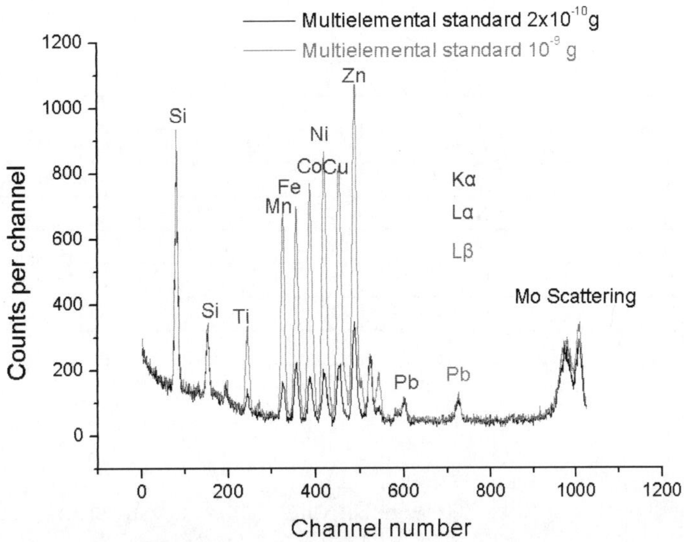

Figure 11: X-ray spectrum of two multielemental standards of $2x10^{-10}$ g and 10^{-9} g, respectively for each.

2.2.3 Sample preparation

Aliquots of 5 uL of each of standard solutions were prepared and were deposited on quartz supports. The Ga element was used as internal standard for all the samples. When the samples deposited on the quartz support are dried under IR light, a thin film is formed. The samples were excited for 300 s and the X-ray spectrum obtained for each one of them were adjusted by AXIL software. Finally the net areas of the characteristics X-ray peaks, for each element present in the sample were evaluated.

2.2.4 Results

For comparative purposes a sample carrier without the sample is measured under the same instrumental conditions, to establish the detection limits of each of the three TRFRX experimental arrangements, by means of the following expression:

$$LD_i(ppm) = \frac{3.\sqrt{BG_i}}{s_i} \tag{6}$$

where BG_i are the counts at the position of the element i in the spectrum and the sensitivity of the I element is given by:

$$S_i = \frac{Area}{Concentration} \tag{7}$$

In table 2 and Figure 12, we report the detection limit obtained for the three TRXRF arrangement. Clearly the new TRXRF Module in Double Reflection exhibits the best results that allow to establish concentrations in the picograms order.

Table 2. Comparative detection limits of the three experimental TRXRF arrangements analyzed (A) Simple reflection, (B) Standard Reflection and (C) new TRXRF Module in Double Reflection (Results in grams x10^{-11})

Z	Elements	Energy (KeV)	(A)	(B)	(C)
20	Ca	3,690	600	65	13
23	V	4,949	300	32	--
24	Cr	5,411	--	--	5
25	Mn	5,894	160	18	4
26	Fe	6,396	--	16	4
27	Co	6,920	160	16	--
28	Ni	7,471	--	--	4,5
28	Cu	8,040	--	--	5,0
30	Zn	8,838	150	15	4,5
34	Se	11,207	--	16	4,3
38	Sr	14,100	150	16	--
39	Y	14,931	150	16	4,1

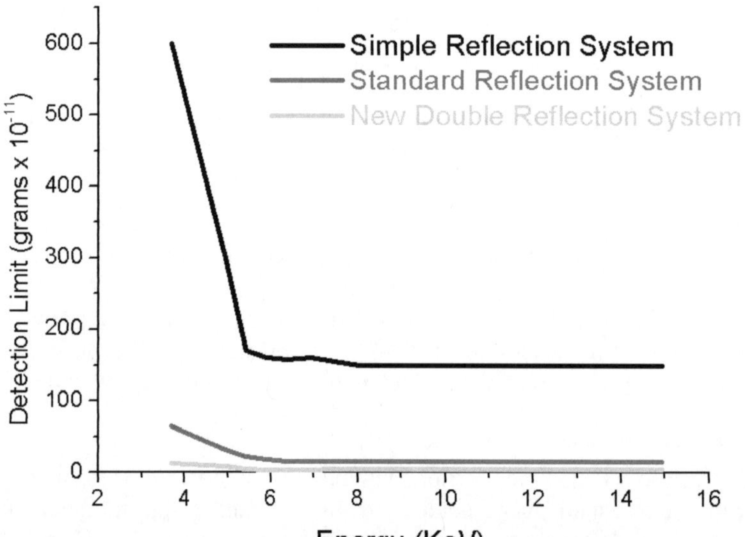

Figure 12: Comparative detection limits of the 3 experimental TRXRF arrangements analyzed (A) Simple reflection, (B) Standard Reflection and (C) new TRXRF Module in Double Reflection

Additionally to verify the applicability of the technique, from the precision and accuracy point of view, two certified rainwater samples were measured for a number of elements. The rain water samples from the Background Air Pollution Monitoring Network (BAPMoN 1, Austria) were measured by using de new TRXRF module in double reflection. Table 3 shows the analytical results.

Table 3. Analysis of certified rainwater simples from de BAPMON by using the new TRXRF Module in Double Reflection

Elements	(ppm)		(ppm)	
	TRXRF	BAPMON-1	TRXRF	BAPMON-2
Mn	40 ± 2	39	14 ± 2	14
Ni	21 ± 1	24	13 ± 2	14
Cu	74 ± 4	73	24 ± 2	22
Pb	110 ± 5	109	33 ± 3	32
Fe	70 ± 2	69	42 ± 2	41
Zn	63 ± 2	64	10 ± 2	9

3. Conclusions

The detection limits obtained for the new prototype exceed 38 to 46 times those obtained by the system in simple reflection and between 4 and 5 times those shown by the TRXRF Standard system, for the energy range studied.

By decreasing the excitation-sample source distance, the intensity of the characteristic radiation increases while the measurement times are reduced.

The adjustment and calibration of the critical angle is considerably simpler than traditional reflection systems, furthermore exhibiting good reproducibility and stability

The results obtained with this new device, project to this technique with applications to the simultaneous analysis of elements, in the order of 10^{-11} grams, for a variety of applications.

The costs involved in the construction of the new system are significantly lower than the conventional systems, because the number of precision mechanical parts is minimized. On the other hand an obsolete X-ray generator connected to a modern X-ray reflection and detection system is reused.

Acknowledgements: Programa CENIM de la Universidad Tecnológica Metropolitana.

References

[1] R Klockenkämper. *Total-Reflection X-Ray Fluorescence Analysis*, (1997). (John Wiley and Sons, New York).

[2] P Wobrauschek, C Streli. in *Encyclopedia of Analytical Chemistry: Applications, Theory and Instrumentation*, ed. R. A. Meyers, Vol.15, p.13384 (2000), (John Wiley and Sons, Chichester).

[3] M Tahri, M Bounakhla, M Zghaïd, A Benchrif, F Zahry, Y Noack, F Benyaïch. *X-Ray Spectrom.*, **42**, 284 (2013).

[4] D.J.Leland. Analysis of aerosols using total reflection X-ray spectrometry. Anal. Chem. **59**, 519(1987).

[5] M A W. Marks, T Wenzel, M J Whitehouse, M Loose, T Zack, M Barth, L Worgard, V Krasz, G N Eby, H Stosnach, G Markl. *Chem. Geol.*, **291**, 241 (2012).

[6] M A Alvarez-Vazquez, C Bendicho, R Prego. *Microchem. J.*, **112**, 172 (2014).

[7] L Telgmann, M Holtkamp, J Künnemeyer, C Gelhard, M Hartmann, A Klose, M Sperling, U Karst. *Metallomics*, **3**, 1035 (2011).

[8] Z Polgári, Z Ajtony, P Kregsamer, C Streli, V G Mihucz, A Réti, B Budai, J Kralovánszky, N Szoboszlai, G Záray. *Talanta*, **85**, 1959 (2011).

[9] V B Yadav, R H Pillay, S K Jha. *J. Radioanal. Nucl. Chem.*, **300**, 57 (2014).

[10] M C Rodriguez Castro, V Andreano, G Custo, C Vázquez. *Microchem. J.*, **110**, 402 (2013).

[11] R Jenkins, R W Gould, D Gedcke. Quantitative X-ray spectrometry. Copyright Maecel Kekker Inc. 1981. U.S.A.

[12] A H Compton, S K Allison. X-rays in theory and experiment. D.Van Nostrand Company. Princeton, N.J.(1935).

[13] P.Wobrauscheck, H.Aiginger, G.Owesny, C.Strelli. J.Trace and Microprobe Techniques. **6**(3), 295-336(1988).

[14] N V Alov. Inorganic Materials, **47**, Nº1, 4-14(2011).

[15] V Poblete. CHI/02/010 Project, Comisión Chilena de Energía Nuclear, Ander auspices of IAEA, (1989 to 1991).

[16] V Poblete, P Castillo, J Gálvez. Analysis of impurities in semiconductor thin films trough X-ray fluorescence spectrometry by total reflection. 42nd annual Denver X-ray conference, Denver, Colorado. August 2-6, 1993.

[17] P Kregsamer. Quantitative X-ray analysis system (version3.2). Under the auspices of IAEA. (1995).

IOP Conf. Series: Journal of Physics: Conf. Series **1043** (2018) 012062 doi:10.1088/1742-6596/1043/1/012062

Monte Carlo modelling of photon transport using Heterogeneous Computing

E. Doerner. C. Rebolledo. V. Gomez

Institute of Physics, Pontificia Universidad Católica de Chile, Vicuña Mackenna 4860, Santiago 7820436, Chile

E-mail: edoerner@fis.puc.cl

Abstract. In this work the use of the OpenCL framework to implement the parallel processing of the Monte Carlo (MC) simulation of the transport of photons on matter is presented. Our platform, named OCL_MC, relies on the physical model for photons implemented in EGSnrc and adapts it for execution across heterogeneous platforms consisting of central processing units (CPUs) and graphics processing units (GPUs). Runtime and accuracy of OCL_MC was tested against a selection of phantoms and several photon beams in the kV range, using different heterogeneous platforms. We could demonstrate that the dose distributions calculated with OCL_MC are in good agreement with EGSnrc. Our implementation scales very well across the different processors available in the studied heterogeneous platforms. Therefore, OCL_MC is capable to exploit all available processing resources inside modern computing systems without sacrificing the accuracy and reliability of the MC method.

1. Introduction

Monte Carlo (MC) simulations are widely used to model the effects of ionizing radiations for applications in medical physics. MC is considered the most precise method for dose distribution calculations, but it usually requires large computation time. To overcome this problem, we have developed the platform OCL_MC, which can efficiently use the different processing units in a heterogeneous computing system through OpenCL [1], which was designed specifically to simplify the coding of applications for heterogeneous systems that incorporate multicore CPUs, GPUs, accelerators, among others.

2. Materials and Methods

OCL_MC implements the photon transport method implemented on EGSnrc [2]. The main purpose of this work is to accelerate the simulation using efficiently all the resources of a system, while not compromising the precision and reliability of the original method. For this purpose, the simulation process, memory-access scheme and data transfer mechanism across computing devices have been optimized in order to exploit the potential of the underlying hardware, without sacrificing the reliability of the original implementation.

To validate our platform performance and precision tests were carried out for a selection of phantoms and three kV photon beams. They were selected based in the geometries typically used in high impact MC studies [3]. All phantoms consist of 61x61x150 voxels with a size of 0.5x0.5x0.2 cm^3. We used a homogenous water phantom and two water phantoms with a layer of bone and lung ranging from z = 5 to 10 cm. The obtained dose distributions were compared with the distributions obtained using the EGSnrc platform. The performance tests were carried out in two systems consisting

IOP Conf. Series: Journal of Physics: Conf. Series **1043** (2018) 012062 doi:10.1088/1742-6596/1043/1/012062

on an Intel Core i5 6500 CPU and an AMD Radeon R9 M380 GPU (PC) and 2x Intel Xeon E5-2620 CPUs and 3x NVIDIA Tesla K40m GPUs (Server), respectively. OC_MC was compiled using the Intel Parallel Studio XE 2017 set of compilers and the Intel OpenCL SDK. For all tests 2.56×10^8 particle histories were simulated.

3. Results and Discussion

The precision tests found a satisfactory conformity between the dose distributions calculated with OCL_MC and EGSnrc. An example of such tests is shown in Figure 1. The performance tests showed OCL_MC being capable of shorter computation times than EGSnrc. OCL_MC allows the user to use all the processing units at the same time significantly reducing computation times without compromising the precision and consistency of the MC method.

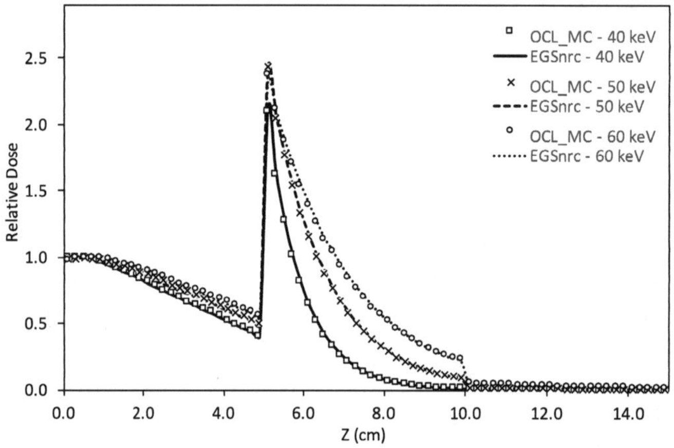

Figure 1. Comparison of the simulation results of OCL_MC and EGSnrc using the bone phantom. The difference between OCL_MC and EGSnrc is within 1.5%.

Table 1. Total simulation time for different cases of OCL_MC and EGSnrc.

System	Phantom	T_{OCL_MC} (s)	T_{EGSnrc} (s)	Speedup
	H_2O	138.5	418.2	3.0
PC	H_2O-Bone-H_2O	113.1	274.0	2.5
	H_2O-Lung-H_2O	130.5	430.9	3.3
	H_2O	22.0	93.9	4.3
Server	H_2O-Bone-H_2O	19.8	66.5	3.4
	H_2O-Lung-H_2O	19.8	90.6	4.6

4. Conclusions

In this work we presented an OpenCL based Monte Carlo platform that is capable of leverage the potential of modern heterogeneous computing systems consisting of different kinds of processors such as CPUs and GPUs. Thanks to this capability, OCL_MC is capable of obtaining a significant speed-up when compared with the original implementation without introducing any approximation to the underlying physics that would compromise the accuracy and reliability of the dose calculations. OCL_MC is a portable code that is able to exploit current and future advances in computing power though the OpenCL API.

References

[1] Khronos OpenCl Working Group. The OpenCL specification, version 1.2, 15 November 2011.
[2] Kawrakow I, Mainegra-Hing E, Rogers D, Tessier F and Walters B 2015 NRCC Report PIRS-701
[3] Ziegenhein P, Pirner S, Kamerling CP, Oelfke U. Fast CPU-based Monte Carlo simulation for radiotherapy dose calculation. Phys. Med. Biol. 60 (2015) 6097.

Atmospheric delay for electromagnetic signals associated with space / time variations of the refractive index in the lower atmosphere: Applications in spatial geodesy.

V. Henrquez[1,2], I. Fustos[3], M. Cordoba[4], C. Aravena[1,2]

[1]Estudiante de Geología, Universidad Católica de Temuco, Chile. [2]Departamento de Ciencias Ambientales, Universidad Católica de Temuco, Chile.[3]Facultad de Ingeniería y Ciencias, Universidad de la Frontera, Chile. [4]Observatorio Volcanológico de los Andes del Sur, Sernageomin, Chile.

E-mail: ivo.fustos@ufrontera.cl

Abstract. The feasibility quantifying atmospheric delay of electromagnetic signals using a mesoscale numerical model (WRF model) is explored. The experiment investigates a new alternative to reduce uncertainty sources in data exploitation for spatial geodesy applications. Our experiment was compared with GPS estimations of stations located in the South-America. The GPS estimations are consistent with values of our proposed methodology. This opens the possibility of using a mesoscale numerical model applied to spatial geodesy. However, these results, despite being promising, need further analysis and research, opening the opportunity to numerical modeling of atmospheric delays using mesoscale models.

1. Introduction

The techniques based on remote sensing have been a data source for volcanic deformation studies during the last years [1, 2]. However, remote measurements like GPS or synthetic aperture radar (SAR) have an intrinsic uncertainty in their measurements. Cheng et al. [3] showed that there are sources of uncertainty in GPS and SAR signals related to the atmospheric component. This is consistent with the basic fundamentals of electromagnetic theory, which depends on the material in which the electromagnetic signal travels [4]. A study case corresponds to the space geodesy data. For space geodesy, one of the most difficult issues corresponds to the atmospheric effect due to the water vapor presence. Previous studies have shown that spatial and temporal changes of 20% in relative atmospheric humidity could lead to 10-14 cm errors in measurements of volcanic deformation [5]. Remy et al. [6] have found that the incorrect elimination of atmospheric uncertainty can lead to misinterpretation in the sign InSAR coming to associate the atmospheric influence to the signs of volcanic deformation. Therefore, the atmosphere is a limiting factor for geodetic applications, such as radar altimeters, SAR and GPS.

To solve the atmospheric delay problem, global products such as the Vienna functions [7] or models derived from global climate models [5] exists. However, these products are too coarse respect to the spatial resolution and are far of the minimum requirements for correcting geodetic data for local studies. This work offers a new alternative of atmospheric correction using numerical models of mesoscale in development. For this, mesoscale weather simulations (Weather Research and Forecasting) were implemented to a theoretical model of atmospheric delays that allowed estimating the values locally. In this way, this work addresses the atmospheric delays with the goal of quantifying their effect reducing the uncertainty associated. We perform atmospheric delay map for 8 GPS stations located in active volcanoes of the Southern Andes. We choose Villarrica (VnV) and Nevados de Chilln (VNC) volcanoes due to the presence of differential GPS stations to validate our work.

2. Study zone

The Villarrica volcano is located in the south of Chile (39.42 S - 71.93 W). Many cities such as Villarrica, Pucón, Coñaripe, and Lican-Ray are emplaced around the volcano. Moreover, the Villarrica volcano is classified as a high-risk volcano due to historical eruptive processes. The crater has a height of 2847 m.a.s.l. with a basal diameter of 12,5 km covering an area of 490 km^2. The main structure is covered by a glacier of approximately 30 km^2. Moreover, the crater has a diameter of 200 m that shows a constant emission of gases. Currently, 5 GPS stations are emplaced near the crater: Llafenco (LLFN), Tralco (TRAL), Vn. Villarrica II (VNV2), Lomo (LOMO) and Vetado (VTDO) (see Map 1 in Annexes), which monitor the volcano, constantly delivering position data.

Moreover, Nevados de Chillán complex (NCC) is located close to Villarrica Volcano (36 51.2'S, 71 23.8' W). NCC volcano is classified as a high-risk volcano due to their proximity to villages and geological characteristics. The volcano height reaches the 3216 m.a.s.l with a basal diameter of 13.8 km. The last major eruption was registered in 1973. The volcano has 5 GPS stations named Chintuco, Fresco (FRSC), Portezuelo (PTZL), Shangrila (SHLA) and Ñuble (NUBL), which are constantly delivering position data for analysis.

3. Methodology

We perform an atmospheric delay mapping in order to understand spatio-temporal changes. We use an integration of the air refraction index as a function of the surface (elevation) and the vertical according to the proposal of [8].

$$\delta = \frac{10^{-6}}{cos(\theta)} \int_{z_o}^{z_{ref}} \frac{k_1 R_d}{g_m}(P(z) - P(z_ref)) + \left(K_2 - \frac{R_d}{R_v}K_1\right)\frac{e}{T} + K_3\frac{e}{T^2} \tag{1}$$

Where z_{ref} was chosen as the height above the troposphere (typically 15,000 m). The value was considered due to this zone does not have larger variations in the time. The first term in equation 1 corresponds to the dry air component of the delay path (hydrostatic term), while the second term is related to the humidity of the air (wet term). We forced the equation 1 through a mesoscale climate model called the Weather and Research Forecast (WRF) [9]. The WRF model allows obtaining atmospheric variables from a dynamic downscaling process [9] based on data from the NCEP reanalysis system [10].

This model provides values of various meteorological parameters with a spatial resolution of 3 km covering the study area from 2012 to the present time scale. Vertical stratification is described in 52 pressure levels, densely spaced at low altitude (25 hPa interval), with the highest level of about 50 km (1 hPa). The total delay maps were combined by stacking to produce time series. These results were compared with GPS measures of the chilean volcanic monitoring network of national agencies (OVDAS-Sernageomin).

4. Results

The results were obtained in a defined temporal baseline according to available GPS data. In VnV, WRF data is in a time series from January 2012 to December 2015, the same for VNC

IOP Conf. Series: Journal of Physics: Conf. Series **1043** (2018) 012063 doi:10.1088/1742-6596/1043/1/012063

Figure 1. Correlation map between atmospheric delay (in meters) and topography for Villarrica Volcano.

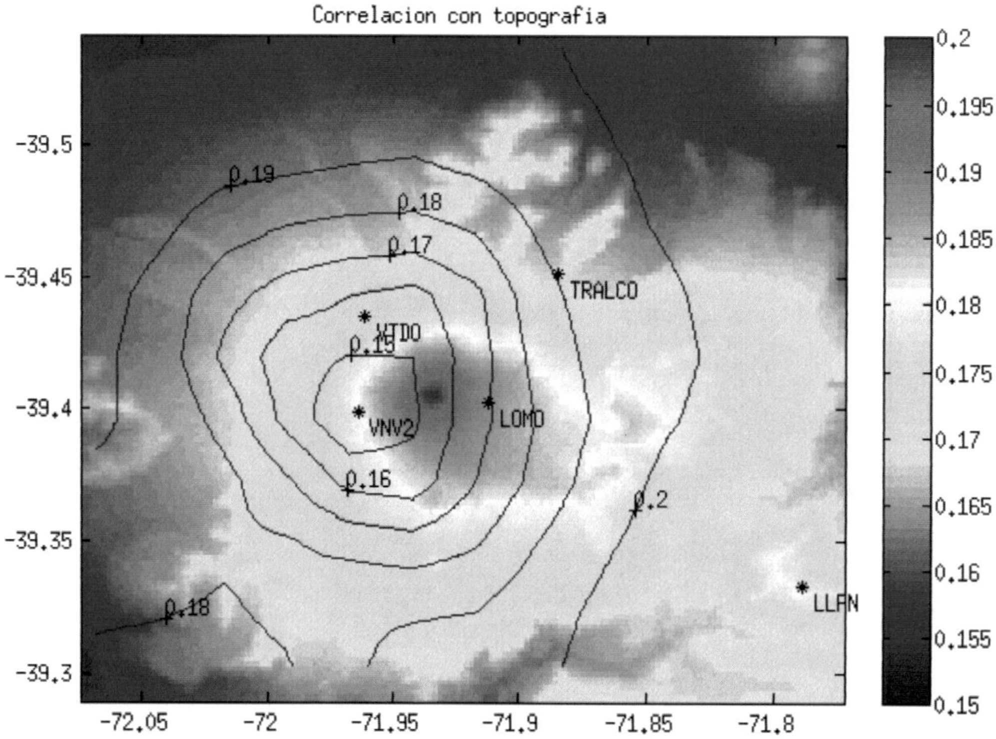

data. However, GPS data for VnV have a time series from December 2013 to December 2015, and for VNC a series from December 2015 to February 2016.

Table 1. Comparison of atmospheric delay (in centimeters) for the proposed methodology and GPS estimations.

GPS station	Min value modeled	Max Min value modeled	Min value mated	GPS esti-	Max value mated	GPS esti-
LOMO	-6.28	7.63	0.08		9.43	
TRAL	-5.23	6.38	0.04		16.05	
VNV2	-6.71	10.04	0.07		15.06	
VTDO	-6.59	9.03	0.15		16.95	
FRSC	-9.49	10.78	0.43		6.95	
NUBL	-9.82	11.95	0.18		9.72	
SHLA	-9.49	10.78	0.19		6.78	
PTZL	-9.60	11.81	0.27		8.49	

Our results show that VnV experiments episodes of atmospheric delay ranging between 0.7 cm and 2.1 cm. These are similar to GPS stations that show temporal variations between 1.5 cm up to 2.7 cm. Moreover, VNC shows temporal variations between 2.5 cm and 2.7 cm for the delay calculated using our atmospheric delay proposed scheme. These values are consistent

Figure 2. Relative atmospheric delay for the Nevados de Chillán volcano. Temporal baseline for four days (01 to 04 January 2015).

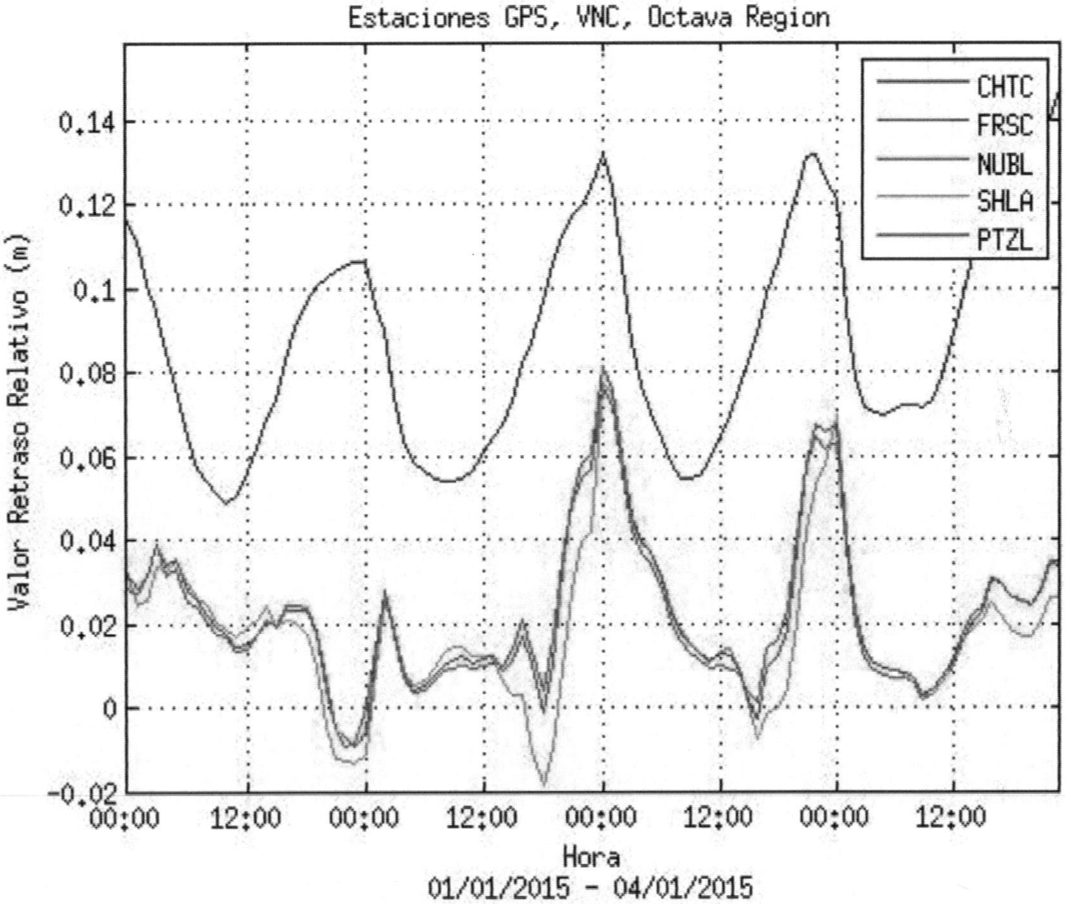

again with GPS measurements (temporal variations of 2 cm up to 2.5 cm). Additionally, a seasonal variability was observed in the values of atmospheric delays in the two zones, which correlates with the local dynamics of the low atmosphere.

5. Discussion and Summary

We propose that the present methodology is able to quantify atmospheric delay. We observe the feasibility to implement an atmospheric delays model at a local-scale as possible in the study zone. Our results show the ability of atmospheric models to estimate atmospheric delays on continental areas. These estimations were demonstrated using a GPS data comparison.

Our product will allow correct geodetic data of satellite measurements like radar altimeters or SAR. The methodology will reduce the uncertainty, improving atmospheric delay estimations associated with potential rates of deformation. For example, our study can be used for volcanic deformation estimations associated with magma intrusion activity. Therefore, new estimations with these corrections will allow taking adequate decisions in volcanic risk assessments. However, the current results are preliminary yet. Moreover, from the literature, we think that the estimation could be less accurate on the coastal zones due to the parameterizations during

Figure 3. Relative atmospheric delay for the Villarrica volcano. Temporal baseline for four days (01 to 04 January 2012).

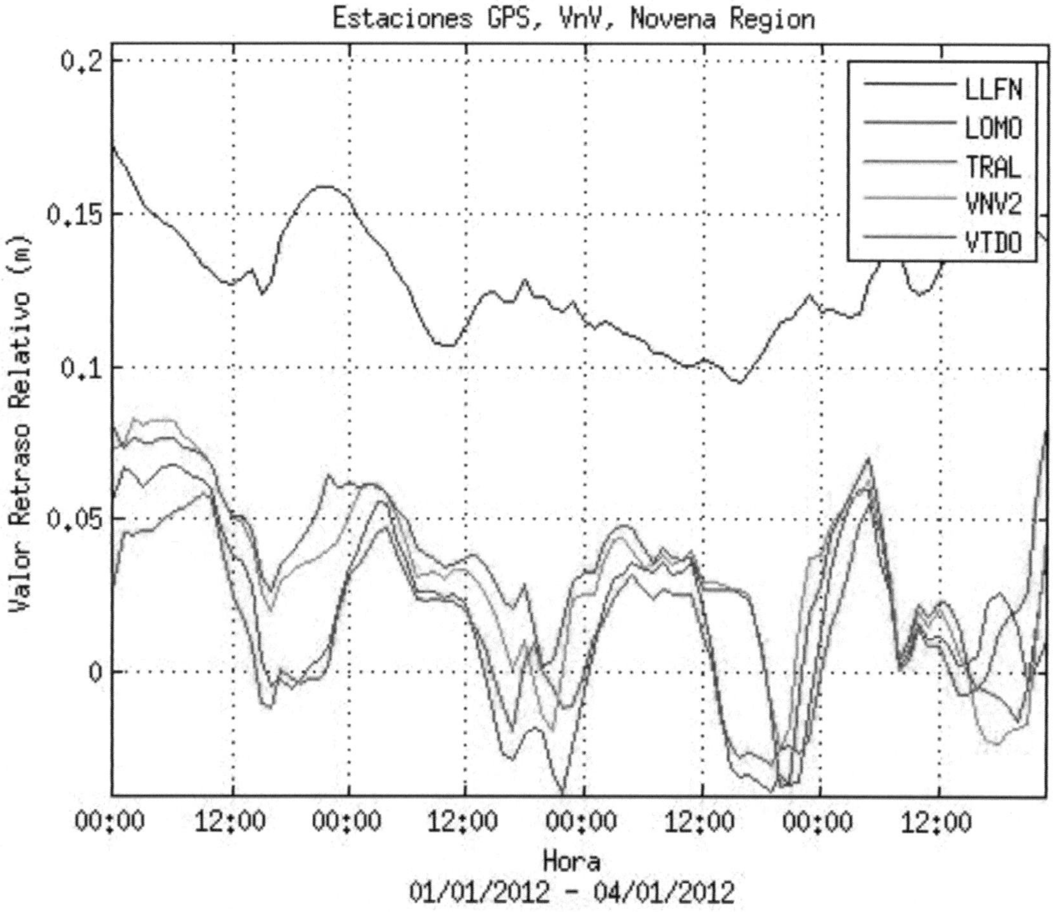

the implementation of the WRF model. Likewise, there may be strong temporal fluctuations in areas with complex topography as shown in Figures 1 and 2. This work corresponds to a preliminary approach within the group of geophysics and remote sensing of the Universidad de la Frontera and Catholic University of Temuco in close collaboration with the OVDAS volcanological observatory.

References

[1] Ebmeier, S. K., Biggs, J., Mather, T. A., Elliott, J. R., Wadge, G., & Amelung, F. (2012). Measuring large topographic change with InSAR: Lava thicknesses, extrusion rate and subsidence rate at Santiaguito volcano, Guatemala. Earth and Planetary Science Letters, 335336, 216225. https://doi.org/10.1016/j.epsl.2012.04.027

[2] Jay, J., Costa, F., Pritchard, M., Lara, L., Singer, B., & Herrin, J. (2014). Locating magma reservoirs using InSAR and petrology before and during the 2011-2012 Cordón Caulle silicic eruption. Earth and Planetary Science Letters, 395(September 2016), 254266. https://doi.org/10.1016/j.epsl.2014.03.046

[3] Shilai Cheng, Daniele Perissin, Hui Lin, Fulong Chen, Atmospheric delay analysis from GPS meteorology and InSAR APS, Journal of Atmospheric and Solar-Terrestrial Physics, Volume 86, September 2012, Pages 71-82, ISSN 1364-6826, https://doi.org/10.1016/j.jastp.2012.06.005.

[4] John David Jackson, 2010. Classical electrodynamics. dition 2, illustre. Wiley Editors, 1975. ISBN 047143132X, 9780471431329. 848 pages.
[5] Jolivet, R., Grandin, R., Lasserre, C., Doin, M.-P., & Peltzer, G. (2011). Systematic InSAR tropospheric phase delay corrections from global meteorological reanalysis data. Geophysical Research Letters, 38(17), n/an/a. https://doi.org/10.1029/2011GL048757
[6] Remy, D., Chen, Y., Froger, J. L., Bonvalot, S., Cordoba, L., & Fustos, J. (2015). Revised interpretation of recent InSAR signals observed at Llaima volcano (Chile). Geophysical Research Letters, 42(10), 38703879. https://doi.org/10.1002/2015GL063872
[7] Boehm, J., and H. Schuh (2004), Vienna mapping functions in VLBI analyses, Geophys. Res. Lett., 31, L01603, doi:10.1029/2003GL018984.
[8] Berrada Baby, H., P. Gol, and J. Lavergnat (1988), A model for the tropospheric excess path length of radio waves from surface meteorological measurements, Radio Sci., 23, 10231038
[9] Skamarock, W. C., J. B. Klemp, J. Dudhia, D. O. Gill, D. M. Barker, W. Wang and J. G. Powers, 2005. A Description of the Advanced Research WRF Version 2
[10] National CentersforEnvironmentalPrediction/NationalWeatherService/NOAA/U.S. Department of Commerce, 2000. NCEP FNL Operational Model Global Tropospheric Analyses, continuing from July 1999, http://rda.ucar.edu/datasets/ds083.2, Research Data Archive at the National Center for Atmospheric Research, Computational and Information Systems Laboratory, Boulder, Colo.

IOP Conf. Series: Journal of Physics: Conf. Series **1043** (2018) 012064 doi:10.1088/1742-6596/1043/1/012064

Filling of Missing Data in Atmospheric Series with Linear Krigeage

E. Mera[1, 4], B. Cordero[2], A. Condal[3], L. Gutierrez[5], C. Ríos[1] and L. Da Silva[5]

[1] Universidad Tecnológica Metropolitana, Physics Department, Box 9845, Santiago, Chile.
[2] Universidad Tecnológica Metropolitana, Statistics Department, Box 9845, Santiago, Chile.
[3] Université Laval, Département des Sciences Géomatiques, G1K 7P4, Canada.
[4] Universidad Metropolitana Cs. Educación, Physics Department, Technology lab, Box 147, Ñuñoa, Chile.
[5] Universidad Tecnica Federico Santa Maria, Box 110-V, Valparaiso, Chile.

E-mail: emera@utem.cl

Abstract. Numerous investigations indicate an important relationship between atmospheric variables and various indicators of environmental contamination that affect public health [1]. Most of the series of meteorological variables used, as independent variables, have defects, Being the most common the lack of specific data or missing data lagoon [2]. There are numerous methods to solve this problem, Most of them require an assumption of normality in the behavior of the data and do not consider their spatial dependence. Through geostatistics, There is an alternative solution to this problem, Making use of regionalized variables, of random functions, of the variogram and Kriging, which is defined as the best linear unbiased estimator [5]. Finalize analysis with method validation, through the cross-validation method “leave one out” [6]. It is concluded that in annual series of atmospheric variables, with time-zone temporal resolution, that the Krigeage presented errors under 5%.

1. Introduction

It is usual to observe, in the course of studies of meteorological parameters and other associated with these, A series of problems caused by the data lagoons existing in the data series, which has been studied by researchers, who have used various methods for their resolution [1][2], among which are the methods of normalized ratio, inverse distance, multiple regression, triangulation, moving averages, auto regressive, logarithmic. These methods are mostly based on normality postulates, which is not always true. Taking into account the above, the fact of being able to define a regionalized variable and to work with random functions, which consider the temporal space dependence of the data, the Geostatistics offers the possibility of estimating the moment of second order, Of any random distribution function, for data series that meet the intrinsic hypothesis [5], Which was later supported by the use of the variogram and the linear Kriging estimator, it is possible to estimate and then fill data lagoons more effectively.

IOP Conf. Series: Journal of Physics: Conf. Series **1043** (2018) 012064 doi:10.1088/1742-6596/1043/1/012064

The objective of the present investigation is fill lagoons in data series of meteorological variables, through the linear Krigeage method, for present complete data series.
Tests performed with meteorological data series of various atmospheric variables such as temperature, pressure and relative humidity allowed to establish the following working protocol:

1. Obtaining data of a meteorological variable, which presents missing data.
2. Exploratory analysis of the data.
3. Reparametrization of the data.
4. Obtention and treatment of experimental variogram and theoretical model.
5. Linear Krigeage Estimation for Missing Data Series.
6. Cross Validation "Leave one out".
7. Error analysis.

2. Results

Based on the steps of the protocol previously established, the results are:

2.1. Obtaining data of a meteorological variable, which presents missing data:

To perform the analysis, was taken as the database, The data series of the MACAM NETWORK of Environmental Health Service, of the year 1999, of relative humidity of the air, registered in 7 monitoring stations in the city of Santiago de Chile.

2.2. Exploratory analysis of the data:

Product of the exploratory analysis of the data were detected missing data lagoons, were eliminated the outlier found and it was verified that the data correspond to a regionalized variable, not normal.

2.3. Reparametrization of the data:

The variogram of the data series was constructed assuming the following reparametrization of the data:

Table 1. Reparametrization of the data.

Data	Reparametrization
X – Geographic	*to ACCUMULATED RUN HOURS*
Y – Geographic	*to DAILY HOURS*
Z – Data	----

It should be mentioned that due to the nature of the data, when they present a variogram of sinusoidal type, the following reparametrization must be assumed:

Table 2. Reparametrization of the data with variogram of sinusoidal type.

Data	Reparametrizatio
X – Geographic	*to DAY*
Y – Geographic	*to DAILY HOURS*
Z – Data	----

2.4. Obtention and treatment of experimental variogram and theoretical model:

In the structural analysis of the data, we defined the random function Z(u), in function of the spatio - temporal parameters. The variogram function was also defined by:

$$2\gamma(h) = V(Z(u) - Z(u+h))$$

Equation 1. Variogram function

Where "h" is the distance to which two observations are separated. Figure 1 shows the graphical variogram, which was obtained through a variogram estimator, defined by:

$$2\gamma(h) = V(Z(u) - Z(u+h))$$

Equation 2. Variogram estimator

To this experimental variogram, which did not present anisotropy, a theoretical model was fitted, formed by a nugget effect and a spherical model, according to the analysis carried out with Surfer 7. It can be appreciated: A discontinuity at the origin expressed by a great nugget effect (Nugget) c_0=350; plateau c_1= 450; range a= 2500; lag h=120.

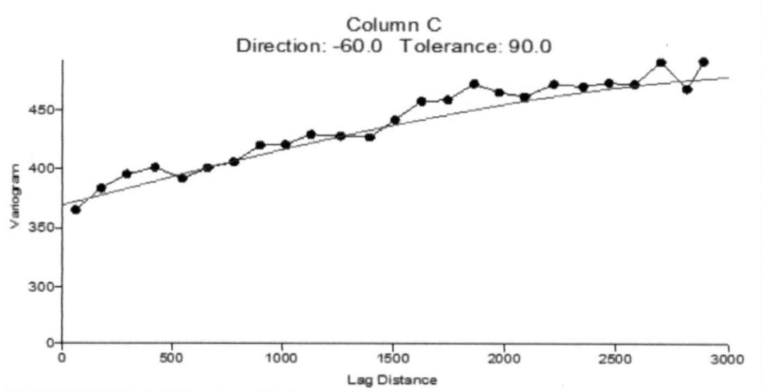

Figure 1. Variogram of the data.

It should be mentioned that when the variograms come from the second reparametrization suggested because the data originally have sinusoidal variograms, daily variograms of the analyzed data should be made, assuming as a final variograptic model the modal variogram model presented by the daily variograms of the data series.

2.5. Linear Krigeage Estimation for Missing Data Series:

Through this method the estimated values Z * (u) are obtained, with the conditions of being an unbiased estimator and minimum variance, which is achieved by assigning weights to each observed value Z(u), with te equation:

$$Z^*(u_0) = \sum_{i=1}^{n} \lambda_i Z(u_i)$$

Equation 3. Linear Krigeage Estimator

Where the "n" coefficients λ_i are determined by solving the system of kriging equations, using the Lagrange multiplier.
In practice, the estimation was performed with linear Krigeage, using the Surfer 7 program, obtaining the estimated values and with them filling the missing data was performed, as shown in the figures 2 and 3.

EOH		F	L	M	N	O	P	Q
16/01/1999	**2400**	84	85	87	83	80	87	85
17/01/1999	**0100**	85	91	89	84	82	88	89
17/01/1999	**0200**	88	98	93	90	85	90	91
17/01/1999	**0300**	95	100	97	93		91	90
17/01/1999	**0400**	96		98	95	85		89
17/01/1999	**0500**	95			94	84		89
17/01/1999	**0600**	94		99	92	85		89
17/01/1999	**0700**	94			92	84	90	94
17/01/1999	**0800**	95	100		92	83	90	97
17/01/1999	**0900**	96	100		90	84	90	94
17/01/1999	**1000**	92			86	81	87	89
17/01/1999	**1100**	85	94		81	78	83	83
17/01/1999	**1200**	79	87	98	75	71	77	77
17/01/1999	**1300**	71	80	88	70	67	72	73
17/01/1999	**1400**	68	76	81	64	62	71	74

Figure 2. Section of the series with data lagoons.

EOH		F	L	M	N	O	P	Q
16/01/1999	**2400**	84	85	87	83	80	87	85
17/01/1999	**0100**	85	91	89	84	82	88	89
17/01/1999	**0200**	88	98	93	90	85	90	91
17/01/1999	**0300**	95	100	97	93	89	91	90
17/01/1999	**0400**	96	93	98	95	85	90	89
17/01/1999	**0500**	95	94	93	94	84	90	89
17/01/1999	**0600**	94	93	99	92	85	90	89
17/01/1999	**0700**	94	92	93	92	84	90	94
17/01/1999	**0800**	95	100	91	92	83	90	97
17/01/1999	**0900**	96	100	90	90	84	90	94
17/01/1999	**1000**	92	89	88	86	81	87	89
17/01/1999	**1100**	85	94	85	81	78	83	83
17/01/1999	**1200**	79	87	98	75	71	77	77
17/01/1999	**1300**	71	80	88	70	67	72	73
17/01/1999	**1400**	68	76	81	64	62	71	74

Figure 3. Section of the series with data lagoons filled with Linear Krigeage.

2.6. Cross Validation "Leave one out":

This process was performed with the "Leave one out" method, obtaining, as shown in figure 4, the dispersion diagram when comparing the actual values with the values obtained with linear Krigeage.

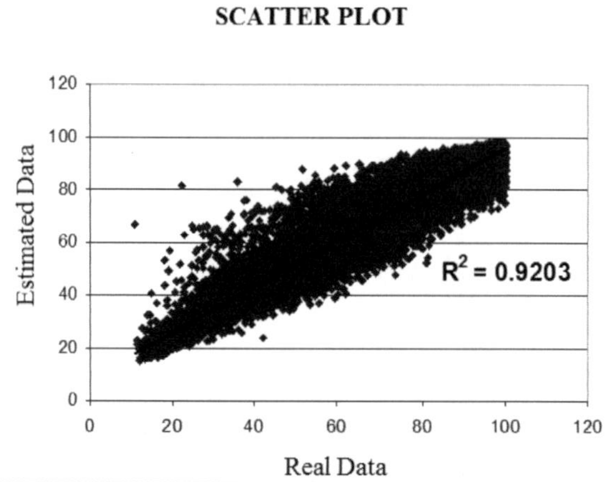

Figure 4: Scatter plot real and estimated data

2.7. Error analysis:

In the present study it was concluded that in the annual series of atmospheric variables, with spatial time resolution, that the Krigeage presented errors under 5%, in figure 5, the effectiveness of the linear Krigeage method is shown, when comparing the series of real data with the estimated data series. It stands out the fact that at the extremes presents a good approximation, because that is where decisions are made on the behavior of the variable.

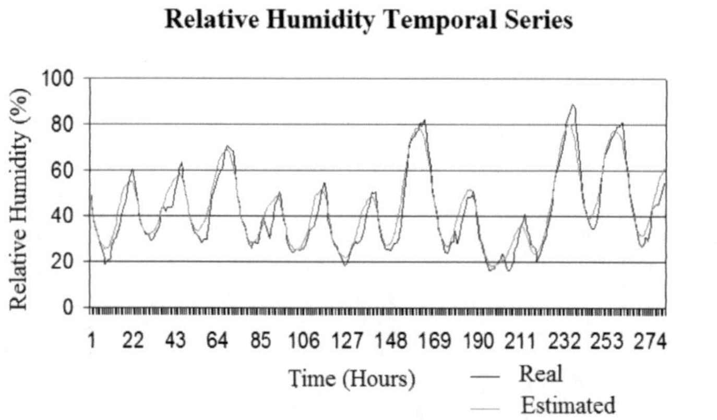

Figure 5: Comparing real data with estimated data with linear Krigeage.

Finally, the data estimates, validated by the "Leave one out" method, obtained a high degree of acceptance, which is shown by the high degree of correlation between the real values and the estimated values obtained, r = 0.96. (Figure 4).

The completed data, reviewed by experts on the meteorological theme, gave 96% approval, which constitutes a high degree of effectiveness of the method. In practice, 3,060 data were filled (approximately 5% of the total data). It was verified that in series of data of atmospheric variables, with temporal and spatial resolution, that the Krigeage method presented a standard deviation of the small error: σe =5.15, Which is a low value to those recorded by other methods of filling data already noted.
The most relevant of the present method of filling data series, is that space-time data is taken into account and the postulate of the normality of data is not required, if other data series filling methods are required.

3. Discussion

Considering that variographic and kriging techniques establish the stationarity constraint, for interpolation conditions to be recorded in a temporal space domain, it is proposed that these techniques can be replicated in other research fields by establishing parameterization modifications, which allow the variographic and krigeaje techniques based on spatial parameters to change to temporal parameters for a determined sector of space.

References

[1] Mirón I et al. 2006. Tratamiento y estudio de series de temperatura para su aplicación en salud públi-ca: El caso de Castilla-La Mancha. *Rev. Esp. Salud Publica.*, Madrid, v. n. **2**.

[2] Cano R et al. 2004. El Clima entre el Mar y la Montaña. Asociación Española de Climatología y Universidad de Cantabria, Serie A, **Nº 4**, Santander.

[3] Eischeid J et al. 1995. The quality control of longterm climatological data using objetive data analysis. *Journal Applied Meteorology*. Rev.**34**:2787-95.

[4] Young K. 1992. A three-way model for interpolating monthly precipitation values. *Mon Weather. Rev.***120**:2561-9.

[5] Matheron G. 1963. Traité de Géostatistique appliquée. Le krigeage. Tome II. Éditions Technip, Paris.

[6] Stone M. 1977. An Asymptotic Equivalence of Choice of Model by Cross-Validation and Akaike's Criterion. *J. R. Stat. Soc.*, **B**, **38**, 44-47.

Acknowledgments

The authors appreciates the support received by the Program of Environmental Applications of Nuclear Technology of the U.T.E.M. One of the authors of the present investigation L. Da Silva dies in February 2017.

IOP Conf. Series: Journal of Physics: Conf. Series **1043** (2018) 012065 doi:10.1088/1742-6596/1043/1/012065

Generation of electricity and waste management by using plasma

Emilio Figueroa and Voltaire Fuentes
Departamento de Física, Facultad de Ciencias Naturales, Matemática y del Medio Ambiente, Universidad Tecnológica Metropolitana, Santiago 7800002

efigueroa@utem.cl

Abstract. Using a plasma torch can intervene practically all types of solid urban waste, residential as well as industrial waste, and obtain electrical energy in the process. The process behind plasma waste treatment is known as plasma gasification, where enough energy is transferred to a solid or liquid substance such that it is transformed into partly in gas, more precisely in synthesis gas (syngas), and another partly in molten lava. The molten lava cools to become an inert vitreous sub product, leaving potentially hazardous products trapped in a crystalline net, such as some heavy metals. Hot gases can be used for electricity production and this can be used both to energize the same treatment plant and to deliver electricity to the external power grid. This is a brief report dedicated to study the viability of implementing the treatment of urban residues by the use of plasma.

1. Introduction

A theme present today in all countries is the amount of garbage per capita produced daily. This amount has reached unmanageable levels that explains the need to implement new and sustainable technologies for the management and treatment of waste. This is a brief report that is part of a more larger study dedicated to assessing the feasibility of implementing the treatment of urban residues by the use of plasma in a small OECD country. Urban solid waste, both from households and industries, can be roughly classified as hazardous and non-hazardous [1], both harmful to the environment or to humans. Among the hazardous ones can be mentioned toxic, combustible, explosive and radioactive waste. Among the non-hazardous are organic, metallic, glass and plastic waste. Due to the increasing amount of waste generated by the population and industry in recent years the waste management is a crucial issue that complies with environmentally responsible standards. The technologies involved in waste management should focus on maximizing material recovery, improving processes that allow us to obtain usable energy from the same waste and above all minimizing environmental impact. By using a plasma torch you can handle all types of waste and get electrical power in the process. Three different types of plasma torches are: the transpiration stabilized arc [2, 3], the arc with stabilized coaxial flow [4] and the stabilized jet with axial symmetry [5]. The process behind plasma waste treatment is known as plasma gasification, where enough energy is transferred to the waste such that it is transformed into partly in gas, more precisely syngas which is basically composed of H_2 and CO, and another partly in molten lava, known as slag. The plasma gasification is a technologically advanced process that in recent years has been reused due to the lack of implementing environmentally friendly techniques to remove wastes and convert them into other usable products and this is a non-incineration thermal process that uses extremely high temperatures. Due to high temperatures, any another more complex molecule is separated into its basic constituents. The molten lava cools to become an inert vitreous sub product, leaving potentially hazardous products, such as some heavy metals, trapped in a crystalline net. Hot gases can be used in conjunction with a turbine system to generate electricity and this electric power can be used both to energize the same treatment plant and to deliver electricity to the external power grid mean an integrated plasma gasification combined cycle system.

2. Electricity Generation mean a Plasma Torch and Financial viability

A plasma is created by applying energy to a gas to rearrange the electronic structure of atoms and molecules. This energy can be of thermal origin or be transported by an electric current or by electromagnetic radiation [6]. Depending on the amount of energy transferred, the properties of the plasma change, that is it changes its electronic density or its temperature [7]. One can distinguish between two classes of laboratory plasmas. The first are high temperature or fusion plasmas, where the electrons, ions and neutral species are in a thermodynamic equilibrium. The second is low temperature plasmas or gas discharges. For the latter, a further distinction can be made between thermal plasmas in which a state of quasi-equilibrium and cold plasmas where a non-equilibrium state occurs [6].

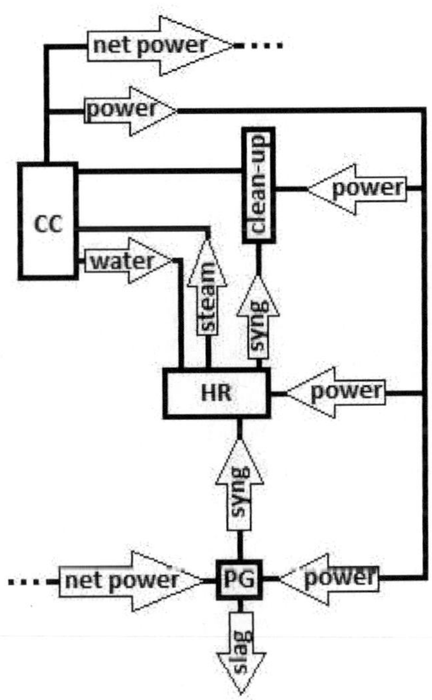

Figure 1. A schematics diagram of a plant. The plasma gasification reactor (PG) is supported by the net power and a by a self-generated power. The gasification resultants sub products are the slag and the syngas. After of a heat recovery (HR) process the syngas is cleaned and sent to the combined cycle (CC) system dedicated to power generation, this power can be used for the different subsystems.

An integrated plasma gasification combined cycle system plant combines the process of plasma gasification with a combined cycle power generation system. Said cycle consists of one or more gas turbines and a steam turbine. The syngas is burned in highly efficient gas turbines to produce electricity. Excess heat from the turbines is used to generate steam which is sent to a steam turbine to produce additional electricity and in this case also using heat from the gasification reaction. The synthesis gas from the plasma gasification must be cooled down before it can be used for the generation of electricity. Besides the gas must be cleaned to comply with the technical specifications of the fuel used by the gas turbine chosen for the treatment plant. In order to drastically reduce air pollution, the cleaning system must achieve the efficient removal of any particulate matter, mercury, Sulphur compound, chlorides and other volatile elements. These gas turbines used in such a plant are similar to jet engines and also very similar to natural gas combined cycle gas turbines but specially adapted for use with syngas. In particular, these gas turbines are adapted to operate with highly hydrogen-rich synthesis gas and this system must include a carbon dioxide capture device.

The Figure 1 shown a schematics diagram of a system of this kind. The plasma gasification reactor (PG) is supported by net power and also can be supported by self-generated power at the same plant. If the self-generated power at the plant is greater than the total net power used net power can be generated. The gasification resultants sub-products are hot slag and the syngas. The slag is cooled down and stored properly. After of a heat recovery (HR) process the syngas is cleaned and sent to one the turbines of the combined cycle (CC) system of two turbines dedicated to the power generation at the same plant, water is converted in steam at the HR system and used by the other turbine. If the self-generated power is greater than the total net power used for the plant net power can be generated. This power also can be used for the different subsystems of the plant: the PG, HR and the clean-up stage.

The amount of electricity that can be generated in this way is directly proportional to the amount of waste that the plant can treat daily, the financial viability of a plant of this type also depends on this amount. The number of tons of waste produced annually by a given urban area is proportional to the number of habitants in it [1]. An estimated price of a plant of this type is MUSD $ 68 [8], the latter is assumed an estimated government contribution of 30% over the total plant cost. A conservative amount of energy that a plant of this type can generate is 0.55 MWh/ton [8], and can be up to 1 MWh/ton [8]. An estimated price in a OECD country (Chile in this example) in the early November of 2016 is USD $ 60 per MWh [9].

3. Conclusions

In areas of very low density of population, such as small islands and rural areas, the financial viability of a plant of this type is questioned, but not the technical viability. In any case, it is concluded that the current form in which the treatment of waste is made must be changed [1], optimized and modernized both in large areas of the population and in areas where it is reduced. As a solution to this dilemma, the following hypothesis is proposed: the creation of sister waste treatment plants, the largest one built in a large urban area and the smallest in an area of few habitants, both financially linked, can make waste treatment viable through the use of plasma in both areas, especially in the area of few population. All this associated to technologies of minimum emission to the atmosphere [10] and extreme care of the environment [10].

References

[1] V. Fuentes, Tesis de Magister, Universidad Tecnológica Metropolitana, Santiago de Chile, 2001.
[2] J. E. Anderson and E. R. G. Eckert, AIAA Journal **vol. 5**, No 4, April 1967.
[3] G. Dyos, United States Patent **S3830428**, Aug 20, 1974.
[4] C. D. Lanzo, Nasa Technical Report Nasa **TN D-4517**, April 1968.
[5] A. Douglas, United States Patent **EP0610177** A1, Aug. 1994.
[6] Tendero C, Tixier C, Tristant P, Desmaison J, Leprince P. Acta B; **61,** 2–30, 2006.
[7] Huang H, Tang L. Energy Convers Manage; **48,** 1331–1337, 2007.
[8] Proceedings of the 18th Annual North American Waste-to-Energy Conference **NAWTEC18-3502**, Orlando, Florida, USA. May 11-13, 2010.
[9] Center of Economic Dispatch of Charge, Central Interconnected System, Chile.
[10] Proceedings of the 14th Annual North American Waste-to-Energy Conference **NAWTEC14-3187**, Tampa, Florida, USA. May 1-3, 2006.

IOP Conf. Series: Journal of Physics: Conf. Series **1043** (2018) 012066 doi:10.1088/1742-6596/1043/1/012066

Effects of ozone layer variation in Ultraviolet solar radiation level received at ground in Arica north of Chile

M Rivas and E Rojas

Laboratorio de Radiación Solar Ultravioleta, Departamento de Física Facultad de Ciencias Universidad de Tarapacá, Arica, Chile

E-mail: mrivas@uta.cl

Abstract. There is an important relationship between overexposure to the sun and various types of skin diseases and sight. Experimental measurements of solar ultraviolet index (UVI) measurements in Arica (ARI) in the period between 2006 to 2015, indicate that during the summer 83.1% was extreme (over 11), when occurs the annual minimum of ozone layer in this latitude. The effects to UVI by the ozone layer variation was study with satellital measurements of thickness of the ozone layer in Arica.

1. Introduction

The electromagnetic radiation emitted by the Sun comprises a broad spectrum. A small part of the ultraviolet solar radiation reaches the ground level after passing through the atmosphere and its different components, including the stratospheric ozone layer. The ultraviolet solar radiation B (UVB) 280 - 315 nm is strongly absorbed by the ozone layer, while the ultraviolet solar radiation A (UVA) 315-400 nm is not absorbed by the ozone layer and its levels are high from dawn until dusk at ground level, whereas (UVC) 100-280 nm is completely absorbed by the ozone layer in the stratosphere not reaching ground level [1].

The effects of ultraviolet solar radiation on the skin depend on factors such as the thickness of the skin penetrated by a specific wavelength and the person's skin type [2].

In the case of the stratospheric ozone layer, the measurement with satellite instruments has reached great precision, allowing a permanent sampling all over the Planet. Although UV solar irradiances and erythema doses have been derived that affect the surface of the Earth from data obtained with satellite, based equipment such as NASA's TOMS instruments [3-4], because the measurements of UVB depend strongly on local conditions, that influence the state of the atmosphere. The amount of UVB received at ground level depends on multiple local factors such as: cloudiness, suspended particles (aerosols), soil reflectivity (albedo), altitude of the place above sea level, stratospheric ozone. This factors makes its study more complex and introduces greater sources of indetermination in this type of measurements being necessary to contrast with calibrated biometers located in the place.

2. Materials and methods

Arica is a city located in the subtropical zone of northern Chile (25 meters above sea level, latitude: 18 ° 49'S; longitude: 70 ° 19'W) which presents a stable micro-climate. The weather conditions throughout the year, such as the lack of rainfall (less than 5 mm per decade), predictable wind regime, high percentage of clear sky days and high soil reflectivity, mainly due to desertification and composition mainly because of very reflective sand. The city is located at sea level; therefore, the population carries out a great number of outdoor activities.

3. Instrumentation

There are essentially two types of physical instruments for measuring solar UV irradiance: spectral, covering a region of the spectrum with very fine wavelength spacing (or measuring at fixed wavelengths with very good spectral resolution) and bandwidth Wide, which measure in a wide range of wavelengths with relative response close to unity in all of them, or approaching the response function of some biological system (usually called biometrics). The widespread use of a standardized metric for reporting UV radiation risk-namely erythemally weighted UV (UVEry, in W / m2) or the UV index (UVI = 40 • UVEry), has also facilitated meaningful comparisons. The most recent of these assessments predicts that although the ozone layer will gradually recover over the next few decades, the outlook for future UV is less certain [5].

The instrument is located in the highest part of the UV solar radiation laboratory with clear view. The measurements were carried out with a YES-UVB broadband biometer [6], calibrated periodically through a cooperation agreement between Tarapacá University (UTA) with the Chilean Meteorological Direction (DMC), by means of which this instrument forms part of the solar ultraviolet measurement network of our country. Calibrations are carried out in accordance with the recommendations of the World Meteorological Organization (WMO). The biometer has been in operation since September 2006.
Stratospheric ozone data are based on satellite measurements by the ozone measuring instrument (OMI / Aura satellite) and TOMS (Spectrometer for total ozone mapping), data are available at http://jwocky.gsfc. Nasa.gov. OMI provides almost global coverage in one day with a spatial resolution of 13 x 24 km. TOMS is the oldest instrument for collecting and analyzing the stratospheric ozone layer and the properties of atmospheric solar UV radiation, this was replaced by OMI after operating between 1996 and 2005 [7]. Measurements of UVB solar irradiance were performed in Arica between September 20th, 2006 and December 31st, 2015, every five minutes.

4. UVI and ozone Arica measurements

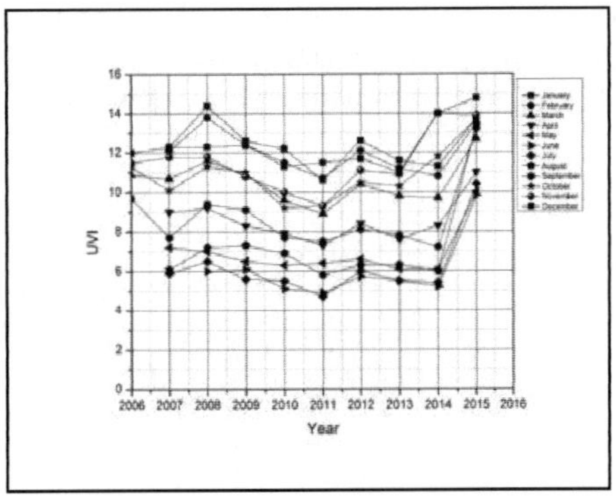

Figure 1. Arica 2006 – 2015. Monthly averages of maximum daily UVI values Arica 2006 - 2015.

Figure 1 shows the monthly averages of daily maximum UVI in Arica between 2006 - 2015. The seasonal variation of UVI is observed, especially with high levels of UVI during the summer months. The data of the monthly averages of UVI measured at the solar midday reached in the summer of 2008 the maximum value of the analyzed period, corresponding to UVI = 14.4 and the minimum value of the average monthly UVI corresponds to July of the year 2011 UVI = 4.7. The minimum values are presented in the winter months of June and July. For most of the year the maximum values measured at solar midday UVI in Arica are over 6.

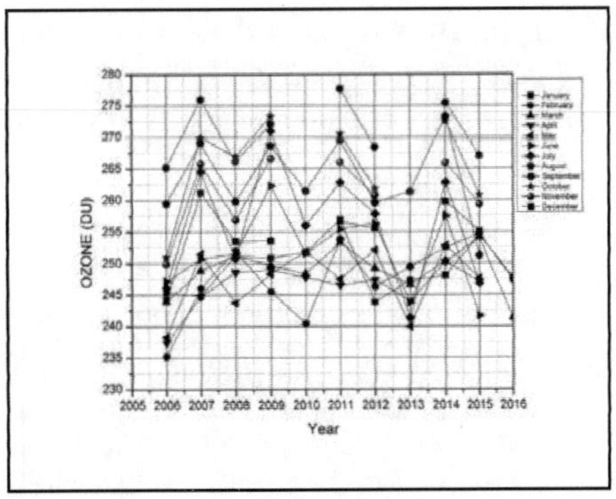

Figure 2. Average monthly values of the thickness of the ozone layer in Arica, 2006-2016.

Figure 2 shows the stratospheric ozone contents in Arica provided by the satellite instrument OMI / NASA between the years 2006 - 2016. The monthly averages of stratospheric ozone measured in the city of Arica range from a minimum value of 235.2 Dobson units (DU) in February 2006 and a maximum value of 277.7 (DU) in September 2011. The maximum thickness of the ozone layer is produced in Arica during the months of September and October and the minimum values are observed during the months of January, February and April. These results show that in this subtropical region the thickness of the ozone layer is lower during the summer when, due to the lower zenith angle, more UVB solar radiation is received and, therefore, more protection is needed, which represents a greater risk to health due the lower thickness of the ozone layer causes an increase in UVB radiation reaching ground level [8-9].

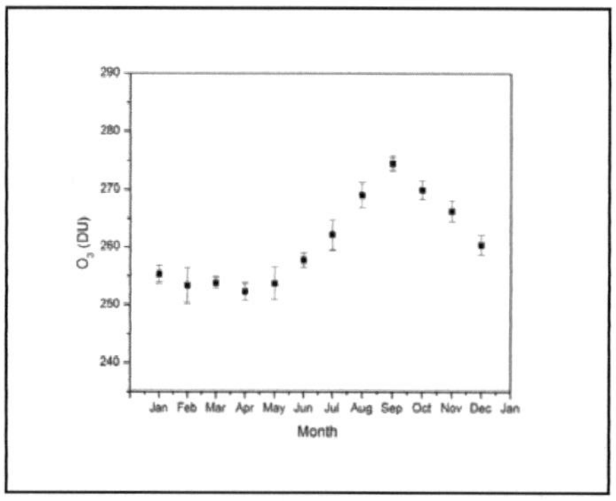

Figure 3. Average monthly values of the thickness of the ozone layer in Arica, 1978-2015

The monthly average of the satellite measurements of ozone are observed in figure 3, the average monthly ozone reaches the highest values during the months of September 274.5 DU and October 269.9 DU, and the minimum values in the months of February 253.4 DU and April 252.3 DU.

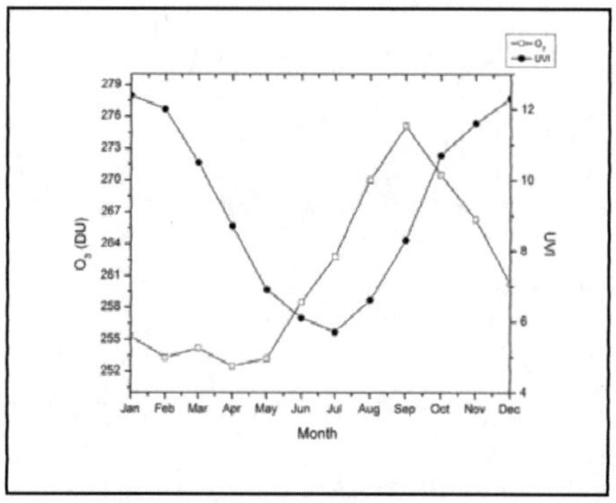

Figure 4. Monthly average values of UVI 2006-2015, and thickness of the ozone layer 1978-2015

Figure 4 shows the monthly averages of the thickness of the stratospheric ozone layer and the ultraviolet solar index in Arica. The UVI measurements indicate that the highest monthly averages of the year are in the months January 12.4, February 12, November 11.5 and December 12.4 all values of UVI above 11 and the minimum values of the year are found on May 6.9, June 6 and July 6.2.

Nevertheless, the minimum values of the thickness of the ozone layer in Arica are found in January 255.3, February 253.4 March 253.8 April 252.3 and May 253.7 DU and the maximum values in August 269, September 274.5 and October 269.9 DU.

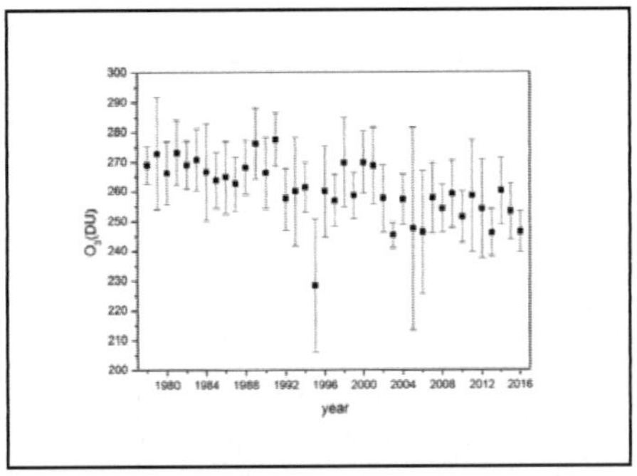

Figure 5. Annual average values of ozone, with their respective standard deviation, Arica 1978-2016

Figure 5 shows the annual averages of the thickness of the ozone layer in Arica provided by different TOMS / NASA and OMI / NASA satellite instruments between 1978 and 2016.

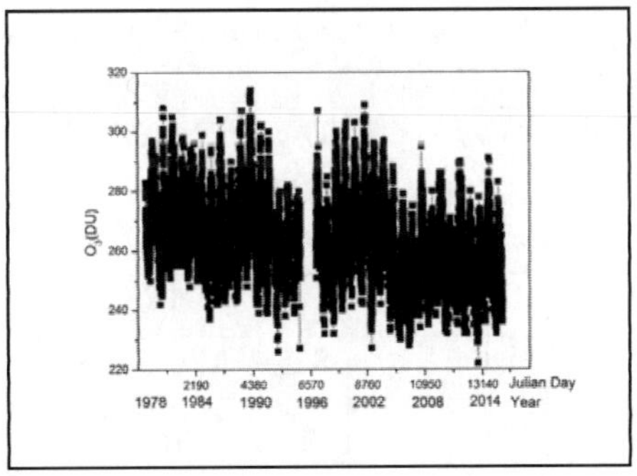

Figure 6. Daily satellite measurements of stratospheric ozone in (DU), Arica (1978 – 2016)

Year	O_{3min} (DU)	O_{3max} (DU)	% ΔO_3	% ΔUVI	$\frac{\Delta UVI}{\Delta O_3}$
2006	235.2	265.2	12.8	16.1	1.26
2007	244.7	276.0	12.8	16.1	1.26
2008	243.7	266.9	9.5	12.0	1.26
2009	245.5	273.3	11.3	14.0	1.24
2010	240,4	261.5	8.8	10.9	1.24
2011	246.5	277.7	12.7	15.9	1.25
2012	243.8	268.4	10.0	12.6	1.26
2013	244.0	261.3	7.1	8.9	1.25

Table 1. Average maximum and minimum monthly values of O3 2006-2013 and percentages of variation of O3. The percentages of variation of UVI were calculated with the TUV program

Table 01 shows the annual maximum and minimum values of the thickness of the ozone layer. The maximum values are measured by satellite during the months of September and October in this latitude and the minimum values are measured during the months of January - February in Arica between the years 2006-2013, and the corresponding annual percentage variation of the thickness of the layer of ozone. Using the TUV program [10]. The corresponding variation in the ultraviolet solar index (UVI) was calculated for the summer months, the time of the year in which the maximum values of UVI are measured, given the annual variation of the thickness of the ozone layer. The percentages of annual variations in the thickness of the ozone layer ranged from 7.1 - 12.8% and the variations of UVI calculated with the TUV program were between 8.9 - 16.1% in the same period. These results are especially relevant at this latitude because for every 1% decrease in the thickness of the ozone layer the UVI increases by 1.26%, which has consequences in the increase of new cases of non-melanoma cancer.

Year	$O_{3\,(DU)}$	UVI Albedo =0.165	UVI Albedo =0.02	ΔUVI	%ΔUVI
2006	235.2	13.3	12.4	0.93	7.5
2007	244.7	9.1	8.5	0.63	7.4
2008	243.7	6.9	6.5	0.47	7.3
2009	245.5	12.6	11.7	0.88	7.5
2010	240,4	12.9	12.0	0.91	7.6
2011	246.5	9.0	8.4	0.62	7.4
2012	243.8	13.0	12.1	0.91	7.5
2013	244.0	9.1	8.4	0.63	7.4

Table 2. Minimum monthly values of O3 2006-2013 and variation of UVI for two values of different albedos corresponding to sand 0.165 and lawn 0.02 and the percentage of variation of corresponding UVI, calculated with TUV program.

Table 2 shows the monthly minimum values of O3 2006-2013 and the variation of UVI for two different albedos values corresponding to the one existing in Arica (corresponding albedo sand = 0.165) and an albedo corresponding to cities in the southern zone (lawn albedo = 0.02). The percentage of variation of corresponding UVI was calculated using the TUV program.

The minimum values of ozone are measured by satellite during the summer months in this latitude between the years 2006-2013. The corresponding annual percentage variation of the UVI calculated for two different albedos (sand and lawn) allows to quantify the percentage of increase of the UVI in Arica due to its lack of vegetation compared to cities of the south zone where the surface of the soil is predominantly covered by Vegetation, calculations were performed using the TUV program. The of UVI increase due to the albedo difference were between 7.3 - 7.6% in the same period.

These results have special relevance in this latitude because people living in this area only by effect of the lack of vegetation, which implies surfaces with a greater albedo the UVI increases by more than 7%, have consequences due to overexposure to UVB solar radiation in the increase of new cases of non-melanoma cancer.

5. Results

The analysis of the results revealed that 16.6% of the maximum daily UVI measured at solar noon during the summer season in Arica are high and very high (8 <UVI <10) and 83.1% to extreme (> 11 UVI) according to the scale used by the World Health Organization (WHO). The maximum solar ultraviolet irradiance in this latitude is received during the summer months January percentages - February, during these months the thickness of the ozone layer has annual minimum values. The maximum values of the thickness of the ozone layer occur during September - October.

The percentage difference between values to the maximum and minimum average annual thickness of the ozone layer in Arica 2006 - 2013 varies 7.1 -12.8%. UVI was calculated using TUV model, reveals that the corresponding differences in the solar ultraviolet index UVI for these variations in the thickness of the ozone layer during the same period, are 8.9 - 16.1%.

6. Acknowledgments

Miguel Rivas and Elisa Rojas, wish to thank to the funding provided by the project UTA-Mayor 2017.

IOP Conf. Series: Journal of Physics: Conf. Series **1043** (2018) 012066 doi:10.1088/1742-6596/1043/1/012066

References

[1] Diffey BL 2002 What is the light? *Photodermatol. Photoimmunol. Photomed.* 18 (2) 68

[2] Sklar LR, Almutawa F, Lim HW and Hamzavi I 2013 Effects of ultraviolet radiation, visible light, and infrared radiation on erythema and pigmentation: a review. *Photocem. Photobiol. Sci.* 12 54

[3] Herman JR, Barthia PK, Ziemke J, Ahmad Z, Larko D 1996 UV-B increases (1979-1992) from decreases in total ozone *Geophysical Research Letters* 23(16) D 10.1029/96GL01958

[4] Herman JR 2010 Global increase in UV irradiance during the past 30 years (1979-2008) estimated from satellite data. *Journal of Geophysical Research 115. D04203. 2010. DOI: 10.1029/2009JD012219.*

[5] Mckenzie L, Liley JB, Bjorn LO 2008 UV Radiation: Balancing Risk and Benefits. *Photochemistry and Photobiology, DOI:10.1111/j.1751-097.2008.00400.x*

[6] Yankee Environmental Systems 2005 Manual Model Ultraviolet-B Broadband Turners Falls, MA 01376, USA.

[7] McPeters RD, Bhartia PK, Krueger AJ, Herman JR, Wellemeyer CG, Seftor CJ, Jaross G,Torres O, Moy L, Labow G, Byerly W, Taylor SL, Swissler T, Cebula RP 2009 Earth Probe Total Ozone Mapping Spectrometer (TOMS) Data Products User's Guide. *National Aeronautics and Space Administration (NASA) Technical Publication 1998-206895, MD. Available at: http://macuv.gsfc.nasa.gov/doc/epusrguide.pdf. Accessed on 23 October 2009.*

[8] Godlee F 1992 Dangers of ozone depletion *BJM* 303 *1326*

[9] Jones RR 1987 Ozone depletion and cancer risk *Lancet* 2 443

[10] Madronich S 2003 *Tropospheric ultraviolet visible TUV version 4.2. Atmospheric Chemistry Division National Center for Atmospheric Research* (NCAR P O Box 3000, Boulder Colorado, USA).